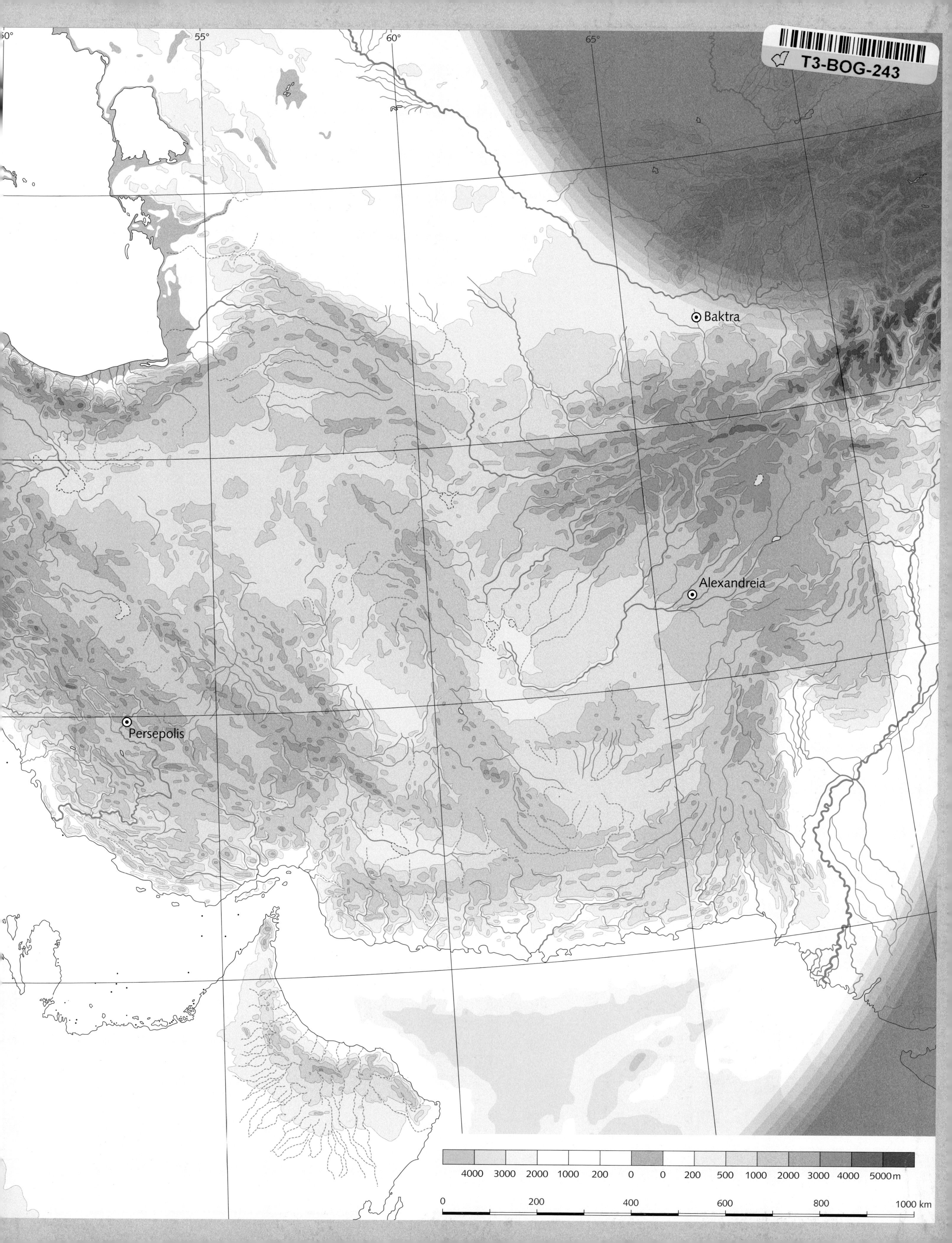

50°
55°
60°
65°
Baktra
Alexandreia
Persepolis
4000
3000
2000
1000
200
0
0
200
500
1000
2000
3000
4000
5000 m
0
200
400
600
800
1000 km

Brill's New Pauly

Supplements 3

HISTORICAL ATLAS OF
THE ANCIENT WORLD

Brill's New Pauly

Supplements

EDITORS

Hubert Cancik
Manfred Landfester
Helmuth Schneider

Brill's

Historical Atlas of the Ancient World

New Pauly

Edited by
Anne-Maria Wittke, Eckart Olshausen
and *Richard Szydlak*

in collaboration with
Vera Sauer and other specialists

English edition by
Christine F. Salazar with *Dieter Prankel,*
Duncan Smart (translation)
and *Corinna Vermeulen* (editing)

LEIDEN - BOSTON
2010

BRILL

Contents

Authors

Walter Eder, Berlin (Maps pp. 88, 91, 93, 99)
Ulrich Fellmeth, Stuttgart (Maps pp. 201, 203, 247)
Klaus Freitag, Münster (Maps pp. 100–103)
Andreas Fuchs, Tübingen (Maps pp. 2, 3 [Map B], 33, 47, 49, 51, 53, 55)
Thomas Hoppe, Stuttgart (Maps pp. 79–81)
Jens Kamlah, Tübingen (Maps pp. 33, 44, 45)
Martin Köder, Tübingen (Map p. 111)
Gerhard Meiser, Jena (Maps pp. 66, 67)
Renate Müller-Wollermann, Tübingen (Maps pp. 3 [Map A], 17, 19, 21, 33, 58, 59)
Hans Georg Niemeyer, Hamburg (†) (Maps pp. 71, 73)
Miroslav Novák, Tübingen (Maps pp. 10–15, 33, 43, 57)
Eckart Olshausen, Stuttgart (*passim*)
Wolfgang Röllig, Tübingen (Map p. 41)
Karl Strobel, Klagenfurt (Map p. 124)
Martina Terp, Tübingen (Map p. 83)
Christian Winkle, Stuttgart (Maps pp. 107, 172, 173, 238)
Anne-Maria Wittke (*passim*)

Cf. the detailed listing of authors and rights on p. 274

Koninklijke Brill NV incorporates the imprints Brill, Hotei Publishing, IDC Publishers, Martinus Nijhoff Publishers and VSP.

Original German language edition:
Anne-Maria Wittke, Eckart Olshausen, Richard Szydlak: Historischer Atlas der antiken Welt (= Der Neue Pauly Supplemente 3) published by J. B. Metzler'sche Verlagsbuchhandlung und Carl Ernst Poeschel Verlag GmbH Stuttgart, Germany.

Cover design: TopicA (Antoinette Hanekuyk)
Front: Delphi, temple area
ISBN 978 90 04 17156 5

This book is printed on acid-free paper.
PRINTED IN THE NETHERLANDS

Preface

With globalization, supposedly, the world is becoming a village. Conversely, it might be said that our horizons are broadening – that we are learning to survey the regions of the world in conscious awareness of the global context. This applies equally to the historical 'backward glance'. In recent decades, the scholarly disciplines studying the ancient world have widened their horizons, and established an understanding of the 'global' contexts of the ancient world, a world that was far more than just the Graeco-Roman sphere of the Mediterranean. The peripheries of the great centres, the indigenous peoples and the great empires and kingdoms of still more ancient civilizations have come back into focus, as have the so-called 'Dark Ages' and Late Antiquity. Moreover, a generally increased awareness of the geographical dimension of historical processes has become apparent.

New conception

The new *Historical Atlas of the Ancient World* takes account of these developments. In several respects, therefore, it breaks new ground.

1. The very scope of the project – with 161 main maps in colour and 44 insets – makes possible a new cartographical portrayal of Antiquity from the 3rd millennium BC (Ancient Near East) to the 15th century AD (Byzantine Empire) that reflects this range, multiplicity of perspective and intensity of new discovery.
2. A particular concern of the Atlas in the sense mentioned above is to focus on the great civilizations of the Ancient Near East, hence shedding light on peoples, kingdoms and epochs that have once more begun to assume a central place in ancient historical studies. At the same time, this perspective sharpens the focus on the conditions for the emergence of Classical Antiquity (as a former 'peripheral culture' of the Near Eastern kingdoms) and the perpetual interactions between West and East. Other themes likewise traditionally neglected, for instance the early phases of Aegean and central and western Mediterranean history, and the history of Late Antiquity, are dealt with in more detail here.
3. The corpus of maps depicts not only political and military narrative history, but also developments in such fields as economic and administrative history and the history of religion and culture. Thus, several pages are devoted to ancient concepts of the world and voyages of exploration. The emergence, evolution and administration of the Roman provinces are comprehensively portrayed in a series of maps. Trading routes and economic relations, languages and cultures – all here find their cartographical presentation here.
4. It is our intention for the maps themselves to be more than mere visual products of lists of historical facts. Their particular cartographic language offers a dynamic interpretation of historical conditions. They establish hierarchies among the various strata of historical evidence, highlighting crucial issues in the process. They also offer a wealth of supplementary information. The cartography of this Atlas demonstrates the advantage of the simultaneous presentation of historical conditions and processes that only the (carto-) graphic image makes possible.
5. Another innovation is the double-page presentation of a map with its attendant commentary opposite. The commentary page supplements the map with brief explanations of the historical background, notes on the conception of the map, general information on the sources of historical information, map insets providing more detailed or additional information, factual information compiled in tabular form, details of sources and bibliographies.

Intended readership

Its conception should make the Atlas attractive to various audiences. It is suitable for teaching and research in the Classics, being based on (and presenting) current scientific and scholarly information. Not least thanks to the commentaries, which facilitate the reading and understanding of the cartographic material, the maps are also ideal for use in schools, and are readily accessible for the general reader with an interest in history.

Enabling conditions

For a cartographic work of this scale to be realized, a supportive environment is needed. This background is provided by *Brill's New Pauly* Encyclopaedia of the Ancient World, as a supplement to which the *Historical Atlas of the Ancient World* is published.

The maps of the *New Pauly* have received a thoroughly positive welcome in the world of scholarship and teaching and in schools. It was therefore only natural to emulate this cartographic concept. All three main authors of the Atlas have already played a central role in the creation of the maps in the *New Pauly*.

Users of the Atlas will therefore continually encounter the substance of the *Pauly* maps. In fact, some 60 percent of maps in the Atlas are derived from the *New Pauly*, with the remainder newly developed. Most of the *Pauly* maps were redesigned in colour and in a more generous atlas format and, to a considerable extent, revised, with a new focus or emphasis.

An indispensable aid in the creation and revision of maps was the topographical *Barrington Atlas of the Greek and Roman World* (2000), edited by R.J.A. Talbert. Information on rulers is mostly taken from Supplement Vol. 1 of the *New Pauly*, W. Eder/J. Renger, *Chronologies of the Ancient World* (2007), and the articles of the *New Pauly* encyclopaedia in general form the constant basis for the commentary texts.

The authors

Remits were divided as follows among the three authors responsible for the entire work:

Anne-Maria Wittke (Department of Classical Studies, University of Tübingen) was responsible for the maps from the great early civilizations up to the Hellenistic period. She also compiled the index, with technical support from Richard Szydlak.

Eckart Olshausen (Department of Ancient History, University of Stuttgart) designed most of the maps from classical Roman antiquity to the end of the Byzantine Period.

Richard Szydlak (cartographer at the University of Tübingen) was responsible for overall cartographic processing. Only through the intensive process of communication between historians and cartographer was the particular quality of the maps as described above made possible. The cartographer is thus a genuine co-author of this work.

Other specialists contributed to the Atlas, as authors either of new maps or of original *New Pauly* maps. Details are found in the lists of authors on pp. IV and 274 ff.

Acknowledgments

Profound thanks are owed to Vera Sauer, who worked on the conception of new maps and the composition of commentaries within the field of Prof. Olshausen, as well as on the final editing of the index and at all stages of the printing process. – Susanne Fischer rendered sterling service to the Atlas by editing almost all maps and commentaries. – The same is true of Christian Winkle, who did extensive work on editing the maps and commentaries within the field of Prof. Olshausen.

The authors are grateful to Oliver Schütze, J.B. Metzler Verlag, for his thorough and patient attention to the editing and publishing process, to Prof. Thomas Schäfer, Director of the Institute of Classical Archaeology at Tübingen, which hosted the project in collaboration with the Historical Institute of the University of Stuttgart, and to many other colleagues, but particularly the late Walter Eder, for expert advice willingly proffered.

Last but not least, thanks are owed to Bernd Lutz (who initiated the project), Johannes Kunsemüller and the late Günter Schweitzer at the publishers for long years of support.

Anne-Maria Wittke, Eckart Olshausen,
Richard Szydlak October 2007

Contents (maps)

Contents (maps)

Systematic list of maps

The list gives abbreviated map titles and page numbers.
Other themes may be derived via the index.

Systematic list of maps

Notes to the user

The series of Supplements to *Brill's New Pauly* are intended to amplify and expand upon volumes 1–15 (Antiquity) of the Encyclopaedia. The Atlas is designed as an independent work, but may readily be used in conjunction with the alphabetical encyclopaedia volumes. We draw particular attention to the geographical reconciliation of modern and ancient place names and the index for locating place names in the Index Volume of *Brill's New Pauly* (forthc.).

Double-page view

The double page allotted to a particular map theme is usually organized so that the commentary is placed on the left-hand page, the coloured main map(s) on the right. The commentary contains supplementary maps and tables as well as references to sources and literature. In many cases, sources, bibliographic references and tables are placed in the appendix for reasons of space. It has not always been possible to include on the corresponding maps all places mentioned in the text. In such cases, we recommend consulting the Atlas index for the item's location.

Spelling and typography

Topographical terms, including names of territories and provinces, are given in their full ancient form in the Atlas. Modern names are indicated in cursive script. All typographical specifications are explained in the pertinent map legend. The transcription of Eastern proper names follows the conventions of the Tübinger Atlas des Vorderen Orients (TAVO). In addition, divergent spellings that are familiar in the English language are retained, where this facilitates location and reference. Cf. the transcription tables below.

Bibliographic references and tables

The references to literature in commentaries and the tables that refer to sources make use of abbreviations for periodicals, series and collections and for ancient authors. These abbreviations follow the conventions of *Brill's New Pauly* (see List of Abbreviations, p. XVI).

Index

The index provides Cartesian coordinates for all names contained in the maps for: places, territories, peoples, kingdoms, provinces, dioceses, bodies of water, mountains and mountain ranges, passes, roads, etc. It also includes historical figures, dynasties, periods, alliances, wars, battles, peace treaties, natural resources, agricultural products and other technical terms mentioned in the commentary pages.

All entries that do not refer to places or people are annotated with one of the following qualifications: body of water, cape, desert, dioecesis, island, kingdom, mountain (range), oasis, pass, people, *praefectura*, province, *regio*, road (in the sense of trade route), strait, territory, theme, *tribus*.

That an unannotated entry refers to a place and not a person can usually be deduced from the fact that it is listed along with map coordinates.

A number of names which incorporate a topographical designation (e.g. *mons*, *lacus*, *mare* etc.) are listed both under the generic term and under the specific name, i.e. both as 'Albanus mons' and 'mons Albanus'. Roman personal names are generally listed under the *nomen gentile*. Exceptions are made for very well-known figures such as Caesar, Cicero and Marcus Antonius (Mark Antony).

Notes to the user

List of transliterations

Transliteration of Ancient Greek

α	a	alpha	ν	n	nu
αι	ai		ξ	x	xi
αυ	au		ο	o	omicron
β	b	beta	οι	oi	
γ	g	gamma; γ before γ, κ, ξ, χ: n	ου	ou	
δ	d	delta	π	p	pi
ε	e	epsilon	ρ	r	rho
ει	ei		σ, ς	s	sigma
ευ	eu		τ	t	tau
ζ	z	z(d)eta	υ	y	upsilon
η	ē	eta	φ	ph	phi
ηυ	ēu		χ	ch	chi
θ	th	theta	ψ	ps	psi
ι	i	iota	ω	ō	omega
κ	k	kappa	ʽ	h	spiritus asper
λ	l	la(m)bda	ᾳ	ai	iota subscriptum (similarly ῃ, ῳ)
μ	m	mu			

In transliterated Greek the accents are retained (acute ´, grave `, and circumflex ˆ). Long vowels with the circumflex accent have no separate indication of vowel length (makron).

Transliteration table of Hebrew consonants

א	a	alef	מ	m	mem
ב	b	bet	נ	n	nun
ג	g	gimel	ס	s	samek
ד	d	dalet	ע	c	ayin
ה	h	he	פ	p/f	pe
ו	w	vav	צ	ṣ	tsade
ז	z	zayin	ק	q	qof
ח	ḥ	khet	ר	r	resh
ט	ṭ	tet	שׂ	ś	sin
י	y	yod	שׁ	š	shin
כ	k	kaf	ת	t	tav
ל	l	lamed			

Transliteration and pronunciation of Modern Greek

Only those sounds and combinations of sounds are listed that differ from Ancient Greek

Consonants

β	v	
γ	gh	(voiced velar fricative) before back vowels
	y	(voiced palatal fricative) before front vowels
δ	dh	as in *the*
ζ	z	as in *zoo*
θ	th	as in *thing*

Consonantal combinations

γκ	ng	g in initial position
μπ	mb	b in initial position
ντ	nd	d in initial position

Vowels

η	i
υ	i

Diphthongs

αι	e	
αυ	av	
	af	before hard consonants
ει	i	
ευ	ev	
	ef	before hard consonants
οι	i	
υι	ii	

The spiritus asper is not sounded. The Ancient Greek accents have generally been preserved in their original positions. However, the distinctions between ´, `, and ˆ have disappeared.

Pronunciation of Turkish

Turkish uses Latin script since 1928. Pronunciation and spelling generally follow the same rules as European languages. Phonology according to G. Lewis, Turkish Grammar, 2000

A	a	French a in *avoir*	L	l	l in *list* or in *wool*
B	b	b	M	m	m
C	c	j in *jam*	N	n	n
Ç	ç	ch in *church*	O	o	French o in *note*
D	d	d	Ö	ö	German ö
E	e	French ê in *être*	P	p	p
F	f	f	R	r	r
G	g	g in gate or in angular	S	s	s in *sit*
Ğ	ğ	lengthens preceding vowel	Ş	ş	sh in *shape*
			T	t	t
H	h	h in *have*	U	u	u in *put*
I	ı	i in *cousin*	Ü	ü	German ü
İ	i	French i in *si*	V	v	v
J	j	French j	Y	y	y in *yet*
K	k	c in *cat* or in *cure*	Z	z	z

Transliteration of Arabic, Persian, and Ottoman Turkish

ء, ا	ʾ, ā	ʾ	ʾ	hamza, alif
ب	b	b	b	bāʾ
پ	–	p	p	pe
ت	t	t	t	tāʾ
ث	ṯ	s̱	s̱	ṯāʾ
ج	ğ	ğ	ğ	ğīm
چ	–	č	č	čim
ح	ḥ	ḥ	ḥ	ḥāʾ
خ	ḫ	ḫ	ḫ	ḫāʾ
د	d	d	d	dāl
ذ	ḏ	ẕ	ẕ	ḏāl
ر	r	r	r	rāʾ
ز	z	z	z	zāy
ژ	ž	ž	ž	že
س	s	s	s	sīn
ش	š	š	š	šīn
ص	ṣ	ṣ	ṣ	ṣād
ض	ḍ	ḍ	ḍ	ḍād
ط	ṭ	ṭ	ṭ	ṭāʾ
ظ	ẓ	ẓ	ẓ	ẓāʾ
ع	ʿ	ʿ	ʿ	ʿain
غ	ġ	ġ	ġ	ġain
ف	f	f	f	fāʾ
ق	q	q	q, k	qāf
ك	k	k	k, g, ñ	kāf
گ	–	g	g, ñ	gāf
ل	l	l	l	lām
م	m	m	m	mīm
ن	n	n	n	nūn
ه	h	h	h	hāʾ
و	w, ū	v	v	wāw
ي	y, ī	y	y	yāʾ

Transliteration of other languages

Akkadian (Assyrian-Babylonian), Hittite and Sumerian are transliterated according to the rules of RLA and TAVO. For Egyptian the rules of the Lexikon der Ägyptologie are used. The transliteration of Indo-European follows RIX, HGG. The transliteration of Old Indian is after M. MAYRHOFER, Etymologisches Wörterbuch des Altindoarischen, 1992ff. Avestian is done according to K. HOFFMANN, B. FORSSMAN, Avestische Laut- und Flexionslehre, 1996. Old Persian follows R.G. KENT, Old Persian, [2]1953 (additions from K. HOFFMANN, Aufsätze zur Indoiranistik vol. 2, 1976, 622ff.); other Iranian languages are after R. SCHMITT, Compendium linguarum Iranicarum, 1989, and after D.N. MACKENZIE, A Concise Pahlavi Dictionary, [3]1990. For Armenian the rules of R. SCHMITT, Grammatik des Klassisch- Armenischen, 1981, and of the Revue des études arméniennes, apply. The languages of Asia Minor are transliterated according to HbdOr. For Mycenaean, Cyprian see HEUBECK and MASSON; for Italic scripts and Etruscan see VETTER and ET.

List of abbreviations

Ancient authors and titles of works

Abd	Abdias
Acc.	Accius
Ach.Tat.	Achilles Tatius
Act. Arv.	Acta fratrum Arvalium
Act. lud. saec.	Acta ludorum saecularium
Acts	Acts of the Apostles
Ael. Ep.	Aelianus, Epistulae
NA	De natura animalium
VH	Varia historia
Aen. Tact.	Aeneas Tacticus
Aesch. Ag.	Aeschylus, Agamemnon
Cho.	Choephori
Eum.	Eumenides
Pers.	Persae
PV	Prometheus
Sept.	Septem adversus Thebas
Supp.	Supplices
Aeschin. In Ctes.	Aeschines, In Ctesiphontem
Leg.	De falsa legatione
In Tim.	In Timarchum
Aesop.	Aesopus
Aet.	Aetius
Aeth.	Aetheriae peregrinatio
Alc.	Alcaeus
Alc. Avit.	Alcimus Ecdicius Avitus
Alex. Aphr.	Alexander of Aphrodisias
Alci.	Alciphron
Alcm.	Alcman
Alex. Polyh.	Alexander Polyhistor
Am	Amos
Ambr. Epist.	Ambrosius, Epistulae
Exc. Sat.	De excessu Fratris (Satyri)
Obit. Theod.	De obitu Theodosii
Obit. Valent.	De obitu Valentiniani (iunioris)
Off.	De officiis ministrorum
Paenit.	De paenitentia
Amm. Marc.	Ammianus Marcellinus
Anac.	Anacreon
Anaxag.	Anaxagoras
Anaximand.	Anaximander
Anaximen.	Anaximenes
And.	Andocides
Anecd. Bekk.	Anecdota Graeca ed. I. Bekker
Anecd. Par.	Anecdota Graeca ed. J.A. Cramer
Anon. De rebus bell.	Anonymus de rebus bellicis (Ireland 1984)
Anth. Gr.	Anthologia Graeca
Anth. Lat.	Anthologia Latina (Riese [2]1894/1906)
Anth. Pal.	Anthologia Palatina
Anth. Plan.	Anthologia Planudea
Antiph.	Antiphon
Antisth.	Antisthenes
Apc	Apocalypse
Apoll. Rhod.	Apollonius Rhodius
Apollod.	Apollodorus, Library
App. B Civ.	Appianus, Bella civilia
Celt.	Celtica
Hann.	Hannibalica
Hisp.	Iberica
Ill.	Illyrica
It.	Italica
Lib.	Libyca
Mac.	Macedonica
Mith.	Mithridatius
Num.	Numidica
Reg.	Regia
Sam.	Samnitica
Sic.	Sicula
Syr.	Syriaca
App. Verg.	Appendix Vergiliana
Apul. Apol.	Apuleius, Apologia
Flor.	Florida
Met.	Metamorphoses
Arat.	Aratus
Archil.	Archilochus
Archim.	Archimedes
Archyt.	Archytas
Arist. Quint.	Aristides Quintilianus
Aristaen.	Aristaenetus
Aristid.	Aelius Aristides
Aristob.	Aristoboulus
Aristoph. Ach.	Aristophanes, Acharnenses
Av.	Aves
Eccl.	Ecclesiazusae
Equ.	Equites
Lys.	Lysistrata
Nub.	Nubes
Pax	Pax
Plut.	Plutus
Ran.	Ranae
Thesm.	Thesmophoriazusae
Vesp.	Vespae
Aristot. An.	Aristotle, De anima (Bekker 1831–70)
An. post.	Analytica posteriora
An. pr.	Analytica priora
Ath. pol.	Athenaion Politeia
Aud.	De audibilibus
Cael.	De caelo
Cat.	Categoriae
Col.	De coloribus
Div.	De divinatione
Eth. Eud.	Ethica Eudemia
Eth. Nic.	Ethica Nicomachea
Gen. an.	De generatione animalium
Gen. corr.	De generatione et corruptione
Hist. an.	Historia animalium
Mag. mor.	Magna moralia
Metaph.	Metaphysica
Mete.	Meteorologica
Mir.	Mirabilia
Mot. an.	De motu animalium
Mund.	De mundo
Oec.	Oeconomica
Part. an.	De partibus animalium
Phgn.	Physiognomica
Ph.	Physica
Poet.	Poetica
Pol.	Politica
Pr.	Problemata
Rh.	Rhetorica
Rh. Al.	Rhetorica ad Alexandrum
Sens.	De sensu
Somn.	De somno et vigilia
Soph. el.	Sophistici elenchi
Spir.	De spiritu
Top.	Topica
Aristox. Harm.	Aristoxenus, Harmonica
Arnob.	Arnobius, Adversus nationes
Arr. Anab.	Arrianus, Anabasis
Cyn.	Cynegeticus
Ind.	Indica
Peripl. p. eux.	Periplus ponti Euxini
Succ.	Historia successorum Alexandri
Tact.	Tactica
Artem.	Artemidorus
Ascon.	Asconius (Stangl Vol. 2, 1912)
Athan. ad Const.	Athanasius, Apologia ad Constantium
c. Ar.	Apologia contra Arianos
Fuga	Apologia de fuga sua
Hist. Ar.	Historia Arianorum ad monachos
Ath.	Athenaeus (Casaubon 1597) (List of books, pages, letters)
Aug. Civ.	Augustinus, De civitate dei
Conf.	Confessiones
Doctr. christ.	De doctrina christiana
Epist.	Epistulae
Retract.	Retractationes
Serm.	Sermones
Soliloq.	Soliloquia
Trin.	De trinitate
Aur. Vict.	Aurelius Victor
Auson. Mos.	Ausonius, Mosella (Peiper 1976)
Urb.	Ordo nobilium urbium
Avell.	Collectio Avellana
Avien.	Avienus
Babr.	Babrius
Bacchyl.	Bacchylides
Bar	Baruch
Bas.	Basilicorum libri LX (Heimbach)
Basil.	Basilius
Batr.	Batrachomyomachia
Bell. Afr.	Bellum Africum
Bell. Alex.	Bellum Alexandrinum
Bell. Hisp.	Bellum Hispaniense
Boeth.	Boethius
Caes. B Civ.	Caesar, De bello civili
B Gall.	De bello Gallico
Callim. Epigr.	Callimachus, Epigrammata
fr.	Fragmentum (Pfeiffer)
H.	Hymni
Calp. Ecl.	Calpurnius Siculus, Eclogae
Cass. Dio	Cassius Dio
Cassian.	Iohannes Cassianus
Cassiod. Inst.	Cassiodorus, Institutiones
Var.	Variae
Cato Agr.	Cato, De agri cultura
Orig.	Origines (HRR)
Catull.	Catullus, Carmina
Celsus, Med.	Cornelius Celsus, De medicina
Celsus, Dig.	Iuventius Celsus, Digesta
Censorinus, DN	Censorinus, De die natali
Chalcid.	Chalcidius
Charisius, Gramm.	Charisius, Ars grammatica (Barwick 1964)
1 Chr, 2 Chr	Chronicle
Chron. pasch.	Chronicon paschale
Chron. min.	Chronica minora
Cic. Acad. 1	Cicero, Academicorum posteriorum liber 1
Acad. 2	Lucullus sive Academicorum priorum liber 2
Ad Q. Fr.	Epistulae ad Quintum fratrem
Arat.	Aratea (Soubiran 1972)
Arch.	Pro Archia poeta
Att.	Epistulae ad Atticum
Balb.	Pro L. Balbo
Brut.	Brutus
Caecin.	Pro A. Caecina
Cael.	Pro M. Caelio
Cat.	In Catilinam
Cato	Cato maior de senectute
Clu.	Pro A. Cluentio
De or.	De oratore
Deiot.	Pro rege Deiotaro
Div.	De divinatione
Div. Caec.	Divinatio in Q. Caecilium
Dom.	De domo sua
Fam.	Epistulae ad familiares
Fat.	De fato
Fin.	De finibus bonorum et malorum
Flac.	Pro L. Valerio Flacco
Font.	Pro M. Fonteio
Har. resp.	De haruspicum responso
Inv.	De inventione
Lael.	Laelius de amicitia
Leg.	De legibus
Leg. agr.	De lege agraria
Lig.	Pro Q. Ligario
Leg. Man.	Pro lege Manilia (de imperio Cn. Pompei)
Marcell.	Pro M. Marcello
Mil.	Pro T. Annio Milone
Mur.	Pro L. Murena
Nat. D.	De natura deorum
Off.	De officiis
Opt. gen.	De optimo genere oratorum
Orat.	Orator
P. Red. Quir.	Oratio post reditum ad Quirites
P. Red. Sen.	Oratio post reditum in senatu
Parad.	Paradoxa
Part. or.	Partitiones oratoriae
Phil.	In M. Antonium orationes Philippicae
Philo.	Libri philosophici
Pis.	In L. Pisonem
Planc.	Pro Cn. Plancio
Prov. cons.	De provinciis consularibus
Q. Rosc.	Pro Q. Roscio comoedo
Quinct.	Pro P. Quinctio
Rab. perd.	Pro C. Rabirio perduellionis reo
Rab. Post.	Pro C. Rabirio Postumo
Rep.	De re publica
Rosc. Am.	Pro Sex. Roscio Amerino
Scaur.	Pro M. Aemilio Scauro
Sest.	Pro P. Sestio
Sull.	Pro P. Sulla
Tim.	Timaeus
Top.	Topica
Tull.	Pro M. Tullio
Tusc.	Tusculanae disputationes
Vatin.	In P. Vatinium testem interrogatio
Verr. 1, 2	In Verrem actio prima, secunda
Claud. Carm.	Claudius Claudianus, Carmina (Hall 1985)
Rapt. Pros.	De raptu Proserpinae
Clem. Al.	Clemens Alexandrinus
Cod. Greg.	Codex Gregorianus
Cod. Herm.	Codex Hermogenianus
Cod. Iust.	Corpus Iuris Civilis, Codex Iustinianus (Krueger 1900)
Cod. Theod.	Codex Theodosianus
Col	Letter to the Colossians
Coll.	Mosaicarum et Romanarum legum collatio
Colum.	Columella
Comm.	Commodianus
Cons.	Consultatio veteris cuiusdam iurisconsulti
Const.	Constitutio Sirmondiana
1 Cor, 2 Cor	Letters to the Corinthians
Coripp.	Corippus
Curt.	Curtius Rufus, Historiae Alexandri Magni
Cypr.	Cyprianus

List of abbreviations

Dan	Daniel
Demad.	Demades
Democr.	Democritus
Dem. Or.	Demosthenes, Orationes
Dig.	Corpus Iuris Civilis, Digesta (Mommsen 1905, author presented where applicable)
Din.	Dinarchus
Diod. Sic.	Diodorus Siculus
Diog. Laert.	Diogenes Laertius
Diom.	Diomedes, Ars grammatica
Dion. Chrys.	Dion Chrysostomus
Dion. Hal. Ant. Rom.	Dionysius Halicarnasseus, Antiquitates Romanae
Comp.	De compositione verborum
Rhet.	Ars rhetorica
Dionys. Per.	Dionysius Periegeta
Dion.	Thrax Dionysius Thrax
DK	Diels/Kranz (preceded by fragment number)
Donat.	Donatus grammaticus
Drac.	Dracontius
Dt	Deuteronomy = 5. Moses
Edict. praet. dig.	Edictum perpetuum in Dig.
EM	Etymologicum magnum
Emp.	Empedocles
Enn. Ann.	Ennius, Annales (Skutsch 1985)
Sat.	Saturae (Vahlen ^{2}1928)
Scaen.	Fragmenta scaenica (Vahlen ^{2}1928)
Ennod.	Ennodius
Eph	Letter to the Ephesians
Ephor.	Ephorus of Cyme (FGrH 70)
Epict.	Epictetus
Eratosth.	Eratosthenes
Esr	Esra
Est	Esther
Et. Gen.	Etymologicum genuinum
Et. Gud.	Etymologicum Gudianum
Euc.	Euclides, Elementa
Eunap. VS	Eunapius, Vitae sophistarum
Eur. Alc.	Euripides, Alcestis
Andr.	Andromache
Bacch.	Bacchae
Beller.	Bellerophon
Cyc.	Cyclops
El.	Electra
Hec.	Hecuba
Hel.	Helena
Heracl.	Heraclidae
HF	Hercules Furens
Hipp.	Hippolytus
Hyps.	Hypsipyle
Ion	Ion
IA	Iphigenia Aulidensis
IT	Iphigenia Taurica
Med.	Medea
Or.	Orestes
Phoen.	Phoenissae
Rhes.	Rhesus
Supp.	Supplices
Tro.	Troades
Euseb. Chron.	Eusebius Chronicon
Dem. evang.	Demonstratio Evangelica
Hist. eccl.	Historia Ecclesiastica
On.	Onomasticon (Klostermann 1904)
Praep. evang.	Praeparatio Evangelica
Eust.	Eustathius
Eutr.	Eutropius
Ev. Ver.	Evangelium Veritatis
Ex	Exodus = 2. Moses
Ez	Ezechiel
Fast.	Fasti
Fest.	Festus (Lindsay 1913)
Firm. Mat.	Firmicus Maternus
Flor. Epit.	Florus, Epitoma de Tito Livio
Florent.	Florentinus
Frontin. Aq.	Frontinus, De aquae ductu urbis Romae
Str.	Strategemata
Fulg.	Fulgentius Afer
Fulg. Rusp.	Fulgentius Ruspensis
Gai. Inst.	Gaius, Institutiones
Gal	Letter to the Galatians
Gal.	Galenus
Gell. NA	Gellius, Noctes Atticae
Geogr. Rav	Geographus Ravennas (Schnetz 1940)
Gp.	Geoponica
Gn	Genesis = 1. Moses
Gorg.	Gorgias
Greg. M. Dial.	Gregorius Magnus, Dialogi (de miraculis patrum Italicorum)
Epist.	Epistulae
Past.	Regula pastoralis
Greg. Naz. Epist.	Gregorius Nazianzenus, Epistulae
Or.	Orationes
Greg. Nyss.	Gregorius Nyssenus
Greg. Tur. Franc.	Gregorius of Tours, Historia Francorum
Mart.	De virtutibus Martini
Vit. patr.	De vita patrum
Hab	Habakkuk
Hagg	Haggai
Harpocr.	Harpocration
Hdt.	Herodotus
Hebr	Letter to the Hebrews
Hegesipp.	Hegesippus (= Flavius Josephus)
Hecat.	Hecataeus
Hell. Oxy.	Hellennica Oxyrhynchia
Hen	Henoch
Heph.	Hephaestio grammaticus (Alexandrinus)
Heracl.	Heraclitus
Heraclid. Pont.	Heraclides Ponticus
Herc. O.	Hercules Oetaeus
Herm.	Hermes Trismegistus
Herm. Mand.	Hermas, Mandata
Sim.	Similitudines
Vis.	Visiones
Hermog.	Hermogenes
Hdn.	Herodianus
Hes. Cat.	Hesiodus, Catalogus feminarum (Merkelbach/West 1967)
Op.	Opera et dies
Sc.	Scutum (Merkelbach/West 1967)
Theog.	Theogonia
Hil.	Hilarius
Hippoc.	Hippocrates
H. Hom.	Hymni Homerici
Hom. Il.	Homerus, Ilias
Od.	Odyssea
Hor. Ars P.	Horatius, Ars poetica
Carm.	Carmina
Carm. saec.	Carmen saeculare
Epist.	Epistulae
Epod.	Epodi
Sat.	Satirae (sermones)
Hos	Hosea
Hsch.	Hesychius
Hyg. Astr.	Hyginus, Astronomica (Le Boeuffle 1983)
Fab.	Fabulae
Hyp.	Hypereides
Iambl. Myst.	Iamblichus, De mysteriis
Protr.	Protrepticus in philosophiam
VP	De vita Pythagorica
Iav.	Iavolenus Priscus
Inst. Iust.	Corpus Juris Civilis, Institutiones (Krueger 1905)
Ioh. Chrys. Epist.	Iohannes Chrysostomus, Epistulae
Hom. ...	Homiliae in ...
Ioh. Mal.	Iohannes Malalas, Chronographia
Iord. Get.	Iordanes, De origine actibusque Getarum
Iren.	Irenaeus (Rousseau/Doutreleau 1965–82)
Is	Isaiah
Isid. Nat.	Isidorus, De natura rerum
Orig.	Origines
Isoc. Or.	Isocrates, Orationes
It. Ant.	Itinerarium, Antonini
Aug.	Augusti
Burd.	Burdigalense vel Hierosolymitanum
Plac.	Placentini
Iul. Vict. Rhet.	C. Iulius Victor, Ars rhetorica
Iuvenc.	Iuvencus, Evangelia (Huemer 1891)
Jac	Letter of James
Jdg	Judges
Jdt	Judith
Jer	Jeremiah
Jer. Chron.	Jerome, Chronicon
Comm.	in Ez. Commentaria in Ezechielem (PL 25)
Ep.	Epistulae
On.	Onomasticon (Klostermann 1904)
Vir. ill.	De viris illustribus
1 – 3 Jo	1st – 3rd letters of John
Jo	John
Jon	Jona
Jos. Ant. Iud.	Josephus, Antiquitates Iudaicae
BI	Bellum Iudaicum
Ap.	Contra Apionem
Vit.	De sua vita
Jos	Joshua
Jud	Letter of Judas
Julian. Ep.	Julianus, Epistulae
In Gal.	In Galilaeos
Mis.	Misopogon
Or.	Orationes
Symp.	Symposium
Just. Epit.	Justinus, Epitoma historiarum Philippicarum
Justin. Apol.	Justinus Martyr, Apologia
Dial.	Dialogus cum Tryphone
Juv.	Juvenalis, Saturae
1 Kg, 2 Kg	1, 2 Kings
KH	Khania (place where Linear B tables were discovered)
KN	Knossos (place where Linear B tables were discovered)
Lactant. Div. inst.	Lactantius, Divinae institutiones
Ira	De ira dei
De mort. pers.	De mortibus persecutorum
Opif.	De opificio dei
Lam	Lamentations
Lex Irnit.	Lex Irnitana
Lex Malac.	Lex municipii Malacitani
Lex Rubr.	Lex Rubria de Gallia cisalpina
Lex Salpens.	Lex municipii Salpensani
Lex Urson.	Lex coloniae Iuliae Genetivae Ursonensis
Lex Visig.	Leges Visigothorum
Lex XII tab.	Lex duodecim tabularum
Lib. Ep.	Libanius, Epistulae
Or.	Orationes
Liv.	Livius, Ab urbe condita
Per.	Periochae
Lk	Luke
Luc.	Lucanus, Bellum civile
Lucian. Alex.	Lucianus, Alexander
Anach.	Anacharsis
Cal.	Calumniae non temere credendum
Catapl.	Cataplus
Demon.	Demonax
Dial. D.	Dialogi deorum
Dial. meret.	Dialogi meretricium
Dial. mort.	Dialogi mortuorum
Her.	Herodotus
Hermot.	Hermotimus
Hist. conscr.	Quomodo historia conscribenda sit
Ind.	Adversus indoctum
Iupp. trag.	Iuppiter tragoedus
Luct.	De luctu
Macr.	Macrobii
Nigr.	Nigrinus
Philops.	Philopseudes
Pseudol.	Pseudologista
Salt.	De saltatione
Somn.	Somnium
Symp.	Symposium
Syr. D.	De Syria dea
Trag.	Tragodopodagra
Ver. hist.	Verae historiae, 1, 2
Vit. auct.	Vitarum auctio
Lucil.	Lucilius, Saturae (Marx 1904)
Lucr.	Lucretius, De rerum natura
Lv	Leviticus = 3. Moses
LXX	Septuaginta
Lycoph.	Lycophron
Lycurg.	Lycurgus
Lydus, Mag.	Lydus, De magistratibus
Mens.	De mensibus
Lys.	Lysias
M. Aur.	Marcus Aurelius Antoninus Augustus
Macrob. Sat.	Macrobius, Saturnalia
In Somn.	Commentarii in Ciceronis somnium Scipionis
1 Macc, 2	Macc Maccabees
Mal	Malachi
Manil.	Manilius, Astronomica (Goold 1985)
Mar. Vict.	Marius Victorinus
Mart.	Martialis
Mart. Cap.	Martianus Capella
Max. Tyr.	Maximus Tyrius (Trapp 1994)
Mela	Pomponius Mela
Melanipp.	Melanippides
Men. Dys.	Menander, Dyskolos
Epit.	Epitrepontes
fr.	Fragmentum (Körte)
Pk.	Perikeiromene
Sam.	Samia
Mi	Micha
Mimn.	Mimnermus
Min. Fel.	Minucius Felix, Octavius (Kytzler 1982, ^{2}1992)
Mk	Mark
Mod.	Herennius Modestinus
Mosch.	Moschus
Mt	Matthew
MY	Mycenae (place where Linear B tables were discovered)
Naev.	Naevius (carmina according to FPL)
Nah	Nahum
Neh	Nehemia
Nemes.	Nemesianus
Nep. Att.	Cornelius Nepos, Atticus

Hann.	Hannibal
Nic. Alex.	Nicander, Alexipharmaca
Ther.	Theriaca
Nicom.	Nicomachus
Nm	Numbers = 4. Moses
Non.	Nonius Marcellus (L. Mueller 1888)
Nonnus	Dion. Nonnus, Dionysiaca
Not. Dign. Occ.	Notitia dignitatum occidentis
Not. Dign. Or.	Notitia dignitatum orientis
Not. Episc.	Notitia dignitatum et episcoporum
Nov.	Corpus Iuris Civilis, Leges Novellae (Schoell/Kroll 1904)
Obseq.	Julius Obsequens, Prodigia (Rossbach 1910)
Opp. Hal.	Oppianus, Halieutica
Cyn.	Cynegetica
Or. Sib.	Oracula Sibyllina
Orib.	Oribasius
Orig.	Origenes
OrMan	Prayer to Manasseh
Oros.	Orosius
Orph. A.	Orpheus, Argonautica
fr.	Fragmentum (Kern)
H.	Hymni
Ov. Am.	Ovidius, Amores
Ars am.	Ars amatoria
Epist.	Epistulae (Heroides)
Fast.	Fasti
Ib.	Ibis
Medic.	Medicamina faciei femineae
Met.	Metamorphoses
Pont.	Epistulae ex Ponto
Rem. am.	Remedia amoris
Tr.	Tristia
P	Papyrus editions according to E.G. Turner, Greek Papyri. An Introduction, 159–178
P Abinn.	Papyrus editions according to H.I. Bell et al. (ed.), The Abinnaeus Archive papers of a Roman officer in the reign of Constantius II, 1962
P Bodmer	Papyrus editions according to V. Martin, R. Kassel et al. (ed.), Papyrus Bodmer 1954ff.
P CZ	Papyrus editions according to C.C. Edgar (ed.), Zenon Papyri (Catalogue général des Antiquités égyptiennes du Musée du Caire) 4 vols., 1925ff.
P Hercul.	Papyrus editions according to Papyri aus Herculaneum
P Lond.	Papyrus editions according to F.G. Kenyon et al. (ed.), Greek Papyri in the British Museum 7 vols., 1893–1974
P Mich	Papyrus editions according to C.C. Edgar, A.E.R. Boak, J.G.Winter et al. (ed.), Papyri in the University of Michigan Collection 13 vols., 1931–1977
P Oxy.	Papyrus editions according to B.P. Grenfell, A.S. Hunt et al. (ed.), The Oxyrhynchus Papyri, 1898 ff.
Pall. Agric.	Palladius, Opus agriculturae
Laus.	Historia Lausiaca
Pan. Lat.	Panegyrici Latini
Papin.	Aemilius Papinianus
Paroemiogr.	Paroemiographi Graeci
Pass. mart.	Passiones martyrum
Paul. Fest.	Paulus Diaconus, Epitoma Festi
Paul. Nol.	Paulinus Nolanus
Paulus, Sent.	Julius Paulus, Sententiae
Paus.	Pausanias
Pelag.	Pelagius
Peripl. m. eux.	Periplus maris Euxini
Peripl. m.m.	Periplus maris magni
Peripl. m.r.	Periplus maris rubri
Pers.	Persius, Saturae
1 Petr, 2 Petr	Letters of Peter
Petron. Sat.	Petronius, Satyrica (Müller 1961)
Phaedr.	Phaedrus, Fabulae (Guaglianone 1969)
Phil	Letter to the Philippians
Phil.	Philo
Philarg.Verg. ecl.	Philargyrius grammaticus, Explanatio in eclogas Vergilii
Philod.	Philodemus
Philostr. VA	Philostratus, Vita Apollonii
Imag.	Imagines
VS	Vitae sophistarum
Phlp.	Philoponus
Phm	Letter to Philemon
Phot.	Photius (Bekker 1824)
Phryn.	Phrynichus
Pind. fr.	Pindar, Fragments (Snell/Maehler)
Isthm.	Isthmian Odes
Nem.	Nemean Odes
Ol.	Olympian Odes
Pae.	Paeanes
Pyth.	Pythian Odes
Pl. Alc. 1	Plato, Alcibiades 1 (Stephanus)
Alc. 2	Alcibiades 2
Ap.	Apologia
Ax.	Axiochus
Chrm.	Charmides
Clit.	Clitopho
Crat.	Cratylus
Crit.	Crito
Criti.	Critias
Def.	Definitiones
Demod.	Demodocus
Epin.	Epinomis
Ep.	Epistulae
Erast.	Erastae
Eryx.	Eryxias
Euthd.	Euthydemus
Euthphr.	Euthyphro
Grg.	Gorgias
Hipparch.	Hipparchus
Hp. mai.	Hippias maior
Hp. mi.	Hippias minor
Ion	Ion
La.	Laches
Leg.	Leges
Ly.	Lysis
Men.	Menon
Min.	Minos
Menex.	Menexenus
Prm.	Parmenides
Phd.	Phaedo
Phdr.	Phaedrus
Phlb.	Philebus
Plt.	Politicus
Prt.	Protagoras
Resp.	Res publica
Sis.	Sisyphus
Soph.	Sophista
Symp.	Symposium
Thg.	Theages
Tht.	Theaetetus
Ti.	Timaeus
Plaut. Amph.	Plautus, Amphitruo (fr.according to Leo 1895 f.)
Asin.	Asinaria
Aul.	Aulularia
Bacch.	Bacchides
Capt.	Captivi
Cas.	Casina
Cist.	Cistellaria
Curc.	Curculio
Epid.	Epidicus
Men.	Menaechmi
Merc.	Mercator
Mil.	Miles gloriosus
Mostell.	Mostellaria
Poen.	Poenulus
Pseud.	Pseudolus
Rud.	Rudens
Stich.	Stichus
Trin.	Trinummus
Truc.	Truculentus
Vid.	Vidularia
Plin. HN	Plinius maior, Naturalis historia
Plin. Ep.	Plinius minor, Epistulae
Pan.	Panegyricus
Plot.	Plotinus
Plut.	Plutarchus, Vitae parallelae (with the respective name)
Amat.	Amatorius (chapter and page numbers)
De def. or.	De defectu oraculorum
De E	De E apud Delphos
De Pyth. or.	De Pythiae oraculis
De sera	De sera numinis vindicta
De Is. et Os.	De Iside et Osiride (with chapter and page numbers)
Mor.	Moralia (apart from the separately mentioned works; with p. numbers)
Quaest. Graec.	Quaestiones Graecae (with chapter numbers)
Quaest. Rom.	Quaestiones Romanae (with ch. numbers)
Symp.	Quaestiones convivales (book, chapter, page number)
Pol.	Polybius
Pol. Silv.	Polemius Silvius
Poll.	Pollux
Polyaenus, Strat.	Polyaenus, Strategemata
Polyc.	Polycarpus, Letter
Pompon.	Sextus Pomponius
Pomp. Trog.	Pompeius Trogus
Porph.	Porphyrius
Porph. Hor. comm.	Porphyrio, Commentum in Horatii carmina
Posidon.	Posidonius
Priap.	Priapea
Prisc.	Priscianus
Prob.	Pseudo-Probian writings
Procop. Aed.	Procopius, De aedificiis
Goth.	Bellum Gothicum
Pers.	Bellum Persicum
Vand.	Bellum Vandalicum
Arc.	Historia arcana
Procl.	Proclus
Prop.	Propertius, Elegiae
Prosp.	Prosper Tiro
Prov	Proverbs
Prudent.	Prudentius
Ps (Pss)	Psalm(s)
Ps.-Acro	Ps.-Acro in Horatium
Ps.-Aristot. Lin. insec.	Pseudo-Aristotle, De lineis insecabilibus
Mech.	Mechanica
Ps.-Sall. In Tull.	Pseudo-Sallustius, In M.Tullium Ciceronem invectiva
Rep.	Epistulae ad Caesarem senem de re publica
Ptol. Alm.	Ptolemy, Almagest
Geog.	Geographia
Harm.	Harmonica
Tetr.	Tetrabiblos
PY	Pylos (place where Linear B tablets were discovered)
4 Q Flor	Florilegium, Cave 4
4 Q Patr	Patriarch's blessing, Cave 4
1 Q pHab	Habakuk-Midrash, Cave 1
4 Q pNah	Nahum-Midrash, Cave 4
4 Q test	Testimonia, Cave 4
1 QH	Songs of Praise, Cave 1
1 QM	War list, Cave 1
1 QS	Communal rule, Cave 1
1 QSa	Community rule, Cave 1
1 QSb	Blessings, Cave 1
Quint. Decl.	Quintilianus, Declamationes minores (Shackleton Bailey 1989)
Inst.	Institutio oratoria
Quint. Smyrn.	Quintus Smyrnaeus
R. Gest. div. Aug.	Res gestae divi Augusti
Rhet. Her.	Rhetorica ad C. Herennium
Rom	Letter to the Romans
Rt	Ruth
Rufin.	Tyrannius Rufinus
Rut. Namat.	Rutilius Claudius Namatianus, De reditu suo
S. Sol.	Song of Solomon
Sext. Emp.	Sextus Empiricus
Sach	Sacharia
Sall. Catil.	Sallustius, De coniuratione Catilinae
Hist.	Historiae
Iug.	De bello Iugurthino
Salv. Gub.	Salvianus, De gubernatione dei
1 Sam, 2 Sam	Samuel
Schol. (before an author's name)	Scholia to the author in question
Scyl.	Scylax, Periplus
Scymn.	Scymnus, Periegesis
Sedul.	Sedulius
Sen. Controv.	Seneca maior, Controversiae
Suas.	Suasoriae
Sen. Ag.	Seneca minor, Agamemno
Apocol.	Divi Claudii apocolocyntosis
Ben.	De beneficiis
Clem.	De clementia (Hosius ^{2}1914)
Dial.	Dialogi
Ep.	Epistulae morales ad Lucilium
Herc. f.	Hercules furens
Med.	Medea
Q Nat.	Naturales quaestiones
Oed.	Oedipus
Phaedr.	Phaedra
Phoen.	Phoenissae
Thy.	Thyestes
Tranq.	De tranquillitate animi
Tro.	Troades
Serv. auct.	Servius auctus Danielis
Serv. Aen.	Servius, Commentarius in Vergilii Aeneida
Ecl.	Commentarius in Vergilii eclogas
Georg.	Commentarius in Vergilii georgica
Sext. Emp.	Sextus Empiricus
SHA Ael.	Scriptores Historiae Augustae, Aelius
Alb.	Clodius Albinus
Alex. Sev.	Alexander Severus
Aur.	M. Aurelius
Aurel.	Aurelianus
Avid. Cass.	Avidius Cassius

Car.	Carus et Carinus et Numerianus
Carac.	Antoninus Caracalla
Clod.	Claudius
Comm.	Commodus
Diad.	Diadumenus Antoninus
Did. Iul.	Didius Iulianus
Gall.	Gallieni duo
Gord.	Gordiani tres
Hadr.	Hadrianus
Heliogab.	Heliogabalus
Max. Balb.	Maximus et Balbus
Opil.	Opilius Macrinus
Pert.	Helvius Pertinax
Pesc. Nig.	Pescennius Niger
Pius	Antoninus Pius
Quadr. tyr.	Quadraginta tyranni
Sev.	Severus
Tac.	Tacitus
Tyr. Trig.	Triginta Tyranni
Valer.	Valeriani duo
Sid. Apoll. Carm.	Apollinaris Sidonius, Carmina
Epist.	Epistulae
Sil. Pun.	Silius Italicus, Punica
Simon.	Simonides
Simpl.	Simplicius
Sir	Jesus Sirach
Socr.	Socrates, Historia ecclesiastica
Sol.	Solon
Solin.	Solinus
Soph.	Aj. Sophocles, Ajax
Ant.	Antigone
El.	Electra
Ichn.	Ichneutae
OC	Oedipus Coloneus
OT	Oedipus Tyrannus
Phil.	Philoctetes
Trach.	Trachiniae
Sor. Gyn.	Soranus, Gynaecia
Sozom. Hist. eccl.	Sozomenus, Historia ecclesiastica
Stat. Achil.	Statius, Achilleis
Silv.	Silvae
Theb.	Thebais
Steph. Byz.	Stephanus Byzantius
Stesich.	Stesichorus
Stob.	Stobaeus
Str.	Strabo (books, chapters)
Suda	Suda = Suidas
Suet. Aug.	Suetonius, Divus Augustus (Ihm 1907)
Calig.	Caligula
Claud.	Divus Claudius
Dom.	Domitianus
Gram.	De grammaticis (Kaster 1995)
Iul.	Divus Iulius
Tib.	Divus Tiberius
Tit.	Divus Titus
Vesp.	Divus Vespasianus
Vit.	Vitellius
Sulp. Sev.	Sulpicius Severus
Symmachus, Ep.	Symmachus, Epistulae
Or.	Orationes
Relat.	Relationes
Synes. epist.	Synesius, Epistulae
Sync.	Syncellus
Tab. Peut.	Tabula Peutingeriana
Tac. Agr.	Tacitus, Agricola
Ann.	Annales
Dial.	Dialogus de oratoribus
Germ.	Germania
Hist.	Historiae
Ter. Maur.	Terentianus Maurus
Ter. Ad.	Terentius, Adelphoe
An.	Andria
Eun.	Eunuchus
Haut.	H(e)autontimorumenos
Hec.	Hecyra
Phorm.	Phormio
Tert. Apol.	Tertullianus, Apologeticum
Ad nat.	Ad nationes (Borleffs 1954)
TH	Thebes (place where Linear B tables were discovered)
Them. Or.	Themistius, Orationes
Theoc. Epigr.	Theocritus Epigrammata
Id.	Idyllia
Theod. Epist.	Theodoretus, Epistulae
Gr. aff. Cur.	Graecarum affectionum curatio
Hist. eccl.	Historia ecclesiastica
Theopomp.	Theopompus
Theophr. Caus. pl.	Theophrastus, De causis plantarum
Char.	Characteres
Hist. pl.	Historia plantarum
1 Thess, 2 Thess	Letters to the Thessalonians
Thgn.	Theognis
Thuc.	Thucydides
TI	Tiryns (place where Linear B tablets were discovered)
Tib.	Tibullus, Elegiae
1 Tim, 2 Tim	Letters to Timothy
Tit	Letter to Titus
Tob	Tobit
Tzetz.	Anteh. Tzetzes, Antehomerica
Chil.	Chiliades
Posth.	Posthomerica
Ulp.	Ulpianus (Ulpiani regulae)
Val. Fl.	Valerius Flaccus, Argonautica
Val. Max.	Valerius Maximus, Facta et dicta memorabilia
Varro Ling.	Varro, De lingua Latina
Rust.	Res rusticae
Sat. Men.	Saturae Menippeae (Astbury 1985)
Vat.	Fragmenta Vaticana
Veg. Mil.	Vegetius, Epitoma rei militaris
Vell. Pat.	Velleius Paterculus, Historiae Romanae
Ven. Fort.	Venantius Fortunatus
Verg. Aen.	Vergilius, Aeneis
Catal.	Catalepton
Ecl.	Eclogae
G.	Georgica
Vir. ill.	De viris illustribus
Vitr. De arch.	Vitruvius, De architectura
Vulg.	Vulgate
Wisd	Wisdom
Xen. Ages.	Xenophon, Agesilaus
An.	Anabasis
Ap.	Apologia
Ath. pol.	Athenaion politeia
Cyn.	Cynegeticus
Cyr.	Cyropaedia
Eq.	De equitandi ratione
Eq. mag.	De equitum magistro
Hell.	Hellenica
Hier.	Hiero
Lac.	Respublica Lacedaemoniorum
Mem.	Memorabilia
Oec.	Oeconomicus
Symp.	Symposium
Vect.	De vectigalibus
Xenoph.	Xenophanes
Zen.	Zeno
Zenob.	Zenobius
Zenod.	Zenodotus
Zeph	Zephania
Zon.	Zonaras
Zos.	Zosimus

Primary source corpora, periodicals, series

ABAW	Abhandlungen der Bayerischen Akademie der Wissenschaften. Philosophisch-historische Klasse
ABSA	Annual of the British School at Athens
AE	L'Année épigraphique
AJA	American Journal of Archaeology
AJPh	American Journal of Philology
ANET	*J.B. Pritchard*, Ancient Near Eastern Texts Relating to the Old Testament, ^{3}1969; Ndr. 1992
ANRW	*H. Temporini, W. Haase* (Hrsg.), Aufstieg und Niedergang der römischen Welt, 1972 ff.
AntAfr	Antiquités africaines
ArchHom	Archaeologia Homerica, 1967 ff.
AS	Anatolian Studies
AU	Der altsprachliche Unterricht
BaM	Baghdader Mitteilungen
BCH	Bulletin de Correspondance Hellénique
BJ	Bonner Jahrbücher des Rheinischen Landesmuseums in Bonn und des Vereins von Altertumsfreunden im Rheinlande
BSTH(B)	Bulletin archéologique du Comité des travaux historiques et scientifiques. (B) Afrique du Nord
Byzantion	Byzantion. Revue internationale des études byzantines
ByzZ	Byzantinische Zeitschrift
CAH	The Cambridge Ancient History, 12 Text- und 5 Tafelbde., 1924–39 (Bd. 1 als 2. Aufl.); Bde. 1–2, ^{3}1970–75; Bde. 3,1 und 3,3 ff., ^{2}1982 ff.; Bd. 3,2, ^{1}1991
CIG	Corpus Inscriptionum Graecarum, 4 Bde., 1828–77
CIL	Corpus Inscriptionum Latinarum, 1863 ff.
CM	Clio Medica. Acta Academiae internationalis historiae medicinae.
CPh	Classical Philology
CRAI	Comptes rendus des séances de l'Académie des Inscriptions et Belles-lettres
CW	The Classical World
DNP	Der neue Pauly. Enzyklopädie der Antike. Band 1–16, hg. von H. Cancik, H. Schneider und M. Landfester, Stuttgart 1996–2003
EA	Epigraphica Anatolica. Zeitschrift für Epigraphik und historische Geographie Anatoliens
EAZ	Ethnographisch-archäologische Zeitschrift
Emerita	Emerita. Revista de linguistica y filologia clasica
EncIr	*E. Yarshater* (Hrsg.), Encyclopaedia Iranica, 1985
Eranos	Eranos. Acta Philologica Suecana
FGrH	*F. Jacoby*, Die Fragmente der griechischen Historiker, 3 Teile in 14 Bden., 1923–58; Teil I: ^{2}1957
FHG	*C. Müller* (Hrsg.), Fragmenta Historicorum Graecorum, 5 Bde., 1841–70
Germania	Germania. Anzeiger der Römisch-Germanischen Kommission des Deutschen Archäologischen Instituts
GGM	*C. Müller* (Hrsg.), Geographi Graeci Minores, 2 Bde., Tabulae, 1855–61
GRBS	Greek, Roman and Byzantine Studies
Hermes	Hermes. Zeitschrift für klassische Philologie
Historia	Historia. Zeitschrift für Alte Geschichte
IEJ	Israel Exploration Journal
IG	Inscriptiones Graecae, 1873 ff.
IGLS	Inscriptions grecques et latines de la Syrie, 1929 ff.
ILCV	*E. Diehl* (Hrsg.), Inscriptiones Latinae Christianae veteres orientis, 3 Bde., 1925–31; Ndr. 1961; *J. Moreau, H.I. Marrou* (Hrsg.), Suppl., 1967
ILS	*H. Dessau* (Hrsg.), Inscriptiones Latinae selectae, 3 Bde. in 5 Teilen, 1892–1916; Ndr. ^{4}1974
JEA	The Journal of Egyptian Archaeology
JNES	Journal of Near Eastern Studies
JÖAI	Jahreshefte des Österreichischen Archäologischen Instituts
JRA	Journal of Roman Archaeology
JThS	Journal of Theological Studies
GGM	*C. Müller* (Hrsg.), Geographi Graeci Minores, 2 Bde., Tabulae, 1855–61
JRS	Journal of Roman Studies
Klio	Klio. Beiträge zur Alten Geschichte
Latomus	Latomus. Revue d'études latines
LMA	*R.-H. Bautier, R. Auty* (Hrsg.), Lexikon des Mittelalters
MBAH	Münstersche Beiträge zur antiken Handelsgeschichte
MDAI(Ist)	Istanbuler Mitteilungen des Deutschen Archäologischen Instituts
MDAI(K)	Mitteilungen des Deutschen Archäologischen Instituts (Abteilung Kairo)
MEFRA	Mélanges d'Archéologie et d'Histoire de l'École Française de Rome. Antiquité
MGH	Monumenta Germaniae Historica inde ab anno Christi quingentesimo usque ad annum millesimum et quingentesimum, 1826 ff.
NC	Numismatic Chronicle
ODB	*A.P. Kazhdan et al.* (Hrsg.), The Oxford Dictionary of Byzantium, 1991 ff.
OGIS	*W. Dittenberger* (Hrsg.), Orientis Graeci inscriptiones selectae 2 Bde., 1903–05; Ndr. 1960
PBSR	Papers of the British School at Rome
PdP	La Parola del Passato
PP	*W. Peremans* (Hrsg.), Prosopographia Ptolemaica (Studia hellenistica), 9 Bde., 1950–81; Ndr. Bd. 1–3, 1977
PropKg	*K. Bittel et al.* (Hrsg.), Propyläen Kunstgeschichte 22 Bde., 1966–80; Ndr. 1985
RAL	Rendiconti della Classe di Scienze morali, storiche e filologiche dell'Academia dei Lincei
RE	*G. Wissowa et al.* (Hrsg.), Paulys Real-Encyclopädie der classischen Altertumswissenschaft, Neue Bearbeitung, 1893–1980; *C. Frateantonio, M. Kopp, D. Sigel* et. al., Gesamtregister I. Alphabetischer Teil, 1997
REL	Revue des études latines
RGA	*H. Beck et al.* (Hrsg.), Reallexikon der germanischen Altertumskunde, ^{2}1973 ff. (1. Lfg. 1968), Ergänzungsbde. 1986 ff.
RhM	Rheinisches Museum für Philologie
RIC	*H. Mattingly, E.A. Sydenham*, The Roman Imperial Coinage, 10 Bde., 1923–94
RLA	*E. Ebeling et al.* (Hrsg.), Reallexikon der Assyriologie und vorderasiatischen Archäologie, 1928 ff.
RPh	Revue de philologie
SEG	Supplementum epigraphicum Graecum, 1923 ff.
SMEA	Studi Micenei ed Egeo-Anatolici
SNG	Sylloge Nummorum Graecorum
StV	Die Staatsverträge des Altertums Bd. 2: *H. Bengtson, R. Werner* (Hrsg.), Die Verträge der griechisch-römischen Welt von 700 bis 338, ^{2}1975; Bd. 3: *H.H. Schmitt (Hrsg.)*, Die Verträge der griechisch-römischen Welt 338 bis 200 v. Chr., 1969
Syll.^{3}	*F. Hiller von Gaertringen et al.* (Hrsg.), Sylloge inscriptionum Graecarum, 4 Bde., ^{3}1915–24; Ndr. 1960
TAVO	*H. Brunner, W. Röllig* (Hrsg.), Tübinger Atlas des Vorderen Orients, Beihefte, Teil B: Geschichte, 1969 ff.
Tyche	Zeitschrift für Epigraphik
TIR	Tabula Imperii Romani, 1934 ff.
WS	Wiener Studien. Zeitschrift für klassische Philologie und Patristik
ZPE	Zeitschrift für Papyrologie und Epigraphik

Maps and Commentaries

The Egyptian and Ancient Near Eastern concepts of the world

I. The world from the point of view of the Egyptian New Kingdom (*c.* 1570–1080 BC) (map A)

The Egyptian world view was Egyptocentric. Reflecting the culture's dualism, it centred on the two halves of the country, Ta-Mehu (Lower Egypt or the Nile Delta) and Ta-Shemau (Upper Egypt, i.e. the actual Nile Valley up to the first cataract at Elephantine). Both parts of the country were continuously reunited in the person of the king. This was symbolically represented by the linking of the armorial plants of the two kingdoms: the Lower Egyptian papyrus and the Upper Egyptian rush. There is no evidence that Egypt really consisted of these two parts before it achieved statehood.

Although Egypt proper never extended beyond this territory, the topos of 'expanding borders' became the dominant theme in the royal texts of the New Kingdom. The Egyptian kingdom did indeed reach its greatest territorial expansion in this age. Thutmosis I (1504–1492 BC) and Thutmosis III (1479–1425 BC) set up boundary stelae at Kurgus in Nubia south of the fourth cataract, about 250 km upstream from Gabal Barkal, as well as in Mesopotamia on the Euphrates near Karkemish; but these marked the furthest points of their advance, not the borders of their actual domain. As a matter of course this growth of territory was accompanied by an increase of knowledge about the conquered countries and peoples. Thus onomastica of geographical names, descriptions of the fates of Egyptian soldiers abroad and cultural and topographical guidelines were compiled at that time. Foreign places were invariably described as unpredictable or dangerous: the terrain was forbidding, wild animals and vermin posed a continuous threat, the water was not fit to drink. There was much longing to return home; above all, the nostalgia for Memphis became a literary topos.

In spite of existing knowledge about foreign countries, the notion of the 'Nine-Bow-Peoples', which stemmed from the Old Kingdom (*c.* 2680–2160 BC) was still alive. Here the bow symbolizes a society of hunters, and the figure nine signifies completeness. The 'Nine-Bow-Peoples' traditionally comprise, apart from
Ta-Mehu (Lower Egypt) and
Ta-Shemau (Upper Egypt),
the Hau-Nebu in the Aegean,
the Mentu-Setet in the Middle-East, outside the Egyptian sphere of influence (i.e. beyond the Euphrates),
the Temehu or Tehenu, who lived in Libya between the western delta and the Cyrenaica and north of the Egyptian oases,
Sechet-Jam, the oases of the Libyan desert,
Pedjtu-Shu, the eastern desert between the Nile and the Red Sea, the Iuntu-Seti in the area of the fourth cataract of the Nile and
Shat, beyond (i.e. south of) Kush, originally Sai Island, covering a region between the second and the third Nile cataracts.

This list of names is found approximately 120 times on various temple walls, pylons and other objects dating from the Middle Kingdom (from *c.* 2060 BC) to the end of Roman rule. One problem is that we do not know whether the 'Nine Bows' refer to countries or peoples; another problem stems from the imprecise use of those names and the change of meaning they underwent in the course of time. The more the Egyptians knew about the surrounding territories, the further away they placed the 'Nine Bows'; conversely, they moved closer to Egypt again as its influence on the surrounding countries began to wane after the New Kingdom.

Apart from designations of regions, the topographical lists quoted above contain individual place names, frequently written into ovals representing city walls. When the toponyms refer to the northern or north-eastern regions, the sites can usually be accurately identified, since here we have access to additional textual sources in local languages. As for the regions in the west and in the south, which have left no textual records of their own, most sites cannot be located nearly as accurately.

Furthermore, other types of texts contain traditionally different names for well-known regions, such as Wawat for Lower Nubia (i.e. below the second cataract) and Kush for Upper Nubia (above the second cataract). To the latter the epithet 'wretched' was frequently added. Retenu referred to Palestine, Charu was Syria and Hatti the term for the Hittite kingdom. Isi or Ires probably referred to Cyprus, although this is still a controversial point. Keftu can clearly be identified as Crete. Concerning the location of Punt, also entitled 'the land of God', a new suggestion has recently been presented: rather than on the coast of Somalia, it is now supposed to have been situated in the Arabian peninsula.

II. The world from the point of view of the Neo-Assyrian kingdom (8th/7th cents. BC) (map B) and the so-called Babylonian world map (*c.* first half of the 8th cent. BC) (supplementary map)

The ancient Mesopotamians did not have a science of geography in the modern sense. A general standard of proficiency in this field can therefore not be assumed. Even the job-related knowledge that merchants, military men and administrators had at their disposal may have overlapped only in certain areas. When scholars who had access to the written tradition were involved, contemporary, historical and mythological conceptions and ideas blended into a colourful, speculative amalgam. Two examples may serve to illustrate the highly divergent conceptions.

Map B includes the most remote countries that have so far been attested to in Assyrian inscriptions and letters from the 8th and 7th cents. BC. As would be expected, the scope of knowledge clearly ranged beyond the regions into which Assyria made military incursions. Countries a long way from the kingdom's border and as such outside its sphere of direct political interest are only mentioned by chance or in passing in the records. Thus the map can reflect only the minimum of geographical knowledge available at the Assyrian court. Actual geographical expertise may have ranged far beyond that.

In ancient Mesopotamia, rivers and the course of the sun were essential for orientation. Anything situated upstream was termed 'upper', anything downstream was seen as 'lower'. Therefore the Mediterranean was called 'Upper Sea' and the Persian Gulf 'Lower Sea'.

With Euphrates and Tigris both flowing from northwest to southeast, the windrose, which was based on this flow, was turned by one eighth to the left compared to ours. Thus it pointed to the north-west, north-east, south-east and south-west. The four directions of the wind were supplemented by the directions of sunrise (east) and sunset (west).

In the supplementary map, the graphic representation is based on the map sketched by a bookish Babylonian scholar, probably of the 8th cent. BC. The scope of knowledge compared to that shown by map B (see green contour line) is shockingly narrow: while the inner circle – consisting of Babylonia and neighbouring Elam (Susa) – is more or less identical (with map B), Assyria and Urartu are remote, nebulous marginal shapes. The biggest surprise is the complete absence of the Syrian region, which had always been in close contact with Mesopotamia. The perfunctory execution of the sketch, however, demonstrates that the author never intended to create an accurate map. His sole aim seems to have been to point out that the circular area surrounded by the ocean should be viewed as the known part of the world. The badly damaged text found on the same tablet treats of the conditions which prevail in the triangular areas bordering on the outside of the Ocean. The author was clearly not interested in passing on geographical knowledge about the known world, but was speculating about the shape of the world as it might appear to an outside observer. The representation as a star in the shape of two concentric circles with triangular shapes attached to its outside essentially matches the contemporary symbols of the gods Sīn (Moon), Šamaš (Sun) and Ištar (Venus). Obviously the world inhabited by humans is seen as one heavenly body among many others. The sketch depicts only one of many competing ancient Mesopotamian structural models of the universe. It is unique in that it did not have any successor. Whereas the concept of a circular world surrounded by an ocean was current in Antiquity and by its influence in the Middle Ages, the shape of the star remained a singular occurrence.

So-called Babylonian World Map (*c.* 1st half of 8th cent. BC)

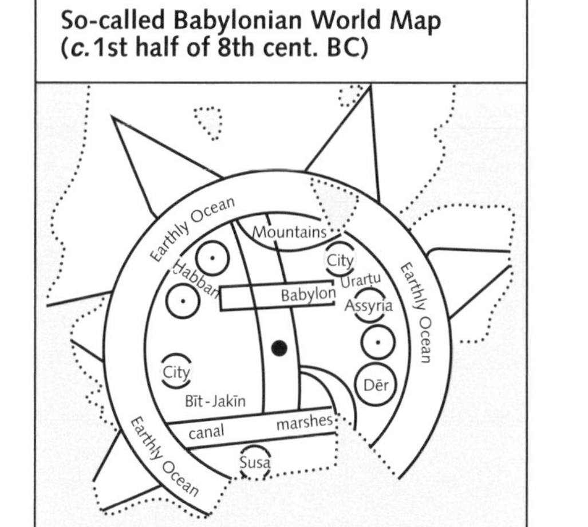

→ Maps pp. 17, 19, 21, 53, 55, 59

Literature

Egyptian 'map of the world': E. Edel (†), M. Görg, Die Ortsnamenlisten im nördlichen Säulenhof des Totentempels Amenophis' III., 2005; N. Grimal, Les listes des peuples dans l'Égypte du deuxième millénaire av. J.-C. et la géopolitique du Proche Orient, in: E. Czerny et al. (eds.), Timelines. Studies in Honour of M. Bietak, I, 2006, 107–119; D. O'Connor, S. Quirke (eds.), Mysterious Lands, 2003; E. Uphill, The Nine Bows, in: Jaarbericht van het Vooraziatisch-Egyptisch Genootschap, Ex Oriente Lux 19, 1965–66, 393–420; D. Valbelle, Les neuf arcs. L'égyptien et les étrangers de la préhistoire à la conquête d'Alexandre, 1990. *The Assyrian world picture and the Babylonian map of the world*: W. Horowitz, Mesopotamian Cosmic Geography, 1998; B. Janowski, B. Ego (eds.), Das biblische Weltbild und seine altorientalischen Kontexte, 2001.

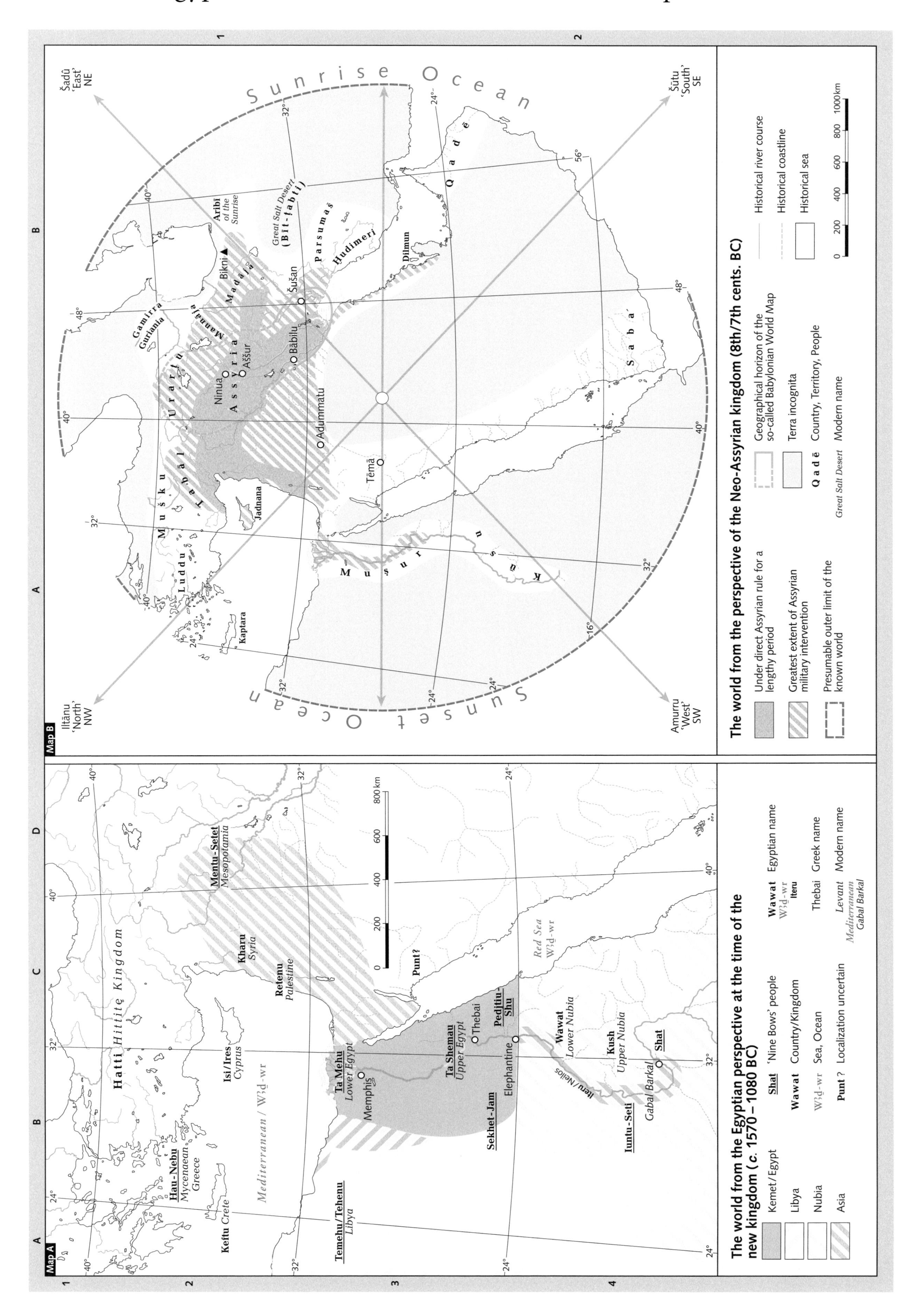
Map A
The world from the Egyptian perspective at the time of the new kingdom (c. 1570 – 1080 BC)
Kemet / Egypt
Libya
Nubia
Asia
Shat 'Nine Bows' people
Wawat Country/Kingdom
Wꜣḏ-wr Sea, Ocean
Punt ? Localization uncertain
Wawat Egyptian name
Wꜣḏ-wr
Iteru
Thebai Greek name
Levant
Mediterranean
Gabal Barkal
Modern name
Hatti Hittite Kingdom
Hau-Nebu
Mycenaean Greece
Keftu Crete
Isi/Ires
Cyprus
Mediterranean / Wꜣḏ-wr
Temehu / Tehenu
Libya
Ta Mehu
Lower Egypt
Memphis
Ta Shemau
Upper Egypt
Thebai
Pedjtiu-Shu
Elephantine
Sekhet-Jam
Wawat
Lower Nubia
Kush
Upper Nubia
Shat
Gabal Barkal
Iuntu-Seti
Iteru / Neilos
Retenu
Palestine
Kharu
Syria
Mentu-Setet
Mesopotamia
Punt?
Red Sea
Wꜣḏ-wr
0 200 400 600 800 km
Map B
The world from the perspective of the Neo-Assyrian kingdom (8th/7th cents. BC)
Under direct Assyrian rule for a lengthy period
Greatest extent of Assyrian military intervention
Presumable outer limit of the known world
Geographical horizon of the so-called Babylonian World Map
Terra incognita
Qadê Country, Territory, People
Great Salt Desert Modern name
Historical river course
Historical coastline
Historical sea
0 200 400 600 800 1000 km
Sunrise Ocean
Sunset Ocean
Iltānu 'North' NW
Šadû 'East' NE
Šūtu 'South' SE
Amurru 'West' SW
Gamirra
Guriania
Mannaja
Madaja
Bikni
Aribi of the Sunrise
Great Salt Desert (Bit-ṭabti)
Parsumaš
Hudimeri
Dilmun
Qadê
Urarṭu
Mušku
Tabāl
Luddu
Assyria
Ninua
Aššur
Bābilu
Šušan
Adummatu
Têmâ
Jadnana
Kaptara
Muṣur
Kūsu
Sabaʾ

The world through the eyes of ancient authors

The world through the eyes of ancient authors

Information about cartography in the Graeco-Roman world has come down to us from a number of ancient writers, but there are only a few rare cases of actual maps surviving, and even those have only been preserved in fragmentary form.

In the present context the ongoing debate about the scope and purpose of these maps will have to be be put aside, as the focus is on the geographical world picture that emerges – i.e. can be reconstructed – from these maps and from what can be derived – i.e. can be reconstructed – from their portrayal in the writings of the ancients.

The use of the term 'reconstruction' could be misleading, as maps were not attached to the works in question, as we are told explicitly e.g. by Hellanicus, an author of ethnographic writings from the 5th cent. BC (Dion. Hal. Rhet. 10,11). Thus when we talk about 'reconstructions' in the cases of e.g. Herodotus and probably Erathosthenes as well, they refer to the respective authors' notions of the world, which were recorded textually, not cartographically.

I. Hecataeus (6th/5th cents. BC, map A)

The first map of the world is believed to have been drawn up by Anaximander of Miletus in the early 6th cent.; its reconstruction draws on surviving fragments of the geographical work of Hecataeus, who probably was one of his students.

Hecataeus enriched the map with information he had collected on his travels. The representation is characterized by rigid geometrical concepts: The perfectly circular earth disk is divided by the Mediterranean into two equally large segments, i.e. Europe and Asia, which encompasses Libya. It is encircled by the currents of Oceanus which is connected with the Mediterranean and the Red Sea as well as the Caspian Sea (*Kaspia thalatta*). It (Oceanus) is also the source of the Nilus/Nile.

While the Ister (Danube) runs down the middle of Europe, the Nile in similar fashion splits Asia into two halves. In the face of the problematic nature of the source material, Felix Jacoby was the first to warn that it would be very difficult to reconstruct Hecataeus' map. Aristagoras may have taken along a bronze copy of that map or a comparable one in 499 BC when as an envoy of his hometown of Miletus he asked the Spartans to support the Ionian uprising against the Great King of Persia (Hdt. 5, 49). Such maps were rather common at the time; the Spartans were probably familiar with them, as Anaximander himself had paid an earlier visit to Sparta. The map offered an impressive visualization both of the size and of the wealth of the Persian empire (see supplementary map). The Spartans were stunned by the vastness of the Persian empire. Marching for three months to cover the distance from the Aegean coast to the Great King's residence at Susa was something even they could not stomach, which is why they expelled Aristagoras from their city the same day.

II. Herodotus (5th cent. BC, map B)

Herodotus' geographical world picture differs markedly from Anaximander's and Hecataeus' version: Both Ister (Danube) and Nilus(Nile) run almost parallel from their respective sources in the west; an array of seas ranging from the Pillars of Hercules to the east coast of Lake Maeotis (Sea of Azov) running parallel to these two rivers separates the two land masses of Europe in the north and Libya in the south with Asia occupying the east.

To the south the African and Asian land masses are bordered by Oceanus which links up with the Mediterranean at the Pillars of Hercules (Strait of Gibraltar). The view still held by Hecataeus that the earth was an island completely encircled by Oceanus was abandoned by Herodotus, who describes the Caspian Sea as an inland body of water, no longer with any connection to Oceanus.

III. Eratosthenes (3rd cent. BC, map C)

In the third volume of his *Geographiká*, fragments of which have survived, this Cyrenaean polymath created a map of the Earth by drawing on his knowledge of mathematics and astronomy. He followed Dicaearchus (4th cent. BC) in that he chose as his line of orientation the parallel which runs through Rhodes along the Taurus to India (≈ 36° northern latitude). The positions of the other six parallels – the southernmost crosses Taprobane (today's Sri Lanka), the northernmost runs through Thule – he worked out by comparing temperatures and types of vegetation. On his map the Indus does not flow from the north-west to the south-east, but from north to south. His Caspian Sea again has a passage opening up to Oceanus. The idea that the Earth is a sphere had been around from the 6th cent. and was generally accepted by the 4th cent. BC. As evidenced in his treatise on the measurement of the Earth, its spherical shape was a given to Eratosthenes, but this is not reflected at all in the design of his map.

IV. Claudius Ptolemaeus (2nd cent. AD, map D)

Marinus of Tyre (2nd cent. AD) and his slightly younger contemporary Ptolemy represent the high point of scientific geography in Antiquity. Both geographers considered all seas as inland seas within a single *oikoumene* (=inhabited world). Marinus' map using transverse cylindrical projection displays a rectangular grid of latitudinal circles running parallel to the latitudinal circle of Rhodes (≈ 36° of northern latitude) and of meridians spaced out eastwards from the prime meridian running through the Islands of the Blessed in the west. In this type of projection the conversion of the meridians as they approach the poles is ignored. In all there are 16 meridians crossing the parallel of Rhodes at a right angle. They are distributed equally at a distance (≈ 15°) that translates into a difference of one hour for sunrise and sunset from one meridian to the next. Making comparatively sparing use of astronomical observations, Marinus gleaned his distance data mainly from travel reports and itineraries recording mileages and numbers of days spent marching or sailing.

Ptolemy's *Geographike Hyphegesis* is the oldest surviving set of instructions for mapping the Earth by applying mathematics and astronomy according to a consistent and systematic plan. His work builds on the foundations laid by Marinus, even though he was critical of some aspects of the latter's work. It is not improbable that Ptolemy included maps in his work. In fact some, 13th-cent. Greek MSS have come down to us with maps attached to them. The authenticity of those maps had been generally denied until recently, when a strong case was made for their being genuine. Ptolemy construed both a simpler projection (line of orientation: the Rhodes parallel; the parallels are drawn concentrically around the pole of orientation with meridians of equal length extending from the pole of projection) and a more complex modified conic one (line of orientation: the Syene parallel, curved meridians – see map D). (Ptol. 1,23 and 24,1-8) The material for this grid is found in books 2-7 where the co-ordinates for places in Europe, Libya and Asia are provided. Book 8 supplies information on the content of 26 section maps.

World Map of Hecataeus

Map A

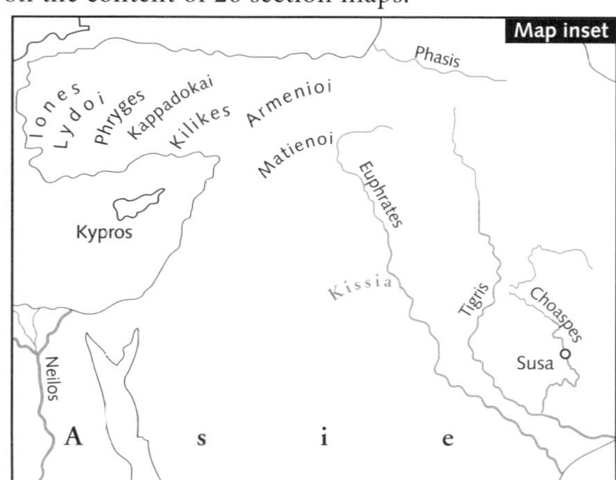

India were shipped down the Oxus to the Caspian Sea, from where they were transported up the river Kyros/Kura to the Black Sea. The Oxus mentioned in this context is hard to identify. One of the rivers emptying into the Caspian near Turkmenbashi may have been mistaken for the river whose upper reaches are known under the name of Amu-Darya, a left branch of which (Wadi Usboi) did in fact still empty into the Caspian Sea in the Neolithic age.

21. Megasthenes, an official of Seleucus I (312–281 BC), was delegated to be ambassador at Palimbothra, the residence of the Maurya ruler Sancracottus (320–298 BC) between 305 or 303/02 and 298 BC. There he collected information about the country and its inhabitants and recorded it in his *Indika,* which has survived in fragmentary form.

22. Daïmachus of Plataeae, an official of Seleucus I (312–281 BC) and Antiochus I (292 or 281–261 BC), was delegated to Palimbothra as an ambassador to Amitrochates after 298 BC. The written memories of his journey and stay in India have come down to us only in fragmentary form.

The Far East

23. It was probably in the first cent. AD that Maës Titianus, a Macedonian trader, travelled along the Silk Road, which had been used since the 4th cent. BC. His eastward route took him from Hierapolis on the Euphrates (Membidj, Syria), Dura-Europus (aṣ-Ṣāliḥiya, Syria), Hecatompylus (Sahr-i Qumis, Iran), Antiochia (Erk Kala/Gyaur Kala, Turkmenistan) and Baktra (Balkh, Afganistan) to Pyrgus Lithinus (in the Pamir mountains; location not identified); there he sought information about the remaining distance to China by questioning travellers who were about to continue to Sera (possibly Luoyang).

The map

In the present context 'long-distance exploration' refers to travels going beyond the geographical framework of the known world at the time in question. This includes exploratory expeditions like Megasthenes' Indian journeys (between 305 or 303/02 and 298 BC), but also odysseys like Colaeus' ending up at Tartessus (7th cent. BC), searches for trading opportunities like Patrocles' sailing across the Caspian Sea (285/282 BC), military campaigns like Alexander's march to India in 326/25 BC (which also had an exploratory angle to it) or the purely military March of the Ten Thousand (402/01 BC).

Repeatedly certain political and social developments (stemming from different motives) either triggered or caused a flurry of such exploratory activities – such as the development of Carthage into a major power in the Western Mediterranean (7th/6th cents. BC) or the efforts made by the Roman emperors from Augustus to Domitian to secure the empire's African border.

The sources

Information about long-distance exploration can be found in Herodotus, who was interested in anything foreign, and in the works of various geographers such as Strabo and Ptolemy, but also in the Elder Pliny's *Natural History.* Of many works by ancient geographers and of the so-called *periploi*, i.e. Greek descriptions of coasts, only fragments or abridged versions survive. They are accessible to today's readers only in a variety of separate and partly quite old, publications.

The *Geographi Graeci Minores* by C. Müller (1882) and the collection of fragments by A. Diller (The Tradition of the Minor Greek Geographers, 1951, 102–146) are still important works of reference. An essential tool still awaiting completion is the collection of fragments by F. Jakoby; it so far covers only three areas (III A: Autoren über verschiedene Städte (Länder), 1940; III B: Autoren über einzelne Städte (Länder), 1950, III C: Autoren über einzelne Länder (Ägypten bis Geten und Illyrien bis Thrakien), 1958). There are also useful collections of fragments of the writings of certain geographers or seafarers like Posidonius (*About the Ocean and People Living on Its Shores*: W. Theiler, ed., Poseidonios, Die Fragmente 1, 1982, 6–82; 2, 1982, 6–78) or Pytheas of Massalia (S. Bianchetti, ed., Pitea di Massalia, L'Oceano, 1998, with Italian translation and commentary).

Sources and literature

General: P. Arnaud, Pouvoir des mots et limites de la cartographie dans la géographie grecque et romaine, in: Dialogues d'histoire ancienne 15, 1989, 9–29; K. Brodersen, Terra Cognita, ²2003; P. Fabre, Les Grecs et la connaissance de l'Occident. Le mythe occidental, 1981; K.M. Girardet, Kontinente und ihre Grenzen in der griechisch-römischen Antike, in: S. Penth et al. (eds.), Europas Grenzen (Limites 1), 2006, 19–65; B. Isaac, The Limits of Empire. The Roman Army in the East, 1990; T. Kotula, Le monde romain et ses périphéries sous la République et sous l'Empire, 2001; J.S. Romm, The Edges of the Earth in Ancient Thought. Geography, exploration and fiction, 1992; H. Sonnabend, Die Grenzen der Welt. Geographische Vorstellungen der Antike, 2007; D. Timpe, Entdeckungsgeschichte, in: RGA 7, 1989, 307–389.

Northern Europe: **1. Colaeus:** Hdt. 4,152; cf. Plin. HN 7,197. – Literature: A. Laronde, Cyrène et la Libye hellénistique. Libykai historiai de l' époque républicaine au principat d'Auguste, 1987, 223f.; A. Tovar, Iberische Landeskunde II.1, 1974, 70. **2. Himilko:** Avien. 113–129; 380–389; 406–413; cf. 91; 96; 154; Plin. HN 2,169. – Literature: K. Geus, Himilko Nr. 1, in: Prosopographie der literarisch bezeugten Karthager (Studia Phoenicia 13; Orientalia Lovaniensia Analecta 59), 1994, 157–159; Huss, 84f. **3. Pytheas:** S. Bianchetti (ed.), Pitea di Massalia, L'Oceano, 1998 (with introduction, Italian translation and commentary); Ead., La geografia di Pitea e la diorthosis di Polibio, in: G. Schepens, J. Ballansée (eds.), The Shadow of Polybius (Studia Hellenistica 42) 2005, 257–270. – Literature: S. Magnani, Il viaggio di Pitea sull'Oceano, 2002.

Northwestern Africa: **4. Hanno:** Aristot. Mir. 833a 11; Arr. Ind. 43,11f.; Hanno, Periplus, ed. by K. Bayer, in: G. Winkler, R. König (eds.), C. Plinius Secundus der Ältere, Naturkunde 5, 1993, 337–353 (text, translation, commentary; bibliography 360–363; cf. GGM I, 1–14); Mela 3,90; 93; Plin. HN 2,169; 5,8. – Literature: K. Geus, Hanno Nr. 3, in: Prosopographie der literarisch bezeugten Karthager (Studia Phoenicia 13; Orientalia Lovaniensia Analecta 59), 1994, 98–105; Huss, 75–83. **5. C. Suetonius Paullinus:** Cass. Dio 60,9,1; Solin. 24,15; Plin. HN 5,11–15. – Literature: B.E. Thomasson, Fasti Africani. Senatorische und ritterliche Amtsträger in den römischen Provinzen Nordafrikas von Augustus bis Diokletian, 1996, 197f. no. 2.

Central Northern Africa: **6. Mago:** Aristot. fr. 103 Rose (= Athen. 2,22). – Literature: K. Geus, Hanno Nr. 3, in: Prosopographie der literarisch bezeugten Karthager (Studia Phoenicia 13; Orientalia Lovaniensia Analecta 59), 1994, 179f.; W. Huss, Die antike Mittelmeerwelt und Innerafrika, in: H. Duchhardt et al. (eds.), Afrika, 1989, 1–29. **7. L. Cornelius Balbus:** Plin. HN 5,36f. – Literature: J. Desanges, Le triomphe de Cornélius Balbus, 19 av. J.-C., in: Revue Africaine 101, 1957, 5–43; P.J. Holiday, Roman Triumphal Paintings, in: The Art Bulletin 79, 1997, 130–147; H. Lhote, L'expédition de Cornélius Balbus au Sahara 19 av. J.-C., in: Revue Africaine 98, 1954, 41–83; B.E. Thomasson, Fasti Africani. Senatorische und ritterliche Amtsträger in den römischen Provinzen Nordafrikas von Augustus bis Diokletian, 1996, 21 no. 4. **8. Septimius Flaccus:** Ptol. 1,8,5f. – Literature: J.L. Berggren, A. Jones, Ptolemy's Geography. An Annotated Translation of the Theoretical Chapters, 2000, 145–147; J. Desanges, Recherches sur l'activité des Méditerranéens aux confins de l'Afrique (VIe siècle avant J.-C. – IVe siècle après J.-C.), 1978, 197–213; Id., Rom und das Innere Afrikas, in: H. Duchhardt et al. (eds.), Afrika, 1989, 31–50.

Northeastern Africa: **9. Hatshepsut:** sources with bibliography in M. Werbrouck, Le temple d'Hatshepsout à Deir el-Bahari, 1949; R. Herzog, Punt, 1968. **10. Necho II:** Diod. Sic. 1,33,9; Hdt. 4,42; cf. 2,158f.; Str. 17,1,25 (construction of the canal). – Literature: W. Huss, Die antike Mittelmeerwelt und Innerafrika bis zum Ende der Herrschaft der Karthager und der Ptolemaier, in: H. Duchhardt et al. (eds.), Afrika, 1989, 1–29; A.B. Lloyd, Herodotus, Book II, Commentary, 1988, 149–158; J. Moje, Die angebliche phönizische Umseglung Afrikas im Auftrag des Pharaos Necho, in: A.I. Blöbaum, J. Kahl, S.D. Schweitzer (eds.), Festschrift E. Graefe, 2003, 197–209. **11. Eudoxus:** Nep. ap. Mela 3,90; 92; Plin. HN 2,169; 6,188; Posidon. ap. Str. 2,3,4. – Literature: W. Otto, O. Bengtson, Zur Geschichte des Niedergangs des Ptolemäerreiches. Ein Beitrag zur Regierungszeit des 8. und des 9. Ptolemäers (ABAW 17, 1938), 194ff.; J.H. Thiel, Eudoxus of Cyzicus, 1966 (English translation of the 1939 Dutch version). **12. Diogenes:** Marinus of Tyre ap. Ptol. 1,9. – Literature: J.L. Berggren, A. Jones, Ptolemy' s Geography. An Annotated Translation of the Theoretical Chapters, 2000, 68 with note 33. **13. P. Petronius:** Cass. Dio 54,5,4f.; Plin. HN 6,181f.; Str. 17,1,54; cf. R. Gest. div. Aug. 26; Prop. 4,6,78. – Literature: A.E.P. Weigall, A Report on the Antiquities of Lower Nubia, 1907; cf. K. Buschmann, TAVO B V 22, 1987, additional map. **14. Nero:** Cass. Dio 63,8,1; Plin. HN 6,181; 6,184f.; 12,19: Aethiopiae forma; Sen. Q. nat. 6,8,3–5 – Literature: R. Hennig, Terrae incognitae 1, ²1944, 356–362.

Arabia: **15. Scylax:** Hdt. 4,44; cf. 3,102; Hecataeus FGrH 1 F 295; the fragments of his *Periplous* are in F. Jacoby, FGrH 709. – Literature: A. Peretti, Dati storici e distanze marine nel Periplo di Scilace, in: Studi Classici e Orientali 38, 1988, 13–137. **16. L. Aelius Gallus:** Plin. HN 6,160; R. Gest. div. Aug. 26,5; Str. 2,5,12; 16,4,22–24. – Literature: H. von Wissmann, Die Geschichte des Sabäerreichs und der Feldzug des Aelius Gallus, in: ANRW II 9,1, 1976, 308–544.

The Near East: **17. March of the Ten Thousand:** Xen. An. – Literature: T. Mitford, Thalatta, Thalatta. Xenophon's view of the Black Sea, in: Anatolian Studies 50, 2000, 127–132.

India: **18. Alexander the Great:** the sources are stated in J. Seibert (cf. below); for Scylax of Caryanda, cf. FGrH 709. – Literature: J. Hahn, Alexander in Indien, 2000; J. Seibert, Die Eroberung des Perserreiches durch Alexander den Großen auf kartographischer Grundlage, Darstellungs- und Kartenband, 1985, 155–184, with maps 26f. **19. Nearchus:** FGrH 133 Nearchos F 1–28. – Literature: E. Badian, Nearchus the Cretan, in: Yale Classical Studies 24, 1975, 147–170; H. Berve, Das Alexanderreich auf prosopographischer Grundlage 2, 1926, 269–272 no. 544; J. Seibert, Das Alexanderreich (336–323 v. Chr.), TAVO B V 1, 1985. **20. Patrocles:** Memnon ap. Phot., Bibl. 224; Plin. HN 6,58; cf. Eratosthenes fr. III B 68 Berger; FGrH 712 Patrokles F 4f. – Literature: H. Berger, Die geographischen Fragmente des Eratosthenes, 1880, 94–97; M.U. Erdsoy, in: R.J.A. Talbert, Barrington Atlas of the Greek and Roman World, 2000, commentary (CD-ROM) on map 8, pp. 77f.; K.J. Neumann, Die Fahrt des Patrokles auf dem Kaspischen Meere und der alte Lauf des Oxos, in: Hermes 19, 1884, 165–185. **21. Megasthenes:** Arr. Anab. 5,6,2; Ind. 5,3; Str. 2,1,9; 15,1,36; cf. FGrH 715 Megasthenes. – Literature: A. Mehl, Seleukos Nikator und sein Reich (Studia Hellenistica 28) 1986, 186–191; A. Kuhrt, S. Sherwin-White, From Samarkhand to Sardis, 1993, 91–113; E. Olshausen, Prosopographie der hellenistischen Königsgesandten 1 (Studia Hellenistica 19) 1974, 172–174 no. 127; D. Panchenko, Scylax' Circumnavigation of India and its interpretation in early Greek geography, ethnography and cosmography 1, in: Hyperboreus 4, 1998, 211–242. **22. Daïmachus:** Str. 2,1,9; cf. FGrH 716 Daïmachos. – Literature: A. Mehl, Seleukos Nikator und sein Reich (Studia Hellenistica 28) 1986, 186–191; E. Olshausen, Prosopographie der hellenistischen Königsgesandten 1 (Studia Hellenistica 19) 1974, 171f.

The Far East: **23. Maës Titianus:** Amm. Marc. 23,6,60; Ptol. 1,11,4 and 7; 1,12,8; 6,13,2. – Literature: J.L. Berggren, A. Jones, Ptolemy's Geography. An annotated translation of the theoretical chapters, 2000, 150–152.

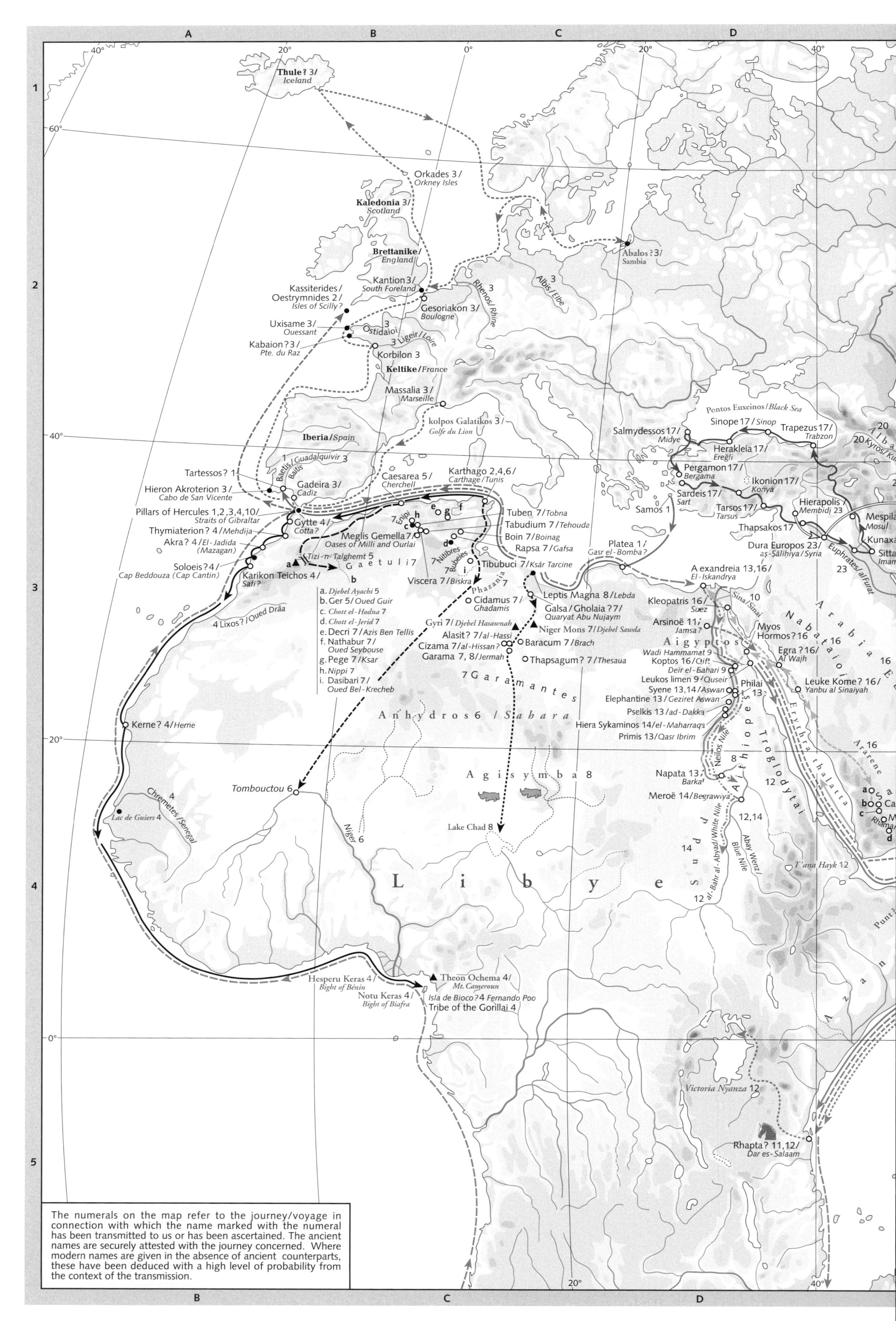

The numerals on the map refer to the journey/voyage in connection with which the name marked with the numeral has been transmitted to us or has been ascertained. The ancient names are securely attested with the journey concerned. Where modern names are given in the absence of ancient counterparts, these have been deduced with a high level of probability from the context of the transmission.

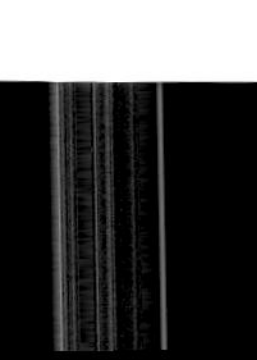

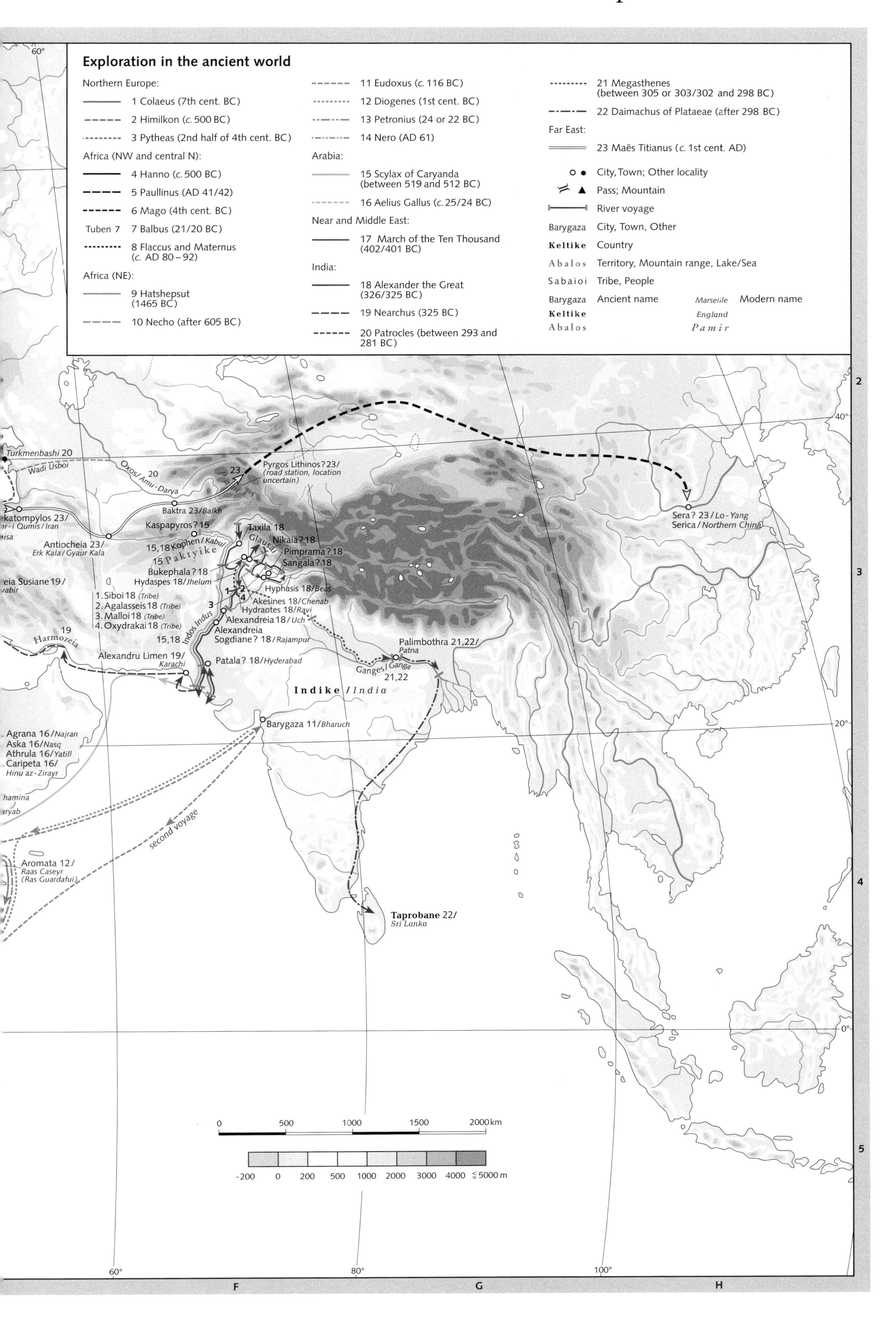
Exploration in the ancient world
Northern Europe:
1 Colaeus (7th cent. BC)
2 Himilkon (c. 500 BC)
3 Pytheas (2nd half of 4th cent. BC)
Africa (NW and central N):
4 Hanno (c. 500 BC)
5 Paullinus (AD 41/42)
6 Mago (4th cent. BC)
Tuben 7 7 Balbus (21/20 BC)
8 Flaccus and Maternus (c. AD 80–92)
Africa (NE):
9 Hatshepsut (1465 BC)
10 Necho (after 605 BC)
11 Eudoxus (c. 116 BC)
12 Diogenes (1st cent. BC)
13 Petronius (24 or 22 BC)
14 Nero (AD 61)
Arabia:
15 Scylax of Caryanda (between 519 and 512 BC)
16 Aelius Gallus (c. 25/24 BC)
Near and Middle East:
17 March of the Ten Thousand (402/401 BC)
India:
18 Alexander the Great (326/325 BC)
19 Nearchus (325 BC)
20 Patrocles (between 293 and 281 BC)
21 Megasthenes (between 305 or 303/302 and 298 BC)
22 Daimachus of Plataeae (after 298 BC)
Far East:
23 Maës Titianus (c. 1st cent. AD)
City, Town; Other locality
Pass; Mountain
River voyage
Barygaza City, Town, Other
Keltike Country
Abalos Territory, Mountain range, Lake/Sea
Sabaioi Tribe, People
Barygaza Keltike Abalos Ancient name
Marseille England Pamir Modern name
Turkmenbashi 20
Wadi Usboi
Oxos/Amu-Darya
20
23
Pyrgos Lithinos? 23/ (road station, location uncertain)
Baktra 23/Balkh
Hekatompylos 23/
Kaspapyros? 15
Antiocheia 23/ Erk Kala/Gyaur Kala
15,18 Kophen/Kabul
15 Paktyike
Taxila 18
Nikaia? 18
Pimprama? 18
Sangala? 18
Bukephala? 18
Hydaspes 18/Jhelum
Hyphasis 18/Beas
Akesines 18/Chenab
Hydraotes 18/Ravi
Alexandreia 18/Uch
Alexandreia Sogdiane? 18/Rajampur
1. Siboi 18 (Tribe)
2. Agalasseis 18 (Tribe)
3. Malloi 18 (Tribe)
4. Oxydrakai 18 (Tribe)
Indos Indus
15,18
19
Harmozeia
Alexandru Limen 19/ Karachi
Patala? 18/Hyderabad
Palimbothra 21,22/ Patna
Ganges/Ganga
21,22
Indike/India
Barygaza 11/Bharuch
Agrana 16/Najran
Aska 16/Nasq
Athrula 16/Yatill
Caripeta 16/ Hinu az-Zirayr
second voyage
Aromata 12/ Raas Caseyr (Ras Guardafui)
Taprobane 22/ Sri Lanka
Sera? 23/Lo-Yang
Serica/Northern China
0 500 1000 1500 2000 km
-200 0 200 500 1000 2000 3000 4000 ≦5000 m
60°
80°
100°
40°
20°
0°
F G H
2 3 4 5

Mesopotamia in the 2nd half of the 3rd millennium BC

In Mesopotamia, the period in question encompasses the later Early Dynastic (*c.* 2500–2200 BC), the Akkadian (ca 2200–2050 BC) and the Neo-Sumerian (*c.* 2050–1950 BC) periods. In northern Syria, it includes the Early Bronze Ages III and IV and the beginning of the Middle Bronze period. The state of the source material on this period shows little homogeneity. For the Neo-Sumerian period we have at our disposal a considerable number of texts from several cities in southern Mesopotamia, whereas there are only isolated data available for the Early Dynastic and Akkadian periods, mostly in the shape of plain royal inscriptions; there are also a few trade-related and literary texts. In northern Syria and northern Mesopotamia, writings from this period have come to light at Ebla (Tall Mardīḫ) and Nabada (Tall Baydar) only recently. They contain a number of place names, but many of these places cannot be located with certainty. On the other hand, a considerable number of Early Bronze Age sites whose ancient names are as yet unknown have been excavated and achaeologically examined in the last decades, especially in northern Mesopotamia. Therefore the map can only present the result of a temporary synthesis of known facts.

The Early Dynastic period

The term 'Early Dynastic period' is a first indication that the world of those 3rd-millennium states was characterized by a pronounced particularism which led to the creation of a multitude of relatively small political units. In most cases these principalities, which are frequently termed 'city-states', comprised towns of varying sizes as well as villages and hamlets, as we can see from the example of Lagaš: apart from the eponymous cult city, which apparently dominated in the early 3rd cent., it had two more 'big cities', namely Girsu, the actual capital, and Nimin. The inscriptions also mention smaller places, some of them with special functions, e.g. the port city of Gu' aba on the Persian Gulf. Scientific research has shown that the Gulf extended much further into the alluvial area of southern Iraq in the 3rd millennium than it does today.

Several other examples make it clear that the situation in Lagaš was not uncommon. Thus the principality of Umma, alongside the capital of the same name, contained at least one other sizeable city named Zabalam. The principalities were permanently at war with each other, constantly forming new coalitions. The aim was to exert hegemonic power over the complete 'country' in order to be entered into the 'Sumerian King List', a compilation originating at the end of the 3rd millennium, which insinuated that the country had always been ruled by one city or by one dynasty. This was achieved by presenting demonstrably contemporaneous dynasties as a succession of ruling families. This fiction and the merging of the individual city panthea into a comprehensive pantheon contributed to creating a southern Mesopotamian *koine* which encompassed Sumerian-speaking groups as well as Semitic population groups.

As far as the texts from Ebla and Nabada indicate, a comparable situation existed in the Levant and in northern Mesopotamia, although the cities there were significantly smaller. Some principalities in those regions seem to have dominated their neighbours for a considerable length of time. Among those powers were Ebla (situated south of Aleppo), Mari on the Euphrates and Nagar (Tall Brak), Urkeš (Tall Mozan) and Abarsal (Tall Chuēra?) in northern Mesopotamia. The armed conflict between the regional powers of Mari and Ebla is fairly well documented. Ultimately, however, they were both conquered by the Akkadian kingdom and later on became dependencies of the Ur III kingdom.

In the area of present-day Iran larger political units existed, probably originating from confederations of tribes or minor principalities, such as Elam, which on several occasions managed to influence political developments in Mesopotamia. It also acted as a trading centre for the exchange of goods between Mesopotamia and the regions of modern eastern Iran and Afghanistan, which are rich in raw materials (lapis lazuli, cornelian, tin?).

The geographical horizon of that period extended from the Indus valley (called Meluḫḫa by the Sumerians) via Oman (Magan) und Bahrain (Dilmun) to Central Anatolia (Purušḫanda) and the Mediterranean.

The Akkadian kingdom (*c.* 2200–2100 BC) and the Third Dynasty of Ur (*c.* 2050–1950 BC) (maps B and C)

Towards the end of the Early Dynastic age, the armed conflict between the southern Mesopotamian principalities came to a head; the Semitic dynasty of the newly founded city of Agade (Akkad) emerged victorious. In this process the first great empire in history was created: it covered nearly the whole of Mesopotamia and intermittently extended into the Levant as far as the coast of the 'Upper Sea' (i.e. the Mediterranean). This was accompanied by attempts at administrative centralization.

The Akkadian rulers were replaced by the once more Sumerian kings of the Third Dynasty (according to the calculation of the Sumerian King List) from the southern Mesopotamian port Ur. Apparently they were much more efficient at building a centralized administration. This kingdom finally collapsed at the transition from the third to the second millennium; the causes were domestic disturbances, the immigration of new ethnic groups (Amurru or Amorites) and an invasion by Elamite troops.

→ Map p. 13

The Kingdom of Agade (*c.* 2200–2100 BC)

Heartland · Maximum extent · Capital · City, localization certain/uncertain · Campaigns · Historical coastline · Historical river course · Simurrum: Ancient name · *Tall Huwaira*: Modern name

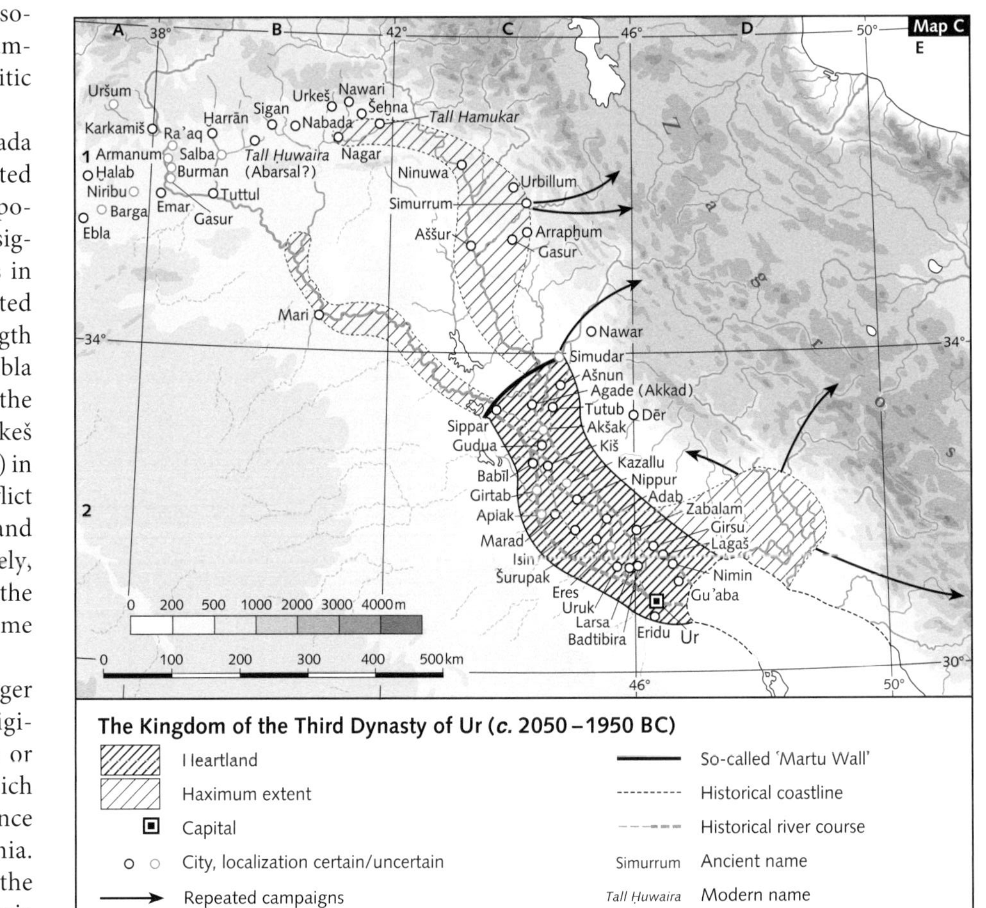

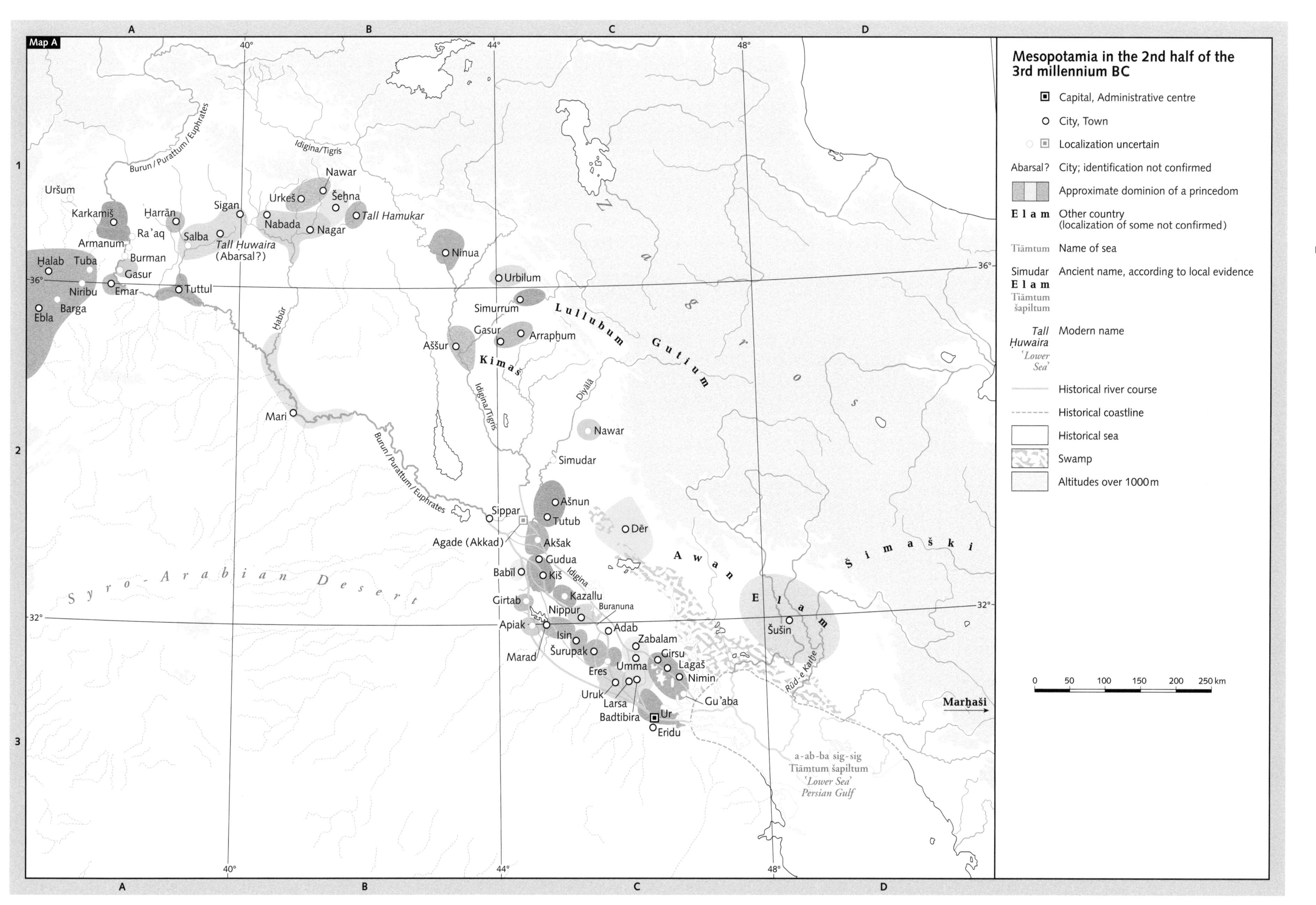
Map A
Mesopotamia in the 2nd half of the 3rd millennium BC
Capital, Administrative centre
City, Town
Localization uncertain
Abarsal? City; identification not confirmed
Approximate dominion of a princedom
Elam Other country (localization of some not confirmed)
Tiāmtum Name of sea
Simudar Elam Tiāmtum šapiltum Ancient name, according to local evidence
Tall Ḫuwaira 'Lower Sea' Modern name
Historical river course
Historical coastline
Historical sea
Swamp
Altitudes over 1000 m
0 50 100 150 200 250 km
Uršum
Karkamiš
Ḫarrān
Ra'aq
Armanum
Ḫalab
Tuba
Niribu
Barga
Ebla
Burman
Gasur
Emar
Tuttul
Salba
Sigan
Tall Ḫuwaira (Abarsal?)
Burun/Purattum/Euphrates
Idigina/Tigris
Nawar
Urkeš
Šeḫna
Nabada
Nagar
Tall Hamukar
Ninua
Urbilum
Simurrum
Gasur
Arrapḫum
Aššur
Kimaš
Lullubum
Gutium
Zagros
Ḫabūr
Mari
Diyālā
Nawar
Simudar
Sippar
Agade (Akkad)
Ašnun
Tutub
Akšak
Gudua
Babīl
Kiš
Idigina
Kazallu
Girtab
Nippur
Buranuna
Apiak
Marad
Isin
Šurupak
Adab
Zabalam
Girsu
Lagaš
Umma
Nimin
Eres
Uruk
Larsa
Badtibira
Ur
Eridu
Gu'aba
Dēr
Awan
Elam
Šušin
Šimaški
Rūd-e Karḫe
Marḫaši
Syro-Arabian Desert
a-ab-ba sig-sig
Tiāmtum šapiltum
'Lower Sea'
Persian Gulf
36°
32°
40°
44°
48°
A
B
C
D
1
2
3

The Ancient Near East in the 17th and 16th centuries BC

The Ancient Near East in the 17th and 16th centuries BC

The two maps highlight the political conditions during the 'Early Babylonian period'. This age, named after a stage in the development of the Babylonian dialect, a variety of the Akkadian language, encompassed the first half of the 2nd millennium BC. It followed on the heels of the fall of the 'Neo-Sumerian' kingdom of the 3rd Dynasty of Ur in the 20th cent. BC and ended with the conquest and destruction of the city of Bābilim/Babylon by the Hittites in the late 16th cent. BC. Generally the Early Babylonian period in Mesopotamia, Syria and Anatolia was characterized by constant change between particularism and centralism. Although there were phases when smaller political units were predominant, there were also several successful attempts at establishing larger, centralized empires. The one fairly consistent entity appears to have been the Elamite kingdom in southwestern and southern Iran.

I. The Ancient Near East in the middle of the Early Babylonian period around 1700 BC (map A)

Map A visualizes the political situation in the middle of the Early Babylonian period (around 1700 BC) shortly before the establishment of a great empire by Ḫammurapi of Bābilim/Babylon (cf. supplementary map). It is mainly based on the large text corpus of the royal archives at Mari on the middle reaches of the Euphrates, with its international correspondence. A few years earlier the Northern or Upper Mesopotamian kingdom of Šamšī-Adad I of Assyria (1769–1712, according to the 'short chronology'), whose territory is marked on the map, had disintegrated. It had extended from the city of Tuttul in the valley of the Baliḫ (the westernmost tributary of the Euphrates) across the area surrounding the Ḫābūr (another tributary to the Euphrates), with its capital of Šubat-Enlil and the region along the middle reaches of the Euphrates (which centred on Mari), as far as Assyria and into the foothills of the Zagros mountain range. At Ekallātum near Aššur and at Mari two sons had been set up as vice-regents of the Great King, who resided in Šubat-Enlil. The elder brother, Išme-Dagan, governed the actual core territory of Assyria and – to its north – the principality of Nurrugum with the cult city of Ninuwa (Ninet, Nineveh). After his father's death, he became the ruler of what was left of the Assyrian state. His brother Yasmaḫ-Addu, who governed in Mari, was ousted by Zimrī-Lim, a descendant of the family that had ruled the city before.

The following decades were marked by the clashes of a number of sizeable kingdoms, all of which were allied with several smaller principalities. For reasons of clarity the map only displays the larger polities and – roughly – their spheres of influence. Because of the permanent state of war between ever-changing coalitions the borders constantly shifted. Therefore the map can only show a state of affairs that has been arbitrarily selected from this period.

The most important kingdoms in Mesopotamia were Bābilim (Babylon), Eshnunna and Larsa. The latter had succeeded in defeating its old rival Isin and in incorporating its territory during the reign of King Rīm-Sîn. The kingdom of Mari extended along the middle reaches of the Euphrates, while what was left of the Assyrian empire was grouped along the middle reaches of the Tigris. In the Ḫābūr region, the territory of the defunct Northern Mesopotamian kingdom was divided up among several smaller principalities. The Levant was dominated by the rivalling empires of Yamḫad (with its capital Ḫalab/Aleppo) in the north and Qatna (Qatanum) in the south, while Yamḫad's power increased steadily. In Southern Anatolia there was the principality of Anum-Hirbi of Mama. Central Anatolia contained several small principalities, which were unified in the 18th cent. by King Anitta of Kuššara. One of the largest and most powerful kingdoms of the age was Elam. It consisted of the regions around the cities of Šušim (Susa) in the lowlands of today's Ḫuzistan (Khuzestan) and Anšan (Anshan) on the plateau of what was to become the Persis.

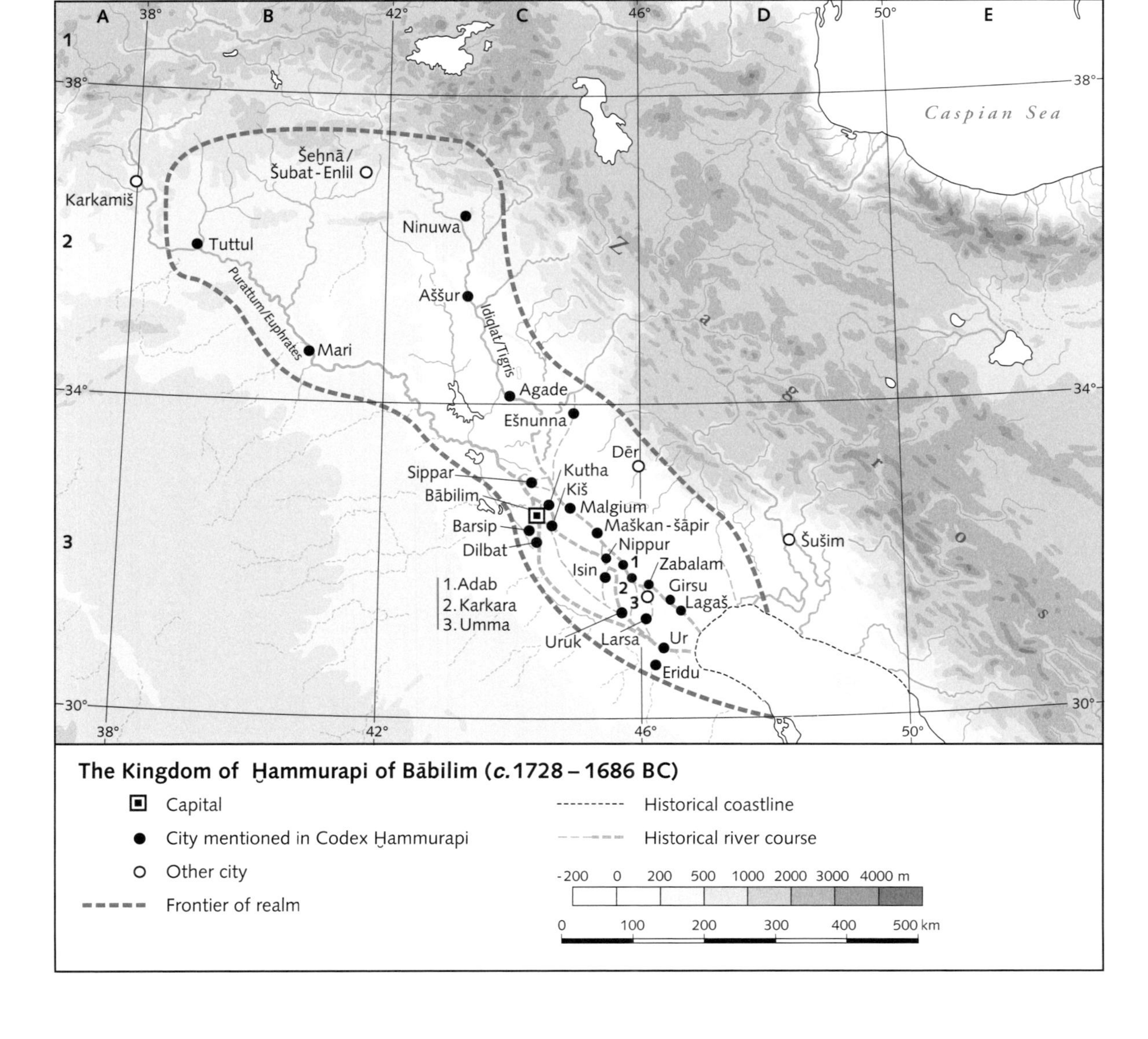

The Kingdom of Ḫammurapi of Bābilim (*c.* 1728 – 1686 BC)

II. The kingdom of Ḫammurapi of Bābilim (*c.* 1728–1686 BC) and the Ancient Near East towards the end of the Early Babylonian period around 1550 BC (supplementary map and map B)

By a rapid succession of victories over Larsa, Elam, Eshnunna and Mari at the beginning of the 17th cent., Ḫammurapi of Bābilim – for the first time in more than 250 years – succeeded in unifying almost all of Mesopotamia. The extent of his kingdom is described in the prologue of his famous law code; it is depicted on the supplementary map.

After only two generations, large parts of the kingdom were lost again. This is the situation illustrated by map B; the limitation of source material, however, which comes from a few smaller local archives, prevents a precise reconstruction. In the south the 'Sea Land' gained independence, while the principality of Ḫana on the middle reaches of the Euphrates and Aššur on the Tigris followed suit. In addition, two political units came into being which were to play a leading role in Mesopotamia during the next period. From the Zagros mountains the Kassites (whose country was called Kaššū) had immigrated along the old military highway (for information on its course see → map on p. 47) and through the area along the Diyālā (a left tributary of the Tigris); they managed to extend their sphere of influence temporarily from western Iran to the lower reaches of the Ḫābūr. On the upper reaches of the Ḫābūr, Hurrian mercenaries, who had probably been deported from the Zagros mountains, founded a principality whose name has not been established (some sources call it Ḫanigalbat, while the Hittites only mention a 'King of the Hurrians'); later on this was to develop into the Mittani kingdom. In addition, the so-called Early Hittite empire (on the map: Ḫattusa) came into being in Central Anatolia under Ḫattusili I (*c.* 1590–1560 BC). Around 1531/30 BC, Ḫattusili's grandson Mursili I (*c.* 1560–1531/30 BC) succeeded in capturing Ḫalab and Bābilim, an event that marked the end of the Early Babylonian period.

Sources

The quality and the amount of sources available for this long period diverge considerably, depending on the region and the phase. This substantially impairs the reconstruction of the political situation. The best-documented phase is that around 1700 BC, which is covered especially by the large archive from the royal palace at Mari on the Euphrates. Furthermore, numerous smaller archives from Babylonia, many royal inscriptions (among others, the stela bearing the Babylonian king Ḫammurapi's law code) and various kinds of economic records provide a wealth of additional multi-facetted documentation. Throughout the Early Babylonian period, the year names supply essential information: every year was named after the previous year's most important historical or religious event. As the lists of year names from various Babylonian cities are extant, numerous events and their backgrounds can be reconstructed.

The source material for the centuries preceding the age of the Mari archives is quite meagre overall. Apart from isolated inscriptions on buildings and literary texts from Babylonia and Assyria it is mainly the letters and commercial documents of Assyrian merchants in Anatolia which pass on historical information regarding the period around 1900 BC and later (the so-called *kārum* period). It is also the first time that we gain insight into the political conditions in Central Anatolia.

For the time after the destruction of Mari we also have to rely on smaller archives, but together with various royal inscriptions they contribute to a fairly coherent picture of conditions in Babylonia.

→ Maps pp. 11, 15, 19 23

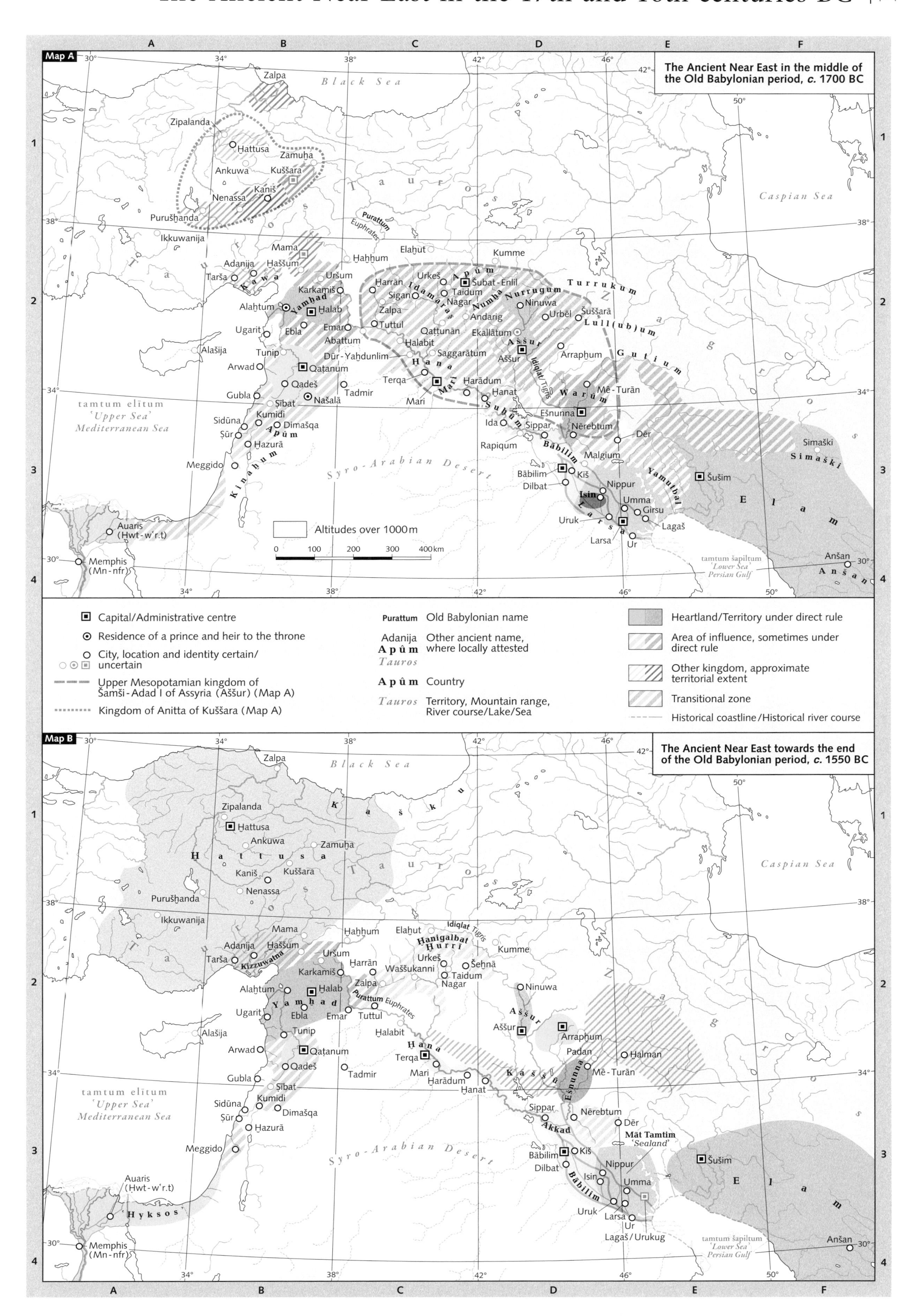
Map A
The Ancient Near East in the middle of the Old Babylonian period, c. 1700 BC
Black Sea
Caspian Sea
Zalpa
Zipalanda
Hattusa
Zamuḫa
Ankuwa
Kuššara
Kaniš
Nenassa
Purušḫanda
Ikkuwanija
Mama
Adanija
Haššum
Tarša
Uršum
Karkamiš
Halab
Alaḫtum
Ugarit
Ebla
Emar
Abattum
Alašija
Tunip
Arwad
Qaṭanum
Qadeš
Gubla
Šibat
Našalā
Tadmir
Kumidi
Sidūna
Dimašqa
Šūr
Hazurā
Meggido
Auaris
(Hwt-w'r.t)
Memphis
(Mn-nfr)
Puratum
Euphrates
Elaḫut
Haḫḫum
Kumme
Harrān
Urkeš
Šubat-Enlil
Sigan
Taidum
Nagar
Zalpa
Tuttul
Andarig
Ninuwa
Urbēl
Šuššarā
Qaṭṭunān
Ekallātum
Halabit
Aššur
Arrapḫum
Dūr-Yaḫdunlim
Saggarātum
Terqa
Harādum
Mari
Hanat
Mê-Turān
Ešnunna
Nērebtum
Dēr
Ida
Sippar
Rapiqum
Malgium
Bābilim
Kiš
Dilbat
Nippur
Isin
Umma
Girsu
Uruk
Lagaš
Larsa
Ur
Šušim
Simaški
Anšan
Elam
Turrukum
Lullubum
Gutium
Idiqlat / Tigris
tamtum elītum
'Upper Sea'
Mediterranean Sea
tamtum šapiltum
'Lower Sea'
Persian Gulf
Syro-Arabian Desert
Altitudes over 1000 m
0 100 200 300 400 km
Capital/Administrative centre
Residence of a prince and heir to the throne
City, location and identity certain/uncertain
Upper Mesopotamian kingdom of Šamši-Adad I of Assyria (Aššur) (Map A)
Kingdom of Anitta of Kuššara (Map A)
Purattum Old Babylonian name
Adanija Other ancient name, where locally attested
Apûm Country
Tauros Territory, Mountain range, River course/Lake/Sea
Heartland/Territory under direct rule
Area of influence, sometimes under direct rule
Other kingdom, approximate territorial extent
Transitional zone
Historical coastline/Historical river course
Map B
The Ancient Near East towards the end of the Old Babylonian period, c. 1550 BC
Hattusa
Kašku
Hanigalbat
Hurri
Kizzuwatna
Yamḫad
Wašsukanni
Šeḫnā
Halman
Padan
Kaššu
Akkad
Mât Tamtim
'Sealand'
Hyksos
Lagaš/Urukug

The Ancient Near and Middle East in the 15th to 13th centuries BC

The two main maps and the supplementary one attempt to reflect the extremely complex political situation in the Eastern Mediterranean region and the Middle East during the Late Bronze Age (*c.* 1500–1200 BC). While map A highlights the circumstances before the destruction of the Mittani kingdom by the Hittites around 1350 BC, map B depicts the constellation which resulted from that event and remained stable until 1200 BC. The subject of the supplementary map is the historical geography of Syria and the Levant, the contact zone between the great empires of the age.

I. The Ancient Near and Middle East in the 15th and 14th centuries BC (map A)

In the early 15th cent. BC, the destructive campaigns undertaken by the rulers of the so-called Old Hittite Empire had caused the fall of the empires of the Early Babylonian period and led to the emergence of two new major powers in Mesopotamia and Northern Syria: the Central Babylonian kingdom (Karduniaš/Bābil) ruled by the Kassites and the Mittani kingdom dominated by the Hurrians.

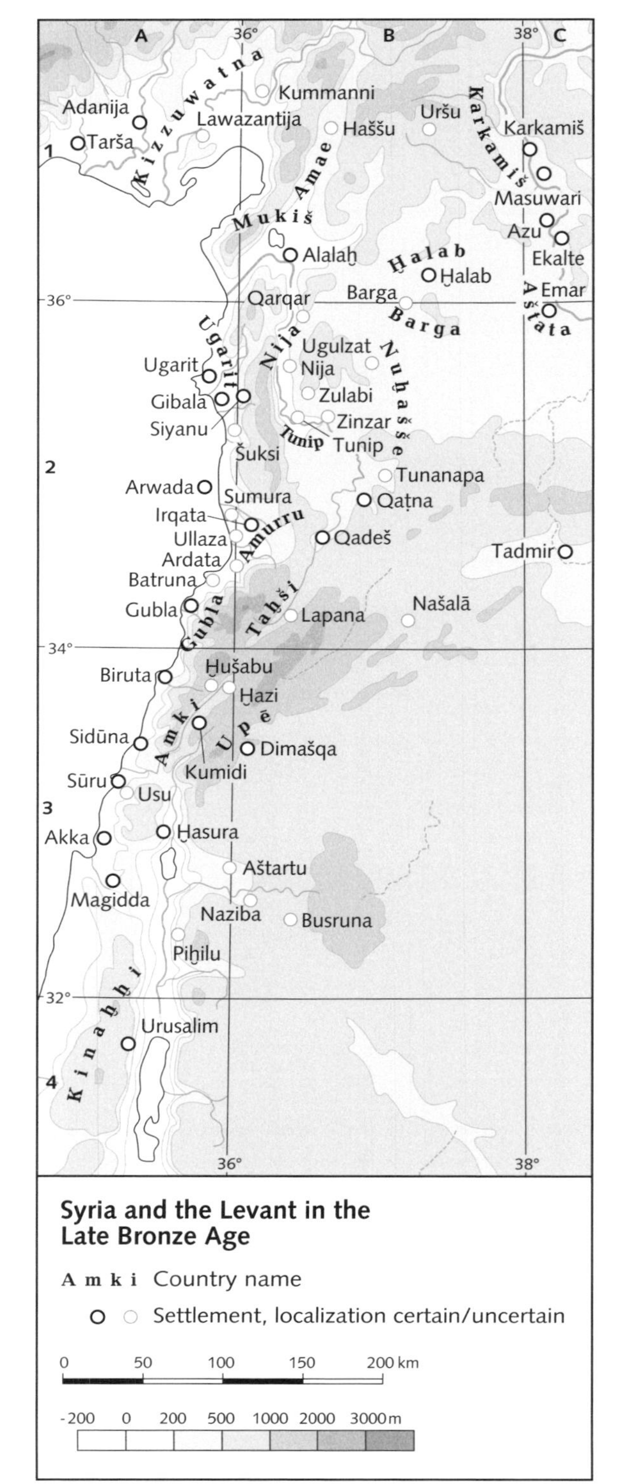

Syria and the Levant in the Late Bronze Age

From the 17th cent. onwards, the Kassites had been immigrating from what is now Iran, following the 'great military highway' (for its course see → map p. 47). They were able to seize power in Southern Mesopotamia after the fall of the Old Babylonian kingdom. Along with Babylonia, their territory included parts of western Iran and – for a time – Dilmun (today's Bahrain on the Persian Gulf). Apart from the names of kings and of a few gods as well as a very small number of technical terms related to horse breeding, no words from the Kassite language have survived. Apparently the Kassites assimilated quickly and thoroughly to Babylonian culture. The Kassite king Kurigalzu I (*c.* 1410–1380 BC) founded Dūr-Kurigalzu, near modern Baghdad, as his new capital.

Without exception the Mittani kings bore Indo-Aryan, not Hurrian names. This suggests that before the founding of the kingdom some cultural contacts must have existed between the Hurrians, who were to form the kingdom's elite, and probably nomadic Indo-Aryan groups, who resurfaced in Northern India towards the end of the 2nd millennium. These contacts may have taken place in the Zagros region, from where the Hurrian groups were apparently deported into the Ḫābūr area by the Assyrians around 1700 BC. From marauding bandits they subsequently rose to join the ranks of upper-class local Hurrians to found what came to be called, from 1450 BC onwards, the Mittani kingdom. Like the Hittite and the Kassite kingdoms, it owed its military success to the wider use – then documented for the first time – of light chariots, which the Hyksos brought to Egypt as well. The Mittani kingdom centred on the region around the headwaters of the Ḫābūr, where the capital cities of Waššukanni (probably today's Tall Faḫariya), Ta'idu (probably present-day Tall Hamidiya) and Irrite (on the Baliḫ river) were situated. At the height of its power around 1400 BC, the kingdom extended from the coast of the Levant to the foothills of the Zagros range. The small kingdoms of Aššur and Arrapḫa were under Mittani control at this point.

The Middle Elamite kingdom managed to establish itself as a further major power, its territory extending from the lowlands around Šušim (Susa, Ḫuzistan) to the plateau of Anšan (Persis).

In Central Anatolia, the Middle Hittite kingdom was able to consolidate after the disturbances that had marked the end of the Old Hittite period, and around 1400 BC it succeeded in extending its territory again. The Cilician and Cappadocian principality of Kizzuwatna acted as a buffer state and as a cultural bridge between the Hurrians and Mittani in northern Syria and Hittite central Anatolia. Around 1350 BC, Suppiluliuma I founded the great Hittite empire.

When the reign of the Hyksos came to an end, the rulers of the Egyptian New Kingdom succeeded in penetrating as far as northern Syria in occasional campaigns. By concluding an agreement with the Mittani kingdom around 1400 BC, Egypt firmly established its domination over the southern Levant. The diplomatic correspondence discovered at Aḫet-Aten (Amarna) highlights the balance of power between two kingdoms of equal status. The small Syrian and Levantine principalities were placed under the occasionally changing suzerainty of either major power (cf. supplementary map).

II. The Ancient Near and Middle East in the 13th century BC (map B)

The situation changed as a result of the expansion of the Hittite empire 'Ḫattusa' from the middle of the 14th cent. BC, to which the Mittani kingdom fell victim. Along with the Hittite empire, the Middle Assyrian empire also benefitted from the situation as it gained independence and went on to conquer the core territory and the east of the former Mittani kingdom. The Assyrians clashed with the temporarily allied kingdoms of the Hittites and Babylonians, but managed to prevail and to extend their power as far as the vicinity of the Hittite fortress of Karkamiš on the Euphrates.

After the battle of Qadeš (fought between Rameses II and Muwatalli in 1275 BC) the great Hittite empire, which had expanded considerably in the 14th cent., concluded a peace treaty with Egypt guaranteeing its hegemony over the northern Levant. At the same time, the Hittites conquered the western Anatolian country Arzawa (cf. map A). In spite of these successes, the Hittites were troubled by internal political strife and by the permanent threat to their core territory posed by the Kaskaeans, apparently a largely nomadic people from the Pontic mountains south of the Black Sea. These problems were exacerbated by the establishment of secundogenitures at Tarḫuntašša in southern Anatolia and at Karkamiş in northern Syria, which were soon granted considerable autonomy. At the beginning of the 12th cent. BC, the end of the world of Late Bronze Age states was brought about by famines, civil wars and large-scale migrations.

Sources

Our sources for the period in question stem from the large royal archives of Ḫattusa (Hittite empire) and Aḫet-Aten/Amarna (Egypt, New Kingdom) as well as from numerous smaller archives (e.g. from Dūr-Katlimmu, Assyria), which contain the correspondence between the major powers as well as their correspondence with their vassals). In addition there are royal inscriptions, reflective 'autobiographies' and historical introductions to state treaties as well as large quantities of commercial records from almost every region of the Orient. The source situation profits from the fact that during this period the Akkadian language, recorded in Akkadian cuneiform script on durable clay tablets, was the *lingua franca* of the entire civilized world.

Another potentially important source for reconstructing the extent of political units is ceramic ware, which became highly standardized at this time. Thus the establishment of the kingdom of the Mittani can be linked with the emergence of the so-called Nuzi ware, which was painted in a specific manner. The same is true of the incorporation of its core territory into the Middle Assyrian empire, indicated by the spread of certain typically Assyrian vase shapes. Furthermore, the expansion of the Kassite Middle Babylonian kingdom along the 'great military highway' far into the Zagros region can be proved from the spread of matching ceramic vase types. Certain luxury goods, on the other hand, especially those from Cyprus, were traded over long distances both in their own right or as containers for perishable goods to be sold. Therefore they can in any case be used as indicators of chronological contexts and trade relations, but not for determining the extent of political units.

→ Maps p. 13, 19, 21, 23, 29, 33, 47; BNP 6, 2005, s.v. Ḫattusa

Literature

T. Bryce, The Kingdom of the Hittites, 1998; W. Eder, J. Renger (eds.), Herrscherchronologien der antiken Welt, 2004; B.I. Faist, Der Fernhandel des assyrischen Reiches zwischen dem 14. und 11. Jahrhundert v. Chr. (thesis, Tübingen) 2001; J. Freu, Histoire du Mitanni, 2003; H.J. Nissen, Geschichte Alt-Vorderasiens, 1998; W. Orthmann (ed.), Der Alte Orient, PropKg vol. 14, 1975; P. Pfälzner, Die Entwicklung der Keramik vom 14. bis zum frühen 11. Jahrhundert v. Chr. im nördlichen Mesopotamien (thesis Berlin) 1991; RLA, 1928ff.; G. Wilhelm, Grundzüge der Geschichte und Kultur der Hurriter, 1982.

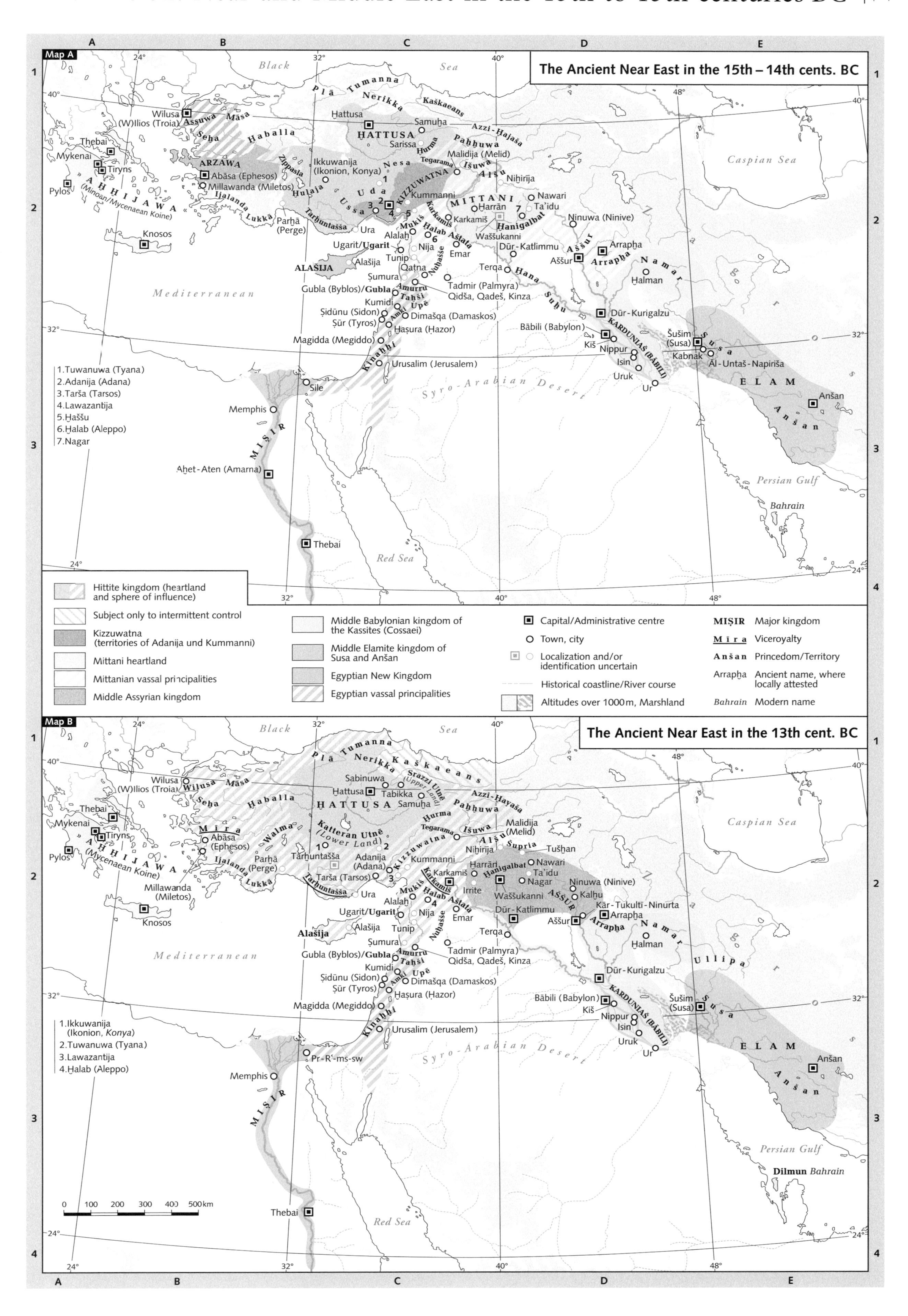
Map A
The Ancient Near East in the 15th – 14th cents. BC
1.Tuwanuwa (Tyana)
2.Adanija (Adana)
3.Tarša (Tarsos)
4.Lawazantija
5.Ḫaššu
6.Ḫalab (Aleppo)
7.Nagar
Hittite kingdom (heartland and sphere of influence)
Subject only to intermittent control
Kizzuwatna (territories of Adanija und Kummanni)
Mittani heartland
Mittanian vassal principalities
Middle Assyrian kingdom
Middle Babylonian kingdom of the Kassites (Cossaei)
Middle Elamite kingdom of Susa and Anšan
Egyptian New Kingdom
Egyptian vassal principalities
Capital/Administrative centre
Town, city
Localization and/or identification uncertain
Historical coastline/River course
Altitudes over 1000 m, Marshland
MIṢIR Major kingdom
Mira Viceroyalty
Anšan Princedom/Territory
Arrapḫa Ancient name, where locally attested
Bahrain Modern name
Map B
The Ancient Near East in the 13th cent. BC
1.Ikkuwanija (Ikonion, Konya)
2.Tuwanuwa (Tyana)
3.Lawazantija
4.Ḫalab (Aleppo)
0 100 200 300 400 500 km

Egypt in the Old Kingdom and the 1st Intermediate Period

I. The Old Kingdom (*c.* 2680–2160 BC)

Roughly speaking, the Old Kingdom extended from the 3rd Dynasty through to the 8th Dynasty; a 7th Dynasty is part of the Egyptian tradition, but did not exist in reality. As usual in the context of Egyptian history, dynasties are not necessarily royal houses defined in terms of kinship, but batches of kings who were posthumously grouped together (e.g. on the so-called Turin king list); a common residence was a defining factor.

The Old Kingdom was preceded by the Thinite period (1st and 2nd Dynasties, encompassing about 300 years, with Abydus as the central town) as well as the Predynastic period, whose final phase is hypothetically classified as the Dynasty 0 in present-day research. The beginnings of statehood and the development of writing in Egypt are continuously retreating further into the past as a result of new archaeological finds. In the course of the Old Kingdom, the Egyptian state developed only gradually; only in the 4th Dynasty was it fully formed to a certain degree. Its centre in this period was situated at Memphis (close to modern Cairo), where the residence as well as the burial sites of the kings and their officials were situated (see map C).

At that time and probably for some time to come, land use in Egypt had not yet been fully developed either in terms of agriculture or settlement. Especially the swampy delta with its numerous watercourses had only been opened up in parts. This is suggested by the fact that so-called foundation estates (established for the funerary cult of deceased kings and officials) were newly set up predominantly in that region. Settlements could only be established on the sand bars framing the watercourses and have so far been hardly explored by archaeologists. The Faiyūm oasis was also largely uninhabited because no hydro-engineering measures had yet been taken to stop the two lakes flanking it – one to the north and one to the south – from merging during the periodic flooding.

The country's uniform administrative division into so-called nomes (*nomoí* = districts) run by a governor is a fictitious concept. Instead we must assume that there were central temple sites which controlled the surrounding areas. The most important provincial capitals were Elephantine as the southern border town, Abydus as a religious centre with the temple of Osiris, and Thebae/Thebes, which was founded in the 6th Dynasty and subsequently became the capital city. Other sites like Coptus or Panopolis/Aḫmīm were important as starting points of caravan trails into the desert. In the Wādī al-Ḥammāmāt, which branches off near Coptus, greywacke was mined (as well as gold in some tributary valleys). It also served as an access route to the Red Sea and on to Punt, from where incense was imported. Punt's location – either Somalia or Arabia – is a point of controversy.

From the 5th Dynasty onwards, the oases played an important part as did the desert trails leading, among others, across the western desert to Nubia. From Lycopolis/Asyūṭ the 'Route of forty days' (Darb al-arbaʿīn) branched off to the south. Nubia was explored in occasional expeditions in this period. Raw materials, cattle and people were taken back to Egypt, but the country was not annexed yet. The Wādī al-Allāqi is considered one of the main regions for the mining of gold, while the mining in the Wādī Maġāra on the Sinai was predominantly for turquoise. The stone blocks used for the pyramids and the Mastaba tombs were mainly quarried at Ṭura on the eastern bank of the Nile opposite Memphis.

II. The 1st Intermediate period (*c.* 2160–2060 BC)

The duration of the 1st Intermediate period is still a matter of controversy, especially as far as the combined stretch of the poorly documented 9th and 10th Dynasties is concerned. The period is characterized by the split of the political unit into two kingdoms: the domain of the Heracleopolitan rulers (representing the 9th and 10th Dynasties) and that of the Thebans (encompassing the 11th Dynasty). The original authority on the schism of the Heracleopolitan kingdom is the Greco-Egyptian writer Manetho (3rd cent. BC), but his version is not confirmed by pharaonic sources. The two separate kingdoms did not come into existence at the same time, the Theban kingdom emerging later than its Heracleopolitan counterpart. The Heracleopolitan domain comprised northern Egypt and closely followed the Memphite traditions of the Old Kingdom. The Theban kingdom encompassed the south; the border lay near Lycopolis/Asyūṭ. Under Mentuhotep II, the second king of the Theban-based 11th Dynasty, Egypt was reunified, although it is not possible to determine the exact date. This reunification appears to have been achieved mainly by military means, as evidenced, among other things, by a report on warlike activities in the tomb of a local ruler at Asyūṭ as well as another one at Thebes.

Apart from the Heracleopolitans and Thebans, there were several other local rulers who strove to expand their sphere of influence. This resulted in a situation similar to civil war. Together with a series of low Nile inundations, it led to poor harvests which, combined with bad storage management, caused food shortages and famines. As a countermeasure the first attempts were made to irrigate fields artificially through basin irrigation. The situation of civil strife also meant that expeditions into the desert to retrieve raw materials more or less came to a halt, and that practically all foreign contacts were broken off. The Nubians who were hired as mercenaries, were an exception; they settled predominantly in Upper Egypt, e.g. at Pathyris/al-Gabalain.

The sources for this period consist almost exclusively of tombs and their inscriptions, most of which were found in Upper Egypt; there is hardly any research on settlements. Recent excavations at Heracleopolis may produce additional insights. Sometimes the dating of archaeological material at the cemeteries still proves difficult; it is expected, however, that dendrochronological examinations of coffins may shed new light on the matter.

III. Pyramid cemeteries from the Old Kingdom to the Middle Kingdom

Map C covers the pyramid cemeteries of the Old Kingdom. It shows the burial sites of the kings and their cult facilities down to the 12th Dynasty (i.e. the Middle Kingdom), as well as the sites of the sun cult complexes of the 5th Dynasty. Moreover, this region also contains the necropoleis of all members of the royal families as well as their high-ranking officials, who wished to be buried close to their kings and were assigned burial sites by them.

→ Maps pp. 19, 21; BNP7, 2005, s.v. Kings' Lists

Literature

Map A: Elephantine: M. Ziermann, Elephantine XXVIII. Die Baustrukturen der älteren Stadt (Frühzeit und Altes Reich), 2003; *Kōm al-Hisn*: A. Cagle, The Spatial Structure of Kom el-Hisn: An Old Kingdom Town in the Western Nile Delta, Egypt, 2003.

Maps A and B: Abydus: M.D. Adams, The Abydos Settlement Site Project. Investigation of a Major Provincial Town in the Old Kingdom and First Intermediate Period, in: C.J. Eyre (ed.), Proceedings of the Seventh International Congress of Egyptologists (Cambridge, 3–9 September 1995), 1998, 19–30.

Map B: F. Gomaà, Ägypten während der Ersten Zwischenzeit, 1980; W. Schenkel, Memphis – Herakleopolis – Theben. Die epigraphischen Zeugnisse der 7. bis 11. Dynastie Ägyptens, 1965; Id., Die Bewässerungsrevolution im Alten Ägypten, 1978; S.J. Seidlmayer, Gräberfelder aus dem Übergang vom Alten zum Mittleren Reich. Studien zur Archäologie der Ersten Zwischenzeit, 1990; *Dandara*: R.A. Slater, The Archaeology of Dendereh in the First Intermediate Period (thesis, University of Pennsylvania), 1974.

Map C: New drawing after: R. Stadelmann, Die ägyptischen Pyramiden. Vom Ziegelbau zum Weltwunder, 1985, 6 (with the author's revisions of the map); M. Bárta, J. Krejčí (eds.), Abusir and Saqqara in the Year 2000, 2000.

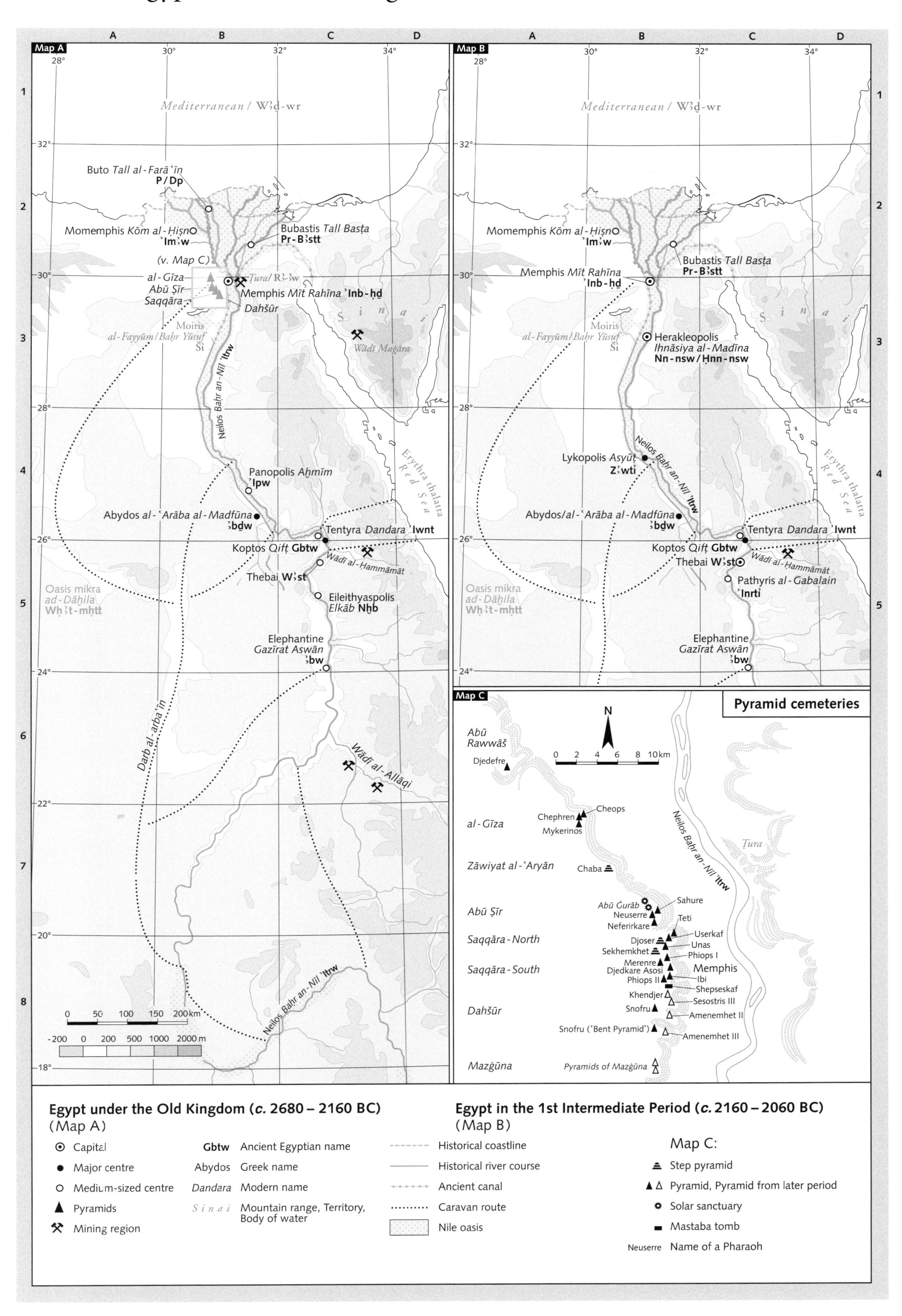

Map A
Map B
Map C
Mediterranean / W3d-wr
Buto Tall al-Farā'īn
P / Dp
Momemphis Kōm al-Ḥiṣn
'Im3w
Bubastis Tall Basṭa
Pr-B3stt
(v. Map C)
al-Gīza
Abū Ṣīr
Saqqāra
Tura/ R3-3w
Memphis Mīt Rahīna 'Inb-ḥd
Dahšūr
Moiris
al-Fayyūm / Baḥr Yūsuf
Š
Wādī Maġāra
Sinai
Neilos Baḥr an-Nīl 'Itrw
Panopolis Aḫmīm
'Ipw
Abydos al-'Arāba al-Madfūna
3bdw
Tentyra Dandara 'Iwnt
Koptos Qifṭ Gbtw
Wādī al-Hammāmāt
Thebai W3st
Eileithyaspolis
Elkāb Nḫb
Oasis mikra
ad-Dāḫila
Wḥ3t-mḥtt
Elephantine
Gazīrat Aswān
3bw
Erythra thalatta
Red Sea
Darb al-arba'īn
Wādī al-Allāqi
Herakleopolis
Ihnāsiya al-Madīna
Nn-nsw / Ḥnn-nsw
Lykopolis Asyūṭ
Z3wti
Pathyris al-Gabalain
'Inrti
Pyramid cemeteries
Abū Rawwāš
Djedefre
al-Gīza
Cheops
Chephren
Mykerinos
Zāwiyat al-'Aryān
Chaba
Abū Ṣīr
Abū Ġurāb
Sahure
Neuserre
Neferirkare
Saqqāra-North
Teti
Djoser
Userkaf
Unas
Sekhemkhet
Phiops I
Saqqāra-South
Merenre
Djedkare Asosi
Memphis
Phiops II
Ibi
Shepseskaf
Khendjer
Sesostris III
Dahšūr
Snofru
Amenemhet II
Snofru ('Bent Pyramid')
Amenemhet III
Mazġūna
Pyramids of Mazġūna
Ṭura
0 50 100 150 200 km
-200 0 200 500 1000 2000 m
0 2 4 6 8 10 km
Egypt under the Old Kingdom (c. 2680 – 2160 BC) (Map A)
Egypt in the 1st Intermediate Period (c. 2160 – 2060 BC) (Map B)
Capital
Major centre
Medium-sized centre
Pyramids
Mining region
Gbtw Ancient Egyptian name
Abydos Greek name
Dandara Modern name
Sinai Mountain range, Territory, Body of water
Historical coastline
Historical river course
Ancient canal
Caravan route
Nile oasis
Map C:
Step pyramid
Pyramid, Pyramid from later period
Solar sanctuary
Mastaba tomb
Neuserre Name of a Pharaoh

Egypt in the Middle Kingdom and the 2nd Intermediate period

Egypt in the Middle Kingdom and the 2nd Intermediate period

I. The Middle Kingdom (*c.* 2060–1800 BC)

We cannot pinpoint an exact date, but the Middle Kingdom started when the fragmented country was reunited under Mentuhotep II in the 11th Dynasty. It came to an end with the demise of the 12th Dynasty, which in turn caused the gradual dissolution of state unity. Some scholars regard the 13th Dynasty as part of the Middle Kingdom. The duration of the Middle Kingdom cannot be determined with precision, not least because there is not enough information about royal co-regencies.

The kings of the 11th Dynasty were the descendants of the Theban local rulers of the 1st Intermediate Period (→ map B on p. 17). After the reunification, which was apparently brought about by military means, their capital therefore was Thebae/Thebes. Amenemhet I, the first king of the 12th Dynasty, however, moved his residence into the region of al-Lišt on the fringe of the Faiyūm depression. Like the time of this move, the exact location of the capital is unknown, but we can assume that it lay near his pyramid.

Older Egyptologist research held that the canonical division of the country into so-called nomes was codified at Karnak (Thebae, east bank) on the White Chapel of Sesostris I, the successor of Amenemhet I. There, alongside the names of the 36 nomes, the major settlements and gods as well as the north-south extension of the individual areas are stated. It seems, however, that in fact this is not a list of administrative units, but of cultic entities; the major sites are not provincial capitals, but major religious centres. In reality there were probaly fewer than 36 provincial governors or nomarchs. It was possible for a major cult centre to be identical with the administrative capital, but this did not follow automatically. The nomarchs were frequently also the high priests of their region's main sanctuary.

Whereas the colonization of the Nile Valley and the Delta seems to have been complete at this point, that of the Faiyūm had only started at the onset of the Middle Kingdom. By building a sluice-gated dam at the entrance to the Faiyūm, the possibility was created to regulate the amount of water fed into the Faiyūm basin from the Baḥr Yūsuf river; thus, part of the region was drained and turned into arable land. This achievement is ascribed to the kings Sesostris I and Amenemhet III, both of whom had pyramids erected at the access point to the Faiyūm. The latter is also known to posterity under his Greek name of Lamares. His funerary temple and the attached settlement near Hawwāra later became known as a labyrinth (for further pyramid complexes of the Middle Kingdom cf. → map C, 'Pyramid cemeteries', p. 17).

Alongside the capitals, the places in the Nile Valley where the caravan routes branched off into the desert were of central importance as well, e.g. Lycopolis/Asyūṭ, Coptus and the border town of Elephantine. Abydus continued to play its time-hallowed part as a religious centre and a cult site of the god Osiris; in the Middle Kingdom it became a pilgrimage shrine, where wealthy believers from all over the country left small chapels and stelae. In the Delta the spacious residence of a 12th-Dynasty provincial governor has been excavated at the site of Bubastis.

Among the mining areas – where kings, officials and labourers left inscriptions and graffiti on the rock – are the Wādī al-Hūdī, where amethyst was mined, and Ḥatnūb with its alabaster quarries. Greywacke and – in some tributary valleys –gold were mined in the Wādī al-Ḥammāmāt. This wādī was also used as the access route to the Red Sea, where a port of unknown location was the departure point for travel to Punt. (The location of Punt, from where incense was imported – either in Somalia or in Arabia – is a point of controversy.) In the region of Gabal az-Zait, galena was mined, which served as the basic ingredient for, among other things, eye make-up; the unlocated Red Sea harbour close by was also a departure point for journeys to Punt. The Wādī al-Allāqi in Nubia is considered one of the prime regions for gold mining. In the Wādī Maġāra on the Sinai, turquoise was the predominant mining product and a temple was erected for the tutelary goddess Hathor. In the reign of Amenemhet IV (*c.* 1800 BC), documentation abruptly ceases.

Asyūṭ and Elephantine were the departure points for overland routes to Nubia, the trail from Asyūṭ being known as the 'Route of forty days' (Arabic: Darb al-arbaʿīn); the main way to Nubia, however, was the Nile. During the Middle Kingdom, this country was increasingly opened up in order to secure its resources. For this purpose fortresses were built on the banks of the Nile, especially at the second cataract south of the 22nd latitude. Buhen (situated north of the second cataract) gradually lost its character as a border fortress and turned into a customs and trading post. Kūbān from the beginning served as a site for the collecting and processing of gold and copper, raw materials from the Wādī el-Allāqi. Some of the fortresses at the second cataract also took on different functions. Thus Mirgissa served predominantly as a trading post; Askūt on the other hand became a depot for further ventures in Upper Nubia. When Semna was built under Sesostris III, all the complexes to the north changed into civilian stations. In order to rule over the surrounding illiterate Nubian population, the so-called C group, ʿAnība was erected in a region of vast tilled fields. At the same time another illiterate civilization emerged in Upper Nubia, which centred on Kerma and is known as 'Kerma Moyen' (Middle Kerma).

Contacts with the Levant resulted from occasional trading journeys or military campaigns. The military enterprise which is documented most thoroughly in the temple of Ptah at Memphis is Amenemhet II's Palestine campaign (*c.* last quarter of the 20th cent. BC), which brought a multitude of people and large quantities of cattle and goods to Egypt; on his way back he even reached Cyprus.

II. The 2nd Intermediate period (*c.* 1800–1570 BC)

The 2nd Intermediate period was characterized by the fragmentation of the country into several regional domains. The number of kings and the dates of their time in office as well as the durations of individual dynasties and that of the entire period are still matters of controversy. The problems are compounded by the fact that the Turin King List, one of the most important Egyptian sources for establishing the order of reigns, has been re-edited for this era – with the results still being discussed.

We can be relatively certain about the 13th Dynasty, although some scholars assign it to the Middle Kingdom. The rulers resided near al-Lišt and were buried in the Memphite region (cf. → map C, 'Pyramid cemeteries', p. 17). It is uncertain whether the last kings ruled in succession or simultaneously at different places. Nor is it clear whether its reign coincided with the 14th Dynasty, which ruled over the Delta with the exception of Bubastis and Athribis. In all, the 13th Dynasty comprised thirty kings and a period of 130 to 155 years. The 14th and 15th Dynasties consisted of Hyksos ('Rulers of foreign countries'), who originally came from Palestine and resided at Avaris in the Eastern Delta (for the region of Avaris cf. → p. 21, map C). Their territory hardly extended beyond the Delta; occasionally Hermupolis is mentioned as it southernmost outpost. The contemporaneous rulers of the 16th and 17th Dynasties resided at Thebae/Thebes with their sphere of influence extending northwards to the mouth of the Faiyūm or to the area south of Hermupolis. The period from the end of the 13th to the beginning of the 18th Dynasty (the latter belonging to the New Kingdom) covered about 100 years. With the campaign led by the Theban ruler Kamose, the last king of the 17th Dynasty, the Hyksos were expelled from the Nile Valley and the unity of the country basically restored. Kamose's campaigns against the Hyksos and their allies, the rulers of Nubia, are documented on two stelae. The consolidation of unity was, however, left to Ahmose I, the first king of the 18th Dynasty and as such of the New Kingdom, who conquered Avaris.

Evidence of Egyptian settlements is mainly archaeological for this period, with ceramic ware providing most of the dating clues. The site of Tall al-Yahūdīya, ancient Leontopolis, is eponymous of a whole type of ceramic ware. Conversely, philological documents are rare and succinct, frequently just mentions of names or titles on seals and scarabs.

Dẹr el- Ballāṣ deserves mention as a newly founded and important town. It comprised a royal residence in the shape of a fortress with a small settlement attached to it and a second fortress which served as an observation post. Both became disused in the early 18th Dynasty.

It appears that most mining activities areas were discontinued after the 13th Dynasty at the latest.

Egyptian control of Nubia equally came to an end. The local fortresses lost their military character and, from the 13th Dynasty onward, were inhabited by Egyptian civilians. In the late 2nd Intermediate Period, a Nubian element was added to the population, before Egyptian suzerainty was restored under Kamose. In southern Upper Nubia a Nubian civilization, still illiterate, emerged, which reached its peak in Kerma ('Kerma Classique'). The capital Kerma with its fortress-like so-called Defūfa encompassed an area of more than 30 ha. The rulers of Kerma and the Hyksos in the Delta kept in touch via the caravan trails with the aim of encircling the rulers of Thebes.

During the 13th Dynasty, connections with the Levant and the Middle East extended as far as Byblos, later only as far as Palestine. It was exclusively the Hyksos – with their original and main site at Sharuhen (present-day Tell el-ʿAǧǧul) 20 km south of Gaza – who kept those relations alive. Apart from contacts with Cyprus there is no further evidence of long-distance trading.

→ Maps pp. 17, 21; map s.v. Thebes, BNP 14, 2009

Literature

F. Gomaà, Die Besiedlung Ägyptens während des Mittleren Reiches, 2 vols., 1986, 1987; P. Lacau, H. Chevrier, Une chapelle de Sésostris Ier à Karnak, 2 vols., 1956, 1969; K.S.B. Ryholt, The Political Situation in Egypt during the Second Intermediate Period c.1800–1550 B.C., 1997. *Askūt*: S.T. Smith, Askut in Nubia. The Economics and Ideology of Egyptian Imperialism in the Second Millennium B.C., 1995. *Buhen*: W.B. Emery (†), H.S. Smith, A. Millard, The Fortress of Buhen. The Archaeological Report, 1979. *Dẹr al-Ballās.*: P. Lacovara, Deir el-Ballas. Preliminary Report on the Deir el-Ballas Expedition (1980–1986), 1990. *Elephantine*: C. von Pilgrim, Elephantine XVIII. Untersuchungen in der Stadt des Mittleren Reiches und der Zweiten Zwischenzeit, 1996. *Gabal az-Zait*: G. Castel, G. Soukiassian, Gebel el-Zeit I. Les mines de galène (Egypte, IIe millénaire av. J.-C.), 1989. *Kerma*: Ch. Bonnet, Le temple principal de la ville de Kerma et son quartier religieux, 2004. *Mirgissa*: J. Vercoutter, Mirgissa I, 1970. *Tall Habwa*: M. Abd el-Maksoud, Tell Heboua (1981–1991). Enquête archéologique sur la Deuxième Période Intermédiaire et le Nouvel Empire à l' extrémité orientale du Delta, 1998. *Wādī al-Hūdi*: I. Shaw, R. Jameson, Amethyst Mining in the Eastern Desert: A Preliminary Survey at Wadi el-Hudi, in: JEA 79, 1993, 81–97.

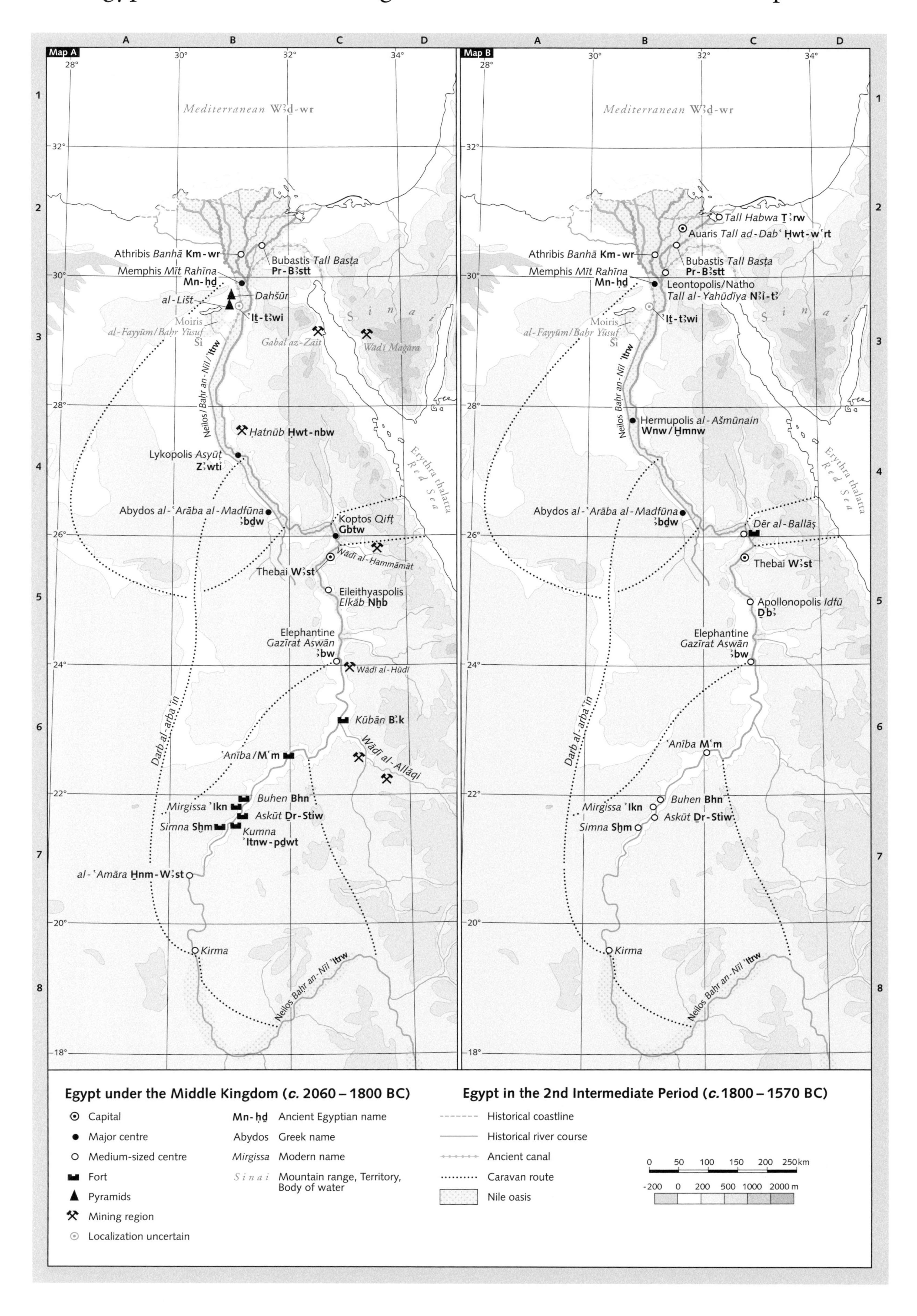
Map A
Mediterranean W3ḏ-wr
Athribis Banhā Km-wr
Bubastis Tall Basṭa
Pr-B3stt
Memphis Mīt Rahīna
Mn-ḥḏ
Dahšūr
al-Lišt
Iṯ-t3wi
Moiris
al-Fayyūm/Baḥr Yūsuf
Sinai
Gabal az-Zait
Wādī Maġāra
Neilos/Baḥr an-Nīl/ ʾItrw
Ḥatnūb Ḥwt-nbw
Lykopolis Asyūṭ
Z3wti
Erythra thalatta
Red Sea
Abydos al-ʿArāba al-Madfūna
3bḏw
Koptos Qifṭ
Gbtw
Wādī al-Ḥammāmāt
Thebai W3st
Eileithyaspolis
Elkāb Nḫb
Elephantine
Gazīrat Aswān
3bw
Wādī al-Hūdī
Kūbān B3k
Wādī al-Allāqi
ʿAnība/Mʿm
Darb al-arbaʿin
Buhen Bhn
Mirgissa ʾIkn
Askūt Ḏr-Stiw
Simna Sḫm
Kumna
ʾItnw-pḏwt
al-ʿAmāra Ḥnm-W3st
Kirma
Neilos Baḥr an-Nīl ʾItrw
Map B
Mediterranean W3ḏ-wr
Tall Habwa Ṯ3rw
Auaris Tall ad-Dabʿ Ḥwt-wʿrt
Athribis Banhā Km-wr
Bubastis Tall Basṭa
Pr-B3stt
Memphis Mīt Rahīna
Mn-ḥḏ
Leontopolis/Natho
Tall al-Yahūdīya N3i-t3
Iṯ-t3wi
Moiris
al-Fayyūm/Baḥr Yūsuf
Sinai
Neilos Baḥr an-Nīl ʾItrw
Hermupolis al-Ašmūnain
Wnw/Ḫmnw
Erythra thalatta
Red Sea
Abydos al-ʿArāba al-Madfūna
3bḏw
Dēr al-Ballāṣ
Thebai W3st
Apollonopolis Idfū
Ḏb3
Elephantine
Gazīrat Aswān
3bw
Darb al-arbaʿin
ʿAnība Mʿm
Buhen Bhn
Mirgissa ʾIkn
Askūt Ḏr-Stiw
Simna Sḫm
Kirma
Neilos Baḥr an-Nīl ʾItrw
Egypt under the Middle Kingdom (c. 2060 – 1800 BC)
Capital
Major centre
Medium-sized centre
Fort
Pyramids
Mining region
Localization uncertain
Mn-ḥḏ Ancient Egyptian name
Abydos Greek name
Mirgissa Modern name
Sinai Mountain range, Territory, Body of water
Egypt in the 2nd Intermediate Period (c.1800 – 1570 BC)
Historical coastline
Historical river course
Ancient canal
Caravan route
Nile oasis
0 50 100 150 200 250 km
-200 0 200 500 1000 2000 m

Egypt in the New Kingdom

In the New Kingdom, Egypt reached its peak and greatest extension beyond the Egyptian core territory, as far as Nubia and the Middle East. Moreover, the New Kingdom is archaeologically and philologically better documented than any other period of pharaonic Egypt.

The New Kingdom covers the timespan from the 18th to the 20th Dynasties; the 19th and the 20th Dynasties are also termed the age of the Ramessides, after the period's most frequent royal name. The 18th Dynasty lasted from *c.* 1570 to 1315 BC, the 19th reigned from *c.* 1315 to 1200 BC and the 20th was in power from *c.* 1200 to 1080 BC. The number, sequence of accession and names of the kings have been firmly established. There were only two co-regencies: of Thutmosis III and Hatshepsut in the 18th, and of Sethos I and Ramesses II in the 19th Dynasty; any other constellations are controversial. The transition periods between the dynasties were characterized by internal strife.

The New Kingdom was initiated by by Ahmose, the first king of the 18th Dynasty, who restored state unity by succesfully campaigning against the Hyksos (originally from Palestine) and expelling them. It ended with the gradual dissolution of the state, caused in part by the weakness of the Ramesside kings and in part by increasing Libyan infiltration. The result was a division of power between two ruling houses, a Tanis-based one and another one ruling in Thebae/Thebes (→ map B on p. 33).

I. Capitals

The site of the capital changed several times. At first it was at Thebes, as this was Ahmose's original residence. Palaces, administrative buildings and living quarters must have been situated on the east bank of the Nile, but there are no remains. Only a palace belonging to Amenophis III (Malqata) was excavated on the west bank. The west bank also contains the graves of the kings and officials along with the Dẹr al-Madīna settlement, where the labourers and artisans were housed. In the fifth year of his reign, Amenophis IV (Echnaton) for reasons of religious politics moved his residence to al-ʿAmārina/Amarna, which had been founded to serve this purpose. An extensive palace complex, administrative buildings, houses and a temple dedicated to Aton were excavated at this site. Echnaton and his officials had themselves laid to rest in the eastern desert, where the labourers' living quarters associated with the project have also been found. Echnaton's successor Tutanchamun, however, moved to Memphis in the third year of his reign, giving up al-ʿAmārina, which subsequently faded away as a city. It was only near the end of the 18th Dynasty, at a date which can no longer be determined with any degree of precision, that the residence was possibly moved back to Thebes.

Sethos I, the second king of the 19th Dynasty, built a new capital at Qanṭīr, close to the old Hyksos residence of Avaris/Tall aḍ-Ḍabʿa (see map C). The tensions *vis-à-vis* the Near East may have been the reason for shifting the capital to the eastern Delta. When the New Kingdom came to an end, this residence was also deserted. The site is located on a sand gezira, i.e. a kind of sand bar, on the east bank of the Pelusian branch of the Nile and covers at least 10 km². The excavations have not yet been completed. So far they have brought to light parts of the palace complex, stables and arsenals, but the actual settlement has not yet been unearthed. The city's main temple, dedicated to the god Amun-Re-Harachte-Atum, is also known.

Although Qanṭīr remained the capital until the end of the New Kingdom, Thebes with its temples at Karnak and Luxor remained the most important religious centre of the country. It was also the place where all kings, with the exception of Echnaton, and a multitude of officials had themselves entombed. Memphis, on the other hand, played the part of a 'second capital' throughout the New Kingdom period, not least because of its cult of Ptah, who rose to the status of being one of the 'state gods'. It was especially in the 18th Dynasty, under the reigns of Thutmosis IV and Amenophis III, that Memphis as well its port of Prw-nfr was an extremely important place. With the exception of the Amarna period, Memphis may even have been the residence from the reign of Thutmosis III to the end of the 18th Dynasty. In any case, the princes were educated at Memphis and numerous officials chose to be buried there from the end of the Amarna period to the early period of the Ramessides. Chaemwese, a son of Ramesses II (19th Dynasty), undertook extensive restoration of ageing building complexes in the Memphite region.

II. The Egyptian province

The canonical division of the country into nomes, if it ever was an administrative reality, became obsolete in the period of the New Kingdom at the latest. The province was governed by mayors, who resided at a central town from where they controlled the surrounding country. Large sections of agrarian land were the property of great temples like the one at Karnak, even if great distances separated them from their proprietors. The Ramesside papyri Harris and Wilbour contain information about the situation of the estates as well as ownership and taxation. A provincial town could achieve importance either because it was the starting point of a caravan trail, or because it was the site of a large temple. Among the first category are Lycopolis/Asyūt with the 'Route of forty days' (Arabic: Darb al-arbaʿīn), which led to Nubia; Coptus, the access point to the Wādī al-Ḥammāmāt; and Elephantine, a traditional border crossing into Nubia. Large temples were mostly erected during the long reign of Ramesses II, like the one for the god Herishef at Heracleopolis, the one for Amun at Hermupolis Amun, for Min at Panopolis/Aḫmīm, for Osiris at Abydus and Amun-Re at Thebes. Towns in the eastern Delta which were more or less situated along the route to the Near East also increased in importance. Examples are Tall Habwa with its fortress and Diospolis inferior/Tall al-Balāmūn with its Ramesside temple of Amun. The oases in the western desert were incorporated into the Egyptian administrative system and acted as suppliers of agrarian goods, e.g. wine.

The traditional mining areas were the Wādī al-Ḥammāmāt, where greywacke was mined, and the Wādī al-Allāqi in Nubia, where gold was extracted. In the Sinai turquoise was mined in the Wādī Maġāra as well as copper at Timnā. There were frequent expeditions to the Sinai under Thutmosis III and Sethos I, but they petered out after the reign of Ramesses VI towards the end of the 20th Dynasty.

III. The colony of Nubia

After the expulsion of the Hyksos and the consolidation of the kingdom, the kings of Egypt attempted to break Nubia's independence. In his third year in power, Kamose, the last ruler of the 2nd Intermediate Period, had already recaptured the fortress of Buhen. Thutmosis I made several military incursions into Nubia, capturing the Kushite capital of Kerma and advancing as far as Kurgus 200 km upstream from Gabal Barkal and had a border stela erected. This is the southernmost point the Egyptians were ever to reach. The actual southern border of the Egyptian kingdom, however, was at Gabal Barkal, where Thutmosis I and Thutmosis III both left inscriptions and built a fortress along with a temple of Amun.

Nubia became a colonial territory and was ruled by a so-called Prince of Kush. At ʿAnība an administrative post was set up, whose main purpose was the requisitioning of goods. Similar posts were created at Buhen and al-ʿAmāra. The former fortresses to the north of ʿAnība, which had turned into predominantly Nubian settlements in the 2nd Intermediate Period, were turned into fortresses again. New fortresses and fortified settlements were built further to the south, along with temples which were mostly dedicated to the cult of the Theban Amun. Abū Simbil represents an exception in that its temple was first of all dedicated to the deified Ramesses II. There is no evidence of the existence of Egyptian settlements between Kawa and Gabal Barkal. This could indicate that Egypt was represented there by local Nubian princes under Egyptian orders; this region, however, has hardly been examined by archaeologists. Tabo originally was a Kushite settlement as well In the 20th Dynasty, Egyptian internal weakness resulted in a decreasing influence on Nubia. The administrative structure collapsed and the Egyptian settlements disintegrated.

IV. External Relations

The contacts with the Near and Middle East, especially Palestine and Syria, are depicted on → p. 13, map B, and on → p. 15, maps A and B.

During the 20th Dynasty, there was increasing infiltration from Libya. This occurred especially in the regions of the western Delta, i.e. Heracleopolis and Thebes. (→ map A on p. 33). For protection against the Libyans, but also against the so-called Sea Peoples, Ramesses II had a chain of small fortresses erected that ran westward along the Mediterranean coast. Temple inscriptions and depictions of Merenptah and Ramesses III immortalized the Egyptian victory over the Sea Peoples.

→maps pp. 13, 15, 19, 23, 29, 33

Literature

Maps A and B: E. Edel (†), M. Görg, Die Ortsnamenlisten im nördlichen Säulenhof des Totentempels Amenophis' III., 2005; *Aḫmīm*: K.P. Kuhlmann, Materialien zur Archäologie und Geschichte des Raumes von Achmim, 1983; *al-ʿAmārina*: B.J. Kemp, S. Garfi, A Survey of the Ancient City of El-ʿAmarna, 1993; *Athribis*: P. Vernus, Athribis. Textes et documents relatifs à la géographie, aux cultes, et à l'histoire d'une ville du Delta égyptien à l'époque pharaonique, 1978; *Bubastis*: Ch. Tietze, M. Abd el-Maksoud, Tell Basta. Ein Führer über das Grabungsgelände, 2004; *Buhen*: W.B. Emery, H.S. Smith, A. Millard, The Fortress of Buhen. The Archaeological Report, 1979; *Diospolis inferior*: A.J. Spencer, Excavations at Tell el-Balamun, 3 vols., 1996–2003; *Eileithyaspolis*: F. Depuydt, S. Hendrickx, D. Huyge, Elkab IV (fasc. 1, 2). Topographie, 1989; *al-Gīza*: Ch.M. Zivie, Giza au deuxième millénaire, 1976; *Hermupolis*: A.J. Spencer, Excavations at el-Ashmunein. I. The Topography of the Site, s.a.; II. The Temple Area, 1989; *Kawa*: M.F. Laming Macadam, The Temples of Kawa. II. History and Archaeology of the Site, 2 vols., 1955; *Memphis*: D.G. Jeffreys, The Survey of Memphis. Part One: The Archaeological Report, 1985; *Mirgissa*: J. Vercoutter, Mirgissa I, 1970; *Nubia*: T. Säve-Söderbergh, L. Troy, New Kingdom Pharaonic Sites. The Finds and the Sites, 2 vols., 1991; *Qanṭīr*: J. Dorner, Die Topographie von Piramesse – Vorbericht, in: Ägypten und Levante 9, 1999, 77–83; *Sinai*: M. Chartier-Raymond, B. Gratien, C. Traunecker, J.-M. Vinçon, Les sites miniers pharaoniques du Sud-Sinaï. Quelques notes et observations de terrain, in: Cahier de Recherches de l'Institut de Papyrologie et d'Égyptologie de Lille 16, 1994, 31–77; *Tall Habwa*: M. Abd el-Maksoud, Tell Heboua (1981–1991). Enquête archéologique sur la Deuxième Période Intermédiaire et le Nouvel Empire à l'extrémité orientale du Delta, 1998.
Map C: new drawing after: J. Dorner, Die Topographie von Piramesse – Vorbericht, in: Ägypten und Levante 9, 1999, Plan 1 (with the author's revisions of the map).

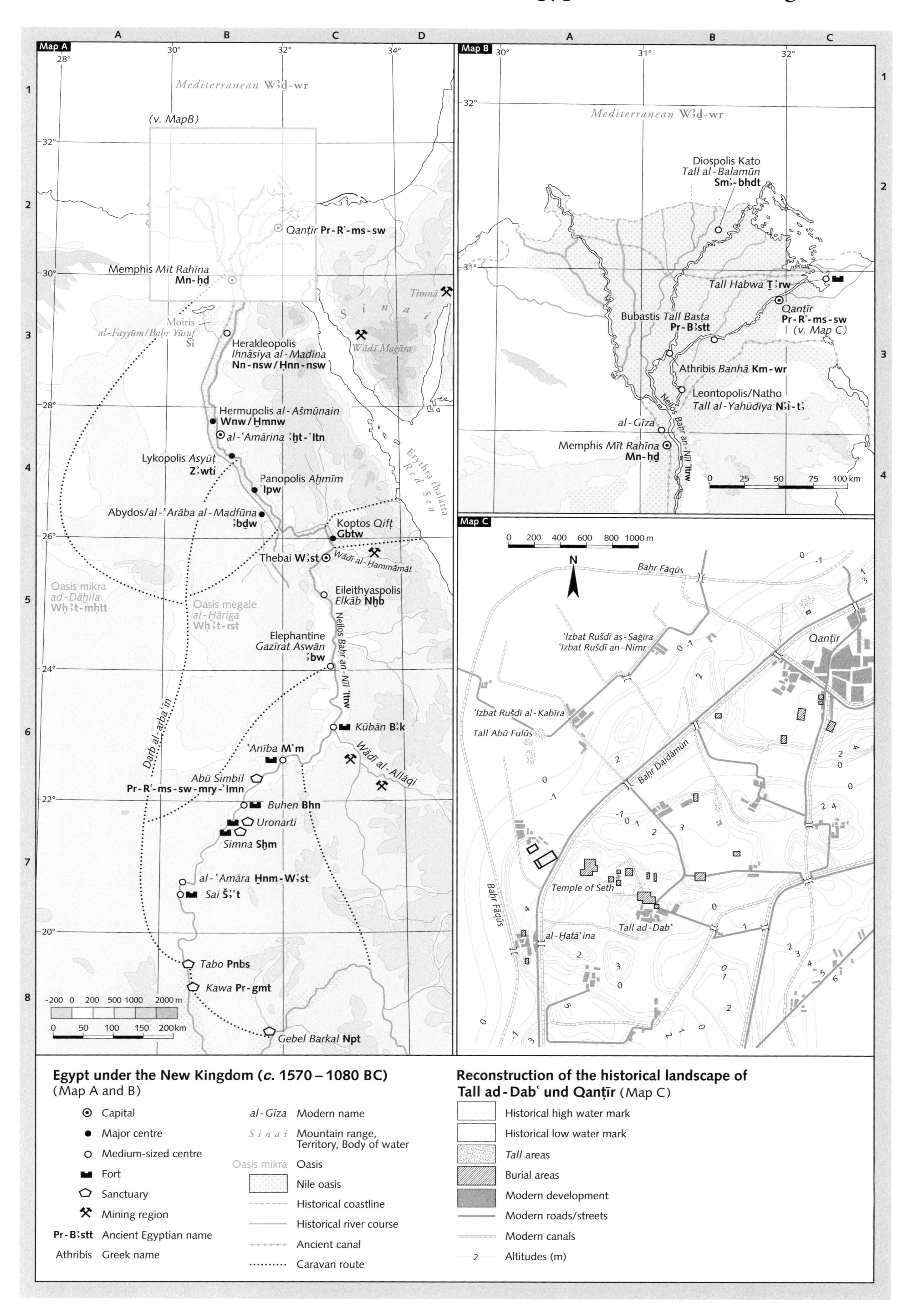
Map A
Mediterranean W3d-wr
(v. MapB)
Qanṭīr Pr-Rʿ-ms-sw
Memphis Mīt Rahīna
Mn-ḥd
Timnā
Sinai
Moiris
al-Fayyūm/Baḥr Yūsuf
Herakleopolis
Ihnāsiya al-Madīna
Nn-nsw/Ḥnn-nsw
Wādī Maġāra
Hermupolis al-Ašmūnain
Wnw/Ḫmnw
al-ʿAmārina 3ḫt-ʾItn
Lykopolis Asyūṭ
Z3wti
Panopolis Aḫmīm
ʾIpw
Erythra thalatta
Red Sea
Abydos/al-ʿArāba al-Madfūna
3bḏw
Koptos Qifṭ
Gbtw
Thebai W3st
Wādī al-Ḥammāmāt
Oasis mikra
ad-Dāḫila
Wḥ3t-mḥtt
Oasis megale
al-Ḫāriga
Wḥ3t-rst
Eileithyaspolis
Elkāb Nḫb
Elephantine
Gazīrat Aswān
3bw
Neilos Baḥr an-Nīl ʾItrw
Darb al-arbaʿīn
Kūbān B3k
ʿAnība Mʿm
Wādī al-ʿAllāqi
Abū Simbil
Pr-Rʿ-ms-sw-mry-ʾImn
Buhen Bhn
Uronarti
Simna Smn
al-ʿAmāra Ḫnm-W3st
Sai Š3ʿt
Tabo Pnbs
Kawa Pr-gmt
Gebel Barkal Npt
-200 0 200 500 1000 2000 m
0 50 100 150 200 km
Map B
Mediterranean W3d-wr
Diospolis Kato
Tall al-Balamūn
Sm3-bḥdt
Tall Habwa Ṯ3rw
Qanṭīr
Pr-Rʿ-ms-sw
(v. Map C)
Bubastis Tall Basṭa
Pr-B3stt
Athribis Banhā Km-wr
Leontopolis/Natho
Tall al-Yahūdīya N3i-t3
al-Gīza
Memphis Mīt Rahīna
Mn-ḥd
Neilos Baḥr an-Nīl ʾItrw
0 25 50 75 100 km
Map C
0 200 400 600 800 1000 m
N
Baḥr Fāqūs
ʿIzbat Rušdī aṣ-Ṣaġīra
ʿIzbat Rušdī an-Nimr
Qanṭīr
ʿIzbat Rušdī al-Kabīra
Tall Abū Fulūs
Baḥr Daidāmūn
Temple of Seth
Baḥr Fāqūs
Tall ad-Dabʿ
al-Ḫatāʿina
Egypt under the New Kingdom (c. 1570 – 1080 BC)
(Map A and B)
Capital
Major centre
Medium-sized centre
Fort
Sanctuary
Mining region
Pr-B3stt Ancient Egyptian name
Athribis Greek name
al-Gīza Modern name
Sinai Mountain range, Territory, Body of water
Oasis mikra Oasis
Nile oasis
Historical coastline
Historical river course
Ancient canal
Caravan route
Reconstruction of the historical landscape of
Tall ad-Dabʿ und Qanṭīr (Map C)
Historical high water mark
Historical low water mark
Tall areas
Burial areas
Modern development
Modern roads/streets
Modern canals
Altitudes (m)

The Hittite Empire, 'Ḫattusa', in the 13th cent. BC

There is demonstrable proof that as early as the 3rd millennium BC, Hittites were living in the area bordering on the inside of the arc formed by the river Halys. From the end of the 18th cent. BC onwards, they succeeded in creating a territorial state which was already characterized by the specifically Hittite organization of government, in which the king and his clan shared an equal degree of responsibility. The Hittite Empire, particularly the so-called Great Empire (14th/13th cents.), represents the only development of an extensive territorial state in Asia Minor before the emergence of the Achaemenid Kingdom in the 6th cent. BC.

Although the rise to dominance was not a continuous process, but was interrupted by several distinct periods of weakness, the kings' efficient political organization at home and purposeful policy of military and non-military expansion established the nation of Ḫattusa as a great power on a par with Egypt and Babylonia.

The main map displays the extent of the empire and its sphere of influence in three different periods: the reign of Ḫattusili I (*c.* 1565–1540), with the first documentary evidence of Ḫattusa as the capital city; secondly, the reign of Tudḫalija I (*c.* 1420–1400), whose campaigns contributed massively to the expansion of the empire; and finally, the period of the Great Hittite Empire (14th/13th cents.) – when in ruling over Asia Minor, Northern Syria and a part of Mesopotamia, Hittite expansion reached its peak – up to its collapse around 1200/1180 BC.

I. Development under Ḫattusili I, Tudḫalija I and their successors

According to the sources (which re-emerge after a gap of approximately 130 years), the Hittite policy of expansion was revived under Ḫattusili I. It was directed towards western Asia Minor and against the great kingdom of Ḫalpa/Aleppo in Northern Syria. His son and successor Mursili I (*c.* 1540–1530) succeeded in capturing Ḫalpa, but a subsquent period of weakness, caused by a dynastic power struggle, resulted in the complete loss of the conquered territories. The kingdom of Mittani and the Kaskaeans, who invaded the Pontus region at about that time (see below), took advantage of the situation. From about 1500, there is a definite shift in the foreign policy of the Hittite kings: in addition to a strategy of military conquest, they began to conclude treaties and set up alliances. Tudḫalija I resumed the struggle for the restoration of Hittite hegemony in Syria and also waged war in Western Asia Minor. His successors, however, had to concentrate mostly on securing the core territory against the increasing danger of Kaskaean incursions after the sack of Ḫattusa.

II. The Great Hittite Empire of Ḫattusa

The rise of Ḫattusa to the status of empire and third great power in the Near East began under Suppiluliuma I (*c.* 1355–1320), who in a single campaign succeeded in conquering (or turning into vassal states, e.g. in the North Syrian Confederation of States) territories in the east (Isuwa), all the small states between the arc of the Euphrates and the Mediterranean which had been controlled by Mittani, and territories which had been under Egyptian rule. A similar policy was subsequently adhered to in Western Asia Minor. After the destruction of Arzawa this resulted in the Confederation of Arzawa (comprising the states of Mirā, Ḫaballa, Sēḫa and later Wilusa) .

The map illustrates the political structure of the state. Spreading out from the Hittite core territory inside the arc of the Halys with the capital of Ḫattusa (Boğazköy), the Upper Country and the Lower Country to the southwest, it consisted of further incorporated and external states as well as vassal states bound to the empire by treaties. To support this system further, secundogenitures were established at Ḫalpa (Aleppo), Karkamissa (Karkamiš) and Tarḫuntassa. They survived the empire's decline for a considerable stretch of time, until the whole region disintegrated again into smaller political units. Whereas particularly Karkamissa had very close links with the capital, there was an independence movement in Tarḫuntassa, which probably contributed to the collapse of Ḫattusa around 1200/1180.

The Kaskaeans, a tribe inhabiting the region north of Ḫattusa, posed a constant threat to the empire that is well documented by Hittite sources. They repeatedly succeeded in invading it and occupying large regions for considerable periods. Like the Chalybes who are documented by later Greek sources, they are linked to metallurgy. Only very little is known, however, about their organization, way of life (nomadic or semi-nomadic?) and material culture. According to recent research, they may have been responsible for the destruction of the Hittite capital, but they may also have resettled it, albeit at a lower level of sophistication.

The greatest uncertainties about the reconstruction of historical geography as well as history concern the western part of Asia Minor. The region was probably divided up into several mostly Luwian-speaking states, which apparently forged occasional alliances or united to form 'leagues'. After repeated conflicts with Ḫattusa, they successively became vassal states under Hittite rule or within the Hittite sphere of influence – the difference is hard to see. Some Hittite place names have by now been identified; among them are Parḫā/Perge and Winuwanda/Oinoanda in the South and Lazba/Lesbos, Abasa/Ephesos and Millawanda/Miletus on the western coast of Asia Minor. The latter belonged to Aḫḫiyawa-land, a country also mentioned in Hittite texts and increasingly identified as Mycenaean Greece by researchers. However, contrary to earlier assumptions it seems only to have been a kind of bridgehead. The issue of Mycenaean-Greek settlements in southwestern Asia Minor is still a controversial point among scholars.

It is difficult to determine to what extent Alasija/Cyprus was controlled by the Hittites. While no clear evidence has been discovered in the material culture of Cyprus, at least there is documented proof of military clashes at sea and on land.

III. The approximate distribution of Anatolian languages in the second millennium BC (*c.* 1580–1200)

The supplementary map reveals that the Hittite language was spoken only in a comparatively small area, mostly inside the arc of the Halys. It was the language of the extended royal clan and very much a language of diplomacy. The indigenously Anatolian language with the broadest range of distribution was Luwian, which encompassed various dialects. There are indications that Luwian was also used in Hittite-speaking areas and probably in northwestern Asia Minor as well. Whereas Hittite has come down to us in Babylonian cuneiform on clay tablets and other text-bearing objects, Palaic and cuneiform Luwian are known to us only from glosses and sayings inserted into Hittite texts. Luwian from the age of the Great Empire has also survived in hieroglyphic Luwian inscriptions, albeit mostly from the southeastern region.

Sources

The source material is relatively rich, with evidence stemming especially from the excavated archives of clay tablets from Ḫattusa/Boğazköy, Tabikka/Maşat and Sabinuwa/Ortaköy, as well as other epigraphic and literary finds. In many cases however the identification and the location of places, countries, rivers and mountains mentioned by their Hittite names is questionable and the subject of scholarly controversy. The map presented here is essentially based on reconstructions by the Hittitologist F. Starke. Apart from the Hittite sources, textual and material ones (from excavations), it is above all the Assyrian-Akkadian and Egyptian sources to which we owe our knowledge of personal and geographical names. Although in a scholarly context it is conventional to use the Assyrian names (e.g. Akkadian Ḫatti instead of Hittite Ḫattusa) – a usage similar to the handling of Greek names –, on the present map we have deviated from this practice by using the Hittite names as far as possible.

› Maps pp. 13, 15, 27, 29, 33, 39, 87;
Ḫattusa BNP 6, 2005

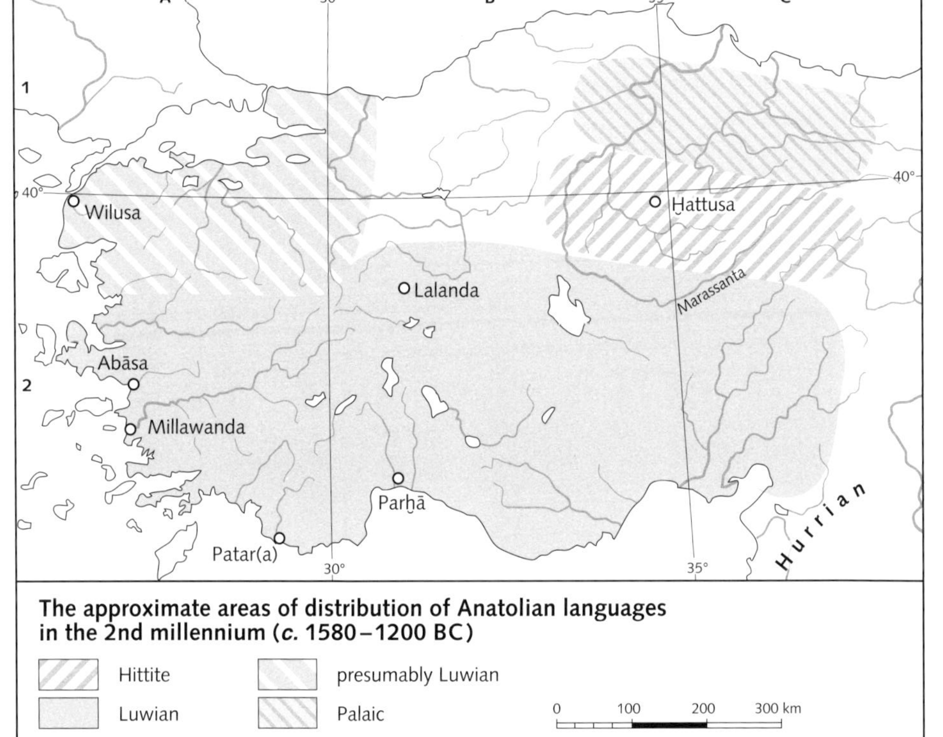

The approximate areas of distribution of Anatolian languages in the 2nd millennium (*c.* 1580–1200 BC)

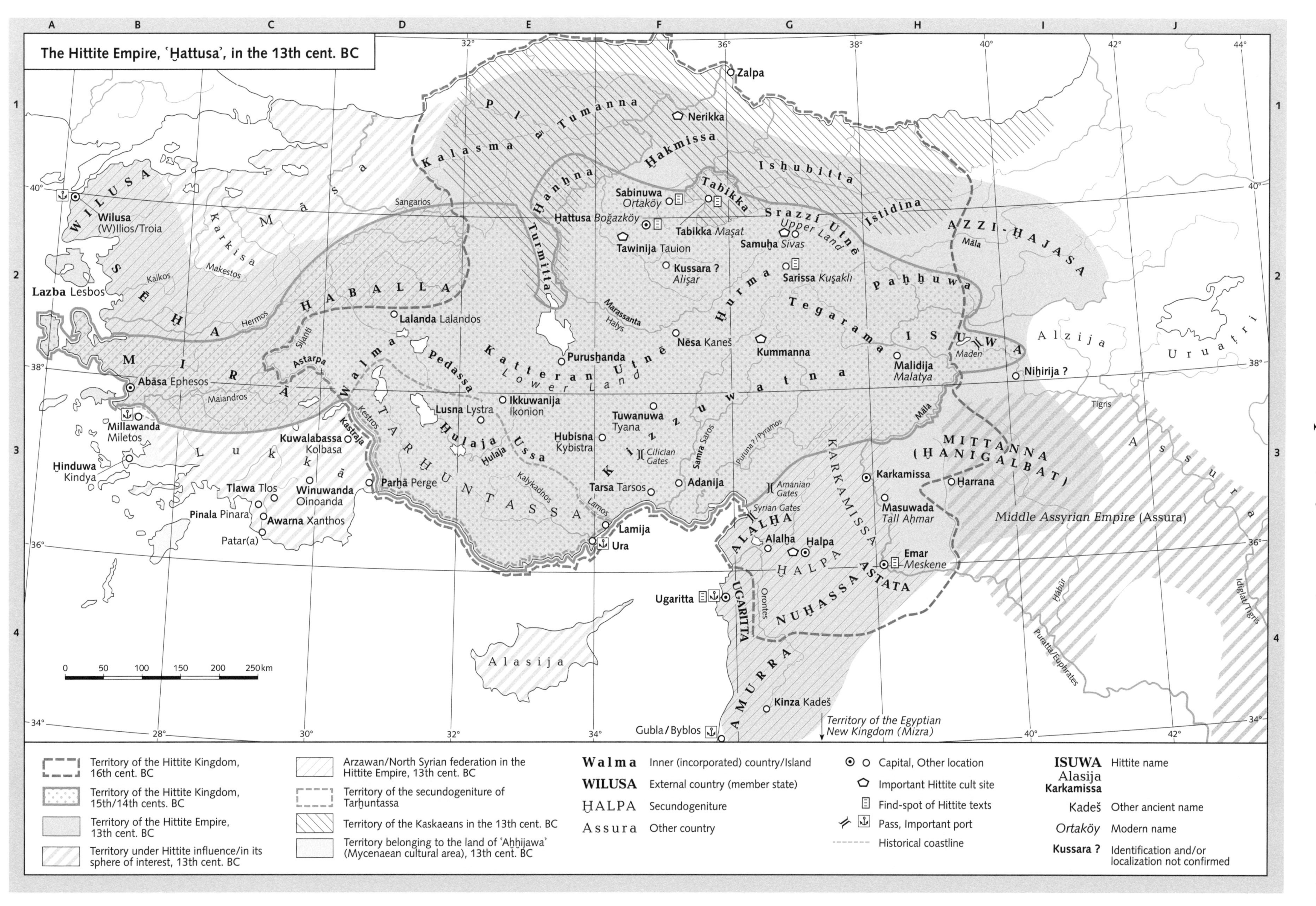
The Hittite Empire, 'Ḫattusa', in the 13th cent. BC
0 50 100 150 200 250km
Territory of the Hittite Kingdom, 16th cent. BC
Territory of the Hittite Kingdom, 15th/14th cents. BC
Territory of the Hittite Empire, 13th cent. BC
Territory under Hittite influence/in its sphere of interest, 13th cent. BC
Arzawan/North Syrian federation in the Hittite Empire, 13th cent. BC
Territory of the secundogeniture of Tarḫuntassa
Territory of the Kaskaeans in the 13th cent. BC
Territory belonging to the land of 'Aḫḫijawa' (Mycenaean cultural area), 13th cent. BC
Walma Inner (incorporated) country/Island
WILUSA External country (member state)
ḪALPA Secundogeniture
Assura Other country
Capital, Other location
Important Hittite cult site
Find-spot of Hittite texts
Pass, Important port
Historical coastline
ISUWA Hittite name
Alasija
Karkamissa
Kadeš Other ancient name
Ortaköy Modern name
Kussara ? Identification and/or localization not confirmed
WILUSA
Wilusa (W)Ilios/Troia
Lazba Lesbos
Abāsa Ephesos
Millawanda Miletos
Ḫinduwa Kindya
Karkisa
Lukkā
Tlawa Tlos
Pinala Pinara
Awarna Xanthos
Patar(a)
Winuwanda Oinoanda
Kuwalabassa Kolbasa
Kastraja
Parḫā Perge
ḪABALLA
Lalanda Lalandos
Pedassa
Ḫulaja
Lusna Lystra
Ikkuwanija Ikonion
TARḪUNTASSA
Ḫubisna Kybistra
Tuwanuwa Tyana
Tarsa Tarsos
Lamija
Ura
Adanija
Kizzuwatna
Alasija
Turmitta
Ḫattusa Boğazköy
Sabinuwa Ortaköy
Tawinija Tauion
Kussara ? Alişar
Nēsa Kaneš
Purusḫanda
Tabikka Maşat
Samuḫa Sivas
Sarissa Kuşaklı
Kummanna
Zalpa
Nerikka
Ḫakmissa
Kalasma
Tumanna
Ishubitta
Istidina
AZZI-ḪAJASA
ISUWA
Malidija Malatya
Tegarama
KARKAMISSA
Karkamissa
Masuwada Tall Aḥmar
Emar Meskene
ASTATA
Ḫarrana
Niḫrija ?
MITTANNA (ḪANIGALBAT)
Middle Assyrian Empire (Assura)
Alzija
Uruatri
Ḫalpa
Alalḫa
NUḪASSA
Ugaritta
UGARITTA
Kinza Kadeš
AMURRA
Gubla / Byblos
Territory of the Egyptian New Kingdom (Mizra)
Puratta/Euphrates
Idiglat/Tigris
Orontes
Halys
Sangarios
Maiandros
Hermos
Kaikos
Kestros
Samra Saros
Puruna ? / Pyramos

The Aegean area in the Bronze Age

The Aegean area in the Bronze Age

Far from separating the nations living on its shores, as is the case today, the Aegean Sea in Antiquity served as a medium of interaction to its coastal communities in Macedonia, Greece and Crete, along the shoreline of Asia Minor and in the Ancient Balkan region in the north. In the Aegean Bronze Age (*c.* 2700–*c.* 1200 BC) and the subsequent Iron Age (*c.* 1200–*c.* 900 BC), cultural assets and ideas spread along the sea routes to such a degree that we can speak of an 'Aegean koine', although interpretation of the surviving material and textual sources shows that the nature of this phenomenon varied. Whereas in the Early Bronze Age (map A) economic relations seem to have prevailed, the so-called Minoan koine (*c.* 2200– *c.* 1400 BC; map B) and the Mycenaean koine (*c.* 1400–*c.* 1200 BC; → map p. 29) may have been driven by expansive power politics as well.

I. The Aegean region in the Early Bronze Age (*c.* 2700–2200 BC) (map A)

The evidence of bronze metallurgy is an important indicator of contacts in the Early Bronze Age, with the Trojan/East Aegean culture group apparently leading the way. This new technology entailed increased mobility in order to safeguard the supply of raw materials and the marketing of the products. It also involved the division of labour in the processing of the raw materials and the emergence of elites, whose existence is documented archaeologically by treasure finds such as 'Priam's treasure' at Troy II). The use of longboats led to a substantial increase in the travelling range and to the addition of new markets and producers. Both innovations – bronze metallurgy and the introduction of the longboat – triggered a surge of cultural advances in the Aegean region: the archaeological remains clearly show that among the elites of the four culture groups which had formed around the Aegean Sea and on its islands, there was a tendency to accumulate wealth and, as a consequence, prestige. Analyses of lead isotopes as well as finds of certain metal objects have revealed that metal was traded among the four culture groups.

The exchange of religious ideas is illustrated by imports and local imitations of marble female figurines of the Cycladic (so-called canonical) type on the Greek mainland and in Crete. The exchange of social practices is reflected by the introduction in the Cyclades and Crete of a certain type of mainland vessel (sauce boat) used for drinking or pouring, and the emergence of drinking vessels (*depas*, stirrup jar) stemming from Asia Minor in the Cyclades and on the Greek mainland. At the same time, the individual regions preserved their own cultural identities, which is documented by numerous finds of objects whose spread is limited.

Towards the end of the 3rd millennium, the first Aegean koine was terminated by a series of destructions. There is archaeological evidence of discontinuities of settlement and changes in the material culture in the Cyclades as well as in central and southern Greece; their causes are a matter of sharp controversy among scholars (migrations or invasions from western Asia Minor and the islands of the eastern Aegean; the collapse of social structures; climate change).

II. The expansion of the Minoan koine in the Middle and Late Bronze Ages (*c.* 2200–1400 BC) (map B)

Crete was largely spared from the upheaval; on the contrary, the Minoan civilization must be considered the first high culture in the Aegean region. It produced the so-called Old Palaces, which were erected in the middle of settlements on Crete and served as centres of administration, trade and religion. Further evidence is provided by Minoan settlements and trading outposts in the Aegean, probably as far as Samothrace. The recent introduction of sailing permitted the extension of Minoan territory or of the Minoan political (or at least cultural) sphere of influence along with the export of Cretan goods. The various sea routes were also used to import finished products as well as raw materials, especially metals such as silver and lead from Siphnos and the Laureum mines (near Thoricus in Attica). From the middle of the 3rd millennium, important sea routes, guarded by bases along the way and in use both in the Old and in the New Palace Periods, led to the southern Peloponnese (via Cythera). Various parts of Asia Minor and Rhodes were accessed via Carpathos from the early 2nd millennium onward. Rhodes in turn was the departure point of two routes. One of these ran along the southern coast of Asia Minor into the Eastern Mediterranean (Kamares ware and precious metal objects were found among others on Cyprus, in the Levant and as far away as Egypt and Mari in Syria). A second route followed the west coast via the islands of Cnidos, Cos and Iasos to Samos, from where it continued to the north-eastern Aegean (seal marks on Samothrace). A third route linked Crete with the Argolid, Attica and Thessaly, running via Thera, Melos and Ceos. These contacts are documented by finds of imported goods on Crete and by written sources.

Around 1750 BC, the so-called Old Palace Age came to an end, probably as a result of the destructions caused by earthquakes. In a seamless transition, the New Palaces were erected; some indicators point to Knossos as taking the leading role among them. In the New Palace Period, the Minoan presence and influence in the Aegean region is documented by even more positive and widespread evidence than before; they are closely linked with the search for raw materials and the trade in finished goods.

Although it is still a moot point among researchers, there are good reasons for assuming the existence of a Minoan maritime empire rather than a network of independent trading entities: purely Minoan outposts (Castri on Cythera, Trainda/Ialysus on Rhodes, Millawanda/Miletus) as well as settlements with a predominantly indigenous population and a Minoan presence (Thera, Melos, Ceos), clues to the active use of Linear A script (Thera, Melos, Ceos, Cythera, Miletus) and of the Minoan system of weights – lead weights whose distribution range is similar to that of the written records. This is complemented by finds of Minoan household wares and loom weights of the discoid Minoan standard type on a number of Cycladic and Dodecanese islands and on the south-western coast of Asia Minor. The perception of Crete as a political unit by the Egyptians is another indicator (Kaftu/Keftu = 'Crete and the islands in the middle of the great green sea'). Depictions of Cretans in Egyptian Thebes and Egyptian written documents are proof of the diplomatic and commercial relations between the two nations. Central and southern mainland Greece were probably part only of the 'cultural' koine of the Minoized Aegean region in the Late New Palace Period (import of Minoan prestige objects, imitation of the Minoan style and of Minoan iconography, ceramics of the New Palace period).

III. Trading contacts in the Bronze Age (supplementary map)

The reconstruction of the trade routes (only diagrammatically shown here) in the Aegean region and far beyond from the Early Bronze Age to the beginning of the Late Bronze Age is based, on the one hand, on finds of imported objects from different regions, which have been excavated on Crete (metal objects, ceramic ware, ivory, stone vessels) and, on the other hand, on the evidence of the transfer of technologies and ideas in connection with the import and processing of metals (e.g. the use of seals to supervise the traffic of goods). Metal was not only extracted in the Aegean region itself, but was also imported from the Levant (as a relay station) and probably the Balkans, as rich finds in graves at Nidri on Leucas indicate. Leucas sat on the Ionian-Adriatic route and must have been part of the trading system. The burial gifts are of Cycladic, East Aegean and Cretan types, whereas the lay-out of the graves is Balkanic. Clues pointing to early contacts between Crete and the western Mediterranean region can be found in Sicily, on Sardinia and possibly even on the Iberian Peninsula. The distribution range of Minoan ceramics ware (e.g. Kamares ware) extended via Cyprus as far as the Levant and Egypt. The archives in Mari on the Euphrates contain records of imports from Crete. Aegean fresco technique wall paintings have been identified in the Levant (Alalaḫ, Qaṭna, Tall Kabri; → map p. 29) and Egypt (Avaris/Tall aḍ-Ḍabʿa; → maps pp. 21, 29).

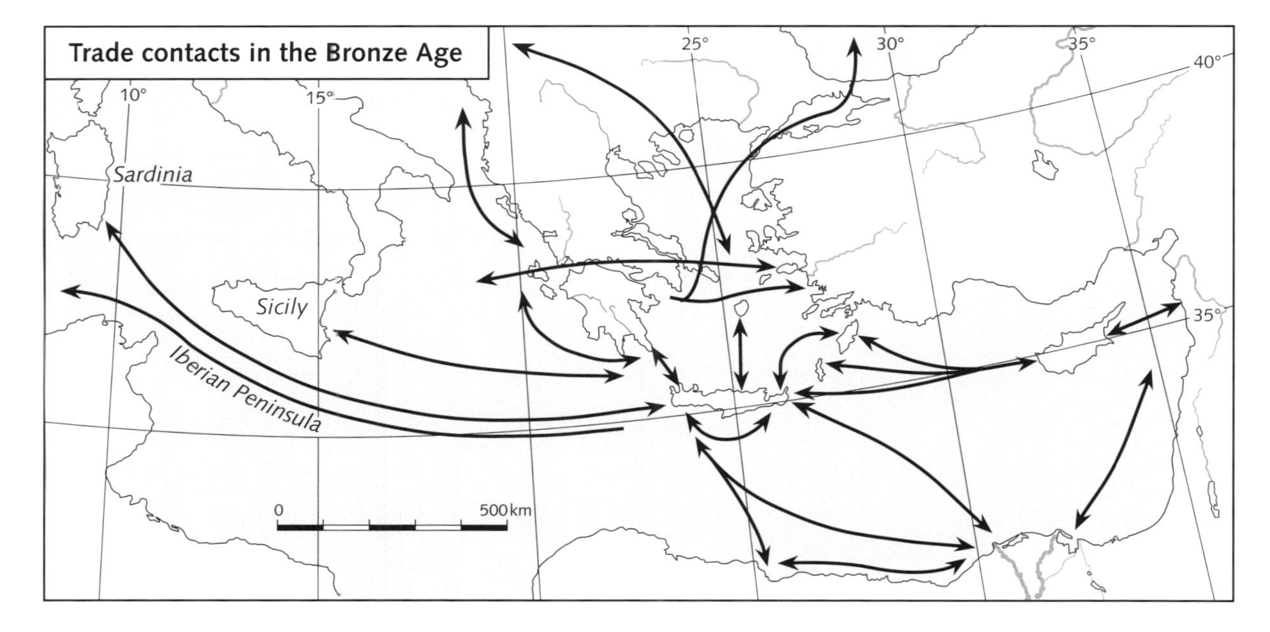

Sources

Whereas our knowledge of the Early Bronze Age is based on finds from archaeological contexts of the various culture groups, the Minoan koine supplies us with a wider spectrum of finds and, in addition, with Linear A texts (which have so far not been decoded). The source material for the Mycenaean koine is even better, since the material evidence is supplemented by decipherable Linear B texts (→ maps pp. 27, 29).

Owing to the political situation prevailing until the end of the Cold War, the relationships between the Ancient Balkanic region and the other Aegean civilizations have not yet been sufficiently explored.

→Maps pp. 3, 15, 17, 19, 21, 23, 27, 29, 33, 35, 37, 39; Aegean koine, BNP 1, 2002, 174–187

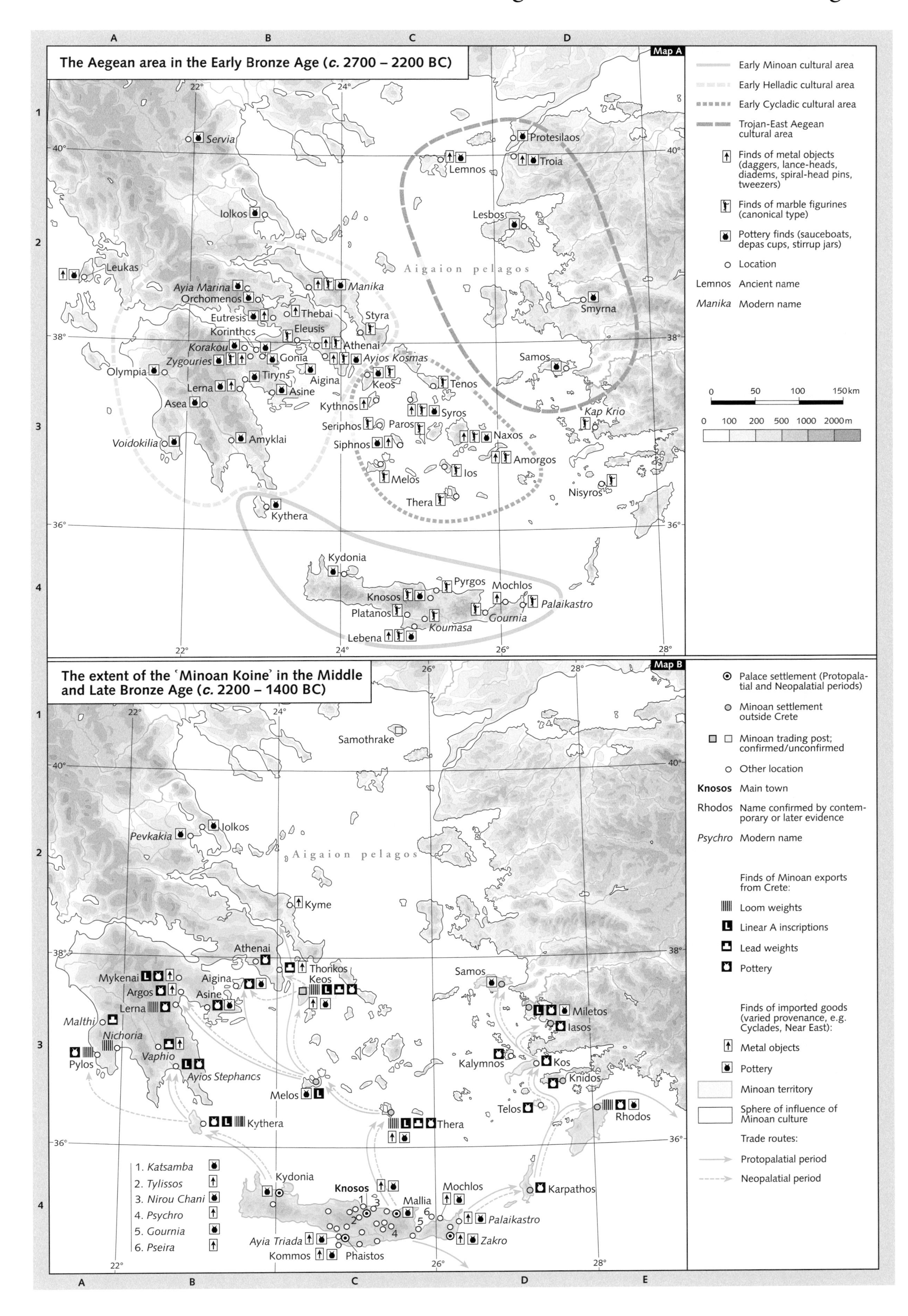
The Aegean area in the Early Bronze Age (c. 2700 – 2200 BC)
Map A
Early Minoan cultural area
Early Helladic cultural area
Early Cycladic cultural area
Trojan-East Aegean cultural area
Finds of metal objects (daggers, lance-heads, diadems, spiral-head pins, tweezers)
Finds of marble figurines (canonical type)
Pottery finds (sauceboats, depas cups, stirrup jars)
Location
Lemnos Ancient name
Manika Modern name
Aigaion pelagos
Servia
Protesilaos
Troia
Lemnos
Lesbos
Iolkos
Leukas
Ayia Marina
Orchomenos
Manika
Eutresis
Thebai
Styra
Korinthos
Eleusis
Korakou
Athenai
Zygouries
Gonia
Ayios Kosmas
Olympia
Tiryns
Aigina
Keos
Tenos
Lerna
Asine
Asea
Kythnos
Syros
Seriphos
Paros
Naxos
Siphnos
Amyklai
Voidokilia
Amorgos
Ios
Melos
Thera
Kythera
Smyrna
Samos
Kap Krio
Nisyros
Kydonia
Pyrgos
Mochlos
Knosos
Palaikastro
Platanos
Gournia
Koumasa
Lebena
The extent of the 'Minoan Koine' in the Middle and Late Bronze Age (c. 2200 – 1400 BC)
Map B
Palace settlement (Protopalatial and Neopalatial periods)
Minoan settlement outside Crete
Minoan trading post; confirmed/unconfirmed
Other location
Knosos Main town
Rhodos Name confirmed by contemporary or later evidence
Psychro Modern name
Finds of Minoan exports from Crete:
Loom weights
Linear A inscriptions
Lead weights
Pottery
Finds of imported goods (varied provenance, e.g. Cyclades, Near East):
Metal objects
Pottery
Minoan territory
Sphere of influence of Minoan culture
Trade routes:
Protopalatial period
Neopalatial period
Samothrake
Iolkos
Pevkakia
Aigaion pelagos
Kyme
Athenai
Thorikos
Mykenai
Aigina
Keos
Argos
Asine
Lerna
Malthi
Nichoria
Vaphio
Pylos
Ayios Stephanos
Samos
Miletos
Iasos
Kalymnos
Kos
Knidos
Melos
Telos
Rhodos
Kythera
Thera
Kydonia
Knosos
Mochlos
Karpathos
Mallia
Palaikastro
Zakro
Ayia Triada
Kommos
Phaistos
1. Katsamba
2. Tylissos
3. Nirou Chani
4. Psychro
5. Gournia
6. Pseira

The extent of Mycenaean culture in the Aegean area (17th to 11th cents. BC)

The Mycenaean civilization has been named after its main site Mykenai/Mycenae, located in the Peloponnesian Argolid. In scholarly usage the term signifies the Late Bronze Age (late 17th–11th cents. BC) civilization of mainland Greece and the continuation of the Minoan palace system across the whole Aegean region under Mycenaean auspices (→ map p. 25).

Its essential manifestations are the finds and findings at the sites immortalized by Homer's *Iliad* and *Odyssey*. They are supplemented by the sites where Linear B tablets and other objects have been found (→ supplementary map) which bear the earliest representations of any Greek dialect. The main significance of the Mycenaean civilization lies in its role as a spatial and chronological intermediary. By integrating the 'Mycenaean koine' into the cultural and political context of the Eastern Mediterranean (→ map p. 29) and extending its sphere of influence into the Western Mediterranean (→ map p. 31), it made a fundamental contribution to the emergence of Greek civilization.

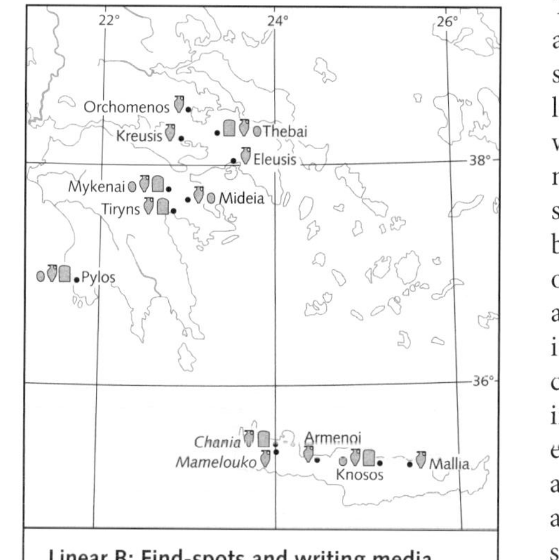

Linear B: Find-spots and writing media (*c.* 1420 – 1180 BC)

I. Chronology

The Mycenaean age can be subdivided into the Early Mycenaean Era (1600–1450), the Palace Era (1450–1200) and the Postpalatial Era (1200–1050), the dates being approximations. The onset of the transition to the Greek Early Iron Age, also known as the 'Dark Ages', cannot be dated accurately as there is no relevant archaeological evidence of settlements. Therefore the 11th cent. is regarded as the gradual decline of the Mycenaean and called the sub-Mycenaean phase.

It should be noted that there are connections to Egypt. Thus the earliest Mycenaean period, the so-called Shaft Grave Era, is considered contemporaneous with the beginning of the New Kingdom in Egypt. Finds of Mycenaean ceramic ware in King Akhenaton's palace at Amarna are considered proof of a flourishing Mycenaean civilization in the mid 14th cent. The time of crisis between 1200 and 1150 BC, which is documented in Egyptian sources and affected a large part of the Eastern Mediterranean, also brought about the collapse of most palaces in the areas settled by Mycenaeans, even though it did not spell the end of their civilization.

1. The Early Mycenaean Era (c. 1600–1450 BC)

The core settlement area of the Mycenaeans comprised the Peloponnese and the southern part of Central Greece. Arable land for producing an agricultural surplus was a rarity in these territories. The only mineral resources were small quantities of silver, lead and copper, while bronze had to be imported. The local trading assets were olive oil and above all ceramic ware, but also finished luxury goods crafted from imported raw materials. While Minoan culture initially was the main source of inspiration to the Early Mycenaean Era, especially with regard to pottery, this situation was clearly reversed later; the causes are not quite clear, but there were obvious social and political changes. They are manifested by the switch from family graves (shaft graves) to elaborate *tholos* tombs, which can only have been constructed in a hierarchical social context and by a division of labour.

It was during this period that Crete was captured by the Mycenaeans; they took over its Aegean hegemony by aggressive expansion. This is evidenced by the spread of Mycenaean settlements and settlements with a significant admixture of Mycenaean components. It has been suggested that this was made possible by a concentration of power in the hands of a group which controlled the import of key raw materials.

2. The Palace Era (c. 1450–1200 BC)

In the so called Palace Era, which followed the Early Mycenaean Era, mighty palaces (some of them archaeologically unverified to date), decorated with frescoes on walls and floors were erected for purposes of representation, central government and the production and distribution of goods. They were headed by a *wa-na-ka* (ruler) and situated at the central settlements of the various economic regions, which – due to the topographical restrictions – can be accurately defined: the Argolid with Mycenae, Midea and Tiryns, Messenia with Pylos, probably Laconia with the Menelaion, Attica with Athens and Eleusis, Boeotia with Thebes, Orchomenus and Gla and presumably the area near Iolcus; furthermore Crete with Knossos and possibly Cydonia and Mallia. Not all of these residences were fortified castles; there are no protective walls around the Laconian and Messenian palaces, for instance. One possible explanation for the enormous walls (at Mycenae and Tiryns) and the extensions of the fortifications in the 13th cent. BC, when there was no recognizable external threat, is rivalry among the rulers of the Argolid – the subject of the myths of Proetus and Acrisius as well as of Heracles and Eurystheus (Tiryns versus Argos).

It was a period of stunning feats of engineering and logistics. At Tiryns, a large dam was raised and the river threatening the city was diverted into a channel. A sizeable water reservoir has been identified at Mycenae, while in Boeotia a big chunk of acreage was gained by draining the Copais basin near Orchomenus. Overall, there is evidence of increasing prosperity: the settlements were growing in size, and a new type of spacious house with integrated workshops and storage rooms emerged in the palaces' vicinity. Apart from the elite's *tholos* tombs, we find sprawling necropoleis comprising mostly chamber tombs, frequently a long way from the respective settlements. Manufactured products such as ceramic ware, terracotta goods, tools and weapons, seals and jewellery are characterized by a uniform development and a large degree of homogeneity, whose effects can even be felt among the local elites of Northern Thessaly and Macedonia (→ map p. 105). The Aegean and western Asia Minor, which were controlled by the Great Hittite Empire or at least under its influence, were also firmly within the range of Mycenaean economic and cultural influence. This is proven by import finds – e.g. at Troy, Larisa, Ephesus and Miletus – and by the existence of Mycenaean trading posts, among others at Miletus and Müskebi. Mycenaean trade in the eastern and western Mediterranean flourished.

3. The Postpalatial Era and the 11th century

The Mycenaean palace system broke down in the decades around 1200 BC. What caused the collapse is still a matter of debate: a case can be made for internal strife, but also for the impact of external forces. The consequences were dramatic in any case. In the so-called Postpalatial Era (*c.* 1200/1150–1050 BC), the system of the centralized palace economies fell apart and dissolved into separate regional units. The skill of writing was obviously lost, for there are no Linear B documents dating from that time. Although there is a continuity of settlement in many places, large buildings were no longer erected, only small one-room or two-room houses, some of them with an apsis. We can also observe a turning towards religion (as manifested by terracotta figurines and indications of worship in small new 'chapels') and the development of distinctive local pottery styles. The last area still identifiable as culturally Mycenaean was the Southern Aegean core territory. A general decrease in population in all the Mycenaean settlement areas – some areas were completely deserted – continued into the sub-Mycenaean period. There is evidence, however, of the presence of Mycenaean groups on Cyprus.

The excavated material from the sub-Mycenaean phase (11th cent.) consists mostly of graves and very few, scarcely verified remains of settlements. We can assume that people lived in simple farmhouses which have left no archaeological traces, and supported themselves by herding cattle in a semi-nomadic way of life. The end of this phase has been linked to the appearance of proto-geometrical pottery, although hard stratigraphical evidence is missing. The breakdown of the palace system and its political functions cleared the way for the development of individual settlements and the *polis* structure.

II. Linear B: sites and media (*c.* 1420–1180 BC) (supplementary map)

Linear B documents are written in Mycenaean, the earliest Greek dialect known to us so far. Their spread is largely coterminous with that of Mycenaean civilization in the Late Bronze Age. There are three distinguishable regional foci: Crete, with finds at Knossos, Khania, Armenoi, Mallia and Mamelouko; the Peloponnese, with finds in the Argolid (Mycenae, Tiryns and Midea) and in Messenia at Pylos; and Central Greece, with finds at Eleusis in Attica and at the Boeotian sites of Thebes, Creusis and Orchomenus. The texts were discovered either in palace archive rooms or at sites pointing towards manufacturing activities.

The oldest texts come from Late Minoan Knossos (*c.* 1420-1400), which proves that Crete must have been under Mycenaean rule from about the middle of the 15th cent. BC. The rest of the texts known from Knossos – which include inscriptions on vases – date from the years between 1375 and 1350. The earliest find at Pylos dates from the same period. All texts from Thebes and Khania and the majority of finds at Mycenae belong to the phase around 1280 (LM III B). The youngest texts were found in mainland Greece (Mycenae, Pylos: *c.* 1220-1180); some inscriptions on Cretan vases date from the same period.

There are about 6,000 Linear B texts, most of which have been recorded on clay tablets. Other materials were used as well, such as storage and transport vessels and so-called nodules. Frequently in the form of succinct notes, they deal exclusively with administrative and practical matters. Nevertheless we can draw conclusions from them about various areas of Mycenaean civilization, such as the administrative and economic systems and the social and political organization of the palaces.

→ Maps pp. 21, 23, 25, 29, 31, 33, 105; Cretan-Mycenaean archaeology, BNP/CT 1, 2006; Linear B, BNP 7, 2005; Mycenaean culture and archaeology, BNP 9, 2006

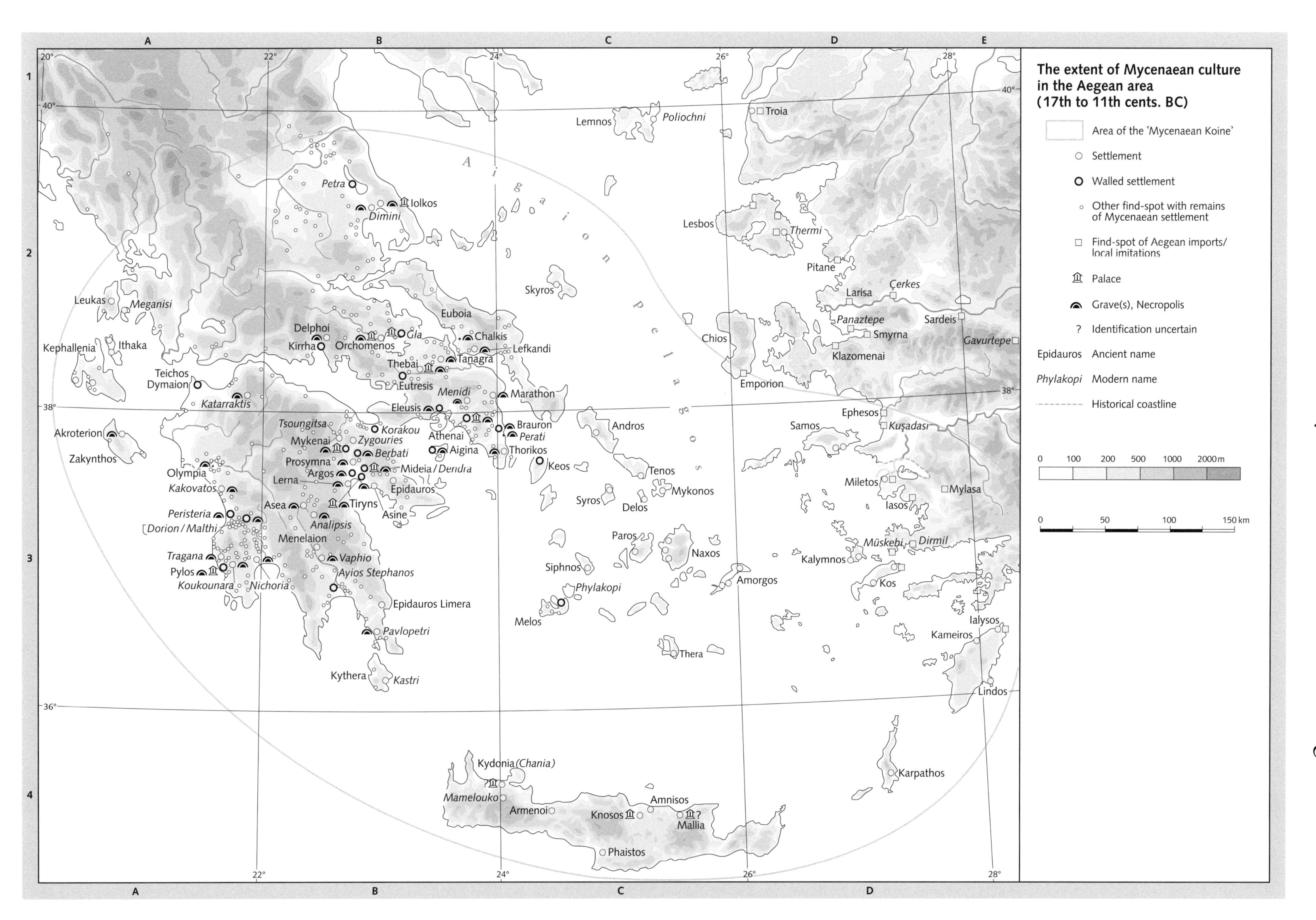
The extent of Mycenaean culture in the Aegean area (17th to 11th cents. BC)
Area of the 'Mycenaean Koine'
Settlement
Walled settlement
Other find-spot with remains of Mycenaean settlement
Find-spot of Aegean imports/ local imitations
Palace
Grave(s), Necropolis
? Identification uncertain
Epidauros Ancient name
Phylakopi Modern name
Historical coastline
0 100 200 500 1000 2000 m
0 50 100 150 km
Aigaion pelagos
Lemnos
Poliochni
Troia
Thermi
Lesbos
Pitane
Larisa
Çerkes
Panaztepe
Smyrna
Klazomenai
Sardeis
Gavurtepe
Chios
Emporion
Samos
Ephesos
Kuşadası
Miletos
Mylasa
Iasos
Müskebi
Dirmil
Kalymnos
Kos
Ialysos
Kameiros
Lindos
Karpathos
Amorgos
Naxos
Thera
Mykonos
Tenos
Delos
Paros
Andros
Syros
Siphnos
Phylakopi
Melos
Keos
Skyros
Euboia
Lefkandi
Chalkis
Tanagra
Marathon
Brauron
Perati
Thorikos
Athenai
Aigina
Menidi
Eleusis
Thebai
Eutresis
Gla
Orchomenos
Delphoi
Kirrha
Iolkos
Dimini
Petra
Korakou
Zygouries
Berbati
Mykenai
Tsoungitsa
Prosymna
Argos
Mideia / Dendra
Epidauros
Tiryns
Asine
Lerna
Analipsis
Asea
Menelaion
Vaphio
Ayios Stephanos
Epidauros Limera
Pavlopetri
Kythera
Kastri
Nichoria
Koukounara
Pylos
Tragana
Peristeria
[Dorion / Malthi]
Kakovatos
Olympia
Katarraktis
Teichos Dymaion
Zakynthos
Akroterion
Kephallenia
Ithaka
Leukas
Meganisi
Kydonia (Chania)
Mamelouko
Armenoi
Knosos
Amnisos
Mallia
Phaistos
20°
22°
24°
26°
28°
36°
38°
40°
A
B
C
D
E
1
2
3
4

The Eastern Mediterranean in the Late Bronze Age (*c.* 1400–1200 BC) – political and cultural interconnections

This map focusses on the interaction between the territories settled and culturally dominated by the Mycenaeans, the so-called Mycenaean koine (*c.* 1400–*c.* 1200 BC), and the system of states in the eastern Mediterranean under the hegemony of Ḫattusa and Egypt. On the basis of archaeological evidence and Linear B inscriptions, it traces the development of the Mycenaean koine in Greece and the Aegean region. It also shows its connections with the eastern Mediterranean as evidenced by Mycenaean imports into Asia Minor, the Levant and Egypt, and also by the corpus of names in Hittite and Egyptian texts which refer to the core region of the Mycenaean koine. Imports into Mycenaean territory from Egypt and the Near East have not been charted on this map.

I. The Mycenaean koine

1. Definition

The term 'Mycenaean koine' refers primarily to the highly homogeneous Mycenaean civilization of the so-called Palace Era (*c.* 1450–1200 BC). The development and expansion of this cultural koine is closely linked to the emergence and history of the Mycenaean 'palace states', centralized theocracies running a redistributive economy, which required a large and complex administration. Research of these contexts relies on the evaluation of the Linear B texts, most of which come from the archives at Mycenae, Pylos, Thebes and Knossos. They are also the earliest existing source of the Greek language. The systems of writing, measures and numbers along with the architecture, furnishings, function and organization of the Mycenaean palaces clearly owe much to Minoan influences. Scholars are in disagreement whether this is also true of the monarchic ideology; the emphatic centralism of governance and administration evokes oriental models as well. There is equal controversy about the question of the existence of a 'Mycenaean empire' ruled by Mycenae. The homogeneous nature of architecture, organization, cult and religion, economy and art in the palace states, combined with the administrative centralism, seems to support this hypothesis, as does the Hittite name 'Aḫḫiyawa' (Achaia). On the basis of the Linear B texts it seems more probable however that the palatial centres were closely connected, but should nonetheless be regarded as independent units.

2. Expansion of the Mycenaean koine; international contacts

The Mycenaean Palace Era saw an increase of Mycenaean influence throughout the Mediterranean. As early as the 15th cent., Minoan settlements on the Cyclades fell to the Mycenaeans; Knossos and large parts of Crete were added around 1400 at the latest. Altogether they took over Crete's former domain and its entire sphere of influence in the Aegean. Minoan outposts became Mycenaean, e.g. on Rhodes. In mainland Greece, Mycenaean influence probably expanded into Phocis, the Spercheius Valley, western Thessaly and areas north of Mount Olympus. It is generally accepted nowadays that – as was the case in Minoan times – there were only scattered Mycenaean trading posts such as Miletus/Millawanda, Iasus or Panaztepe on the west coast of Asia Minor, whereas the aforementioned Aḫḫiyawa (or a part of it) definitely was not located here; both in Ḫattusa and in western Anatolia, this term was reserved for mainland Greece and the islands.

The incorporation of all these areas into the Mycenaean sphere of influence was initiated by the palaces, who used them as trading posts. The Mycenaeans inherited the Minoan trading network in the eastern Mediterranean. It is a proven fact that they maintained trading and diplomatic contacts with the Egyptian New Kingdom and its areas of influence in the southern Levant as well as with the northern Levant (especially Ugarit; cf. supplementary map), which was part of the Hittite empire. Although we cannot exclude the possibility of direct contacts with the Hittites, there is little probability of such contacts. It is likely, however, that there was a Mycenaean trade route touching the coast of southern Anatolia (Tarsus, Mersin) and Cyprus. The Mycenaeans also maintained trade relations with places to the north and in the western Mediterranean.

Ancient oriental and Mycenaean Linear B texts (archive tablets, inscriptions), pictorial evidence and archaeological finds serve to document above all the trade with the East, which culminated between *c.* 1340 (the period of Akhnaton and Amarna) and 1250 BC (Ramesses II). On the Mycenaean side, the import trade in metals, oil, wine, grain and commodities – including luxury goods – was monopolized by the palaces, whereas provincial areas had only indirect access to it. Vice versa, Mycenaean export goods – mostly ceramic ware – originated from the palace centres. Such objects have been found in Upper and Lower Egypt, throughout the Levant, in northern Syria, at scattered sites in central Anatolia (inner arc of the river Halys) and southern Anatolia (e.g. Tarsus) and more frequently in western Anatolia. There is evidence of Mycenaean cultural influence on Cyprus, but (at least during that period) no trading posts seem to have been set up of the type found e.g. in the western Mediterranean along the Gulf of Taranto and in Sicily (→ map p. 31).

II. The topography of the Kingdom of Ugarit (14th cent. to *c.* 1180 BC) (supplementary map)

The site of the town of Ugarit, present-day Ra's Šamra, on the coast of Syria, was permanently inhabited from *c.* 6500 to *c.* 1180 BC, flourishing from the 14th cent. to its demise in 1180. It owed it position as a metropolis of trade to its favourable strategic situation at the intersection of important sea and land trade routes. This is documented by archaeological finds and written sources.

The city of Ugarit – a site excavated uninterruptedly since 1929, under French direction – was the centre and residence of a city state typical of the region of north ern Syria and the Levant. On the so-called acropolis a vast royal palace was situated as well as extensive temple sites and archives withs clay tablets in eight different languages. Among these are Ugaritic (also found on other sites in the kingdom and e.g. on Cyprus), Akkadian, Hurritic, Hittite, Egyptian and Cypro-Minoan. Half of the tablets (*c.* 2,100) are inscribed in Mesopotamian cuneiform, the other half (*c.* 2,000) in an alphabetic cuneiform script.

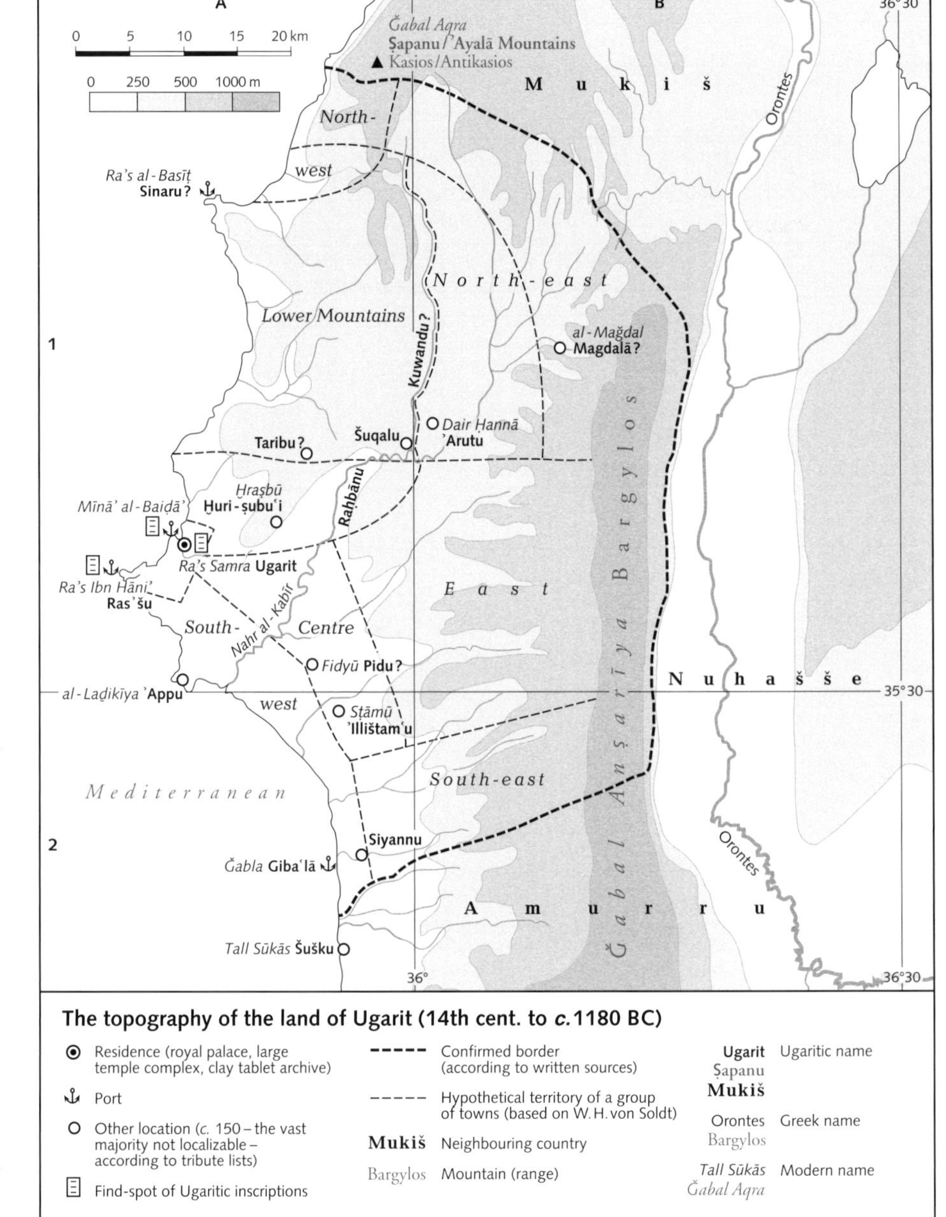

They bear witness to the city's extensive international contacts.

Apart from the capital, the Kingdom of Ugarit comprised another 150 settlements. Their existence can be derived from tribute lists, even though most of the settlements themselves cannot be located. Among them are several sea ports for the powerful trade and war fleet, which according to Hittite sources were called on to support Suppiluliuma II of Ḫattusa in his battle against Cyprus. The country's borders can be reconstructed only by matching topographical features with references in written sources. The country was probably subdivided into groupings of settlements with (hypothetically outlined) territories.

→ Maps pp. 13, 15, 23, 25, 27, 31, 33; Aegean Koine B., BNP 1, 2002; Ugarit, BNP 15, 2009

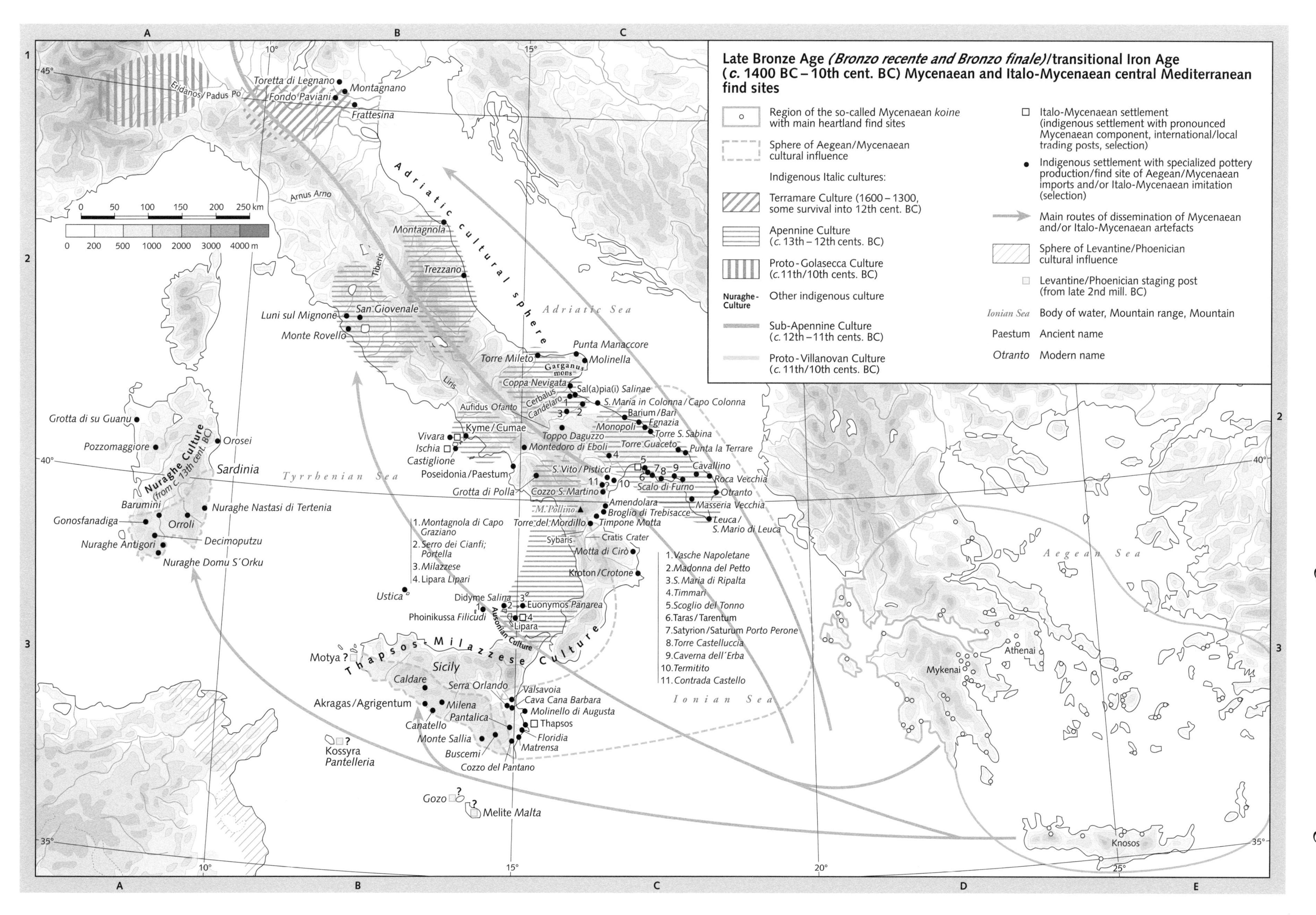
Late Bronze Age (Bronzo recente and Bronzo finale)/transitional Iron Age (c. 1400 BC – 10th cent. BC) Mycenaean and Italo-Mycenaean central Mediterranean find sites
Region of the so-called Mycenaean koine with main heartland find sites
Sphere of Aegean/Mycenaean cultural influence
Indigenous Italic cultures:
Terramare Culture (1600 – 1300, some survival into 12th cent. BC)
Apennine Culture (c. 13th – 12th cents. BC)
Proto-Golasecca Culture (c. 11th/10th cents. BC)
Nuraghe-Culture Other indigenous culture
Sub-Apennine Culture (c. 12th – 11th cents. BC)
Proto-Villanovan Culture (c. 11th/10th cents. BC)
Italo-Mycenaean settlement (indigenous settlement with pronounced Mycenaean component, international/local trading posts, selection)
Indigenous settlement with specialized pottery production/find site of Aegean/Mycenaean imports and/or Italo-Mycenaean imitation (selection)
Main routes of dissemination of Mycenaean and/or Italo-Mycenaean artefacts
Sphere of Levantine/Phoenician cultural influence
Levantine/Phoenician staging post (from late 2nd mill. BC)
Ionian Sea Body of water, Mountain range, Mountain
Paestum Ancient name
Otranto Modern name
0 50 100 150 200 250 km
0 200 500 1000 2000 3000 4000 m
Eridanos/Padus Po
Toretta di Legnano
Montagnano
Fondo Paviani
Frattesina
Arnus Arno
Adriatic cultural sphere
Montagnola
Tiberis
Trezzano
Luni sul Mignone
San Giovenale
Monte Rovello
Adriatic Sea
Punta Manaccore
Torre Mileto
Molinella
Garganus mons
Liris
Coppa Nevigata
Sal(a)pia(i) Salinae
Cerbalus
Candelaro
Aufidus Ofanto
S. Maria in Colonna/Capo Colonna
Barium/Bari
Egnazia
Monopoli
Torre S. Sabina
Kyme/Cumae
Vivara
Ischia
Castiglione
Toppo Daguzzo
Montedoro di Eboli
Torre Guaceto
Punta la Terrare
Cavallino
Poseidonia/Paestum
S. Vito/Pisticci
Roca Vecchia
Scalo di Furno
Otranto
Grotta di Polla
Cozzo S. Martino
Amendolara
Masseria Vecchia
M. Pollino
Broglio di Trebisacce
Torre del Mordillo
Timpone Motta
Leuca/S. Mario di Leuca
Cratis Crater
Sybaris
Motta di Cirò
Kroton/Crotone
Grotta di su Guanu
Pozzomaggiore
Orosei
Sardinia
Nuraghe Culture (from c. 13th cent. BC)
Tyrrhenian Sea
Barumini
Nuraghe Nastasi di Tertenia
Gonosfanadiga
Orroli
Decimoputzu
Nuraghe Antigori
Nuraghe Domu S'Orku
1. Montagnola di Capo Graziano
2. Serro dei Cianfi; Portella
3. Milazzese
4. Lipara Lipari
Ustica
Didyme Salina
Euonymos Panarea
Phoinikussa Filicudi
Lipara
Ausonian Culture
1. Vasche Napoletane
2. Madonna del Petto
3. S. Maria di Ripalta
4. Timmari
5. Scoglio del Tonno
6. Taras/Tarentum
7. Satyrion/Saturum Porto Perone
8. Torre Castelluccia
9. Caverna dell'Erba
10. Termitito
11. Contrada Castello
Motya ?
Thapsos-Milazzese Culture
Sicily
Caldare
Serra Orlando
Akragas/Agrigentum
Milena
Valsavoia
Cava Cana Barbara
Molinello di Augusta
Pantalica
Thapsos
Canatello
Monte Sallia
Floridia
Matrensa
Buscemi
Cozzo del Pantano
Kossyra Pantelleria
Ionian Sea
Gozo
Melite Malta
Aegean Sea
Athenai
Mykenai
Knosos

The eastern Mediterranean and Near East (12th – mid 10th cent. BC)

Between *c.* 1200 and 1150, the Late Bronze Age civilizations and political units in the Aegaean region, in Anatolia, Syria and Palestine were obliterated. The collapse of the political, economic and cultural systems had many causes and happened in different ways in different regions. The evidence is problematic: substantial written sources are available only for Egypt and Assyria. Apart from those we have to rely on linguistic research, later texts and archaeological appraisals.

I. The eastern Mediterranean region and the Near East (12th cent. BC; map A)

Around 1200 BC, the palace system in Greece collapsed (among other things owing to harvest failures, minor immigrations from the Adriatic and Balkan regions as well as local natural disasters). Most palaces were destroyed (with the exception of Athens) without being rebuilt. In the Postpalatial Era (*c.* 1200–1050), there was a decline in population, accompanied by a breakdown of the infrastructure, but also cultural continuity and continuity of settlement at numerous sites – even in places of retreat such as Cos, Rhodes and Cyprus. At some locations even population growth (Euboea, the Ionian Islands, Achaea) or the emergence of new centres (Attica) can be observed. There were still supraregional contacts with Crete, Rhodes, Cos, Cyprus, Syria and Egypt, but at a lower level of intensity.

In western Asia Minor, imperial structures probably managed to persist for quite a while (Arzawian states). At the same time population shifts among indigenous or immigrant groups occurred in northern and central Anatolia. In the long term only the so-called Hittite successor states Tarḫuntassa and Karkamiš turned out to be stable entities.

As early as the 13th cent., cuneiform texts record incursions and then immigrations – perhaps from Arabia – by Aramaean nomads targeting areas along the middle reaches of the Euphrates. This over time resulted in the destruction of numerous cities in the area and in the central Levant. There are also indications that formerly sedentary Semitic farmers from northern Mesopotamia were forced by the unstable economic situation to take up a nomadic or transhumance type of cattle farming. In the course of time they may have developed into a separate ethnic group. This could provide an explanation for the decline of population figures in the towns and their simultaneous increase among the nomads. A combination of the two models also seems possible.

In Palestine the Bronze Age system of 'city states' collapsed between 1200 and 1100 BC, as did supraregional trade and Egyptian suzerainty. Around 1200 BC, numerous cities ceased to exist. In many of them burnt layers have been found, which were not always the result of warlike activities (nowhere can they be ascribed to a specific group). At many places the urban culture survived (Dor, Bẹt Šəān), while some sites actually began to flourish (Kinneret). Some areas were dominated by non-sedentary groups such as the Shasu nomads. For the southern coastal plain, Egyptian documents and archaeological finds bear witness to the recent immigration of a new group, the Philistines (pəlištīm).

In Egypt, a gradual decline of governmental power set in after the end of Ramesses III's reign. Through a combination of internal (e.g. frequent changes on the throne) and external factors, this led to the collapse of the New Kingdom: the rule over the extraterritorial areas in Nubia (decline of Egyptian settlements) and the Near East (only bases in Palestine were left) came to an end. In Egypt, Nubia and the Sinai, mining activities came to a standstill. At the same time Libyan tribes increasingly infiltrated the western delta and the areas surrounding Heracleopolis and Thebes in particular. Under Ramesses III, victories over the Libyans and the so-called Sea Peoples, who were partly allied to the Libyans, were immortalized in pictures and inscriptions on the outer walls of temples.

The murder of Tukultī-Ninurta I by one of his sons (1197 BC) triggered a crisis in Assyria which lasted until the end of the 12th cent. Under Aššur-dān I (1168–1133), Assyria lost its influence over the northern mountain areas, where the Mušku people established themselves around 1165, while further to the east the warrior bands of the Papḫu people (Paphaeans) stirred up trouble as mercenaries or allies of local princes. Apart from occasional encounters with dispersed Hittite army units (Šubari), Assyria was not affected by the events in the Mediterranean region. In fact Tiglatpileser I (1114–1076) quickly managed to pacify the northern periphery of his kingdom.

Adad-šuma-uṣur, who expelled the Assyrians from Babylon in 1182, once more restored the Kassite dynasty in Babylonia. It came to a final close when the Elamite king laid claim to the Babylonian throne on the basis of centuries of intermarriage between his family and Kassite royals, a claim that he backed up with violence (military campaign, sacking of Babylon by his son *c.* 1150). The Elamite kingdom subsequently reached the peak of its territorial expansion, but Babylonia made a rapid recovery and took its revenge under Nebuchadnezzar I (1126-1105), who after defeating the Elamite king Ḫutelutuš-Inšušinak pillaged Šušan/Susa.

II. The eastern Mediterranean and the Near East (11th cent. to mid 10th cents.; map B)

Around 1100 BC, during a phase of relative prosperity, local principalities with residences and fortified centres (Mycenae, Tiryns, Achaea, Arcadia, Laconia, Euboea, Paros) established themselves in the Greek region. This was accompanied by a revival of fresco painting, the re-use of Mycenaean tombs and the manufacturing of ornamental ceramic ware, albeit at a more modest level. Instead of agriculture, animal husbandry became the dominant branch of farming. The few known settlements from this phase were frequently short-lived, were were temporarily or permanently deserted or relocated. In the middle of the 11th cent., there were further destructions, general impoverishment, declining populations, the final decay of the material culture and invasions by new population groups.

The Greek (re-)settlement of the coast of Asia Minor (Ionians, Dorians, Aeolians) began around 1050 BC. Western, central and eastern Anatolia were inhabited by autochthonous Anatolian peoples (some of them probably Luwian-speaking), with an archaeologically and linguistically verified admixture of ancient Balkanic, Caucasian and East-Anatolian components, which explains e.g. the ethnogenesis of the Phrygians. The settlement structure typically consisted of numerous small settlement cells associated with a centre (Sardis, Gordium, Ancyra/Ankara, Iconium, Boğazköy, Kaman Kalehöyük). There were cultural and perhaps political links with the Luwian principalities in south-east Anatolia and northern Syria, where Tarḫuntassa (documented as Tabal or Que from the 9th cent., → map p. 43) and Karkamiš had fallen apart. Cultural and political traditions can be guessed at, but the artwork is patterned on Hittite models; the kings of Que (Adanija) claimed descent from Mopsos/Mukšu, probably a ruler of Tarḫuntassa. The political clout of the principality of Karkamiš – culturally in the Hittite mould – decreased continually into the 8th cent.: the former secundogeniture Malida gained independence in the 11th cent.; the territories along the middle reaches of the Euphrates were lost to Aramaeans, who extended their sphere of domination across large parts of the northern and central Levant as well. Individual cities remained under a strong cultural Hittite influence, which was probably kept alive by Luwian elites (e.g. the realm of Taitas of P/Watasatini).

Towards the end of the 12th cent., the incursions from the Syrian and Mesopotamian steppes into settled areas by groups of Aramaean nomads intensified ominously, probably as the result of climatic change (crop failures). The disastrous combination of military incursions and famines plunged the established Mesopotamian kingdoms into a series of struggles for survival, which lasted a century and a half and led to enduring changes in the composition of the Mesopotamian population. In Assyria, Tiglatpileser I chased his Aramaean enemies along the Euphrates into the Syrian steppe almost every year from 1111 onwards, whereas Aššur-bēl-kala (1073–1056) was forced to take on Aramaean raiders in every part of his realm. The Babylonian empire was spared at first; Marduk-nādin-aḫḫē (1100–1083) waged an unsuccessful war against Assyria. Towards the end of his reign, Aramaean raiders began to target Babylonia as well, and in the reign of Adad-apla-iddina (1069–1048) important cities were sacked. During the 11th and 10th centuries the weakened empires suffered substantial losses of territory: around 1000 BC, numerous small Aramaean kingdoms were founded on former Assyrian teritory, and tribal Chaldaean kingdoms (Kaldu) emerged along the Euphrates, just south of Babylon. Their territory included the sites of ancient Babylonian cult centres. The cores of both the Assyrian and the Babylonian empires survived, although the Assyrian position was much more comfortable in the 10th cent. The Elamite kingdom collapsed shortly after 1100, leaving a gap in the records of more than 350 years. Its fate was linked to events on the Iranian plateau; owing to a lack of sources however the immigration or ethnogenesis of those speakers of Indo-European languages who are identified as Medes and Persians in Assyrian inscriptions from the 9th and 8th cents. remains in the dark.

Between 1100 and 1000, Palestine saw the beginning of a phase of consolidation; the newly created ethnic and social structures began to stabilize. Numerous small settlements had sprung up in the mountainous regions. The typical Early Iron Age village culture was characterized by a marked subsistence economy embedded in a tribal society. The emerging tribes of the Israelites, Judaeans, Ammonites and Moabites were mostly (unlike the Philistines) the descendants of the autochthonous rural population. At some central sites in the area of Early Iron Age village culture (e.g. Ḥâṣor, Šəkem, Rabbat Bənẹ Ammon and Dibon) a process of transition towards a town culture can be observed around 1000 BC that was to determine further development.

Power in the 21st Dynasty in Egypt was divided between a royal dynasty residing at Tanis and a dynasty of high priests ruling from Thebes: both were of Libyan descent and linked by family ties. The border between their dominions was situated north of Heracleopolis. Their style of government evinced feudal characteristics. The Theban high priests were also warriors who maintained fortresses in central Egypt. The capital of the royal dynasty was moved from Qanṭīr, where the Nile was silting up, to Tanis and many building components were taken there from Qanṭīr. Tanis was not only the kings' residence, but also their burial place, just as the high priests at Thebes were buried there as well. There are no sources at all documenting external relations.

→ Maps pp. 23, 27, 29, 34f., 45, 47

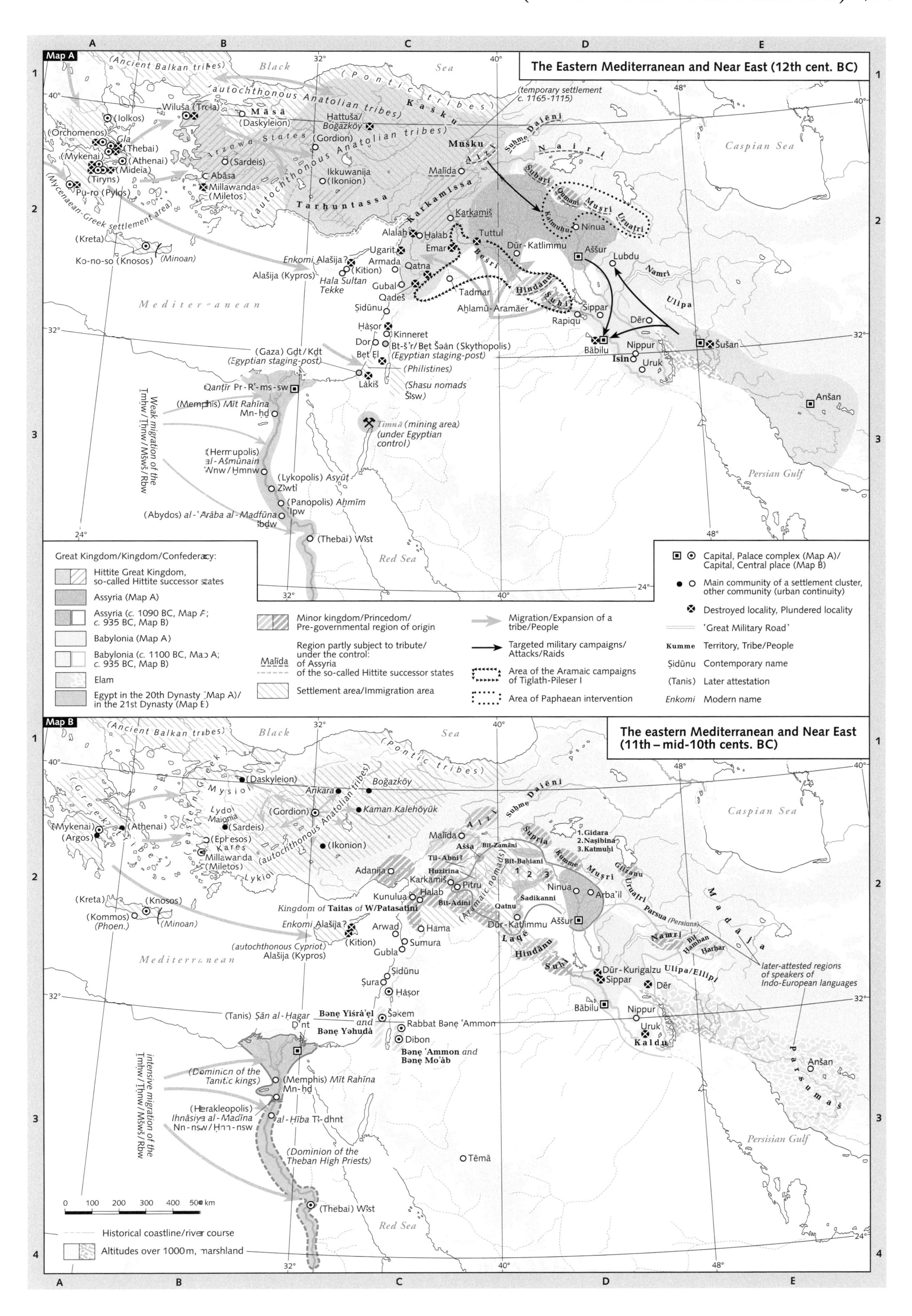
Map A
The Eastern Mediterranean and Near East (12th cent. BC)
Map B
The eastern Mediterranean and Near East (11th – mid-10th cents. BC)
Great Kingdom/Kingdom/Confederacy:
Hittite Great Kingdom, so-called Hittite successor states
Assyria (Map A)
Assyria (c. 1090 BC, Map A; c. 935 BC, Map B)
Babylonia (Map A)
Babylonia (c. 1100 BC, Map A; c. 935 BC, Map B)
Elam
Egypt in the 20th Dynasty (Map A)/ in the 21st Dynasty (Map B)
Minor kingdom/Princedom/ Pre-governmental region of origin
Region partly subject to tribute/ under the control: of Assyria of the so-called Hittite successor states
Settlement area/Immigration area
Migration/Expansion of a tribe/People
Targeted military campaigns/ Attacks/Raids
Area of the Aramaic campaigns of Tiglath-Pileser I
Area of Paphaean intervention
Capital, Palace complex (Map A)/ Capital, Central place (Map B)
Main community of a settlement cluster, other community (urban continuity)
Destroyed locality, Plundered locality
'Great Military Road'
Kumme Territory, Tribe/People
Ṣidūnu Contemporary name
(Tanis) Later attestation
Enkomi Modern name
Historical coastline/river course
Altitudes over 1000m, marshland
0 100 200 300 400 500 km

Greece, the Aegean and western Asia Minor, 12th–9th cents. BC

Greece, the Aegean and western Asia Minor from the 12th to the 9th cents.

The Postpalatial period (*c.* 1200–1050 BC, subdivided chronologically into the Late Helladic and the sub-Mycenaean or sub-Minoan phases) after the decline of the Mycenaean palace culture, did not entail a radical culture change nor the definitive end of Mycenaean civilization. It was followed by the so-called Dark Ages (*c.* 1050–800 BC) covering the Protogeometric period (*c.* 1050–900) and part of the Geometric period (*c.* 900-700), which is subdivided into the Early Geometric (*c.* 900–850) and the Geometric phase (*c.* 850–750). For the Postpalatial period as well as the 'Dark Ages', we have to rely mostly on the findings of settlement archaeology and linguistic deduction for insights into the events in Greece and the Aegean, since there are no written records for the transitional period from the Bronze Age to the Iron Age.

I. The most important sites (main map)

The excavated settlements and necropoleis in mainland Greece – e.g. Athens and Perati (Attica), Lefkandi (Euboea), Mycenae, Tiryns and Argos (Argolid), Dyme (north-western Peloponnese), Nichoria (Messenia) – as well as on the Aegean islands (e.g. Zagora on Andros, Koukounaries on Paros, Emporium on Chios, Serraglio on Cos) and in Greek Asia Minor (Ancient Smyrna) reveal the emergence of local 'baronial' rulers, a hierarchical society and an *oikos* economy largely based on animal husbandry and probably involving transhumance. Furthermore there are proto-urban centres or fortified residences and precursors of later temple buildings, but no palaces. The finds suggest internal Aegean contacts, but also a limited amount of long-distance marine trade; there are verified import finds from Cyprus and the Near East, but also from Italy and central Europe. They attest to technological innovations, among others, in the production of of iron objects (Argos) and large ceramic vessels decorated with geometric and figurative ornaments (Athens), which must be considered aristocratic status symbols.

The fading of Mycenaean traditions and the emergence of regional cultures did not happen in a uniform manner in the various regions, nor at the same time and speed (cf. the chronology of individual sites). There was also a varying degree of social and economic differentiation, ranging from simple, village-oriented animal husbandry culture to areas with (proto-)urban centres, monarchical governance and far-ranging international relations (e.g. Athens with Attica, Lefkandi with Euboea, Knossos with Crete). It should also be borne in mind that on the periphery of the Mycenaean culture (e.g. in Achaea, Arcadia, Aetolia, Acarnania and Phocis), even while it flourished, social structures were fairly straightforward; the predominant type of regional rulership would centre on fortified manorial settlements, which also served as refuges.

The massive upheavals and the lack of stability throughout this period can be deduced from the numerous destruction layers found at almost every site; most of them were the result of warlike conflicts, some were caused by earthquakes, both of which led to population decline, general poverty and the deterioration of the material culture.

II. Greek ethnogenesis and early forms of organization

It must have been during the last phase of the Late Bronze Age and the 'Dark Ages' that in Greece and in the Aegean the complex process of ethnogenesis took place from which originated – through the fusion of already resident Greeks and newcomers – the Greek tribes and states with their different languages and forms of social organization.

It is a matter of scholarly controversy when and how tribes organized themselves, e.g. as personal associations or 'tribal states': perhaps already during the migrations or in the settling phase, in part by peaceful infiltration and acculturation, in part by force. An early form of organization which preceded the consolidation of *poleis* and tribal states was represented by certain cult leagues of equal and independent members (tribes or subtribes, groupings of individual settlements such as the Doric Hexapolis, inhabitants of certain regions or islands). These so-called *amphictyonic* leagues probably came into being in the 9th cent. They centred on a supra-local shrine of a deity whose cult was practised collectively. The most important and politically influential early *amphictyony was apparently the 'Pylian Amphictyony' of Anthela' (near Thermopylae) with its central shrine of Demeter* (Hdt. 7,200); in the early 6th cent. it was expanded to become the Pylian-Delphic *Amphictyony, supported by twelve tribes (éthnē):* Thessalians, Perrhaebi, Magnesians, Dolopians, Malieis, Aenianes and Phthiotes in Thessaly, Locrians, Ionians (from Euboea and Attica), Boeotians, Phocians and Dorians in central Greece.

III. The distribution of Greek dialects (supplementary map)

To identify the main routes of the numerous migrations during this period in general and the geographic distribution of the Greek tribes in particular, we have to rely on the linguistic reconstruction of dialect areas. Since the results of the population shifts did not stabilize fully until the middle of the 8th century, conclusions about any earlier linguistic states should be treated as informed hypotheses. Speakers of the North-West Greek dialect occupied Thessaly (which like Boeotia spoke Aeolian), all of central Greece and the northwestern Peloponnese. The Dorians, who were close linguistic relatives, took over the remainder of the Peloponnese (except for Arcadia), the southern Aegean islands and the southern part of the Greek settlement area in western Asia Minor. Some of the previous residents were integrated, others began to move elsewhere from the middle of the 11th cent.: the Thessalian Aeolians migrated to north-western Asia Minor; the Ionians settled on the central Cyclades, Chios and Samos and on the central coast of Asia Minor; the pre-Dorian 'Achaeans' retreated to Arcadia or moved to Cyprus.

→ Maps pp. 27, 33, 37; Dark Ages, BNP 4, 2004

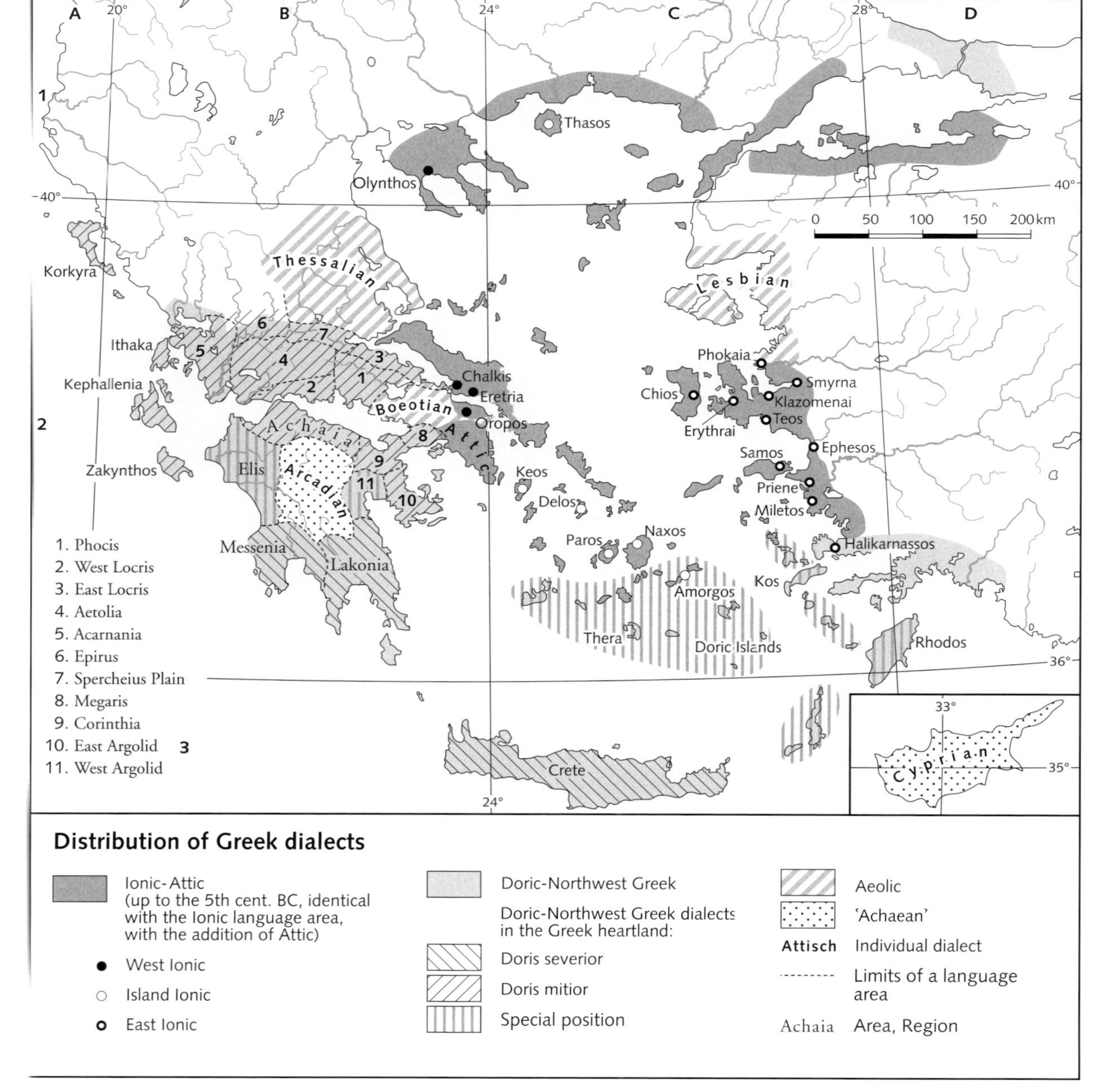

Literature

J. Boardman, Kolonien und Handel der Griechen, 1981; W. Eder, K.-J. Hölkeskamp (eds.), Volk und Verfassung im vorhellenistischen Griechenland, 1997; H.-J. Gehrke, H. Schneider (eds.), Geschichte der Antike, 2006; J. Latacz (ed.), Zweihundert Jahre Homerforschung, 1991; I. Morris, Archaeology as Cultural History. Words and Things in Iron Age Greece, 2000; D. Musti et al. (eds.), La transizione dal Miceneo all'alto Arcaismo, 1991; F. de Polignac, Cults, Territory, and the Origins of the Greek City-State, 1995; G. Roux, L'Amphictionie, Delphes et le temple d'Apollon, 1979; P. Siewert, L. Aigner-Foresti, Föderalismus in der griechischen und römischen Antike, 2005; K. Tausend, Amphiktyonie und Symmachie, 1992; C. Ulf (ed.), Wege zur Genese griechischer Identität, 1996; K.-W. Welwei, Die griechische Frühzeit 2000 bis 500 v. Chr., 2002.

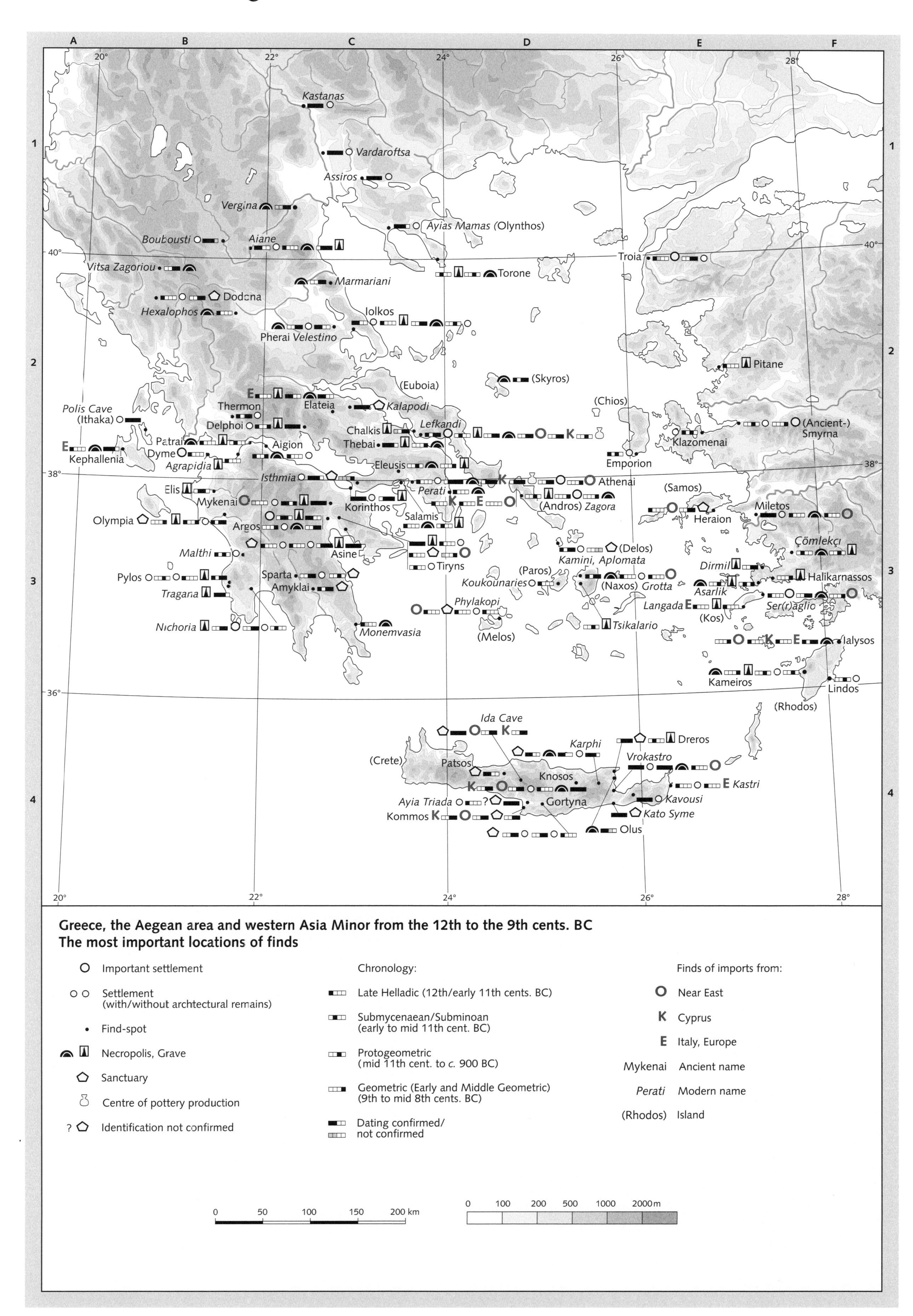

Kastanas
Vardaroftsa
Assiros
Vergina
Ayias Mamas (Olynthos)
Boubousti
Aiane
Vitsa Zagoriou
Torone
Troia
Marmariani
Dodona
Hexalophos
Iolkos
Pherai Velestino
Pitane
(Skyros)
(Euboia)
Polis Cave
(Ithaka)
Thermon
Elateia
Kalapodi
(Chios)
Delphoi
Lefkandi
Chalkis
Thebai
Klazomenai
(Ancient-)
Smyrna
Kephallenia
Patrai
Dyme
Aigion
Agrapidia
Eleusis
Emporion
Isthmia
Athenai
Elis
Perati
(Samos)
Mykenai
Korinthos
(Andros) Zagora
Miletos
Olympia
Argos
Salamis
Heraion
Malthi
Asine
Tiryns
(Delos)
Kamini, Aplomata
Çömlekçı
Pylos
Sparta
Amyklai
(Paros)
Koukounaries
(Naxos) Grotta
Dirmil
Halikarnassos
Tragana
Asarlik
Phylakopi
Langada
Ser(r)aglio
Nichoria
Monemvasia
(Melos)
(Kos)
Tsikalario
Ialysos
Kameiros
Lindos
(Rhodos)
Ida Cave
Dreros
Karphi
(Crete)
Patsos
Vrokastro
Knosos
Kastri
Ayia Triada
Gortyna
Kavousi
Kommos
Kato Syme
Olus
Greece, the Aegean area and western Asia Minor from the 12th to the 9th cents. BC
The most important locations of finds
Important settlement
Settlement (with/without archtectural remains)
Find-spot
Necropolis, Grave
Sanctuary
Centre of pottery production
Identification not confirmed
Chronology:
Late Helladic (12th/early 11th cents. BC)
Submycenaean/Subminoan (early to mid 11th cent. BC)
Protogeometric (mid 11th cent. to c. 900 BC)
Geometric (Early and Middle Geometric) (9th to mid 8th cents. BC)
Dating confirmed/ not confirmed
Finds of imports from:
O Near East
K Cyprus
E Italy, Europe
Mykenai Ancient name
Perati Modern name
(Rhodos) Island
0 50 100 150 200 km
0 100 200 500 1000 2000 m

Greece, the Aegean and western Asia Minor, late 9th – *c.* mid 6th cents. BC

Greece, the Aegean and western Asia Minor, late 9th – *c.* mid 6th cents. BC

The main map is concerned with Greece, the Aegean and Asia Minor in the period following the 'Dark Ages', from the late 9th cent. to the middle of the 6th cent., a period which in art encompasses the Geometric (*c.* 900–700 BC) and the Archaic (*c.* 700–500 BC) periods. This era has not yet attracted sufficient attention from archaeological and historical research. Its politically most important events were the development of the *polis* structure in the Greek cultural area and the so-called Great Greek Colonization, which started around 750 BC. In Asia Minor, Greek cities had established themselves and were taking an active part in the process of colonization, with Miletus playing a leading role. They were repeatedly targeted by the expanding Lydian kingdom, which becomes historically verifiable during the 7th-cent. reign of King Gyges. The formation of this kingdom, its relations with the Phrygian kingdom centring on Gordium and with autochthonous Anatolian ethnic groups – as yet identifiable as cultures only – such as the Carians in the south and the Mysi in the north are not yet fully understood. Another research gap concerns the Ancient Balkans region in the Iron Age, which was probably subdivided along tribal lines, and its outside contact zones in northern Greece and the Propontis region.

I. The Greek cultural area

It is difficult to identify the early political structure in mainland Greece, on the islands and in Asia Minor. From the late 9th cent. onward, politically independent units of equal status associated in supra-local cult leagues (amphictyonic leagues) centring on their respective shrines. Since they agreed on certain rules in case of a conflict between members, the leagues had a politically stabilizing effect.

Oil, grain and probably wine and other agrarian products were the most important economic assets from the early 8th cent. onward. Accordingly, land and soil, fixed settlements and the settling of larger groups played an increasing role. This is illustrated by storage structures e.g. in Nichoria (Messenia), in early settlements on the Cyclades and at Smyrna as well as by models of granaries from Athens. This process overlapped with the so-called Great Colonization starting around 750 BC. Both developments influenced the emergence of the *polis* as a socio-political principle of organization and its spreading as the dominant form of organization within the Greek area of settlement in the Archaic period. The development of the *polis* took place in the time between the 8th and the 6th cents. *Polis* is defined as a state in the sense of a self-governing association of citizens; it might or might not include urban settlements or an urban core settlement with associated surroundings (as supported by archaeological and/or literary evidence). So-called city states were particularly common in a part of the Peloponnese, Attica, Euboea and on the coast of Asia Minor. When the colonization movement got under way, a distinction was made between *metropolis* (mother-city) and *apoikia* (colony; → map p. 69). Simultaneously, tribal associations developed into tribal states and later on also federal states (*ethne*/*koina*), especially in north-western Greece, in regions without any or hardly any urbanization, where pasture farming, transhumance and a semi-nomadic lifestyle were predominant.

The adoption of the alphabet in the 8th cent. and its fast spread in many varieties (initially only attested by proprietary, dedicatory and funerary inscriptions; 'individual' verses) are proof of the existence of a tight network of contacts, exchange and mutual influence. The same is true of the spread – as far as Asia Minor, the Levant and distant colonies in the west – of painted Euboean, Corinthian or Attic ceramic ware, bearing no longer exclusively geometric, but also (once more) figurative decorations. Shared dialects and local styles of vase painting can be seen as expressions of local or regional identity at centres such as Argos, Laconia, Boeotia, Thessaly, Crete, the Cyclades and eastern Greece. A key part in this cultural as well as political web of internal Greek and far-reaching international relations was played by the religious shrines with their by now monumental edifices: as religious centres and venues of great festivals Olympia, Delphi and the Heraea of Perachora and Samos achieved supra-regional, even Panhellenic status (as attested by a soaring increase of votive offerings from inner Asia Minor, the Near East and Italy). These innovations have been documented archaeologically, but are also reflected in the *Iliad* and the *Odyssey*, which were written down towards the end of the 8th cent.

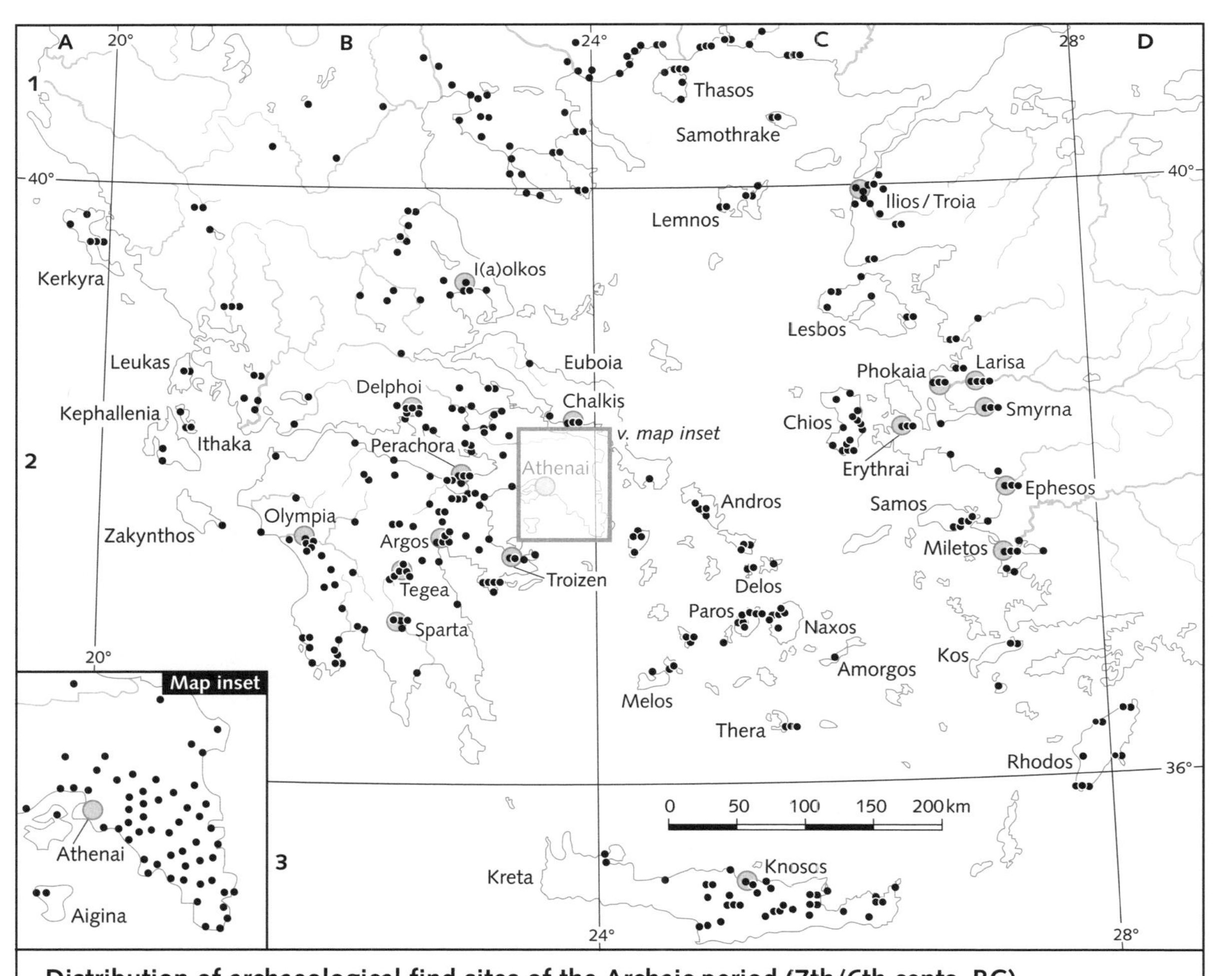

Distribution of archaeological find sites of the Archaic period (7th/6th cents. BC)

- Find site: settlement and/or necropol(e)is and/or sanctuary (selection, after F. Lang)
- Location (for orientation)

II. The autochthonous Anatolian cultural region

Among the autochthonous Anatolian cultures are the Lydian and the Carian ones. The location of the latter has been mainly reconstructed on the basis of topographical facts, of the Carian characteristics of the settlements (according to archaeological and literary evidence; e.g. Halicarnassus) and of the distribution of Carian finds, especially ceramic ware (from *c.* 1050 to 550 BC). Documents written in Carian, which began to emerge in the 7th and 6th cents., are also among the resources, but margins of error are still considerable. On Lydian soil the Lydian kingdom centring on the capital of Sardis came into existence in the 7th cent. at the latest; it lasted until 546 BC, when it was conquered by the Persians. In analogy to the Carian region, Lydia's cultural and even more its political boundaries cannot be determined with any degree of precision. The origin of the Lydians, for example, is still a matter of scholarly dispute; literary sources offer contradictory information as to which territories were ruled over by the Lydians or which relations of dependence existed (for tribute payments by Greek coastal cities, cf. Hdt. 1,6). Archaeological research on these questions has not yet been evaluated sufficiently (e.g. for Dascylium) or has not even been undertaken so far.

The Lydian kingdom was a feudal state with a class of aristocrats breeding horses on large estates. There is also proof of merchants' and artisans' activities: in Sardis e.g. workshops where gold from the Pactolus river, from mines in Mysia, near Pergamum (cf. Aristot. Mir. 834a) and from the Pontus area (probably imported by Milesians, cf. Aristeas FGrH 35 F 4) was worked as well as ivory, leather and wool. The Lydian kings cultivated international contacts (Assyria: Assurbanipal's royal inscriptions; for Media see Hdt. 1,73,4). From the second half of the 7th cent., Lydia repeatedly attacked Greek cities (→ map p. 39), although there is evidence of positive contacts with Greek shrines, e.g. consultations of the Delphic oracle and votive offerings, big monetary deposits in Didyma and the assignment of confiscated enemy estates to the temple of Artemis at Ephesus (Hdt. 1,92,2ff.; Nicolaus of Damascus, FGrH 90 F 65), who was also worshipped in an affiliated temple at Sardis.

III. The diffusion of archaeological sites (7th/6th cents. BC) (supplementary map)

The archaeological sites from the Archaic period, which are charted in this map (without making any distinction between settlements, necropoleis and shrines) provide the material foundation for the above conclusions on the cultural and political developments. The map is based on the results of F. LANG's studies. This apparently homogeneous picture presents problems caused by the rather uneven state of research and the range of materials examined. Thus perishable materials are very difficult, if not impossible to detect, which in particular bars us from using written sources whose former existence must be assumed. The problem is compounded by the fact that some sites are covered by modern buildings.

→Maps pp. 35, 39, 69, 91, 101, 103, 105, 127; Lydia, BNP 8, 2006; Polis, BNP 11, 2007

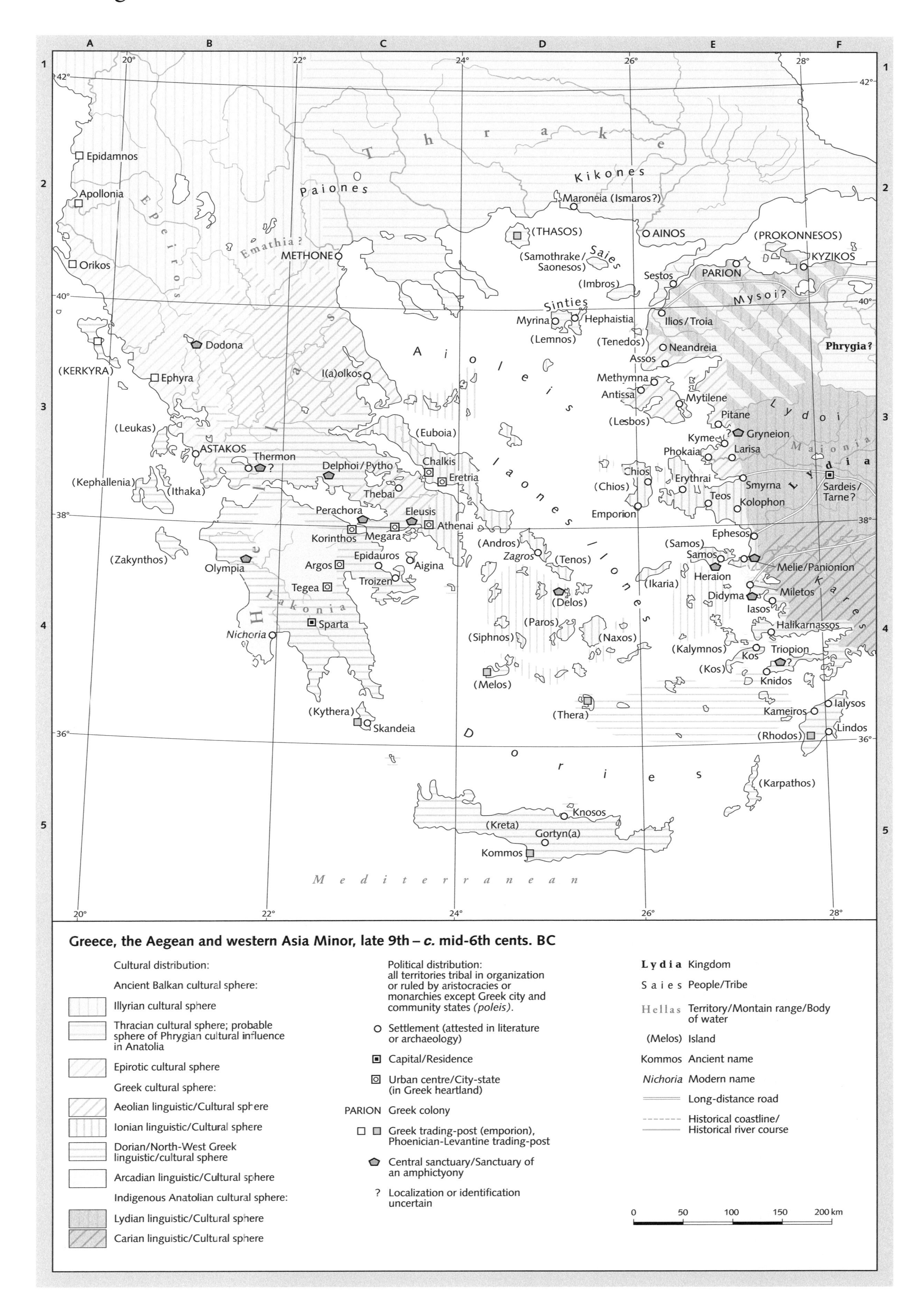
Thrake
Epidamnos
Apollonia
Orikos
Epeiros
Emathia?
Paiones
METHONE
Kikones
Maroneia (Ismaros?)
(THASOS)
AINOS
(PROKONNESOS)
KYZIKOS
(Samothrake/ Saonesos)
Saies
(Imbros)
Sestos
PARION
Mysoi?
Sinties
Myrina
Hephaistia
(Lemnos)
(Tenedos)
Ilios/Troia
Neandreia
Assos
Phrygia?
Dodona
(KERKYRA)
Ephyra
I(a)olkos
Aiolees
Methymna
Antissa
Mytilene
(Lesbos)
Pitane
Lydoi
Gryneion
Kyme
Larisa
Phokaia
Maionia
Lydia
Sardeis/ Tarne?
(Leukas)
ASTAKOS
Thermon
Delphoi/Pytho
(Euboia)
Chalkis
Eretria
Thebai
(Kephallenia)
(Ithaka)
Chios
(Chios)
Erythrai
Smyrna
Teos
Kolophon
Emporion
Iaones
Perachora
Eleusis
Athenai
Korinthos
Megara
Ephesos
(Samos)
Samos
(Andros)
Zagros
(Tenos)
(Zakynthos)
Olympia
Epidauros
Aigina
Argos
Troizen
Heraion
Melie/Panionion
Miletos
Didyma
(Ikaria)
Iasos
Tegea
Lakonia
Hellas
(Delos)
Karees
Sparta
Nichoria
(Paros)
(Naxos)
Halikarnassos
(Siphnos)
(Kalymnos)
Kos
Triopion
(Kos)
Knidos
(Melos)
(Thera)
Ialysos
Kameiros
Lindos
(Rhodos)
(Kythera)
Skandeia
Dories
(Karpathos)
Knosos
(Kreta)
Gortyn(a)
Kommos
Mediterranean
Greece, the Aegean and western Asia Minor, late 9th – c. mid-6th cents. BC
Cultural distribution:
Ancient Balkan cultural sphere:
Illyrian cultural sphere
Thracian cultural sphere; probable sphere of Phrygian cultural influence in Anatolia
Epirotic cultural sphere
Greek cultural sphere:
Aeolian linguistic/Cultural sphere
Ionian linguistic/Cultural sphere
Dorian/North-West Greek linguistic/cultural sphere
Arcadian linguistic/Cultural sphere
Indigenous Anatolian cultural sphere:
Lydian linguistic/Cultural sphere
Carian linguistic/Cultural sphere
Political distribution: all territories tribal in organization or ruled by aristocracies or monarchies except Greek city and community states (poleis).
Settlement (attested in literature or archaeology)
Capital/Residence
Urban centre/City-state (in Greek heartland)
PARION Greek colony
Greek trading-post (emporion), Phoenician-Levantine trading-post
Central sanctuary/Sanctuary of an amphictyony
? Localization or identification uncertain
Lydia Kingdom
Saies People/Tribe
Hellas Territory/Montain range/Body of water
(Melos) Island
Kommos Ancient name
Nichoria Modern name
Long-distance road
Historical coastline/ Historical river course
0 50 100 150 200 km

Anatolia, 10th–7th cents. BC

The map covers the so-called Dark Ages, a time during which – as far as we know – at least western and central Asia Minor were scriptless until the 8th cent. This period saw the emergence of the Phrygian and Lydian kingdoms, the only verified major kingdoms in the western half of Asia Minor before the rise of the Persian Achaemenids. For Anatolia the two empires also marked the transition from the so-called prehistoric era to the historic one; they are mentioned in Assyrian source texts as well. The territories of the so-called Luwian and Aramaean principalities as well as Assyria and Urarṭu are not considered here (→ maps p. 43 and 47, 49 and 51, 41 and 53).

I. The territorial divisions in Asia Minor in the Early Iron Age

Research is still in its infancy concerning what used to be the 'Arzawian federation of states' in the 13th and possibly right into the 10th cent., i.e. the largely Luwian-speaking Early Iron Age territories in western Asia Minor such as the Troad, Mysia, Lydia (see below) and Caria (Hittite sources and Luwian inscriptions, → map p. 33). It is believed that Luwian states, e.g. in the shape of relatively small principalities, continued to survive into the 8th cent.; Homer's *Iliad* provides evidence of the existence of specifically Anatolian institutions. The toponyms can even be traced back to the Hittite era.

If we exclude the areas where states such as the Phrygian and Lydian kingdoms were formed, we can generally assume that the other autochthonous Anatolian populations were probably organized in small units, e.g. in south-western and southern Asia Minor, the Black Sea area and inside the arc of the Halys river. They were presumably oriented towards some local settlement cell and comprised one or several central towns. It is also likely that throughout this settlement area a (semi-)nomadic way of living was practised.

The west coast of Asia Minor and its off-shore islands, once part of the Mycenaean sphere of influence, were once again targeted by the Greeks from the 11th cent. onward. This time it was the tribes (who were to form federations later) of the Ionians in the central section, the Aeolians to the north and the Dorians to the south who constituted a distinct cultural community. By the 8th cent., the establishment of the so-called city-states had been completed. From the middle of the century they participated in the 'Great Greek Colonization', with Miletus being particularly active. Their sense of a common identity found its expression in the cult practices at the religious centres.

II. The growth of state-like entities in Asia Minor

1. Phrygia

Scholars would frequently conclude that in the wake of the demise of the Great Hittite Empire (1200/1180 BC) another great territorial state emerged in Asia Minor during the Early Iron Age (*c.* 1200 to 700 BC): the Phrygian kingdom (11th/10th to 7th cents. BC), whose size and political power in Asia Minor supposedly rivalled that of the Hittite empire. Although according to recent research the claims concerning its territorial size have to be downgraded considerably, there is no gainsaying the fact that King Midas of Phrygia was a historical figure. Both contemporary late 8th-cent. ancient oriental sources as well as slightly later Greek ones testify that Midas (Assyrian: Mitā of Mušku) was a contemporary of Sargon II of Assyria (722–705 BC), Rusas I of Urarṭu (verified *c.* 720 BC) and several rulers of Luwian and Aramaean principalities. It is also documented that the influence of Phrygian culture extended far beyond its territory and that the Phrygian language was still spoken in many parts of Asia Minor as late as the 3rd cent. BC (→ supplementary map p. 116). Phrygian is not an Anatolian language. It seems to have originated in the region of the Ancient Balkans and Thrace, from where the (proto-) Phrygians probably immigrated in the 2nd millennium (Hdt. 7,73; Str. 7,3,2; 14,5,29). However, it is very likely that their ethnogenesis took place in northwestern Asia Minor.

Phrygia's cultural and political centre must have been Gordium (Assyrian: Mušku?), a border town in the Hittite age, situated on the right bank of the river Sangarius. It boasted an acropolis housing palatial *megaron* buildings and vast necropoleis with large tumulus structures. It was also the hub of Anatolia's most important overland routes.

For the development – probably starting in the 11th/10th cents. – and the political organization of the kingdom as well as its relations with the Great Kingdom of Mirā to the west (around Midas City), which seems to have outlasted the Hittite empire for some time, we have to rely on informed guesses, because the Phrygian inscriptions cannot be translated with any satisfactory degree of certainty. Therefore scholars subdividing Phrygia and the attached cultural centres, some of which may have been politically independent, have to make do with regional terms such as central (Gordium and Ancyra), western (Midas City, Dorylaeum?) and eastern Phrygia (Kaman Kalehöyük and Boğazköy); these terms are based on topography, archaeology and art history. Whether the southern Propontis around Dascylium (which was part of Hellespontic Phrygia in a later age) can really be considered part of early Phrygia, as is suggested by first probings, depends on the final outcome of ongoing archaeological investigations. There are no known Phrygian toponyms. Early contacts, at least cultural ones, with the Luwian and Aramaean region have been verified. With Urarṭu, which probably emerged at the same time, no direct exchange seems to have taken place.

During Midas' reign the expansion of the kingdom presumably reached its peak, although we only have reasonably reliable information (which comes from Assyrian sources) about the borders with the Luwian principalities and the Assyrian empire. It suggests that the border ran roughly along the Taurus mountain range. In the other regions all we have to go by for cautious conclusions about zones of cultural influence is the spread of Phrygian inscriptions and material culture such as pottery, bronze objects and stepped altars. The territories to the west of the Luwian states and south of Iconium (Konya), probably predominantly Luwian-speaking areas, are particularly difficult to categorize culturally.

Alongside Gordium, Celaenae is named in (admittedly rather late) Greek and Latin sources (Curt. 3,1,1ff. referring to Xen. An. 1,2,7; also Liv. 38,13,5) as Phrygia's central settlement, with a royal family of its own. Under Persian rule it remained an important site with a palace and a *paradeisos* and retained this status in Hellenistic times, when it was known as Apamea. This seems to indicate that it was a small local kingdom.

The line of Phrygia's western border and thus its relationship with Lydia (with which according to later Greek sources there were manifold contacts) are questions still waiting to be answered by new research. It is therefore marked on the map as a contact zone. The northern teritorry extending to the Pontic Mountains can only be described as *terra incognita*.

2. Lydia

Concerning the Lydian royal dynasties of the rather mythical Heraclidae and the historical Mermnadae established by Gyges (Assyrian: Guggu), the state of the sources is far better than in the case of the Phrygians, but there are also numerours open questions. They regard the origins of the Lydians, among other things. One theory (based on a Luwian source) assumes inner-Anatolian roots, possibly from the area around Dascylium. Later Greek sources, however, classify that very area as territory acquired by northward expansion from Sardis.

We are also in the dark about the beginnings and the growth of the Lydian kingdom, which emerged in territory that was part of the former federation of Arzawa (see above and p. 33) and whose existence can be verified in the 8th/7th cents. Greek sources from the 8th cent. (Homer, *Iliad*) mention the land of Maeonia in the same context as the Tmolus range, Lake Gyges (*Gygaia limne*) and the city of Tarne, which might be identical with Sardis as the cultural and political centre (a place with an acropolis, an ample urban area and large necropoleis with tumulus structures). The Lydian core territory can be reconstructed from its topography, but its external relationships are far from clear. This goes for Phrygia, where literary and archaeological sources have revealed a considerable cultural overlap, as well as other adjoining territories, especially Caria, parts of which must be considered as areas of Lydian expansion. It also applies to the Greek cities on the west coast of Asia Minor, which were repeatedly attacked by Lydian kings and seem to have been under Lydian suzerainty from time to time, and to Assyria, whose sources document the country under the name of Luddu and the capital Sardis as Šibartu.

Sources

So far, any statements on areas which were not covered by ancient oriental written documents are restricted to the sparse results of archaeological and – sporadically – scientific research. The difficulty is compounded by the fact that in this region only very few significant excavations of settlements have been carried out and their findings published. The spread of inscriptions is also of limited significance: it was not before the 8th cent. that the first (mainly undecipherable) Phrygian inscriptions emerged in central Anatolia, contemporaneous with mostly brief Greek ones on the west coast of Asia Minor, which are of limited or no use at all as historical documents. There are Urarṭian inscriptions dating from approx. the same time, but none of these were found on the western periphery of that kingdom. Lydian (6th cent.), Carian (7th/6th cents., translatable only since 1996) and Lycian (from the 5th cent. onward) written sources appeared even later. The Luwian-Aramaean region in southeastern Anatolia and Assyria are the only areas where the use of writing continued without interruption.

→ Maps pp. 23, 27, 33, 37, 41, 43, 47, 49, 51, 53, 69, 87; Lydia, BNP 8, 2006 (bibliography)

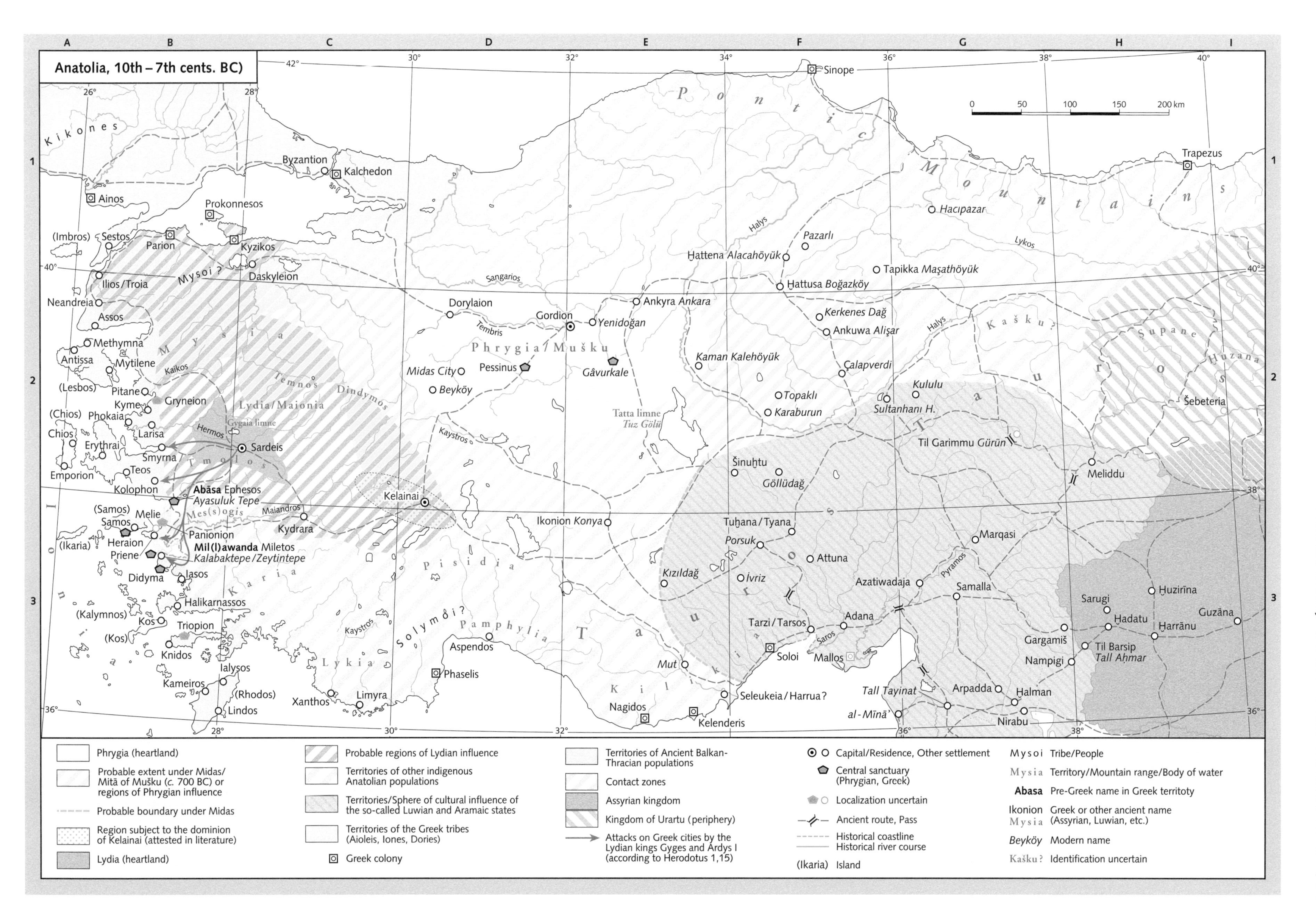

Anatolia, 10th – 7th cents. BC)
Phrygia (heartland)
Probable extent under Midas/Mitā of Mušku (c. 700 BC) or regions of Phrygian influence
Probable boundary under Midas
Region subject to the dominion of Kelainai (attested in literature)
Lydia (heartland)
Probable regions of Lydian influence
Territories of other indigenous Anatolian populations
Territories/Sphere of cultural influence of the so-called Luwian and Aramaic states
Territories of the Greek tribes (Aioleis, Iones, Dories)
Greek colony
Territories of Ancient Balkan-Thracian populations
Contact zones
Assyrian kingdom
Kingdom of Urartu (periphery)
Attacks on Greek cities by the Lydian kings Gyges and Ardys I (according to Herodotus 1,15)
Capital/Residence, Other settlement
Central sanctuary (Phrygian, Greek)
Localization uncertain
Ancient route, Pass
Historical coastline
Historical river course
(Ikaria) Island
Mysoi Tribe/People
Mysia Territory/Mountain range/Body of water
Abasa Pre-Greek name in Greek territoty
Ikonion Greek or other ancient name (Assyrian, Luwian, etc.)
Beyköy Modern name
Kašku? Identification uncertain
0 50 100 150 200 km

Urarṭu and eastern Anatolia *c.* 700 BC

The map presents the political situation in eastern Anatolia and western Iran a few decades before the fall of the Assyrian Empire (614/12 BC). It centres on the kingdom of Urarṭu, even though – in terms of world history – Urarṭu only played a minor role as an opponent of Assyria.

I. Geopolitical situation

Urarṭu was flanked by Qulḫā/Colchis in the north, an area only sparingly mentioned in ancient sources, and by the territory of the Mannāya/Mannaeans in the southeast, a tribal association of Indo-Aryan descent only known from Assyrian sources. The rugged mountains of the Taurus and *Tur Abdīn* to the south acted as a buffer zone against Assyria which according to the available sources was populated by the Nairi people (→ map p. 33). It ran over into territories under Aramaean rule like Bīt-Baḫiāni, Katmuḫḫu and Amedi, which were situated in the triangle between Tigris and Euphrates. West of the Euphrates there were further Aramaean areas centring on Ḫalman/Aleppo and Ḫamat/Hama. The northern part of this region was dominated by so-called Late Hittite principalities such as Gargamiš/Karkamiš, Miliṭi/Melitene and Que/Cilicia. In Cappadocia they bordered on a league of states called Tablāne/Tabāl. It seems that there was no contact with the peripheral parts of the Phrygian kingdom inside the arc of the Halys. On the other hand, it is still unclear whether the Kaškaeans/Kašku, who had made incursions from the Pontus area as early as the age of the Great Hittite Empire, were present on the upper reaches of the Halys.

II. Core territory and expansion

The core territory of Urarṭu (originally called Biaine, hence modern 'Van'), centred on Lake Van/Thospitis Limne, a non-draining body of water. On its shore the castle of the capital Ṭurušpā/Tušpa (Van Kale) was built on a steep rock by Sarduri I (*c.* 840 – 830/25 BC). Later Rusa II (documented for 673/672 BC) had the new residence Rusaḫinele (Toprakkale) built – somewhat to the northeast, but still close to Lake Van – along with a reservoir, today's Keşiş Gölü.

Argišti I (*c.* 785/80–760 BC), Sarduri II (*c.* 760–730 BC) and Rusa I (*c.* 730–714/13 BC) were particularly active expanding their territory: in the north-west as far as Qulḫā/Colchis, in the north-east to Lake Sevan/Lychnitis Limne, in the south-east to Lake Urmia/Matiane Limne, while in the west Urarṭu's territory was expanded across the Euphrates, if only for a short period. Both the central territory, which comprised Great Ararat and Little Ararat, as well as the surrounding areas are extremely rugged mountain terrain, including the Zagros Range, Hakkari, the Lesser Caucasus with the Ararat Massif, the Karasu Mountains and the eastern Taurus Range. As a result, the clusters of settlements were very small, which limited the central government's influence. On the other hand, the terrain offered protection against enemy incursions and could be easily defended by large castles in advanced positions.

Sources

Only a small portion of the source material for the history of Urarṭu can be gleaned from the inscriptions of the Urarṭian kings, which are texts hewn into rocks in Assyrian cuneiform, predominantly in Urarṭian (rarely in Assyrian). Examples are Sarduris I's Assyrian foundation inscription in Tušpa, the bilingual Kelišin stele of Išpuini (*c.* 830/25–820/10 BC) on a pass in the Zagros mountains at an altitude of *c.* 3000 m, south-west of Wiše/Uēsi on today's Iraqi-Iranian border, and the Urarṭian annals of Argišti I on the Rock of Van. They are mostly brief and easy to decipher, and the Urarṭian language has been largely decoded by now. However, apart from formulaic genealogical data they rarely offer historically useful information. This is provided more often by Assyrian campaign reports, as the Neo-Assyrian rulers from *Shalmaneser* III (858–824 BC) to Assurbanipal (669 to *c.* 635 BC) were frequently involved in fights with their northern neighbours. The Urarṭians may have been considered a threat because of their expansionist tendencies. But they were also such skilful smelters of ores, that they could have guaranteed the supply of bronze and iron to the Assyrian army. Especially Sargon II (721-705 BC) made his mark, when in 714 BC he plundered and destroyed among others the important temple of the god Ḫaldi in Muṣaṣir in his thoroughly documented eighth campaign (→ map p. 51). The chronology of the Urarṭian kings relies almost completely on Assyrian reports.

These sources, however, are supplemented by archaeological evidence, most of which comes from Turkish excavations at Altıntepe, Toprakkale (Rusaḫinele), Van and Ayanis, from Soviet sites at Karmirblur (Teišebai) and German excavations at Bastam (Rusai patari). A typical type of pottery with brown glazing serves as an indicator of the location of Urarṭian settlements. Even more noticeable and often a still identifiable feature of the terrain are the canals, which were frequently dug to divert the water of the mountain streams into the settlements and onto the fields; some of these canals are still in use today. Semiramis, the legendary Assyrian queen, is often credited with ordering their construction. Furthermore the territory was sprinkled with fortresses like Cavuš Tepe, Turki Tepe, Haftavan Tepe, Qalat, Qaleh Ismail Agha etc. (not charted on the map). Most of them were placed in the border areas, but also on the sites of selected settlements, where they were to serve as refuges. The Urarṭians were obviously excellent stonemasons. To lay the foundations of their castles they cut into the bedrock – frequently in the form of steps – until they had created a solid rest for the blocks, which were then stacked up, without the use of mortar, to serve as foundations for the walls. On these foundations or semi-basements of sometimes smoothed and almost jointless staggered stonework arose the walls of mudbricks. Some of these fortresses contained vast complexes of stables, storage buildings, utility rooms, as well as rooms for official occasions and temples. They bear witness to the economic strength of the state, which was probably to a large extent fuelled by the mining of copper and tin and probably iron (e.g. at Qulḫā). In any case, Urarṭu's crafts are largely represented by bronze cauldrons, furniture fittings, weapons, belts, etc. They were apparently manufactured for the cult of the god Ḫaldi in the major temples of the country, but also for export. Their distinctive style of decoration was clearly inspired by Assyrian court art.

Although Urarṭu's history and culture had already been discovered more or less simultaneously with that of Mesopotamia by F.E. Schulz (1799–1829), it was not before the 'Armenian expedition' (1898/99) by W. Belck and W.C. Lehmann-Haupt that academically useful materials, e.g. numerous rubbings of rock inscriptions, were brought back to Europe. The inscriptions, then still named Khaldic after the Urarṭian supreme god Ḫaldi, are in a simplified version of Neo-Assyrian cuneiform. They were easily deciphered, but the texts are not fully understood to this day, because Urarṭian does not belong to any of the known language families. It is, however, related in some degree to Hurritic, a language that was documented in northern Mesopotamia and Ḫattusa in the 2nd millennium BC. Like e.g. Hurrian and Sumerian and among modern languages Bask, it belongs to the ergative languages. They shift the load of signalling transitivity to the subject of the sentence, which is marked by the suffix "-she", while the verb and the "direct object" remain unmarked. The remainder of the syntactic relationships is handled by chains of suffixes added to the unchanged root of the verb. The building and consecration inscriptions are frequently quite formulaic and contain very little information that is useful to the historian or the historian of religion. A limited number of Urarṭian 'hieroglyphics' seem to have been used exclusively in economic contexts.

→ Maps pp. 23, 33, 43, 47, 49, 51, 53, 55, 61

Literature

N.V. Arutjunjan, Korpus urartskich klinoobraznych nadpisej, 2001 (s. M. Salvini, SMEA 43, 2001, 241–267); R. Biscione, S. Hmayakyan, N. Parmegiani (eds.), The North-Eastern Frontier. Urartians and Non-Urartians in the Sevan Lake Basin, 2002; A. Çilingiroğlu, Urartu Kralliği: Tarihi ve Sanati, 1997; Id., M. Salvini, Ayanis I: Ten Years' Excavations at Rusahinili Eiduru-kai, 2001; I.M. Diakonoff, S.M. Kashkai, Geographical Names According to Urartian Texts. Répertoire Géographique des Textes Cuneiformes 9, 1981; W. Eder, J. Renger (eds.), Herrscherchronologien der antiken Welt, 2004; G. Garbrecht, Historische Wasserbauten in Ost-Anatolien – Königreich Urartu, 9.–7. Jahrhundert v. Chr., in: Ch. Ohlig (ed.), Schriften der Deutschen Wasserhistorischen Gesellschaft vol. 5, 2004, 1–103; W. Kleiss, H. Hauptmann et al., Karte von Urartu. Verzeichnis der Fundorte und Bibliographie. Archäologische Mitteilungen aus Iran, suppl. 3, 1976; W. Kleiss, Größenvergleiche urartäischer Burgen und Siedlungen, in: R.M. Boehmer, H. Hauptmann (eds.), Beiträge zur Altertumskunde Kleinasiens. FS K. Bittel, 1983, 283–290; F.W. König, Handbuch der chaldischen Inschriften, 1955/57; C.F. Lehmann-Haupt, Armenien einst und jetzt, 3 vols., 1910–1931; G.A. Melikišvili, Urartskie klinoobraznye nadpisi (I), 1960; II as: Otkritija i publikacii 1954–1970, in: Vestnik Drevnej Istorii 1971/3, 229–255; 4, 267–293; M. Salvini, Geschichte und Kultur der Urartäer, 1995; U. Seidl, Bronzekunst Urartus, 2004; R.-B. Wartke, Urartu. Das Reich am Ararat, 1993; P. Zimansky, Ecology and Empire: The Structure of the Urartian State, 1985; Id., Ancient Ararat. A Handbook of Urartian Studies, 1998.

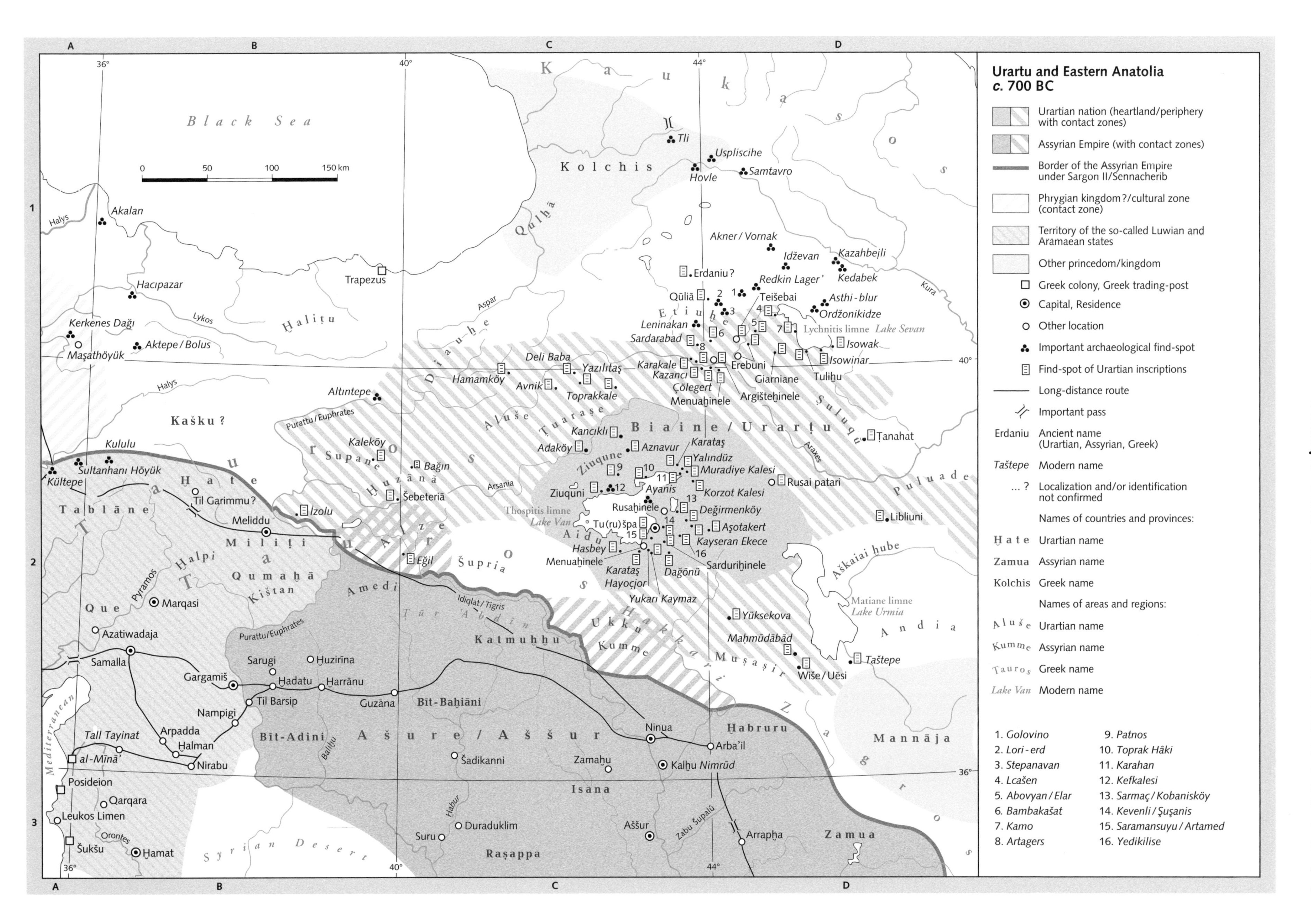
Urartu and Eastern Anatolia c. 700 BC
Urartian nation (heartland/periphery with contact zones)
Assyrian Empire (with contact zones)
Border of the Assyrian Empire under Sargon II/Sennacherib
Phrygian kingdom?/cultural zone (contact zone)
Territory of the so-called Luwian and Aramaean states
Other princedom/kingdom
Greek colony, Greek trading-post
Capital, Residence
Other location
Important archaeological find-spot
Find-spot of Urartian inscriptions
Long-distance route
Important pass
Erdaniu Ancient name (Urartian, Assyrian, Greek)
Taštepe Modern name
... ? Localization and/or identification not confirmed
Names of countries and provinces:
Ḫate Urartian name
Zamua Assyrian name
Kolchis Greek name
Names of areas and regions:
Aluše Urartian name
Kumme Assyrian name
Tauros Greek name
Lake Van Modern name
1. Golovino
2. Lori-erd
3. Stepanavan
4. Lcašen
5. Abovyan / Elar
6. Bambakašat
7. Kamo
8. Artagers
9. Patnos
10. Toprak Hâki
11. Karahan
12. Kefkalesi
13. Sarmaç / Kobanisköy
14. Kevenli / Šuşanis
15. Saramansuyu / Artamed
16. Yedikilise
Black Sea
0 50 100 150 km
Kolchis
Kaukasos
Qulḫā
Ḫaliṭu
Trapezus
Kaškus ?
Biaine / Urarṭu
Etiuḫe
Diauḫe
Aluše
Tuaraše
Ziuqune
Aidu
Šupria
Thospitis limne
Lake Van
Lychnitis limne Lake Sevan
Matiane limne Lake Urmia
Andia
Mannāja
Zamua
Ḫabruru
Zagros
Ukku
Kumme
Hakkari
Muṣaṣir
Puluade
Katmuḫḫu
Amedi
Ṭūr ʿAbdīn
Išuwa
Tablāne
Ḫate
Tauros
Quē
Ḫalpi
Qumaḫā
Kištan
Miliṭi
Supane
Ḫuzānā
Bīt-Adini
Bīt-Baḫiāni
Aššure / Aššur
Isana
Raṣappa
Syrian Desert
Mediterranean
Halys
Lykos
Aspar
Kura
Araxes
Arsania
Purattu / Euphrates
Idiqlat / Tigris
Pyramos
Orontes
Baliḫu
Ḫabur
Zabu Šupalû
Akalan
Hacıpazar
Kerkenes Dağı
Maşathöyük
Aktepe / Bolus
Kululu
Sultanhanı Höyük
Kültepe
Til Garimmu ?
Meliddu
Altıntepe
Kaleköy
Bağın
Šebeteriā
Izolu
Eğil
Hamamköy
Deli Baba
Yazılıtaş
Avnik
Toprakkale
Kancıklı
Adaköy
Aznavur
Karataş
Yalındüz
Muradiye Kalesi
Ayanis
Ziuquni
Korzot Kalesi
Rusaḫinele
Değirmenköy
Aşotakert
Tu(ru)špa
Kayseran Ekece
Hasbey
Menuaḫinele
Sarduriḫinele
Dağönü
Karataş
Hayocjor
Yukarı Kaymaz
Rusai patari
Libliuni
Ṭanahat
Yüksekova
Maḫmūdābād
Wiše / Uēsi
Taštepe
Tli
Usplisciche
Hovle
Samtavro
Akner / Vornak
Idževan
Kazahbejli
Kedabek
Erdaniu ?
Redkin Lager'
Qūliā
Teišebai
Asthi-blur
Ordžonikidze
Leninakan
Sardarabad
Isowak
Isowinar
Karakale
Kazancı
Erebuni
Giarniane
Tuliḫu
Çolegert
Menuaḫinele
Argištehinele
Marqasi
Azatiwadaja
Samalla
Gargamiš
Sarugi
Ḫuzirīna
Ḫadatu
Ḫarrānu
Til Barsip
Nampigi
Guzāna
Arpadda
Ḫalman
Nirabu
Tall Tayinat
al-Mīnā'
Posideion
Qarqara
Leukos Limen
Šukšu
Ḫamat
Šadikanni
Zamaḫu
Ninua
Arba'il
Kalḫu Nimrūd
Duraduklim
Suru
Aššur
Arrapḫa
36°
40°
44°
A
B
C
D
1
2
3

The Luwian-Aramaic principalities *c.* 900 BC

I. Antecedents and genesis

The system of Late Bronze Age states that had developed over centuries had collapsed in the 12th cent. BC. Numerous political units like the Great Hittite Empire, the New Kingdom in Egypt and the Middle Babylonian Empire failed to survive the turmoil. Small principalities, too, lost large parts of their territories or disappeared completely. Only Assyria managed to hold on to its status as a great power until the early 11th cent. Then, towards the end of the Middle Assyrian Period, it, too, lost vast tranches of its Upper Mesopotamian possessions along the Baliḫ and the Ḫābūr, two left-hand tributaries of the Euphrates, to Aramaean tribes who had first infiltrated the area as nomads. Only sections of the lower Ḫābūr near the provincial towns of Dūr-Katlimmu and Šadikanni, the 'Ḫābūr Triangle' with the city of Kaḫat, as well as sections of the upper Tigris near the city of Tušḫan remained under what was probably a loose kind of Assyrian control.

In Northern Syria the city of Gargamiš, a secundogeniture of the Hittite royal dynasty, exercised hegemony over the former provinces of the Great Hittite Empire until the 11th cent. BC.

The main beneficiaries of the changing political situation were the Semitic Aramaeans (Aramū). Where they originated from is a matter of controversy among scholars. Some clues seem to indicate that they infiltrated from Arabia into the marginal zones of the Fertile Crescent; thus an important tribal group, which was to settle in northern Mesopotamia, is called 'Temanids' (Temanayū, cf. map B on p. 33), a hint at origins from the important oasis of Temā in the north of today's Saudi-Arabia. According to other indicators there is some probability that unstable economic conditions forced formerly sedentary Semitic farmers from the northern Mesopotamian region to switch to a nomadic or transhumant way of life as cattle breeders, over time turning into a discrete ethnic group. Incursions by Aramaic nomads are repeatedly mentioned in cuneiform texts as early as the 13th and 12th cents. In the 11th cent., the Assyrian kings had to fight hard to cope with the increasing pressure the Aramaeans were exerting on their core territory. No later than the early 10th cent., numerous small Aramaean principalities, which were named after an ethnicon following the Bīt-PN pattern meaning 'House (= dynasty) of PN'. The most important of these small realms were the Upper Mesopotamian principalities of Bīt-Baḫiāni and Naṣibīna, founded by the Temanayū tribal group, Bīt-*Adini*, situated on the middle reaches of the Euphrates, the northern Levantine principalities of Bīt-Agūsi and Bīt-Gabbār as well as southern Levantine Aram (Damascus).

II. The establishment of the Luwian and Aramaic principalities

Towards the end of the 11th cent. at the latest, several principalities established themselves in northern Syria and southern Anatolia. Due to the influence of their Luwian-Hittite elites, they retained essential elements of the Late Bronze Age culture of the Hittites, especially in religion, art and architecture, but also in their administrative organization and their ideology of government. Thus this age has been termed the 'Late Hittite Period' in the history of art and culture, whereas the political term 'Hittite successor states' (F. Starke), which is also used by some scholars, applies only to a handful of principalities, among these are Gargamiš and Melid, which explicitly wished to connect, ideologically as well as genealogically, with the Great Hittite Empire.

Moreover, Aramaic rulers managed to seize power in some of the Luwian principalities, e.g. in Ḥamat. In a variety of ways the new rulers adopted elements of Neo-Hittite culture, in several cases even taking Luwian names (e.g. Panamuwa and Kilamuwa of Sam'al/Bīt-Gabbār).

The multi-layered, complex and regionally varying processes of transculturation between the established Luwian, Canaanite and Hurrite populations and the recent Aramaean arrivals resulted in a new, multifaceted cultural mix. This makes it difficult to distinguish clearly between 'Luwian' and 'Aramaean' principalities and to define their respective characteristics. The principality of Bīt-Gabbār with its capital Sam'al is a case in point: several of its rulers bore Luwian names in spite of their indubitably Aramaean origins, while others had Aramaic names. The local artwork and architecture was so closely modelled on Gargamiš that it can be seen as a provincial variant of that Neo-Hittite metropolis, but the inscriptions left by the rulers of Sam'al are almost excusively in an Aramaic dialect, which identifies its inhabitants as Aramaeans. Therefore it comes as no surprise that in archaeological research Sam'al is classed as 'Neo-Hittite', while historic philologists term it an 'Aramaic' principality.

A striking trait of numerous Aramaean principalities is the foundation of new residential cities close to the old regional capitals, which had remained inhabited. The latter contained the shrines – some of which had been in existence for more than a thousand years – of the regional deities, who continued to be revered by the new Aramaean rulers. For this reason, many Aramaic principalities are characterized by two main settlements situated close to each other: the new political metropolis and the old cult centre.

III. The Levant

On the coast of the Levant, most of the old seaports had survived the destructions of the Late Bronze Age; only Ugarit in the north had been obliterated. These towns, the most important of which were Ṣūr/Tyre, Ṣidūna/Sidon, Gubla/Byblos and Arwad/Arados are now classified as 'Phoenician' in scholarship. They ruled their respective hinterlands right into the Lebanon range, amassing vast fortunes from their maritime trade, which reached as far as the western Mediterranean.

IV. Assyria

From around 900 BC, the rejuvenated Assyrian Empire began to expand under King Adad-nẹrārī II (911–891 BC). Assyria began by exacting tribute from the Aramaic principalities in Upper Mesopotamia or conquering them and from the reigns of Assurnasirpal II (883–859 BC) and Shalmaneser III (858–824 BC) onward exerted an ever-increasing pressure on the principalities of northern and western Syria. Analogously the culture of Luwian and Aramaic principalities is marked by the reception and adaptation of Assyrian elements, especially in their artwork and style of dress.

Sources

The map approximately reflects the political landscape in Syria and southern Anatolia just before the resumption of an expansive policy by the Assyrian Empire in the early 9th century. The exceedingly fragmentary state of the sources however makes it very difficult to draw such a map: from the Luwian and Aramaean principalities themselves, only scattered bits of information about the extent of their territories have come down to us in the shape of simple royal inscriptions; the Assyrians on the other hand only recorded those situations in their annals which were relevant to their own policy of expansion. Thus, concerning the late 10th cent., there are only records about the principalities east of the Euphrates like Naṣibīna or Bīt-Baḫiāni, whereas the small western realms were ignored by the Assyrians until the middle of the 9th cent. This means that any kind of reconstruction runs the risk of being ahistoric or containing anachronisms.

→ Maps pp. 33, 45, 47, 63, 69, 71

Literature

G. Bunnens (ed.), Essays on Syria in the Iron Age, 2000; J.D. Hawkins, Corpus of Hieroglyphic Luwian Inscriptions, 2000; Id., Die Erben des Großreiches, in: Kunst- und Ausstellungshalle der BRD GmbH (ed.), Die Hethiter und ihr Reich, 2002, 56–61, 264–273; G.B. Lanfranchi, The Ideological and Political Impact of the Neo-Assyrian Imperial Expansion on the Greek World in the 8th and 7th Centuries B.C., in: S. Aro, R. Whiting (eds.), The Heirs of Assyria, 2000, 7–34; E. Lipinski, The Aramaeans. Their Ancient History, Culture, Religion, 2000; H.C. Melchers (ed.), The Luwians, 2003; M. Novák, Die Religionspolitik der aramäischen Fürstentümer im 1. Jt. v. Chr., in: M. Hutter, S. Hutter-Braunsar (eds.), Offizielle Religion, lokale Kulte und individuelle Religiosität, 2004, 319–346; Id. et al. (eds.), Die Außenwirkung des späthethitischen Kulturraumes, 2004; Id., Luwians and Arameans. Processes of an Acculturation, in: W. van Soldt (ed.), Ethnicity in Ancient Mesopotamia, 2005, 252–266; W. Orthmann, Untersuchungen zur späthethitischen Kunst, 1971; Id., Kontinuität und neue Einflüsse, in: Kunst- und Ausstellungshalle der BRD GmbH (ed.), Die Hethiter und ihr Reich, 2002, 274–281; K.A. Raaflaub, Aramäer und Assyrer: Die Schriftzeugnisse bis zum Ende des Assyrerreiches, in: G. Bunnens (ed.), Essays on Syria in the Iron Age, 2000, 177–186; H. Sader, Les États Araméens de Syrie, 1987; R.B. Wartke, Sam'al. Ein aramäischer Stadtstaat des 10. bis 8. Jh. v. Chr., 2006.

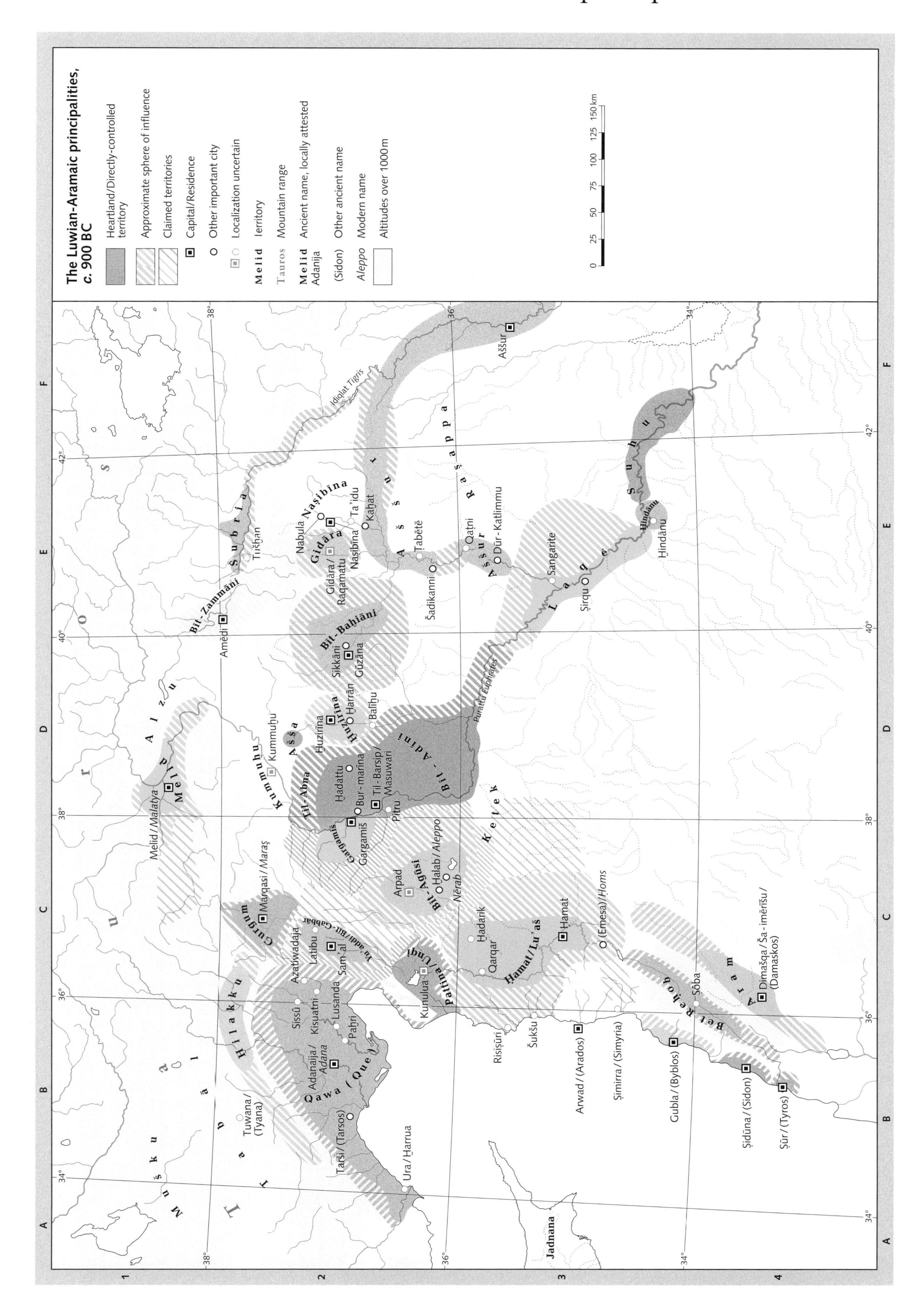
The Luwian-Aramaic principalities, c. 900 BC
Heartland/Directly-controlled territory
Approximate sphere of influence
Claimed territories
Capital/Residence
Other important city
Localization uncertain
Melid Territory
Tauros Mountain range
Melid Adanija Ancient name, locally attested
(Sidon) Other ancient name
Aleppo Modern name
Altitudes over 1000m
0 25 50 75 100 125 150 km

Palestine from the 10th to the 6th cents. BC

Maps A and B illustrate the emergence and development of the states on the Syro-Palestinian land bridge: from the chaos and disruptions of the transition from the Late Bronze Age to the Early Iron Age (12th to 10th cent. BC) to the period when Nebuchadnezzar II's Neo-Babylonian empire was the predominant power (587/86 BC).

I. The topographical relief of Palestine and its impact on cultural history

In Antiquity, Palestine's geographical situation and characteristics exerted a decisive influence on its history. This small area is a mosaic of different climates and landscapes.

1) The long stretch of coastline does not encourage the establishment of ports' facilities, in that it lacks natural bays and inlets. The fertile coastal plain, which was traversed by an important interregional trade route, had been the location of urban centres since the Bronze Age. Unlike the Palestinian coast, the headlands and bays of the Phoenician stretch of the coast provide ideal conditions for building harbours (Ṣor/Tyre, Ṣidon, Beirūt/Biruta/Berytos, Gəbal/Biblos). – 2) Originally, the central Palestinian mountain range (Har Efrayim) with its peaks Har ʿEbal and Har Gərizim had been wooded like its northern continuation, the Lebanon range. Places such as Ḥebron and Yərūšālayim/Jerusalem were located on its ridge. The west-facing slopes of the range receive sufficient precipitation to sustain an intensive terrace agriculture (predominantly olives and wine) in Antiquity, while the east-facing slopes lie in the rain shadow. Thus the route from Jerusalem to Yəriḥo/Jericho leads through steppe terrain. – 3) The Jordan trench extends from the sources of the Yardēn/Jordan to the Yam ha-Melaḥ/Dead Sea, dropping to *c.* 400 metres below sea level. – 4) The steep ascent of the trans-Jordan highlands caught enough rain from the residual clouds to permit agriculture; in Antiquity, this area was also intensively used for terrace agriculture. On the plateau of the mountain range east of the Jordan, another important trade route connected the southern part of the country with Syria. – 5) In the south of the west-Jordan territory, the cultivated land on the heights of Bəʾer Šeba was bordered by the Negev desert, which marked the south-western end of the 'Fertile Crescent'. – 6) North of Mount Carmel lies the small plain of Yizrəʿel, where – because of its fertile soil and rich water resources – important cities had sprung up (Məgiddo, Taʿnak and Bet Šəan).

The Syro-Palestinian land bridge connects Asia Minor with Egypt as well as the Mediterranean region with Arabia. In Antiquity, Palestine was a transit region, which – unlike the Phoenician cities – lacked harbours and thus was not focussed on naval trade, but remained a country of farmers and shepherds. Its fragmented topography made it difficult to establish large political units and promoted the independence of individual subregions. For this reason, Palestine in the course of history was home to an endless diversity of social and ethnic groups.

II. Israel and Judah in the context of the history of Palestine until 722 (map A)

The Israelites and Judaeans emerged as groups within the Early Iron Age village culture on Palestinian soil. The majority of them were descended from the established indigenous population. The Pəlištīm/Philistines on the southern coastal plain, on the other hand, were a group of recent immigrants (first mentioned in Egyptian sources in Madinat Hābū, in the context of a combined land-and-sea battle in 1177 BC as prst/pw-r+a-s+a-*t). The five most important Philistine towns – ʿAzza/Gaza, Ašqəlon, Ašdod, ʿEqron and Gat (unlocalized) –, like the Phoenician cities to the north, retained the political system of independent cities governed by kings which had been predominant in Bronze Age Syria and Palestine.

In the interior part of the country, a new political system came to the fore. There several small regional kingdoms successively emerged and shaped Palestine's history during Iron Age II (*c.* 1000-586 BC): Ăram Dammeśeq, Yiśraʾʾel/Israel, Yəhudaʾ/Judah, ʿAmmon, Moʾaʾb and Edom. About the origins and the early stages of the monarchy in Israel and Judah we are largely in the dark. The first king about whose reign historians can feel confident is Solomon, who ruled in Jerusalem around the middle of the 10th cent. BC. The exact area covered by his kingdom is unknown, but he founded the central shrine of this young kingdom: the temple of Jerusalem, the location of which on the Temple Mount at the site of today's Dome of the Rock is considered an established fact (cf. map C: Jerusalem). Immediately next to it stood another building that the Old Testament ascribes to Solomon: the palace of Jerusalem's royal dynasty. Its members traced their ancestry back to David, which is why they were known as 'House of David' (e.g. in the Hasael inscription from the middle of the 9th cent.)

The records documenting the time after Solomon mention two kingdoms in the land west of the Jordan: Israel in the north and Judah in the south. They existed alongside each other until the Assyrians conquered the Northern Kingdom in 722. Relations between the two were sometimes friendly, sometimes competitive and sometimes hostile. The same can be said of their relationships with the other neighbours: Philistines, Phoenicians, Aramaeans, Ammonites, Moabites and Edomites. King Hasael's inscription from Dammeśeq (middle of the 9th cent.) which was found at Dan, bears witness to the fact that the borders between the two kingdoms were often disputed.

Israel's royal residence was moved several times in the early stages (Šəkem, Pənuʾẹl, Tirṣā). Šoməron/Samaria, founded by King Omri in the first quarter of the 9th cent., was the first capital to retain this status permanently. Dynastic rule was the exception. Omri's son Ahab (871–852) was married to Isebel, who came from a royal family of Phoenicians. On the whole, the Northern Kingdom was more closely bound up in interregional relationships than its southern counterpart. Thus it was first and foremost the Northern Kingdom which was affected by the westward expansion of the Assyrian empire, which from the 9th cent. swept into the Syro-Palestinian land bridge in a succession of waves. This development reached a first peak when Tiglatpileser III in 732 carved large chunks of Israel from the territory of the Northern Kingdom (cf. Map B). In 722, the Assyrians under the command of

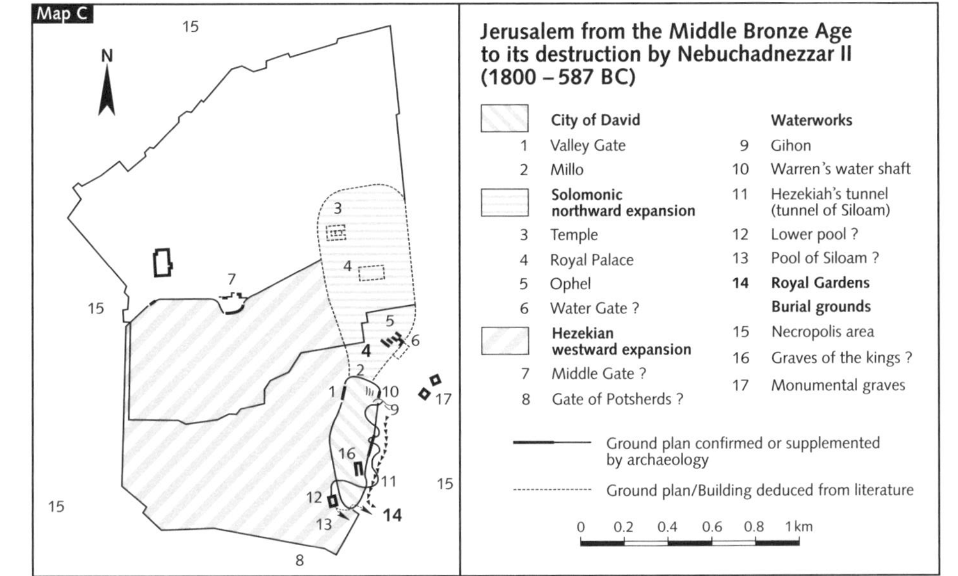

Salmanassar V finally conquered what was left of the Northern Kingdom. It became the province of Sāmerīna and part of the ruling class was deported to Mesopotamia. Another part of the population fled to the Southern Kingdom of Judah.

III. The Southern Kingdom of Judah before 587/86 BC (map B)

After 722, large numbers of refugees from the former Northern Kingdom flocked to Judah, especially to the capital of Jerusalem, whose fortifications were considerably extended under King Hiskia (725–679) in order to protect large areas in the west of the city (cf. map C: Jerusalem). In 701, Hiskia's attempts to escape Assyrian suzerainty led to a military expedition into Palestine by the Assyrian ruler Sennacherib, in the course of which numerous cities in the coastal plain (e.g. ʿEqron and Laʾkiš) were destroyed, while Jerusalem was spared. Among the king's foremost duties in Iron Age Palestine were the care for, and protection of, the central shrines of the kingdom's most important deity (cf., for instance, the temple inscriptions of Bodaštart, King of Ṣidon, and ʾAkyiš/Ikausu, King of ʿEqron). The fall of the Northern Kingdom sharpened the focus on Jerusalem, which became the prime cult centre; this furthered the development of the worship of Yahweh into monolatry and later on into fully fledged monotheism.

In 605, the Babylonians achieved suzerainty over Palestine. Nebuchadnezzar II captured Jerusalem in 598, exiling King Jehoiachin and numerous members of the ruling class to Babylonia and enthroning Zedekiah. When Zedekiah revolted against him, Nebuchadnezzar II in 586 had Jerusalem conquered for the second time and razed to the ground. This was the final blow to Judah as a state in its own right.

Sources

The accounts of the Old Testament, which constitute our main source, were mostly composed at a much later date. This means that they have to be dated and analyzed by philological methods first, and, in a second step, analyzed and evaluated in terms of their historical reliability before they can be used as historiographic source material: Apart from the biblical texts, we have at our disposal a very small number of inscriptions (e.g. the royal inscriptions, cf. maps A and B) and, above all, Assyrian sources. Maps A, B and C are also based on the findings of archaeology, which provide an essential source for the reconstruction of the living conditions in Palestine during the Iron Age.

→ Maps p. 33, 43, 47, 49 and 51, 53, 55; Jerusalem, BNP 6, 2005

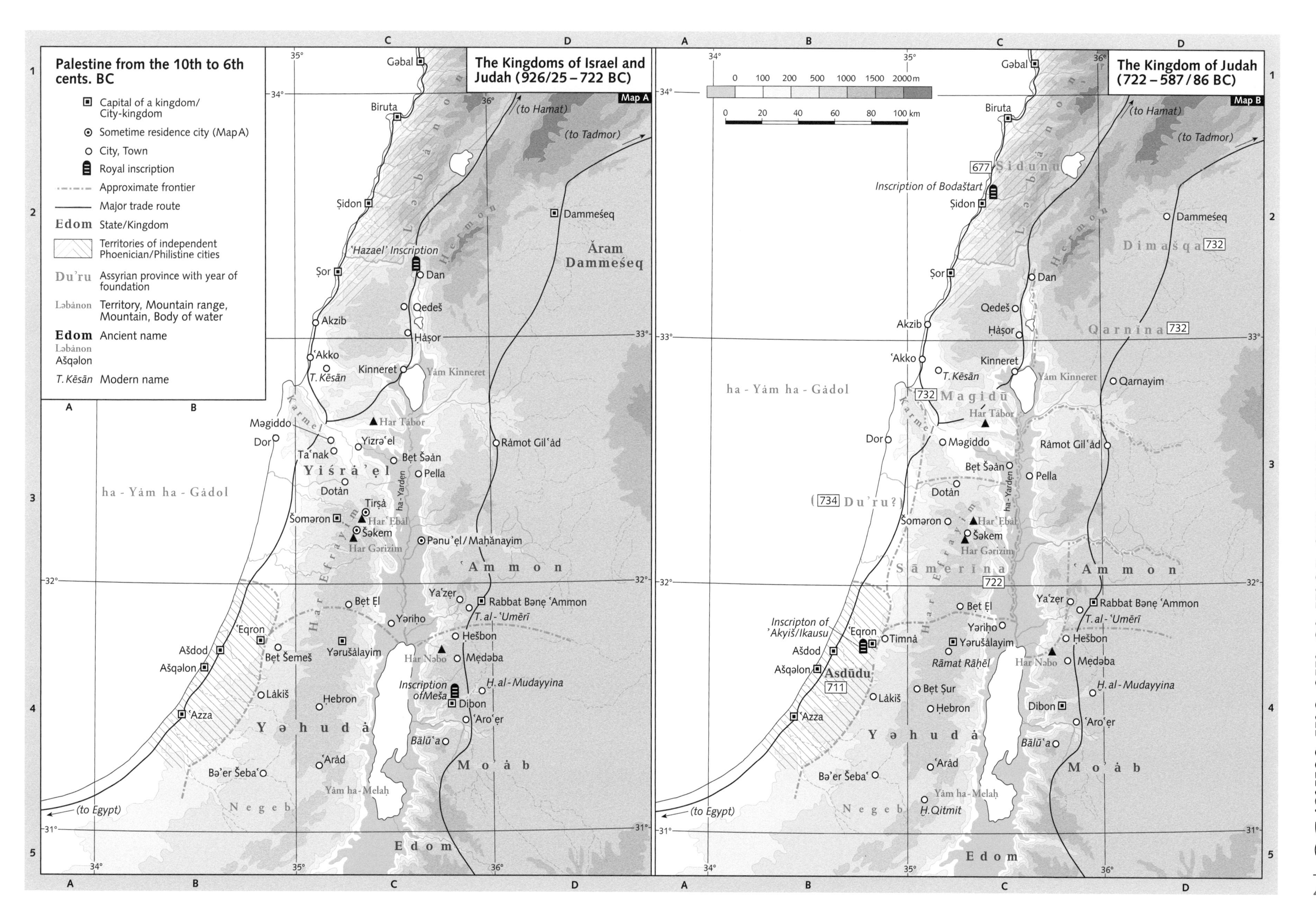

Palestine from the 10th to 6th cents. BC
Capital of a kingdom/ City-kingdom
Sometime residence city (Map A)
City, Town
Royal inscription
Approximate frontier
Major trade route
Edom State/Kingdom
Territories of independent Phoenician/Philistine cities
Du'ru Assyrian province with year of foundation
Ləbānon Territory, Mountain range, Mountain, Body of water
Edom Ləbānon Ašqəlon Ancient name
T. Kēsān Modern name
The Kingdoms of Israel and Judah (926/25 – 722 BC)
Map A
Gəbal
Biruta
Ṣidon
Ṣor
Akzib
ʿAkko
T. Kēsān
Kinneret
Yām Kinneret
'Hazael' Inscription
Dan
Qedeš
Ḥaṣor
Dammeśeq
Ăram Dammeśeq
Məgiddo
Dor
Taʿnak
Yizrəʿel
Har Tābor
Bet Šəān
Pella
Yiśrāʾel
Dotān
Tirṣā
Šoməron
Har ʿEbāl
Šəkem
Har Gərizim
Panuʾel / Maḥănayim
Rāmot Gilʿād
ʿAmmon
ha - Yām ha - Gādol
ha-Yarden
Har Efrayim
Yaʿzer
Rabbat Bənę ʿAmmon
T. al-ʿUmērī
Beṯ El
Yəriḥo
ʿEqron
Bet Šemeš
Yərušālayim
Ḥešbon
Har Nəbo
Mędəba
Ašdod
Ašqəlon
Lākiš
Ḥebron
Inscription of Meša
Ḥ. al-Mudayyina
Dibon
ʿAroʿer
ʿAzza
Yəhudā
Bālūʿa
ʿArād
Bəʾer Šebaʿ
Mōʾāb
Yām ha-Melaḥ
Negeb
Edom
(to Egypt)
(to Hamat)
(to Tadmor)
The Kingdom of Judah (722 – 587/86 BC)
Map B
0 100 200 500 1000 1500 2000 m
0 20 40 60 80 100 km
677 Ṣidunu
Inscription of Bodaštart
Dimašqa 732
Qarnina 732
Qarnayim
732 Magidū
(734 Du'ru?)
Sāmerīna 722
Inscripton of ʿAkyiš/Ikausu
Timnā
Asdūdu 711
Rāmat Rāḥēl
Beṯ Ṣur
Ḥ. Qiṭmit

Mesopotamia and the Levant in the 10th and 9th cents. BC

Map A illustrates how the Assyrian Empire slowly regained its strength. Its rulers focussed all their energies on retrieving the territories it had lost in the turmoil of the 11th and 10th cents. The starting point of map B is the foundation of the new Assyrian capital of Kalḫu; it topicalizes the subsequent phase of Assyrian power politics, which – after continually expanding its scope – came to a halt with the wars for the succession of Shalmaneser III.

I. Assyria (maps A and B)

Around the middle of the 10th cent., the Assyrian Empire had been reduced to the core area around the cities of Aššur, Ninua/Ninive and Arba'il as well as some isolated territories and outposts within a patchwork of small kingdoms run by an Aramaean elite.

The Assyrian kings Aššur-dān II (935–912 BC), Adad-nārāri II (912–891 BC), Tukultī-Ninurta II (891–884 BC) and Assurnasirpal II (884–859 BC) set about to reduce the separate political units and small kingdoms that had sprung up among their scattered territories to the status of vassal states or simply annex them. Assurnasirpal II 's inscriptions record in detail the atrocities committed in the process. As there were no rivals to match its strength, Assyria's unhindered expansion resulted in its regaining control of Upper Mesopotamia, making it the supreme power in the Near East.

The restored Assyrian sense of power manifested itself in Assurnasirpal II's magnificent residential city of Kalḫu. Its construction was begun in 879 BC, but it was not officially consecrated before 866 BC. It was, however, considered Assyria's official capital from 878 BC onwards, a status it was to retain until the end of the 8th cent.

In the second half of Assurnasirpal II's reign (map B) the three objectives were taking shape which Assyria was to pursue for the next century and a half, i.e. to expand its empire in two directions simultaneously: towards the east into the Iranian region, where Assyria obtained the horses that were so vital to its armies, and towards the east into the highly developed and prosperous territories of the Syrian region. Assurnasirpal still brought the conquest of Zamua to a successful conclusion in the east. In the west, he reached the shore of the Mediterranean around 870 BC during one of his campaigns.

The third objective was to prevent any further expansion of the Urarṭian Empire. Shalmaneser III (859–824 BC) achieved a string of impressive successes at the beginning of his reign. He managed to bring the conflict with King Aramu of Urarṭu to a temporary conclusion with a victorious campaign right across Urarṭu (856 BC) and by conquering the realm of Bīt-Adini (858–855 BC) he not only brushed aside the last obstacle on the route to the west, but with the capital Til-Barsip he also acquired a useful base for further campaigns into Syria. From that time on, the scope of Assyria's activities expanded continuously. To the west, the complete Syrian region was affected by Assyria's military campaigns (858–829 BC). The same goes for the small realms situated along the 'Great Military Highway' in the east (843 BC) and Qawa in the north-west (839 BC). The continuous and considerable efforts however failed to achieve anything at all in the long run, despite spectacular military feats. Victories no longer translated into lasting results, and the defeated opponents were only prepared to pay temporary tributes. In Syria, the main target of Shalmaneser's expansion policy, the strongly fortified centres of the local realms turned out to be unconquerable. Ša-imērīšu/Dimašqa emerged as his main opponent. It had at first just been part of an alliance of central Syrian kingdoms led by Adad-idri, which repelled the Assyrians (853–845 BC, battle of Qarqar 853 BC). After the dissolution of the alliance under Hazael, Ša-imērīšu was on its own, but managed to fend off Shalmaneser's attacks (841–837 BC). When the Assyrians subsequently concentrated on conquering the north-west, i.e. Tabāl, Melidu and Qawa (836–831 BC), they did not make any lasting gains either. Simultaneously Shalmaneser lost his influence on the region along the 'Great Military Highway.'

His life ended in tragedy: outshone by his general Dajjān-Aššur from 830 onwards, he died witnessing a long war of succession waged between his sons (826–820). Dajjān-Aššur's emergence marked the beginning of an era in Assyrian history which was to last until 746 BC. It was characterized by a shift of power towards the Assyrian aristocracy at the expense of the royal dynasty.

II. Urarṭu (maps A and B)

Separated from Mesopotamia by high mountain ranges, the Urarṭian empire at first developed in isolation from Assyria. After 866 BC, the growing threat it posed to the states of Šupria and Gilzanu within the Assyrian sphere of interest, led to a war, which in 856 BC ended in defeat for King Aramu of Urarṭu. In the subsequent decades, when he avoided conflict with Assyria, not even retaliating after an Assyrian incursion in 844 BC, Urarṭu must have undergone essential structural changes. When Sarduri I resumed hostilities against Shalmaneser III in 830 BC, his empire was stronger than ever before. The Assyrian succession war presented an opportunity to him and his successor Išpuini to annex the fertile areas to the west of Lake Urmia.

III. Babylonia

The kingdom of Babylonia had also survived the turmoil of the 11th and 10th cents., but unlike Assyria it only managed to retrieve parts of its former territory with the tribal areas of the Kaldu (Chaldaeans) in southern Babylonia remaining almost completely beyond the king's control. Babylonia never pursued the Assyrian type of expansion policy.

Around 905 BC, Assyria's king Adad-nārāri II had wrested areas east of the Tigris away from his weaker Babylonian neighbour, but before 891 BC he concluded a pact of mutual assistance with the Babylonian king Nabū-šuma-ukīn. Apart from a brief conflict in the Euphrates valley in 878 BC, this pact was respected by both sides and renewed a number of times. Shalmaneser III put an end to a quarrel about the succession to the Babylonian throne in 851/850 BC by supporting Marduk-zākir-šumi, who reciprocated this good turn by successfully supporting Šamšī -Adad V's claim to the throne in the war of the Assyrian succession in 820 BC.

Sources

As the political history of the period in question can only be reconstructed from Assyrian sources, our perspective is bound to be slanted. Royal inscriptions from Assyria are available throughout but it is only from the reigns of Tukultī-Ninurta II, Assurnasirpal II and Shalmaneser III that they offer detailed and coherent accounts.

Lists of eponyms make it possible to determine dates more precisely from 910 onwards, while from 840 BC eponym chronicles complement or correct the data given by the inscriptions. In addition there are pieces of information from later chronicles. In marked contrast Babylonia offers only a handful of contemporary source texts for this period.

In the Anatolian and Syrian regions as well as in Upper Mesopotamia there are Luwian and Aramaean inscriptions providing isolated pieces of information on the time before the Assyrian reconquest. Initially Urarṭu seems to have been without writing until the Assyrian cuneiform script was adopted, at first together with the Assyrian language, toward the end of the 9th cent. (inscriptions of Sarduri I). From the reign of Išpuini onwards, inscriptions in the Urarṭian language have also come down to us.

→Maps pp. 33, 41, 43, 45, 49

Literature

E. Cancik-Kirschbaum, Die Assyrer, 2003; E. Ebeling et al. (eds.), RLA, 1928ff.; W. Eder, J. Renger (eds.), Herrscherchronologien der antiken Welt, 2004; A.K. Grayson, Royal Inscriptions of Mesopotamia. Assyrian Periods, vols. 2 and 3, 1991, 1999; J.D. Hawkins, Corpus of Hieroglyphic Luwian Inscriptions, 2000; E. Lipinski, The Aramaeans. Their Ancient History, Culture, Religion, 2000; H.C. Melchers (ed.), The Luwians, 2003; H.-J. Nissen, Geschichte Altvorderasiens, 1998; M. Salvini, Geschichte und Kultur der Urartäer, 1995; M.W. Waters, A Survey of Neo-Elamite History, 2002.

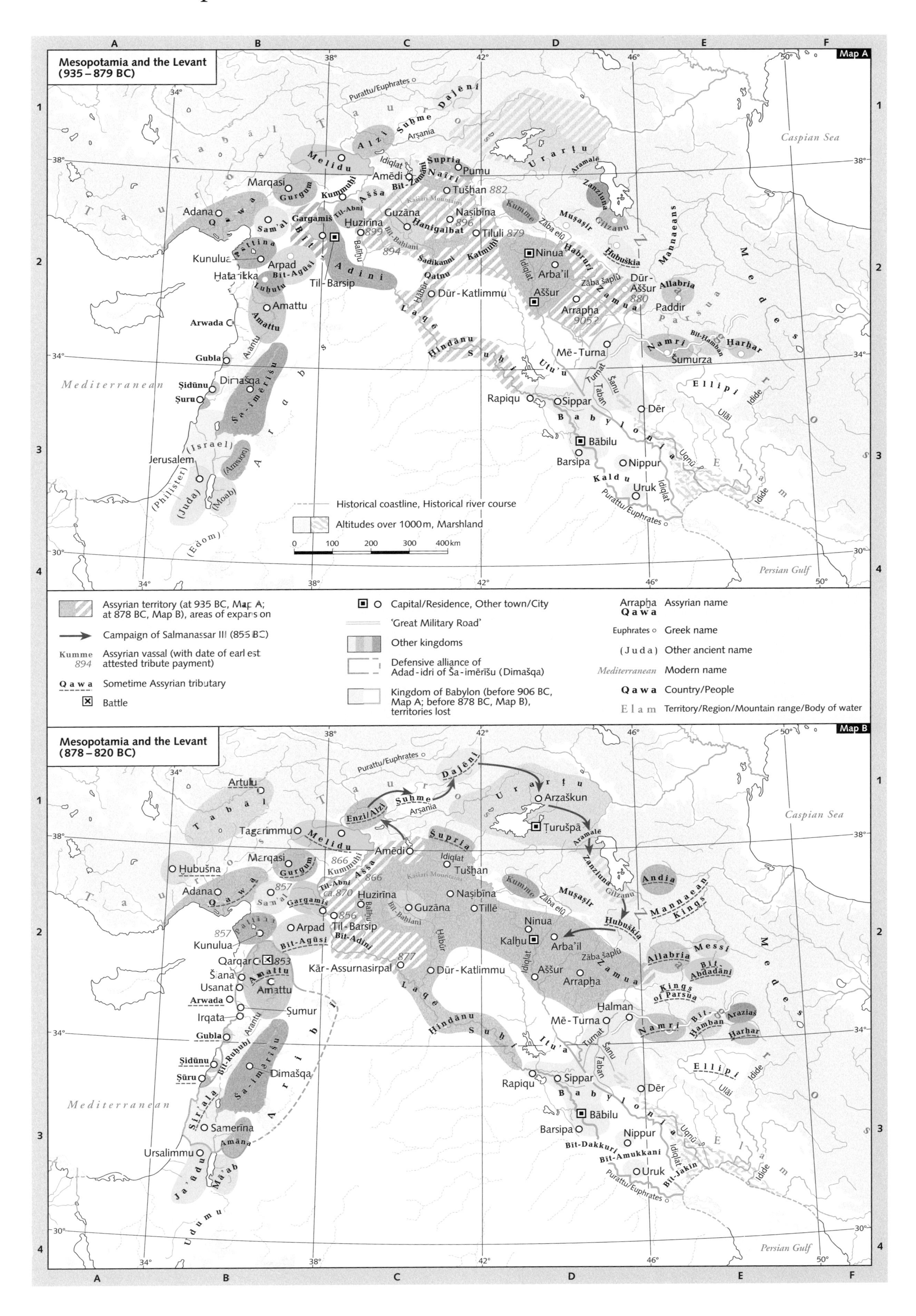
Mesopotamia and the Levant (935 – 879 BC)
Map A
Caspian Sea
Mediterranean
Persian Gulf
Purattu/Euphrates
Idiqlat
Marqasi
Adana
Kunulua
Ḫatarikka
Arpad
Til-Barsip
Amattu
Arwada
Gubla
Ṣidūnu
Ṣuru
Dimašqa
Jerusalem
(Israel)
(Ammon)
(Moab)
(Juda)
(Philister)
(Edom)
Amēdi
Pumu
Tušḫan 882
Guzāna
Huzirīna
Naṣibīna
Tīluli 879
Dūr-Katlimmu
Ninua
Arba'il
Aššur
Arrapḫa
Dūr-Aššur
Paddir
Mē-Turna
Šumurza
Rapiqu
Sippar
Dēr
Bābilu
Barsipa
Nippur
Uruk
Urarṭu
Mannaeans
Medes
Ellipi
Historical coastline, Historical river course
Altitudes over 1000 m, Marshland
0 100 200 300 400 km
Assyrian territory (at 935 BC, Map A; at 878 BC, Map B), areas of expansion
Campaign of Salmanassar III (855 BC)
Kumme 894 Assyrian vassal (with date of earliest attested tribute payment)
Qawa Sometime Assyrian tributary
Battle
Capital/Residence, Other town/City
'Great Military Road'
Other kingdoms
Defensive alliance of Adad-idri of Ša-imērīšu (Dimašqa)
Kingdom of Babylon (before 906 BC, Map A; before 878 BC, Map B), territories lost
Arrapḫa Qawa Assyrian name
Euphrates Greek name
(Juda) Other ancient name
Mediterranean Modern name
Qawa Country/People
Elam Territory/Region/Mountain range/Body of water
Mesopotamia and the Levant (878 – 820 BC)
Map B
Artulu
Tagarimmu
Ḫubušna
Arzaškun
Ṭurušpā
Tušḫan
Naṣibīna
Tillē
Guzāna
Kalḫu
Qarqar 853
Šiana
Usanat
Irqata
Ṣumur
Kār-Assurnasirpal
Samerīna
Ursalimmu
Ḫalman
Bit-Dakkuri
Bit-Amukkani
Bit-Jakin

Mesopotamia and neighbouring regions (819–746 BC)

This map covers the long period from the end of the succession war between the sons of Shalmaneser III to Tiglatpileser III's accession to the throne. It is marked by the temporary demise of the Babylonian Kingdom and the dynamic territorial expansion of Urarṭu. Assyria, although shaken by the war of succession, resumed its policy as a great power, albeit on a smaller scale compared to Shalmaneser III's time. In addition it was greatly hampered by internal strife from 764 BC onward.

I. Babylonia

The Babylonian king Marduk-zākir-šumi (*c.* 855/851–819? BC) had helped to end the Assyrian war of succession between the sons of Shalmaneser III by intervening to support Šamšī-Adad V. He had, however, linked his support to conditions the new Assyrian ruler was bound to find humiliating. Apparently there were further provocations, for it led to a sea change in Assyria's policy towards Babylon. For almost the entire 9th cent., it had been characterized by mutual respect and support, but now Šamšī-Adad shattered the Babylonian Kingdom in an unbroken series of campaigns which lasted from 814 to 811 BC and cost the lives of two of Marduk-zākir-šumi's successors. The cult centres of Babylonia, which were also sacred to the Assyrians, remained untouched, but in annexing the territories north of the Tigris, Šamšī-Adad deprived the Babylonian Kingdom of its most important resources and consequently its political clout. As a result Babylonia disintegrated into a multitude of effectively independent cities and tribal associations. The beneficiaries were the large Chaldaean associations of Bīt-Dakkuri, Bīt-Amukkani and especially Bīt-Jakīn. They emerged strengthened from the turmoil. Their leaders were therefore able to compete with the urban elites for the throne of Babylon when the kingdom began its slow recovery from about 770 BC.

II. Assyria

The war of accession from 826 to 820 BC was the first existential crisis of the Neo-Assyrian Empire, which only managed to survive because there was no opponent strong enough to exploit that long period of weakness. Nevertheless, Assyria's influence in Syria and western Iran had largely evaporated and the vassal realm of Gilzanu had been vanquished and destroyed by Urarṭu. Šamšī-Adad V (824–811 BC) during his reign was unable to undo the consequences of the crisis; his campaigns against Urarṭu and into western Iran ended in failure. In 817 and 816 an uprising in Tillē required his full attention, while from 814 BC he was exclusively focussed on Babylonia. Adad-nārāri III (811–783 BC) resumed the campaigns in Syria and western Iran, dividing his attention equally between these two theatres of war. West of the Euphrates, he restricted himself to arbitrating conflicts among the minor Syrian kingdoms and receiving their tribute payments. In the Iranian region the conquest of the kingdom of Namri (797 BC) tightened his control of the 'Great Military Highway', whose starting point had already been in Assyrian hands since Šamšī-Adad's conquests and Babylon's downfall. At this time the Assyrian army began to penetrate into Median territory.

The confli ct about the succession had not only damaged Assyria's international standing, it had also weakened the king's position at home: Šamšī-Adad V retained an unusual number of his father's powerful officials instead of replacing them with his own followers, as was the custom. The people who apparently largely determined Adad-nārāri III's actions openly took centre-stage alongside the king in an unprecedented manner: among them were first his mother Šammuramat and then Nergal-ēriš, a high official (at least 803–775 BC), who by accumulating offices controlled a considerable portion of the power structure. Nevertheless, the dominating personality of that era was the eunuch Šamšī-ilu, who rose to the exalted position of supreme commander (*c.* 786 BC) in the reign of Adad-nārāri III. To this post, which also included the governorships of Til-Barsip and Namri, he managed to hold on in a succession of reigns until at least 752 BC. He was in charge of the war against Urarṭu during Shalmaneser IV's reign (783–773 BC); the kings Aššur-dān III (773–756 BC) and Aššur-nirari V (756–746 BC) probably were little more than his puppets. Šamšī-ilu, however, was not the only powerful Assyrian official at that time and his rivals were not prepared to put up with his dominance forever. Thus in 764, Assyria was plunged into its second serious crisis, during which it was almost continuously paralyzed by rivalries within its elite. A renewed civil war (763-758 BC), in the course of which the capital itself was fought over as well as Arrapḫa and Guzāna, was followed by a period in which the lords of the empire instead of co-operating simply eyed each other suspiciously. It was another crisis Assyria survived. It is true that its hold over the tributaries on its periphery weakened again, but even the most powerful governors, who had practically become independent, did not secede from the empire. Yet above all, there was again no serious threat from an external enemy, who might have exploited the crisis to destroy the empire. In 746 BC a palace revolution swept Tiglatpileser III into power, whose undisputed claim to the throne put an end to what had basically been a crisis of legitimacy.

III. Urarṭu

After the failure of his campaign of 819, Šamšī-Adad V gave up his attempts to attack Urarṭu in the area around Lake Urmia. Urarṭu's king Išpuini (*c.* 830/25–820/10 BC) had thus decided the second great clash between Assyria and Urarṭu in his favour. In the following decades, he and his son Minua (*c.* 810–785/80 BC) achieved a considerable expansion of Urarṭian territory: in the west they reached the Euphrates, in the north they began the conquest of the Araxes valley (foundation of Minuaḫinili). The two great powers at first avoided any direct confrontation, but it became inevitable when in the reign of Argišti I (785/80–760/56 BC) Urarṭu began to penetrate northern Syria and western Iran, advancing into the territory around the 'Great Military Highway'. The control of these two regions was a central concern to the kings of Assyria. The war lasted almost as long as Shalmaneser IV's reign (783-773 BC), but the outcome was inconclusive. Argišti's subsequent attempt to subjugate the territory of the Mannaeans caused the Mannaeans to unite in defence and led to the birth of the Mannaean kingdom, which successfully resisted Urarṭian expansion. At the same time in the north Urarṭu's territorial gains in the Araxes region were shored up by the foundation of Erebuni and Argištiḫinili. Sarduri II (760/56 to *c.* 730 BC) initially benefited from the continuing crisis paralyzing Assyria. Thus he managed in 755 or 754 BC to defeat Aššur-nārāri V's army. It is remarkable, though, that even then he did not have the military resources for a direct assault on Assyria.

Sources

Events in this period are much more difficult to trace than earlier ones. There are very few Assyrian royal inscriptions giving detailed reports. For long stretches of time the only source available are the meagre eponym chronicles. The Urarṭian royal inscriptions, which emerged at that point in time, are very brief and have so far only been partly deciphered. The situation inside Assyria is reflected in the inscriptions of Assyrian officials, who immortalized themselves in a manner which was normally reserved for kings. From various kingdoms and principalities in Syria and Anatolia, Phoenician, Luwian and Aramaic inscriptions have come down to us which offer isolated pieces of information.

→ maps pp. 41, 47, 50

Literature

E. Cancik-Kirschbaum, Die Assyrer, 2003; E. Ebeling et al. (eds.), RLA, 1928ff.; W. Eder, J. Renger (eds.), Herrscherchronologien der antiken Welt, 2004; A.K. Grayson, Royal Inscriptions of Mesopotamia. Assyrian Periods, vol. 3, 1999; J.D. Hawkins, Corpus of Hieroglyphic Luwian Inscriptions, 2000; F.W. König, Handbuch der chaldischen Inschriften, 1955–1957; E. Lipinski, The Aramaeans. Their Ancient History, Culture, Religion, 2000; H.C. Melchers (ed.), The Luwians, 2003; H.-J. Nissen, Geschichte Altvorderasiens, 1998; M. Salvini, Geschichte und Kultur der Urarṭäer, 1995.

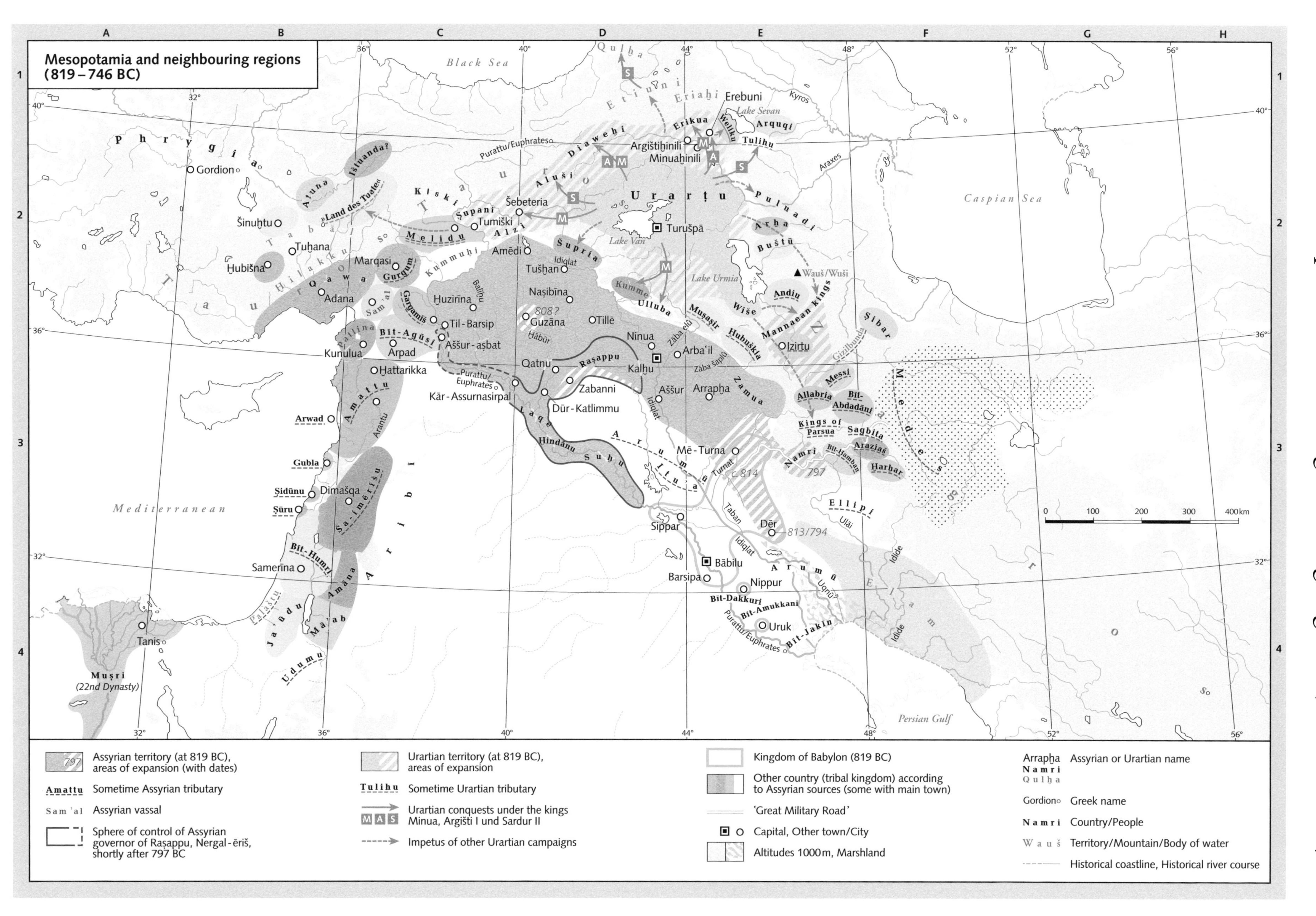

Mesopotamia and neighbouring regions (819 – 746 BC)
Assyrian territory (at 819 BC), areas of expansion (with dates)
Sometime Assyrian tributary
Assyrian vassal
Sphere of control of Assyrian governor of Raṣappu, Nergal-ēriš, shortly after 797 BC
Urartian territory (at 819 BC), areas of expansion
Sometime Urartian tributary
Urartian conquests under the kings Minua, Argišti I und Sardur II
Impetus of other Urartian campaigns
Kingdom of Babylon (819 BC)
Other country (tribal kingdom) according to Assyrian sources (some with main town)
'Great Military Road'
Capital, Other town/City
Altitudes 1000m, Marshland
Assyrian or Urartian name
Greek name
Country/People
Territory/Mountain/Body of water
Historical coastline, Historical river course

Mesopotamia and neighbouring regions (745–711 BC)

The map covers the phase in Assyrian history when its kings Tiglatpileser III, Shalmaneser V and Sargon II realized Shalmaneser III's long-standing plans of conquest in Syria and western Iran by prevailing against the Urarṭian Empire for good in both regions.

I. Assyria

Tiglatpileser III (745–727 BC) succeeded in reunifying the empire's fragmented powers and using them to full effect, which meant that for several generations further conflict among the Assyrian elite was either avoided or at least quickly resolved: when Sargon II (722–705 BC) ousted his brother Shalmaneser V (727–722 BC) in a *coup d'état*, he was largely in control of the situation within a year. Having regained their capacity to take the initiative and enjoying sustained internal stability, the Assyrian kings were in a position to pursue a policy of unprecedented expansion. Thus in 745 BC the Neo-Assyrian Empire entered the phase of the greatest expansion in its entire history, which was to raise it to superpower status within the Ancient Near and Middle East. Between 745 and 708 BC, the Assyrians achieved what had been the three central aims of their policy. The first one, the rule over the rich Syrian territories, first envisaged in the middle of the 9th cent., was essentially achieved when Tiglatpileser III seized and destroyed the two main centres of resistance Bīt-Agūsi (743–740 BC) and Dimašqa (Damascus) (733–732 BC), as this left the remaining Syrian kingdoms without a chance of serious resistance. Many of their rulers re-oriented themselves radically by becoming vassals to the Assyrian superpower, which alienated their subjects and destroyed the internal unity of their kingdoms so vital to the will to survive. Almost without exception they were annexed by Tiglatpileser III (Kunulua), Shalmaneser V (Sam'al, Quwe, Amattu, Samerīna) and Sargon II (Gargamiš, Gurgum, Melidu, Til-Garimmu, Kummuḫi). The last uprising, which involved large portions of Syria, was suppressed in 720 BC. Just a few peripheral areas, most of them under Assyrian tutelage, were able to retain a modicum of independence. Only the impregnable Phoenician island cities of Ṣūru/Tyros and Arwad remained capable of open resistance to Assyria.

The second long-term objective, securing access to the horse-breeding regions on the Iranian plateau, was achieved by gaining complete control of the 'Great Military Highway'. Along the highway and in the surrounding areas, Bīt-Ḫamban and Parsua (744 BC), Kišesim and Ḫarḫar (716 BC) were incorporated into the Assyrian system of provinces. Even fairly small Assyrian armies could move across the vast, but hardly developed territory of Media to the east without encountering any threat. Nevertheless, Assyria was only able to set up a very loose type of suzerainty over the Median princes, who were locked in chaotic internal strife. The third objective, the thwarting of Urarṭian ambitions, was realized in several stages: by defeating Sardur II in the area of Kummuḫi (743 BC, battle of Kištan and Ḫalpi), Tiglatpileser III managed to expel the Urarṭians from the Syrian region. He follwed this up by reducing their sphere of influence in the mountainous border regions (Ulluba 739, Alzi 735). In western Iran the kingdom of Mannaea, founded in the first half of the century and an ally of Assyria since 744 BC, was protecting Assyrian estates from Urarṭian incursions. In the reign of Sargon II, however, the rivalry between Mannaea and Zikirtu led to another clash of the two great powers from 719 onward. For Urarṭu's king Rusa I it ended in a heavy defeat in the battle of Mt. Uauš (714), in the devastation of his south-eastern border areas and in the sack of Muṣaṣir, a shrine most sacred to the Urarṭians.

II. Urarṭu

Thus between 743 and 714 BC Assyria got the better of Urarṭu both in Northern Syria and in western Iran. Dramatic as Sardur II's (*c.* 760/56–730 BC) and Rusa I's (*c.* 730–714/13 BC) defeats may appear in the inscriptions of their Assyrian enemies, yet they simply confirmed the stalemate between the two empires, which had already been indicated at the beginning of the century: in normal circumstances the Urarṭians were not capable of defeating the Assyrian army, while the Assyrians were no real threat to the Urarṭian Empire, which was protected by mountains and guarded by fortresses. Whereas the Assyrians vanquished the Urarṭian forces on the battlefield in 743 and 714, the Urarṭians in 735 successfully defended their royal castle Ṭurušpa when Tiglatpileser III laid siege to it. In spite of being humiliated on several occasions, Urarṭu did not suffer any territorial losses. It even managed to expand its territory in the northern part of western Iran and near Lake Sevan, both of which were outside the Assyrian sphere of interest. After 714 BC, both empires finally accepted the situation and were henceforth anxious to avoid further conflict.

III. Egypt

Assyria's territorial expansion brought it into contact with ever more distant regions. Tiglatpileser III had already tried to control the trade with Egypt by decreeing in 734 BC that it would have to be routed through Ḫazzutu (Gaza). At that stage, disagreements with Egyptian rulers were not yet settled by direct intervention; instead, the Assyrian kings resorted to trade boycotts. In Egypt, which was torn by political strife, the Assyrians at first had to deal only with numerous small kingdoms in the Nile delta. Sargon II, for instance, from 716 to 711 BC cultivated friendly contacts with Osorkon IV, the ruler of Tanis and Bubastis, but at that time the rulers of Kuš had already started to extend their kingdom in several surges from Nubia across the whole of Egypt.

IV. Anatolia

In the north-west, Tiglatpileser III managed to establish a loose kind of suzerainty over the kings of Tabāl, but it was undone again by the emergence of a rival: King Mitā of Mušku (Midas of Phrygia), whose policy aimed at the incorporation of Quwe and the expulsion of Assyrian power from Tabāl and northern Syria. He was at war with Assyria at the latest in the reign of Sargon II. In the course of the conflict, the Taurus range became the border between the two spheres of control: to the north of it, Mitā thwarted any Assyrian efforts to tighten their control over the kings of Tabāl by destroying within a year the province Sargon II had installed in Bīt-Purutaš in 713 BC. South of the Taurus, however, the king of Assyria prevailed. He annexed all vassal states on simple suspicion of being in league with Mitā and in 715 BC cut off his enemy's access to the Mediterranean by seizing Ḫarrua. In 711 BC, Sargon II effectively recognized the Taurus border when he strenghtened the defences of Til-Garimmu to prevent further incursions by Mitā.

V. Babylonia and Elam

Assyria was keeping a close watch on the situation in Babylonia, especially the developments in its temple cities, as these were also sacred to the Assyrians. Although the Babylonian Empire was only a shadow of its former self, being the ruler of the legendary Babylonian Kingdom carried tremendous prestige. At first Tiglatpileser III supported King Nabonassar of Babylonia (747–734 BC), but when the political situation began to destabilize after the latter's death and the Chaldaean prince Mukīn-zēri of Bīt-Amukanni usurped the throne of Babylon, Tiglatpileser III intervened, subjugated northern Babylonia, and in 729 BC mounted the Babylonian throne himself. Shalmaneser V continued the dual Assyrian-Babylonian kingship, but Marduk-aplu-iddina, the Chaldaean king of Bīt-Jakīn and, up to that point, Assyria's ally, exploited the temporary weakness caused by Sargon II's *coup d'état* to expel the Assyrians and entered Babylon in 721 BC as its king. The Chaldaean-Babylonian Kingdom managed to assert itself for the time being, as it had found a powerful ally in the kingdom of Elam, which was making a comeback on the stage of history. For the time being, Sargon II, who failed to defeat the Elamite army in the battle of Dēr (720 BC), had to defer the reconquest of Babylon.

Sources

The Assyrian kings Tiglatpileser III and Sargon II left a vast corpus of inscriptions, on which the reconstruction of the historical developments is essentially based. The years from 728 to 722 are the only period not covered by inscriptions. These were the final years of Tiglatpileser's reign, when he apparently did not commission any major new inscriptions, and the reign of Shalmaneser V, whose memory was obliterated by his brother Sargon. Both Tiglatpileser III and Sargon II had their achievements immortalized in relief sculptures. Complementary information about events in Assyria and Babylonia may be gleaned from eponym lists, eponym chronicles and the Babylonian chronicle, which was composed in a later period. Direct insights can be gained from the partly surviving correspondence among members of the Assyrian elite and numerous administrative documents. Royal inscriptions have also been left by the Urarṭian kings of that period, whereas Assyrian sources are all we have to go by for the events in Babylon, in Elam and on the Iranian plateau. Isolated pieces of information are provided by the inscriptions of the local rulers of Tabāl, the realm of Adana (Quwe) and the Syrian region.

→ Maps pp. 39, 41, 43, 53, 59

Literature

E. Cancik-Kirschbaum, Die Assyrer, 2003; E. Ebeling et al. (eds.), RLA, 1928ff.; W. Eder, J. Renger (eds.), Herrscherchronologien der antiken Welt, 2004; A. Fuchs, Die Inschriften Sargons II. aus Khorsabad, 1994; J.D. Hawkins, Corpus of Hieroglyphic Luwian Inscriptions, 2000; F.W. König, Handbuch der chaldischen Inschriften, 1955–1957; E. Lipinski, The Aramaeans. Their Ancient History, Culture, Religion, 2000; H.C. Melchers (ed.), The Luwians, 2003; H.-J. Nissen, Geschichte Altvorderasiens, 1998; S. Parpola, The Correspondence of Sargon II, 1987; M. Salvini, Geschichte und Kultur der Urarṭäer, 1995; M.W. Stolper, Elam, 1984; H. Tadmor, The Inscriptions of Tiglath-Pileser III, King of Assyria, 1994; M.W. Waters, A Survey of Neo-Elamite-History, 2002; M. Zick, Im Land der Königin von Saba, in: Antike Welt 4, 2006, 33–41 (with further literature).

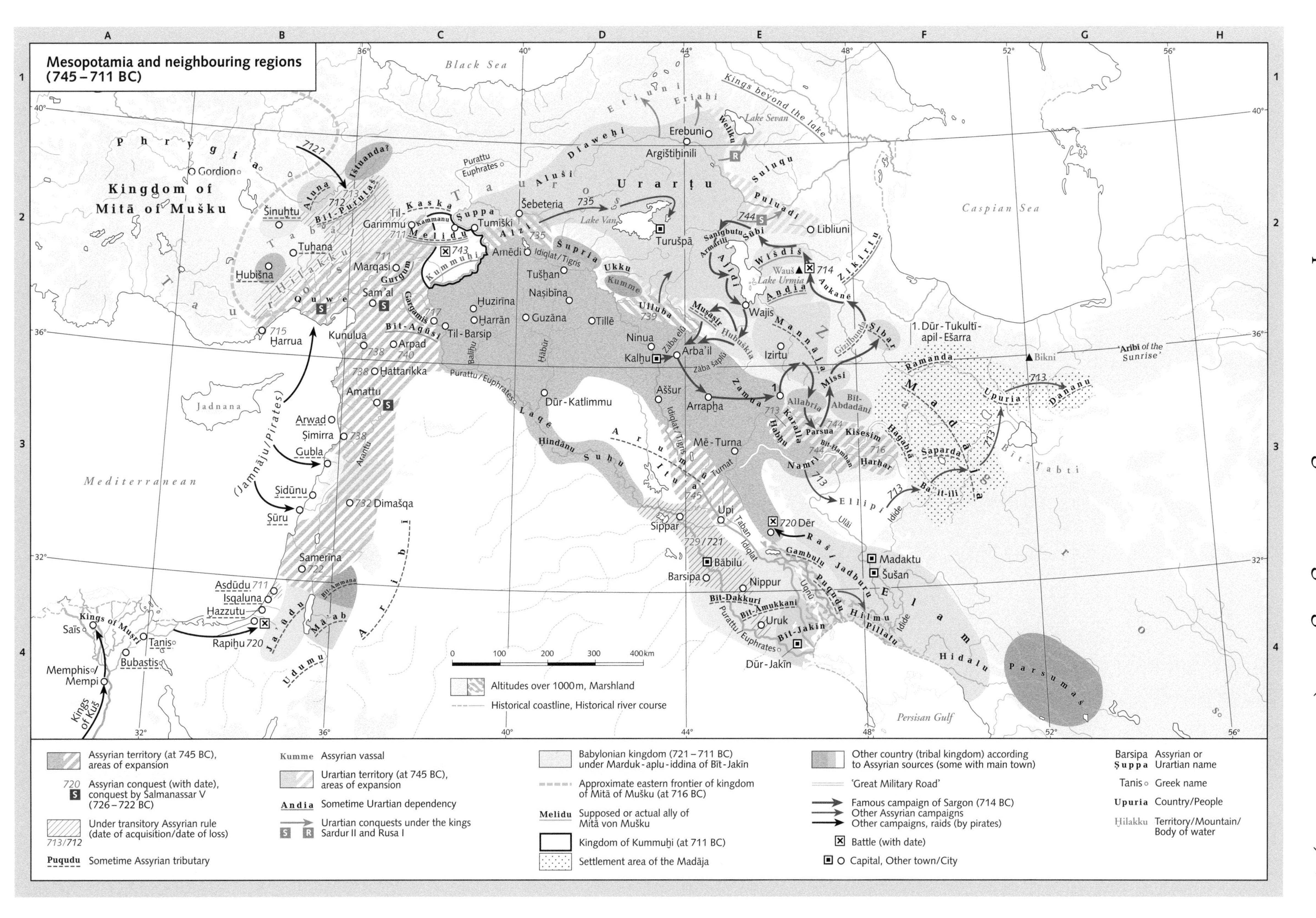
Mesopotamia and neighbouring regions (745–711 BC)
Black Sea
Caspian Sea
Mediterranean
Persian Gulf
Kingdom of Mitā of Mušku
Phrygia
Urarṭu
Gordion
Erebuni
Argištiḫinili
Turušpā
Libliuni
Šebeteria
Tumiški
Amēdi
Tušḫan
Naṣibina
Guzāna
Tillē
Ḫuzirina
Ḫarrān
Til-Barsip
Arpad
Ḫattarikka
Amattu
Arwad
Ṣimirra
Gubla
Ṣidūnu
Ṣūru
Dimašqa
Samerīna
Asdūdu
Isqaluna
Hazzutu
Rapiḫu
Tanis
Bubastis
Saïs
Memphis/Mempi
Kings of Muṣri
Kings of Kūš
Ḫarrua
Kunulua
Marqasi
Sam'al
Ṣinuḫtu
Tuḫana
Ḫubišna
Garimmu
Dūr-Katlimmu
Ninua
Kalḫu
Arba'il
Aššur
Arrapḫa
Mē-Turna
Izirtu
Wajis
Upi
Sippar
Bābilu
Barsipa
Nippur
Uruk
Dūr-Jakīn
Dēr
Madaktu
Šušan
1. Dūr-Tukultī-apil-Ešarra
Bikni
'Aribi of the Sunrise'
Lake Van
Lake Urmia
Lake Sevan
Kings beyond the lake
Jadnana
(Jamnāju/Pirates)
Madāja
Elam
Parsumaš
Hidalu
Ellipi
Namri
Missi
Zamua
Mannāja
Ulluba
Kumme
Ukku
Šupria
Alzi
Meliddu
Kummuḫi
Gurgum
Gargamiš
Bīt-Agūsi
Que
Ḫilakku
Tabal
Atuna
Istuandaʔ
Bīt-Purutaš
Kaska
Ṣuppa
Alusi
Diaweḫi
Etiuni
Eriaḫi
Weliku
Suluqu
Puluadi
Šubi
Wišdiš
Andia
Aukanē
Zikirtu
Sibar
Gizilbunda
Musaṣir
Ḫubuškia
Allabria
Karalla
Parsua
Bīt-Abdadāni
Kišesim
Ḫarḫar
Bīt-Ḫamban
Ḫagabtā
Ṣaparda
Bā'it-ili
Upuria
Dananu
Ramanda
Bīt-Tabti
Laqē
Ḫindānu
Suḫu
Aribi
Arumu
Rašî
Jadburu
Gambulu
Puqudu
Ḫilmu
Pillatu
Bīt-Dakkūri
Bīt-Amukkāni
Bīt-Jakīn
Ammana
Ma'ab
Ja'udu
Udumu
Purattu/Euphrates
Idiqlat/Tigris
Zāba elû
Zāba šaplû
Ḫābūr
Baliḫu
Turnat
Taban
Uqnû
Ulāi
Idide
0 100 200 300 400km
Altitudes over 1000m, Marshland
Historical coastline, Historical river course
Assyrian territory (at 745 BC), areas of expansion
720 Assyrian conquest (with date), conquest by Salmanassar V (726–722 BC)
Under transitory Assyrian rule (date of acquisition/date of loss)
713/712
Puqudu Sometime Assyrian tributary
Kumme Assyrian vassal
Urartian territory (at 745 BC), areas of expansion
Andia Sometime Urartian dependency
Urartian conquests under the kings Sardur II and Rusa I
Babylonian kingdom (721–711 BC) under Marduk-aplu-iddina of Bīt-Jakīn
Approximate eastern frontier of kingdom of Mitā of Mušku (at 716 BC)
Melidu Supposed or actual ally of Mitā von Mušku
Kingdom of Kummuḫi (at 711 BC)
Settlement area of the Madāja
Other country (tribal kingdom) according to Assyrian sources (some with main town)
'Great Military Road'
Famous campaign of Sargon (714 BC)
Other Assyrian campaigns
Other campaigns, raids (by pirates)
Battle (with date)
Capital, Other town/City
Barsipa Assyrian or
Ṣuppa Urartian name
Tanis Greek name
Upuria Country/People
Ḫilakku Territory/Mountain/Body of water

Mesopotamia and neighbouring regions in the late 8th and 7th cents. BC

Maps A and B show the Assyrian Empire at its height as a world power. Map A covers the period when Assyria subjugated first Babylonia and then Egypt while practising peaceful coexistence with Urarṭu. A new phase began around 660 BC (map B), but the radical changes in the Near East caused by Egypt's emergence as a great power, the disintegration of Urarṭu and Elam and the havoc wrought in Anatolia by the Cimmerians, failed to affect Assyria, which even benefited from these events.

I. Babylonia and Elam

From 710 to 689 BC, Sargon II (722-705) and Sennacherib (705-681) were predominantly occupied establishing Assyrian control over Babylonia. In this politically fragmented country Assyrian rule was not accepted in the long run, neither among the Chaldaean and Aramaean tribes nor in the cities, whose ancient culture saw itself at the centre of the universe and despised the Assyrian upstarts. Although the Assyrian kings endowed the Babylonian shrines with similar privileges and lavish gifts as the Assyrian ones, these very places turned increasingly into centres of resistance. All this was compounded by the fact that the kings of Elam tried to keep the threat of Assyrian power at arm's length by actively supporting the striving for independence in nearby Babylon and granting asylum to rebel refugees to that end.

The Babylonian ruler Marduk-aplu-iddina was driven out of Babylon in 710 and in 709 he was expelled from Bīt-Jakīn as well. After Sargon's death, he regained his throne (705-704) with Elamite support, just to be driven back into Elam again by Sennacherib. As the refugees continued to resist with pinprick tactics against southern Babylonia, Sennacherib decided in 694 to get on top of the situation by making an amphibious landing on the Elamite coast. This sparked the fiercest war Assyria had been forced to wage so far: an instant Elamite counterattack, which triggered a general uprising in Babylonia, resulted in the overthrow of Assyrian rule and the death of Sennacherib's son, who had been put on the Babylonian throne. The ferocious fighting came to a head in 691, when the Elamite king Ḫuban-menanu led a massive coalition force of Elamite, Iranian and Babylonian armies in an attack on Assyrian territory, which Sennacherib only just managed to bring to a halt in the hard-fought battle of Ḫalule. As a consequence of this defeat the coalition fell apart and – after Elam's retreat – Assyria quickly gained the upper hand. Siege was laid to Babylon in 690 and after its capture in 689, the city was ruthlessly destroyed. This was followed up by mass deportations. Only now, as much of the country lay in ruins and was in a state of complete exhaustion, was it possible for the Assyrians to rule the country for a longer spell.

Sennacherib pursued Babylon with increasing hatred and even initiated religious reforms of the common Assyro-Babylonian pantheon, aiming to replace, wherever possible, Babylonian gods with Assyrian ones. After his murder however, Asarhaddon (681–669), an admirer of Babylonian culture, ascended the throne. He sought to reconcile the Babylonians to Assyrian rule, had Babylon rebuilt and strove for an understanding with Elam, whose internal fabric had been badly damaged by the stresses of waging war against Sennacherib. In the reign of Assurbanipal (669–631), the enmity between Assyria and Elam, which had again been smouldering since 664, finally flared up into open warfare, which ended in the battle of Til-Tuba (653) with the annihilation of the Elamite army and the death of the Elamite king Te'umman.

In spite of this, Assurbanipal's brother Šamaš-šum-ukīn, who according to Asarhaddon's will had been established as King of Babylon, headed an uprising in 652, which spread across the whole of Babylonia. It took Assurbanipal's troops until 648 before it was put down for good (siege of Babylon 650–648). Badly weakened Elam's policy was oscillating between submissiveness to Assurbanipal and active support for the Babylonian rebels, while its hapless kings were overthrown in quick succession. The progressive decay of royal power and worsening rivalries within the Elamite elite brought about the disintegration of the kingdom. The campaigns which Assyria began to undertake in 647, met with no organized resistance whatsoever and compounded the chaos (sack of Susa *c.* 646). The main beneficiary of Elam's fall was to be the kingdom of Parsumaš, whose king Kuraš (Cyrus) contacted Assurbanipal around 640.

II. Assyria

Throughout this period the Assyrian core territory did not have to face any external threat. The fact that Sargon II, in spite of giving permanent proof of his piety, shamefully lost his life campaigning against one of the minor kings of Tabāl in 705, disquietened the Assyrian elite, but did not pose a threat to the empire. Equally the court intrigues concerning the succession of Sennacherib, which culminated in his murder in 681, did not have any serious repercussions, because his son Asarhaddon succeeded in quickly defeating the regicides. Apart from a wave of purges among the high officials (670), life at the centre of the empire continued undisturbed. Not the spoils of war which flowed there in the wake of royal campaigns were the greatest benefit, but to a much greater extent the steady stream of people and goods from the provinces and the vassal states. This immense accumulation of wealth enabled the Assyrian rulers to realize extensive programmes to build temples, palaces, water works, parks and city walls. At Dūr-Šarrukīn (Sargon's fortress, founded 717, inaugurated in 707), Sargon II commissioned a new capital which was created on the drawing board, but lost its status when he died. Sennacherib and Assurbanipal developed Ninua/Ninive into an even more splendid metropolis. This is where Assurbanipal had an important library collected, which comprised a considerable portion of Ancient Mesopotamia's written culture.

III. Egypt

From 707 at the latest, Egypt was ruled by the Nubian Kushite kings, whose attempts at winning over Assyria's vassals in southern Syria and Phoenicia forced the Assyrian kings repeatedly to intervene (Jerusalem 701, Sidon 677). Asarhaddon attempted from 673 onwards to wrest Egypt itself away from the Kushite king Taharka. He succeeded in 671 for the first time, but subsequently the Nile valley changed rulers several times. It was not before 664 that Assurbanipal managed to expel Taharka's son Tantamun for good (sack of Ne'/Thebes). The attempt to integrate the remote, densely populated, culturally different Nile valley into the Assyrian Empire was doomed to failure, for Assyria depended completely on the support of indigenous princes. The one on whose support Assurbanipal had mainly relied from 664 – Psammetichus, King of Saïs – up to 656 unified all of Egypt under his rule and soon afterwards disengaged himself from Assyria's suzerainty.

IV. Cimmerians and Scythians

In this era, the Near East was harrowed for the first time by the mounted warriors from the steppes of Eurasia, whose marauding incursions towards the end of the 8th cent. threatened the northern border of Urarṭu, extending their range over the whole northern periphery of Assyria in the 7th century. Bands of Cimmerians (Gīmirāja) and Scythians (Iškuzāju) partly allied with the Mannaeans roamed across the Iranian plateau; in the west the kingdom of Mušku, which in 709 had started peace negotiations with Sargon II, probably fell victim to their onslaught. When around 666 BC the Cimmerians shifted their activities completely to Central Anatolia, Gyges of Lydia (Luddu) and Mugallu, who in 676 had carved out his realm at Assyria's expense, made unsuccessful appeals for help to Assurbanipal: Gyges fell in battle against the Cimmerian prince Tugdamme (Lygdamis), who after dispatching Mugallu's kingdom, attacked Assyrian territory around 640 BC. To the great powers, these raids were more of a nuisance than a threat: Assyria had already repelled a Cimmerian advance near Hubusnu in 679 and now managed to defeat Tugamme as well, but Cimmerian incursions continued.

V. Urarṭu

The defeat which the Cimmerians inflicted on Argišti II around 710 BC did not have any consequences, because Assyria and Urarṭu stuck firmly to their policy of avoiding war and not retaliating even in cases of clear provocation: neither the fact that Urarṭu granted asylum to Sennacherib's murderers nor the annexation of the buffer state of Šubria by Asarhaddon (672) resulted in further armed conflict. Free from such disturbance, the Urarṭian Empire reached the peak of its internal development during the reigns of Rusa II (attested 673/72) and Rusa III (attested 655/54 or 652). New, splendidly appointed castles and the expansion of the empire's administration attested to the power of the king. After 653, however, the Urarṭian Empire decayed rapidly, in all likelihood as a consequence of interior strife and new incursions by Cimmerians or Scythians. A long time before the end of the 7th cent., the term Urarṭu had come to describe no more than a mere geographical area.

Sources

This period is covered almost completely by – again – an abundance of Assyrian royal inscriptions. As these are without exception inscriptions dedicating buildings, this type of source dries up in the year after the building plans of the ruler in question have been realized; thus the last years of both Sennacherib (688–681) and Assurbanipal (637–631) remain in the dark. Assyria's other variegated written records from this era are put in the shade by the literary compositions from Assurbanipal's library. On the other hand, the eponym lists are missing for the age of Assurbanipal. This makes the exact dating of historical events before 648 difficult, while its becomes an impossibility for the time after that date. A wealth of pictorial sources is offered by the relief sculptures which decorated the palaces of Sargon II, Sennacherib and Assurbanipal. In Urarṭu there was a change in the way the kings presented themselves: Argišti II, Rusa II and Rusa III no longer left extensive war reports; their inscriptions almost exclusively describe the royal building activities.

→ Maps pp. 39, 41, 49, 87

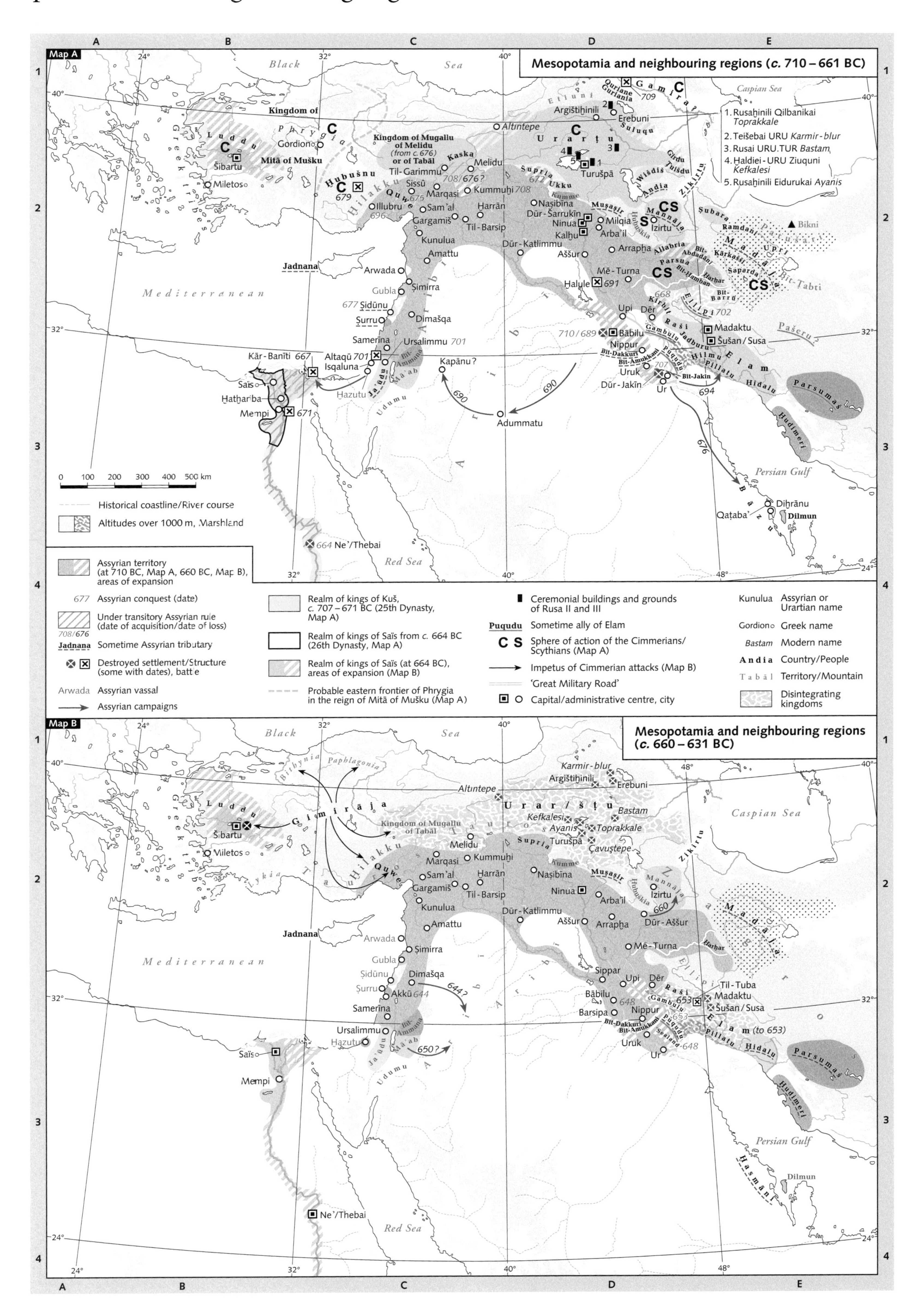
Map A
Mesopotamia and neighbouring regions (c. 710 – 661 BC)
1. Rusaḫinili Qilbanikai Toprakkale
2. Teišebai URU Karmir-blur
3. Rusai URU.TUR Bastam
4. Ḫaldiei-URU Ziuquni Kefkalesi
5. Rusaḫinili Eidurukai Ayanis
Historical coastline/River course
Altitudes over 1000 m, Marshland
Assyrian territory (at 710 BC, Map A, 660 BC, Map B), areas of expansion
Assyrian conquest (date)
Under transitory Assyrian rule (date of acquisition/date of loss)
Sometime Assyrian tributary
Destroyed settlement/Structure (some with dates), battle
Assyrian vassal
Assyrian campaigns
Realm of kings of Kuš, c. 707 – 671 BC (25th Dynasty, Map A)
Realm of kings of Saïs from c. 664 BC (26th Dynasty, Map A)
Realm of kings of Saïs (at 664 BC), areas of expansion (Map B)
Probable eastern frontier of Phrygia in the reign of Mitā of Mušku (Map A)
Ceremonial buildings and grounds of Rusa II and III
Sometime ally of Elam
Sphere of action of the Cimmerians/Scythians (Map A)
Impetus of Cimmerian attacks (Map B)
'Great Military Road'
Capital/administrative centre, city
Assyrian or Urartian name
Greek name
Modern name
Country/People
Territory/Mountain
Disintegrating kingdoms
Map B
Mesopotamia and neighbouring regions (c. 660 – 631 BC)

Mesopotamia and neighbouring regions in the late 7th and 6th cents. BC

Mesopotamia and neighbouring regions in the late 7th and 6th cents. BC

Map A describes the decline and ultimate fall of the Assyrian Empire after the death of Assurbanipal. Map B shows the brief intermediate phase between the empire of the Assyrians and that of the Persians, a period characterized by equilibrium among several major powers in Asia Minor.

I. Assyria (Map A)

At the time of the death of Assurbanipal (631 BC), the empire was in a more favourable position than ever before, since it no longer faced any serious foreign opponents. However, because the king had not succeeded in satisfactorily ordering the succession, the kingdom again fell into a crisis similar to that of the mid 8th cent. This time, however, internal disputes and civil war would mean its ruin.

Assurbanipal's son and successor, the minor Aššur-etel-ilāni (631–627 BC), was entirely under the influence of the eunuch Sīn-šum-lēšir, whose dominant position the Assyrian elite no more accepted than they had that of Šamšī-ilu a century and a half before. Sīn-šum-lēšir successfully fought off at least one attempt at usurping his ward's throne, but both died in unexplained circumstances in 627 BC, after the eunuch had had himself proclaimed king in the end.

Another son of Assurbanipal, Sīn-šar-iškun (627–612 BC), now ascended the throne at Niniveh, but he was not unopposed: strong opposition forces were gathering in the area west of the Euphrates, apparently under the pretender Itti-ili, who from there advanced on Ninua/Niniveh in 623 BC. However, although Sīn-šar-iškun initially focussed his resources on this foe, whom he judged more dangerous than the rebellion now also developing in Babylonia, it only proved possible to put down the Syrian centre of resistance with Egyptian help, and at the cost of relinquishing all territories west of the Euphrates to Pharaoh Psamtik/Psammetichus. By the time Sīn-šar-iškun eventually turned his attention to his Babylonian opponent, Nabopolassar, the latter had become too strong. In spite of Egyptian and Mannaean help, the already weakened Assyria was quickly put on the defensive, especially with the emergence on the scene of another enemy, the Median king Cyaxares/Umakištar, who advanced as far as Arrapḫa in 615 BC. The Median-Babylonian alliance concluded in 614 BC meant the end for Assyria. Aššur fell in 614, and Sīn-šar-iškun met his death during the fall of Niniveh in 612. Finally, in 610, Aššur-uballiṭ II, Assyrian king only by the grace of Egypt, was driven from Ḫarrān.

II. Babylonia (maps A and B)

A revolt against Assyrian rule began in Uruk amidst the confusion of the year 627 BC. During the following year, it spread to the city of Babylon. The coronation as King of Babylon of Nabopolassar (626–605 BC), the leader of the rebellion, marked the foundation of the Neo-Babylonian Empire. Because Assyria was preoccupied with internal matters and in the first years was only able to muster inadequate force to counter him, not only could Nabopolassar hold his own, but he succeeded in expelling the Assyrian overlords from Babylonia by 619 BC, and ultimately, with Median help, in destroying the Assyrian Empire by 610 BC.

The fall of Assyria was followed by decades of dispute between the Neo-Babylonian Empire and Egypt over ownership of Syria and Palestine. Nabopolassar's son, Nebuchadnezzar II (605–562 BC), defeated the Egyptians at Galgameš and Ḫamātu in 605 BC, but his assault on Egypt was repelled with heavy losses in 601 BC. Although Nebuchadnezzar was able to hold on to the disputed territories, the constant Egyptian attempts to incite the local Palestinian and Phoenician powers against him compelled him to intervene repeatedly (capture of Urusalimmu/Jerusalem in 597 and 587 BC, deportation of the Jews to Babylon; thirteen-year blockade of Ṣurru/Tyre). In 568 BC, he had to repel an invasion by Pharaoh Amasis.

Babylonia itself, lastingly freed from unrest, war and foreign rule for the first time in centuries after the expulsion of the Assyrians, experienced a period of economic prosperity that was reflected in countless major royal construction projects in all cities of the country. Babylon in particular was developed into a 'Wonder of the World' under Nebuchadnezzar II.

Following Nebuchadnezzar's death, the dynasty of Nabopolassar dissolved in internal power struggles. Nebuchadnezzar's son Amīl-Marduk (562–560 BC) was deposed by Neriglissar (560–556 BC), whose own son Lābāši-Marduk (556 BC) fell victim to another usurper by the name of Nabonid (556–539 BC). This mysterious and eccentric last Neo-Babylonian king, who withdrew for ten years (553–543 BC) to the oasis of Temā in the middle of the Arabian desert, forfeited the support of the populace of Babylon and thus laid the ground for the Persian conquest.

III. Anatolia and Iran (maps A and B)

The situation in Elam during this period is unclear. The kingdom of Mannāya, one of the last allies of the doomed Assyrian Empire, probably fell victim to its arch-enemy, the kingdom of Zikirtu (Sagartāya).

Even of more significant powers, little is known. The nature and spatial extent of the Median kingdom can only be guessed at. It is known that the Median king Cyaxares (Umakištar in Babylonian sources) played a vital part in the fall of Assyria, that the capital of the kingdom was Agamtanu (Ekbatana) and that another Median king called Astyages (Babylonian: Ištumegu) finally succumbed to the Persians. Otherwise, Media is an enigma.

Information about Lydia (Luddu/Lūdu) is almost equally scarce: its kings finally eliminated the Cimmerian threat before extending their realm across large expanses of western and central Anatolia.

The rise of the Persian Empire is also entirely obscure before the mid 6th cent. Nothing is known of how, in spite of its rather modest territorial foundations, it attained the enormous military power that enabled Cyrus II to conquer first Media (553–550 BC), then shortly afterwards Lydia (*c.* 545 BC) and finally Babylonia (539 BC), in a triumphant campaign as surprising as it was unprecedented in world history. When Cambyses conquered Egypt in 525 BC, all the great civilizations of the Near and Middle East were united in a single empire for the first time.

Sources

After the interruption of the narrative Assyrian sources (*c.* 638 BC), events can generally only be followed sporadically and for the most part sketchily. Babylonian chronicles give descriptions of the Babylonian revolt and the fall of Assyria, but with a lacuna in the years 622-617 BC. The information contained in contemporary documents from various Babylonian cities, showing how Assyria was gradually forced back, has proved to be a valuable supplement.

Babylonian chronicles cover the years 626–623, 616–594, 557 and 555–539 BC. Other than these, there are only sporadic individual pieces of information of varying origin. The Neo-Babylonian rulers left an epigraphic corpus of enormous extent, but of only very limited value in respect of narrative history, because, unlike the Assyrian inscriptions, they contain almost no references to foreign policy, but only describe the construction of major royal buildings. The Babylonian sources give no information about the world outside the kingdom's own borders.

Except for Judaea, the tragic fate of which is recorded in the historical books of the Bible, the only accounts of the other regions of the Near and Middle East in that period are from Greek sources, written at a considerable temporal, spatial and cultural distance.

→ Maps pp. 37, 39, 53, 59, 87

Literature

S. Aro, R.M. Whiting (eds.), The Heirs of Assyria. Melammu Symposia 1, 2000; E. Cancik-Kirschbaum, Die Assyrer, 2003; E. Ebeling et al. (eds.), RLA, 1928ff.; W. Eder, J. Renger (ed.), Herrscherchronologien der antiken Welt, 2004; D.O. Edzard, Geschichte Mesopotamiens, 2004; F. Joannès, La Mésopotamie au 1er millénaire avant J.-C., 2000; M. Jursa, Die Babylonier, 2004; G.B. Lanfranchi, M. Roaf, R. Rollinger (eds.), Continuity of Empire (?), Assyria, Media, Persia, 2003; H.-J. Nissen, Geschichte Altvorderasiens, 1998; M.W. Stolper, Elam, 1984; M. Streck, Assurbanipal und die letzten assyrischen Könige bis zum Untergang Ninivehs, reprint 1975; M.W. Waters, A Survey of Neo-Elamite-History, 2002.

Mesopotamia and neighbouring regions in the late 7th and 6th cents. BC

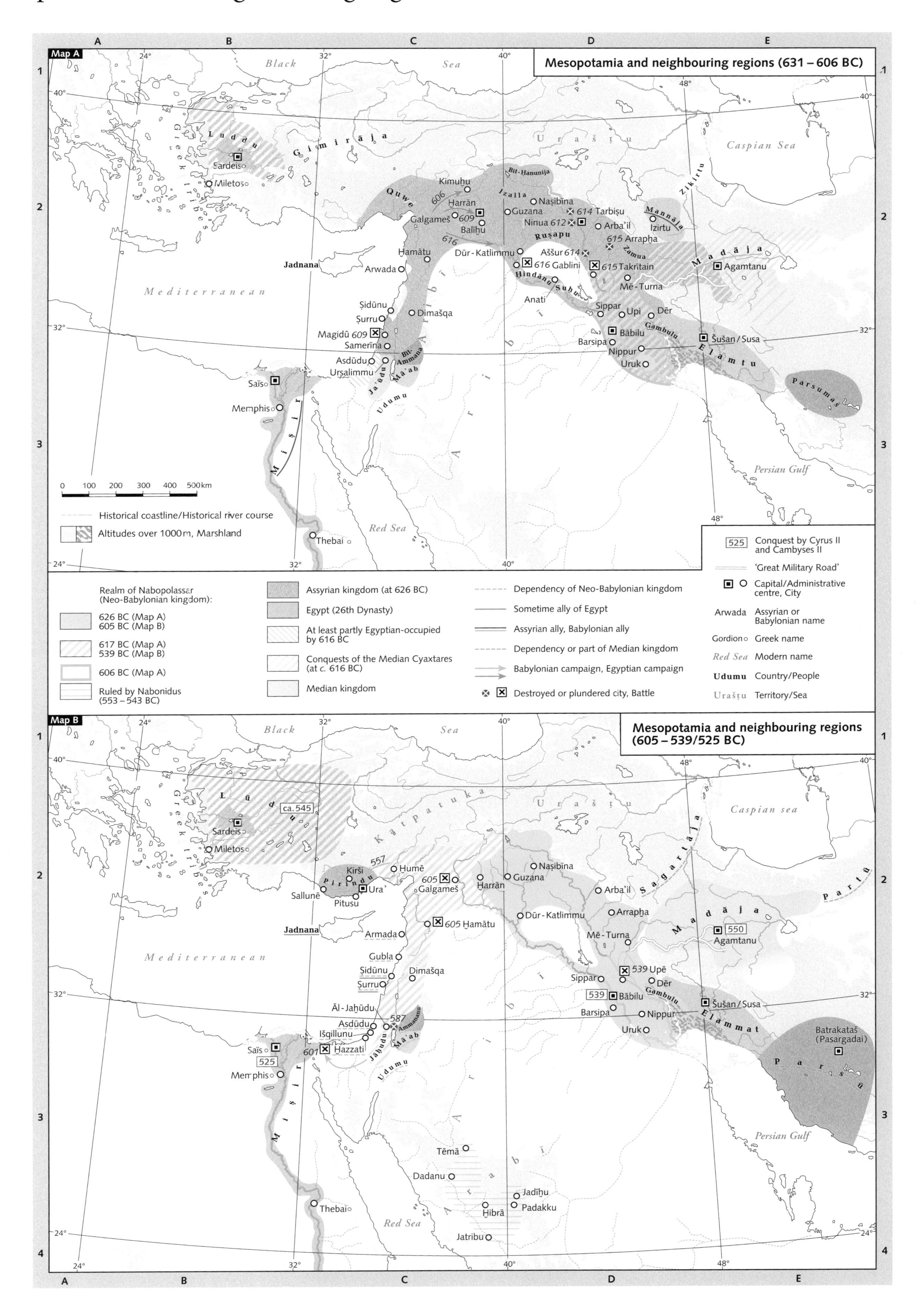

Bābilu ('Gate of God')/Babylon at the time of the Neo-Babylonian Empire (7th/6th cents. BC)

The Mesopotamian metropolis of Bābilu with its 2,200 acres (890 hectares) of enclosed space was one of the largest cities of Antiquity as well as the cultural, economic and at times also political centre of the Ancient Middle East. The city's ruins are situated in the centre of the southern Mesopotamian alluvial plain on a branch of the Euphrates that was called Araḫtu in Antiquity. The city existed at least from the 3rd millennium; its original name was *Ba(b)bal, which in folk-etymology was interpreted as Bāb ilim, i.e. 'God's Gate' (Sumerian: KÁ.DINGIR.RA; OT: Babel; Greek: Babylon).

I. Architecture

1. Fortifications
Babylon's rectangular inner city lay on both banks of the Euphrates. It was enclosed by a defensive wall at a total length of 8.05 km, with the northern and southern walls being *c.* 2,650 respectively *c.* 2,500 metres long and the eastern and western walls extending over *c.* 1,650 and *c.* 1,500 metres. On the east bank there was another wall, called the Eastern Hook. At a total length of 10.6 km, it integrated the elevated 'Summer Palace' in the north along with the outer city into the urban ensemble (inner city: 965 acres (390 ha); outer city: another 1,235 acres (500 ha)). The inner city was protected by a fortification system of extraordinary strength. It consisted of two walls running parallel at a distance of 12 meters, as well as a stone-banked moat. The outer wall, Nimitti-Enlil, was about 7,80 m thick with parts of it consisting of fired bricks, while the inner wall, Imgur-Enlil, was a little wider and consisted of mud bricks. Both of them were equipped with rectangular, regularly shaped towers. A quay wall, which was repeatedly renewed, ran along the river bank.

2. The inner city
The inner city's structure was determined by the most important intra-urban watercourse (Araḫtu), which split it into an eastern and a western segment. The remains of a stone pontoon bridge have been uncovered close to the Ziggurat.

The sacred places: The centre of the city complex, right next to the eastern bank of the Araḫtu, was the location of the city's foremost sanctum: the ziggurat *Etemenanki* (the biblical Tower of Babel) and the ground-level temple *Esangila*, both of which were dedicated to Babylonia's national god Marduk. They are, however, separate buildings, which were erected as independent structures and were separated by the thoroughfare running towards the bridge across the Euphrates. The centre of this most important sacred area of the city had been designed according to traditional Babylonian concepts. It can probably be traced back to the Early Babylonian (20th–16th cents.) plan, when under King Ḫammurapi Bābilu for the first time rose to become the centre of Mesopotamia. Several other temples have been excavated within the area of the city (e.g. the Išḫara temple; the Ninurta temple; the temple to Ištar of Agade/Akkad; the Nabû temple). All of them were laid out according to the standard design with houses surrounding a central courtyard. Sitting across the central axis of this courtyard was a broad antecella (anteroom) which provided access to the identically aligned cella containing the cult niche situated opposite the entrance. The inner sanctum of such temples was usually separated from the building's outside wall by a narrow encircling corridor.

The palace buildings (citadel and 'Summer Palace'): The palace complex was another landmark of Babylon. It consisted of several areas: the 'Southern Castle' situated just inside the inner city wall between the Procession Way and the Euphrates; the 'outwork' pushing out into the river (its function is still unknown); the main palace north of the Southern Castle and thus outside the actual city; and the 'eastern outer defence', a walled-in area east of the Procession Way outside the Ištar Gate. It is now believed to have been a water reservoir.

In untypical fashion, but obviously based on the original plan, the whole complex including the Ištar Gate was elevated in several stages by raising the walls as foundations and then filling up the walled-in spaces successively with debris. Thus the complete area was steadily raised. This emphatically marked out the palace area, the Procession Way and the city gate as a detached citadel system separated from the city by the fortifications. In spite of integrating a section of the city wall, it disrupted the rectangular layout of the city by jutting out into the landscape to the north. Here the influence of Assyrian town planning becomes obvious, where protruding royal palaces and elevated citadels were a distinctive feature. The external plan of the palace area was, in keeping with all buildings of Bābilu, oriented towards the cardinal directions (→map p. 3). The internal plan, however, was aligned towards the north, i.e. oriented along a different axis. The reason for this discrepancy seems to have been that the external layout had to be integrated into the totality of the city's grid pattern, while the internal structure was determined by the orientation of the throne room. This was aligned so that the axis between the main entrance and the throne niche would extend straight towards the *Etemenanki* temple, the Ziggurat of Marduk. Through the open gate the gaze of any visitor standing in the courtyard and looking at the king seated on his throne in the reception area would simultaneously take in the sight of the Ziggurat dominating the skyline. Thus in Babylonia for the first time a conscious connection between the architecture of palace and temple was established, retaining at the same time the apparently required rectangularity of the building, but sacrificing its alignment along the city grid (see below). The western part of the palace in all likelihood contained the royal family's quarters. It was either here, in the vicinity of the outwork, or to the west and northwest of the 'main palace' that the royal gardens were located, which came to be known via later traditions as the 'Hanging Gardens of Semiramis'. It seems that they were not planted areas sitting on top of vaults, but landscape gardens laid out on artificial hills in the tradition of Assyrian parks. The 'Summer Palace', in the northern corner of the outer city (the direction of the 'auspicious wind') also seems to have been artificially elevated.

3) The street system
The street system can be reconstructed with comparative ease. Running generally in a straight line, the main axes connected the city gates with the centre and the sacred area dedicated to Marduk. The main thoroughfares (e.g. the Procession Way running along the palace from the Ištar Gate to the cult centre) were lavishly paved and of generous and regular width. The thoroughfare which crossed the city from the Marduk Gate widened like a funnel and crossing the Procession Way opened onto a vast open square, almost 100 metres wide, which spread out before the main entrance to the large courtyard enclosing the ziggurat. As far as can be told from the findings from the excavation area called Merkez, there were smaller, unpaved and less regular secondary streets branching off the main axial streets, more or less at a right angle. They provided access to particular residential neighbourhoods, where small alleys, frequently cul-de-sacs, would lead to individual houses or groups of houses.

II. Babylon's cosmological importance (cf. supplementary map)

In Babylonian cosmology, the city was regarded as the centre of the world, at whose centre lay the shrine of Marduk, also known as Bēl (Lord), the father of the gods. The vertical axis of the world, which linked Heaven, Earth, the subterranean freshwater ocean *apsû* and the Underworld, was believed to run through this point. The vertical axis consisted in the ziggurat, the 'Sacred Mountain', whose name É.TEMEN.AN.KI ('House of the foundations of Heaven and Earth') encapsulated that very concept.

The 'horizontal axis of the world', which took its course across the terrestrial world, centred on the city, the home of civilization. It acted as the counterpoint to the wild, uncivilized and dangerous 'steppe', a place populated by ferocious beasts, uncivilized nomads and restless spirits. The city wall represented the transition between the two domains, thus offering physical, but also spiritual protection to the city dwellers. The rectangular, geometric layout of the city seems to symbolize the totality of a world order whose navel (centre) and physical representation Bābilu was considered to be. The Babylonian concept of the 'four corners of the world' was reflected in the city's shape as it had been before in the plans of the older Assyrian metropoleis.

With the shrine of Marduk in the centre and the palace citadel, the ruler's seat, on the periphery, Babylon was a bipolar city. A connection between the two was established when the royal palace was aligned differently from the predominant axial system in order to orient it towards the cult centre (see above).

Although the realization of ideological concepts through city planning devices was partially inspired by the spatial patterns of Assyrian models, the ruler's different position vis-à-vis Marduk and his temple in the middle of the city and at the centre of the universe resulted in a different intra-urban structure.

Sources

From 1899 to 1917, the ruins of the ancient city were archaeologically examined by the Deutsche Orient-Gesellschaft (German Oriental Society). Later excavations were undertaken by the German Archaeological Institute in 1962 and from 1970 onward by the Iraqi Directorate-General of Antiquities. An Italian campaign was also undertaken in the area of the Uraš Gate in 1987 and 1989.

Due to the high water-table it has so far only been possible to examine the city of the first cent. BC, albeit on a large scale. Apart from the archaeological findings, 'Tintir' (= Bābilu), a series of contemporary texts, serves as a main source for the reconstruction of Bābilu's urban structure in the Neo-Babylonian Age (10th–6th cents.). Although it was conceived as a paean to the city, information about the topographical struture of Bābilu can also be gleaned from it. The clay tablets contain the following enumerations: the 51 hymnic names and epithets of Bābilu (Tintir I), a list of *šubtus* (shrines and pedestals; Tintir II), a list of the 43 main cult centres of Bābilu, structured by neighbourhood (Tintir IV), and a list of *parakkus* (enumerating the names of the city's gates, walls, water courses and streets as well as the names and positions of neighbourhoods; Tintir V); Tintir III has not yet been identified.

→ Maps pp. 3, 11, 13, 15, 47, 49, 51, 53, 55

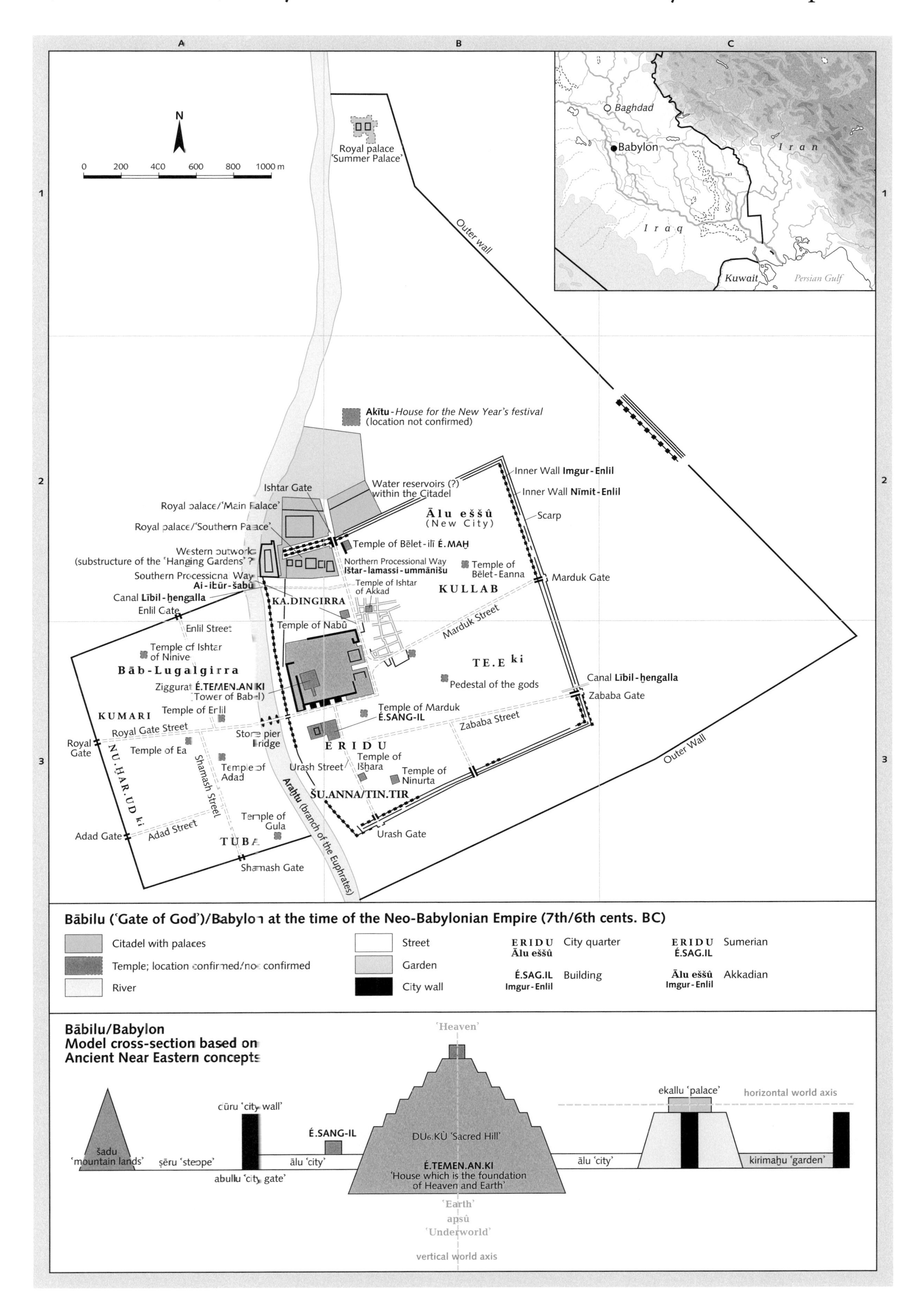
Baghdad
Babylon
Iran
Iraq
Kuwait
Persian Gulf
Royal palace 'Summer Palace'
Outer wall
Akītu-House for the New Year's festival (location not confirmed)
Inner Wall Imgur-Enlil
Inner Wall Nīmit-Enlil
Scarp
Water reservoirs (?) within the Citadel
Ishtar Gate
Royal palace/'Main Palace'
Royal palace/'Southern Palace'
Western outworks (substructure of the 'Hanging Gardens'?)
Southern Processional Way Ai-ibūr-šabû
Canal Lībil-ḫengalla
Enlil Gate
Enlil Street
Ālu eššû (New City)
Temple of Bēlet-ilī É.MAḪ
Northern Processional Way Ištar-lamassi-ummānišu
Temple of Bēlet-Eanna
Temple of Ishtar of Akkad
KULLAB
Marduk Gate
KA.DINGIRRA
Temple of Nabû
Marduk Street
Temple of Ishtar of Ninive
Bāb-Lugalgirra
TE.E ki
Ziggurat É.TEMEN.AN.KI 'Tower of Babel'
Pedestal of the gods
Canal Lībil-ḫengalla
Zababa Gate
Temple of Enlil
KUMARI
Royal Gate Street
Temple of Marduk É.SANG.IL
Zababa Street
Royal Gate
Temple of Ea
Stone pier bridge
ERIDU
NU.ḪAR.UD ki
Shamash Street
Temple of Adad
Urash Street
Temple of Išḫara
Temple of Ninurta
Outer Wall
ŠU.ANNA/TIN.TIR
Araḫtu (branch of the Euphrates)
Temple of Gula
Adad Gate
Adad Street
TUBA
Urash Gate
Shamash Gate
Bābilu ('Gate of God')/Babylon at the time of the Neo-Babylonian Empire (7th/6th cents. BC)
Citadel with palaces
Temple; location confirmed/not confirmed
River
Street
Garden
City wall
ERIDU Ālu eššû City quarter
É.SAG.IL Imgur-Enlil Building
ERIDU É.SAG.IL Sumerian
Ālu eššû Imgur-Enlil Akkadian
Bābilu/Babylon
Model cross-section based on Ancient Near Eastern concepts
'Heaven'
šadu 'mountain lands'
šēru 'steppe'
cūru 'city wall'
abullu 'city gate'
ālu 'city'
É.SANG-IL
DU6.KÙ 'Sacred Hill'
É.TEMEN.AN.KI 'House which is the foundation of Heaven and Earth'
'Earth'
apsû
'Underworld'
vertical world axis
ekallu 'palace'
horizontal world axis
kirimaḫu 'garden'

Egypt in the 3rd Intermediate Period and the Late Dynastic Period (*c.* 1080–332 BC)

The maps show Egypt in periods of foreign rule (Libyans, Nubians) or suzerainty (Assyrians, Achaemenids, Macedonians), but also in times when the unity of the kingdom was restored, territorial gains were made in the Syro-Palestinian region and campaigns were conducted in the south. The contacts with the Mediterranean region, which were recorded in Greek sources as well (Herodotus), were becoming more intensive (Greek trading posts, mercenaries).

The toponyms in the maps are primarily the place names as used in Classical Antiquity, followed by the modern Arabic version, which is used exclusively whenever the classical variety is unknown; the Egyptian names are then added in scholarly transcription, complemented in map A by the surviving Assyrian names.

I. Egypt in the 3rd Intermediate Period (*c.* 1080-664 BC) (map A)

1. Political developments

In the 3rd Intermediate Period (*c.* 1080-664 BC), the political situation was characterized by the country's fragmentation into local political units governed by individual rulers. Egypt was no longer a political entity. In the time from the 21st to the 24th Dynasties, the rulers in the Delta and central Egypt, including the Ma chiefs in the Delta, were of Libyan extraction. The kings of the 25th Dynasty (mid 8th to the mid 6th cent.) originated from Nubia, retaining their residence at Npt/Napata.

The 728 BC campaign by the Nubian ruler Piye, who moved northward from Napata (Gabal Barkal) into the region of Athribis in the Delta, failed to create a lasting unified state, one of the causes being Piye's counterproductive return to Nubia at the end of the campaign. As only a limited number of places can be identified as belonging to specific dominions, it is impossible to determine the frequently shifting political boundary lines. We can, however, assume that in the majority of cases they were identical with geographical barriers and landmarks like branches of the Nile in the Delta which were difficult to cross, or promontories in the Nile Valley which extended as far as the river.

During the period of the Assyrian conquests in Egypt from the first campaign in 674 to their retreat in 655 BC (→ map A on p. 53), numerous places were renamed; attributing the names in the surviving Assyrian records to Egyptian sites is largely a hopeless task. Apart from the absence of etymological relationships, the Assyrian texts also frequently lack any information about geographical positions and were not found at Egyptian locations. Thus cases like the trilingual identification of Egyptian Mn-ḥd/Assyrian Mempi/Greek Memphis and ZÂw/Sai/Saïs are rare exceptions.

The map captures the political situation before Piye's campaign in 728 BC. The local areas of government are marked out by different colours. Their names are mostly derived from the ruling families' ancestral seats (e.g. Heracleopolitans, Thebans).

At the beginning of the 3rd Intermediate Period (21st Dynasty), the kings resided at Tanis, where they also chose to be laid to rest. The local buildings were predominantly constructed from the spolia of the nearby former Ramesside capital. The central site of the territory governed by the 22nd Dynasty (Bubastides) was Bubastis, whose temple was substantially extended by Osorkon II in the middle of the 9th cent. The government seat of Tefnakht (24th Dynasty) was Sais.

2. Sources

The sources predominantly consist of archaeological finds from settlements; in the Delta, the development of archaeological sites has recently been intensified. Inscriptions from that period are comparatively rare and brief. Administrative papyri and royal inscriptions are also few and far between, especially for the 21st Dynasty.

II. Egypt in the Late Dynastic Period (*c.* 664–332 BC) (maps B and C)

1. Political developments

The Late Dynastic Period (*c.* 664–332 BC) was characterized by the restoration of political unity by Psammetichus I (664–610 BC, of Libyan extraction), the first king of the 26th Dynasty, who had liberated Egypt from Assyrian rule. This unity largely remained intact to the end of the Pharaonic Era and beyond. It was only interrupted when the Egypt of the 27th Dynasty was incorporated into the Achaemenid Empire as a satrapy (→ map p. 87); after its reconquest in the 28th Dynasty, it was not until the 31st Dynasty that Egypt again came under Persian suzerainty for a short spell, until Alexander the Great conquered the country in 332 BC.

The capital was Memphis. This was also the case during the Persian rule, when it was the satrap's residential city. Sais functioned alongside Memphis as the site of royal burials. The importance of Thebae/Thebes as a religious and administrative centre as well as the burial site of high-ranking officials and priests was also restored.

The whole era was characterized by close economic and cultural contacts with the Greek world. The Delta, whose settlements were partly inhabited by Greeks, played an increasingly pivotal role. The port of Naucratis was the leading customs and shipment centre for the import of Greek goods. Darius I, King of Persia (521-486 BC), for the first time connected the Nile with the Red Sea by a canal running through the Wādī aṭ-Ṭumilāt.

Nubia was no longer a major force, but in marked contrast to earlier periods the oasis settlements are well documented. In spite of Herodotus' interest in Lake Moeris, there are hardly any records about the Fayyūm, which was only systematically re-colonized in Ptolemaeic times (→ map p. 121). Elephantine was a town with a high concentration of Aramaic-speaking Jews, whose temple is mentioned in textual sources.

Because of its importance the Delta region is the subject of an enlarged map segment (map C). The capital Memphis is displayed on the supplementary map, although very little has been excavated so far; the foreign neighbourhoods can be clearly identified by their names.

2) Sources

The sources are predominantly archaeological. Extensive burial sites have been preserved in Thebes. The development of archaeological sites in the Delta, especially in Tall al-Balamūn and in the area around Tall al-Mashūṭa, has been recently intensified, as ancient sites are being increasingly overbuilt by modern settlements. In contrast to the 3rd Intermediate Period, more numerous and longer texts have survived; these are complemented by papyri in demotic script.

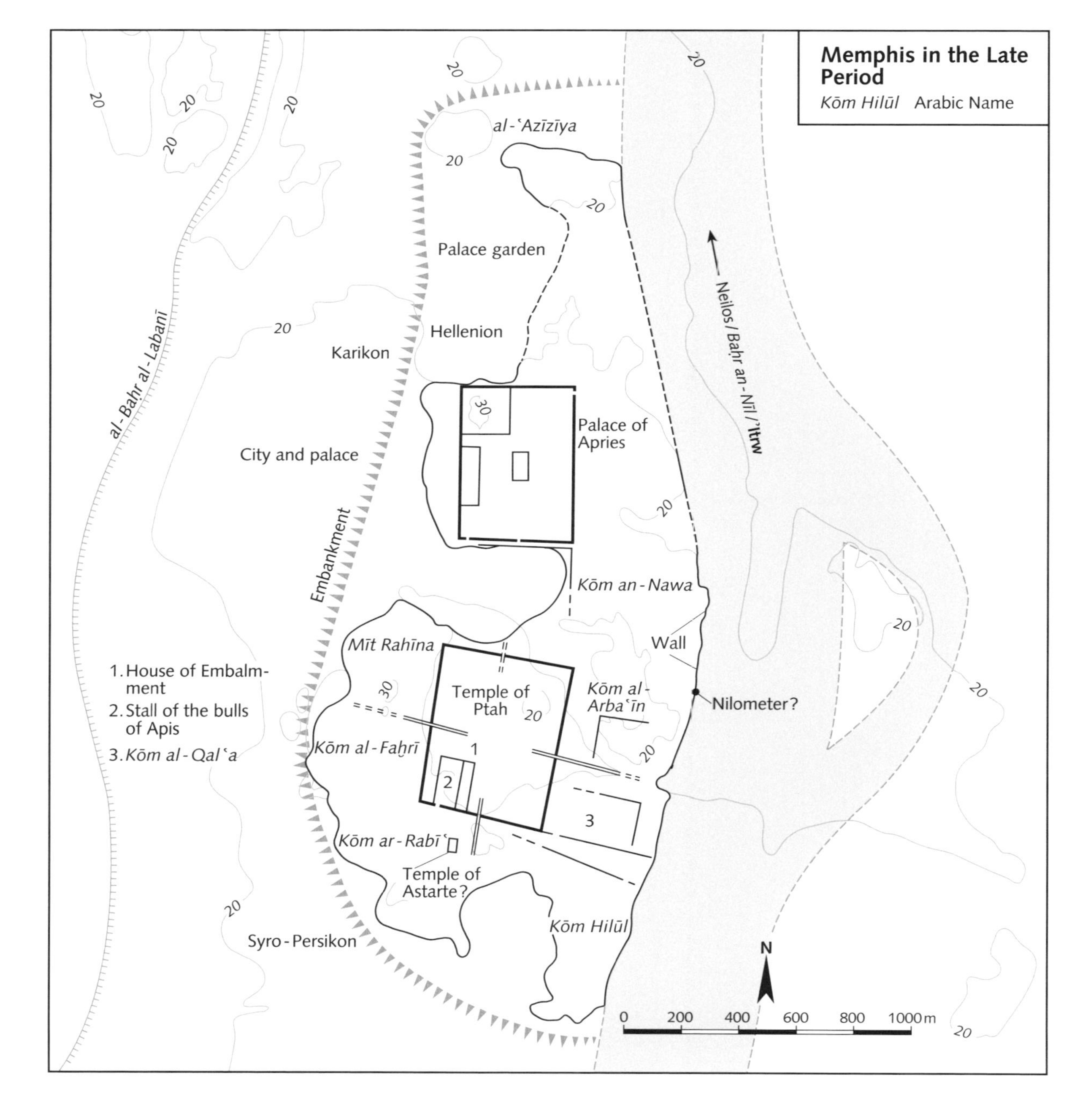

→ Maps pp. 53, 55, 87, 113, 121; Naucratis, map 'Naucratis: archaeological site-map', BNP 9, 2006; Thebes [1], for Wāst/Njwt – 'hundred-gated Thebes', BNP 14, 2009

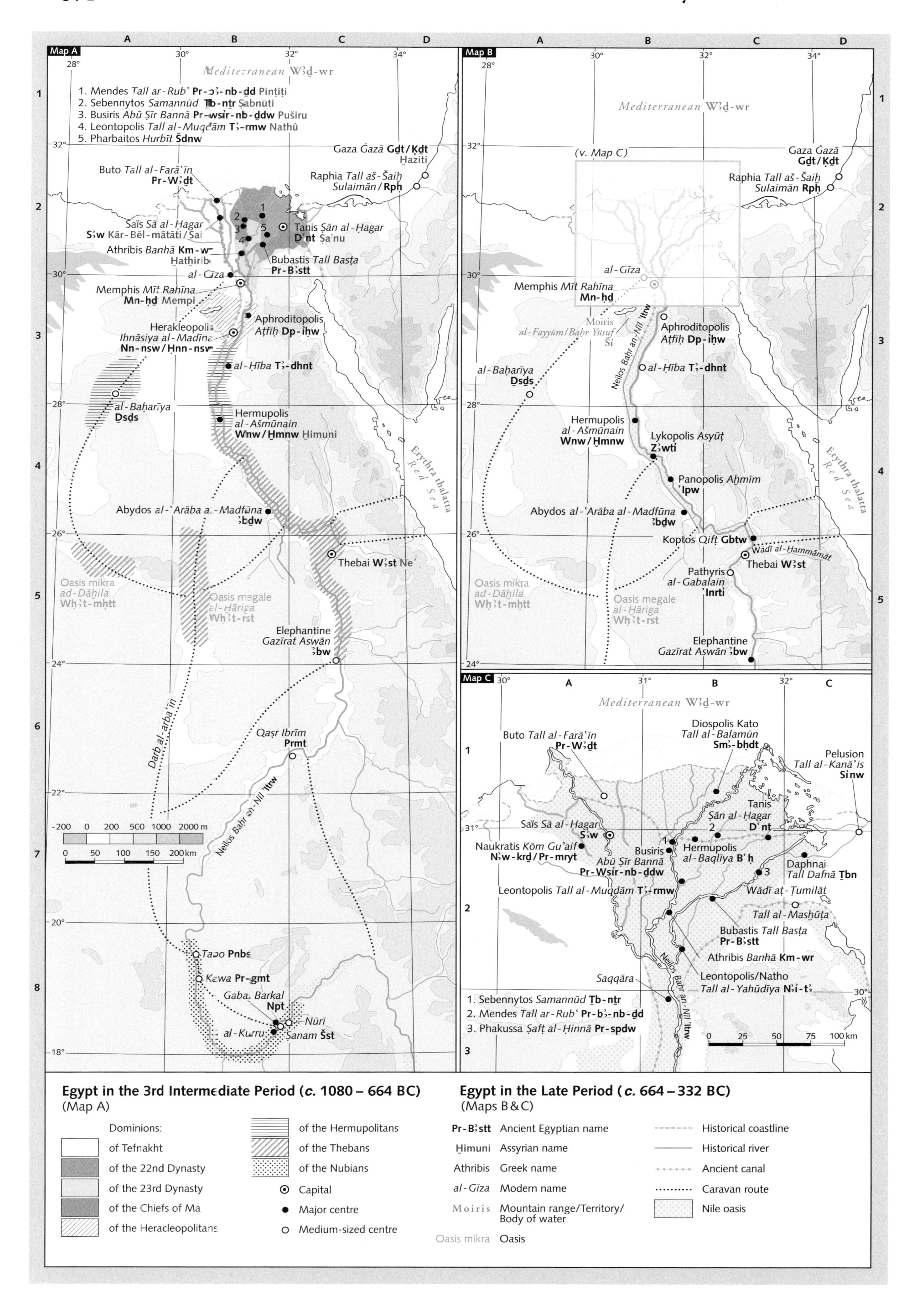
Map A
Mediterranean W3d-wr
1. Mendes Tall ar-Rubʿ Pr-b3-nb-ḏd Pinṭiṭi
2. Sebennytos Samannūd Ṯb-nṯr Šabnūti
3. Busiris Abū Ṣīr Bannā Pr-wsir-nb-ḏdw Puširu
4. Leontopolis Tall al-Muqdām T3-rmw Nathû
5. Pharbaitos Hurbīt Šdnw
Gaza Ġazā Gḏt/Ḳdt Ḫaziti
Raphia Tall aš-Šaiḫ Sulaimān / Rph
Buto Tall al-Farāʿīn Pr-W3dt
Saïs Sā al-Ḥagar S3w Kār-Bēl-mātāti / Sai
Tanis Ṣān al-Ḥagar Dʿnt Ṣaʿnu
Athribis Banhā Km-wr Ḫathiribi
Bubastis Tall Basṭa Pr-B3stt
al-Gīza
Memphis Mīt Rahīna Mn-ḥd Mempi
Aphroditopolis Aṭfīḥ Dp-iḥw
Herakleopolis Ihnāsiya al-Madīna Nn-nsw/Ḥnn-nsw
al-Hība T3-dhnt
al-Baḥarīya Dsds
Hermupolis al-Ašmūnain Wnw/Ḫmnw Ḫimuni
Abydos al-ʿAraba al-Madfūna 3bdw
Thebai W3st Neʾ
Oasis mikra ad-Dāḫila Wḥ3t-mḥtt
Oasis megale al-Ḫāriga Wḥ3t-rst
Elephantine Gazīrat Aswān 3bw
Erythra thalatta Red Sea
Qaṣr Ibrīm Prmt
Darb al-arbaʿīn
Neilos Baḥr an-Nīl ʾItrw
Tabo Pnbs
Kawa Pr-gmt
Gabal Barkal Npt
Nūrī
al-Kurru
Sanam Šst
-200 0 200 500 1000 2000 m
0 50 100 150 200 km
Map B
Mediterranean W3d-wr
(v. Map C)
Gaza Ġazā Gḏt/Ḳdt
Raphia Tall aš-Šaiḫ Sulaimān Rpḥ
al-Gīza
Memphis Mīt Rahīna Mn-ḥd
Moiris al-Fayyūm/Baḥr Yūsuf Š i
Aphroditopolis Aṭfīḥ Dp-iḥw
al-Hība T3-dhnt
al-Baḥarīya Dsds
Hermupolis al-Ašmūnain Wnw/Ḫmnw
Lykopolis Asyūṭ Z3wti
Panopolis Aḫmīm ʾIpw
Abydos al-ʿAraba al-Madfūna 3bdw
Koptos Qifṭ Gbtw
Wādī al-Hammāmāt
Thebai W3st
Pathyris al-Gabalain ʾInrti
Oasis mikra ad-Dāḫila Wḥ3t-mḥtt
Oasis megale al-Ḫāriga Wḥ3t-rst
Elephantine Gazīrat Aswān 3bw
Erythra thalatta Red Sea
Neilos Baḥr an-Nīl ʾItrw
Map C
Mediterranean W3d-wr
Buto Tall al-Farāʿīn Pr-W3dt
Diospolis Kato Tall al-Balamūn Sm3-bḥdt
Pelusion Tall al-Kanāʾis Sinw
Tanis Ṣān al-Ḥagar Dʿnt
Saïs Sā al-Ḥagar S3w
Naukratis Kōm Guʿaif N3w-krḏ/Pr-mryt
Busiris Abū Ṣīr Bannā Pr-Wsir-nb-ḏdw
Hermupolis al-Baqlīya Bʿḥ
Daphnai Tall Dafnā Ṯbn
Leontopolis Tall al-Muqdām T3-rmw
Wādī aṭ-Ṭumilāt
Tall al-Mašḫūṭa
Bubastis Tall Basṭa Pr-B3stt
Athribis Banhā Km-wr
Leontopolis/Natho Tall al-Yahūdīya N3i-t3
Saqqāra
Neilos Baḥr an-Nīl ʾItrw
1. Sebennytos Samannūd Ṯb-nṯr
2. Mendes Tall ar-Rubʿ Pr-b3-nb-ḏd
3. Phakussa Ṣaft al-Ḥinnā Pr-spdw
0 25 50 75 100 km
Egypt in the 3rd Intermediate Period (c. 1080 – 664 BC) (Map A)
Dominions:
of Tefnakht
of the 22nd Dynasty
of the 23rd Dynasty
of the Chiefs of Ma
of the Heracleopolitans
of the Hermupolitans
of the Thebans
of the Nubians
Capital
Major centre
Medium-sized centre
Egypt in the Late Period (c. 664 – 332 BC) (Maps B&C)
Pr-B3stt Ancient Egyptian name
Ḫimuni Assyrian name
Athribis Greek name
al-Gīza Modern name
Moiris Mountain range/Territory/Body of water
Oasis mikra Oasis
Historical coastline
Historical river
Ancient canal
Caravan route
Nile oasis

Core areas of distribution of hieroglyphic, cuneiform, alphabetic and syllabic scripts in the Eastern Mediterranean area

The approximate core areas of distribution of hieroglyphic, cuneiform, alphabetic and syllabic scripts in the Eastern Mediterranean area (*c.* 12th to 7th cents. BC)

The approximate core distribution areas of hieroglyphic, cuneiform, alphabetic and syllabic scripts in the eastern Mediterranean region (*c.* 12th–7th cents. BC)

In the history of writing systems, three different fundamental methods of recording the spoken word have been established: these are – in order of emergence – ideographic, syllabic and phonetic scripts. Any known writing system on this planet, including those of the Ancient World, is bound to use one of these methods or a combination of them, with the phonetic script – including the special case of the mostly oriental consonant script (e.g. Phoenician) – best suited to represent combinations of sounds.

The Early Iron Age (*c.* 12th to 7th cents.) was chosen as the most suitable era for mapping the subject, although some scripts had emerged a considerable time before that period or continued to exist beyond the 7th cent. It was, however, a time when a particularly large number of writing systems existed side by side. It also offers an opportunity to highlight both where in the so-called Dark Ages (*c.* 12th to 9th/8th cents.) writing still continued to be practised and textual sources have been discovered.

I. Hieroglyphic scripts

The Egyptian hieroglyphic writing system had served as a monumental script from the 3rd millennium. Alongside there was the hieratic script, which developed from a cursive variety of the hieroglyphic script. From the 7th cent. onward, the demotic cursive script emerged, which was used for everyday purposes.

Possibly inspired by Egyptian or Levantine models, the Minoans on Crete also developed a hieroglyphic-pictographic script found in 34 locations on Crete, predominantly at Knossos and Mallia, but also on the islands of Cythera and Samothrace. It includes syllabic and logographic characters along with measure and number symbols, but has not been decoded so far. Most of the finds date from the time between 1900 to 1700 BC, i.e. the First Palace Period (not included on p. 61 → see map B, p. 25). In the 17th cent. BC, there was an overlap between the use of hieroglyphics and Linear A script.

The Luwians in southeastern Asia Minor/northern Syria were another people who created hieroglyphics to record their language in monumental inscriptions on rocks, orthostates, stelae and statues at Karkamiš, Azatiwadā/Karatepe, Tuwana/Tyana. They can be dated from the 15th to the 7th cents. In the first millennium, they were also used for administrative purposes, as documented by finds from Kululu at the centre of the southern arch of the Halys and Aššur. At least for a brief period (from the 15th to the 13th cents.) and on a limited scale Luwian hieroglyphics were also used on seals by the Hittites, and in the 13th cent. also to represent the Urarṭaic language in inscriptions at Boğazköy/Ḫattusa, among others. In Urarṭu — admittedly rudimentary — samples of an indigenous hieroglyphic script have been discovered as well: at Toprakkale near Tušpa/Van, Karmir-blur, and Baṣṭām (→ map p. 41).

II. Cuneiform scripts

At the end of the 4th millennium BC, the cuneiform script, which combined logographic and syllabic elements, was invented in Sumerian southern Mesopotamia. It remained in use for approximately 3000 years and was adopted by numerous non-Sumerian-speaking nations like the Hittites (until *c.* 1200 BC, → maps pp. 22/23), the Assyrians (Aššur, Ninua/Ninive), the Syrians (until *c.* 600 BC) and the Urarṭians (Tušpa/Van, → map p. 41). This caused the script to become regionally and chronologically diversified in some cases; in others, e.g. in the Neo-Assyrian Empire, a standardization of the characters can be observed. There was no monumental script for exclusive use in legal texts or royal inscriptions.

This map does not include: a) the alphabetic left-to-right cuneiform, which was used almost exclusively in Ugarit (→ map p. 28) and thus was obliterated around 1190 with the fall of the city of Ugarit, and b) the ancient Persian cuneiform, which was developed in the reign of Darius I. Its use was restricted to royal edicts and consequently it vanished after the fall of the Achaemenid Empire (→ map p. 87)

III. Linear and syllabic scripts

Another syllabic writing system, the so-called Linear B script, which also included ideographs and special characters for numbers, weights and measures, has come down to us from the 2nd millennium from Mycenaean Greece (until *c.* 1200, → maps pp. 26/27). It was also found on Crete, where its predecessor, the undecoded Linear A script, can be documented from 1650-1450. Linear A was probably also largely of the syllabic type, as were the inscriptions on the 'Disc of Phaistos' (created between *c.* 1650-1550). Linear B records have come to light on Cyprus as well, where from about the mid 11th cent. it took the form of the Cypriot script (Paphos, Citium), which presumably developed from the Late Bronze Age Cypro-Minoan scripts (Cypro-Minoan 1–3; 1: 16th–11th cents. (all of Cyprus), 2: 13th/12th cents. (Enkomi) resp. 3: 14th/13th cents. Ugarit) and was in use into the Hellenistic Age. It served to transcribe both the Eteo-Cypriot language (Amathus) and the Cypriot-Greek dialect.

IV. Alphabetic scripts

The Ancient Semitic consonantal alphabetic scripts can be documented in strongly varying local forms from the 17th cent. onwards. Presumably spreading from Gubla/Byblos, a standardized Phoenician variety consisting of 22 characters made slow but steady progress in Syria and Palestine from the 11th cent. In the 9th/8th cents., a cursive variety was developed which diverged into local variants like Aramaic and Hebrew.

Phrygian, which is counted among the Indo-European, not the Anatolian languages, has only survived in the shape of fragmented remains. There are *c.* 250 inscriptions in Ancient Phrygian alphabetic script, which have come down to us e.g. from Gordium, Ayazın, Boğazköy. According to current research they date from the 8th cent. onward, but have remained practically undecoded. The alphabet is close to the Greek one, but researchers assume that there was no direct influence, but rather a very similar process of adapting Semitic models.

For the time after the adoption of the Semitic system, which probably happened around 800 BC, it is possible to reconstruct a type of Greek proto-alphabet, which redefines certain superfluous consonant characters as representations of the vowels a, e, i, o and adds an additional character to represent the 'u' sound to complete the first fully phonetic script. Greek model alphabets within the various dialect groups such as Ionic-Attic, Arcadian-Cypriot (Achaean), Aeolian and Doric-North-West Greek have been found in many local, clearly distinguishable script variants, e.g. from Athens, Euboea (found at Marsiliana d'Albegna in Etruria), Boeotia, Corinth, Achaia and eastern Ionia (found on Samos). They were used alongside each other until the so-called East-Ionian standard alphabet displaced all other varieties (*c.* 400 BC). The local or epichoric alphabets led to the creation of further alphabets to transcribe other languages, e.g. Lydian, Carian, Lycian and Sidetic. These alphabets emerged in areas adjacent to the territories settled by the Greeks in the mainland, in western Asia Minor and in the colonies. Since this happened outside the selected time frame, they have not been included in this map.

→ Maps pp. 11, 21, 25, 29, 35, 41, 59, 87; Greek, map 'Greek-speaking areas before Hellenism', BNP 5, 2004; Asia Minor[C.], map 'Successor states of the Hittites (12th-8/7th cents. BC)', BNP 6, 2005; Cypriot script, BNP 3, 2003; Writing [II.], BNP 15, 2009; Writing [III.], BNP 15, 2009

Literature

C. Brixhe, M. Lejeune, Corpus des inscriptions paléo-phrygiennes, 1984; H. Günther, O. Ludwig (eds.), Schrift und Schriftlichkeit – Writing and its Use, 3 vols., 1994–1996; J.D. Hawkins, Corpus of Hieroglyphic Luwian Inscriptions, 2000; A. Heubeck, Schrift, ArchHom III, ch. X, 1979; F. Prayon, A.-M. Wittke, Kleinasien vom 12. bis 6. Jh. v. Chr. TAVO suppl. B 82, 1994; M. Salvini, Geschichte und Kultur der Urartäer, 1995; A. Schlott, Schrift und Schreiber im Alten Ägypten, 1989.

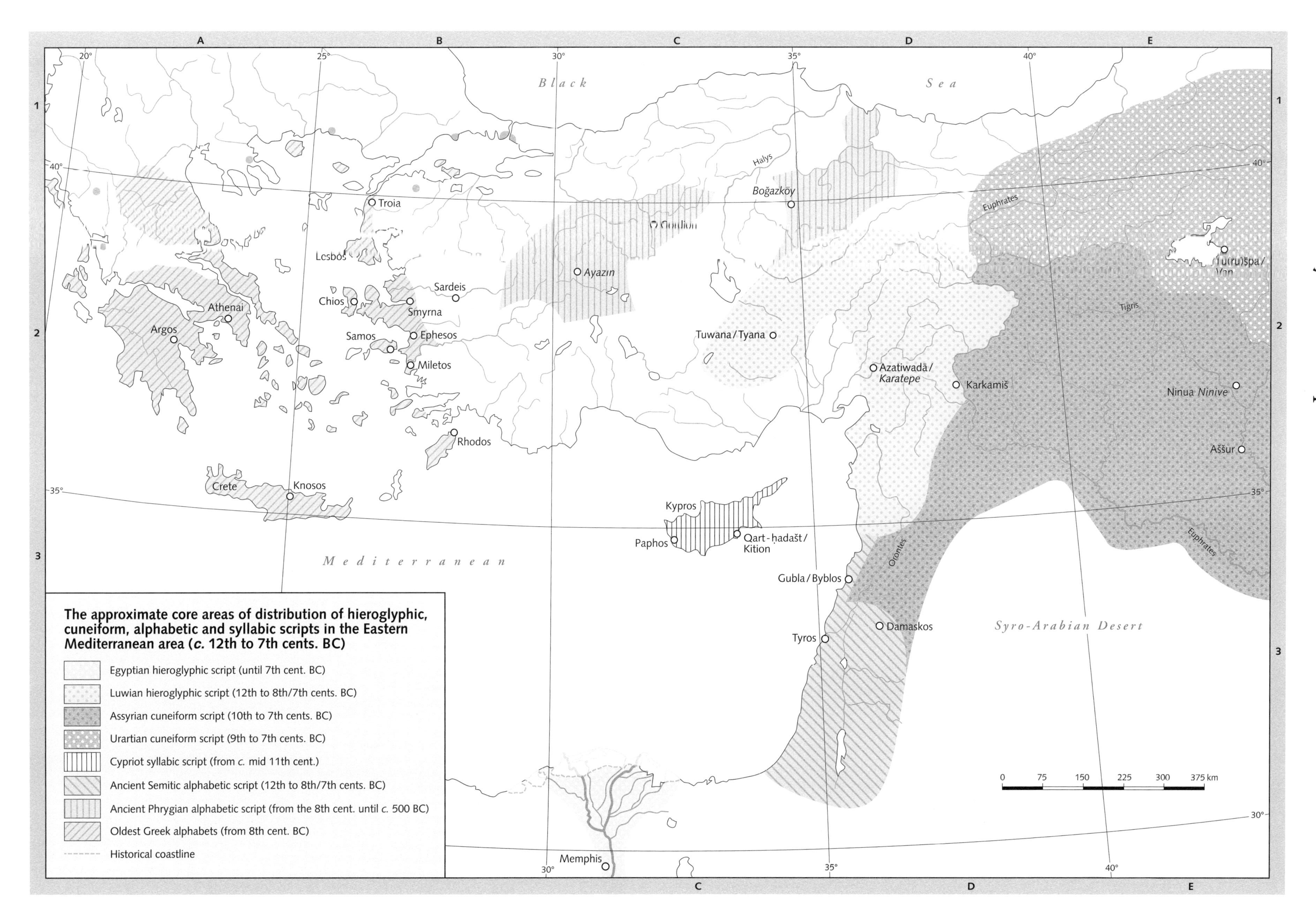
The approximate core areas of distribution of hieroglyphic, cuneiform, alphabetic and syllabic scripts in the Eastern Mediterranean area (c. 12th to 7th cents. BC)
Egyptian hieroglyphic script (until 7th cent. BC)
Luwian hieroglyphic script (12th to 8th/7th cents. BC)
Assyrian cuneiform script (10th to 7th cents. BC)
Urartian cuneiform script (9th to 7th cents. BC)
Cypriot syllabic script (from c. mid 11th cent.)
Ancient Semitic alphabetic script (12th to 8th/7th cents. BC)
Ancient Phrygian alphabetic script (from the 8th cent. until c. 500 BC)
Oldest Greek alphabets (from 8th cent. BC)
Historical coastline
Black Sea
Mediterranean
Syro-Arabian Desert
Troia
Lesbos
Chios
Samos
Smyrna
Sardeis
Ephesos
Miletos
Rhodos
Knosos
Crete
Athenai
Argos
Ayazın
Boğazköy
Halys
Tuwana/Tyana
Azatiwadā/Karatepe
Karkamiš
Euphrates
Tigris
Ninua Ninive
Aššur
Orontes
Damaskos
Gubla/Byblos
Tyros
Qart-ḥadašt/Kition
Kypros
Paphos
Memphis
0 75 150 225 300 375 km

The Iberian Peninsula and its contacts in the Late Bronze Age and in the Early Iron Age

The Iberian Peninsula and its contacts in the Late Bronze Age and in the Early Iron Age (*c.* 13th to 7th/6th cents. BC)

The Iberian peninsula and its contacts in the Late Bronze, Final Bronze and Early Iron Ages (*c.* 13th–7th/6th cents. BC)

As reflected in Greek myth (the Pillars of Hercules), the cultures in the southwest and along the Mediterranean coast of the Iberian Peninsula and its 200 km wide hinterland strip had been trading partners of the contemporaneous high civilizations of the eastern Mediterranean region from the Final Bronze Age. Nevertheless their cultures and the prehistoric cultures of the central plateau and the north-west have only in recent times been accorded appropriate interest in research. The choice of period here, with its focus on the Early Iron Age and the onset of the Phoenician colonization (of the trading post type) is based on the chronology of the eastern Mediterranean region, especially the Aegean. Thus it does not correspond with the greatly varying regional chronologies of the linguistically and otherwise differentiated Late Bronze and Early Iron Age cultures.

I. Bronze and Iron Age cultures (main map)

The main settlement areas of the autochthonous Late and Final Bronze Age cultures (16th-11th cents., regionally lasting into the 9th or 8th cent.), which are only partly identical with those of the later Iberians and Celtiberians, have been mapped only diagrammatically. The territory can be roughly divided into three Early Iron Age autochthonous cultural regions. The earliest is the so-called Tartessian culture (*c.* 8th to 6th cents.), for which both a core territory and an area of influence have been identified. Situated in the southeast of the Iberian Peninsula, they are crossed by the Baetis/Guadalquivir, the Rio Tinto, the Anas/Guadiana and the Segura rivers. Its northern boundary is formed by the course of the Tagus/Tejo. This region is rich in a great variety of minerals. The Iberian and Iberianized cultural region, which was dominated by non-Indo-European Iberian tribes, can be traced back to the 7th cent. It extended along the east coast, in the south-east, south, and southwest of the peninsula. The peninsula's western half, however, was inhabited by Indo-Europeans; here it is difficult to distinguish between Pre-Celtic indigenous and Celtic or Celticized (Celtiberian/Hispano-Celtic) cultural areas.

In the Atlantic west, with its climatically more moderate mountain zone ('ore-bearing crescent') gold, copper and tin were mined from early times. This attracted Phoenician and later on West-Phoenician seafarers to the coasts of the Atlantic and the southern Mediterranean as well as to the region called Tartessus, later known as Turdetania, which is problably identical with biblical Taršiš, even though this is still a matter of controversy among scholars.

Early Greek finds in the southern peninsula – the sources point to the Phocaeans, keen travellers and explorers (Hdt. 1,163–165) – indicate that the Greeks were also attempting to establish a foothold to enlarge their area of settlement and economic influence. As elsewhere, they were foiled for the time being by Phoenician/West-Phoenician resistance. Their venture of setting out from their base in Massalia/Massilia to found colonies in the northeast (Emporium/Emporiae, founded *c.* 520 BC) was more successful, especially after the Etruscans had been driven from the western Mediterranean upon their defeat at Alalia (Corsica) *c.* 540/535 BC. This opened the sea and land routes to the Spanish north-east and, via the valleys of the Rhodanus/Rhône and the Garumna/Garonne, to the north-west of the Iberian Peninsula, giving access to the coveted metal resources.

The international interest in the Iberian Peninsula is reflected in the various appellations for the complete area or parts of it. Thus the abovementioned Old Testament Phoenician toponym Taršiš probably signifies the Tartessian region in the south-west, where written documents were produced between the 7th and the 5th cents. The term Hispania goes back to the peninsula's West-Phoenician/Punic name, while 'Iberia' is likely to be derived from the Greek name for the river Ebro: Iber (Lat. Iberus), which seems to indicate an approach from the north as documented in the literature about the colonization, which originated from Massilia.

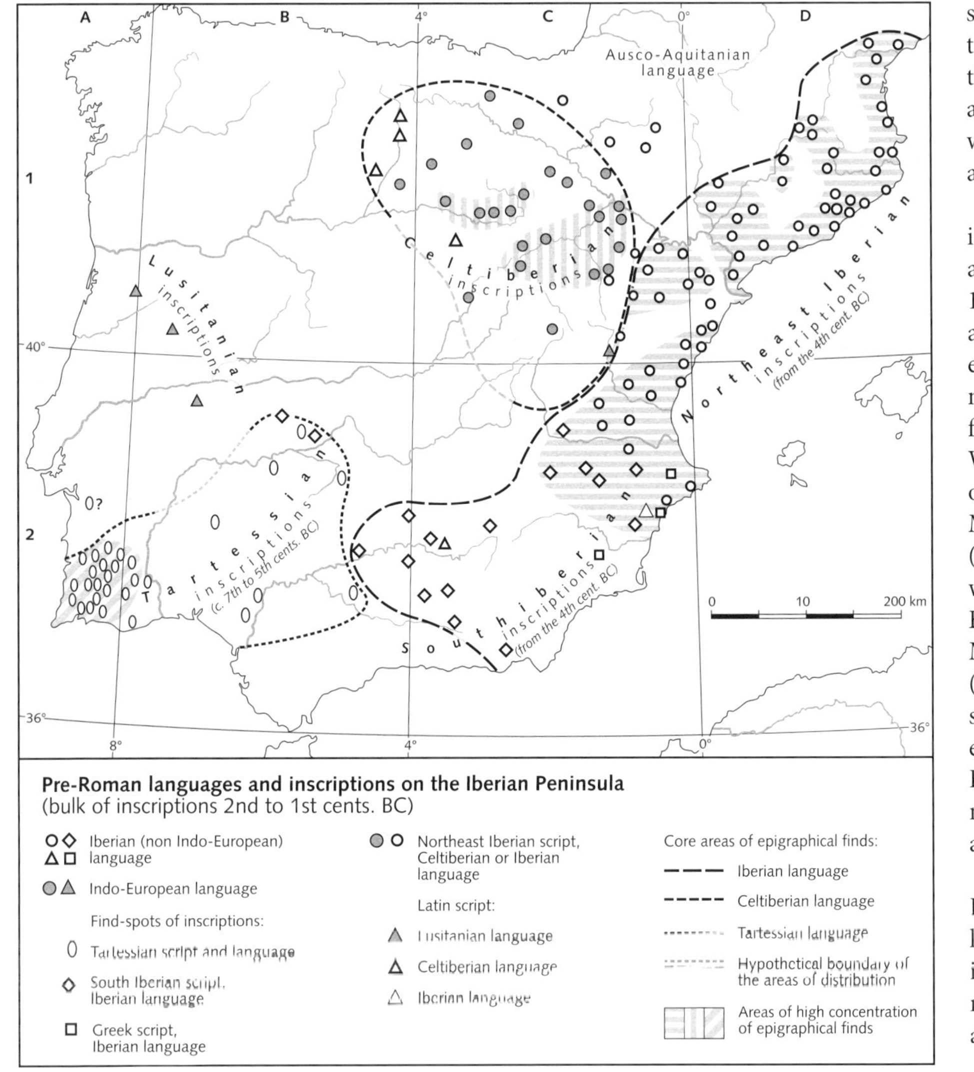

II. Pre-Roman languages and inscriptions on the Iberian Peninsula (supplementary map)

The distribution of languages (and scripts) across the Iberian Peninsula supports the definition of the different cultural areas, even if the written documents are of a rather late date. There are inscriptions in non-Indo-European Iberian as well as in several Indo-European languages – e.g. Celtiberian and Lusitanian –, which were transcribed in various scripts such as Tartessian, South-Iberian, North-Iberian as well as Greek and Latin.

Sources

Our knowledge is predominantly based on the findings of Celtiberian and Iberian archaeology, so named to reflect the settlement areas and linguistic characteristics of the Indo-European Celtiberian and the non-Indo-European Iberian tribes. On the central plateau, there was also a Pre-Celtic indigenous contact zone between Iberian and 'Celtiberian' cultural areas. A patchwork quilt of larger and smaller, more or less homogeneous cultures is typical of the whole of the first millennium, of which the wealth of tribal names documented in Antiquity provides evidence, thus supporting the findings of archaeology.

The finds from the first half of the first millennium bear witness to an amazing intensity of long-distance communication in the tradition of the Late Bronze Age, both inside the pensinsula and Trans-Pyrenaean, West-Mediterranean and Atlantic.

Literary documents dealing with the Iberian Peninsula are, however, of a relatively late date and come either from the Old Testament (referring to Phoenician journeys to Taršiš) or from Greco-Roman sources starting approx. in the 6th cent. The bulk of Pre-Roman inscriptions (cf. supplementary map) dates from the 2nd and 1st cents.

→ Maps pp. 31, 65, 69, 71, 83, 85

Literature

M. Blech, M. Koch, M. Kunst (eds.), Hispania antiqua. Denkmäler der Frühzeit, 2 vols., 2001; D. Brandherm, Zur Datierung der ältesten griechischen und phönizischen Importkeramik auf der Iberischen Halbinsel. Bemerkungen zum Beginn der Eisenzeit in Südwesteuropa, in: Madrider Mitteilungen 47, 2006, 1–23; Les Ibères, exhibition catalogue, Paris, Barcelona, Bonn, 1997.

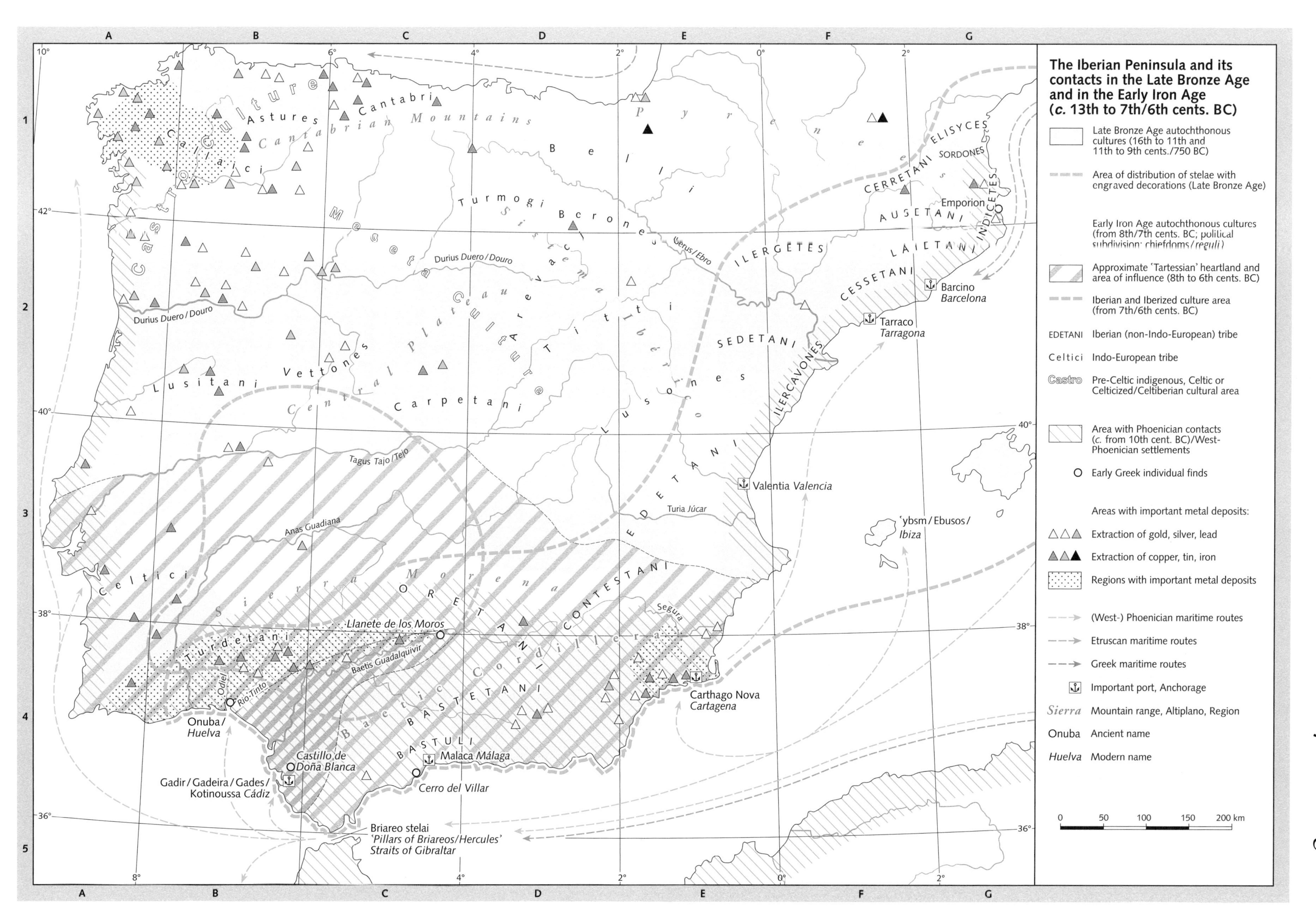
The Iberian Peninsula and its contacts in the Late Bronze Age and in the Early Iron Age (c. 13th to 7th/6th cents. BC)
Late Bronze Age autochthonous cultures (16th to 11th and 11th to 9th cents./750 BC)
Area of distribution of stelae with engraved decorations (Late Bronze Age)
Early Iron Age autochthonous cultures (from 8th/7th cents. BC; political subdivision: chiefdoms/reguli)
Approximate 'Tartessian' heartland and area of influence (8th to 6th cents. BC)
Iberian and Iberized culture area (from 7th/6th cents. BC)
EDETANI Iberian (non-Indo-European) tribe
Celtici Indo-European tribe
Castro Pre-Celtic indigenous, Celtic or Celticized/Celtiberian cultural area
Area with Phoenician contacts (c. from 10th cent. BC)/West-Phoenician settlements
Early Greek individual finds
Areas with important metal deposits:
Extraction of gold, silver, lead
Extraction of copper, tin, iron
Regions with important metal deposits
(West-) Phoenician maritime routes
Etruscan maritime routes
Greek maritime routes
Important port, Anchorage
Sierra Mountain range, Altiplano, Region
Onuba Ancient name
Huelva Modern name
0 50 100 150 200 km
Pyrenees
Cantabrian Mountains
Castro Culture
Emporion
Barcino Barcelona
Tarraco Tarragona
'ybsm / Ebusos / Ibiza
Valentia Valencia
Carthago Nova Cartagena
Malaca Málaga
Cerro del Villar
Briareo stelai
'Pillars of Briareos/Hercules'
Straits of Gibraltar
Gadir / Gadeira / Gades / Kotinoussa Cádiz
Onuba / Huelva
Castillo de Doña Blanca
Llanete de los Moros
Baetis Guadalquivir
Anas Guadiana
Tagus Tajo / Tejo
Durius Duero / Douro
Iberus / Ebro
Turia Júcar
Segura
Odiel
Rio Tinto
Sierra Morena
Turdetani
Oretani
Contestani
Bastetani
Bastuli
Celtici
Lusitani
Vettones
Carpetani
Callaici
Astures
Cantabri
Turmogi
Berones
Belli
Edetani
Sedetani
Ilercavones
Ilergetes
Cessetani
Laietani
Ausetani
Indicetes
Cerretani
Elisyces
Sordones

Italy from the 10th to the 8th cents. BC

The map illustrates the transition period bridging the gap from the Late Bronze Age, when the Mycenaeans were present in Italy, to the completion of the Etruscan ethnogenesis and the emergence of their cities in the northern part of central Italy as well as renewed Greek settlement activities in southern Italy and Sicily in the context of the 'Great Greek Colonization' in the middle of the 8th cent. At first there were no identifiable centres of power. It was only when supra-regional trade relations were intensified, when both Phoenician and Greek settlements were founded and the Etruscan cities took shape, that the political structures in the western Mediterranean, now fed by a multitude of different influences, underwent a noticeable change. The prehistoric period in hand must still be regarded as comparatively unexplored, although (particularly in recent times) the excavations and the examinations of individual cultures have been supplemented by comprehensive projects like a comparison of urbanization phenomena or a programme dealing with metal working and trading, the results of which have yet to be published.

I. Cultures

In the Early Iron Age, whose onset was subject to great local differences, ranging from the 12th to the 9th cents., various cultures in Italy can be distinguished archaeologically. Some of them are linked by scholars to tribes which developed through the centuries and were only later (from *c.* 6th/5th cents.) documented in literature, mostly by Greek historians.

The transition from the Bronze Age to the Iron Age is characterized by increasingly closer relationships among the different cultural groups and by the replacement at many locations of the traditional burial (inhumation) by cremation, a feature which is not restricted to the territory of the Villanova culture.

1. The Villanova culture
The Villanova culture (early period: *c.* 900-820, transition period: *c.* 820-770, final period: 770-730 BC) is one of the most thoroughly researched cultures of the Early Iron Age. Although it primarily originated from the Proto-Villanovan culture, it also displays features stemming from eastern Mediterranean, central and eastern European and Danubian sources. Although Villanova, its eponymous site, was located in the neighbourhood of Bologna, the culture's actual core territory encompasses ancient Etruria, its offshore islands and northern Latium. Expansion in the early phase took it northward beyond the Apennine into areas around Bologna in the Po Plain and the surroundings of Rimini. Features very similar to those of the Villanova culture – above all its tradition of cremating the dead – can be found near Fermo on the Adriatic coast as well as in southern Italy (Sala Consilina, Pontecagnano). In Etruria, this period brought fundamental changes, not least because of the mining of its resources (iron ore). The newly built settlements were to become the nuclei of the later Etruscan cities, just as the Villanova culture must be viewed as the socio-cultural and economic predecessor – or early phase – of the Etruscan civilization. There were close contacts with the Greek world and Sardinia from the 8th cent. onward.

2. Cultures in northern Italy
The Golasecca culture in northern Italy (proto-GC: 12th/11th cents.; GC: 11th–4th cents. BC) can be divided into three groups, ranging from the source area of the Ticino/Ticinus to Lago Maggiore/Verbanus lacus and Lago di Como/Larius lacus as far as the Po/Padus. Their cremation customs and lavishly appointed warrior tombs point to connections with the Hallstatt culture; the Golasecca culture is linked with the Celtic tribes of the Insubres, Orobii and Lepontii, which are documented in later sources, but there are also ties with the Este culture and Etruria.

The relatively homogeneous Proto-Venetic and Este culture, located between the Po/Padus, Lake Garda/Benacus lacus, the northeastern arch of the Alps and the Adriatic Sea, with its important centres at Este, Padua and Frattesina, has provided us with the material legacy of the Venetic tribes; dating from the period between 1000 and 300 BC, it also displays links with the Hallstatt and Golasecca cultures and Etruria.

3. Adriatic cultures
The Picenian culture (9th-3rd cents. BC) is one of the so-called Adriatic cultures. It is also the most important central Adriatic culture and had proven links with Etruria (metal objects). Finds such as Baltic amber suggest far-reaching trade contacts. To its south lay the territory of the Iapygian culture, which emerged in the 11th or 10th cent. According to Greek sources (Hecataeus, Herodotus, Aristotle, Strabo and others), the term 'Iapyges' was usually applied to the complete population of Apulia (Dauni, Peucetii, Messapii and the eponymous Iapyges). There is no solid evidence supporting their Illyrian origins, but they did practise a continuous exchange of goods with the tribes on the opposite coast of the Adriatic as well as the Greek cities on the south coast of Italy.

4. Indigenous Italian cultures
The territories of the remaining Italian cultures, some of whom went through a Bronze Age phase as well as an Iron Age one and whose inhabitants are usually classified by the umbrella term 'Italic peoples', cover the centre and the southwest of the peninsula. The resident cultures include the Terni culture, but also the Latial culture, which in its early stages is closely related to the Villanova culture, the Fossa Grave culture, which extended from Latium to the south, and the Oenotrian culture in the interior (Dion. Hal. Ant. Rom. 1,11,2–4; 12,1). On the mainland, the distribution area of the so-called Ausonian culture partly overlaps with the territory of the Fossa Grave culture and later Greek and Etruscan settlement areas. It also extends across the Lipari Islands and on to Sicily.

5. Sicilian and Sardinian cultures
On Sicily, the Pantalica (northern and southern) and Cassibile cultures at the important sites of Leontini, Punta Castelluzzo and Morgantina continue to display some Aegean features. The Sant'Angelo-Muxaro culture in the neighbourhood of Agrigento must be set apart as a separate development. Alongside further cultures there were the Siculi, whom scholars generally view as an autochthonous group, and whose existence is documented in non-contemporaneous written documents from the 6th and 5th cents. (→ map p. 67). They displaced the Sicani, who moved to the western part of the island (Thuc. 6,2,4f.; Diod. Sic. 5,2,6). Another group are the Elymi, documented in writing from the 8th cent., mostly in the shape of graffiti and 5th-cent. coin legends, but their origins and their migrations across Sicily are still rather obscure. Alongside these cultures, there is evidence of early (Western) Phoenician influence in western Sicily. The same is true of Sardinia, which otherwise was dominated by the Nuraghic Iron Age culture.

II. Travel routes

From the 8th cent. onward, the Greeks established in their search for new territories their *apoikiai* (colonies) predominantly in those parts of Italy where there was little urban development, such as the coasts of Sicily, Calabria and Lucania and the Gulf of Ionia, but avoided southern Etruria and the areas around Capua, Pontecagnano and Sala Consilina as well as Apulia. The settlements along the Tyrrhenian and Ionian shorelines up to Sybaris were not linked by a coastal road. Most of the settlements on the fertile plains were located some distance away from the coast. The rivers, however, served as communication and traffic arteries. The topographical structure created by the Apennine favoured the north-to-south flow of communication over the west-to-east one. In the south, communication ran from the eastern Lucanian plateau via Matera to the Apulian plains and on to the coast. Further routes ran from Salerno across the plain surrounding Posidonia and along the Silarus valley to Sala Consilina, from where they probably crossed the Passo di Conza to follow the Aufidus valley to the Adriatic or in the direction of Sybaris. South of the Silarus (Sele), the north-to-south course of the rivers connected the region with Calabria and the Ionian Gulf (Aciris/Agri valley). The northern section of the route followed the Sacco/Liris valley between Campania and Latium. In central Italy there was a west-east axis in the shape of the route from southern or central Etruria along the valleys of the Tiber, the Nerina and the Tenna to Fermo, the embarkation point for crossing into Istria or Illyria.

→ Maps pp. 31, 67, 75 and 77, 79 and 81, 83; Este Culture, BNP 5, 2004; Golasecca Culture, BNP 5, 2004; Villanova Culture, BNP 15, 2009; Iapyges, Iapygia, BNP 6, 2005.

Literature

G. Bartoloni, F. Delpino (eds.), Oriente e Occidente: metodi e discipline a confronto. Riflessioni sulla cronologia dell'età del ferro italiana. Atti dell'incontro di studi (Roma, 30-31 ottobre 2003), Mediterranea I, 2005; S. Haynes, Kulturgeschichte der Etrusker, Mainz 2005; S. Moscati, Storia degli Italiani dalle origini all'età di Augusto, 1999; G. Nenci, G. Vallet, Bibliografia topografica della colonizzazione greca in Italia, Scuola Normale Superiore – École Française de Rome – Centre Jean Berard, vols. I–XX, 1977–2007; M. Pallottino, Storia della prima Italia, 1984; L. Franchi dell'Orto (ed.), Die Picener. Ein Volk Europas, exhibition catalogue 1999; G. Pugliese Carratelli (ed.), Italia omnium terrarum alumna. La civiltà dei Veneti, Reti, Liguri, Celti, Piceni, Umbri, Latini, Campani e Iapigi, 1988; Id. (ed.), Italia omnium terrarum parens. La civiltà degli Enotri, Choni, Ausoni, Sanniti, Lucani, Brettii, Sicani, Siculi, Elimi, 1989; M. Torelli (ed.), Gli Etruschi 2000; K. v. Welck, R. Stupperich (eds.), Italien vor den Römern, 1996.

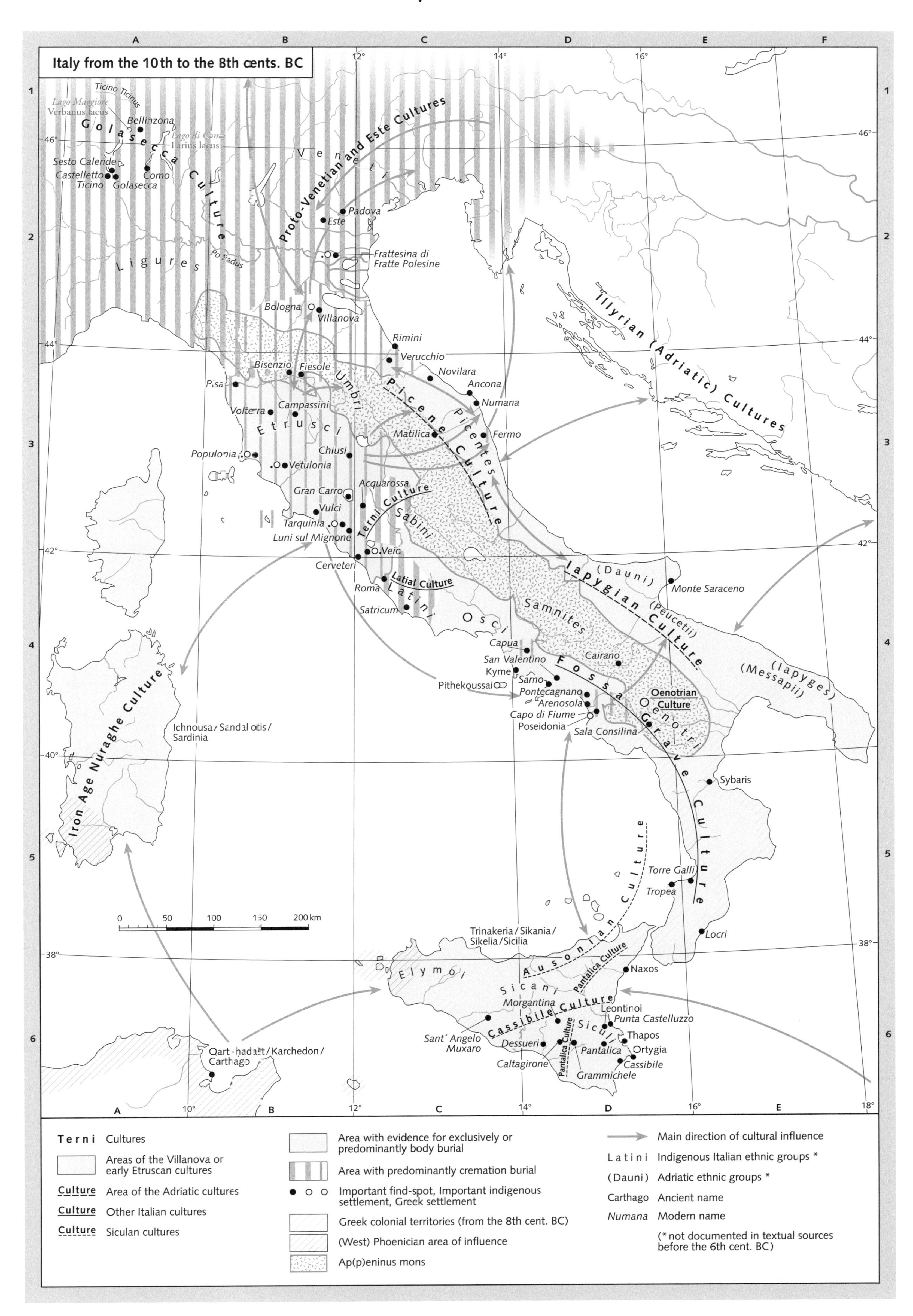
Italy from the 10th to the 8th cents. BC
Golasecca Culture
Proto-Venetian and Este Cultures
Veneti
Ligures
Illyrian (Adriatic) Cultures
Etrusci
Umbri
Picene Culture
Picentes
Terni Culture
Sabini
Latial Culture
Latini
Osci
Samnites
Iapygian Culture
(Dauni)
(Peucetii)
(Iapyges)
(Messapii)
Fossa Grave Culture
Oenotrian Culture
Oenotri
Iron Age Nuraghe Culture
Ausonian Culture
Pantalica Culture
Cassibile Culture
Elymoi
Sicani
Siculi
Bellinzona
Sesto Calende
Castelletto Ticino
Golasecca
Como
Padova
Este
Frattesina di Fratte Polesine
Bologna
Villanova
Rimini
Verucchio
Novilara
Ancona
Numana
Fermo
Matilica
Bisenzio
Fiesole
Pisa
Volterra
Campassini
Populonia
Vetulonia
Chiusi
Acquarossa
Gran Carro
Vulci
Tarquinia
Luni sul Mignone
Veio
Cerveteri
Roma
Satricum
Monte Saraceno
Capua
San Valentino
Kyme
Sarno
Pithekoussai
Pontecagnano
Arenosola
Capo di Fiume
Poseidonia
Sala Consilina
Cairano
Sybaris
Torre Galli
Tropea
Locri
Ichnousa / Sandaliotis / Sardinia
Trinakeria / Sikania / Sikelia / Sicilia
Naxos
Morgantina
Leontinoi
Punta Castelluzzo
Thapos
Ortygia
Cassibile
Pantalica
Grammichele
Caltagirone
Dessueri
Sant´ Angelo Muxaro
Qart-hadašt / Karchedon / Carthago
Lago Maggiore
Verbanus lacus
Lago di Como
Larius lacus
Ticino
Ticinus
Po
Padus
0 50 100 150 200 km
Terni Cultures
Areas of the Villanova or early Etruscan cultures
Culture Area of the Adriatic cultures
Culture Other Italian cultures
Culture Siculan cultures
Area with evidence for exclusively or predominantly body burial
Area with predominantly cremation burial
Important find-spot, Important indigenous settlement, Greek settlement
Greek colonial territories (from the 8th cent. BC)
(West) Phoenician area of influence
Ap(p)eninus mons
Main direction of cultural influence
Latini Indigenous Italian ethnic groups *
(Dauni) Adriatic ethnic groups *
Carthago Ancient name
Numana Modern name
(* not documented in textual sources before the 6th cent. BC)

Languages in ancient Italy before the spread of Latin

Due to numerous, partly overlapping migrations from the second millennium onward, the linguistic map of pre-Roman Italy is characterized by great diversity. Both archaeological evidence and historical sources regarding these migrations are available in exceptional cases only, such as the infiltration of the Padana by the Celts, which roughly happened from the 6th to the 4th cents. (Liv. 5,34) or the southward expansion of the Samnites (Strab. 5,242). On the other hand, almost a score of languages and their dialect varieties have come down to us directly in the shape of inscriptions as a result of the early introduction of alphabetic scripts (before 700 to 6th cent. BC at the latest), whereas we know hardly more than the names of some of the other languages. New finds of inscriptions have been continually adding corrective details to our understanding of the early history of languages in Italy. Thus texts from the 6th cent. BC which were found in recent years in the Gulf of Naples (e.g. at Vico Equense) point to a closer link of the local language variety with the South Picenian language of eastern central Italy before the Samnites moved into Campania in the 5th cent. BC.

The relationships of ancient Italian languages (map and stemma)

At least four Indo-European language branches can be identified:
(a) Italic or Italo-Venetic (Latin: the city of Rome, Latium); Faliscan: north of Rome, around Falerii; Osco-Umbrian/Sabellan: central eastern Italy (from the 7th cent.); Venetic: *c.* 300 inscriptions from the 6th to the 1st cents.);
(b) Celtic (Lepontic: *c.* 100 inscriptions, 6th–1st cents., around Lugano; Gaulish: few inscriptions, the bilingual Todi Stone, 3rd/2nd cent., from the left bank of the Padus to the boundary of the Lepontic territory; by the 4th cent., the entire Padana was Celtic-speaking);
(c) Messapian (Illyrian?, *c.* 300 inscriptions, 6th–1st cents.);
(d) Greek (in the areas of Greek settlement).
The relationship of the indigenous languages of Sicily (Siculan, Elymian) and probably Northern Picenian with other languages has so far not been established.
The non-Indo-European languages are represented by:
(a) the Tyrsenian group (Raetic: verified both north and south of the Brenner Pass; Etruscan: *c.* 9,000 inscriptions from before the 7th cent. onward in the Etruscan core territory, in the Padana, in Campagna and probably in the Etruscan settlement area on Corsica; the epichoric dialect preserved in a small number of inscriptions on the Greek island of Lemnos belongs to the same group);
(b) the Semitic group (Punic: *c.* 170 inscriptions from archaic trading colonies on Sicily, Sardinia and at Pyrgi as well as in the Western Phoenician settlement area outside Italy.

As to the affiliations of the Camunian (*c.* 70 inscriptions, 5th to 1st cents., east of the Lepontic language area), the Ligurian (Riviera) and the Daunian (northern Apulia) languages, scholars are still in the dark.

Although the findings of linguistics have established with a high degree of probability that the immigration of the so-called Italic peoples happened in the second millennium BC, possibly in two waves (the first one bringing the Latins and the second one the Sabelli), there is as yet no archaeological evidence to back this up. The funeral practice of cremation is usually treated as a significant feature in archaeological contexts in that it was preferred by the Etruscans and Latins, whereas the Sabelli continued to practise inhumation. The southward movement of cremation in the middle of the second millennium, however, does not necessarily indicate a migration of populations, which severely limits the scope of any historical and cultural conclusions.

In the case of the Etruscan language, epigraphic documents and an evaluation of onomastic findings demonstrate that the Etruscans advanced from the Tyrrhenian coast into the interior, which up to then had been inhabited by Umbro-Sabinians and, in its southern areas, by Latins; by the 7th cent., this process had been largely concluded. It is equally possible to work out from what time Etruscan was actually spoken in the core territory and in the 'colonial areas' of northern Italy and Campania. Thus the findings of linguistics support and supplement the information about pre-Roman times provided by literary sources and the results of archaeological research.

The classificatory division of the indigenous population of Sicily into Elymi, Sicani and Siculi (partly based on cultural differences), made in Antiquity and accepted by modern scholarship, is a different matter. Any attempt to use linguistic and epigraphical findings (*c.* 7th – 5th cents.) to provide support for this theory is problematic, because it has not been possible to establish any close linguistic connection with 'the Italic language family'.

→ Maps pp. 65, 75

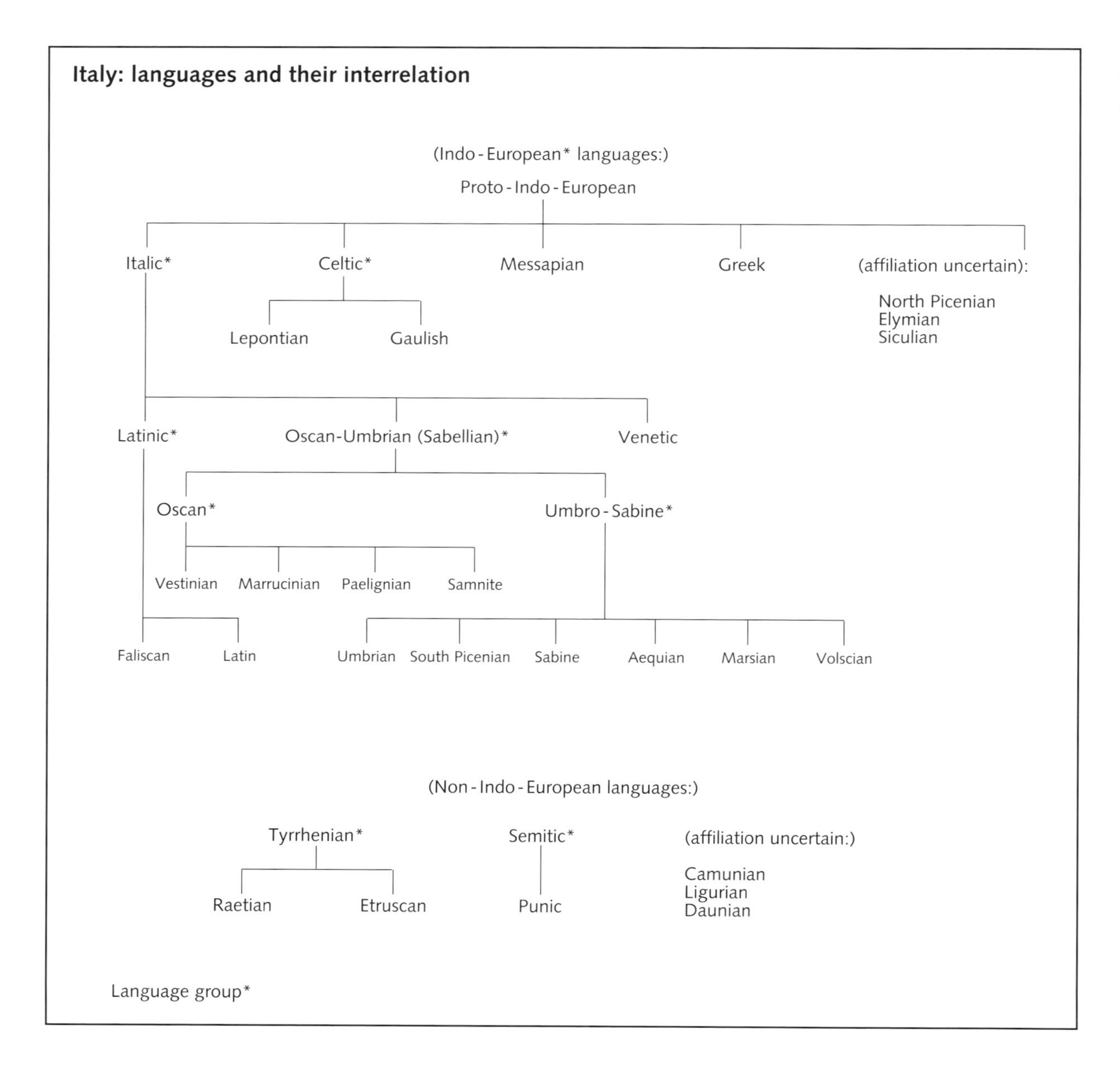

Literature

L. Agostiniani, Le iscrizioni anelleniche di Sicilia, 1977; G. Fogolari, A.L. Prosdocimi, I Veneti antichi. Lingua e cultura, 1988; R. Guerra, Antiche popolazioni dell' Italia preromana, 1999; M. Lejeune, Lepontica, 1971; Id., Recueil des Inscriptions Gauloises. 2,1: Textes Gallo-étrusques ..., 1988; M. Pallottino, G. Mansuelli, A.L. Prosdocimi, O. Parlangeli (eds.), Popoli e civiltà dell' Italia antica, vol. 6 (A.L. Prosdocimi, Lingue e dialetti), 1978; H. Rix, Etruskische Texte, 2 vols., 1995; Id., Sabellische Texte. Die Texte des Oskischen, Umbrischen und Südpikenischen, 2002; C. Santoro, Nuovi Studi Messapici, 3 vols., 1982–1984; S. Schumacher, Die rätischen Inschriften, 1992.

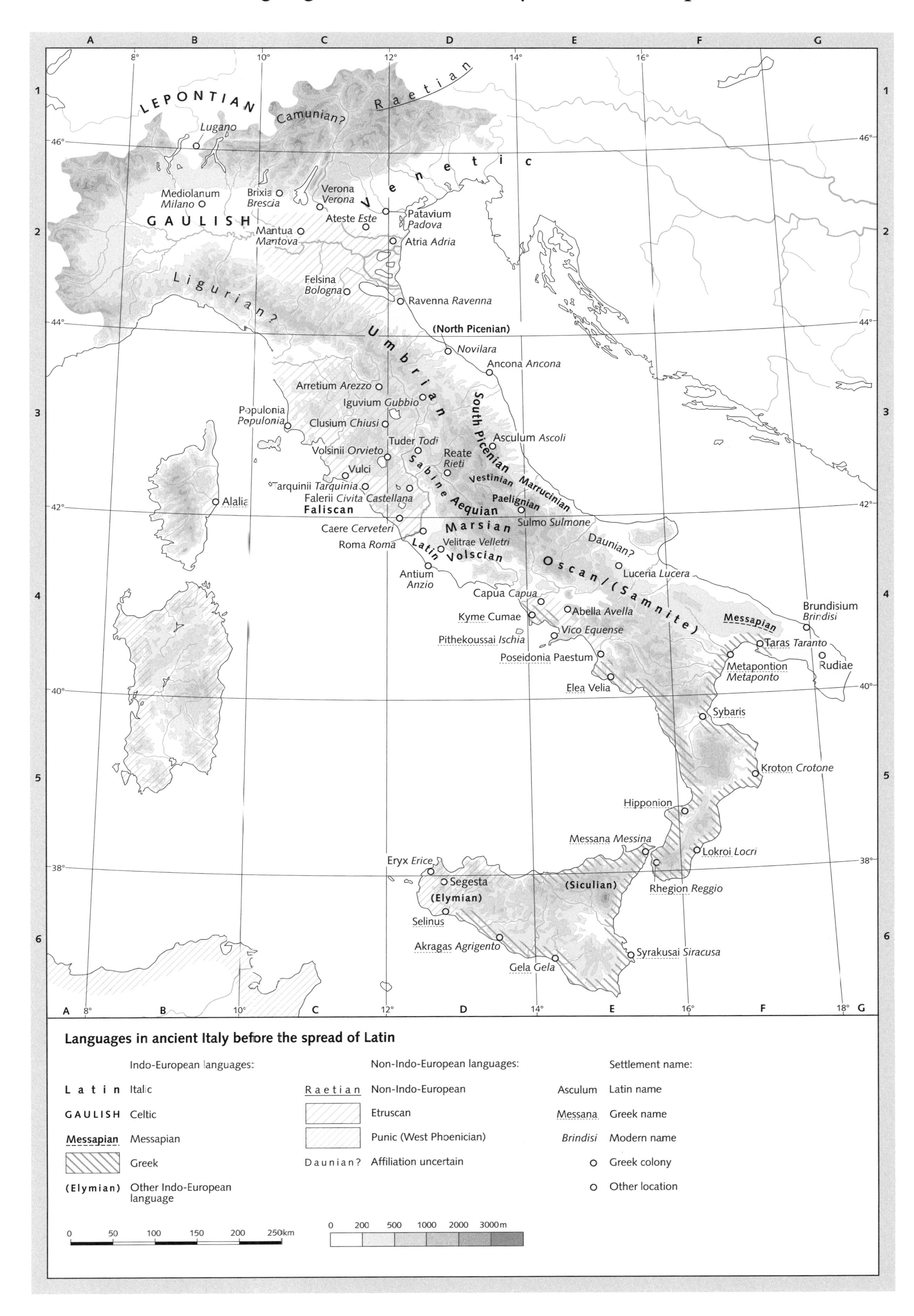

A B C D E F G
1 2 3 4 5 6
8° 10° 12° 14° 16° 18°
46° 44° 42° 40° 38°
LEPONTIAN
Camunian?
Raetian
Lugano
Venetic
Mediolanum Milano
Brixia Brescia
Verona Verona
GAULISH
Ateste Este
Patavium Padova
Mantua Mantova
Atria Adria
Ligurian?
Felsina Bologna
Ravenna Ravenna
(North Picenian)
Umbrian
Novilara
Ancona Ancona
Arretium Arezzo
Iguvium Gubbio
Populonia Populonia
Clusium Chiusi
South Picenian
Asculum Ascoli
Tuder Todi
Volsinii Orvieto
Reate Rieti
Sabine
Vulci
Vestinian
Marrucinian
Tarquinii Tarquinia
Falerii Civita Castellana
Faliscan
Aequian
Paelignian
Alalia
Sulmo Sulmone
Marsian
Caere Cerveteri
Daunian?
Roma Roma
Latin
Velitrae Velletri
Volscian
Oscan/(Samnite)
Luceria Lucera
Antium Anzio
Capua Capua
Abella Avella
Kyme Cumae
Messapian
Brundisium Brindisi
Pithekoussai Ischia
Vico Equense
Taras Taranto
Rudiae
Poseidonia Paestum
Metapontion Metaponto
Elea Velia
Sybaris
Kroton Crotone
Hipponion
Messana Messina
Lokroi Locri
Eryx Erice
Segesta
(Elymian)
(Siculian)
Rhegion Reggio
Selinus
Akragas Agrigento
Syrakusai Siracusa
Gela Gela
Languages in ancient Italy before the spread of Latin
Indo-European languages:
Latin Italic
GAULISH Celtic
Messapian Messapian
Greek
(Elymian) Other Indo-European language
Non-Indo-European languages:
Raetian Non-Indo-European
Etruscan
Punic (West Phoenician)
Daunian? Affiliation uncertain
Settlement name:
Asculum Latin name
Messana Greek name
Brindisi Modern name
Greek colony
Other location
0 50 100 150 200 250km
0 200 500 1000 2000 3000m

Colonization: Phoenicians, Greeks and Etruscans in the Mediterranean area (*c.* 11th–6th cents. BC)

The term 'colonization' here encompasses various settlement movements in the Mediterranean region from the 11th cent. on, which had a significant impact on the political geography and history of the Mediterranean world. The peoples involved were the Phoenicians, Greeks and Etruscans. This definition does not include, for example, immigrations during the 3rd/2nd millennia, the Minoan and Mycenaean expansions, or the Celtic incursions into the Mediterranean region.

In spite of the sometimes weak basis of literary and archaeological sources and considerable differences in the courses, causes and purposes of colonial foundations, some common characteristics can be observed, such as springing from individual communities and trading interests. Colonization movements in the western Mediterranean began to interweave from the 6th cent., which increasingly led to conflict between Greeks, Phoenicians/Western Phoenicians/Carthaginians and Etruscans.

I. Phoenician colonization

The western voyages embarked upon, probably from the early 12th cent. BC, from the individual Phoenician city-states (esp. Tyre, Sidon and Byblus) were primarily aimed at the metal resources of Cyprus, the Iberian Peninsula, Sardinia and Etruria, for the supply of domestic industry. However, they also served the well-attested intermediary trade. An early foundation date (12th cent. BC) has been calculated for many settlements (Qart-ḥadašt/Citium, Ityke/Utica, 'Gdr/Gades) on the basis of the testimony of Greek and Latin sources (Aristotle, Flavius Josephus, Pliny, Virgil, Justin, Vellius), although archaeological evidence from the Early Iron Age is lacking. This must in principle be assessed as a sign that Bronze Age contacts survived the disturbances around 1200 BC. The main source for the voyages to Taršiš, probably to be identified with Tartessus in southern Spain, is the Old Testament (1 Kgs 10: 22; Ezek 27: 12).

Scholars in fact reserve the title 'colony' (*apoikia*) for the Tyrian foundations of Qart-ḥadašt/Citium on Cyprus and Qart-ḥadašt/Carchedon/Carthago. All other Phoenician settlements of the first phase (*c.* 11th-9th cents.) in the 'regions of expansion' (Cyprus and the Aegean, central Mediterranean (i.e. Italy with Sicily, Sardinia), African Mediterranean and Atlantic coasts, Iberian Peninsula), for which the archaeological evidence is only sporadic, are classified as trading posts and workshops (*enoikismoi*). Numerous permanent, archaeologically verifiable emporia, factories and sanctuaries then arose in the second phase (7th-5th cents.), some of which rapidly developed to become fortified urban centres with harbours and necropoleis (e.g. Motya on Sicily). However, Phoenicians (and probably other Levantines) also sometimes lived as *metoikoi* (e.g. at Pithecussae).

II. The so-called Great Greek Colonization

The most substantial colonization movement was the so-called 'Great Greek Colonization', in which numerous Ionian, Dorian and Aeolian and some Achaean cities (apart from Athens) took part, and which probably led to a doubling in size of the Greek commonwealth in the Mediterranean and Black Sea regions during the heyday between *c.* 750 and 580 BC. It lasted until around 500 BC at varying degrees of intensity, and included the foundation by original *apoikiai* of their own daughter-cities in turn, sometimes with the participation of the original mother-city. The points of origin were some twenty Greek communities on the mainland and Aegean islands and in western Asia Minor. The preferred destination regions (in approximate chronological order, cf. also the chronological synopsis) were, in the 8th and 7th cents., Sicily, southern Italy, the Chalcidice of the northern Aegean, the sea routes to the west (via Corcyra) and to the Black Sea (Propontis). The settlement of the Black Sea region began in the second half of the 7th cent. BC. In North Africa, Cyrene was founded (subsequently with its own colonies), and in the far west, Massalia (with its own colonies and trading posts). As well as on archaeological finds, our knowledge is based mostly on the works of Herodotus and Thucydides, and on occasional references in Strabo.

III. Etruscan colonization

The northward and southward expansion from the Etruscan heartland between the Arnus and the Tiber began as early as the Villanova phase (9th/8th cents. BC), reaching Felsina/Bononia/Bologna, Fermo and Verucchio in the north, and Pontecagnano and Sala Consilina by sea in the south (on locations→→ map p. 65). The far more extensive colonization that started in the 7th cent. led to foundations attested in archaeology and literature (Diodorus Siculus, Livy, Pliny: e.g. Marzabotto, Mantua, Bononia/Bologna, Atria/Adria, Spina) north of the Apennines and in Campania (e.g. Nola, Capua, Pompeii), perhaps organized in the form of twelve-city leagues as in the heartlands. The identity of the founders, however, can only be conjectured from the alphabets used and from finds. The Greeks (Massalia) put a stop to the Etruscan expansion along the Tyrrhenian coast into Ligurian territory (Genua) in the 6th cent.

→ Maps pp. 25, 27, 29, 31, 33 (on the so-called 'Ionian Colonization'), 35, 63, 65, 67, 69, 75, 77, 81, 167; Colonization [III]–[V], DNP 3, 2003 (with stemmata 'Ionian Colonization', 'Dorian Colonization').

Literature

J. Boardman, The Greeks Overseas. The Archaeology of Their Early Colonies and Trade 1964; La colonisation grecque en Méditerranée occidentale. Actes de la rencontre scientifique en hommage à G. Vallet, 1999; J.-P. Decoeudres (ed.), Greek Colonists and Native Population, 1990; F. Krinzinger (ed.), Die Ägäis und das westliche Mittelmeer. Beziehungen und Wechselwirkungen 8. bis 5. Jahrhundert v. Chr., 2000; D. Mertens, Städte und Bauten der Westgriechen. Von der Kolonisationszeit bis zur Krise um 400 v. Chr., 2006; A. Möller, Naukratis, 2000; R. Rolle, K. Schmidt, R. F. Docter (eds.), Archäologische Studien in Kontaktzonen der antiken Welt. FS H.G. Niemeyer, 1998; W. Schuller, Griechische Geschichte, [5]2002; E. Stein-Hölkeskamp, Im Land der Kyklopen. Immigranten und Indigene in den südi talischen Siedlungen des 8. und 7. Jahrhundert v. Chr., in: Klio 88, 2006, 311–327; G.R. Tsetskhladze (ed.), The Greek Colonisation of the Black Sea Area, 1998; Id., F. de Angelis (eds.), The Archaeology of Greek Colonisation, Festschrift J. Boardman, 1994.

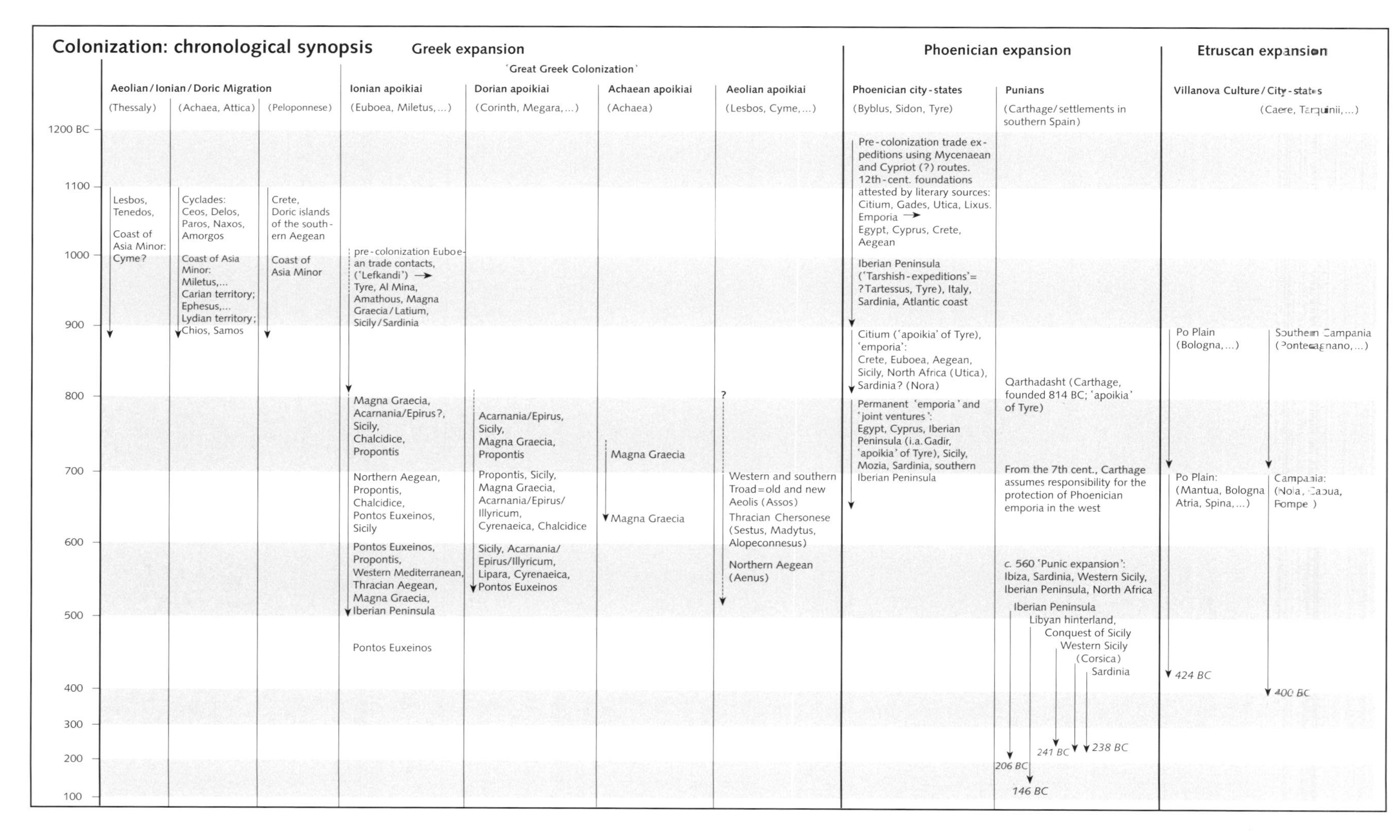

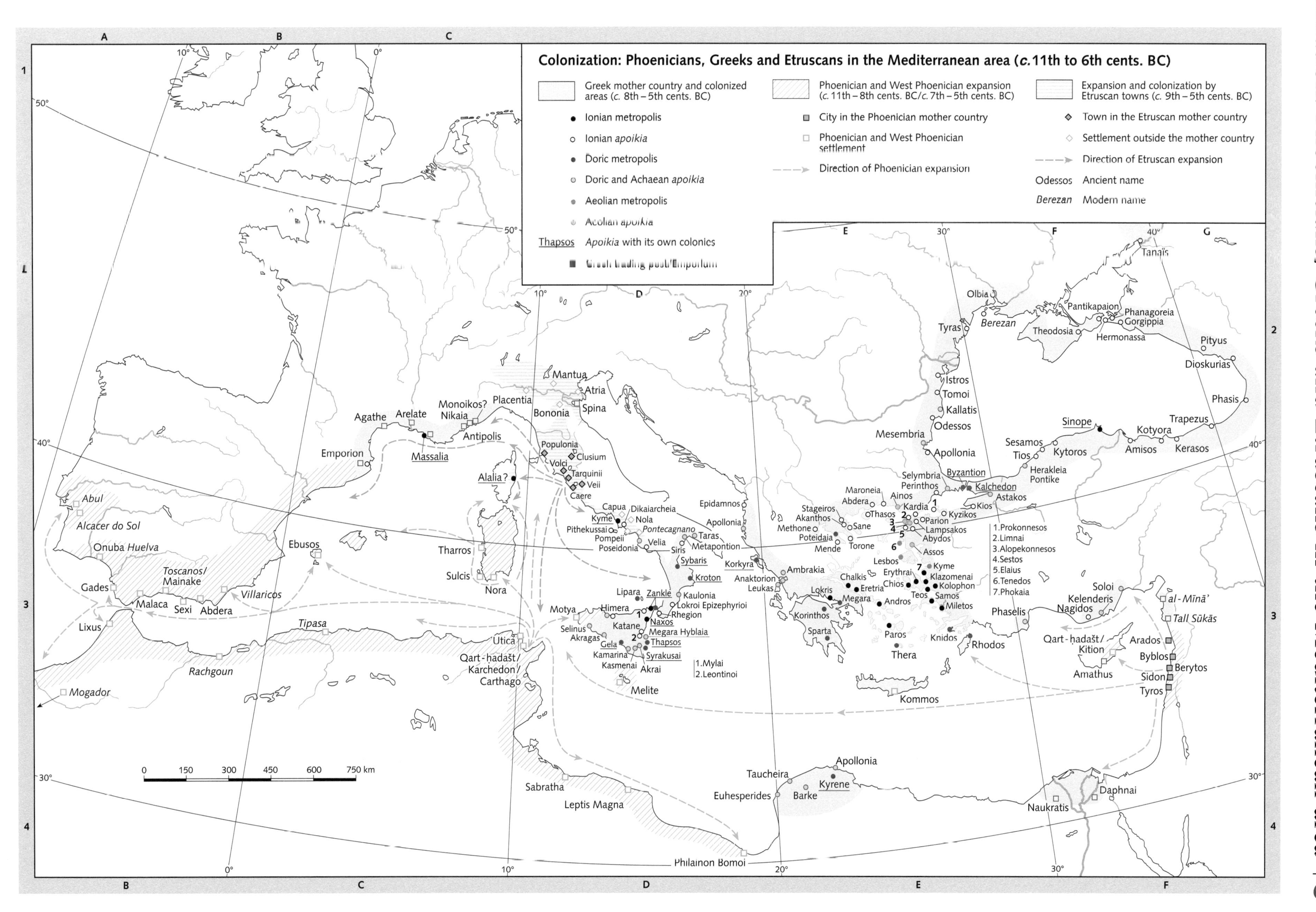
Colonization: Phoenicians, Greeks and Etruscans in the Mediterranean area (c. 11th to 6th cents. BC)
Greek mother country and colonized areas (c. 8th – 5th cents. BC)
Ionian metropolis
Ionian apoikia
Doric metropolis
Doric and Achaean apoikia
Aeolian metropolis
Aeolian apoikia
Thapsos Apoikia with its own colonies
Phoenician and West Phoenician expansion (c. 11th – 8th cents. BC/c. 7th – 5th cents. BC)
City in the Phoenician mother country
Phoenician and West Phoenician settlement
Direction of Phoenician expansion
Expansion and colonization by Etruscan towns (c. 9th – 5th cents. BC)
Town in the Etruscan mother country
Settlement outside the mother country
Direction of Etruscan expansion
Odessos Ancient name
Berezan Modern name
1.Prokonnesos
2.Limnai
3.Alopekonnesos
4.Sestos
5.Elaius
6.Tenedos
7.Phokaia
1.Mylai
2.Leontinoi
0 150 300 450 600 750 km

The Phoenician and West Phoenician/Carthaginian world in the western Mediterranean area (9th–2nd cents. BC)

The Phoenician and Western Phoenician/Carthaginian world in the western Mediterranean (9th-2nd cents. BC)

Not only does the map cover a lengthy period (9th-2nd cents. BC), but it also brings together two separate phases of Phoenician presence in the western Mediterranean. Firstly, there is the period of early Phoenician settlements, trading posts and agrarian communities initiated and controlled by the Levantine mother-cities, notably Tyre (9th-7th/6th cents.). Secondly, what is conventionally referred to as Punic or Carthaginian foundations, are all the Western Phoenician settlements on the African and Iberian coasts, on Sardinia and later on Sicily which were created (7th/6th-2nd cents.) independently of the Eastern Phoenicians in the wake of the rise to predominance of the Tyrian colony of Qart-ḥadašt/Carchedon/Carthago (Carthage) in the 6th/5th cents.

I. Eastern Phoenicians in the West

The Phoenician cities, initially gbl/Byblus (cf. itinerary of the Egyptian Wenamun, ANET 25-29), ṣdn/Sidon (Hom. Il. 6,291; 23,743) and ṣr/Tyros (Tyre), were probably already sending out trading expeditions in the 12th/11th cents., but certainly at least from the 10th cent. BC (OT, 1 Kgs 10: 14–29), exploring westward via Cyprus (Citium), Crete (Kommos) and Malta. The objectives were especially the metal deposits of Etruria, Sardinia and above all the Iberian Peninsula, but also the Atlantic coast of what is now Morocco. Phoenician settlements were established in the course of these 'Taršiš voyages' (which are not uncontroversial among scholars), some as *enoikismoí* (trading offices and workshops) within indigenous settlements. The Phoenician imports found locally also demonstrate contacts sometimes reaching far into the hinterland, esp. into the so-called 'Tartessian' region of southern Spain, which may be the 'Taršiš' of the Old Testament tradition (1 Kgs 10: 22; Ezek 27: 12).

The 'new city' of Qart-ḥadašt/Carchedon/Carthago is regarded as the sole Phoenician colony founded in the west. Along with 'j/Ityke/Utica, Carthage was also one of the first settlements, dated to 814/13 from the literary tradition (Timaeus). It was also supposedly in the 9th cent. that Auza (African coast, not localized) and perhaps Nora (Sardinia) were founded. Although archaeological findings and finds mostly elicit later dates, it may be that Sardinia and Sicily were also being regularly visited by Phoenicians from the 9th/8th cents., to set up 'emporia' there. There was probably some Levantine presence on the Italian coast from the 8th/7th cents. onward, not necessarily all from Phoenician city-states. Permanent settlements were also established in this period on the Mediterranean and Atlantic coasts of Africa. The settlement pattern is the same in almost all locations exploited, for some of which we know the Phoenician names: these were sheltered and easily-defended seaside locations, often on capes, with good harbours and/or on river mouths, with fertile hinterland and/or good access to natural resources. The settlement layouts show tendencies towards an orthogonal street pattern, special zones for workshops, and shipbuilding facilities.

The hinterland of these settlements was generally not politically and administratively subordinate or subjected. Rather, there was a co-operative relationship with the indigenous aristocracy.

II. Western Phoenician ascendancy in the western Mediterranean

The rise of Qart-ḥadašt, popularly Carthage, and the associated hegemony of this city-state in the western and central Mediterranean from the 7th/6th cents. (from the late 6th cent., the term Carthaginian 'empire' is used) are closely connected with the increasing pressure being exerted on the Eastern Phoenicians by the Assyrians, Neo-Babylonians and later Persians. The city rapidly rose to become a leading maritime and commercial power. At its greatest extent, Carthage's realm or sphere of influence reached from Arae Philaenorum on the Gulf of Sidra (Syrtis maior) in the east to Mogador on the African Atlantic coast, and included the south-west (to the mouth of the Calipus/Sado (Abul)) and south of the Iberian Peninsula, the Balearic Islands and the 'granaries' of Corsica (?), Sardinia, and western Sicily. These three islands the Carthaginians lost to Rome as a result of the First Punic War (264–241 BC), which meant that the Iberian Peninsula was thereafter systematically pervaded with military and diplomatic activity.

Carthage consolidated its power on the one hand with trading agreements and friendship pacts (e.g. with the Etruscans in 535 BC; with Rome in 509, 348, 343, 306, 279/78), which laid down respective interests and spheres of interest. Often, though, it pressed its interests, sometimes defensively, sometimes offensively, by military means (e.g. against the Sicilian Greeks from 480 BC). Through the second half of the 4th cent. BC, its position mostly depended on the wealth it had accrued by commerce, the superiority of its fleet, the army, now mostly composed of mercenaries, the Carthaginian sovereign territory and its alliances with numerous cities of North Africa – formal autonomy from Carthage is here attested for some Western Phoenician cities and autonomous administration for cities in Carthaginian territory, e.g. Utica and Hadrumetum. Corsica (?) and Sardinia, and probably the cities of the southern Iberian Peninsula, were treated like the Carthaginian African region. The situation was different in the case of the Carthaginian *epikráteia* on Sicily (important Phoenician settlements: Motya, Solus, Panormus), which was destabilized by frequent conflicts with the Greeks. Local coin minting developed here. The conquered Greek cities also kept their own administrative systems.

The relationship with the tribes of the hinterland, some of which were nomadic, some sedentary, remains unclear. Around 450 BC, the Libyan tribes were said to have been conquered, and the tribute payments formerly made to them cancelled. According to the sources, so-called 'Libyan Phoenicians' lived in many coastal cities.

Sources

Our knowledge of the Eastern Phoenicians in the west and of the Western Phoenicians, notably Carthage, is mainly based on archaeological research (settlement excavations, surveys) and on indirect sources. Although there are a great many inscriptions, esp. simple funerary and votive inscriptions, there are no Phoenician literary texts from the western Mediterranean. In the epigraphic record, scholars make an expedient distinction between Phoenician (i.e. all linguistic evidence of the Eastern Phoenician area and the Phoenician diaspora throughout the Mediterranean, to *c.* 500 BC), Punic (i.e. texts from the western Mediterranean from *c.* 500 BC) and Neo-Punic (i.e. all 'Punic' texts from the western Mediterranean that were written after the destruction of Carthage in 146 BC). We mostly have authors writing in Greek and Latin to thank for information assisting the reconstruction of Phoenician and Carthaginian history (Polybius, Livy); among their sources were the works of historians such as the bilingual (Greek/Punic) Philinus of Acragas (used by Polybius) or works like the *Historiae Poenorum* and the *Punica historia* (Pseudo-Aristotle; Servius). In addition, references in Flavius Josephus (esp. Jos. Ant. Iud., e.g. 8,144; 9,283 = FGrH 783 T 3: Menander of Ephesus translated the Tyrian archives into Greek; Dios was a Hellenistic author of a Phoenician history, cf. Jos. Ap. 1,112 = FGrH 785 F) and in the OT indicate that Tyre had its own historiographic activity during the 10th-8th and 6th cents., although all the texts produced were presumably recorded on perishable material and are now lost. For the history of Carthage and its 'empire', too, the very fragmentary source material for the 5th and 4th cents. improves only in conjunction with the military confrontations with the expanding Rome from the 3rd cent. BC (the so-called Punic Wars), but all information on this period is written from a Roman standpoint.

→ Maps pp. 43, 45, 67, 73, 85, 127, 139; Commerce II., BNP 3, 2003; Phoenicians, Poeni [III. B.], BNP 11, 2007 (with sources)

Literature

M. Blech, Tartessos, in: M. Blech, M. Koch, M. Kunst (eds.), Hispania antiqua. Denkmäler der Frühzeit, 2 vols., 2001, 305–348; C.R. Krahmalkov, A Phoenician-Punic Grammar, 2001; K. Momrak, The Phoenicians in the Mediterranean: Trade, Interaction and Cultural Transfer, in: Altorientalische Forschungen 32, 2005, 168–181; S. Moscati, Die Karthager, 1996; H.G. Niemeyer, Die Phönizier auf dem Weg nach Westen, in: M. Blech, M. Koch, M. Kunst (eds.), Hispania antiqua, 2001, 275–282; Id., Die Punier auf der Iberischen Halbinsel, Ibid., 413–422; R. D'Oriano, Olbia: ascendenze puniche nell' impianto urbanistico romano, in: A. Mastino (ed.), L' Africa romana. Atti del VII convegno di studio su »L' Africa romana«. Sassari, 15–17 dicembre 1989, 1990, 487–495; H. Schubart, Die Phönizier an den Küsten der Iberischen Halbinsel, in: M. Blech, M. Koch, M. Kunst (eds.), Hispania antiqua, 2001, 283–304; M. Sommer, Phönizier, 2005; Id., Europas Ahnen, 2000; N.C. Vella, Landscape, and Territory. Phoenician and Punic Non-Funerary Religious Sites in the Mediterranean: An Analysis of the Archaeological Evidence (thesis, Bristol), 1998.

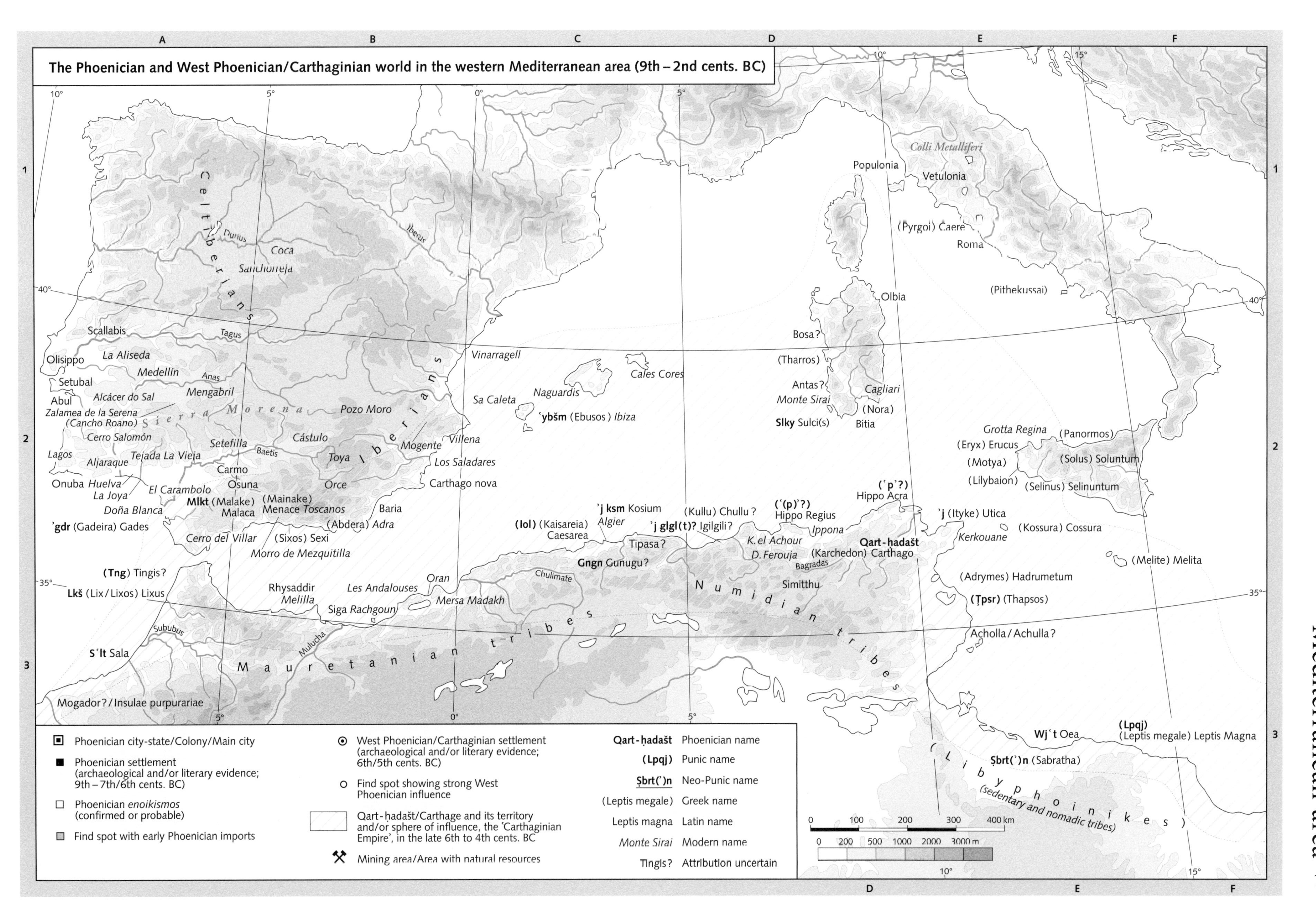
The Phoenician and West Phoenician/Carthaginian world in the western Mediterranean area (9th – 2nd cents. BC)
Celtiberians
Durius
Coca
Sanchorreja
Iberus
Tagus
Scallabis
Olisippo
La Aliseda
Medellín
Anas
Setubal
Alcácer do Sal
Mengabril
Abul
Zalamea de la Serena (Cancho Roano)
Sierra Morena
Pozo Moro
Cerro Salomón
Setefilla
Cástulo
Lagos
Aljaraque
Tejada La Vieja
Baetis
Carmo
Toya
Iberians
Mogente
Villena
Los Saladares
Onuba Huelva
La Joya
El Carambolo
Osuna
Orce
Carthago nova
Mlkt (Malake) Malaca
(Mainake) Menace Toscanos
Doña Blanca
Baria
'gdr (Gadeira) Gades
(Abdera) Adra
Cerro del Villar
(Sixos) Sexi
Morro de Mezquitilla
(Tng) Tingis?
Lkš (Lix/Lixos) Lixus
Rhysaddir
Melilla
Les Andalouses
Oran
Siga Rachgoun
Mersa Madakh
Sububus
Mulucha
S'lt Sala
Mauretanian tribes
Mogador?/Insulae purpurariae
Vinarragell
Cales Cores
Naguardis
Sa Caleta
'ybšm (Ebusos) Ibiza
'j ksm Kosium
Algier
(Iol) (Kaisareia) Caesarea
(Kullu) Chullu?
'j glgl(t)? Igilgili?
Tipasa?
Gngn Gunugu?
Chulimate
Numidian tribes
('(p)'?) Hippo Regius
Ippona
K. el Achour
D. Ferouja
Bagradas
Simitthu
Qart-ḥadašt (Karchedon) Carthago
('p'?) Hippo Acra
'j (Ityke) Utica
Kerkouane
(Kossura) Cossura
(Melite) Melita
(Adrymes) Hadrumetum
(Ṭpsr) (Thapsos)
Acholla/Achulla?
Wj't Oea
(Lpqj) (Leptis megale) Leptis Magna
Ṣbrt(')n (Sabratha)
(Libyphoinikes)
(sedentary and nomadic tribes)
Populonia
Colli Metalliferi
Vetulonia
(Pyrgoi) Caere
Roma
(Pithekussai)
Olbia
Bosa?
(Tharros)
Antas?
Monte Sirai
Cagliari
(Nora)
Slky Sulci(s)
Bitia
Grotta Regina
(Panormos)
(Eryx) Erucus
(Motya)
(Solus) Soluntum
(Lilybaion)
(Selinus) Selinuntum
Phoenician city-state/Colony/Main city
Phoenician settlement (archaeological and/or literary evidence; 9th – 7th/6th cents. BC)
Phoenician enoikismos (confirmed or probable)
Find spot with early Phoenician imports
West Phoenician/Carthaginian settlement (archaeological and/or literary evidence; 6th/5th cents. BC)
Find spot showing strong West Phoenician influence
Qart-ḥadašt/Carthage and its territory and/or sphere of influence, the 'Carthaginian Empire', in the late 6th to 4th cents. BC
Mining area/Area with natural resources
Qart-ḥadašt Phoenician name
(Lpqj) Punic name
Ṣbrt(')n Neo-Punic name
(Leptis megale) Greek name
Leptis magna Latin name
Monte Sirai Modern name
Tingis? Attribution uncertain
0 100 200 300 400 km
0 200 500 1000 2000 3000 m

Carthage

I. Qart-ḥadašt/Carchedon (8th–2nd cents. BC; map A)

According to the report by Timaeus (FGrH 566 F 60; another detailed account in Iustinus, *Elissa*), Carchedon (Phoenician: Qart-ḥadašt/'New Town', Latin: Carthago, commonly Carthage) was founded in 814/13 or 813/12 BC by colonists from Tyre, whose leader was Princess Elissa/Dido. Situated on a peninsula on the African coast of the Mediterranean Sea, it was a perfect site for building port facilities and fortifications. On its northern flank, it was bounded by the Gulf of Ityke/Utica, in the east it bordered on the Gulf of Tunes/Tunis and in the south, where it was linked to the mainland by an isthmus, it adjoined the (landlocked lagoon) lake of Tunes. The earliest archaeological finds date from the 8th cent. BC.

The decisive reason for choosing this site for the settlement was probably its situation as a potential staging post for trading with the Iberian Peninsula (Taršiš/Tartessus), but also the fertility of its hinterland and its suitability as a base for setting out into the African interior. Once Carthage had achieved economic and political stability, its far-reaching and intensive trading activities led to a rapid rise in prosperity, as evidenced not least in the setting up of trading posts and colonies.

Carthage was increasingly successful in co-ordinating the different Phoenician expansion activities and – due to the involvement of the Levantine Phoenician cities (e.g. Tyre) in military clashes with Assyrians and Neo-Babylonians –managed step by step to take over as the leading power in the western Mediterranean region from the 7th/6th cents. onward.

In the context of the conflicting expansion drives of Carthage and the Sicilian Greeks starting in the 5th cent., Carthage and its political system came into the mainland Greeks' view. Aristotle discussed its aristocratic and oligarchic constitution at some length in his *Politics,* comparing it with the Spartan system. (Aristot. Pol. 1272b). According to him, Carthage, like the Greek *poleis,* had magistrates, a council and a popular assembly, which elected the higher magistrates. Whereas a 'vice-regent' (*skn*), acting on behalf of the king of Tyre, was in charge in the early days, he was replaced by a 'king' (*mlk*) as emancipation proceeded. The king in turn lost this position at the top to one or two 'judges' (*špṭ(m)*) in the course of the 6th cent.

From 264 BC, the city – which had been prospering until then – was increasingly drawn into conflict with the expanding might of Rome and was razed to the ground by the Romans in 146 BC.

The topography of the city has come down to us in the shape of descriptions, but has also been partially reconstructed by intensive excavations which were begun in 1973. As early as the 8th cent. BC, i.e. very shortly after its founding phase, Carthage was an extensive settlement of urban character, one of the largest cities in the Mediterranean region, covering 50 to 60 hectares (123.5 to 148.2 acres). In archaic times (8th/7th cent. BC), Carthage was laid out on a fixed grid covered by buildings of the *insula* type (tenement blocks), as was revealed by the archaeological findings underneath the Roman *decumanus maximus.* In the west and the north it was bounded by the large *necropoleis* (nos. 1 and 4) which were located on Byrsa hill as well as on the seaward slopes of the hills called Junon, Douimès and Dermech. In this area the city spread radially down the terraced southeastern slope of Byrsa hill, spilling out into the coastal plain.

Probably from the foundation of the city, the Byrsa housed the acropolis and the the main shrine of Ešmūn (no. 8). On the slope, the remains of another important temple (between nos. 9 and 10) along with some smaller 'chapels' were discovered in the midst of residential neighbourhoods. Proof of the existence of smithies and potteries has been found in the low-lying coastal plain. In the 5th cent., i.e. the period of the Magonids, this plot was built over with large villa-type buildings and integrated into the growing settlement (no. 6). Around the same time the monumental sea wall (no. 7) was constructed, hugging the ancient shoreline, which in Roman times receded eastward. It is assumed that in the 5th cent. there was a harbour north of the Sea Gate (no. 6), which was rebuilt toward the city in the 3rd cent.

In the 4th/3rd cents. BC, the city consisted of three districts: the lower city by the harbour, the Acropolis (Byrsa) and the spacious suburbs of Megara and Mapalia in the north. In the 3rd/2nd cents. BC, the south-eastern slope of the Byrsa, formerly a commercial area, was urbanized by creating a grid of *insulae* of standardized tenement buildings (no. 16).

The older port facilities consisted of hithes (landing stages). The harbours that have survived as lagoons (nos. 11–13) only date from the 3rd cent. BC. The rectangular one was used by merchant ships, the circular one was a naval base equipped with docks and wharves. It was laid out around an 'admiralty island' (App. Lib. 96).

West of the ports we find the so-called Tophet (no. 14), a sacred district which had been in use from the 8th cent. BC. By the 4th cent., it covered an area of 6000 m^2 (1.5 acres) of graves containing the cremated ashes of still-born babies, of children who died in early infancy and probably also of sacrificed infants. The area has yielded more than 20,000 urns and several thousand grave stelae bearing useful iconographic information. South-west of it a commercial area has been discovered that was associated with purple dyeing (no. 17).

Carthaginian iconography, life-style and home furnishings were highly developed. From the 5th cent. BC, Carthaginians increasingly modelled themselves on trends set by the Greek Mediterranean Koine, which they enriched by adding (Western) Phoenician elements.

II. Carthage in Roman times (late 2nd cent. AD) (map B)

In 44 BC, the city was refounded by the Romans as Colonia Iulia Concordia Carthago. Before that, Italic colonists had occasionally been allotted Carthaginian land. In 40 or 39 BC, it seems to have been raised to the status of capital of the Africa Nova province. In 29 BC, another 3,000 colonists were settled, giving the city a new impetus. The emperors Hadrian, Antoninus Pius, Marcus Aurelius and Commodus accorded great importance to Carthage, which by the middle of the 2nd cent. AD had become one of the most important centres of the Empire, reaching a peak under the Severan dynasty.

Christianity also found a timely foothold in Carthage with the earliest evidence dating from the 2nd cent. AD (Christian cemeteries). The numerous churches (a-j), however, all of them built in the regular basilica style, were only erected in the wake of the Edict of Milan in 312 AD, when Carthage became the primatial see of the African Diocesis. These churches have been only partially excavated so far.

The Roman city was newly laid out following a regular pattern. The orientation of the street grid, with the *cardo maximus* running appr. from north to south and the *decumanus maximus* extending from west to east, centred on a survey point on the dome of Byrsa Hill.

The bulk of the architectural remains dates from the 1st to the 3rd cents. AD. Among them are villas, frequently decorated with mosaics, cisterns, temple compounds, but also the usual prestige architecture such as the Forum (no. 9), the Odeion (no. 1), the theatre (no. 2), the amphitheatre (no. 10) and a circus (no. 11). The monumental original 'Imperial Forum' (no. 9) was located on the spacious Byrsa plateau, an artificially flattened space created in Augustan times. After a conflagration in the late Antonine Period it was – along with other large structures – restored on top of its Augustan predecessor, now incidentally the site of the largest North-African basilica (the late-19th-cent. Saint Louis Cathedral, now a museum building). The imperial baths of Antoninus Pius (no. 8), situated by the coast, were also built in the same period. The aqueduct (no. 16) was probably a gift from the emperor Hadrian. Traces of the city wall (no. 17) survived until the 12th cent. AD in the south (old wall) and in the north (Theodosian land wall, 4th cent.) – the remaining course of the wall has been reconstructed deductively.

Sources

There is no surviving Phoenician or West-Phoenician narrative of Carthaginian history. Thus all our knowledge stems from contemporary reports and observations by Greek and Roman writers about the culture and society, the politics, army and economy of the city or the so-called Carthaginian Empire, with the authors, especially the Roman ones, unsurprisingly providing outright hostile interpretations of everything (among others: Aristotle, 4th cent. BC; Timaeus, mid 4th to mid 3rd cent. BC; Livy, 59 BC to AD 17; Flavius Josephus, 1st cent. AD; Appianus, 2nd cent. AD; Justinus, around AD 390). This information is supplemented by (Western) Phoenician, Punic and Neo-Punic inscriptions (dedicatory, votive and commemorative inscriptions cut into stone, metal and clay, amphora stamps, clay bullae, stamp seals, ostraka) of a highly formulaic kind, as well as numismatic finds (coin legends) and the interpretations of the material culture. There is a sharp contrast between the views of earlier and present-day researchers: while the former tended to treat Carthage as a special case, not least because of the myths of Carthage's horrendous mercantile wealth as well as its repulsive religious practices, the latter usually view Carthage as a normal ancient-Mediterranean large *polis*.

From 1973 into the 1990s, twelve international project groups representing different branches of classical scholarship from among others Canada, France, Germany, the UK and the USA went to Carthage, a world heritage site since 1979, to contribute to the UNESCO campaign 'Pour sauver Carthage'. They succeeded in establishing that the site of the city's first foundation in the middle of the 8th cent. was indubitably located at the foot of the eastern slope of Byrsa Hill. They were also able to sketch the urban development of the huge Western Phoenician-Punic metropolis from the archaic period to its destruction in 146 BC and they managed to reconstruct the monumental design of the *splendidissima urbs* in the age of imperial Rome and Late Antiquity.

→ Maps pp. 67, 71, 85, 93, 135, 139, 147; Carthage, BNP 2, 2003

Literature

M.G. Amadasi Guzzo, revised ed. of J. Friedrich, W. Röllig, Phönizisch-punische Grammatik, [3]1999; J. Holst et al., Die deutschen Ausgrabungen in Karthago, 3 vols., 1991–97; W. Huss, Geschichte der Karthager, 1985; Id. (ed.), Karthago (Wege der Forschung 654), 1992; S. Moscati, Die Karthager, 1996; H.G. Niemeyer, R.F. Docter, K. Schmidt (eds.), Karthago. Die Ergebnisse der Hamburger Grabung unter dem Decumanus maximus, 2007; W. Röllig, Das Punische im Römischen Reich, in: G. Neumann, J. Untermann (eds.), Die Sprachen im römischen Reich der Kaiserzeit, 1980, 285–299; W. Schumacher, P. Wülfing, Karthago und die Römer, 1998.

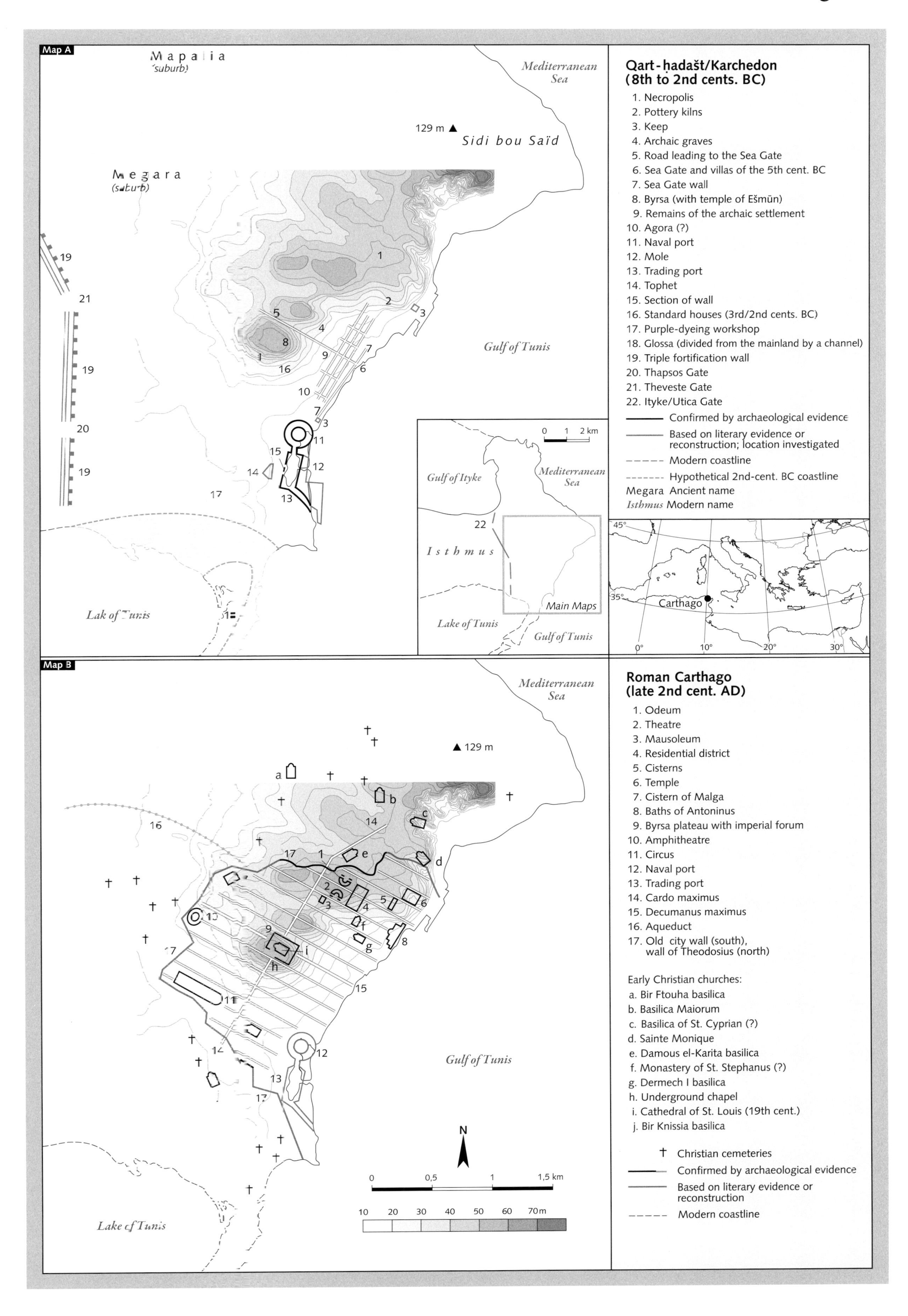
Map A
Mapalia
(suburb)
Megara
(suburb)
Mediterranean Sea
129 m
Sidi bou Saïd
Gulf of Tunis
Lake of Tunis
Qart-ḥadašt/Karchedon (8th to 2nd cents. BC)
1. Necropolis
2. Pottery kilns
3. Keep
4. Archaic graves
5. Road leading to the Sea Gate
6. Sea Gate and villas of the 5th cent. BC
7. Sea Gate wall
8. Byrsa (with temple of Ešmūn)
9. Remains of the archaic settlement
10. Agora (?)
11. Naval port
12. Mole
13. Trading port
14. Tophet
15. Section of wall
16. Standard houses (3rd/2nd cents. BC)
17. Purple-dyeing workshop
18. Glossa (divided from the mainland by a channel)
19. Triple fortification wall
20. Thapsos Gate
21. Theveste Gate
22. Ityke/Utica Gate
Confirmed by archaeological evidence
Based on literary evidence or reconstruction; location investigated
Modern coastline
Hypothetical 2nd-cent. BC coastline
Megara Ancient name
Isthmus Modern name
0 1 2 km
Gulf of Ityke
Mediterranean Sea
Isthmus
Main Maps
Lake of Tunis
Gulf of Tunis
45°
35°
Carthago
0°
10°
20°
30°
Map B
Mediterranean Sea
▲ 129 m
Gulf of Tunis
Lake of Tunis
Roman Carthago (late 2nd cent. AD)
1. Odeum
2. Theatre
3. Mausoleum
4. Residential district
5. Cisterns
6. Temple
7. Cistern of Malga
8. Baths of Antoninus
9. Byrsa plateau with imperial forum
10. Amphitheatre
11. Circus
12. Naval port
13. Trading port
14. Cardo maximus
15. Decumanus maximus
16. Aqueduct
17. Old city wall (south), wall of Theodosius (north)
Early Christian churches:
a. Bir Ftouha basilica
b. Basilica Maiorum
c. Basilica of St. Cyprian (?)
d. Sainte Monique
e. Damous el-Karita basilica
f. Monastery of St. Stephanus (?)
g. Dermech I basilica
h. Underground chapel
i. Cathedral of St. Louis (19th cent.)
j. Bir Knissia basilica
Christian cemeteries
Confirmed by archaeological evidence
Based on literary evidence or reconstruction
Modern coastline
N
0 0,5 1 1,5 km
10 20 30 40 50 60 70 m

The Etruscan core territory: emergence of the Etruscan cities (8th to 7th/6th cents. BC), as well as their heyday and crisis (6th to 4th/3rd cents. BC)

Among the peoples of ancient Italy it was the Etruscans whose highly developed culture attracted the attention of travellers and scholars at an early stage. They were the only Italic people to be actively involved in overseas trade and despite the problematic state of the sources, their history can still be retraced more clearly than that of the other peoples.

The question of their origin and immigration has been the subject of controversy ever since Antiquity (among others Hdt. 1,94; Dion. Hal. Ant. Rom. 1,30). The same is true of their ethnogenesis, which according to recent research seems to have been completed at the beginning of the Villanova Period (*c.* 900/9th cent. BC). They called themselves *Rasna* (Graecized: *Rasenna*), but the following names to identify them have also come down to us: Umbrian *Turskus*, Greek *Tyrrhenoi* or *Tyrsenoi* and Latin *Etrusci*, *Tusci*, rarely also *Lydii* (due to the theory of their immigration from Lydia found in Hdt. 1,94). They settled in western central Italy between the rivers Arnus and Tiber (core territory) and proceeded to colonize the Padana and Campania. Their cultural influence extended to the surrounding areas, especially to the Faliscans, Latium and early Rome. Their material culture, which has been a matter of interest to researchers from the Renaissance onward, encompasses the final stage of the Bronze Age (Proto-Villanova: 11th/10th cents. BC), the Iron Age periods of the Villanova culture (9th/8th cents.), the so-called orientalizing phase (*c.* 800–650 BC) and the Etruscan Period (archaic: *c.* 650 into the 5th cent., classical: from *c.* 500/5th cent. to *c.* 300, and Hellenistic: *c.* 300–100 BC) to the completion of Romanization in the early 1st cent. BC.

Etruria can be divided into several different zones. One of them consists of the southern and nothern coastal areas and their cities with their various crafts and businesses such as trading and the mining and working of metal (Populonia, Vetulonia, Volaterrae, Monti Metalliferi; the island of Ilva/Elba, Monte Argentario, Monte Tolfa) or the production of ceramic ware (Caere, Tarquinii, but also Vetulonia and Populonia). Their links with southern and northern interior Etruria, predominantly agricultural and densely forested, partly downright backward (Villanova phase extending into the 6th cent.) areas, consisted of trails following the valleys of the Umbro, Orcia, Albinia or Marta rivers from east to west, and of navigable rivers – Tiber, Pallia and Clanis forming a north-south axis. Sections of Etruscan roads, e.g. between Pyrgoi/Pyrgi and Caere, have been discovered and identified by archaeologists. Regarding the script, a distinction can be made between the South-Etruscans' alphabet and inscriptions and the North-Etruscan ones.

I. The origins of the Etruscan cities (8th–7th/6th cents. BC)

The majority of the towns that were to develop into the great cities of southern, central and northern Etruria can be traced back seamlessly to Late Bronze Age (so-called Proto-Villanova) or Early Iron Age (Villanova) villages by evaluating the archaeological finds. The subsequent 'orientalizing' phase was characterized by early Levantine/eastern Mediterranean and Greek imports, their imitations and concomitant influences. It also saw the onset of urbanization, when first in the volcanic landscape of the southern coast (7th cent., with the port cities increasingly evolving into economic centres) and shortly afterwards in the hilly north (6th cent.) within a predominantly agrarian area of small settlements an unmistakable trend towards urbanization began to emerge at some central locations, i.e. the budding city states of the 7th and 6th cents. In the coastal settlements, local kings or so-called tyrants (e.g. at Caere) were in charge by dint of their economic power. In the interior, the power structure was characterized by the aristocratic system of clan loyalty.

The bulk of our knowledge, however, is not gleaned from settlements, but from the necropoleis such as the monumental 7th-cent. tombs at Tarquinii and Vetulonia, which offer us clues about the Etruscans' cultural, social (emergence of a gentilitial or clan structure) and economic development (internal and overseas trade).

From the late 8th cent./*c.* 700 BC, the historicity of the Etruscans is proved by their written documents (Veii, Tarquinii), recorded in their own distinctive alphabetic script, which had been inspired by the presence of Greeks and Levantines, especially Phoenicians, on the coast of Campania and southern Etruria (Pithecussae, Cyme, Pyrgi) and went on to evolve into various local and regional variants (model alphabets). More than 10,000 Etruscan inscriptions have been preserved, mostly names and dedications, also glosses. Among them are several lengthy texts, which, however, are practically incomprehensible. There are very few double inscriptions (Phoenician; Latin), while no exactly equivalent bilingual text has surfaced so far. Sadly all Etruscan historiographic and literary writings were recorded on perishable materials and have thus been lost without exception. This means that the bulk of information, especially for the early periods, has come down to us via archaeological finds or had to pass through the filter of later, mostly Greek or Roman textual sources – rather meagre ones at that, because many of those texts have been lost as well.

Foreign traders voyagers were attracted to Etruria from the Late Bronze Age by its abundance of natural resources (metals: iron, copper, tin, zinc, lead; minerals, salt, timber) and agricultural products. It seems, however, that the Etruscans very early made sure that the small number of natural harbours (near Telamon and Populonia) and landfall areas (today's coastline is considerably different due to silting up and the formation of swamps) remained in their hands. They restricted the location of Phoenician and Greek emporia to certain districts of their port cities (Phoenician and Greek shrines at Pyrgi and Graviscae).

Archaeological finds also prove a certain degree of Etruscan cultural influence on neighbouring peoples speaking other languages, especially the Faliscans (around Falerii), the Umbrians and Sabines, the Latins (Rome, Praeneste) and in the north-west – at least for a certain period – the Ligurians. The approximate extent of the Etruscan core territory and their neighbouring Etruscanized spheres of influence ('contact zones') have also been established by archaeological finds. Etruscan expansion drives to the south (Campania, from the 8th cent.?), to the north (Padana, 6th cent.) and to the east (Firmum) – all of them tellingly former Villanova areas – as well as Corsica (especially when Populonia was at the peak of its power from the 6th cent.) were without exception economically motivated.

II. The Etruscan cities in their heyday and period of crisis (6th–4th/3rd cents. BC)

From the 7th cent. onward, increasing production led to the development of a growing class of specialized artisans and small traders and in its wake to a kind of 'democratization' at the expense of the aristocratic elite. The shift of social power in the cities led to a reorientation of political, economic and military interests. It is reflected e.g. in the taking of decisions affecting the whole community: measures like building city walls, an orthogonal street grid, a sewage system, the layout of the necropoleis, were no longer carried out for the benefit of the privileged few but to further the common good by everybody contributing to the communal effort.

From the second half of the 6th cent., there is increasing archaeological evidence of destructions and the desertions of minor towns (e.g. Viterbo), which probably led to a redistribution of territory in favour of the larger cities. There have been attempts to reconstruct the territories of the most important city states, but – in spite of a few local successes – most of this is still largely hypothetical.

As the cities on principle acted autonomously in all political, economic and military matters, it is a contentious issue among scholars if there ever was such a thing as a league of Etruscan city states and, if there was one, if it was of a religious nature (amphictyony, with the 'Fanum Voltumnae' as its federal shrine) and/or the political kind (possibly even fluctuating between the two). One result of the battle of Alalia off the east coast of Corsica in 540 BC (Hdt. 1,166f.), in which a coalition of Etruscan and Western Phoenicians defeated a fleet of Phocaean Greeks, thus decisively changing the balance of power in the western Mediterranean (→ map p. 77), and of the Etruscan defeat at the hands of a Greek force in the battle of Cyme *c.* 524 BC, was an increased Etruscan orientation towards the northern interior (Marzabotto, Padana), central Europe (trading with the Hallstatt and La Tène cultures) as well as direct trading with the Greeks across the Adriatic. The losses of Rome (*c.* 509 BC), Latium (*c.* 504 BC) and later Campania (*c.* 424 BC) also entailed the loss of political and economic power of the South-Etruscan cities of Caere, Tarquinii and Volci. On the other hand Populonia was beginning to flourish. Veii, Falerii, Volsinii, Clusium, the agriculture-based cities in the hinterland along the valley of the navigable Tiber and Clanis rivers also benefited from the changed situation, as did Cortona and Arretium, which had access to the northbound overland routes.

The impact of the continuous Roman conquests in the 4th cent. seems to have contributed to the emergence of a new rural aristocracy in Southern Etruria (Caere, Tarquinii, Volci), which showed a renewed interest in the more thinly populated hinterland (the necropoleis of Tuscana, Blera/Bieda, Castel d'Asso, San Giovenale, San Giuliano, Suana) This was also an age when numerous cities renovated their fortifications or even built new ones, preparing for the wars with Rome that have survived in literary narratives (396 BC conquest of Veii, followed by the fall of Capena, Falerii and Sutri/Sutrium).

→ Maps pp. 31, 65, 67, 77, 83, 93, 111; The Etruscan heartland: settlements and production centres (8th – 2nd cents. BC), BNP 5, 2004, s.v. Etrusci, Etruria II.; Commerce III., BNP 3, 2003; Etruria II., BNP 5, 2004

Literature

L. Aigner-Foresti, Die Etrusker und das frühe Rom, 2003; Ead., P. Siewert (eds.), Entstehung von Staat und Stadt bei den Etruskern, 2006; B. Andreae, H. Spielmann (eds.), Die Etrusker, exhibition catalogue 2004; G. Camporeale, Die Etrusker, 2003 (Italian 2004); M. Cristofani (ed.), Die Etrusker, 2004; Die Etrusker und Europa, exhibition catalogue 1993; F. Falchetti, A. Romualdi, Die Etrusker, 2001; K. Geppert, Studien zu Aufnahme und Umsetzung orientalischer Einflüsse in Etrurien und Mittelitalien vom Ende des 8. bis Anfang des 6. Jahrhunderts v. Chr., (thesis, Tübingen) 2006; S. Haynes, Kulturgeschichte der Etrusker, 2005; F. Prayon, Die Etrusker, [3]2003; Id., Die Etrusker. Jenseitsvorstellungen und Ahnenkult, 2006; Id., W. Röllig (eds.), Der Orient und Etrurien, 2000; P. Siewert, L. Aigner-Foresti, Föderalismus in der griechischen und römischen Antike, 2005; D. Steiner, Jenseitsreise und Unterwelt bei den Etruskern, 2004; K. v. Welck, R. Stupperich (eds.), Italien vor den Römern, 1996.

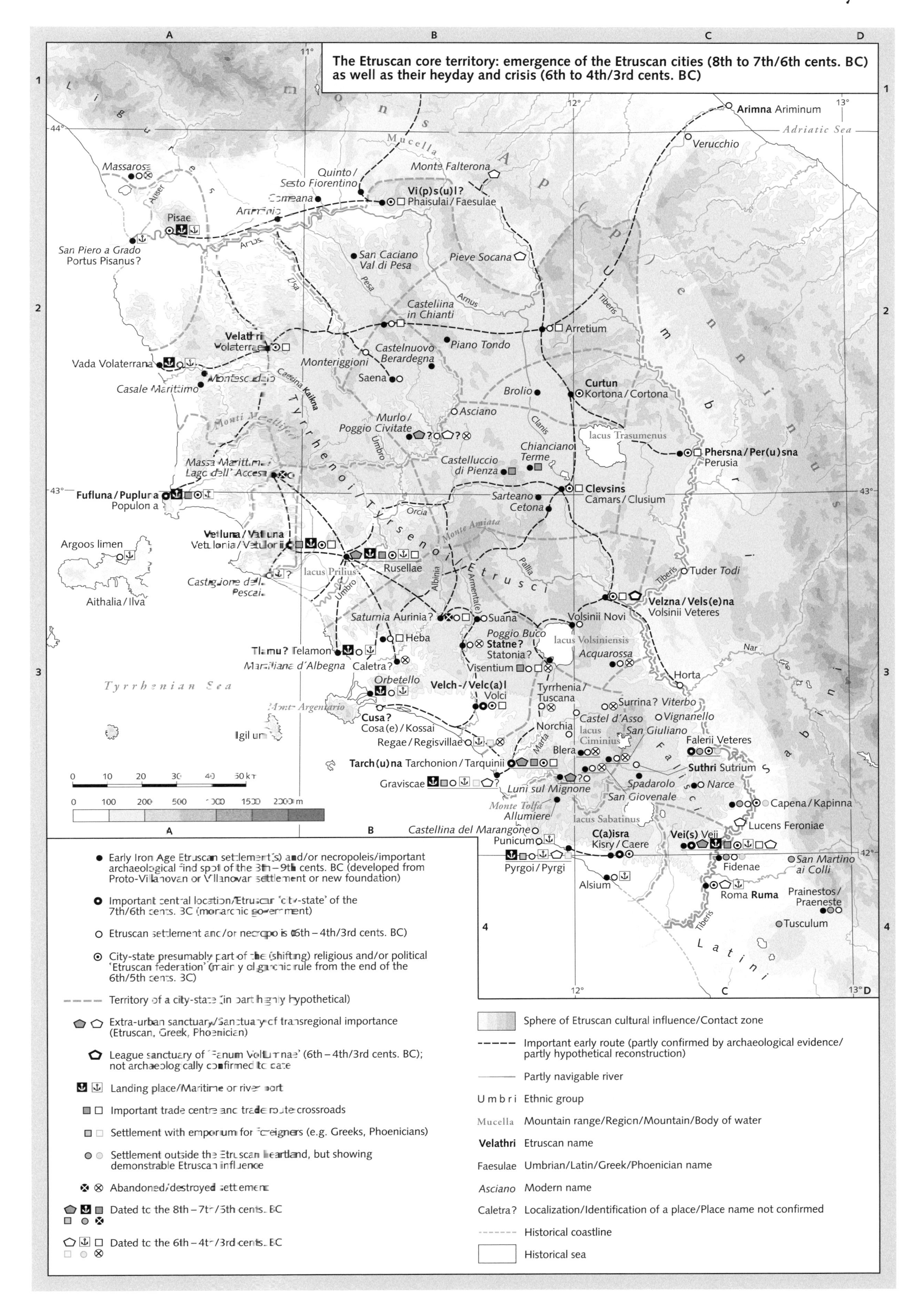
The Etruscan core territory: emergence of the Etruscan cities (8th to 7th/6th cents. BC) as well as their heyday and crisis (6th to 4th/3rd cents. BC)
Arimna Ariminum
Adriatic Sea
Verucchio
Massarosa
Quinto / Sesto Fiorentino
Monte Falterona
Vi(p)s(u)l? Phaisulai / Faesulae
Comeana
Artiminio
Pisae
San Piero a Grado Portus Pisanus?
Arnus
San Caciano Val di Pesa
Pieve Socana
Castellina in Chianti
Arretium
Velathri Volaterrae
Piano Tondo
Castelnuovo Berardegna
Vada Volaterrana
Monteriggioni
Saena
Montescudaio
Casale Marittimo
Curtun Kortona / Cortona
Brolio
Asciano
Murlo / Poggio Civitate
lacus Trasumenus
Phersna / Per(u)sna Perusia
Chianciano Terme
Castelluccio di Pienza
Massa Marittima Lago dell' Accesa
Clevsins Camars / Clusium
Sarteano
Cetona
Fufluna / Pupluna Populonia
Monte Amiata
Argoos limen
Vetluna / Vatluna Vetulonia / Vatluni
Rusellae
lacus Prilius
Castiglione della Pescaia
Aithalia / Ilva
Tuder Todi
Velzna / Vels(e)na Volsinii Veteres
Saturnia Aurinia?
Suana
Volsinii Novi
Heba
Poggio Buco
Statne? Statonia?
lacus Volsiniensis
Tlamu? Telamon
Acquarossa
Marsiliana d'Albegna
Caletra?
Visentium
Horta
Tyrrhenian Sea
Orbetello
Velch- / Velc(a)l Volci
Tyrrhenia / Tuscana
Surrina? Viterbo
Monte Argentario
Cusa? Cosa(e) / Kossai
Castel d'Asso
Vignanello
Igilium
Norchia
lacus Ciminius
San Giuliano
Regae / Regisvillae
Falerii Veteres
Blera
Tarch(u)na Tarchonion / Tarquinii
Suthri Sutrium
Graviscae
Spadarolo
Narce
Luni sul Mignone
San Giovenale
Capena / Kapinna
Monte Tolfa
Allumiere
lacus Sabatinus
Lucens Feroniae
Castellina del Marangone
Punicum
C(a)isra Kisry / Caere
Vei(s) Veii
San Martino ai Colli
Pyrgoi / Pyrgi
Fidenae
Alsium
Roma Ruma
Prainestos / Praeneste
Tusculum
Latini
Etrusci
Umbri
Sabini
Tyrrhenoi / Tyrsenoi
Ligures
Apenninus mons
Tiberis
Clanis
Umbro
Albinia
Pallia
Nar
Marta
Arnus
Pesa
Elsa
Cecina
Orcia
Auser
0 10 20 30 40 50 km
0 100 200 500 1000 1500 2000 m
Early Iron Age Etruscan settlement(s) and/or necropoleis/important archaeological find spot of the 8th – 9th cents. BC (developed from Proto-Villanovan or Villanovan settlement or new foundation)
Important central location/Etruscan 'city-state' of the 7th/6th cents. BC (monarchic government)
Etruscan settlement and/or necropolis (6th – 4th/3rd cents. BC)
City-state presumably part of the (shifting) religious and/or political 'Etruscan federation' (mainly oligarchic rule from the end of the 6th/5th cents. BC)
Territory of a city-state (in part highly hypothetical)
Extra-urban sanctuary/Sanctuary of transregional importance (Etruscan, Greek, Phoenician)
League sanctuary of 'Fanum Voltumnae' (6th – 4th/3rd cents. BC); not archaeologically confirmed to date
Landing place/Maritime or river port
Important trade centre and trade route crossroads
Settlement with emporium for foreigners (e.g. Greeks, Phoenicians)
Settlement outside the Etruscan heartland, but showing demonstrable Etruscan influence
Abandoned/destroyed settlement
Dated to the 8th – 7th/5th cents. BC
Dated to the 6th – 4th/3rd cents. BC
Sphere of Etruscan cultural influence/Contact zone
Important early route (partly confirmed by archaeological evidence/ partly hypothetical reconstruction)
Partly navigable river
Umbri Ethnic group
Mucella Mountain range/Region/Mountain/Body of water
Velathri Etruscan name
Faesulae Umbrian/Latin/Greek/Phoenician name
Asciano Modern name
Caletra? Localization/Identification of a place/Place name not confirmed
Historical coastline
Historical sea

Etruscans, Western Phoenicians and Greeks (6th cent. to *c.* 400 BC)

The conflicts of the Etruscans and West Phoenicians with the Greeks (6th cent. to *c.* 400 BC)

The conflicts of the Etruscans and Western Phoenicians with the Greeks (6th cent. to *c.* 400 BC)

While (Western) Phoenicians, Greeks and Etruscans had arrived at an understanding about their areas of settlement and economic spheres of influence in the western and central Mediterranean in the 7th cent. BC, this balance was increasingly disturbed from the 6th cent. onward. The map shows the central Mediterranean region (Tyrrhenian Sea), i.e. the area of contact between the three most powerful trading nations and the theatre of war for the essential military operations from *c.* 540 BC onward.

I. Causes of the conflicts

One cause of the conflicts was the expansion of the Greek area of settlement and economic activities by setting up colonies on the northern shore of the western Mediterranean. It was there that settlers from Phocaea in eastern Greece founded Massalia around 600 BC. Taking advantage of the abundance of harbours, the fertile hinterland, the closeness of the Rhône/Rhodanus (route to the territories of the Western Hallstatt and LaTène cultures) and the perfect location as a starting point for journeys throughout the Mediterranean, they set up additional trading posts and colonies (Emporium on the Iberian peninsula, Antipolis, Nicaea; Monoecus (Monaco): already mentioned in Hecataeus FGrH 1 F 57) and increased the scope of their trading activities (finds of amphorae and coins). It was there and above all by founding the stage post of Alalia/Aleria in the plain of the Rhotanus on the east coast of Corsica in 565 BC, that the Massaliotes were both encroaching upon the Etruscan sphere of trade and economic interests and moving ominously close to the Western Phoenician/Carthaginian sphere. The issues for all of the western Greek cities were: expansion on the northern shore, preservation of the present settlement areas in southern Italy and on Sicily and the control of their trading routes and areas of economic influence, which were threatened by conflict with the Western Phoenicians from the 5th cent. at the latest. In Sicily (Syracusan kingdom), but also in southern Italy (e.g. Cumae) the threats and conflicts were instrumental in the emergence of *tyrannis* types of government (→ map p. 93).

Another cause of conflict stems from the successive spread of the Sabellian tribes (late 6th. to 4th cents.) and the concomitant changes in Latium, but also in Campania, where (e.g. in 424 BC) clashes occurred between the Etruscans and Campanians on the one hand and Samnites who were advancing towards the coastal areas on the other hand. The beginning expansion of Rome, from where the Tarquinian kings had been expelled around 509, also played a part. In addition, Celts (Latin Galli) had been causing unrest since the 6th cent. by seizing new settlement areas in the Padana – their earliest archaeological traces in northern Italy date from that time – and later launching hostile incursions as far as the Adriatic coast, Etruria and Rome.

II. The course of the conflict

The Etruscan cities and the Western Phoenicians, especially Carthage, were increasingly compelled to defend themselves against the Greeks, but also against the Latins, the rising city of Rome and the Celts. Around 540/535 BC, Carthage and some Etruscan cities (Hdt. 1,166f.), which were possibly led by Caisri/Caere, made an alliance to secure their spheres of influence and the routes to Corsica and Sardinia in the Tyrrhenian Sea. Although the naval battle of Alalia resulted in the Etruscans' and Western Phoenicians' defeat at the hands of the Phocaean Massaliotes, the Phocaeans lost so many ships that they gave up Alalia to resettle at Elea/Velia (Hdt. 1,166f.). The Etruscans subsequently lost control of the northern sea routes to the Massaliotes, which heralded the decline of 'maritime dominance' by the coastal Etruscans. Both the intensified colonization of the Padana by the cities in the interior (Chiusi/Clusium; foundation of Marzabotto in the last quarter of the 6th cent.) and finally the loss of the Campanian territories must be seen as parts of the same chain of events. The Etruscans made up for the loss by opening up new northward trade routes across the Apennine as well as founding settlements such as Marzabotto and Felsina and reaching out into the Padana. Thus they gained access to the Alps and beyond and to the east coast (Atria, Spina and Firmum), where they began to trade directly with Greece without Magna Graecia acting as an intermediary.

This was followed by further clashes between Etruscans and Greeks or Latins, but also with the rising city of Rome (archaeological evidence of 6th-cent. urban structures), which was also involved in military struggles with the Latin League (1st Latin War, battle of Lake Regillus, 488 BC). According to Timaeus (in Dion. Hal. Ant. Rom. 7,3–4) there was a military expedition on land and by sea in 524/23 BC by the Padanian Etruscans ('Tyrrheni of the Ionian Gulf') and their Italic allies (Ombrikoi/Umbri, Daunioi/Dauni and others) to subdue the important city of Cumae, an important trading centre and the northernmost of the Greek colonial towns (in Italy). The Greeks, led by Aristodemus, emerged victorious from the fighting, but the Etruscan defeat did not yet spell the end of Etruscan domination in Campania (there is archaeological evidence of a new flourishing of Etruscan centres like Capua), even though it was a painful blow. It was followed by another one when in 510 BC Sybaris, a close ally of Etruria, was destroyed by neighbouring Croton. The expulsion of the Tarquinian kings from Rome is generally dated at the same period (*c.* 509 BC). It encouraged Lars Porsenna, king of Clevsin/Clusium, to undertake a raid targeting Rome, which he conquered and then handed over to his son Arruns. In this context Rome probably also served as a jump-off base for the conquest of Latium, for Arruns was expected to set up his own kingdom in Aricia. He was, however, beaten in the battle of Aricia (505/04 BC) by a coalition of the Latin cities of Aricia, Antium and Tusculum (cf. Dion. Hal. Ant. Rom. 36,1f.) and their Greek allies from Cumae under the command of Aristodemus. According to Liv. 2,14,5 the war was fought to stop the Etruscan domination of Latium and to interrupt the Etruscan land route to Campania (Liv. 2,9,1–14,9, Dion. Hal. Ant. Rom. 5,21,1–34,5 and others give a slightly different version.) The occurrence of additional military conflicts in that period, probably within Etruria as well, can be deduced archaeologically from the numerous destruction horizons and the desertion of settlements (→ map p. 75).

The growing strength of the Greek cities in Sicily, especially Syracuse, which was governed by *tyrannoi* from 485 onward, brought a new phase in the conflicts between the Etruscans, Western Phoenicians and Greeks.

To the Western Phoenicians/Carthaginians, Sicily was a strategic and economic key area, which they defended again and again by military means against the expansionism of the Sicilian Greeks. Around the middle of the 6th cent., this resulted in the war against Selinus, towards the end of the cent. it meant war against Heraclea at the foot of Mount Eryx. As the Carthaginians wanted to avoid Roman interference in their Sicilian affairs, they concluded their first treaty with Rome in 509 BC (Pol. 3,22–25). Almost simultaneously with Xerxes' offensive in the Aegean and mainland Greece, they tried to gain further ground in Sicily, making the most of a call for help by the tyrant of Himera. Gelon I, the Greek tyrant of Gela, (491/90–478), who at the beginning of his reign already was the ruler of practically the whole of eastern Sicily, managed to seize Syracuse in 485 and continued to pursue the expansive policy of his predecessors. Together with his father-in-law, Theron of Acragas (489–474/73), he succesfully waged war from 485 onward against the Carthaginians in the west, who, possibly together with Etruscan allies, were defeated by the Greeks in the naval battle of Himera in 480 BC (cf. Hdt. 7,165f.). Gelon's brother Hieron I, tyrant of Gela from 485 onward, afterwards tyrant of Syracuse from 478–466, maintaining this expansionist course, intervened in southern Italy (Locri, Sybaris) and in 474 BC in the naval battle of Cumae fought alongside Cumae against the coalition of Etruscan cities and Western Phoenicians. Pithecussae became a Syracusan base at this point.

To the Etruscans, their defeat at Cumae not only meant the loss of naval supremacy, they also lost most of Campania and of their control over the land route to that region, which weakened their influence on Rome and Latium. In 453, Syracuse succeeded in devastating Etruscan Corsica, occupying Ilva/Elba for a short period and striking a successful blow against Etruscan pirates in the North-Tyrrhenian coastal area. It also controlled the sea route to Greece. The Etruscans one last time contributed three ships to fight this enemy (Thuc. 6,88; 7,53–57) during the Athenians' 'Sicilian expedition' of 413 BC. When in 384 Dionysius II of Syracuse attacked and sacked Pyrgi, the port of Caere, while he was on his way to Corsica, the former maritime power Caere and the neighbouring cities were unable to prevent this raid.

From the late 6th cent., but mainly in the 5th and as late as the 4th cent., all coastal settlements were increasingly forced to defend themselves against Italic peoples. These Sabellian tribes from various Apennine areas advanced into the fertile coastal regions in several waves. It is suspected that there was a military coalition of Samnites in the year 438 (?). Around 424, Capua was seized by Samnites, while around 400 Cumae, Pompeii and Posidonia were taken by Lucanians. For the Etruscans the defeat of Capua brought with it the definitive loss of Campania. From now on they had to come to terms with Rome, both peacefully by means of agreements and by using military force, as exemplified by the struggle for Veii, which originated in the mid-420s. Around 406 BC, Rome laid a siege to the city which was to last ten years. It was finally conquered and destroyed in 396 and its territory incorporated into the *ager Romanus*. Around 400 BC, the Etruscans also lost the Padana to the Celts, after temporarily making tribute payments to them (Pol. 1,6; 2,17–18; Diod. Sic. 14,113–117).

The Western Phoenicians, however, succeeded in retaining their sphere of power: they managed to keep Sardinia, made several attempts to wrest Sicily away from the Greeks by warlike means, and struck treaties with Rome, while increasingly turning their attention to the Iberian Peninsula.

Sources

Occasional mentions and references in Greek and Latin authors about the factual and cultural history of the Etruscans are only sparsely supplemented by Etruscan inscriptions, although they are a little more informative about the Etruscan religion. What we know we owe mostly to archaeological investigations. The amount of scholarly literature on the subject, starting from the Renaissance, is hardly manageable. The state of the sources about the history of the Western Phoenicians/Carthage is not much better; again, no indigenous historiography is available (→ map p. 71 on research). The history of the Western Greeks was usually treated within the context of colonization, with a comparative abundance of material about the tyrants of Sicily.

→ Maps pp. 65, 67, 69, 71, 75, 81, 93, 109

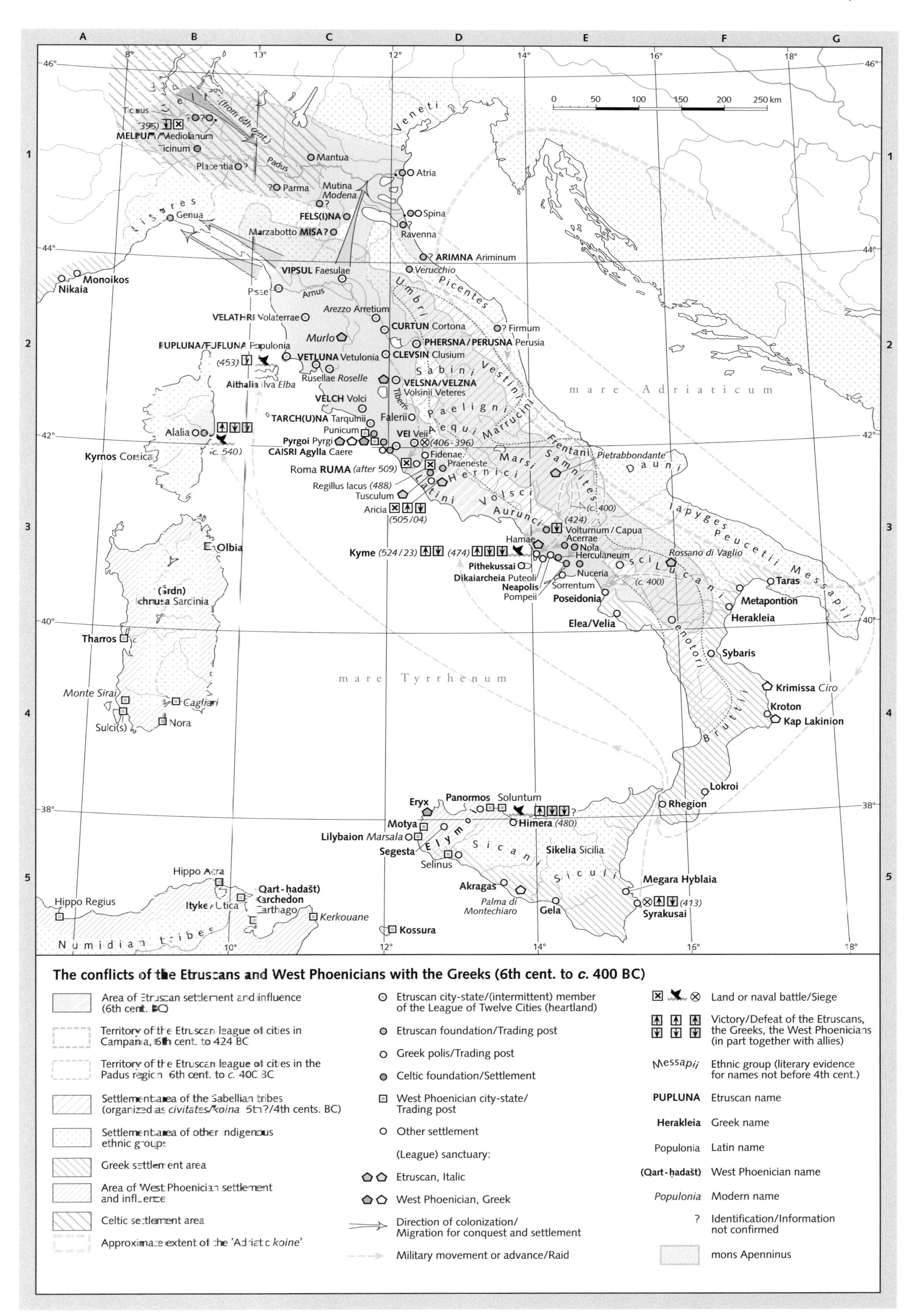
A
B
C
D
E
F
G
1
2
3
4
5
0
50
100
150
200
250 km
mare Adriaticum
mare Tyrrhenum
Veneti
Picentes
Umbri
Sabini
Vestini
Paeligni
Aequi
Marrucini
Frentani
Marsi
Samnites
Hernici
Volsci
Latini
Aurunci
Dauni
Iapyges
Peucetii
Messapii
Osci
Lucani
Oenotori
Brutti
Elymoi
Sicani
Siculi
Numidian tribes
Padus
Arnus
Tiberis
MELPUM/Mediolanum
Ticinum
Placentia
Parma
Mantua
Mutina Modena
FELS(I)NA
Marzabotto MISA ?
Atria
Spina
Ravenna
ARIMNA Ariminum
Verucchio
Genua
Monoikos
Nikaia
VIPSUL Faesulae
Pisae
Arezzo Arretium
VELATHRI Volaterrae
CURTUN Cortona
Firmum
Murlo
PHERSNA/PERUSNA Perusia
CLEVSIN Clusium
VETLUNA Vetulonia
Rusellae Roselle
VELSNA/VELZNA Volsinii Veteres
Aithalia Ilva Elba
VELCH Volci
TARCH(U)NA Tarquinii
Falerii
Punicum
Pyrgoi Pyrgi
CAISRI Agylla Caere
VEI Veii
(406-396)
Fidenae
Praeneste
Roma RUMA (after 509)
Regillus lacus (488)
Tusculum
Aricia
(505/04)
Alalia
(c. 540)
Kyrnos Corsica
Pietrabbondante
(c. 400)
(424)
Volturnum/Capua
Acerrae
Nola
Herculaneum
Hamae
Kyme (524/23)
(474)
Pithekussai
Dikaiarcheia Puteoli
Neapolis
Pompeii
Sorrentum
Nuceria
Poseidonia
Elea/Velia
Rossano di Vaglio
(c. 400)
Taras
Metapontion
Herakleia
Sybaris
Krimissa Ciro
Kroton
Kap Lakinion
Lokroi
Rhegion
Olbia
Tharros
Monte Sirai
Cagliari
Nora
Sulci(s)
Eryx
Panormos
Soluntum
Himera (480)
Motya
Lilybaion Marsala
Segesta
Selinus
Sikelia Sicilia
Akragas
Palma di Montechiaro
Gela
Megara Hyblaia
Syrakusai
(413)
Hippo Acra
Hippo Regius
Qart-hadašt)
Karchedon
Carthago
Ityke/Utica
Kerkouane
Kossura
The conflicts of the Etruscans and West Phoenicians with the Greeks (6th cent. to c. 400 BC)
Area of Etruscan settlement and influence (6th cent. BC)
Territory of the Etruscan league of cities in Campania, 6th cent. to 424 BC
Territory of the Etruscan league of cities in the Padus region 6th cent. to c. 400 BC
Settlement area of the Sabellian tribes (organized as civitates/koina 5th?/4th cents. BC)
Settlement area of other indigenous ethnic groups
Greek settlement area
Area of West Phoenician settlement and influence
Celtic settlement area
Approximate extent of the 'Adriatic koine'
Etruscan city-state/(intermittent) member of the League of Twelve Cities (heartland)
Etruscan foundation/Trading post
Greek polis/Trading post
Celtic foundation/Settlement
West Phoenician city-state/ Trading post
Other settlement
(League) sanctuary:
Etruscan, Italic
West Phoenician, Greek
Direction of colonization/ Migration for conquest and settlement
Military movement or advance/Raid
Land or naval battle/Siege
Victory/Defeat of the Etruscans, the Greeks, the West Phoenicians (in part together with allies)
Messapii
Ethnic group (literary evidence for names not before 4th cent.)
PUPLUNA
Etruscan name
Herakleia
Greek name
Populonia
Latin name
(Qart-hadašt)
West Phoenician name
Populonia
Modern name
?
Identification/Information not confirmed
mons Apenninus

The Hallstatt Culture (*c.* 800-450 BC)

At the beginning of the first millennium, vast parts of Europe were characterized by the Late Bronze Age urnfield culture. From it sprang the cultures of the Iron Age, as reflected in the chronological terminology. The Early Iron Age culture, which left a formative imprint on central Europe, is generally divided into two time horizons. Stages Ha A and B, which were still Late Bronze Age, were followed by the Early (stage Ha C; 8th to 7th cents. BC) and the Late Hallstatt cultures (Ha D; 7th to 5th cents.). The use of iron, which from the 8th cent. onward spread westward from the eastern Mediterranean, brought about essential economic changes, dissolved the traditional systems of goods distribution and social organization, and helped create new local elites. In the Early Hallstatt culture the custom of cremation was retained, but unlike in the Late Bronze Age, the dead were now buried in sometimes huge tumuli. A widespread common characteristic was the new elites' need for conspicuous displays of status and prestige. Throughout the Hallstatt region, some of the towns developed a central function in the settlements structure. On the basis of archaeological findings, the culture can be subdivided into a western and an eastern Hallstatt circle. These eastern and western types come together in the eponymous grave field of Hallstatt in Upper Austria. The problem of distinguishing the eastern and western Hallstatt circles is particularly common in a wide belt of mixed culture which at times would shift from east to west and vice versa.

I. The western Hallstatt circle

From the 8th/7th cents. onward, the western Hallstatt circle extended from eastern France across northern Switzerland and as far as Bohemia and Upper Austria. The predominant funeral type was the burial of cremated bodies in sometimes vast fields of tumuli. From the late 7th cent. (Ha D), a profound and diversified structural change came about. Late Bronze Age traditions receded as southern influences began to dominate. In funerary cults, burial of the body replaced cremation, and metal objects were favoured over ceramic ware as grave goods. At the same time certain towns developed into centres of supraregional importance, the so-called princely residences. In the region between eastern France, northern Switzerland and the river Main these fortified hilltop castles were frequently built at sites located at geographically crucial points, especially close to rivers, which served as the main highways of traffic (→ map p. 83). Among the best-known residences are the Heuneburg on the Upper Danube, Mont-Lassois, the Münsterberg at Breisach, the Marienberg at Würzburg and Závist south of Prague. There was a conspicuous concentration of ceremonial 'princely tombs' in the vicinity of the central settlements. The hallmark objects of the local elites were status symbols like golden torques, four-wheeled chariots or bronze kitchen and table ware imported from the south. The 'princely residences' flourished well into the 5th cent., but collapsed without warning within the course of a generation. In the eastern regions of the western Hallstatt circle, i.e. southeastern Bavaria, particularly in the area between the valley of the Danube and the Alpine foothills, numerous small(ish) moated and fortified 'manorial halls' sprang up, which can be interpreted as the seats of a gentry class. It was a remarkably long-lived phenomenon, lasting from the 8th cent. BC to *c.* 400 BC. The map demonstrates that the distribution of 'princely residences' and 'manor houses' was more or less mutually exclusive.

From about 500 BC onward, Greek authors (Hecataeus, Herodotus) reported about the 'Keltoi' – a collective term denoting a multitude of tribal communities living north of the Greek colonies in southern France and along the upper course of the Danube. Today – in spite of the sparse content of these written sources – it is commonly assumed that the Celtic tribes originated in the western areas of the Hallstatt culture. There is, however, a lively debate debate going on about the exact point in time, from which the prehistoric cultures of central Europe can be justly termed 'Celtic'. Some scholars reserve the first use of the term Celts for the members of the Late Iron Age La Tène culture from 450 BC onward.

II. The eastern Hallstatt circle

The eastern Hallstatt circle evolved from local Late Bronze Age roots in the 8th cent. Its spread can be traced by finds of the typical bull's head pottery. A number of its centres were situated on the very edge of the southeastern Alpine region. Sites yielding abundant finds and vast fortified castles were particularly frequent between Ljubljana and Novo Mesto in the hilly territory flanking the Sava river. A hallmark of the eastern Alpine Hallstatt circle was constituted by the figuratively decorated 'situla art' bronze vessels, inspired by North-Italic models (Etruria, Este culture).

On the whole the eastern Hallstatt circle was a conglomerate of heterogeneous groups, whose funerary customs and archaeological finds were fundamentally different in some cases. What they did share, however, was the custom of burying an elite group of people in splendid 'princely tombs' covered by big tumuli. The most elaborate examples of this phenomenon were found among the Sulmtal group in Styria, in some parts of Slovenia and in northeastern Croatia. Its most outstanding site, which was probably a junction controlling important traffic routes, is Kleinklein to the southwest of Graz. Closely related to it are the Kaptol group to the south and the Transdanubian group. The Kalenderberg group is a special case. Its funerary goods differ from those of the other groups in that they consist mainly of ceramic ware. The Lower Carniola Hallstatt group was distinguished by a number of sprawling hilltop settlements and their adjoining necropoleis. In the 8th cent. their dead were still cremated, as was the general custom, but then there was a switch to body burial that was unique for this group. The south-western border of the eastern Hallstatt circle was formed by the Frög group in Carinthia, while the southernmost groups were the Histrian and the Liburnian groups.

While in the western areas (i.e. Carinthia, Lower Carniola and Slovenia) eastern Hallstatt traits persisted into the 5th cent. BC, they disappeared everywhere else around the middle of the 6th cent. 'Princely tombs' burials were discontinued, the hilltop settlements were deserted. It may have been due to Scythian incursions that the Hallstatt culture came to an end in this area.

III. The Hunsrück-Eifel culture

The regionally diversified Hunsrück-Eifel culture on the northern periphery of the western Hallstatt circle evolved continuously from Late Bronze Age traditions. Its types of ceramic ware and decorations, jewelry and apparel differ substantially from those of its southward Hallstatt neighbours. The Neuwied Basin and the Hunsrück range were the most densely populated areas. One of the most striking features was the custom of building tumulus graves. In the Early Hunsrück-Eifel culture (*c.* 600–475 BC) the funeral rites had already changed from cremation to body burial. Its economy was based on the production of iron and an in part intensive exchange of goods with the Mediterranean region. From the 5th cent. onward, the Late Hunsrück-Eifel culture (475–250 BC) played a role in the development of the La Tène culture.

IV. The Laugen-Melaun group

It was due to the influence of the North-Tyrolean urnfield culture that the Inner-Alpine Laugen-Melaun group, named after two sites in the Adige (Etsch) valley, came into being in the 12th cent. BC. The predominant funerary custom was the use of urn graves, a surviving Late Bronze Age tradition. From the differing types of pottery a tripartite chronology could be derived (stages A–C). Whereas a North-Alpine influence is in evidence at the beginning (12th to 10th cents.), e.g. in the type of bronze vessels, the Laugen-Melaun B phase (10th to 9th cents. BC) pottery and bronze ware is characterized by an unmistakable orientation towards the Este culture and the East-Alpine region. The final phase of the group (stage C) coincides with the Hallstatt culture (7th–5th cent. BC). Its spread – with its southernmost offshoots bordering on the Este culture region – covers an area whose inhabitants were summarily called 'Raeti' in Roman times.

V. The Golasecca culture

Between the 11th and the 4th cents., there existed an archaeologically clearly defineable zone called the Golasecca culture, after a grave field near the exit of the Ticinus river from the Lago Maggiore. The principal impulses for its emergence can be traced to the groups of Late Bronze Age cultures located northwest of the Alpine main ridge. On the basis of archaeological findings the Golasecca culture can be divided into a northern and a southern circle of forms, but the cremation rites were shared within the whole area. From the mid 6th cent. BC, it received important impulses from Etruria (alphabet, art) as well as from the Este territory and the eastern Alpine region, but thanks to its sparsely populated border areas (very few finds) it seems to have retained a high degree of independence. Being part of a supraregional web of relationships, the Golasecca culture played an important role in the transalpine exchange between the cultures of northern Italy and the western Hallstatt circle. Although comparative linguists by studying inscriptions from the 6th and 5th cents. have succeeded in establishing links with the 'Lepontic' dialect, a subspecies of the Celtic language family, it is impossible to determine the ethnicity of the Golasecca culture.

VI. The Este culture

In the course of the 10th cent. BC, a cultural group emerged which can be neatly defined archaeologically. It is called the Palaeo-Venetic or Este culture – after the site of the richest finds – and somewhat surprisingly remained intact until the 3rd cent. BC. In spite of its closed character, the Este culture has clear links with Etruria and the Golasecca culture, especially from the 6th cent. onward. It also exerted a significant influence on the extended area around the northern Adriatic and the south-eastern Alpine Hallstatt regions, by dint of its typical bronze vessels with their figurative decorations, also known as 'situla art'.

→ Maps pp. 31, 65, 67, 81, 83

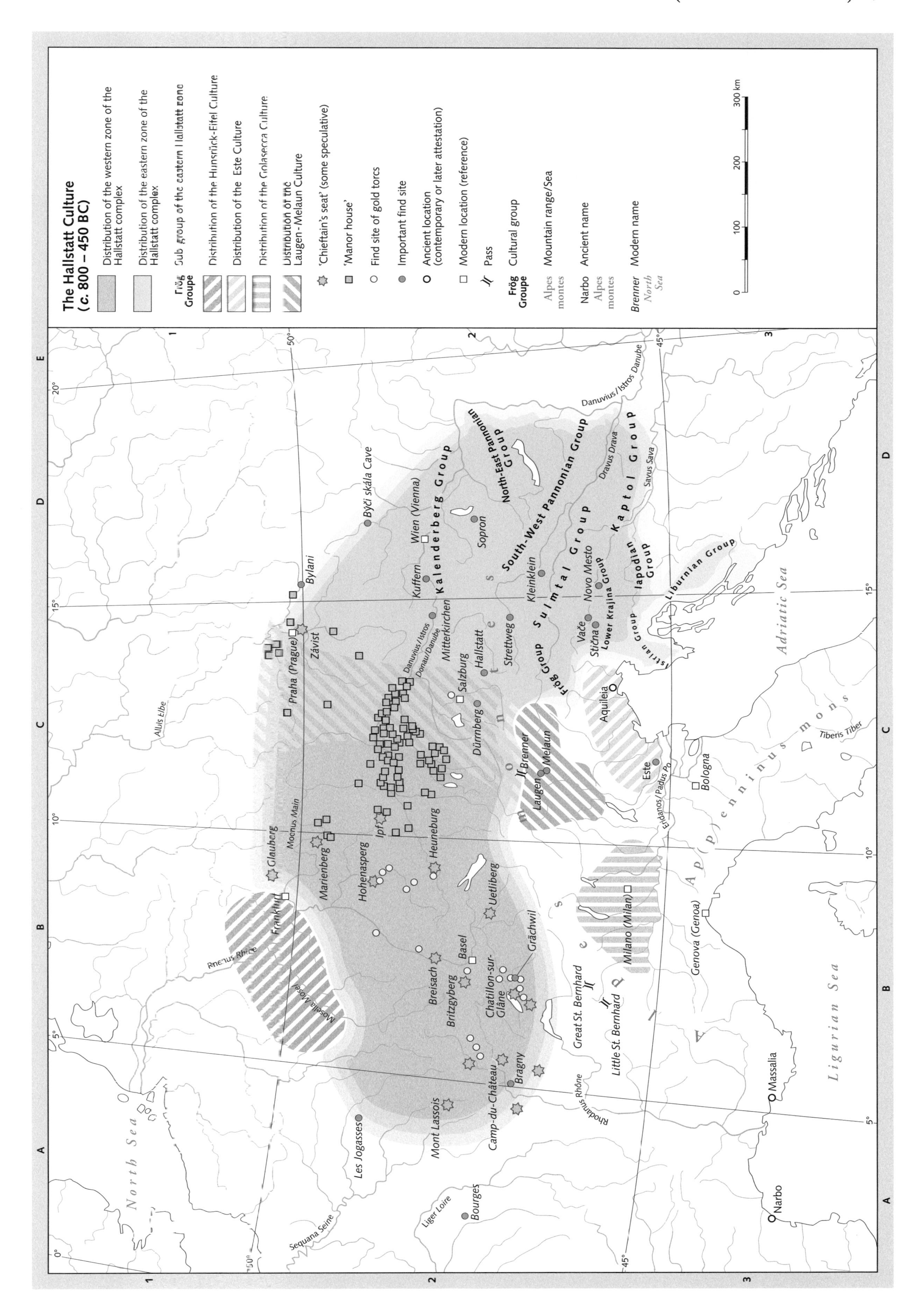
The Hallstatt Culture
(c. 800 – 450 BC)
Distribution of the western zone of the Hallstatt complex
Distribution of the eastern zone of the Hallstatt complex
Frög Groupe
Sub group of the eastern Hallstatt zone
Distribution of the Hunsrück-Eifel Culture
Distribution of the Este Culture
Distribution of the Golasecca Culture
Distribution of the Laugen-Melaun Culture
'Chieftain's seat' (some speculative)
'Manor house'
Find site of gold torcs
Important find site
Ancient location (contemporary or later attestation)
Modern location (reference)
Pass
Cultural group
Alpes montes
Mountain range/Sea
Narbo
Ancient name
Brenner
North Sea
Modern name
0
100
200
300 km
North Sea
Albis Elbe
Rhenus Rhine
Mosella Mosel
Moenus Main
Sequana Seine
Liger Loire
Rhodanus Rhône
Danuvius/Istros Donau/Danube
Danuvius/Istros Danube
Dravus Drava
Savus Sava
Eridanos/Padus Po
Tiberis Tiber
Bourges
Les Jogasses
Mont Lassois
Camp-du-Château
Bragny
Britzgyberg
Breisach
Basel
Châtillon-sur-Glâne
Grächwil
Uetliberg
Heuneburg
Hohenasperg
Marienberg
Glauberg
Frankfurt
Ipf
Praha (Prague)
Závist
Bylani
Býčí skála Cave
Wien (Vienna)
Kuffern
Kalenderberg Group
Mitterkirchen
Salzburg
Hallstatt
Dürrnberg
Strettweg
Kleinklein
Sopron
North-East Pannonian Group
South-West Pannonian Group
Sulmtal Group
Frög Group
Kaptol Group
Novo Mesto
Vače
Stična
Lower Krajina Group
Iapodian Group
Liburnian Group
Istrian Group
Brenner
Laugen
Melaun
Aquileia
Este
Bologna
Milano (Milan)
Genova (Genoa)
Great St. Bernhard
Little St. Bernhard
Alpes
Appenninus mons
Adriatic Sea
Ligurian Sea
Massalia
Narbo

The La Tène Culture (*c.* 450 BC – *c.* AD 1)

In the 5th cent. BC, the cultural structure of the Hallstatt period in eastern France and southwest Germany underwent a fundamental change. The majority of the Hallstatt settlement centres, the so-called princely residences, lost their status, being either destroyed or deserted, while formerly subordinate areas north of the previous centres suddenly rose to prominence: the regions along the Marne and the middle reaches of the Rhine, along with Bohemia. Splendid warrior graves, frequently containing two-wheeled chariots as well as imported goods from the south, were tell-tale signs of this change. At the same time there appeared a radically new style of art featuring life-like plants and circular ornaments, an independent development of Mediterranean impulses. This was the central trait of the Later Iron Age 'La Tène culture', named after an important site of finds in western Switzerland. It can be subdivided into four material and chronological groups, LT A to D, according to the prevalent types of fibulae, ring jewelry and weapons. Stages LT A and B constitute the Early La Tène Period (*c.* 450-250 BC), LT C represents the middle stage (*c.* 250–150 BC) and LT D the Late Period (*c.* 150 BC to the beginning of the CE). Over time, the archaeological spread of the La Tène culture encompassed large parts of western, central as well as south-eastern Europe and Northern Italy. In a switch from Hallstatt funerary customs, the dead were increasingly interred in fields of flat graves instead of tumuli. The weapons in men's graves, esp. lances and swords, indicate the presence of a class of warriors. The women's graves predominantly feature collections of sometimes lavish jewelry. The custom of cremation began to appear in the 4th cent. to become predominant in the 3rd cent. and standard practice in the Late La Tène Period.

The most important sources on the La Tène culture are archaeological finds in the ground, but attempts to harmonize them satisfactorily with the accounts of ancient authors (Herodotus, Posidonius, Strabo, Caesar, Livy and Pliny) and the results of linguistics have only met with partial success.

Due to the ancient Graeco-Roman ethnographers, who used the terms 'Keltoi' or 'Galli' to describe the contemporary transalpine population, the La Tène culture is still viewed as the archaeological heritage left by the 'Celts'. These names, however, simply served as collective terms covering a multitude of diverse tribal formations, who did not share a sense of common identity. In the eastern Mediterranean region the appellation 'Galatai' was predominant. Whereas the westward and south-westward expansion of La Tène elements from the 5th cent. BC onwards is described as 'the prehistoric expansion' of the Celts, the expansion of the La Tène culture to Italy, to the Balkans and down the Danube is linked with the 'Celtic migrations', which were recorded in writing. They were caused both by the dramatic social and economic changes at the end of the Hallstatt period and a considerable deterioration of the climate around the year 400 BC.

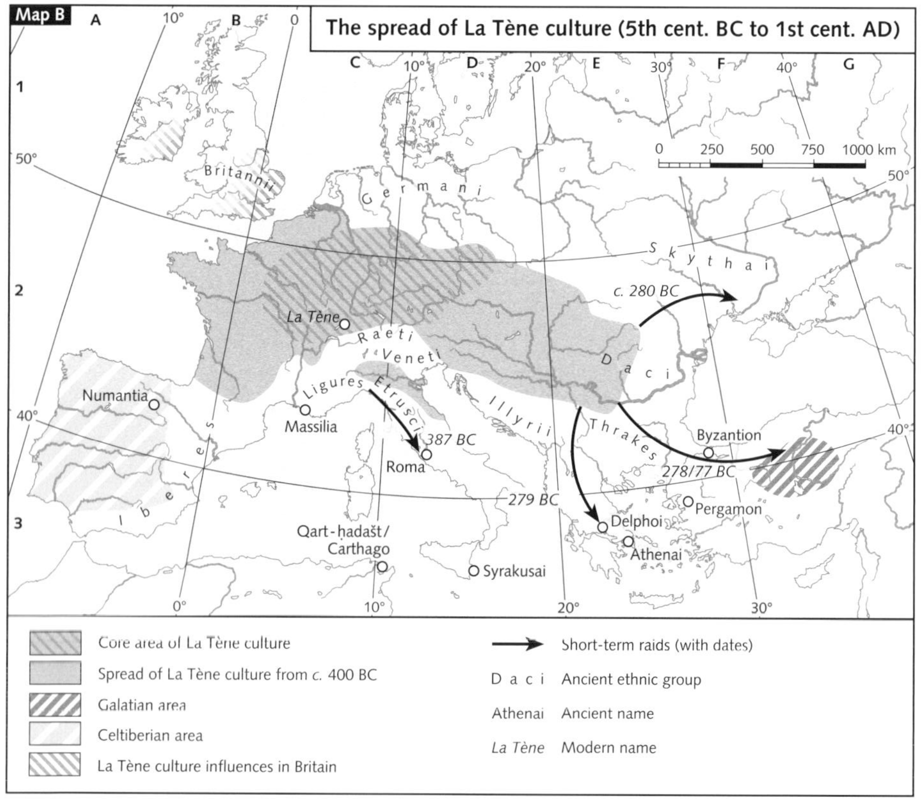

Map B — The spread of La Tène culture (5th cent. BC to 1st cent. AD)

Core area of La Tène culture
Spread of La Tène culture from c. 400 BC
Galatian area
Celtiberian area
La Tène culture influences in Britain
Short-term raids (with dates)
D a c i — Ancient ethnic group
Athenai — Ancient name
La Tène — Modern name

I. The southward expansion

The immigration of Celtic tribes into northern Italy can be verified from written records as well as archaeological finds from *c.* 400 BC. The Insubres seized Melpum (Mediolanum/Milan) and the western Padana, while the Boi(i) took the Etruscan city of Felsina (Bononia/Bologna) along with the area stretching from the river Padus/Po to the Apennines, with the Cenomani settling in the territory between Lake Garda (Benacus lacus) and the Po. The Lingones took possession of the region south of the Po estuary as far as Spina, whereas the Senones occupied the *ager Gallicus* down to Ancona and the Picenum. Judging from the finds in the ground they also took possession of previously unoccupied land. A typical indicator of Celtic conquest is formed by the grave fields of the 4th and 3rd cents. BC, part of whose gifts are very closely related to transalpine finds from the Marne region or Bohemia. This influence is particularly striking in the warrior graves with specific types of helmets and swords or women's graves with corresponding ring jewelry. There was a marked difference between the ways the immigrants dealt with the Italic inhabitants. Whereas, e.g., a distinct La Tène culture evolved in the region encompassing Verona, Brixia/Brescia and Mantua, a culture of mixed Celtic and Etruscan elements developed in the settlement area of the Boi(i). The Lingones left even fewer archaeological traces. Short-term incursions took Celtic tribes like the Senones even further south: thus Rome was sacked in 387 BC, but the Romans in their turn advanced to northern Italy from the 3rd cent. BC onward (battles of Sentinum, Telamon, Clastidium). At the beginning of the 2nd cent. BC, the Roman conquest of northern Italy had been completed. This was followed by a Celtic return migration to the north and into the Danube region, an event that had considerable impact on the Late La Tène culture in those areas.

II. The expansion toward the southeast

On the basis of archaeological finds it can be safely assumed that around the beginning of the 4th cent. BC there were migrations from the core area of the La Tène culture into the north-western Carpathian Basin and parts of Bohemia and Moravia. The Hungarian lowlands seem to have been the target of another Celtic advance taking place around the midddle of the 4th cent. BC, as evidenced by grave fields whose structure as well as material finds correspond to equivalents in the western La Tène region. In the course of the 4th/3rd cents., new population groups from the west immigrated into the Carpathian Basin in successive waves (finds of disk torques, types of pottery). There was a noticeable blending with local cultural elements, which accounts for the formation of a specific set of characteristis in the eastern Celtic La Tène culture. In the course of the 3rd cent., it encompassed the whole Carpathian Basin.

Predatory incursions which cannot be traced by archaeological means, but have been recorded in history, culminated in an attempt to seize the shrine of the Delphic Apollo in 279 BC. At around the same time, Celtic tribes advanced into the region north of the Black Sea. Others crossed the Hellespont (Dardanelles) in 278/277 BC. These so-called Galatai were then settled in central Anatolia (near today's Ankara) by the Ptolemies to serve as mercenaries (→ map p. 124).

There is no proof of the presence of Celtic grave fields after the end of the 2nd cent. BC, while typically Dacian materials are found in settlements dating from the same time. In the 1st cent. BC, the Dacian expansion into the Carpathian Basin was gathering strength (Kingdom of Burebista). As a consequence, the Celtic Scordici, Boii and Taurisci suffered defeats between 60 and 44 BC.

III. The *oppida* civilization

From the 3rd/2nd cents. BC, the reflux of Celtic tribes (e.g. the Boii) from Italy created an economic and social system in western and central Europe as well as the Danubian region which can be justly termed 'early urban'. Its specific features were vast, centrally situated fortifications, which Caesar called *oppida*, coinage and the use of writing. In spite of certain regional differences (types of fortification, size of settlements) the Late La Tène *oppida* civilization ranging from western Europe to Hungary displayed considerable similarities. Whereas most settlements in southern Germany had already been deserted by the middle of the 1st cent. BC, the French ones in many cases turned into the nuclei of Gallo-Roman town development.

IV. Regions influenced by the La Tène culture

Based on the study of the names of water bodies and places, linguists have reconstructed a presence of 'Celtic' populations both in parts of the Iberian Peninsula and the British Isles, which is not fully borne out by the findings of archaeology. Although influences from the La Tène areas were at work in Spain and Portugal, especially in the working of metal objects, they were not sufficiently pervasive to justify the inclusion of the 'Celtiberian culture' in the La Tène group. A similar point can be made about the British Isles. From the 4th cent. BC, especially the style of the ornaments on weapons and metal jewelry betrayed impulses from the La Tène culture. In the South-West of England, burials were also discovered which are reminiscent of sites in the Marne region (Arras culture). According to recent British research, however, these influences cannot be regarded as evidence of immigration from the central European La Tène group. The first case to be corroborated by archaeological findings was the immigration of Belgian tribes in the first cent. BC, which was also reported by Caesar (Aylesford culture, cremation burials, coinage).

→ Maps pp. 63, 79, 83, 85, 124, 165, 167

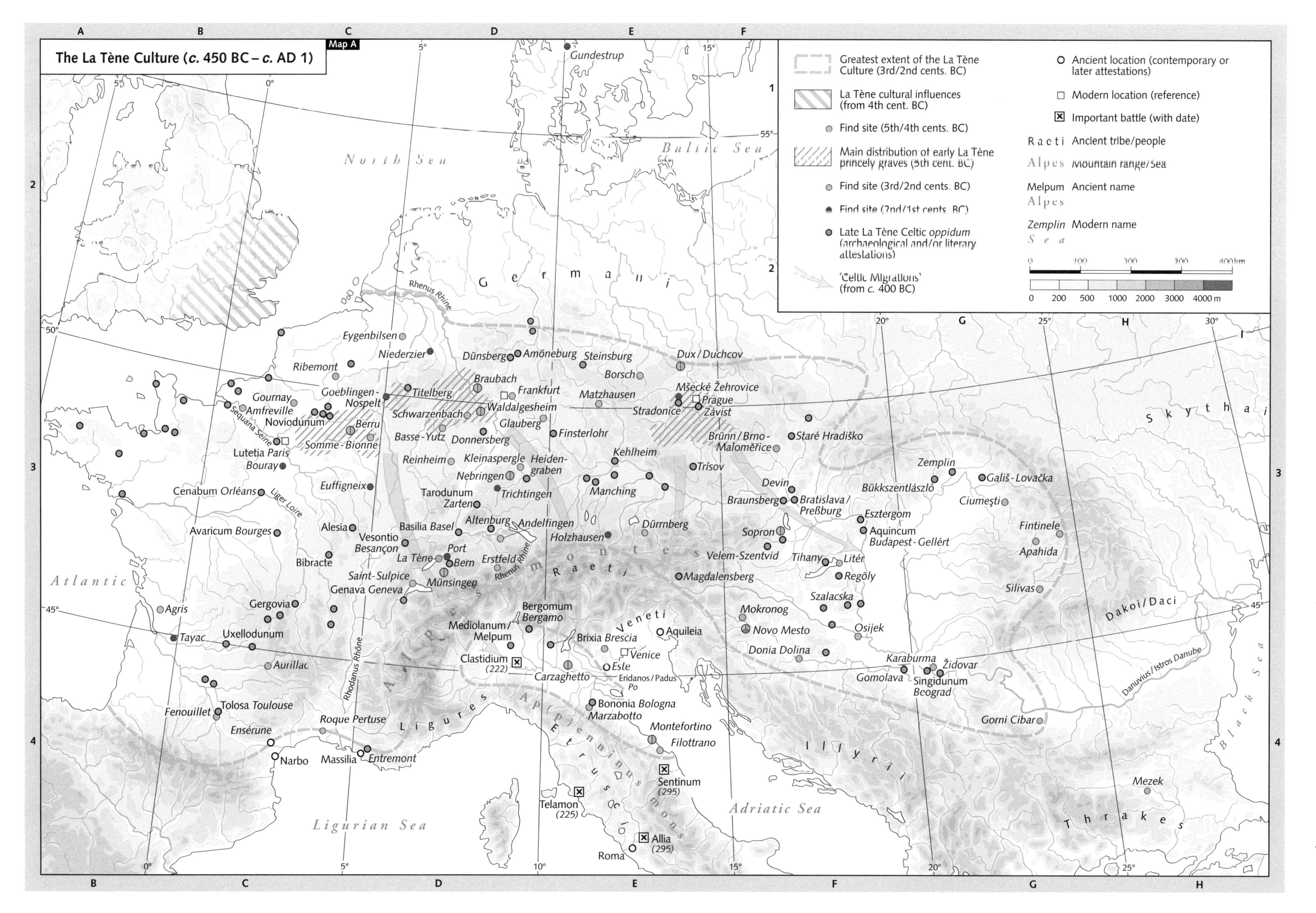
The La Tène Culture (c. 450 BC – c. AD 1)
Map A
Greatest extent of the La Tène Culture (3rd/2nd cents. BC)
La Tène cultural influences (from 4th cent. BC)
Find site (5th/4th cents. BC)
Main distribution of early La Tène princely graves (5th cent. BC)
Find site (3rd/2nd cents. BC)
Find site (2nd/1st cents. BC)
Late La Tène Celtic oppidum (archaeological and/or literary attestations)
'Celtic Migrations' (from c. 400 BC)
Ancient location (contemporary or later attestations)
Modern location (reference)
Important battle (with date)
Raeti Ancient tribe/people
Alpes Mountain range/sea
Melpum Ancient name
Zemplin Modern name
0 200 500 1000 2000 3000 4000 m
North Sea
Baltic Sea
Atlantic
Ligurian Sea
Adriatic Sea
Black Sea
Germani
Skythai
Dakoi/Daci
Thrakes
Illyrii
Raeti
Veneti
Ligures
Etrusci
Alpes
Apenninus mons
Rhenus Rhine
Rhodanus Rhône
Liger Loire
Sequana Seine
Danuvius/Istros Danube
Eridanos/Padus Po
Gundestrup
Eygenbilsen
Niederzier
Ribemont
Goeblingen-Nospelt
Titelberg
Gournay
Amfreville
Noviodunum
Berru
Somme-Bionne
Lutetia Paris
Bouray
Euffigneix
Cenabum Orléans
Avaricum Bourges
Alesia
Bibracte
Vesontio Besançon
Genava Geneva
Saint-Sulpice
Gergovia
Uxellodunum
Aurillac
Agris
Tayac
Tolosa Toulouse
Fenouillet
Ensérune
Narbo
Roque Pertuse
Massilia
Entremont
Schwarzenbach
Basse-Yutz
Donnersberg
Reinheim
Dünsberg
Braubach
Amöneburg
Frankfurt
Waldalgesheim
Glauberg
Kleinaspergle
Heiden-graben
Nebringen
Trichtingen
Tarodunum Zarten
Basilia Basel
La Tène
Port
Bern
Münsingen
Altenburg
Andelfingen
Erstfeld
Holzhausen
Steinsburg
Borsch
Matzhausen
Finsterlohr
Kehlheim
Manching
Dürrnberg
Dux/Duchcov
Mšecké Žehrovice
Prague
Závist
Stradonice
Brünn/Brno-Maloměřice
Třísov
Staré Hradiško
Devin
Bratislava/Preßburg
Braunsberg
Sopron
Velem-Szentvid
Magdalensberg
Tihany
Litér
Regöly
Szalacska
Mokronog
Novo Mesto
Donia Dolina
Osijek
Esztergom
Aquincum
Budapest-Gellért
Bükkszentlászló
Zemplin
Gališ-Lovačka
Ciumeşti
Fintinele
Apahida
Silivas
Karaburma
Židovar
Singidunum Beograd
Gomolava
Gorni Cibar
Mezek
Mediolanum/Melpum
Bergomum/Bergamo
Clastidium (222)
Carzaghetto
Brixia Brescia
Este
Venice
Aquileia
Bononia Bologna
Marzabotto
Montefortino
Filottrano
Sentinum (295)
Telamon (225)
Allia (295)
Roma

Trading relations between the Mediterranean world and north of the Alps (8th–5th cents. BC)

From about the 8th cent., contacts between the Mediterranean region and the area north of the Alps can be observed, as documented by finds of Etruscan and Greek provenance in Hallstatt territory. Both the large number of sites and the presence of certain significant types of artefacts like metal objects and pottery from Etruscan or Greek settlements and cultural areas as well as the increasing volume of goods in the course of the centuries make it highly likely that these goods were not exclusively spoils or gifts, but reached their destination by direct or indirect trading. This observation naturally raises the following questions: Which goods from the north interested the Greeks and Etruscans? Which trade routes did they use?

I. Trading partners and their most important routes of communication

Judging from archaeological and historial sources, the period from *c.* 800 to 600 BC or the 6th cent. BC was one of unrestricted trade between northern and southern Europe involving Phoenicians/Western Phoenicians, Greeks (among others the Phocaeans: Hdt. 1,163) and Etruscans as well as representatives of the western and eastern Hallstatt circles. In this context the western Alps and the Cevennes Range formed a clear barrier between the Mediterranean world and central Europe. In the west this barrier could be by-passed by following the passage of the Rhodanus/Rhône-Arar/Saône-Sequana/Seine valleys, while in the east the Alps had to be crossed to reach the Moravian Gate (outside the map section) and the Vistula/Weichsel river, along a section of which ran the Amber Road. The valley of the river Danuvius/Istrus/Danube provided a central east-west connection with both a navigable river as well as accompanying land routes reaching the Black Sea. Forming a third axis, it linked the two aforementioned routes, with the route across the eastern Alps having several alternative routes branching off towards the Danube on their way north. Over time, passages across the western and central Alps came into use, whose existence can be reconstructed archeologically from the remains of staging posts along the pass routes. Access to the Atlantic and the North Sea was provided by the courses of the rivers Liger/Loire, Sequana/Seine and Rhenus/Rhine.

As long as the Etruscans controlled marine trade in the Tyrrhenian Sea, along the coast of Liguria (Genoa) and the westward route towards the Iberian Peninsula, only very small quantities of pottery from the south were exported into the zone to the north of the Alps, as exports from Etruscan areas consisted largely of bronze objects directly or indirectly associated with the transport and consumption of wine, or of jewelry, fibulae and small statues. However, the battle of Alalia (540/35 BC), the rise of Carthage and the concomitant blockade of the Straits of Gibraltar drastically reduced the hegemonial role the Etruscans had played in the western Mediterranean. Thus the Greek city of Massalia used its newly acquired power to secure the use of the south-north trade route along the Rhodanus/Rhône for the transport of its export goods and the overland import of the much-needed metals. Massalia traded with Hallstatt settlements, whose geographical situation and supraregional economic as well as socio-political status had made them important trade centres. The sudden appearance of great quantities of Massaliote wine amphorae and Attic pottery at sites like the Late Celtic trading centre of Vienna/Vienne, in Bragny and on the Mont Lassois suggests the existence of a well-developed and secure network of long-distance trade routes served by an infrastructure of hubs.

The Etruscans on their part switched to the eastern trade routes across the Alps and the Adriatic region. From Bologna/Felsina, the south-east-north-west route ran as far as Bragny. From the Ticino, a northward route climbed several passes to home in on the Rhine – either across the western Alps (Simplon Pass/Valais/western Swiss lakes) or via the central Alps (Bernardino Pass into the Alpine Rhine valley or alternatively along the Lake Walen-Lake Zurich route). The Golasecca culture, situated between Verbanus lacus/Lago Maggiore and Padus/Po (→ map p. 65), played an important mediating role here. A third eastern passage sprouting several branches opened the path to north-eastern Europe along the rivers At(h)esis/Adige/Etsch–Isarcus/Eisack ('Brenner Pass route', which was to become part of the *via Claudia Augusta*) and to the east the 'Tauern route' via Iuvavum/Salzburg. It is difficult to obtain a clear picture of the various routes of the so-called Amber Road, which used to connect the Baltic Sea, especially the territory of today's Baltic states, with the Adriatic region. These routes are mentioned in early sources, but are still waiting to be defined more precisely by conclusive finds (not on map p. 83; → map p. 85).

The river systems played a vital part in shaping trade and migration patterns. This holds true of the Celtic region, called Gallia by the Romans, served by the Rhône, Garonne, Loire and Seine, and equally of the region between the Padus and the eastern Alpine area. Such favourable natural topographical features were largely missing on the Illyrian coast, whereas the Pontus region was provided with trade routes by the Istrus/Donube and the Borysthenes/Dnieper.

II. Traded raw materials and goods

In addition to the yields of the tin and lead mines in Spain, on the Atlantic coasts of Gaul and the coasts of Britain and in Scandinavia, the southern traders imported copper and iron ore in the shape of cake, ring and axe ingots or even scrap metal. They were also interested in silver and gold. The list was completed by salt and amber, mainly from the coast of the Baltic Sea (Greek: *ēlektron*, Lat.: *sucinum*; according to Pytheas of Massalia, 3rd cent. BC, fr. 8, there was an amber island called Abalus, situated in the North Sea).

The trade in 'natural produce' like iris roots, imported from the Illyrian south coast to be processed into salves and perfumes, features less prominently in ancient sources. It can, however, be safely assumed that beside the trade in iris roots, there was one in ready-made foodstuffs, animal skins, furs as well as slaves.

The products exported to the north by the cultures hugging the Mediterranean have been mainly identified by finds of their packaging, i.e. transport amphorae, and the surviving organic remains contained in them: among these were wine, olive oil, hazelnuts, grapes, stone fruit, but also mussels, crustaceans, meat and fish and furthermore minerals, plants and specific textiles. Throughout the period in question, the finds in the north also prove a demand for metal and ceramic tableware as well as jewelry, fibulae and figurines.

Sources

The information on which the map is based stems from excavation reports and detailed research on specific regions or trade goods. Taking into account different phases of development, it is possible to prove from the distribution of raw materials, traded raw materials, pre-manufactured products and finished products that their distribution was connected to specific routes, which in Antiquity (in this context: Late Hallstatt and Early La Tène Period) were apparently used as principal highways. Although instances of positive archaeological proof are rare (only a few indications of embankments and road surfacing, e.g. in Etruscan areas), but the assumption makes sense, in that these routes were oriented towards natural topographical features like rivers or natural passes (e.g. passages across the Alps). Additional, if not quite contemporary, clues are given by ancient authors providing the names of inhabited places and other geographical features, although some of them have not yet been identified. It is especially from the finds, but also from ancient written sources, that we can assume, that both the amount of traffic and the identity of the users of individual routes could be subject to change. Thus the Rhodanus/Rhône route was first used by the Etruscans, who were displaced primarily by Greeks after the foundation of Massalia.

The evaluation of the finds at different excavation sites and surface finds enable us to take stock – of necessity in a simplifying manner – of the imported finds (especially bronze objects and ceramic ware) in the Alpine region and the zones to its north. The concentration of one or several particular goods in specific cultural areas proves that in certain cases we can even speak of 'market-oriented' production. Like the literary sources, the numismatic and epigraphic materials are of a later date, the latter only from the 3rd cent. BC onwards, apart from a few possible references to the regions north of the Alps. The long-distance trade in general was first represented in literature by accounts of sea journeys – among them Homer's mythical-historical *Odyssey* (late 8th cent.) – and in the *periploi*, Greek descriptions of sea routes and coast lines recorded in logbooks or manuals (cf. also Hdt. 4,42; → map pp. 6f.). Another important source is Festus Rufus Avienus (4th cent. AD), whose so-called *Ora maritima*, the free adaptation of one of the oldest Greek coastal descriptions, possibly by a Massaliote author, has survived to this day. It includes a description of the stretch from Massalia to Tartessus/Tarsis. According to this author, northern central Europe was of interest to the Mediterranean nations primarily because of its abundance of metal deposits.

→ Maps pp. 6f., 63, 65, 69, 71, 75, 79, 81, 85; Etrusci, Etruria II., map 'Etruscan Exports (7th – 5th cents. BC)', BNP 5, 2004.

Literature

G. Bartoloni, Archäologische Untersuchungen zu den Beziehungen zwischen Altitalien und der Zone nordwärts der Alpen während der frühen Eisenzeit Alteuropas, 1998; J. Biel, Die Kelten in Deutschland, 2001; B. Bouloumié, Der Seehandel der Etrusker in Südfrankreich, in: Die Etrusker in Europa, exhibition catalogue, 1992, 168–173; K. Düwel (ed.), Methodische Grundlagen und Darstellungen zum Handel in vorgeschichtlicher Zeit und in der Antike, 1985; M.A. Guggisberg (ed.), Die Hydria von Grächwil. Zur Funktion und Rezeption mediterraner Importe in Mitteleuropa im 6. und 5. Jahrhundert v. Chr., 2004; W. Kimmig (ed.), Importe und mediterrane Einflüsse auf der Heuneburg, 2000; M. Kuckenburg, Die Kelten in Mitteleuropa, 2004; L. Leegaard, The Mediterranean and Central Europe in the 6th and 5th Centuries B.C. The Trade-Route through the Rhône Valley in the Light of Discoveries of Local Plain Wares, in: Acta hyperborea. Danish Studies in Classical Archaeology 9, 2002, 145–168; S.W. Meier, Blei in der Antike, 1995; D. Nash Briggs, Metals, Salt, and Slaves. Economic Links between Gaul and Italy from the Eighth to the Late Sixth Centuries BC, in: Oxford Journal of Archaeology 22, 2003, 243–259; J. Pape, Die attische Keramik der Heuneburg und der keramische Südimport in der Zone nördlich der Alpen während der Hallstattzeit, in: W. Kimmig (ed.), Importe und mediterrane Einflüsse auf der Heuneburg, 2000, 71–175; D. Tiempe, Griechischer Handel nach dem nördlichen Barbaricum, nach historischen Quellen, in: K. Düwel (ed.), Untersuchungen zu Handel und Verkehr der vor- und frühgeschichtlichen Zeit in Mittel- und Nordeuropa, 1. Methodische Grundlagen und Darstellungen zum Handel in vorgeschichtlicher Zeit und in der Antike, 1985, 181–213; S. Zimmer (ed.), Die Kelten – Mythos und Wirklichkeit, 2004.

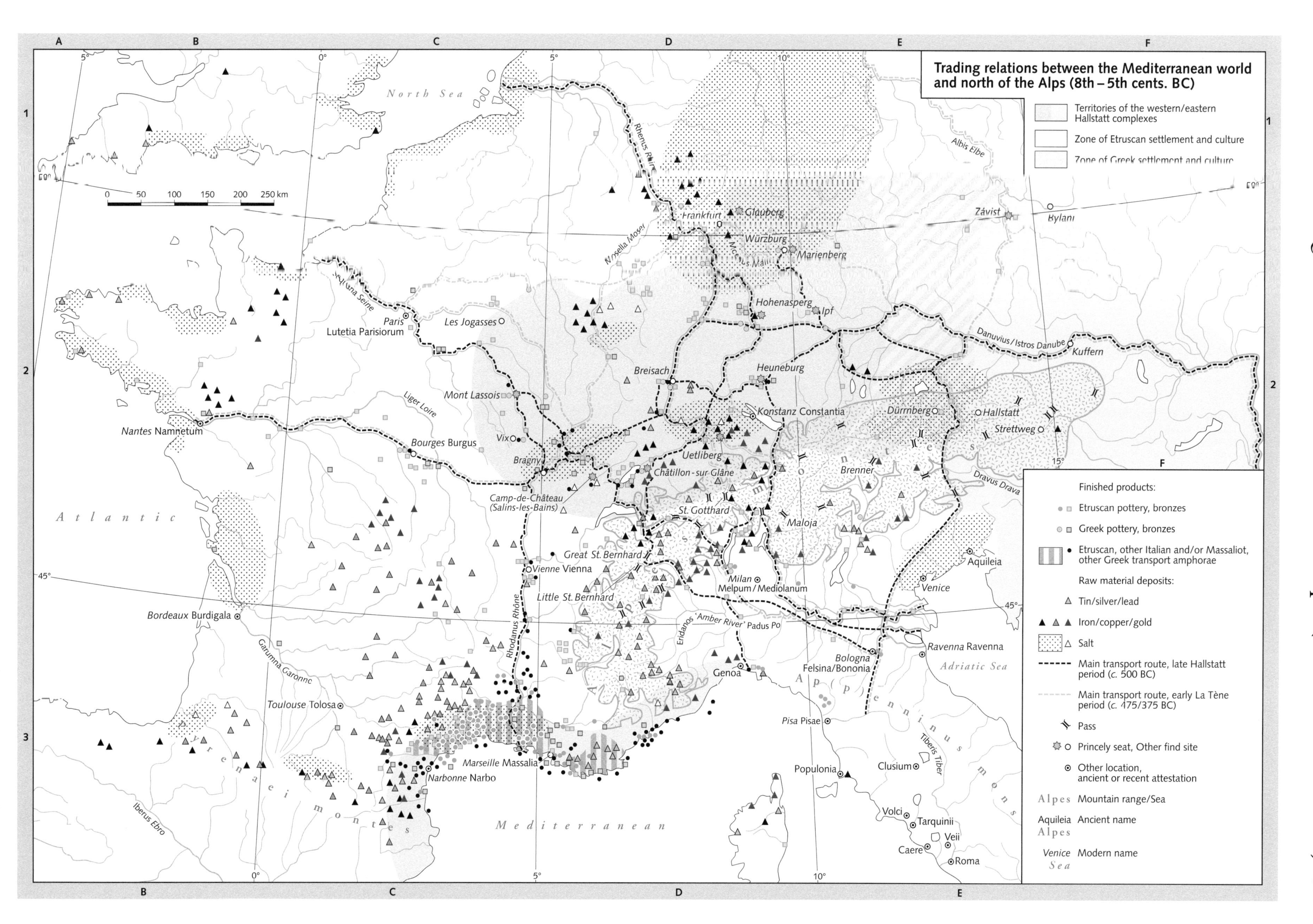
Trading relations between the Mediterranean world and north of the Alps (8th – 5th cents. BC)
Territories of the western/eastern Hallstatt complexes
Zone of Etruscan settlement and culture
Finished products:
Etruscan pottery, bronzes
Greek pottery, bronzes
Etruscan, other Italian and/or Massaliot, other Greek transport amphorae
Raw material deposits:
Tin/silver/lead
Iron/copper/gold
Salt
Main transport route, late Hallstatt period (c. 500 BC)
Main transport route, early La Tène period (c. 475/375 BC)
Pass
Princely seat, Other find site
Other location, ancient or recent attestation
Alpes Mountain range/Sea
Aquileia Ancient name
Venice Modern name
North Sea
Atlantic
Mediterranean
Adriatic Sea
Paris
Lutetia Parisiorum
Nantes Namnetum
Bordeaux Burdigala
Toulouse Tolosa
Narbonne Narbo
Marseille Massalia
Genoa
Pisa Pisae
Populonia
Bologna
Felsina/Bononia
Ravenna Ravenna
Venice
Aquileia
Milan
Melpum/Mediolanum
Vienne Vienna
Bourges Burgus
Mont Lassois
Vix
Bragny
Les Jogasses
Heuneburg
Breisach
Hohenasperg
Ipf
Marienberg
Würzburg
Glauberg
Frankfurt
Konstanz Constantia
Uetliberg
Châtillon-sur-Glâne
Camp-de-Château (Salins-les-Bains)
Great St. Bernhard
Little St. Bernhard
St. Gotthard
Maloja
Brenner
Dürrnberg
Hallstatt
Strettweg
Závist
Kuffern
Clusium
Volci
Tarquinii
Veii
Caere
Roma
Rhenus Rhine
Danuvius/Istros Danube
Dravus Drava
Rhodanus Rhône
Liger Loire
Garumna Garonne
Iberus Ebro
Tiberis Tiber
Eridanos
'Amber River' Padus Po
Albis Elbe
0 50 100 150 200 250 km

Commerce and trade in the Mediterranean world, 7th/6th cents. – 4th cent. BC

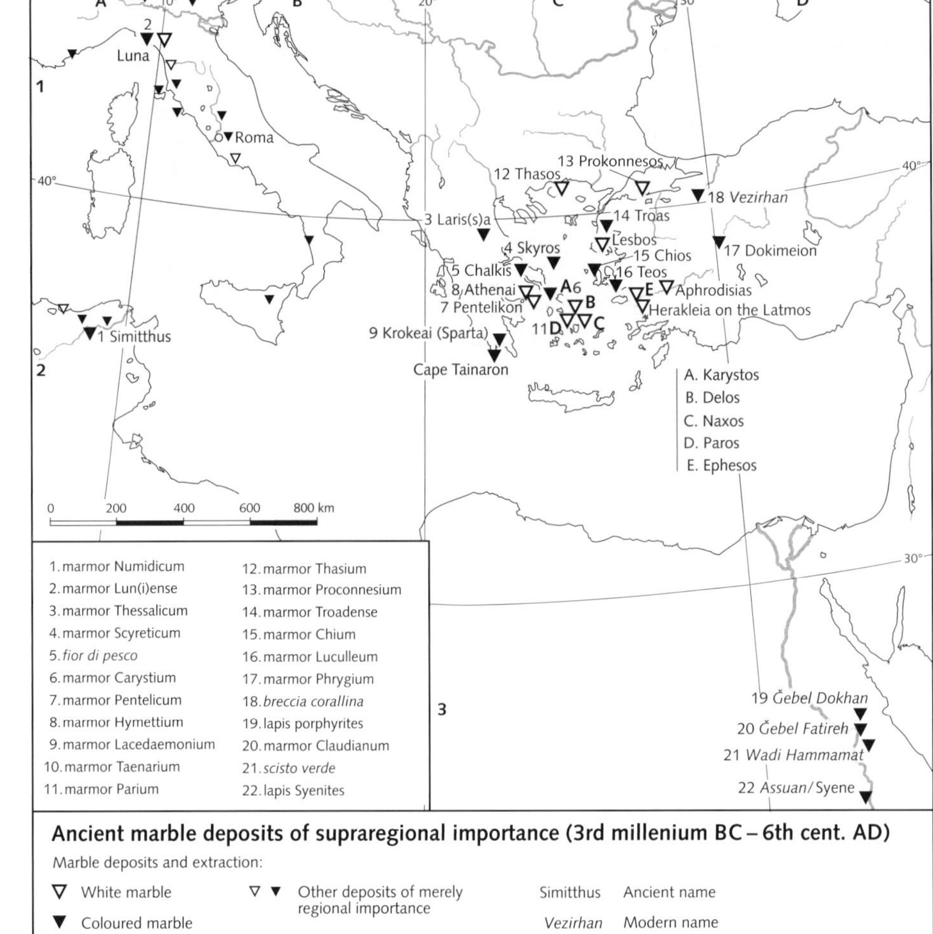

Ancient marble deposits of supraregional importance (3rd millenium BC – 6th cent. AD)

Commerce and trade in the Mediterranean world, 7th/6th cents. – 4th cent. BC

Economy and trade in the Mediterranean region from the 7th/6th cents. to the early 4th cent. BC

In Antiquity, the most important sector of the economy was a type of agriculture which was essentially based on the social organization and labour form of subsistence farming, with crop farming and animal husbandry being largely practised as separate branches. The Mediterranean region is characterized by favoured areas suitable for rain-fed agriculture, situated mostly in the coastal zones and on certain basinal plains, both of which were targeted by Phoenician and Greek colonists. The stable climate in the summer months was a boon for sea travel. Settlements of varying sizes and large cities (e.g. Sidon, Miletus, Athens, Caisri/Caere, Qart-ḥadašt/Carchedon/Carthago) acted as economic centres both for the local markets and supraregional trade. They were also manufacturing sites. Many coastal settlements thus built the necessary port facilities. With agriculture restricted to suitable stretches of land in the plains, frequently supplemented by parcelled terraces on suitable hillsides, the higher grounds could be used for cattle-farming, hunting and poaching. Forests for the production of timber were found at higher altitudes and in areas with higher levels of precipitation, i.e. on the western slopes of mountain ranges and locations close to the coast (→ map p. 201). Monetary and banking transactions (credits, loans, sea loans) did not play a part in the Greek world before the 6th cent.; in Rome and Italy this was only the case from the late 4th/3rd cent. onward. Early coins have been found in just a few Etrurian places. However, even in pre-monetary times access to precious metal deposits (Taršiš?/Tartessus on the Iberian Peninsula, Pangaeum in Thrace, Laurium in Attica) was an important economic, but also political asset.

I. Trade routes

By the 8th/7th cents., some real long-distance trading was in place again. Apart from the Greeks (among others Corinthus, Miletus, Phocaea; later on Athens, Rhodes), it was especially the Phoenicians (among others Tyre, from the 11th/10th cents., Qart-ḥadašt/Carthago, Gades) and other Levantines, who took part in this enterprise. So did the Etruscans (e.g. Caisri/Caere, Tarquinii, Populonia, later Atria, Spina), who in addition to their maritime trade also had early trade contacts with the peoples north of the Alps. There were trading posts to be found in and around the whole Mediterranean Sea, especially wherever either the hinterland was not interested in seafaring activities, wherever there was an absence of stable political structure or wherever a central power either invited or at least tolerated the establishment of trading posts (Greeks at Naucratis). Up to the 6th cent., the Phoenicians/Western Phoenicians (among others Carthage), Greeks and Etruscans adhered to an agreement about the extent of their respective economic spheres of influence, especially in the western Mediterranean.

The traded goods were raw materials of all kinds as well as pre-manufactured and finished products. Among these were grain (from the 5th cent. onward, grain imports became increasingly important for Greek cities: cf. Hdt. 7,147,2; Dem. Or. 20,30–33), wine and olive oil, other foodstuffs, timber (especially for shipbuilding), metals and textiles, but also slaves. It has been scientifically proved that even blocks of marble were shipped across considerable distances (e.g. Proconnesian marble was identified at Didyma; cf. supplementary map).

Apart from the established sea routes, which can also be reconstructed from the locations of shipwrecks (e.g. Igilium/Giglio off the Etruscan coast, Bon-Porté in southern France, Gela on Sicily, Porticello at the Straits of Messina, Cyrenia on Cyprus), information about important overland long-distance trade routes has also come down to us. Among these was the so-called Royal Road, including its predecessors and offshoots, running from Sardis to Susa, the routes from the Levant to Mesopotamia continuing into central Asia along the Grand Military Highway, the Incense Route on the Arabian peninsula and its continuation to Syria on another Royal Road, the southward routes along the Nile and their offshoots to the oases in the west and to the Red Sea, the routes into the area north of the Alps (→ map p. 83), the important east-west connection along the Danube – one of many examples of large rivers, navigable or not, serving as essential axes – and the so-called Amber Road, which over time was characterized by a variety of alternative routes.

II. Sources

Although its importance for grasping historical developments was recognized very early on, economic history is still often neglected by scholars, mainly because of the state of the sources: ancient writers, as a rule, either did not deal with economic questions in their own right (cf. e.g. Pl. Leg. 677a-682e; Str. 2,5,26), or the requisite specialist sources were lost. On the subject of agriculture and (barter) trading or the non-mutual transfer of goods (one-sided presents, but also spoils of war, piracy, robbery) we find references in Homer, Hesiod (*Erga kai hemerai*) and Xenophon (*Oikonomikos*). Aristotle quotes lost specialist writings and discusses zoology, while Theophrastus discourses on botany. Varro mentions Greek specialist writers (Varro, Rust. 1,1,7ff.).

The OT alludes to the early Phoenician trading expeditions undertaken by the city of Tyre. The lost Tyrian historiography was partly reproduced by Flavius Josephus. The Carthaginian writer Mago, a widely respected agricultural specialist whose work was translated into Latin and Greek and widely quoted by Pliny the Elder and Cicero, belongs to the 2nd cent. BC. There are no direct literary sources from the Etruscan context.

Thus, apart from isolated pieces of information on e.g. the organization of transport or the traded products, which can be gleaned from literary sources, it is frequently observations linked to archaeological projects like the identification of emporia or to significant types of ceramic finds (such as transport amphorae) which contribute to our understanding of the economic interplay, especially of trade relationships. These sources are supplemented by epigraphic and numismatic evidence.

III. Ancient marble deposits of supraregional significance (supplementary map)

From the Bronze Age onward, marble was used to create architectural structures, sculptures and vessels. In Greece, white marble was used for such purposes from the 7th cent., while in Rome the first use of white marble can be documented for the 2nd cent. BC, with coloured marble following in the 1st cent. BC. In Late Antiquity and Early Byzantine times both types of marble were still used, in fact frequently re-used, in architecture and sculpture. In all cultures marble was considered a highly prestigious material. The most important quarries exporting white marble were located in Greece, Asia Minor and Italy. This was supplemented by coloured marbles and other coloured rocks, which at times in the age of Imperial Rome were widely used. The numerous local deposits have hardly been researched, in spite of the fact that they made a vital contribution to the infrastructural development of their respective urban environments.

→ Maps pp. 53, 55, 63, 69, 71, 75, 83, 121, 201; Commerce II–IV, BNP 3, 2003

Literature

M. Austin, P. Vidal-Naquet, Gesellschaft und Wirtschaft im alten Griechenland, 1984; R. Bogaert, Grundzüge des Bankwesens im alten Griechenland, 1986; P. Dupont, Trafics mediterranéens archaïques: quelques aspects, in: Krinzinger, 445–460; A. Eich, Die politische Ökonomie des antiken Griechenlands (6.-3. Jh. v. Chr.), 2006; V.D. Hanson, The Other Greeks, 1995; R.J. Hopper, Trade and Industry in Classical Greece, 1979; H. Klees, Sklavenleben im klassischen Griechenland, 1998; F. Krinzinger (ed.), Die Ägäis und das westliche Mittelmeer. Symposion 1999, 2000; S. Lauffer, Die Bergwerkssklaven von Laureion, [2]1979; D. Mertens, Städte und Bauten der Westgriechen, 2006; A. Möller, Naukratis, 2000; C. Neri, Il marmo nel mondo Romano, 2002; P. Pensabene (ed..), Marmi antichi II: cave e technica di lavorazione, provenienza e distribuzione, 1998; S. von Reden, Exchange in Ancient Greece, 1995; R. Rollinger, C. Ulf (eds.), Commerce and Monetary Systems in the Ancient World: Means of Transmission and Cultural Interaction, 2004; R. Sallares, The Ecology of the Ancient Greek World, 1991; I. Scheibler, Griechische Töpferkunst, [2]1995.

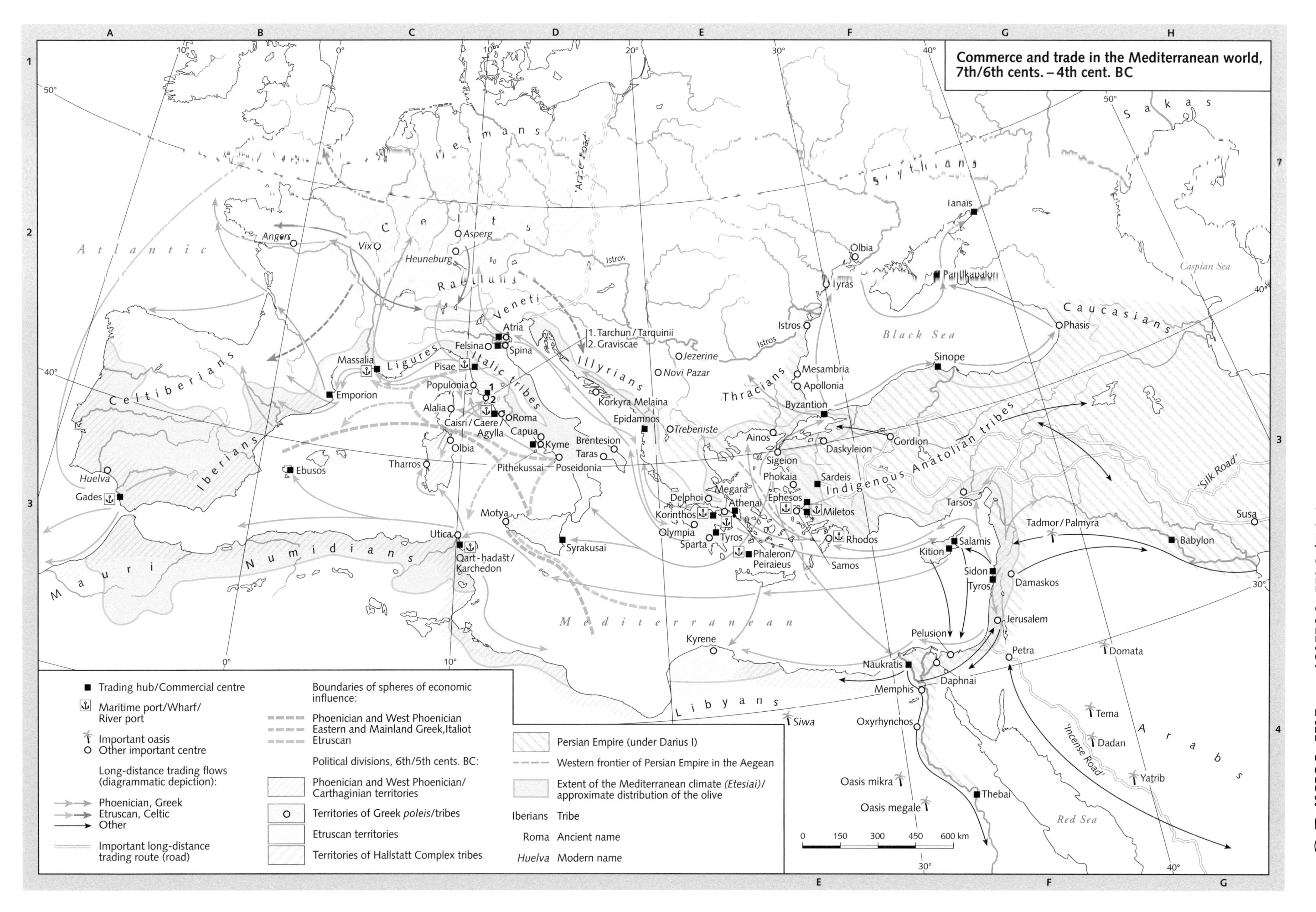
Commerce and trade in the Mediterranean world, 7th/6th cents. – 4th cent. BC
Trading hub/Commercial centre
Maritime port/Wharf/River port
Important oasis
Other important centre
Long-distance trading flows (diagrammatic depiction):
Phoenician, Greek
Etruscan, Celtic
Other
Important long-distance trading route (road)
Boundaries of spheres of economic influence:
Phoenician and West Phoenician
Eastern and Mainland Greek, Italiot
Etruscan
Political divisions, 6th/5th cents. BC:
Phoenician and West Phoenician/Carthaginian territories
Territories of Greek poleis/tribes
Etruscan territories
Territories of Hallstatt Complex tribes
Persian Empire (under Darius I)
Western frontier of Persian Empire in the Aegean
Extent of the Mediterranean climate (Etesiai)/approximate distribution of the olive
Iberians Tribe
Roma Ancient name
Huelva Modern name
1. Tarchun/Tarquinii
2. Graviscae
Black Sea
Caspian Sea
Red Sea
Mediterranean
Atlantic
'Silk Road'
'Incense Road'
Celts
Celtiberians
Iberians
Ligures
Italic tribes
Veneti
Illyrians
Thracians
Indigenous Anatolian tribes
Caucasians
Sakas
Arabs
Libyans
Numidians
Mauri
0 150 300 450 600 km

The Achaemenid Kingdom (6th to 4th cents. BC)

The Achaemenid Empire was more than just the first great Persian empire; for under the rule of the Teispids (i.e. the kings before Darius I) and the Achaemenids (from Darius I), it unified for the first time all of the Near and Middle East and then kept it together, mostly in peace. It lasted from *c.* 550 (conquest of the Median kingdom) until 330 BC, when it was itself conquered by Alexander the Great. The Persians created an exemplary infrastructure comprising a network of roads, a uniform currency, an official language and an efficient and durable administrative system.

I. The expansion of the empire under the Teispids

Like the Medes, the nomadic Persians probably immigrated onto the Iranian High Plateau in the late 2nd/early 1st millennium. Historically they first emerge in Assyrian textual sources of the 9th cent., while their exact location is known from Elamite 7th-cent. sources as Pārsa, later known as Persis. By then it may already have been organized as a kingdom in its own right or still been a patchwork of Persian tribal areas, partly under Elamite, partly under Median rule.

Already in the period stretching from the time of the Assyrian Empire (2nd half of the 7th cent.) to the reign of Cambyses I (600-559), the Persians managed to expand their core territory of Pārsa considerably. Under Cyrus II (559–530), the King of Anšan, the Medes were defeated in the battle of Pasargadae (550), their capital Ekbatana was conquered, their territories were annexed and a twin monarchy was installed.

According to Hdt. 1,73, the Lydian and the Median royal families were dynastically related. This may have been the reason why the Persians next turned to Asia Minor. Another reason could have been a call for help by the 'East Phrygians', whose centre Pteria – newly built in the early 6th cent. inside the arc of the river Halys and now beyond doubt identified as the short-lived structure on the Kerkenes Dağı – had been destroyed by King Croesus in 547/6 (Hdt. 1,76). After a battle in the Pteria region, the Persians pursued their Lydian opponents and seizing their capital Sardes/Sparda in the same year in which they subjugated the Lydian territories (residences at Sardis and Dascylium; Persian palace at Celaenae) and subsequently the Greek coastal cities as well.

Still in the reign of Cyrus II, the formerly allied Neo-Babylonian Empire was conquered from 540 to 539, an event which ended the Babylonian exile of the Jews. Cyrus had himself crowned king to rule over three realms simultaneously. The northern and eastern border areas in Central Asia were, as had been the case in Median times, threatened by Sacaean and Scythian tribes from the steppes. In the years between 547 and his death in 530, Cyrus succeeded in advancing the border as far as the Iaxartes river. It was then secured by fortifications, and his son Bardiya was installed as satrap of that part of the empire.

Cyrus' successor Cambyses II (530-522) added 26th-Dynasty Egypt to the empire. With the exception of events at Pelusium, hardly any resistance was offered to the Persians, as the country had been weakened by succession conflicts. The neighbouring desert nations also succumbed to the Persians, while the kingdom of Meroe in Nubia presumably became a tributary state, but was not subjugated. The plan of a military expedition against the Western Phoenician Carthaginian territories never materialized.

II. The expansion of the empire under the Achaemenids (522/21-330 BC)

After the death of Cambyses II, the 'usurper' Darius I (522-486) had to work hard to quell unrest and secession attempts, which he finally accomplished in 520 BC. He was probably the first Achaemenid, as he invoked his descent from Achaemenes, whereas his predecessors had based their claim to the throne on the house of Teispes. From 519 he undertook several military campaigns in central Asia, subjugated the valley of the Indus, had the sea route from India to Egypt explored by Scylax and reconquered renegade Egypt. He also added territories as far as the Cyrenaea and restored the Nile canal. Possibly to meet the continuing threat from Sacaeans or Scythians (a generic term for tribes from the Asian steppes) by attacking them in their western flank in the northern Black Sea area, he undertook a military expedition to south-eastern Europe in 513/12. In the process a satrapy was set up in the conquered Thracian terriories and the border was advanced as far as the Danube. It is a matter of controversy if Macedonia was conquered or made a pact with the Persians and made tribute payments; in any case, in 500 BC the Persian Empire had reached the borders of the Greek mainland, thus achieving its largest territorial expansion ever.

After the so-called Persian Wars (*c.* 500–478/449, → map p. 89), the Persians lost their European territories. Asia Minor and the Greek cities on its coast forthwith represented the westernmost part of the empire. Egypt seceded in 402 and was only subjugated again in 343. There were repeated troubles and uprisings in Asia Minor, but also in other parts of the empire (Babylon 479; cf. Xenophon, *Anabasis*, 401–400/399; so-called Great Satraps' Revolt around 360). The Indus Valley and the Sacaean areas were lost. Under Darius III (336-330), Alexander the Great invaded the Achaemenid Empire, which ceased to exist as an intact administrative entity with the death of Darius III.

III. The organization of the empire

Under Cyrus II and Cambyses II the empire was structured along federalist lines: the Persian King was simultaneously the King of the Median, the Babylonian and later the Egyptian kingdoms as well. There were also parts of the Empire where the monarchy had been abolished (e.g. Lydia) without destroying the administrative structures, as Persian satraps replaced the former rulers.

Even after comprehensive reforms had been carried out by Darius I, the empire was still characterized by a high degree of local autonomy, a strong central power (the Great King) to whom everybody was responsible, and regional authorities. The 'King's Land' was subdivided into tax and administrative districts (provinces), which – following the terminology used by Herodotus and Diodorus (18,5,3–6,4) – are called satrapies. They were not necessarily identical with the formerly conquered or incorporated territories. This resulted in a difference between the toponyms used in the Persian and Greek literatures, a fact, which caused confusion among scholars right into recent times, because in most cases the adminstrative units (satrapies, Greek tradition) were equated with countries/regions and peoples (Persian tradition). Reconstruction is also made more difficult by the uneven state of the sources: the satrapies in the western half of the empire are much better documented than the eastern ones, because of the Greek source material. However, the overall number of documented cases is rather small by any standard. It has been discovered that under Persian rule some local rulers were allowed to remain in office (the city kings in Phoenicia and on Cyprus; dynasts in Caria, Lycia, Cilicia). Therefore, in contrast to other mappings, the provincial borders have not been charted, the more so because they were continuously redrawn, which apart from very rare cases makes an exact identification impossible. According to recent research, the empire comprised not only well-defineable satrapies, but also semi-autonomous areas or purely military buffer zones as well as independent areas inhabited by nomads or mountain tribes, who were affiliated with specific satrapies without being subject to their administration. They were granted a special status because they played an important role in matters of trade, military liaisons and intelligence. The central administration was located in Susa; Persepolis and Ekbatana housed residences.

The administrative structures were shored up by a finely tuned infrastructure. The most important traffic routes (e.g. the so-called Royal Road from Sardis to Susa, actually from Ephesus to Persepolis; probably also the Silk Road) were developed into a stable network of roads, which criss-crossed the whole empire and connected even the most distant provinces. It featured hostelries and, to guarantee a safe journey, military garrisons which were spaced at regular intervals. From the reign of Darius I, a uniform currency system existed (*daricus*). Until the reign of Artaxerxes I the official administrative language was Elamite, which was then replaced by the imperial Aramaic language. Almost everywhere agriculture was the economic backbone of the Achaemenid Empire. In the remote margins of the empire, especially in central Asia, a nomadic way of life and animal husbandry were predominant. Relatively uniform economic regulations facilitated trade and commerce (cf. clay tablets from Persepolis and relief representations of tributary peoples). The big cities were the most important trade centres, e.g. Babylon (banking), Susa or Tyre (the most important Mediteranean commercial port for trade with the Phoenician colonies and trading posts in the west); Sidon was a navy base. The western part of the empire was integrated into the Greek economic region (money economy), whereas in the east the economy was based on barter trading.

Sources

The amount of sources covering the Achaemenid Empire is limited. The sources can be roughly divided into three main groups: archaeological and epigraphical materials and Greek historiography. The first originate mostly from the large royal residences of Susa, Persepolis and Pasargadae, whereas Ekbatana is largely built over by present-day Hamadan. There are isolated records from Asia Minor, Mesopotamia, Egypt and the eastern parts of the empire, especially findings from palace sites (new excavations e.g. at Celaenae in Asia Minor promise to yield important insights) as well as numerous individual finds. The epigraphic sources are mainly royal inscriptions of limited information value, although there are exceptions like the trilingual inscription of Bīsutūn, containing a detailed report of the achievements of Darius I. There are also administrative records from Persepolis which have not yet been completely sifted. The best-known and, apart from the biblical texts (books of Daniel, Ezra, Esther), only type of source before the advent of archaeological research in the 19th cent. is Greek historiography. There are many authors who deserve a passing mention, but it is Herodotus who must be given pride of place, although using him as a source requires careful analysis and cross-referencing with other sources, non-Greek as well. Other important sources are Xenophon (*Anabasis* and *Cyropaedia*), Plutarch (Life of Artaxerxes II) and the historians of Alexander the Great: Quintus Curtius Rufus, Plutarch, Arrian, Justin, Diodorus and Strabo. Several works which explicitly deal with the Achaemenid Empire, such as the *Persica* by Ctesias, have only survived in fragmentary or secondary form. Archaeological research has revealed that many descriptions given in the historiographic, but also in the literary, texts are distorted or incorrect, which means that balancing the information with Persian sources is an absolute must.

→ Maps pp. 6f., 33, 51, 53, 55, 59, 71, 89, 105, 113, 215, 217

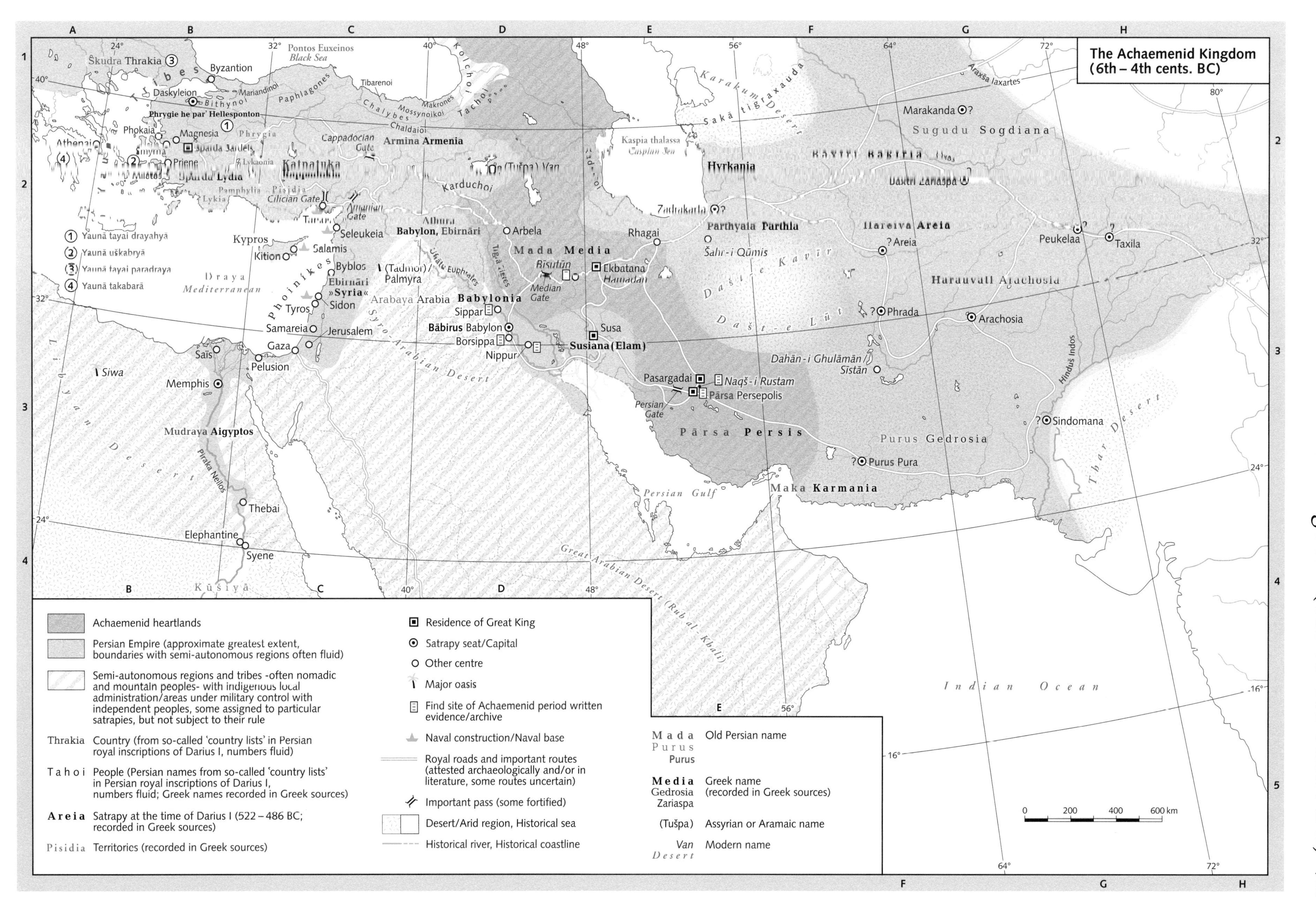
The Achaemenid Kingdom
(6th – 4th cents. BC)
Škudra Thrakia 3
Byzantion
Daskyleion
Bithynoi
Mariandinoi
Paphlagones
Pontos Euxeinos
Black Sea
Tibarenoi
Chalybes
Mossynoikoi
Makrones
Chaldaioi
Kolchoi
Tachoi
Phrygie he par' Hellesponton
Phokaia
Magnesia
Smyrna
Priene
Miletos
Lydia
Phrygia
Lykaonia
Pamphylia
Pisidia
Lykia
Cilician Gate
Cappadocian Gate
Armina Armenia
Amanian Gate
Karduchoi
(Tušpa) Van
Kaspia thalassa
Caspian Sea
Karakum Desert
Sakā tigraxauda
Araxša Iaxartes
Marakanda ?
Sugudu Sogdiana
Hyrkania
Parthyaia Parthia
Šahr-i Qūmis
Areia
Peukelaa
Taxila
Harauvati Arachosia
Arachosia
Phrada
Dašt-e Kavīr
Dašt-e Lūt
Dahān-i Ghulāmān
Sīstān
Hindus Indos
Sindomana
Thar Desert
Purus Gedrosia
Purus Pura
Maka Karmania
Pārsa Persis
Pasargadai
Naqš-i Rustam
Pārsa Persepolis
Persian Gate
Persian Gulf
Susa
Susiana (Elam)
Rhagai
Ekbatana
Hamadan
Bisutūn
Median Gate
Mada Media
Arbela
Tigrā Tigres
Ufrātu Euphrates
Babylon, Ebirnāri
Babylonia
Sippar
Bābirus Babylon
Borsippa
Nippur
Arabaya Arabia
(Tadmor) / Palmyra
Seleukeia
Kypros
Kition
Salamis
Byblos
Ebirnāri
»Syria«
Sidon
Tyros
Phoinikes
Samareia
Jerusalem
Gaza
Pelusion
Syro-Arabian Desert
Draya
Mediterranean
1 Yaunā tayai drayahyā
2 Yaunā uškahyā
3 Yaunā tayai paradraya
4 Yaunā takabarā
Athenai
Saïs
Siwa
Memphis
Mudraya Aigyptos
Libyan Desert
Piraka Neilos
Thebai
Elephantine
Syene
Kūšiyā
Great Arabian Desert (Rub al-Khali)
Indian Ocean
Achaemenid heartlands
Persian Empire (approximate greatest extent, boundaries with semi-autonomous regions often fluid)
Semi-autonomous regions and tribes -often nomadic and mountain peoples- with indigenous local administration/areas under military control with independent peoples, some assigned to particular satrapies, but not subject to their rule
Thrakia Country (from so-called 'country lists' in Persian royal inscriptions of Darius I, numbers fluid)
Tahoi People (Persian names from so-called 'country lists' in Persian royal inscriptions of Darius I, numbers fluid; Greek names recorded in Greek sources)
Areia Satrapy at the time of Darius I (522 – 486 BC; recorded in Greek sources)
Pisidia Territories (recorded in Greek sources)
Residence of Great King
Satrapy seat/Capital
Other centre
Major oasis
Find site of Achaemenid period written evidence/archive
Naval construction/Naval base
Royal roads and important routes (attested archaeologically and/or in literature, some routes uncertain)
Important pass (some fortified)
Desert/Arid region, Historical sea
Historical river, Historical coastline
Mada Old Persian name
Purus
Purus
Media Greek name (recorded in Greek sources)
Gedrosia
Zariaspa
(Tušpa) Assyrian or Aramaic name
Van Modern name
Desert
0 200 400 600 km

The Persian Wars
(*c.* 500–478/449 BC)

'Persian Wars' is a modern umbrella term denoting the attacks by the Achaemenid kings Darius I (522–486 BC) and Xerxes I (486–465 BC) on Greece in the period between the Ionian Revolt (*c.* 500–494 BC) and the middle of the 5th cent. The expansion of Persian rule to the European continent (from 513/12 BC) and the concomitant Achaemenid claim to world dominion had put an end to the hitherto peaceful nature of Greco-Persian contacts. The defence of 'Greek freedom' against the 'barbarians', which was launched by the Ionian Revolt, a revolt of the Greek cities in Asia Minor, and ended in Alexander's military expedition, became the predominant issue for two centuries to come.

Cause and course of the campaigns (cf. supplementary map)

The immediate reason for the Persians to intervene was the support given by Eretria (Euboea) and Athens to the Ionian Revolt. Spreading along the whole coastal area and to the offshore islands, then to the Propontis, Caria, Lycia and Cyprus, the revolt seems to have taken the Persians completely by suprise. The crews of 20 Athenian ships participated in an expedition to Sardis (498), which resulted in the destruction of that city. However, as they retreated, the insurgents suffered a heavy defeat. There was no involvement of combatants from mainland Greece in the naval battle (494) off the island of Lade, north of Miletus, when the Persians destroyed the Greek fleet and went on to conquer and destroy Miletus.

It is still an open question to what extent the incursion by the Persian general Mardonius in 492 was part of the preparations for the conquest of Greece. Combining sea and land operations, the campaign shored up Persian rule in Asia Minor and the northern Aegean region (including Thasos), forcing Macedonia to renew its recognition of Persian suzerainty. However, the complete Persian fleet was wrecked by a tempest off Mount Athos.

Following a final ultimatum by the Great King of Persia in 491 BC, numerous neutral Greek states in central and northern Greece succumbed; only Sparta, its Peloponnesian allies and Athens remained defiant. In 490, Persian elite units led by Datis and Artaphernes crossed the Agean Sea and restored Persian rule over Naxos and some other Cyclades. After attacking and destroying Carystus and Eretria, they landed at Marathon to set out for Athens; but the Athenians and Plataeans, who had advanced to Marathon, inflicted a surprise defeat on them. After initially setting course for Phaleron, the Persian fleet retreated to Asia Minor.

To counter the expected return of the Persians, who wanted to break the resistance of the Greek mainland states, Athens at once strove to gain a foothold in the Aegean (Paros and other Cyclades) and in the Saronic Gulf. Its highly active interior policy was equally driven by fear of the Persians, oligarchic infighting, political reforms and the forced build-up of a modern fleet by Themistocles. Darius' death in 486 BC and uprisings in the Persian Empire delayed the Persian onslaught, until, from 484/83, Xerxes began systematic preparations for a campaign of co-ordinated sea and land operations, with the aim of achieving the permanent subordination of Athens and Greece. Two pontoon bridges were constructed to span the Hellespont. A canal was dug across the Athos peninsula to create a safer naval passage and provision depots were set up in Thrace. Most of northern and central Greece, the majority of the islands and even some anti-Spartan regions of the Peloponnese, as well as the Greeks of Sicily and southern Italy (due to a simultaneous Western Phoenician offensive that may have been agreed upon with the Persians) accepted Persian suzerainty or promised to remain benevolently neutral after a Persian ultimatum. A little earlier an Athenian initiative had led to the formation, in Corinth, of the anti-Persian Hellenic League, a coalition of 31 mostly unimportant cities. The supreme command over land and sea forces was handed to Sparta.

The Persian campaign started in 480 BC. A first Greek line of defence in the Tempe Valley was given up as strategically pointless, because it could have been bypassed. Instead a second line was set up at Thermopylae, while the Greek fleet was waiting for the Persian ships at Cape Artemisium (Euboea). Allegedly by treason, the Persian army managed to bypass Thermopylae; they encountered only a rearguard unit, which they annihilated to a man. The Persian fleet, however, had been weakened by storms. Thus, in the first naval encounters the Greek ships were gaining the upper hand, but then retreated to Salamis, where they completed the evacuation of the city of Athens. The subsequent destruction of Athens made it advisable to retreat behind the Isthmus in this desperate situation, but the consequent sacrifice of the *poleis* to its north would have meant the end of the Hellenic League. Under these circumstances, Themistocles persuaded the Spartans to agree to a naval engagement. With summer drawing to a close, the Persians needed to force a decision at any rate and Themistocles allegedly tricked them into launching an attack in the narrow Sound of Salamis. Thus, late in September 480, *c.* 400 Greek ships defeated Xerxes' somewhat stronger fleet. The king returned to Asia Minor with what was left of his armada. Mardonius and his army spent the winter in Thessaly, only to be defeated by the Greek army in the battle of Plataeae in 479 BC. At about the same time, a fleet of the Hellenic League destroyed Persian ships off the Mycale peninsula north of Miletus.

The changes brought about in the Aegean by the defeat of the Persians enabled the Hellenic League to shift the Greek line of defence a long way to the east and to pursue an aggressive strategy against Persia to secure the freedom of the Greeks in Asia Minor. Numerous Aegean and Anatolian *poleis* seceded from Persia, seeking membership of the Hellenic League. However, this was only granted to the Aegean island states, leaving the problem of Asia Minor unsolved for the time being. After Sparta had resigned from the leadership of the Hellenic League at a conference on Samos, Athens became the hegemonial power in the Aegean, when the Delian League was founded in 478 BC. It is still a matter of controversy, in what manner and at what time the Persian Wars were officially ended with a treaty, possibly the 'Peace of Callias' in 449 BC.

Sources

The almost exclusive ancient source for the Persian Wars is the account by Herodotus, who presents the events from a Greek point of view. Concerning the Persian attitude, we can suspect from a late note in Dio Chrysostom (11,148f.) that the first expedition was meant to be a punitive operation to preserve the naval supremacy the Persians had built up since the reign of Cambyses (530-522), while only the second attempt aimed at shifting the western boundary of the Persian Empire from the Aegean to the Adriatic Sea, with Greece becoming another satrapy or a vassal state. There are no surviving Persian sources. (Further sources: Aesch. Pers.; Plut. Themistocles; Diod. Sic. 11,17–19.)

→ Maps pp. 71, 93, 95, 105; Persian Wars (1), BNP 10, 2007

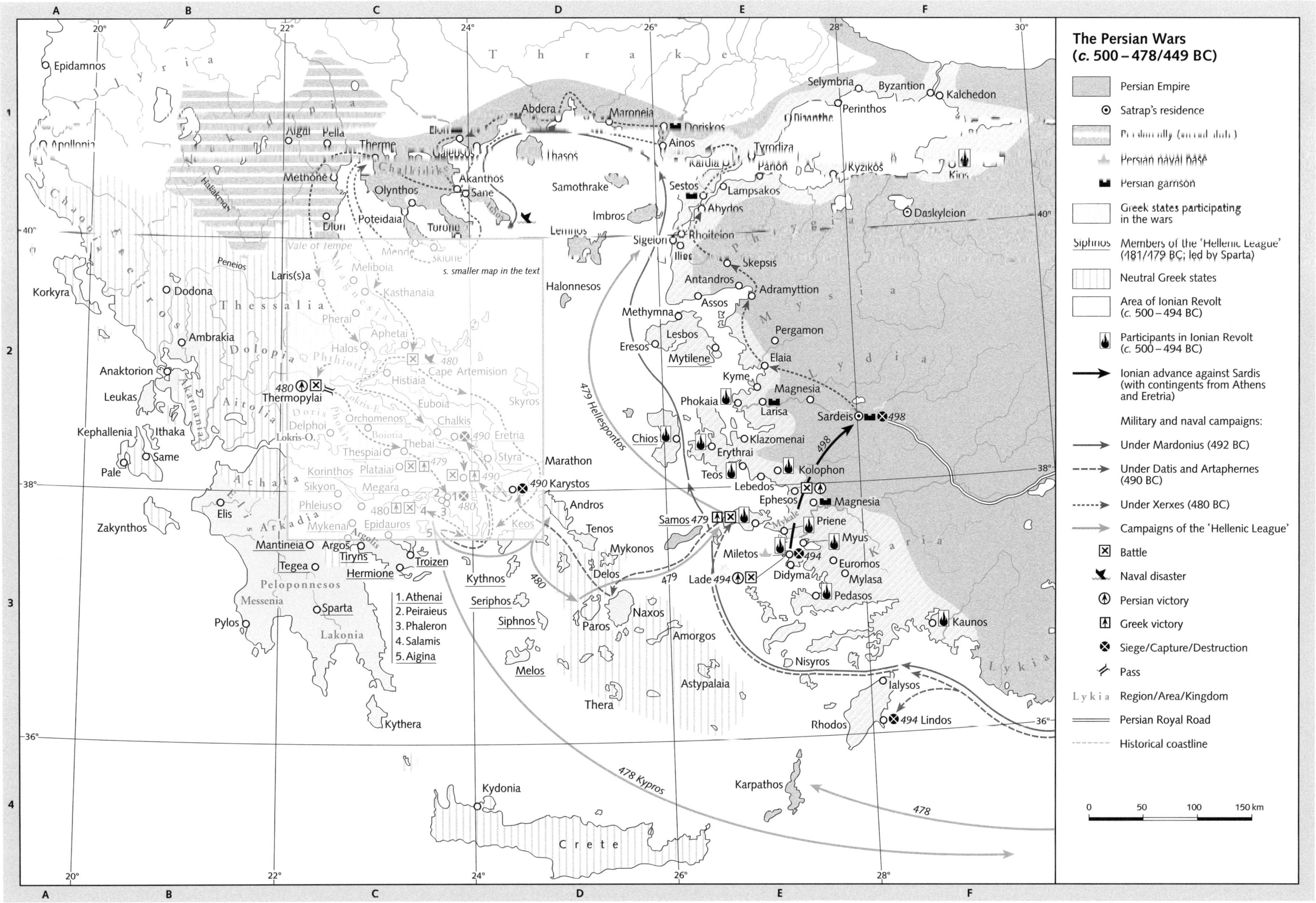

The Persian Wars
(c. 500 – 478/449 BC)
Persian Empire
Satrap's residence
Persian naval base
Persian garrison
Greek states participating in the wars
Siphnos Members of the 'Hellenic League' (481/479 BC; led by Sparta)
Neutral Greek states
Area of Ionian Revolt (c. 500 – 494 BC)
Participants in Ionian Revolt (c. 500 – 494 BC)
Ionian advance against Sardis (with contingents from Athens and Eretria)
Military and naval campaigns:
Under Mardonius (492 BC)
Under Datis and Artaphernes (490 BC)
Under Xerxes (480 BC)
Campaigns of the 'Hellenic League'
Battle
Naval disaster
Persian victory
Greek victory
Siege/Capture/Destruction
Pass
Lykia Region/Area/Kingdom
Persian Royal Road
Historical coastline
0 50 100 150 km
s. smaller map in the text
Thrake
Illyria
Epidamnos
Apollonia
Korkyra
Dodona
Ambrakia
Anaktorion
Leukas
Kephallenia
Ithaka
Same
Pale
Zakynthos
Elis
Pylos
Messenia
Peloponnesos
Lakonia
Sparta
Kythera
Mantineia
Tegea
Argos
Tiryns
Hermione
Troizen
1. Athenai
2. Peiraieus
3. Phaleron
4. Salamis
5. Aigina
Thessalia
Laris(s)a
Peneios
Dolopia
Akarnania
Aitolia
Thermopylai
480
Aigai
Pella
Methone
Therme
Dion
Poteidaia
Olynthos
Chalkidike
Akanthos
Sane
Torone
Abdera
Thasos
Samothrake
Maroneia
Doriskos
Ainos
Imbros
Lemnos
Halonnesos
Skyros
Marathon
490 Karystos
Andros
Tenos
Mykonos
Delos
Kythnos
Seriphos
Siphnos
Melos
Paros
Naxos
Amorgos
Thera
Astypalaia
Kydonia
Crete
Karpathos
Rhodos
Ialysos
494 Lindos
Nisyros
Kaunos
Pedasos
Mylasa
Euromos
Myus
Priene
Didyma
494
Miletos
Lade 494
Samos 479
Mykale
Ephesos
Magnesia
Kolophon
Lebedos
Teos
Erythrai
Klazomenai
Chios
Phokaia
Larisa
Kyme
Elaia
Pergamon
Lesbos
Mytilene
Eresos
Methymna
Assos
Antandros
Adramyttion
Skepsis
Ilion
Sigeion
Rhoiteion
Abydos
Sestos
Lampsakos
Parion
Kardia
Tyrodiza
Kyzikos
Daskyleion
Selymbria
Perinthos
Byzantion
Kalchedon
Sardeis
498
Mysia
Lydia
Karia
Lykia
Phrygia
479 Hellespontos
478 Kypros
478
480
479

City development and town planning in Greece

The layouts of Athens (Athenae) and Miletus typify two different types of Greek city structure, which would today be described as 'naturally grown' in the case of Athens, and as 'planned' to characterize Miletus.

I. Athens (maps A and B)

The overall map of ancient Athens (map A, overlaying the plan of today's Athens) shows the ring-wall (no. 20) erected directly after the destruction of Athens by the Persians (480 BC), the connection with the 'Long Walls' (to the southwest, no. 23) which around the middle of the 5th cent. were built to provide a corridor leading to Piraeus (constructed around the same time), and the positions of the functional zones within the ring-wall. These were – in the south – the religious centre located on the Acropolis from archaic times, a Mycenaean castle from the 2nd millennium BC (no. 1; see also map B), and – in the western part – the political and commercial centre with the *Agora* (no. 4, the site for meetings, court sessions, markets and celebrations), the *Areopagus* (see no. 2) and the *Pnyx* (no. 6, in its monumental 4th-cent. outline), where from *c.* 500 BC the popular assembly held its sessions. The buildings from the Roman era are very close to the centre: the Roman *agora,* a market place from the Augustan age (no. 8), meant to take the pressure off the Athenian Agora, which was cluttered by buidings, and the Library of Hadrian (AD 117–138; no. 12), who as an admirer of Greece made a point of bestowing favours on Athens (see also no. 13) and completed the construction of the temple to the Olympian Zeus (Olympieum, no. 14) begun by the Pisistratids as early as the 6th cent. At the same time the stadium (no. 17) was built on the far side of the Ilissus river (no. 26). It was a gift from the Athenian Herodes Atticus, who als had the *Odeion* erected at the soutwestern foot of the Acropolis (see map B, z).

The irregular orientation of the buildings shows that after its destruction Athens, unlike Miletus, returned to the lay-out of the 'naturally grown' city. As was generally the case in mainland Greece (e.g. in Argos, Corinth, Thebes) after the fall of the Mycenaean civilization in the 12th cent. BC, rural types of settlements had begun to appear on mountainsides and at the feet of fortified hills, where springs usually surfaced. From the 8th cent., these settlements grew and developed the community life of the emerging polis. From the clustering of old shrines south of the Acropolis, Thucydides (2,15) drew conclusions about the extent of the early settlement area, but the shrines located along the *Peripatos* (map B, y; m, n, o) would point to an even distribution of housing around the Acropolis, which was possibly still inhabited. Whenever the settlement grew, it was the course of the traffic routes that suggested the distribution patterns of living quarters, temples and public buildings, while the terrain determined the location and orientation of theatres and stadia. The merger of all Attic communities (*synoikismós*, Thuc. 2,15,2) attributed to mythical times (Theseus) – which in all likelihood did not happen before the 8th cent. – turned Athens into a political and religious centre, but, for the time being, failed to create an accelerated urbanization of either the Acropolis or the Agora, because the inhabitants of Attica remained in their communities.

Acropolis (map B)

It was only in the 6th cent. that alongside earlier shrines the first monumental temples went up: the 'Proto-Parthenon' (*c.* 580, see b); the old Pisistratid Temple of Athena (see e); and around 500 (?) the Pre-Parthenon, the first marble-built version, intended to replace the Proto-Parthenon but never completed; and the first Propylaea (next to c). All of these were destroyed by the Persians in 480 BC. Map B shows the Acropolis and its surroundings. The buildings in the map were erected on the hilltop, whose area had been significantly enlarged by the use of retaining walls (a), and also on the southern slope, from as late as 448 BC, when work on the Parthenon started, to the age of Imperial Rome (2nd cent. AD: z): the small theatre (q) built around 530 at the foot of the hill was extended considerably to the north in 330 (as shown in the map), but limited by the neighbouring Odeion of Pericles (p). Buildings a-d and f-h also stem from the 5th cent.; e was abandoned and functionally/religiously integrated into f. After that there was very little building activity, most of it in connection with dedications (cf. i). The temple for Roma and Augustus (k) was put up to establish the cult of the Roman Emperor in the place where traditionally the altar of the Parthenon was supposed to stand. In the Agora, Augustus made a similar shift of emphasis (see below).

Agora (→ map in BNP 1, 2002)

In spite of attempts from the 3rd cent. onward to redesign the space in the Hellenistic manner (by encasing it with peristyles), the original layout ('ancient style') is still shining through: individual houses and halls framing a square criss-crossed by streets. As late as the 7th cent., sections of the Agora were still used as burial sites, i.e. they were not yet part of the settlement area. It was not before the mid 6th cent. that a central public space serving the needs of Attica as a whole was created, with sites for theatres (orchestra), sports competitions and celebrations (Panathenaeic procession to the Acropolis); the altar of the Twelve Gods, which had been erected next to an ancient sacrificial structure (Eschara) around 520 BC, served as a central reference point for distances within the Attic network of roads. (The importance the Agora held for Attica lends weight to the highly controversial hypothesis of the existence of an earlier 'archaic Agora' [cf. map A, no. 10], if its function was restricted to serving the 9th- to 7th-cent. hillside settlements.) Apart from the cult shrines (for Zeus, Apollo, Patrous, Meter) on the west side and a fountain in the southeastern corner there is no recognizable architectural design. It was only in the 5th cent., especially in its second half, that the western flank was furnished with public buildings. In the south an architectural boundary was created by the South Stoa, several fountains and the mint, whereas the northeastern corner accommodated a law court. Apart from that the north and east sides remained largely undeveloped. The Temple of Hephaestus, begun in 449, dominated the west side and the complete Agora. In the 2nd cent. BC, the Agora was divided by the Middle Stoa; to the south a market area came into being, the northern half received an impressive eastern boundary by the addition of the Stoa of Attalus. In the reign of Augustus, the Odeion of Agrippa was raised (to be converted into a gymnasium in the 2nd cent. AD) and the classical Temple of Ares (the counterpart of the Mars Ultor Temple in Rome) was transferred from the demos of Pallene to the Agora, thus complementing the Roman 'occupation' of the Acropolis with that of the Agora.

II. Miletus (map C)

The map essentially represents the city, which had been destroyed by the Persians in 494, as it was rebuilt from 479/78 onwards and continued to be developed until Late Antiquity (latest buildings: nos. 11, 16, 27) with its 'Ionic-Hippodamian' street grid, buildings and open spaces and the ring-wall, which loosely encompassed the city. It was not, as was the case with all 'planned' cities, oriented towards the street-grid, but followed the terrain to achieve optimum defensibility. In spite of its impressive size the city did not match the spatial dimensions of the 7th and 6th cents. BC, when it extended from the northern tip of the peninsula (no. 1) to the Kalabaktepe (no. 33). In the 9th and 8th cents., the settlement had stretched from the archaic Temple of Athena (no. 26) to the Kalabaktepe, before it was destroyed at the end of the 8th cent. The earliest architectural finds provide evidence of a Minoan settlement from the 16th cent. BC and of two phases of Mycenaean settlement (14th and 13th/12th cents.), all at the site of the Athena Temple.

Unlike Athens, Miletus was rebuilt (after the destruction of 494) according to a new plan – possibly making use of the expertise acquired by Milesian colonists. The debris was methodically removed and dumped at Kalabaktepe (north of no. 33 and west of no. 32), which by then had been largely deserted. The plan was not based on a uniform module, nor was there an uncompromising adherence to axiality: in the hilly northern section (approx. bordered by the line from no. 20 to no. 21) a longitudinal rectangular module was used while an almost square one was chosen for the flat southern part. In addition, the north orientation of the northern grid is by 1.5 degrees more accurate than that of its southern match, which is why left of no. 22 the dotted extended lines of the streets do not meet accurately. The reason is presumably the orientation towards existing structures, most likely the shrine of Apollo (Delphinium, no. 3) for the northern grid and the temple of Athena (no. 26) for the southern one. It is remarkable that no highly visible spaces (special lines of sight and space designs) were chosen to highlight shrines and temples; temples (nos. 3, 11, 22, 26) and heroa (nos. 7, 17) 'disappear' into the street grid.

It seems that the hierarchic division into wide thoroughfares and minor streets to be found in other 'Hippodamian' cities was not applied to Miletus; only the street from the city centre to the classical/Hellenistic Sacred Gate (no. 28) was definitely a little wider than the others. In the town centre, wide spaces in the neighbourhood of the ports were left uncluttered by buildings: a broad strip stretching southward from the Lions' Harbour to what was to become the South Market (no. 19) intersected there with an even wider strip running from the Theatre Harbour to the East Harbour (?). Although these 'reserved spaces' were used for markets and other public purposes from the 5th cent. onward, it took centuries before they were slowly developed architecturally by the addition of Hellenistic and Imperial market buildings and stoai (nos. 5, 8, 9, 12, 19, 20, 24) as well as an impressive city hall (14) and were used to put up a stadium (no. 23) and vast baths (no. 21). The central square (next to nos. 12 and 15), from where the annual procession on the Sacred Road (no. 34) to Didyma set out, had its architectural design rounded off no less than 600 years after the master plan was drawn up, when a fountain (no. 15) and the gate to the South Market (south of no. 15) were completed. The reason why this 'grown' planned city nevertheless creates an impression of unity is to be found in its unchanging commitment to the orientation and size of the module. Even though the Baths of Faustina (no. 21) were oriented to the south – ignoring the grid pattern – to make better use of the sun for their caldaria, on their eastern flank they still comply with the northern grid pattern, although this entailed architectural complications.

→ Maps pp. 23, 25, 27, 29, 69, 87; Hippodamus, BNP 6, 2005; Town planning [B., C.], BNP 14, 2009

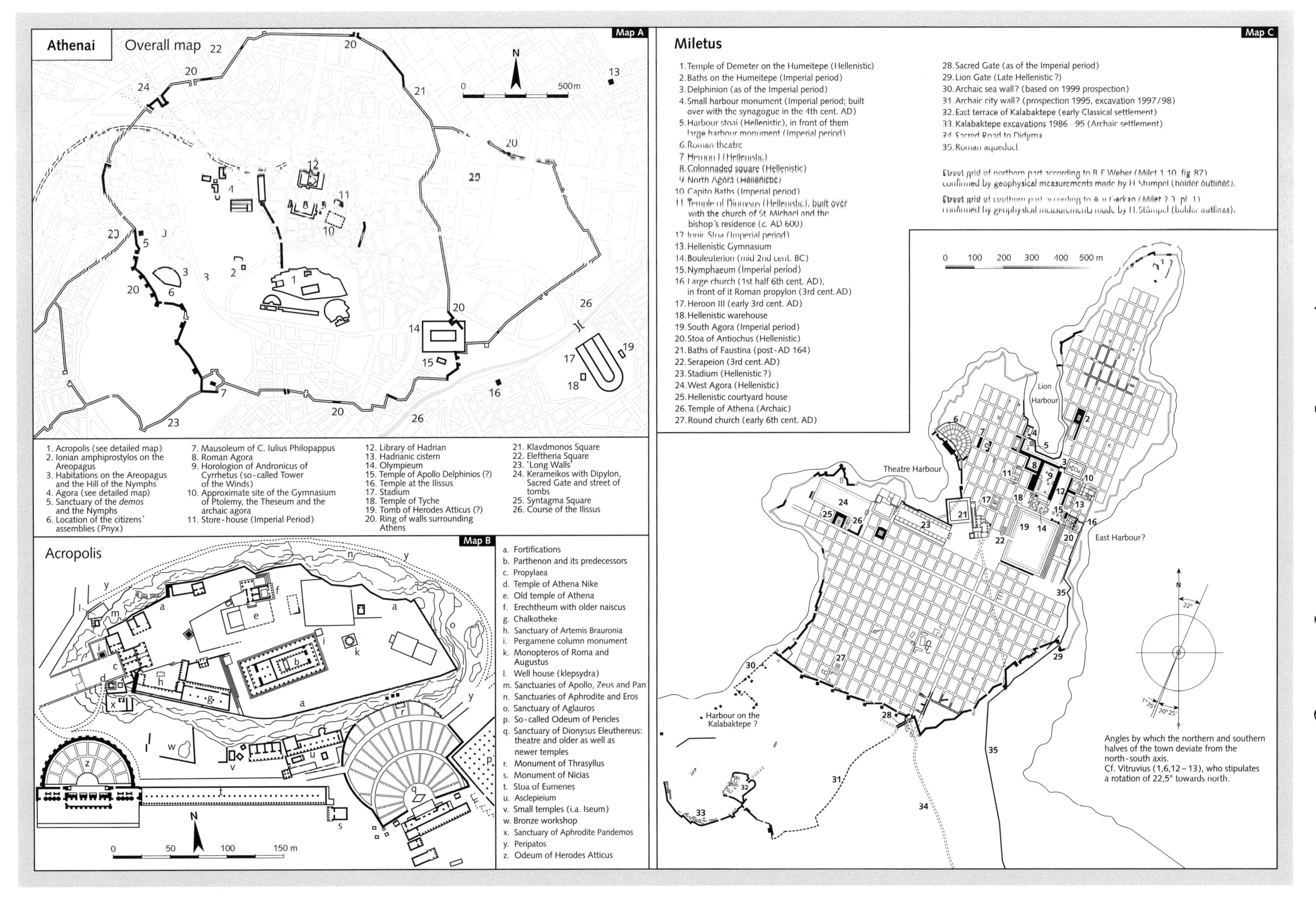
Athenai
Overall map
Map A
0 500m
N
1. Acropolis (see detailed map)
2. Ionian amphiprostylos on the Areopagus
3. Habitations on the Areopagus and the Hill of the Nymphs
4. Agora (see detailed map)
5. Sanctuary of the demos and the Nymphs
6. Location of the citizens' assemblies (Pnyx)
7. Mausoleum of C. Iulius Philopappus
8. Roman Agora
9. Horologion of Andronicus of Cyrrhetus (so-called Tower of the Winds)
10. Approximate site of the Gymnasium of Ptolemy, the Theseum and the archaic agora
11. Store-house (Imperial Period)
12. Library of Hadrian
13. Hadrianic cistern
14. Olympieum
15. Temple of Apollo Delphinios (?)
16. Temple at the Ilissus
17. Stadium
18. Temple of Tyche
19. Tomb of Herodes Atticus (?)
20. Ring of walls surrounding Athens
21. Klavdmonos Square
22. Eleftheria Square
23. 'Long Walls'
24. Kerameikos with Dipylon, Sacred Gate and street of tombs
25. Syntagma Square
26. Course of the Ilissus
Acropolis
Map B
0 50 100 150 m
N
a. Fortifications
b. Parthenon and its predecessors
c. Propylaea
d. Temple of Athena Nike
e. Old temple of Athena
f. Erechtheum with older naiscus
g. Chalkotheke
h. Sanctuary of Artemis Brauronia
i. Pergamene column monument
k. Monopteros of Roma and Augustus
l. Well house (klepsydra)
m. Sanctuaries of Apollo, Zeus and Pan
n. Sanctuaries of Aphrodite and Eros
o. Sanctuary of Aglauros
p. So-called Odeum of Pericles
q. Sanctuary of Dionysus Eleuthereus: theatre and older as well as newer temples
r. Monument of Thrasyllus
s. Monument of Nicias
t. Stoa of Eumenes
u. Asclepieium
v. Small temples (i.a. Iseum)
w. Bronze workshop
x. Sanctuary of Aphrodite Pandemos
y. Peripatos
z. Odeum of Herodes Atticus
Miletus
Map C
1. Temple of Demeter on the Humeitepe (Hellenistic)
2. Baths on the Humeitepe (Imperial period)
3. Delphinion (as of the Imperial period)
4. Small harbour monument (Imperial period; built over with the synagogue in the 4th cent. AD)
5. Harbour stoai (Hellenistic), in front of them large harbour monument (Imperial period)
6. Roman theatre
7. Heroon I (Hellenistic)
8. Colonnaded square (Hellenistic)
9. North Agora (Hellenistic)
10. Capito Baths (Imperial period)
11. Temple of Dionysus (Hellenistic), built over with the church of St. Michael and the bishop's residence (c. AD 600)
12. Ionic Stoa (Imperial period)
13. Hellenistic Gymnasium
14. Bouleuterion (mid 2nd cent. BC)
15. Nymphaeum (Imperial period)
16. Large church (1st half 6th cent. AD), in front of it Roman propylon (3rd cent. AD)
17. Heroon III (early 3rd cent. AD)
18. Hellenistic warehouse
19. South Agora (Imperial period)
20. Stoa of Antiochus (Hellenistic)
21. Baths of Faustina (post-AD 164)
22. Serapeion (3rd cent. AD)
23. Stadium (Hellenistic ?)
24. West Agora (Hellenistic)
25. Hellenistic courtyard house
26. Temple of Athena (Archaic)
27. Round church (early 6th cent. AD)
28. Sacred Gate (as of the Imperial period)
29. Lion Gate (Late Hellenistic ?)
30. Archaic sea wall? (based on 1999 prospection)
31. Archaic city wall? (prospection 1995, excavation 1997/98)
32. East terrace of Kalabaktepe (early Classical settlement)
33. Kalabaktepe excavations 1986 – 95 (Archaic settlement)
34. Sacred Road to Didyma
35. Roman aqueduct
Street grid of northern part according to B. F. Weber (Milet 1.10, fig. 87) confirmed by geophysical measurements made by H. Stümpel (bolder outlines).
Street grid of southern part according to A. v. Gerkan (Milet 2.3, pl. 1) confirmed by geophysical measurements made by H. Stümpel (bolder outlines).
0 100 200 300 400 500 m
Lion Harbour
Theatre Harbour
East Harbour?
Harbour on the Kalabaktepe ?
N
22°
1°35'
20°25'
Angles by which the northern and southern halves of the town deviate from the north-south axis.
Cf. Vitruvius (1,6,12 – 13), who stipulates a rotation of 22,5° towards north.

Tyrannis in the Archaic and Classical Periods

The maps show the (geographical) spread of the Greek *tyrannis* in the Archaic (map A) and Late Classical (map B) periods. This division (first made by Plass) is justified both because of the proven absence of tyrannies in the phase between the end of the earlier *tyrannis* (461 BC, end of Sicilian *tyrannis*) and the later ones (405 BC, Dionysius I of Syracuse) and because of the different political circumstances leading to their emergence.

I. The Archaic *tyrannis* (mid 7th cent. to early 5th cent.; map A)

The archaic tyrannies emerged in a phase of poorly developed statehood as a special form of aristocratic rule. It was a common Greek phenomenon with shared basic features: growing differences in aristocratic wealth because of agriculture, trade or piracy; colonization: port cities/'emigration ports' invariably became the seats of tyrannies; tyrants were members of respected aristocratic families; articulated aristocratic self-representation; it was a short-lived type of government. However, there were also local peculiarities. In Greece, *tyrannis* emerged early (Corinth 657), its objective was to keep down aristocratic competitors and despite widely spread connections, its rule was restricted to the polis in question. In Sicily (from *c.* 600/570) it tended to aspire to territorial gains, especially when it flourished (Gela, *c.* 505-498/7), and to grant a share of the power to aristocratic families.

On the coast of Asia Minor and on the islands, *tyrannis* was also a product of aristocratic infighting from the 7th cent. onward (e.g. Samos, Miletus). From the end of the 6th cent. onward, most of the tyrants there owed their power to a pro-Persian stance or to their appointment by the King of Persia ('vassal tyrants' with military obligations; cf. Hdt. 4,138). There is also documentary evidence of such dependent city overlords ruling over Cos (*c.* 500), Phocaea (late 6th cent.), Mytilene (late 6th cent.) and Halicarnassus (late 6th cent.). There were exceptions: Miltiades the Younger on the Thracian Chersonesus, Hegesistratus at Sigeum and Polycrates of Samos (governing from 538/7 with his brothers, from 533/2 on his own) retained a degree of independence. The latter got involved with Delos (Rheneia) and created a network of relationships with Lygdamis of Naxos, Pharaoh Amasis and Arcesilaus of Cyrene. It was only after his death in 522/21 that his successors were increasingly drawn into the Persian sphere of influence.

Corinth: The thirty years of *tyrannis* under Cypselus were probably made possible by the consent of numerous aristocrats ('popular' tyrant) after the overthrow of the ruling Bacchiadae clan (657/6), who sought refuge in places where they had friends (e.g. Corcyra, Tarquinii: Demaratus with artisans, Cic. Rep. 2,34). Cypselus and his son Periander (627/26–587/86) continued the Bacchiad policy of trade and colonization, exerting unusually close control over the newly founded settlements (e.g. Ambracia, Epidamnus und Potidaea) by setting up their sons as local rulers. Corcyra was taken over by military force and dynastic ties were forged with Athens and Epidaurus. Proven links with Miletus (Thrasybulus), Lydia (King Alyattes) and Egypt (Psammetichus) point to an expansion of trade relations.

Sicyon was the home of the most long-lived line of tyrants: 656/5 to 556/5. Their rule was dominated by wars against Pellene, Argos and Crisa, probably to fund their retinue, but there are also the typical phenomena of self-promotion like Olympic victories, buildings and dedications at Olympia and Delphi alongside dynastic links with Athens.

Cylon's failed attempt in 636 or 632 to secure the Athenian *tyrannis* for himself with the help of his father-in-law, Theagenes, the tyrant of Megara, was seen as a timely warning by the Athenian aristocracy, who were now prepared to scale down their infighting (Draconian Laws, Solon's reforms). After two failures (561/0) and 556/5), Pisistratus had to wait until 546/5 before he succeeded – with external military support (Argos, Eretria, Naxos, Thebes, Thessaly) – in defeating his aristocratic competitors. He passed power on to his sons in 528/7, who were removed by Sparta at the instigation of the Alcmaeonids with support from Delphi (511/10). The map focusses on the web of dynastic relationships and mutual aristocratic obligations of guest-friendship cultivated by Pisistratus and his sons with Argos, Lampsacus (marriages), Eretria (refuge in 555), Rhaecelus (where he joined the colonists), Sparta, Thebes, Thessaly, Macedonia as well as with Lygdamis, who helped Pisistratus seize power and had the favour returned when he aspired to the *tyrannis* of Naxos. Sigeum was conquered; on the Pangaeum, Pisistratus had the rights to mine the precious metals. Unlike other tyrants, Pisistratus and his sons used the shrines at Delos, Athens and Attica to present themselves on the Panhellenic stage.

The living conditions of the colonists in Sicily and southern Italy, a potentially hostile environment (Sicani, Siculi), precluded any interior wrangling for power and promoted homogeneity in the ruling class. For these reasons *tyrannis* only made a late entry (Cleander of Gela *c.* 505–498, his brother Hippocrates 498/97–491/90; historically hardly documented: Leontini around 600, Acragas *c.* 570–555). The usual patterns of aristocratic conduct were now supplemented by aristocratic power-sharing and an emphatically expansionist policy. Thus Gelon (491/90–478) managed to seize almost the whole of eastern Sicily, to take Syracuse in 485 and from that year to wage a successful war against the Carthaginians in the west of the island, in collaboration with his father-in-law Theron of Acragas (*c.* 489–474/73) (in 480 BC he won a battle at Himera, whose tyrant Terillus had called on Carthage for help after being deposed by Theron). His brother Hieron, tyrant of Gela from 485, tyrant of Syracuse from 478-466, continued the policy of expansion. He intervened in southern Italy (Locri, Sybaris) and in a naval battle off Cumae defeated the Etruscans in 474. Thrasybulus, Hieron's brother and successor (466/65), was almost immediately expelled. This also meant the end of Syracusan rule.

An equally expansionist political philosophy, although rarely crowned by success, characterized the *tyrannis* of Anaxilaus of Rhegium (494–476): occupying Zancle/Messana *c.* 492 and thus gaining control of the straits until 461, he was defeated in 480 together with his father-in-law, Terillus of Himera, in the battle of Himera. Regency for his sons from 476 to 467; they were overthrown in 461.

In 504, Aristodemus of Cumae established an apparently harsh *tyrannis*, albeit one which seemed to lack any expansionist traits. He was murdered in 491/90.

II. The Classical-Period *tyrannis*

The later *tyrannis*, which required the tyrant to impose his will on fully developed political entities, is mainly found in three regions – all of them rather close to the margins of the Greek world: Sicily, Thessaly and the eastern Aegean region, but the causes behind its emergence range from external pressure (Carthaginians in Sicily), a retarded development of the *polis* (Thessaly) to a shift in the balance of power (decline of Athenian and Spartan power in the Aegean).

In Sicily in 405, the *tyrannis* of Dionysius I of Syracuse evolved from a democracy and a general's cunning. He had been given plenipotentiary powers to fight the Carthaginians, who had already taken Acragas. Using his personal guards he became a *de facto* tyrant and held on to this position for almost forty years, by taking populist measures like confiscations and the distribution of land, by keeping an army of mercenaries and by staffing all key positions with friends and relations. In spite of inauspicious beginnings – the peace treaty of 405 gave western Sicily as well as Gela and Camarina to Carthage, the cities of eastern Sicily were given guarantees of independence and there was a revolt of the citizens' army in 404 – Dionysius extended his rule to eastern Sicily after 404, allied himself with Locri in southern Italy in 399 and advanced to the western tip of Sicily in 398. The Carthaginian counter-attack nullified all these successes in 397. Syracuse was besieged, but an epidemic among the besieging army saved Dionysius, so that the peace treaty of 392 only left the west with Motya, Panormus and Solus along with their surroundings in Carthaginian hands, while a tight network of military colonies and allies kept a close watch on the bulk of Sicily's territory on the tyrant's behalf. Next Dionysius widened the scope of his activities to include southern Italy (388 victory over the Italiote League on the Elleporus, 387/86 conquest of Rhegium/Straits of Messina) and the Adriatic (385/84 foundation of colonies at Lissos and (L)issa, possibly also Ancona and Atria) established relations with the Molossi, Illyrians and Celts, raided the Etruscan port of Pyrgi and crossed over to Corsica from there. Around 382, he pre-empted an alliance between Carthage and southern Italy with another war on the Carthaginians, conquered Croton and defeated a Carthaginian army. After suffering a defeat, however, he was forced in 374 to recognize the Halycus (Platani) boundary. He was prevented from correcting this in another war (from 368) by his death in 367. With the exception of the period from 355–347, his son Dionysius II, whose succession was confirmed by the populace (!), governed the shrinking territory until 344, when Timoleon put an end to this Sicilian *tyrannis*. Nevertheless, tyrannies, especially Syracusan ones, continued to spring up and govern vast territories (Agathocles 316/5–289/8, Hieron II 275/4–215/4) until Syracuse was conquered by Rome in 212.

In Thessaly, *tyrannis* was a result of the traditional rivalry between the ancient aristocratic families of Pherae, Larisa and Pharsalus, who exploited their leadership role in the Thessalian League or called on external powers for help (Sparta, Thebes, Macedonia, Phocaea). Lycophron of Pherae founded the *tyrannis* in 404 after a victory over Thessalian cities, among them Larisa, which, however, managed to regain its freedom with Persian help and also overcame Pharsalus in 395. Lycophron's son(-in-law) Iason became his successor in 390. From 380 onward, he took the power of Pherae to its peak: commanding an army of mercenaries and from 375 as leader/*tagos* of the Thessalian League, he succeeded in becoming the ruler of the whole of Thessaly and the areas bordering it in the west and the north. Allied to Macedonia, Thebes and Athens (?), he aimed at maintaining a balance of power in Greece. A nephew of Iason's, who succeeded him both as *tagos* and tyrant (369-358), however, was only able to retain the rule over Pherae und Pagasae. A call for help by the Aleuads of Larisa gave Philip II of Macedonia a foothold in Thessaly (352 BC).

The tyrannies in the eastern Aegean, the Propontis and the Black Sea region did not develop a distinctive pattern. This was due to the unstable balance of power ('Great Satraps' Revolt' *c.* 370–350; Philip II's expansionist schemes; Alexander's military expedition; Wars of the Diadochi). Power-hungry individuals either exploited an anarchic period to establish a *tyrannis* or became tyrants with the support of the rulers at the time.

The map also includes cities where a *tyrannis* – mostly brief and frequently not reliably dateable – has been documented, although the term seems to have been used somewhat arbitrarily.

→ Maps pp. 69, 77, 89, 95, 97, 105; Tyrannis, Tyrannos, DNP 15, 2009

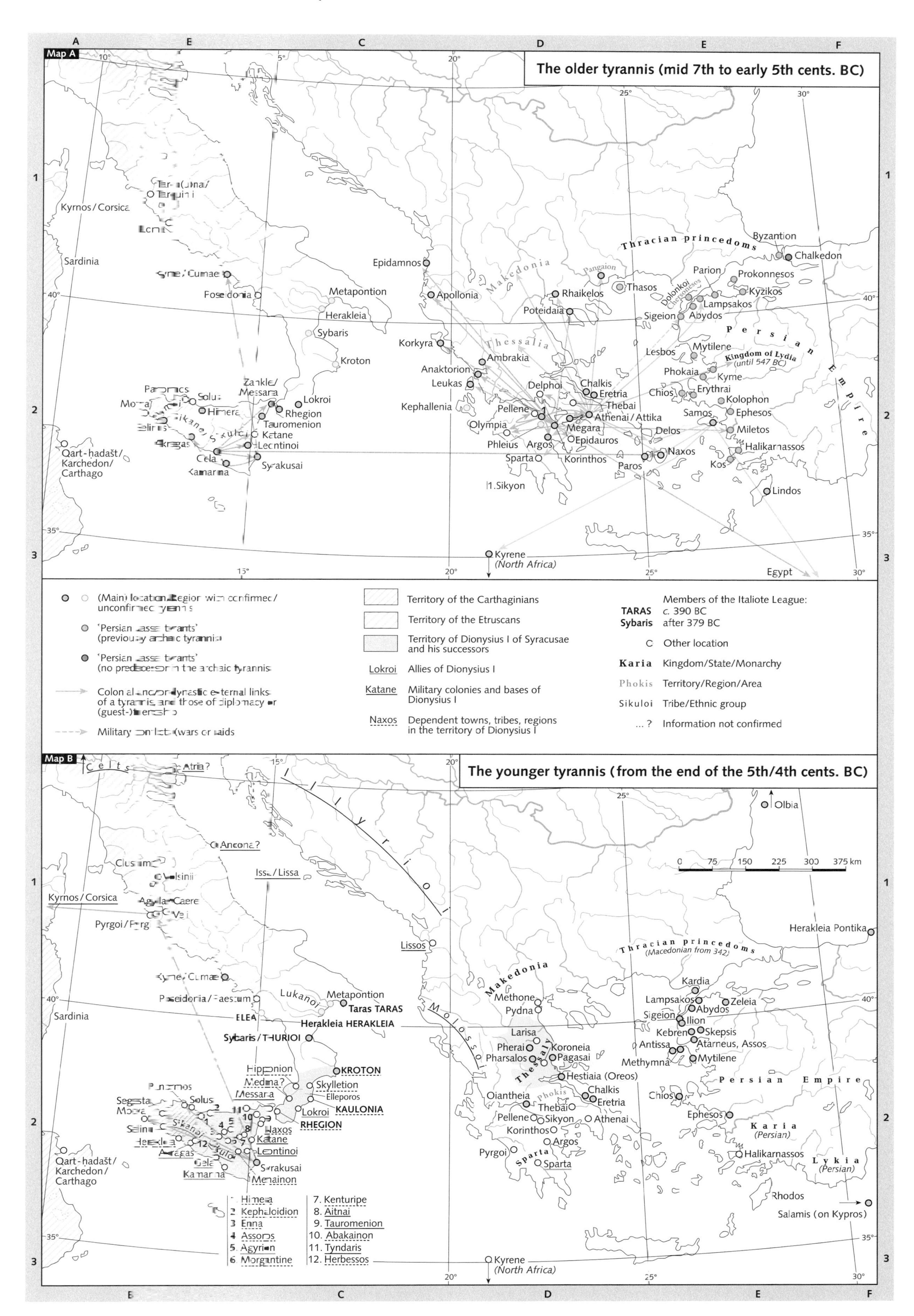
Map A
The older tyrannis (mid 7th to early 5th cents. BC)
Kyrnos/Corsica
Sardinia
Epidamnos
Apollonia
Metapontion
Herakleia
Sybaris
Kroton
Lokroi
Rhegion
Tauromenion
Syrakusai
Korkyra
Ambrakia
Anaktorion
Leukas
Kephallenia
Olympia
Delphoi
Chalkis
Eretria
Thebai
Athenai/Attika
Megara
Epidauros
Korinthos
Phleius
Argos
Sparta
1.Sikyon
Pellene
Rhaikelos
Poteidaia
Thasos
Sigeion
Abydos
Lampsakos
Kyzikos
Prokonnesos
Parion
Byzantion
Chalkedon
Thracian princedoms
Makedonia
Thessalia
Lesbos
Mytilene
Kingdom of Lydia
(until 547 BC)
Phokaia
Kyme
Chios
Erythrai
Kolophon
Samos
Ephesos
Miletos
Delos
Naxos
Paros
Halikarnassos
Kos
Lindos
Persian Empire
Qart-hadašt/
Karchedon/
Carthago
Kyrene
(North Africa)
Egypt
Territory of the Carthaginians
Territory of the Etruscans
Territory of Dionysius I of Syracusae and his successors
Lokroi Allies of Dionysius I
Katane Military colonies and bases of Dionysius I
Naxos Dependent towns, tribes, regions in the territory of Dionysius I
Members of the Italiote League:
TARAS c. 390 BC
Sybaris after 379 BC
C Other location
Karia Kingdom/State/Monarchy
Phokis Territory/Region/Area
Sikuloi Tribe/Ethnic group
... ? Information not confirmed
Map B
The younger tyrannis (from the end of the 5th/4th cents. BC)
Celts
Atria?
Ancona?
Issa/Lissa
Illyrioi
Olbia
Kyrnos/Corsica
Pyrgoi
Sardinia
Lissos
Herakleia Pontika
Thracian princedoms
(Macedonian from 342)
Makedonia
Methone
Pydna
Kardia
Lampsakos
Zeleia
Abydos
Sigeion
Ilion
Kebren
Skepsis
Antissa
Atarneus, Assos
Methymna
Mytilene
Poseidonia/Paestum
Metapontion
Taras TARAS
ELEA
Herakleia HERAKLEIA
Sybaris/THURIOI
KROTON
Skylletion
Elleporos
Lokroi
KAULONIA
RHEGION
Naxos
Katane
Leontinoi
Syrakusai
Kamarina
Menainon
Larisa
Pherai
Pharsalos
Koroneia
Pagasai
Hestiaia (Oreos)
Chalkis
Eretria
Thebai
Athenai
Oiantheia
Pellene
Sikyon
Korinthos
Argos
Pyrgoi
Sparta
Chios
Ephesos
Karia
(Persian)
Halikarnassos
Lykia
(Persian)
Rhodos
Salamis (on Kypros)
Persian Empire
Qart-hadašt/
Karchedon/
Carthago
Kyrene
(North Africa)
0 75 150 225 300 375 km
7. Kenturipe
8. Aitnai
9. Tauromenion
10. Abakainon
11. Tyndaris
12. Herbessos
3 Enna
5 Agyrion
6 Morgantine

The Delian League (478-404 BC)

'Delian League' (also 'Attic Symmachy', in Antiquity: 'the Athenians and their allies') is a scholarly term signifying the standing alliance Athens called into being after the Persian Wars of the 5th cent. BC. Nominally it was an Athenian-led alliance with the Ionians, though not exclusive of others, and a coalition of – at least for the time being – free and independent members (Thuc. 1,96–97, etc.). It was designed to keep the Persians at bay and to defend the freedom of the Greeks, especially that of their cities in Asia Minor. At that time Sparta, which had already handed the leadership of the Hellenic League to Athens, was busy stabilizing its rule over the Peloponnese (Messenia) and its hegemony within the Peloponnesian League (Elis, Arcadia). Thus neither Sparta nor its allies joined the Delian League. Some Greek states remained neutral.

I. The structure of the League

This system of military alliances was based on bilateral, mutual support treaties, for an unlimited period, which Athens, the *hegemon* and owner of he largest fleet, concluded in 478/488 with numerous island and coastal *poleis.* The basic structures of the system were similar to those of the Peloponnesian League, but its organizational arrangements were distinctly more close-knit. The allies either put ships at the League's disposal or paid contributions (*phoroi*), which reflected the individual members' financial potential. However, these had to be paid regularly, not just in the event of a war as in the Peloponnesian League. In essence the money was used to build and maintain a common fleet. The League's war chest was managed exclusively by Athenian treasurers. Reviving the ancient Ionian amphictyony of the Ionian islands, they kept it at the Delian shrine of Apollo, which also served as the convention venue of the League's assembly. There every member was entitled to cast its own vote, but Athens was as dominant as Sparta was within the Peloponnesian League.

II. The League's (i.e. Athens') political objectives

Especially in the Cimonian era of the 470s and 460s, top priority was accorded to the struggle against Persia. Thus around 476 the Persians were ejected from Eion, their last outpost on European soil, and around 466 defeat on land and sea was inflicted on the Persian forces at the mouth of the river Eurymedon in Pamphylia. However, the very first activities also clearly betrayed the pursuit of Athenian self-interest as it strove for the expansion of its sphere of influence, albeit without risking conflict with Sparta. The settlement of Athenians at Eion, the conquest of the island of Scyros in 475, where subsequently a cleruchy was founded, and the forced accession of Carystus (475/470?) to the Delian League were all steps to protect the sea routes to the Black Sea, but also the ones to the northern and the northwestern Aegean region, all of which were of vital importance to Athens. Any resistance from allies met with harsh retribution: the revolts by Naxos (467) and Thasos (465) were punished by confiscating their fleets and collecting huge tribute payments. In 461 relations with Sparta broke down and in the following years Athens attempted to expand its sphere of power in mainland Greece at Sparta's expense. Practising a skilful policy of alliances, Athens drew Thessaly, Megara and Argos, later on also Troezen, to its side; Aegina (456), Zacynthos, Cephallenia and coastal settlements (455) in the northern Gulf of Corinth were either incorporated into the League or conquered. This was a massive encroachment on the interests of both Sparta and Corinth. In the same period, Athens also tried to extend its sphere of power in the Aegean and the eastern Mediterranean region at the expense of Persia. In 460 (and again around 450) it attacked Cyprus and the Levantine coast with a large fleet and subsequently supported an anti-Achaemenid rebellion in Egypt (the 'Egyptian expedition'). However, in doing so it had overextended its resources, so that the Delian League suffered defeats in Thessaly as well as in Egypt (454).

III. The road to the 'Athenian Empire'

These defeats were probably used as a pretext for overhauling the organizational structures of the Delian League in 454. The League's funds were transferred to Athens, its assembly was dissolved and the exclusive right to make decisions on matters concerning the League handed to the popular assembly of Athens. Athens legislated for the League, decided how the tribute payments (which had replaced the voluntary *phoroi*) were to be spent and became the venue for all trials of capital crimes; the introduction of standard weights and measures was enforced throughout the League, its territory was divided up into (tax) districts and – to secure Athenian rule – the allied cities were classified as colonies (*apoikiai)* and tied to Athens by religious rites. Ironically the Delian League seems to have played a decisive part in the breakthrough of democracy in Athens and its spread – sometimes against the recipients' wishes – from there to other cities. In the following years, Athens was mainly concerned with preserving what it had achieved: in 449/8, an agreement with the Persians (Peace of Callias) was arrived at, in effect removing the reason for the League's existence and the justification for collecting *phoros* payments; but the League was not dissolved, because it was doubtful if the Persians would actually honour the agreement.

In 446/5, a peace agreement between Athens and Sparta, meant to last thirty years, terminated almost two decades of a conflict which in present-day scholarship is termed the First Peloponnesian War (460/58/57–446). Due to its hegemonial position as a naval power, which by then had also been recognized by Sparta, Athens was able to restore order to the crumbling system (revolts, withdrawals) of the Delian League and stabilize it. Athens expanded, wherever this was possible, its sphere of power – even entering alliances with some Greek cities in Sicily. Both the chronology and the objectives of these activities are a matter of controversy among scholars; it seems plausible, however, that Athens was planning to reach out for Sicily. Athens also increasingly interfered with the interior policy of its allies, thus further increasing tensions within the League which threatened to put the League itself at risk. The contrast between the aggressive policy pursued by the Athenians both within the Delian League and beyond its boundaries and the rather cautious attitude displayed by Sparta also created an imbalance which threatened to destabilize the agreement between the Delian and the Peloponnesian Leagues. The 'Samian War' of 440, when Samos and Byzantium defected, illustrates the problems within the Delian League, while the conflict about Epidamnus (naval battle (433) off the Sybota islands south of Corcyra, involving Corcyra, allied with and supported by Athens, and Corinth, allied with, but not actively supported by Sparta, cf. → map p. 97) might have had different consequences if both Sparta and Athens had handled their allies differently. The precarious peace with Sparta was put at risk by Athens even more when towards the end of the 430's Megara, a close ally of Corinth and also a member of the Peloponnesian League, was barred from all ports in the Delian League's domain. At the same time Potidaea, a member of the Delian League, was called upon to break off all contact with its mother city of Corinth. These actions finally triggered another Peloponnesian War. After taking over from Sparta as the leading power of the Hellenic League in 479, Athens had successively strengthened its position of dominance within the Delian League, originally a federalistic organization, and established itself as the central power by measures such as setting up military bases and colonies (cleruchies) within the domain of the League, thus transforming it into the territory of the 'Athenian Empire'. Towards the end of the Peloponnesian War, this naval empire rapidly fell to pieces and was terminated by the Athenians' defeat and capitulation in 404.

Sources

The main source for the history of these events is Thucydides' (*c.* 460 to early 3rd cent.) somewhat summary account in the first book of his *History of the Peloponnesian War*, which is supplemented by isolated additional references. There is also a wealth of inscriptions containing e.g. Athenian tribute lists (from 454 onwards), treaties between Athens and individual members of the Delian League and Athenian legislation concerning the League.

→ Maps pp. 89, 97, 127; Delian League, BNP 4, 2004

Literature

C. Habicht, Athen, 1995; S. Hornblower, The Greek World 479–323 B.C., [3]2002; D.B. Meritt et al., The Athenian Tribute Lists, 1939–1953; M.C. Miller, Athens and Persia in the Fifth Century B.C., 1997; P.J. Rhodes, The Athenian Empire, [2]1993; Id., A History of the Classical Greek World: 478–323 B.C., 2006; N. Salomon, Le cleruchie di Atene, 1997; Ch. Schubert, Athen und Sparta in klassischer Zeit, 2003; R. Schulz, Athen und Sparta, 2003; P. Siewert, L. Aigner-Foresti, Föderalismus in der griechischen und römischen Antike, 2005; M. Stahl, Gesellschaft und Staat bei den Griechen, vol. 2, 2003; M. Steinbrecher, Der Delisch-Attische Seebund und die athenisch-spartanischen Beziehungen in der Kimonischen Ära (ca. 478/7–462/1), 1985; L. Thommen, Sparta. Verfassungs- und Sozialgeschichte einer griechischen Polis, 2003; K.-W. Welwei, Das klassische Athen, 1999; Id., Sparta, 2004; W. Will, Thukydides und Perikles, 2003.

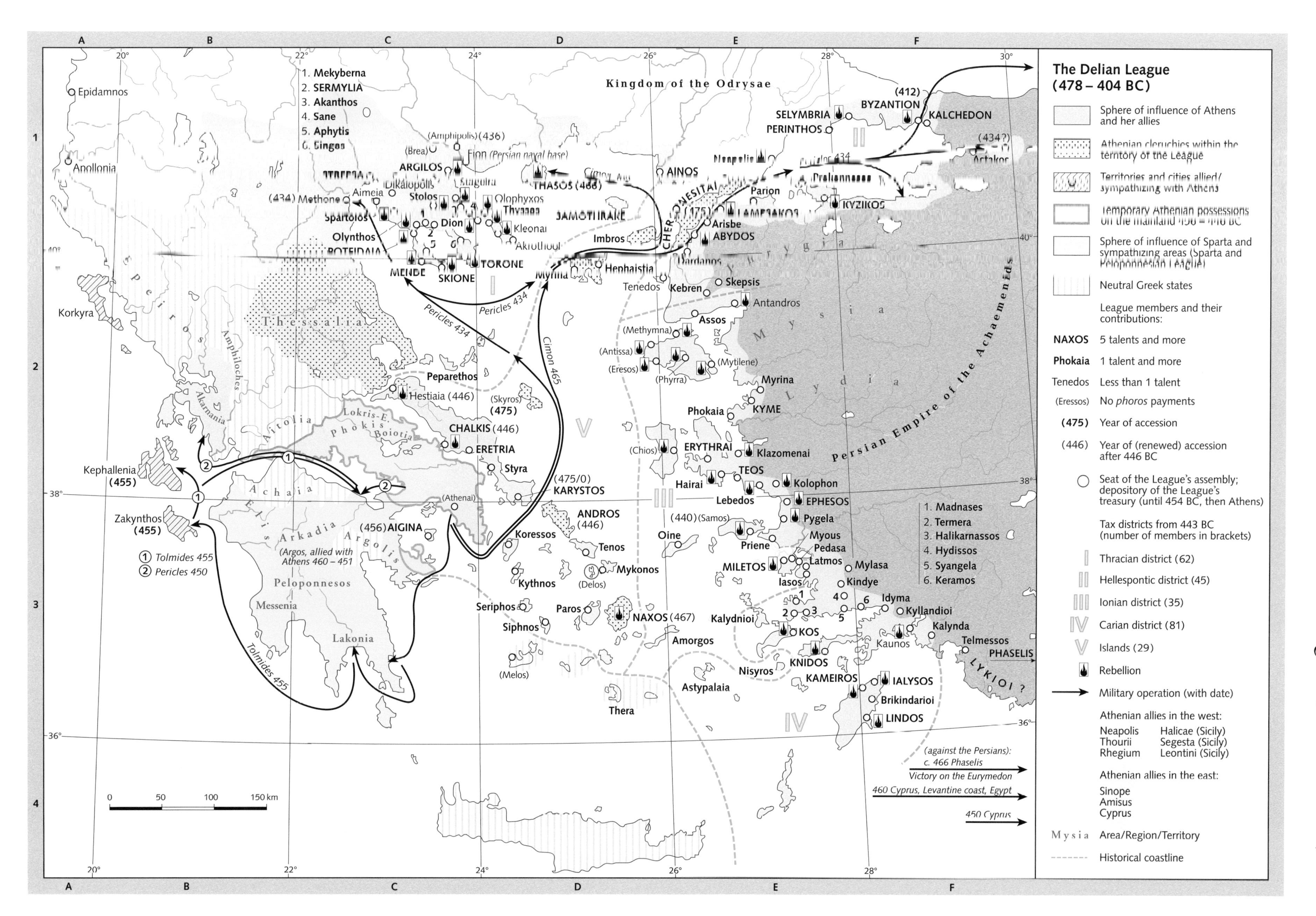
The Delian League (478 – 404 BC)
Sphere of influence of Athens and her allies
Athenian cleruchies within the territory of the League
Territories and cities allied/sympathizing with Athens
Temporary Athenian possessions on the mainland
Sphere of influence of Sparta and sympathizing areas (Sparta and Peloponnesian League)
Neutral Greek states
League members and their contributions:
NAXOS 5 talents and more
Phokaia 1 talent and more
Tenedos Less than 1 talent
(Eressos) No phoros payments
(475) Year of accession
(446) Year of (renewed) accession after 446 BC
Seat of the League's assembly; depository of the League's treasury (until 454 BC, then Athens)
Tax districts from 443 BC (number of members in brackets)
Thracian district (62)
Hellespontic district (45)
Ionian district (35)
Carian district (81)
Islands (29)
Rebellion
Military operation (with date)
Athenian allies in the west:
Neapolis Halicae (Sicily)
Thourii Segesta (Sicily)
Rhegium Leontini (Sicily)
Athenian allies in the east:
Sinope
Amisus
Cyprus
Mysia Area/Region/Territory
Historical coastline
Kingdom of the Odrysae
Persian Empire of the Achaemenids
1. Mekyberna
2. SERMYLIA
3. Akanthos
4. Sane
5. Aphytis
1. Madnases
2. Termera
3. Halikarnassos
4. Hydissos
5. Syangela
6. Keramos
Cimon 465
Pericles 434
Tolmides 455
① Tolmides 455
② Pericles 450
(against the Persians):
c. 466 Phaselis
Victory on the Eurymedon
460 Cyprus, Levantine coast, Egypt
450 Cyprus
0 50 100 150 km

The Peloponnesian War (431– 404 BC)

The theme of the map is 'the war between the Peloponnesians and the Athenians' (Thuc. 1,1,1) which was also called 'the Peloponnesian War' in subsequent ancient writings (Ephorus, the Hellenistic chronographers, Diod. Sic. 12,37,2 etc.; Cic. Rep. 3,44). Among scholars this term is also used for the so-called First Peloponnesian War (460/58/57–446), which means that this one would have to be termed the Second Peloponnesian War. It covers the military conflict between Athens and its Delian League allies (see Delian League, → map p. 95) on the one hand and Sparta, the Peloponnesian League and further *symmachoi* (military allies) on the other in the period from 431 to 404 BC.

I. The Peloponnesian League

After Sparta had expanded its domain in the Peloponnese (by annexing Messenia) up to the middle of the 6th cent. and extended its political influence across all of Greece, it shored up and expanded its sphere of power by creating the Peloponnesian League (ancient term: 'the Lacedaemonians and their allies') a hegemonial symmachy (military alliance) which in the 5th cent. comprised almost all Peloponnesian states – although memberships sometimes vacillated, and with the exception of Argos and Achaea. The supposed political independence and equality of its members was quickly revealed as purely formal, as the structure and loose organization (lack of regular assemblies, one vote per *polis* regardless of its size, majority decisions – which could be overruled by Sparta on religious grounds –, membership tribute payments only to be made in the event of war) guaranteed Sparta unrivalled domination. The alliance pursued predominantly foreign policy objectives, for the most part the struggle against Sparta's enemies. (Sparta had the right to levy troops; any military undertaking had to be agreed to by Sparta.) Nevertheless, Messenia, which had been annexed, remained volatile and Sparta was repeatedly forced to suppress moves towards independence by individual members (Arcadia, Elis). It was also an age in which new urban centres emerged in the Peloponnese. Not only did this bring about a drastic change of settlement patterns, it also spelled the end of the looser tribal organization of the preceding period. The League survived until 365 BC and at some stages even encompassed vast areas of central Greece. After the Peloponnesian War, Sparta also increasingly interfered in the affairs of its allies.

II. Origins and causes of the Peloponnesian War

The origins and causes of the Peloponnesian War are complex, going back to the early years of the 5th cent., a point already emphasized by Thucydides. The Athenian-Spartan peace treaty of 446/5 had failed to overcome the rivalry and mutual suspicions. Sparta was increasingly worried by the expansion of Athenian power and its Delian League under Pericles, which was accompanied by a loss of its authority within its own sphere of power, the Peloponnesian League. The provocative, indirectly anti-Spartan policy practised by Athens in the conflicts between Corcyra and Corinth over Epidamnus (433), towards Potidaea (432–429) and Megara (433/2) resulted in a declaration by Sparta and its allies in 432 that the peace agreed with Athens in 446/5 had been broken. A peaceful settlement was made impossible not only by the promises of military support given to opposing sides by both Athens and Sparta, but also by the efforts of Corinth, Thebes and Perdiccas II of Macedonia to use local conflicts to their own advantage.

III. The Archidamian War (431–421)

In the first part of the war, the so-called Archidamian War (431–421), the Peloponnesians made several incursions into Attica (431, 430, 428, 427 and 425). The Athenians reacted defensively and attempted to decide the war by regular naval attacks on the Peloponnese (e.g. 425 Sphacteria). In 427, Potidaea on the Chalcidice was forced to capitulate. Mytilene, which had defected, was defeated in an operation lasting from 428–27. In 427, Athens also sent ships to Sicily against Sparta's ally Syracuse. The Athenian attempt made in 426 to regain a foothold in central Greece through a combination of land and sea operations ended in failure. In late 424, Athens suffered defeat at Delium in Boeotia. In a partially successful countermove on the Chalcidice and in Thrace, a Spartan army attempted from 424 onward to lure or force Athens' *symmachoi* out of the alliance (conquest of Amphipolis). It was only after the death of the Spartan leaders in a battle near Amphipolis that a general mood of war-weariness set in. Athens, too, had exhausted its resources in its extensive military operations.

IV. The Peace of Nicias and the Interwar Period (421–414/3)

Thus in the spring of 421 a peace agreement for a term of fifty years was concluded (the so-called Peace of Nicias), which essentially restored pre-war conditions. It was unstable from the start. The Interwar Period (421–414/3) was a time of changing coalitions. In particular members of the Peloponnesian League (Corinth, Boeotia) refused to recognize the peace treaty. Sparta was isolated and tried to establish contacts with Athens. This led Argos to set up a rival alliance. When Sparta reacted by concluding an alliance with Boeotia, Athens entered a coalition with Argos, Mantinea and Elis. This, however, was a short-lived affair, as Sparta managed to recover its dominating position in the Peloponnese with a victory in the battle of Mantinea. In 416, Athens conquered the neutral island of Melos and turned it into a cleruchy. Athens again dispatched a fleet to Sicily in 415 in response to a call for help by Segesta. This time the operation was not simply directed against Syracuse, but aimed at subjugating all of Sicily. It all ended in 413 with the complete armada being destroyed off Syracuse. In spite of these commitments, from 414 Athens carried out operations on other fronts as well, which led to a breach with the Persians. From 412 onward, the Achaemenids gave their support to the Spartans, who in return recognized Persian suzerainty over the Greek cities in Asia Minor.

V. The Decelean (414–404) and Ionian Wars (412–405)

In 414, Sparta declared the peace broken and advanced into Attica in 413 setting up a permanent base at Decelea, thus marking the beginning of the so-called Decelean War. Almost simultaneously the so-called Ionian War broke out in 412. Miletus and Chios broke away from Athens. Although Lesbos was retaken (412), numerous allies in Asia Minor seized the opportunity to defect, with Athens being unable to do much about it. Sparta on the other hand, thanks to Persian support, had a fleet at its disposal and made efficient use of it.

Athens' difficulties were compounded when from 411 the Sicilian defeat, the military setbacks in the Aegean and the enduring presence of the Spartans in Attica led to internal strife. The oligarchic forces defeated the democrats and rescinded the constitution. If only because of that, the end of the war seemed near, but democracy was restored and the battle of Cyzicus was won almost simultaneously in 410. The year 407 brought the Athenian defeat at Notium, in 406 the Spartans blockaded the Athenian fleet off Mytilene and even the victory won by a newly built Athenian fleet at Arginusae east of Lesbos couldn't halt the Spartan advance: in 405 the Spartans gained the decisive naval victory at Aigos Potamos on the coast of the Thracian Chersonesus, when they annihilated the last Athenian fleet. The 'Athenian Naval Empire' rapidly fell apart. All naval supply routes (from the Black Sea, Propontis, western Asia Minor) were blockaded by the Spartans, while a Spartan fleet was patrolling off the coast of Attica. On land the siege of the city was completed by the garrison of Decelea and a Spartan army camped to the northwest. Athens was forced to capitulate in the early summer of 404. It had to recognize Spartan hegemony, forfeit all its claims to foreign possessions, hand over what was left of its fleet and have its fortifications razed to the ground. Sparta vetoed further demands by its allies, as it considered Athens in combination with the Lacedaemonians a suitable counterweight to check the aspirations to independence of middle-sized powers like Corinth. These are clear indications that Sparta wanted to increase its domain by taking over the former Athenian territories, because in the last resort it aspired to hegemony over all of Greece (→ map p. 99).

Sources

The main source for the Peloponnesian War up to 411 is Thucydides, with the accounts of Xenophon (Hell. 1,1–2,2) and Diodorus Siculus (13) covering the final years. The division of the war into several stages goes back to Thucydides, who, nevertheless, emphasizes that it must be seen as a single event.

Our knowledge about the Peloponnesian League stems mainly from Herodotus, Thucydides and Xenophon. In contemporary scholarship the Peloponnesian War has recently been the subject of several – mostly English – monographs (among others Bagnall, Kagan, Lazenby, Tritle), with an increasing tendency to cast a critical look at traditional assumptions about origins, lifespan and organization of the Peloponnesian League.

→ Maps pp. 89; 95; 99; 105; Peloponnesian League, BNP 10, 2007; Peloponnesian War, BNP 10, 2007

Literature

N. Bagnall, The Peloponnesian War. Athens, Sparta, and the Struggle for Greece, 2004; B. Bleckmann, Athens Weg in die Niederlage. Die letzten Jahre des Peloponnesischen Krieges, 1998; D. Kagan, The Peloponnesian War, 2003; J.F. Lazenby, The Peloponnesian War. A Military Study, 2003; Ch. Schubert, Athen und Sparta in klassischer Zeit, 2003; L. Thommen, Sparta. Verfassungs- und Sozialgeschichte einer griechischen Polis, Stuttgart 2003; L.A. Tritle, The Peloponnesian War, 2004.

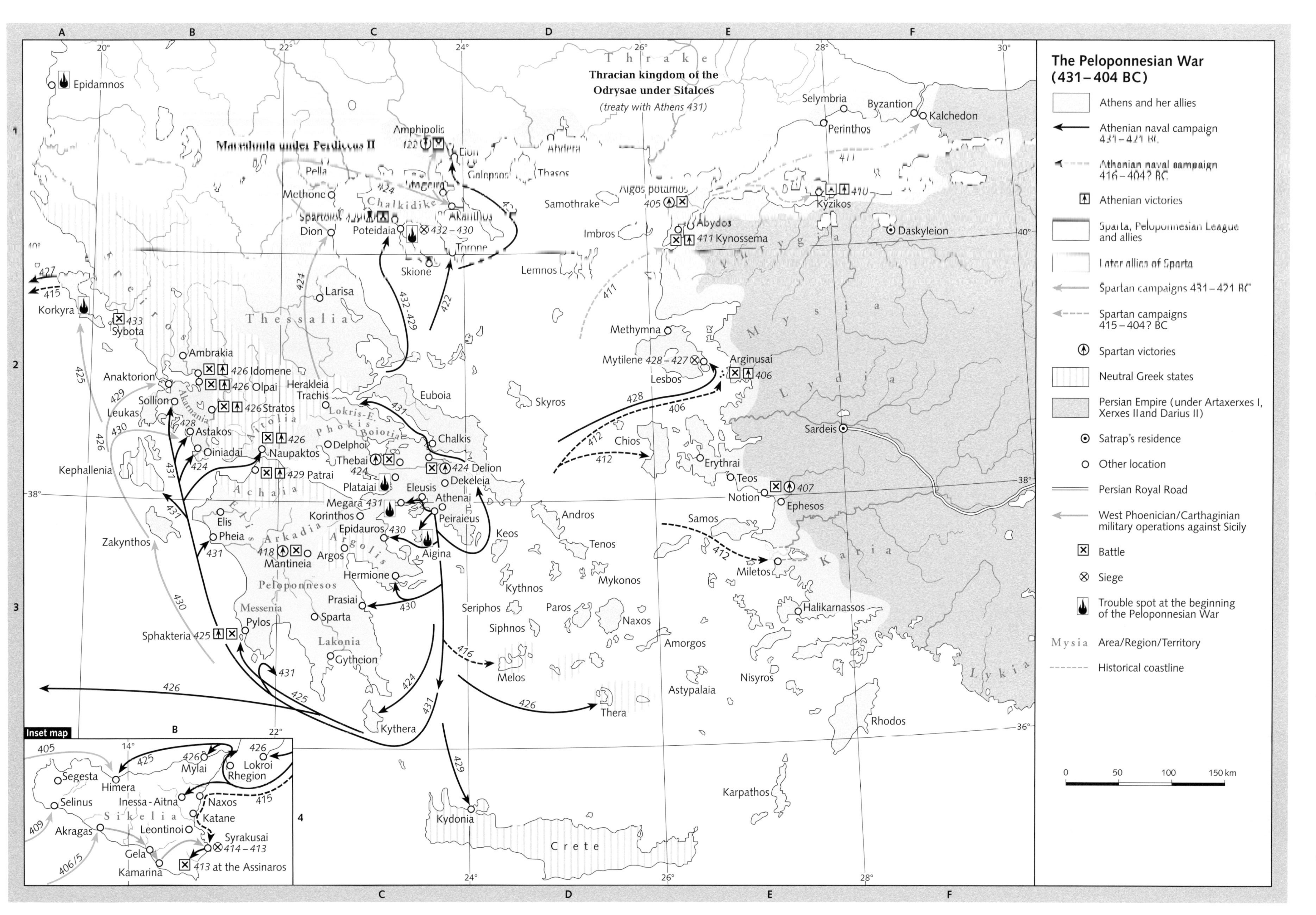
The Peloponnesian War (431–404 BC)
Athens and her allies
Athenian naval campaign 431–421 BC
Athenian naval campaign 416–404? BC
Athenian victories
Sparta, Peloponnesian League and allies
Later allies of Sparta
Spartan campaigns 431–421 BC
Spartan campaigns 415–404? BC
Spartan victories
Neutral Greek states
Persian Empire (under Artaxerxes I, Xerxes II and Darius II)
Satrap's residence
Other location
Persian Royal Road
West Phoenician/Carthaginian military operations against Sicily
Battle
Siege
Trouble spot at the beginning of the Peloponnesian War
Mysia Area/Region/Territory
Historical coastline
0 50 100 150 km
Thracian kingdom of the Odrysae under Sitalces
(treaty with Athens 431)
Macedonia under Perdiccas II
Inset map

League formations with hegemonic tendencies in the Greek world, 4th cent. BC

League formations with hegemonic tendencies in the Greek world, 4th cent. BC

The hegemonial leagues of the 4th cent. shared two features: orientation towards a leading *polis* or person (*hegemon*) within the League and the foreign policy objective of expanding their sphere of action by treaties and/or military means in order to achieve hegemony over neighbouring territories or ideally over Greece and the Aegean region. Some of them were organized as a league of states, i.e. a federation of autonomous *poleis* led by a state empowered with executive foreign policy rights (*hegemon)* and a council of member states (*synhedrion*); others were structured as a territorial union of federal states ('tribal states'; *koina)* sharing a common civic right (alongside the civic rights of the individual states), a primary assembly, central coinage and a federal army led by an elected leadership.

The first type is represented by the Peloponnesian League under Sparta's leadership and the (so-called Second) Delian League with Athens acting as *hegemon*, but also the Theban-led Boeotian League of the 370s, although it did not have a *synhedrion*. The second type was very common in the 4th cent. (e.g. the Leagues in Achaea, Aetolia, Acarnania, Arcadia, the Chalcidice and Epirus). but in the period in question only Thessaly and Phocis displayed expansionist hegemonial tendencies. The Thessalian League, documented from the 5th cent., although chronically weakened by the influence of the powerful aristocratic clans, still managed – under the 4th-cent. leadership of the *tyrannoi* of Pherae – to generate considerable expansionist energies. Using the temple treasures raided from Delphi, the Phocian *koinon*, which had streamlined its military leadership (*strategoi autokratores*), took military action far beyond the boundaries of Phocis between 356 and 346. The Corinthian League founded by Philip II of Macedonia was a unique hegemonial alliance to which all Greeks (except the Spartans) swore allegiance at Corinth in the summer of 337, making him *hegemon* for life.

Neither hegemonial alliances nor expansionist policies were unique to the 4th cent.; however, the rise of small and medium-sized alliances to the status of active foreign policy players and the short duration of their hegemony were without precedent. The causes are to be found in the consequences of the Peloponnesian War (431–404): Athens, for the time being, had lost its position of leadership, while Sparta neither knew how to institutionalize conflict management nor had a strong navy at its disposal. The Persian king used his chance to interfere politically and financially (the 'King's Peace' of 386 BC being the first instance). The construction of durable hegemonial systems was also hampered by the tenet of the freedom and autonomy of the *poleis* which had emerged in the Peloponnesian War and was still cherished in the 4th cent. This meant that there was a systemic contradiction between the members' claims to autonomy and the *hegemon's* entitlement to leadership.

The hegemonies in chronological order

1. Sparta (404-371 BC)
Sparta, the only great power left after 404, was caught up in a war on two fronts. From 400 BC and with increased intensity after 396, it felt the need to protect the Greek cities in the Aegean and on its shores, but was forced from 394 onwards to defend its hegemony against a coalition of Thebes, Athens, Corinth and Argos, which was supported by the king of Persia. This 'Corinthian War' ended with the 'King's Peace' (also known as the Peace of Antalcidas) in 386. This first general peace agreement had been negotiated by Sparta and the king of Persia and been sworn to by the Greek states. The Persians were handed the Greek cities on the coast of Asia Minor, offshore Clazomenae and Cyprus. All other Greek cities were guaranteed autonomy. The map illustrates the situation of the Spartan hegemony right after 386. The cities of Asia Minor had been given up, while what was left of the 5th-cent. Athenian sphere of power (Aegean and northern coastal area including Chalcidice, the northern shore of the Propontis including Byzantium, parts of Thessaly, the islands west of Greece) and mainland Greece from the Peloponnese to southern Macedonia were under Spartan suzerainty. However, its repressive methods of government provoked resistance. In 379, the Thebans expelled the Spartan garrison from their castle (Cadmea) and revived the Boeotian League under Theban leadership. In 377, Athens issued a call to join the newly founded Athenian League, which was openly anti-Spartan. In the north the Chalcidian League, which had been dissolved in 379, was refounded in 376. Since 380, Jason of Pherae, the leader (*tagos*) of the Thessalian League, had been busy conquering Thessaly and parts of northern Greece. As a result Sparta and Athens, at the suggestion of the king of Persia, worked out an agreement based on the autonomy principle of the King's Peace. The Theban refusal to dissolve its League resulted in a war leading to the Spartan defeat at Leuctra in 371 and the reduction of Sparta to a local power in the Peloponnese.

2. Thebes (371–362 BC)
Thebes used the instant disintegration of the Peloponnesian League to extend its hegemony to most of the Peloponnese. In the north, the death of Jason of Pherae (370) made it easier for Thebes to incorporate Thessaly. An understanding with Macedonia had already been achieved and Philip, the future king of Macedonia, was held hostage at Thebes from 368. With backing from the Persian king, the Theban leader Epaminondas, using a newly built fleet, prised Byzantium, Rhodes and Chios (364) away from the Athenian League, which had alread lost Euboea in 370. Thebes, however, like Sparta failed to translate a hegemony achieved by military prowess into a stable political system. Therefore its hegemony collapsed when in 362 its architect, Epaminondas, fell in the battle of Mantinea. (Mantinea was supported by Sparta and Athens.) Yet as hegemon of the Boeotian League, Thebes continued to play an important role in Central-Greek politics.

3. Athens (377–357/338 BC)
After 404, Athens had quickly regained stability. In the Corinthian War it had found a new role on the stage of power politics. After the Persian fleet, led by Conon of Athens, had defeated the Spartans at Cnidos in 394, Athens had also regained its freedom of action in the Aegean (from 393 alliances with island states and Byzantium) and had, in 377, pounced at the chance of founding the Athenian League. It was joined by quite a few members of the 5th-cent. Delian League, because it avoided making the same mistakes. Keeping within the framework of the King's Peace it respected the allies' autonomy, did not collect compulsory contributions (*phoroi*) or establish settlements (*cleruchies*) within the League's territory and it set up a permanent League council (*synhedrion*) at Athens. With Sparta tacitly recognizing it as the second-ranking hegemonial power in 375, this organization had great potential for the stabilization of the Greek political world, especially when Athens was left as the sole great power after Thebes had been weakened by the battle of Mantinea. However, Athens was induced by its expansionist instincts to intervene in various conflicts (e.g. in the Peloponnese and central Greece to prevent Theban expansion, conflicts of succession in Macedonia and Thrace, siding with rebelling satraps against Persia), thus frittering away its military resources. Although in the northern Aegean it managed to secure the grain supply route from the Black Sea and regain Euboea (357), it failed to prevent important allies from 'jumping ship' in the Social War (357–355). Nor could it give adequate support to the Phocians in their struggle with Thebes and Macedonia (so-called Third Sacred War, 356–345). The Phocians were awash with funds (stolen Delphic treasures), and with 10,000 men under arms temporarily maintained the strongest land force in Greece, but their attempt to strengthen their position remained a footnote in history, although it had grave consequences: the Macedonian sphere of influence expanded up to Thermopylae. After 357, the Athenian League came under increasing pressure from Macedonian expansion, losing among others Amphipolis, Pydna and the Chalcidice. Furthermore, it was no longer able to provide efficient protection to the grain route. The Athenian League was dissolved in 338/7.

4. Macedonia/Philip II/Corinthian League (337–322 BC)
All Greek attempts at establishing a stable hegemonial order came to an end with Philip II's victory at Chaeronea. The League founded at Corinth in 337 in the form of a general peace agreement comprised all Greek states except Sparta, as shown on the map. It included the island states; probably not the Greek cities in Asia Minor, although for the first time they were not explicitly excluded. Thus the League's territory exceeded those of all its predecessors. The treaty decreed that the Greeks were to be free and autonomous, that they were entitled to sail the sea without let or hindrance, and that their respective constitutions and territorial possessions would be protected. A *synhedrion*, granting proportional representation to all members, would determine if the peace had been disturbed (internally or externally). If this was the case the assembly would declare war on the perpetrator and authorize the *hegemon* either to conduct the war in person or to delegate the task. Philip was appointed *hegemon* for life. His position in the League was taken over by Alexander the Great in 336.

Sources

The tendency among scholars, noticeable for the past twenty years, to look at the 4th cent. BC in its own right instead of dismissing it as an age of decline or a prelude to the Hellenistic period has led to a growing interest in the hegemonial leagues and their potential function as stabilizing elements in a world that was politically highly volatile – a view shared by the contemporaries. Scholarship focusses on a) the survival of the *polis* as the autonomous organization of a civic community. While some deny this (Runciman), others confirm its existence right into the Late Hellenistic Age (e.g. Habicht); b) the tension between the organization of the Leagues and the concept of *polis* independence, i.e. hegemony versus autonomy. This ties in with the 4th-cent. instrument of the General Peace (*koine eirene*), which was intended to secure peace and autonomy. There is a tendency to look upon the particularistic refusal by the individual *polis* to part with any element of autonomy as the decisive barrier on the road to creating viable supralocal organizations and thus establishing the political framework to secure and sustain general peace (see e.g. Jehne).

→Maps pp. 93, 95, 97, 101 and 103, 113: Athenian League, BNP 2, 2003 (map).

League formations with hegemonic tendencies in the Greek world, 4th cent. BC

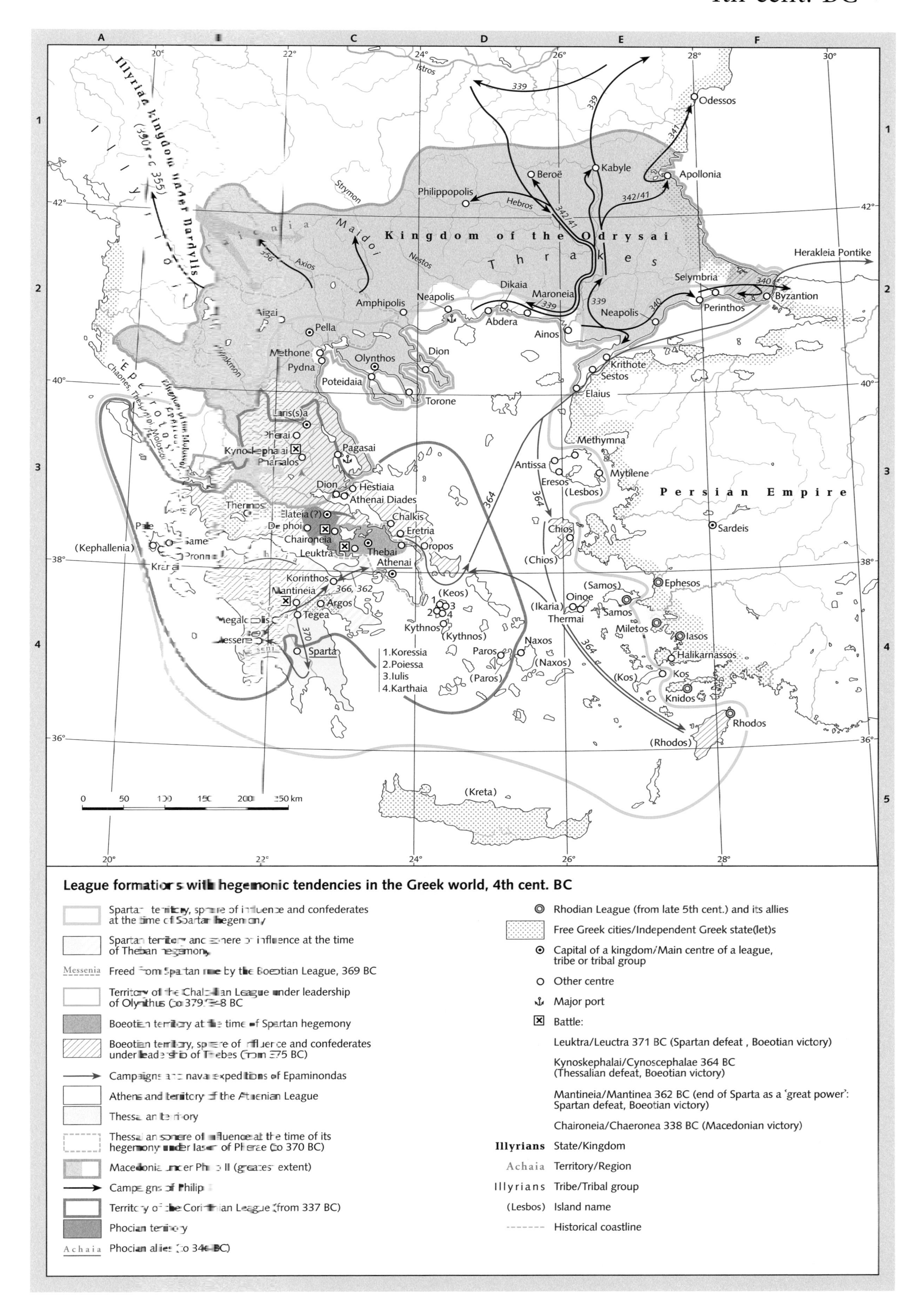
A
C
D
E
F
1
2
3
4
5
20°
22°
24°
26°
28°
30°
42°
40°
38°
36°
Istros
Strymon
Hebros
Axios
Nestos
Illyrian kingdom under Bardylis
(390s–c. 355)
Maidoi
Kingdom of the Odrysai
Thrakes
Odessos
Kabyle
Beroë
Apollonia
Philippopolis
339
341
342/41
342/41
356
340
Herakleia Pontike
Selymbria
Byzantion
Perinthos
Dikaia
Maroneia
Neapolis
Abdera
Ainos
Amphipolis
Aigai
Pella
Methone
Pydna
Olynthos
Poteidaia
Dion
Torone
Krithote
Sestos
Elaius
Epeirotai
Chaones, Thesprotoi, Molossoi
Laris(s)a
Pherai
Kynoskephalai
Pharsalos
Pagasai
Dion
Hestiaia
Athenai Diades
Thermos
Elateia (?)
Delphoi
Chaironeia
Leuktra
Thebai
Chalkis
Eretria
Oropos
Athenai
(Kephallenia)
Same
Krane
Korinthos
366, 362
Mantineia
Argos
Tegea
370
Sparta
(Keos)
Kythnos
(Kythnos)
1.Koressia
2.Poiessa
3.Iulis
4.Karthaia
Paros
(Paros)
Naxos
(Naxos)
364
Methymna
Antissa
Eresos
Mytilene
(Lesbos)
Chios
(Chios)
Persian Empire
Sardeis
(Samos)
Oinoe
Samos
(Ikaria)
Thermai
Ephesos
Miletos
Iasos
Halikarnassos
(Kos)
Kos
Knidos
Rhodos
(Rhodos)
(Kreta)
0
50
100
150
200
250 km
League formations with hegemonic tendencies in the Greek world, 4th cent. BC
Spartan territory, sphere of influence and confederates at the time of Spartan hegemony
Spartan territory and sphere of influence at the time of Theban hegemony
Messenia
Freed from Spartan rule by the Boeotian League, 369 BC
Territory of the Chalcidian League under leadership of Olynthus (to 379/8 BC)
Boeotian territory at the time of Spartan hegemony
Boeotian territory, sphere of influence and confederates under leadership of Thebes (from 375 BC)
Campaigns and naval expeditions of Epaminondas
Athens and territory of the Athenian League
Thessalian territory
Thessalian sphere of influence at the time of its hegemony under Iason of Pherae (to 370 BC)
Macedonia under Philip II (greatest extent)
Campaigns of Philip II
Territory of the Corinthian League (from 337 BC)
Phocian territory
Achaia
Phocian allies (to 346 BC)
Rhodian League (from late 5th cent.) and its allies
Free Greek cities/Independent Greek state(let)s
Capital of a kingdom/Main centre of a league, tribe or tribal group
Other centre
Major port
Battle:
Leuktra/Leuctra 371 BC (Spartan defeat , Boeotian victory)
Kynoskephalai/Cynoscephalae 364 BC (Thessalian defeat, Boeotian victory)
Mantineia/Mantinea 362 BC (end of Sparta as a ‘great power’: Spartan defeat, Boeotian victory)
Chaironeia/Chaeronea 338 BC (Macedonian victory)
Illyrians
State/Kingdom
Achaia
Territory/Region
Illyrians
Tribe/Tribal group
(Lesbos)
Island name
Historical coastline

The Greek leagues

Forms of political organization transcending the individual *polis* were already developing in ancient Greece during the Archaic period, *c.* 700-500 BC, and these are today commonly referred to as 'leagues'. However, the terms used in ancient sources to refer to these organizations, such as *koinon* ('commonwealth'), *ethnos* ('people') or *sympoliteia* ('confederacy', 'union with interchange of civic rights'), are less specific, and were also used of other ancient forms of association (peoples, tribes, societies).

The origins of the leagues are obscure. They are discussed by scholars under the heading of 'ethnogenesis'. In this context, the theory was formerly proposed that the *polis* should be seen as the most developed phase of statehood in Greece, while the *ethnos* represented the more primitive, aboriginal form, out of which the leagues then emerged under the influence of the processes of state development in the *poleis*, and in imitation of these. Today, a much more differentiated view is taken of these constellations and transitional processes, both chronologically and structurally. Like the geographically widely-dispersed intra-Hellenic *ethne*, such as the Achaei, Iones and Dorieis, the regional *ethne* from which leagues emerged in many places, e.g. the Arcadians (map E), Boeotians (map B), Achaeans (map D), Acarnanians (map C), Aetolians (map A), Thessalians, etc., were probably never primordial tribes in the sense of communities of common descent whose origins can be traced to a pre-migratory period outside Greece. Their genesis is now dated to the 8th/7th cents. BC, although questions of how their sense of affiliation and joint political action developed and when the transition from loose co-operation among territorial associates to actual league took place have not been conclusively answered. Some of the aforementioned regional *ethne* are already mentioned in Homer, esp. in the Catalogue of Ships (Hom. Il. 2,494–758), but we do not know to what extent that collective identification of Aetolians, Boeotians, etc. already refers to political co-operation in the strict sense.

Another matter of dispute is whether the co-operation among groups of people who found themselves part of the same *ethnos* was originally cultic in nature. The conception of the early *ethne* as cultic associations brings them close to amphictyonies, one of the earliest of which, the Delphic-Pylian Amphictyony, was centred on the sanctuary of Demeter at Anthela and later also that of Apollo at Delphi. Its members were Central-Greek *ethne*, such as Thessalians, Boeotians, Phocians etc. The co-operation of these *ethne* in this amphictyony in turn had an impact on the intensification of communal identity and internal institutionalization.

I. The formation of leagues in the Archaic and Classical Periods (*c.* 700-330 BC)

Like the *poleis*, the *ethne*, too, underwent a political process of differentiation from the Archaic Period on. The 'segments' joining in federations differed greatly in size, layout and structure. As recent research has shown, settlement structures were also anything but uniform. While the subsidiary units remained largely autonomous, the political structures at league level became ever more distinct. However, evidence does not show this process happening simultaneously in all leagues, or at the same level of intensity. Citizens within a league had double citizenship, of their member state and of the league. Scholars dispute the respective significance of other 'intermediate structures', such as districts, municipalities, etc., between these main member-state and federal levels. The transition from *ethnos* to league did not proceed simultaneously or according to the same pattern everywhere.

II. The leagues

Among the *Aetolians* (map A), a transformation took place from a tribal state to a league in the Classical Period (5th cent.). Before this, the coastal region round Calydon and Pleuron had separated from the Aetolian entity mentioned in the Homeric Catalogue of Ships (Hom. Il. 2,638–644, cf. Str. 10,2,3). The old structure of three great tribes (Thuc. 3,94,4–5; only the Eurytanes have been localized) – Apodotae, Ophiones, Eurytanes –, subdivided in turn into many smaller entities, was abandoned. In its place emerged a league which over the ensuing centuries succeeded in uniting almost the whole of central and north-western Greece. The repelling of the Galatians outside Delphi in 279 was of particular importance to this process for the Aetolians (Paus. 1,4,2; 16,2; 10,19–2). Subsequently, the Aetolians controlled Delphi, the sanctuary of Apollo and the Delphic-Pylian amphictyony until 189.

In *Boeotia* (map B), a league of complex structure had already formed in the Classical Period, as the description of the constitution in the so-called *Hellenica Oxyrhynchia* shows (papyrus fragments of the 2nd cent. AD recording part of a work of history from the first half of the 4th cent. BC; Hell. Oxy. 19,2–4, 374–404). The cities were merged into eleven commensurate districts and had shares in tax payments, federal offices and federal institutions according to the principle of proportionality. The history of the *koinon* was determined by the relationship between its largest city, Thebes, and the other members. At times in the 4th cent., Thebes was able to control the league, and its federal character was largely lost. As well as the core of the Boeotian *poleis*, the *poleis* of Chalcis, Eretria and Megara and parts of eastern Locris, which were not strictly part of the *ethnos*, also at times belonged to the league.

The *Acarnanians* (map C) only formed a league in the 5th cent. BC; its centre was the *polis* of Stratos (Thuc. 2,80; IG IX 12,2, 390; cf. also Thuc. 3,107,2; Xen. Hell. 4,6,4; Aristot. fr. 474 A; Pol.). During the Peloponnesian War, the Acarnanians succeeded in incorporating coastal Corinthian colonies (Anactorium, Alyzea and Astacus) and cities under Corinthian influence (Oeniadae) into their league. Lists of *theorodokoi* reveal the membership of the Acarnanian League in the 4th cent. (IG IV,1^2,95; SEG 36,331). The subsequent history of the Acarnanians was primarily determined by their latent conflict with the Aetolians. Only at times did the River Achelous form the frontier between the two leagues (IG IX 1^2,1 3A). The Acarnanian territory was partitioned in the mid 3rd cent., the west falling to Epirus and the eastern lands becoming Aetolian (Pol. 2,45,1; 9,34,7). Stratos would never again be Acarnanian. A new Acarnanian League formed *c.* 230 around Leucas (Liv. 33,17,1), which had joined the league before (cf. esp. IG IX 1^2,2, 583; SEG 18,261), and this league is attested until the 1st cent. BC. Rome forced Leucas to leave the league after 167. Oeniadae, which the Romans had conquered in 212 (Liv. 26,24,15; Pol. 9,39,2) and which they had subsequently handed over to the Aetolians, was re-incorporated into the Acarnanian League in 189 (Liv. 38,4f.; 11,9; Pol. 21,32,14).

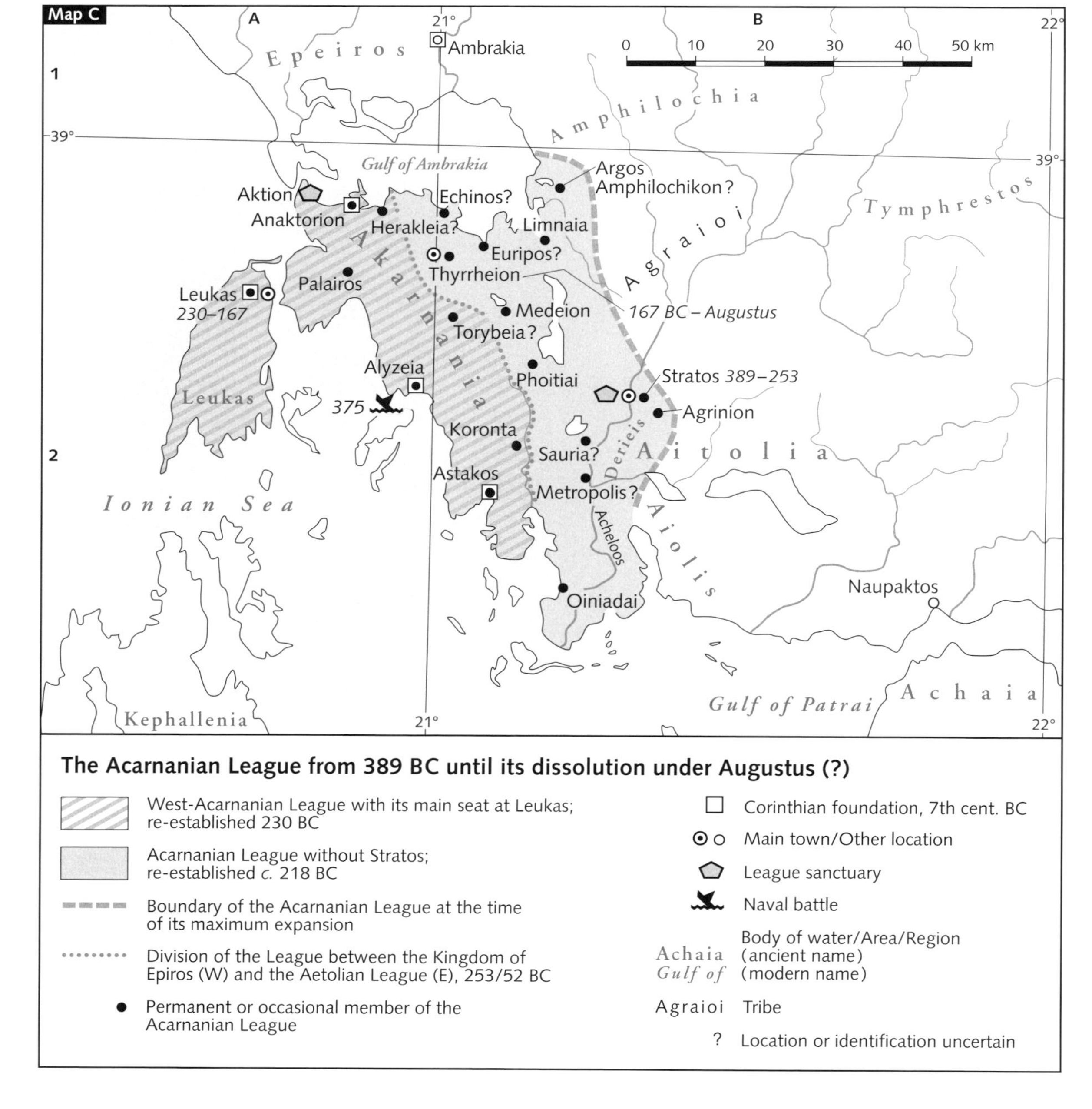

The Acarnanian League from 389 BC until its dissolution under Augustus (?)

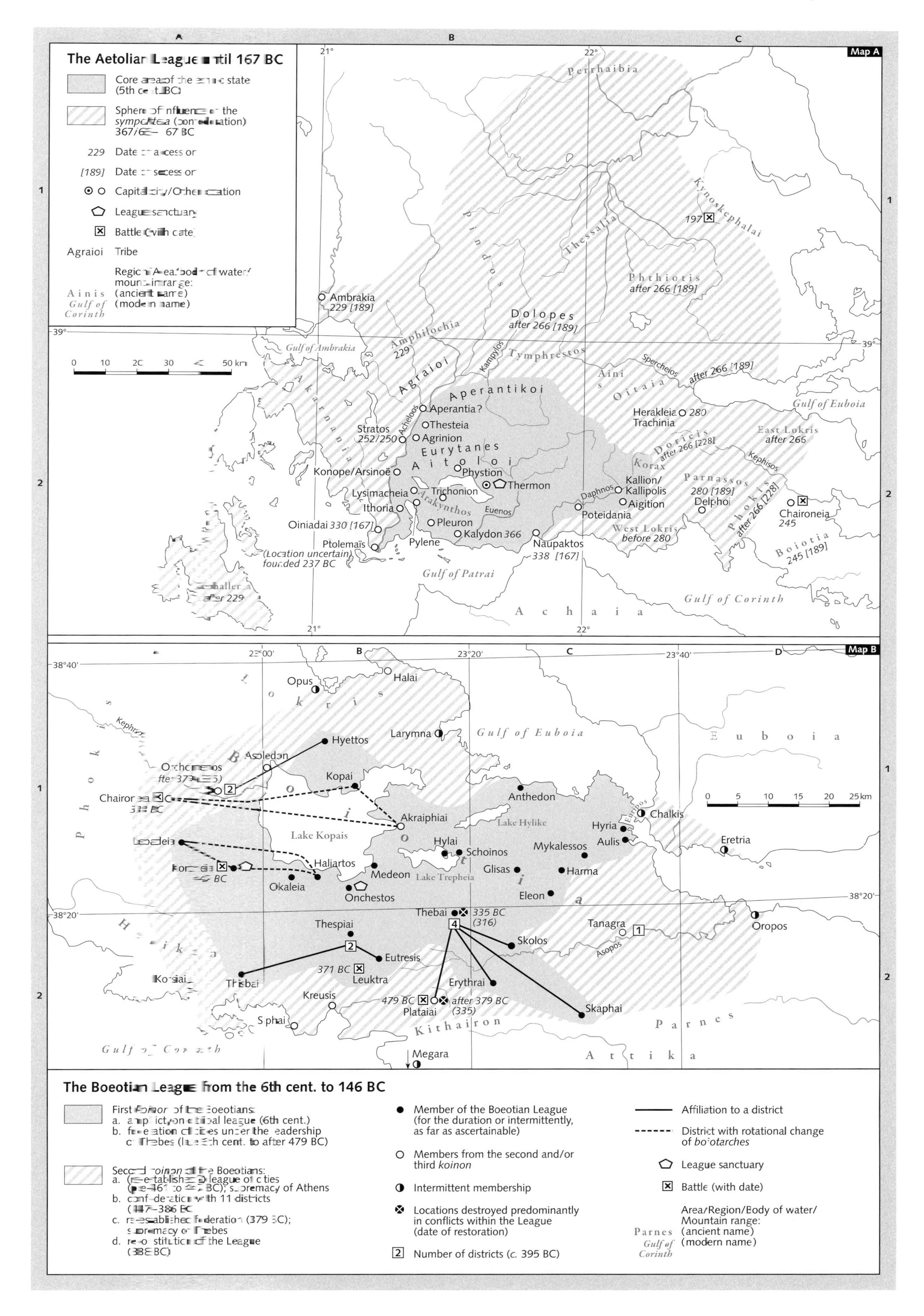
Map A
The Aetolian League until 167 BC
229 Date of accession
[189] Date of secession
League sanctuary
Agraioi Tribe
Ainis (ancient name)
Gulf of Corinth (modern name)
0 10 20 30 50 km
Perrhaibia
Kynoskephalai
197
Thessalia
Pindos
Phthiotis
after 266 [189]
Ambrakia
229 [189]
Dolopes
after 266 [189]
Amphilochia
229
Gulf of Ambrakia
Tymphrestos
Kampylos
Agraioi
Spercheios
after 266 [189]
Ainis
Oitaia
Akarnania
Aperantikoi
Aperantia?
Thesteia
Agrinion
Acheloos
Stratos
252/250
Herakleia 280
Trachinia
Gulf of Euboia
East Lokris
after 266
Eurytanes
Aitoloi
Doris
after 266 [228]
Korax
Kephisos
Konope/Arsinoë
Phystion
Thermon
Kallion/Kallipolis
Aigition
Parnassos
280 [189]
Delphoi
Phokis
after 266 [228]
Chaironeia
245
Lysimacheia
Trichonion
Arakynthos
Ithoria
Daphnos
Poteidania
Euenos
Pleuron
Kalydon 366
Oiniadai 330 [167]
Pylene
Naupaktos
338 [167]
West Lokris
before 280
Boiotia
245 [189]
Ptolemaïs
(Location uncertain)
founded 237 BC
Gulf of Patrai
after 229
Gulf of Corinth
Achaia
21°
22°
39°
Map B
38°40'
23°00'
23°20'
23°40'
38°20'
Opus
Halai
Lokris
Kephisos
Hyettos
Larymna
Gulf of Euboia
Euboia
Boiotia
Kopai
Anthedon
0 5 10 15 20 25 km
Chalkis
Euripos
Akraiphiai
Lake Hylike
Hyria
Aulis
Lake Kopais
Hylai
Schoinos
Mykalessos
Eretria
Phokis
Haliartos
Medeon
Lake Trepheia
Glisas
Harma
Okaleia
Onchestos
Eleon
Thebai
335 BC
(316)
Tanagra
Oropos
Thespiai
Skolos
Asopos
Eutresis
371 BC
Leuktra
Erythrai
Kreusis
479 BC
after 379 BC
(335)
Plataiai
Skaphai
Kithairon
Parnes
Megara
Attika
The Boeotian League from the 6th cent. to 146 BC
Member of the Boeotian League (for the duration or intermittently, as far as ascertainable)
Members from the second and/or third koinon
Intermittent membership
Locations destroyed predominantly in conflicts within the League (date of restoration)
Number of districts (c. 395 BC)
Affiliation to a district
District with rotational change of boiotarches
League sanctuary
Battle (with date)
Area/Region/Body of water/Mountain range:
Parnes (ancient name)
Gulf of Corinth (modern name)

The *Achaeans* (map D) formed a league in the Classical Period; it included twelve cities on the shores of the Gulf of Corinth (Hdt. 1,145; Str. 8,7,4). To these they added the formerly Aetolian cities of Calydon and Pleuron in the first half of the 4th cent. After long periods of political instability after the death of Alexander the Great, an Achaean League was refounded on the initiative of the eastern Achaean *poleis* around Patrae (281/0 BC; Pol. 2,41), and all the former Achaean *poleis* joined it; soon Sicyum and Corinth did so as well, along with many Arcadian *poleis* and, in the early 2nd cent., Elis, Sparta and Messene. The Achaeans' goal of the political union of the entire Peloponnese (cf. Pol. 2,37–41) was only briefly achieved. The smouldering conflict between Sparta and the Achaeans led to repeated quarrels with Rome, which in 146 (Achaean War) dissolved the league and annihilated Corinth.

Strictly speaking, the *Arcadians* (map E) only formed a league during the relatively short period from 371 to 362, founding Megalopolis as its capital. The federation split in two after 362, and the two halves only co-operated politically until the end of the Classical Period. However, the end of the Arcadian League did not mean the end of the *ethnos*. There are suggestions in the sources that some kind of shared Arcadian sense of identity survived in spite of the fact that many of their cities went their own way from the early Hellenistic Period and almost without exception joined the Achaean League. This sense of commonality seems to have been expressed above all in the cult sites of Zeus Lykaios (Paus. 8,38,2–7) and the festivals celebrated together there, and it remained palpable into the Roman Imperial Period.

III. The leagues in the Hellenistic-Roman Period

In the Hellenistic Period in particular, many *koina* also succeeded in integrating regions not belonging to the *ethnos*, the Achaean League in particular proving especially efficient in this respect in the Peloponnese (map D), and the Aetolian League (map A) in north-western and central Greece.

After the collapse of Macedonian rule after 196, the so-called *periokoi* tribes of central Greece (e.g. the Thessalians, Magnetes and Perrhaebi) in particular joined forces as well, following the by now so long-established league structure. Other leagues, such as the Aetolian (map A) and Boeotian (map B), entered long, bitter political processes of disintegration as a result of Roman intervention in the affairs of the Greek states, and in many cases these processes meant the end of the old, traditional federal structures.

Some leagues continue to be attested well into the Imperial Period, although their structure and remit often changed dramatically. Some *koina* were even accorded functions beyond the religious sphere in the organizational structures of the Roman province of Achaia.

→ Maps pp. 35, 37, 81, 97, 187

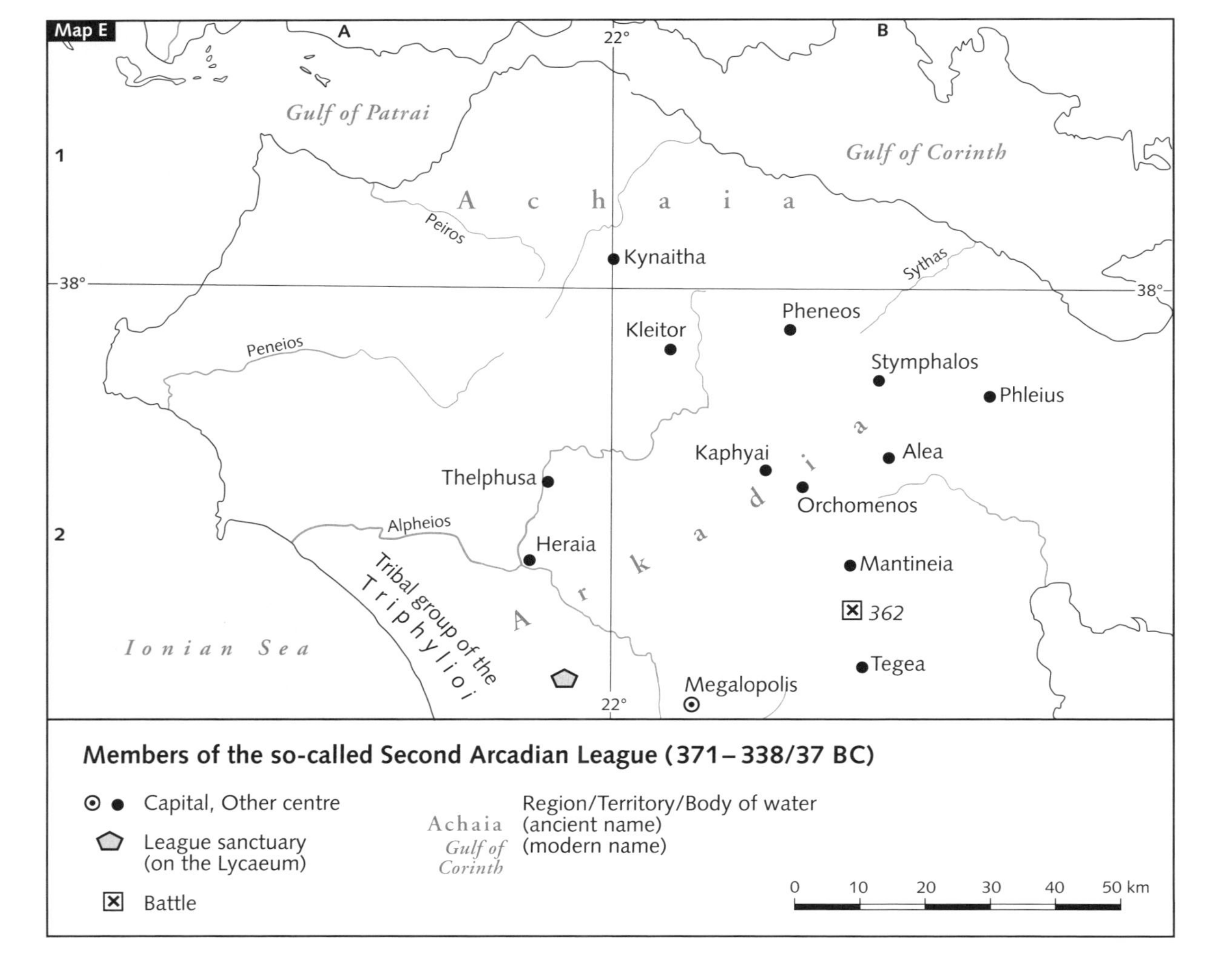

Members of the so-called Second Arcadian League (371–338/37 BC)

Literature

General:
H. Beck, Polis und Koinon, 1997; Id., New Approaches to Federalism in Ancient Greece: Perceptions and Perspectives, in: K. Buraselis, K. Zoumboulakis (eds.), The Idea of European Community in History, vol. 2: Aspects of Connecting Poleis and Ethne in Ancient Greece, 2003, 177–190; T. Corsten, Vom Stamm zum Bund. Gründung und territoriale Organisation griechischer Bundesstaaten, 1999; P. Funke, Staatenbünde und Bundesstaaten. Polis-übergreifende Herrschaftsorganisationen in Griechenland und Rom, in: K. Buraselis (ed.), Unity and Units of Antiquity, 1994, 125–136; Id., Stamm und Polis. Überlegungen zur Entstehung der griechischen Staatenwelt in den Dunklen Jahrhunderten, in: Colloquium aus Anlaß des 80. Geburtstages von A. Heuss, 1993, 29–43; H.-J. Gehrke, Ethnos, Phyle, Polis, in: P. Flensted-Jensen, T.H. Nielsen, L. Rubinstein (eds.), Polis and Politics. Studies in Ancient Greek History, 2000, 159–176; F. Gschnitzer, Stammes- und Ortsgemeinden im alten Griechenland, in: Id. (ed.), Zur griechischen Staatskunde, 1969, 271–297; J. Hall, Hellenicity. Between Ethnicity and Culture, 2001; Id., Ethnic Identity in Greek Antiquity, 1997; J.A.O. Larsen, Greek Federal States, 1968; G.A. Lehmann, Ansätze zu einer Theorie des griechischen Bundesstaates bei Aristoteles und Polybios, 2001; I. Malkin (ed.), Ethnicity and the Construction of Ancient Greek Identity, 2001; C. Morgan, Early Greek States beyond the Polis, 2003; R. Parker, Cleomenes on the Acropolis, 1998; P. Siewert, L. Aigner-Foresti, Föderalismus in der griechischen und römischen Antike, 2005; C. Tanck, Arche – Ethnos – Polis, 1997; C. Ulf, Griechische Ethnogenese versus Wanderungen von Stämmen und Stammstaaten, in: Id. (ed.), Wege zur Genese griechischer Identität, 1996, 240–280; F.W. Walbank, Hellenes and Achaians: 'Greek Nationality' Revisited, in: P. Flensted-Jensen (ed.), Further Studies in the Ancient Greek Polis, 2000, 19–33.

Further literature on the leagues:
A. Aetolian: P. Funke, Polisgenese und Urbanisierung in Aitolien im 5. und 4. Jahrhundert v. Chr., in: M.H. Hansen (ed.), The Polis as an Urban Centre and as a Political Community, 1997, 145–188; J.D. Grainger, The League of the Aitolians, 1999; J.B. Scholten, The Politics of Plunder: The Aitolians and their Koinon in the Early Hellenistic Era, 279–217 B.C., 2000; Id., The Internal Structure of the Aitolian Union: A Case-study in Ancient Greek Sympoliteia, in: K. Buraselis, K. Zoumboulakis (eds.), The Idea of European Community in History, vol. 2, 2003, 65–80.
B. Boeotian: H. Beck, Thebes, the Boiotian League, and the 'Rise of Federalism' in Fourth-Century Greece, in: P. Angeli Bernardini (ed.), Presenza e funzione della città di Tebe, 2000, 331–344; R.J. Buck, The Hellenistic Boiotian League, in: Ancient History Bulletin 7, 1993, 100–106; Id., Boiotia and the Boiotian League, 432–371 B.C., 1994; A. Schachter, Gods in the Service of the State: the Boiotian Experience, in: L. Aigner-Foresti, A. Barzanò, C. Bearzot, L. Prandi, G. Zecchini (eds.), Federazioni e federalismo nell'Europa antica, 1994, 57–86.
C. Acarnanian: O. Dany, Akarnanien im Hellenismus. Geschichte und Völkerrecht in Nordwestgriechenland, 1999; K. Freitag, Der Akarnanische Bund im 5. Jh. v. Chr. in: P. Berktold et al. (eds.), Akarnanien, 1996, 75–85; H.-J. Gehrke, Die kulturelle und politische Entwicklung Akarnaniens vom 6. bis zum 4. Jahrhundert v. Chr., in: Geographia antiqua 3/4, 1993/1994, 41–47; M. Schoch, Beiträge zur Topographie Akarnaniens in klassischer und hellenistischer Zeit, 1997.
D. Achaean: E. Greco (ed.), Gli Achei e l'identità etnica degli Achei d'Occidente, 2002; K. Harter-Uibopuu, Das zwischenstaatliche Schiedsverfahren im achäischen Koinon, 1998; G.A. Lehmann, Erwägungen zur Struktur des achaiischen Bundesstaates, in: ZPE 51, 1983, 237–261; H. Nottmeyer, Polybios und das Ende des Achaierbundes, 1995; A.D. Rizakis, Achaïe. 1. Sources textuelles et histoire régionale, 1995; J. Roy, The Achaian League, in: K. Buraselis, K. Zoumboulakis (eds.), The Idea of European Community in History, vol. 2, 2003, 81–96; T. Schwertfeger, Der Achaiische Bund von 146 bis 27 v. Chr., 1974; R. Urban, Wachstum und Krise des Achäischen Bundes, 1978.
E. Arcadian: M. Jost, Sanctuaires et cultes d'Arcadie, 1986; T.H. Nielsen, Arkadia and its Poleis in the Archaic and Classical Periods, 2002; Id., J. Roy (eds.), Defining Ancient Arkadia, 1999.

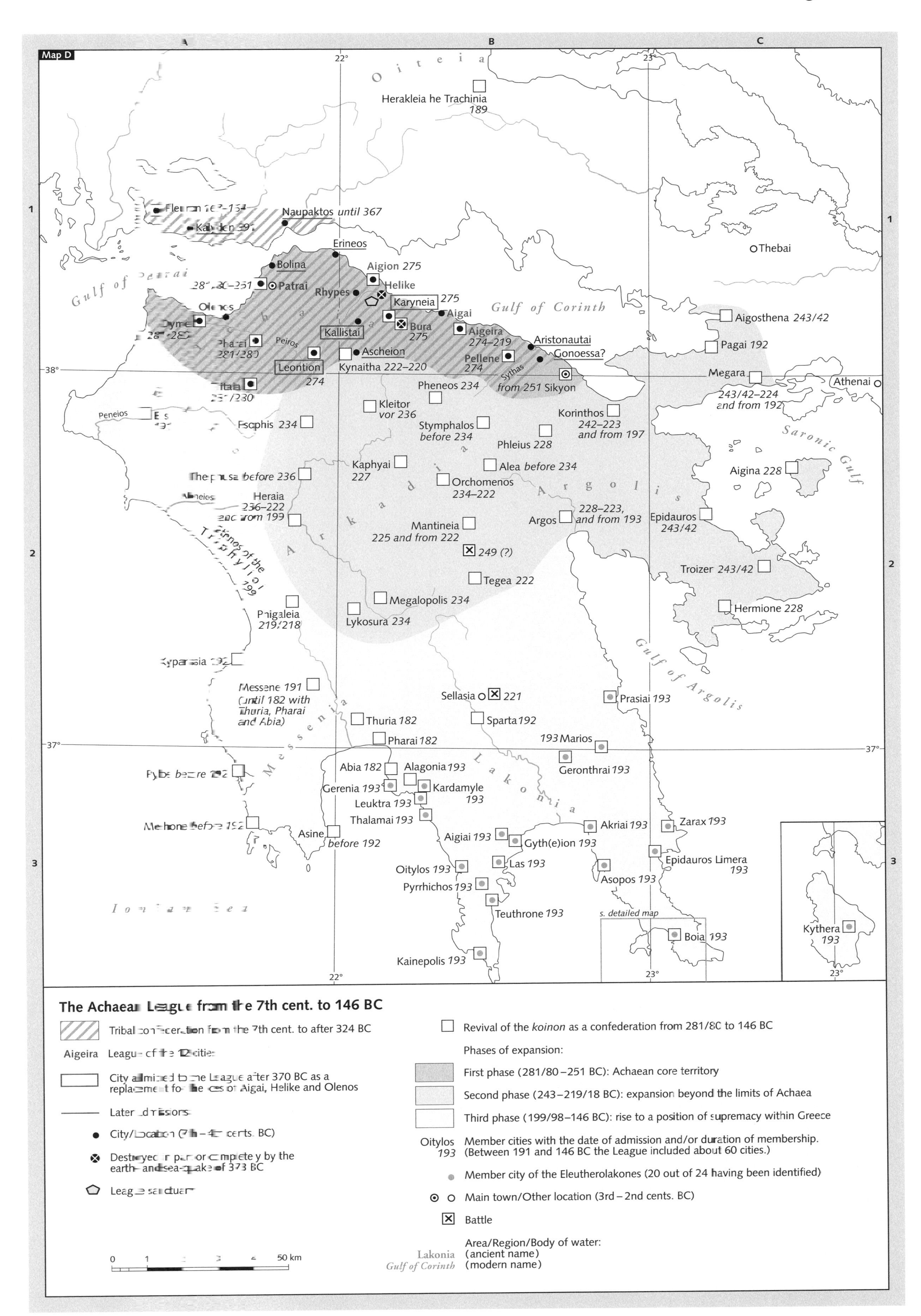
Map D
A
B
C
1
2
3
22°
23°
38°
37°
O i t e i a
Herakleia he Trachinia 189
Naupaktos until 367
Erineos
Bolina
Aigion 275
Patrai
Helike
Rhypes
Karyneia 275
Gulf of Corinth
Aigai
Olenos
Bura 275
Kallistai
Aigeira 274–219
Aristonautai
Gonoessa?
Pellene 274
Leontion 274
Ascheion
Kynaitha 222–220
Sythas
from 251 Sikyon
Peiros
Peneios
Pheneos 234
Kleitor vor 236
Stymphalos before 234
Korinthos 242–223 and from 197
Phleius 228
Psophis 234
Kaphyai 227
Alea before 234
Orchomenos 234–222
Heraia 236–222
Mantineia 225 and from 222
Argos 228–223, and from 193
Epidauros 243/42
249 (?)
Tegea 222
Megalopolis 234
Lykosura 234
Phigaleia 219/218
Troizen 243/42
Hermione 228
A r k a d i a
A r g o l i s
Aigosthena 243/42
Pagai 192
Megara 243/42–224 and from 192
Athenai
OThebai
Saronic Gulf
Aigina 228
Gulf of Argolis
Messene 191 (until 182 with Thuria, Pharai and Abia)
Sellasia 221
Sparta 192
Thuria 182
Pharai 182
Prasiai 193
193 Marios
Geronthrai 193
Abia 182
Alagonia 193
Gerenia 193
Kardamyle 193
Leuktra 193
Thalamai 193
Akriai 193
Zarax 193
Asine before 192
Aigiai 193
Gyth(e)ion 193
Epidauros Limera 193
Oitylos 193
Las 193
Asopos 193
Pyrrhichos 193
Teuthrone 193
s. detailed map
Boia 193
Kythera 193
Kainepolis 193
Messenia
Lakonia
The Achaean League from the 7th cent. to 146 BC
Tribal confederation from the 7th cent. to after 324 BC
Aigeira League of the 12 cities
City admitted to the League after 370 BC as a replacement for the cities of Aigai, Helike and Olenos
Later divisions
City/Location (7th–4th cents. BC)
Destroyed in part or completely by the earth- and sea-quake of 373 BC
League sanctuary
Revival of the koinon as a confederation from 281/80 to 146 BC
Phases of expansion:
First phase (281/80–251 BC): Achaean core territory
Second phase (243–219/18 BC): expansion beyond the limits of Achaea
Third phase (199/98–146 BC): rise to a position of supremacy within Greece
Oitylos 193 Member cities with the date of admission and/or duration of membership. (Between 191 and 146 BC the League included about 60 cities.)
Member city of the Eleutherolakones (20 out of 24 having been identified)
Main town/Other location (3rd–2nd cents. BC)
Battle
Area/Region/Body of water:
Lakonia (ancient name)
Gulf of Corinth (modern name)
0 1 50 km

The development of the Macedonian Kingdom from the 7th cent. until 336 BC

The development of the Macedonian Kingdom from the 7th cent. until 336 BC

Until its final political union under Philip II (360/59–336 BC), the region of Macedon was the contact zone between the non-Greek, Illyrian tribes on the Adriatic coast to the west (separated from Macedon by mountain ranges), the Thracian tribes to the east, the northern regions as far as the Istrus/Danube, and the Greek *poleis* to the south (also separated by high mountains). Politically speaking, Macedonia rose within just two generations from being little more than a scarcely-noticed peripheral phenomenon in the Greek north to becoming the leading Greek power and, with Alexander the Great, the dominant power in the eastern Mediterranean and western Asia.

The heartland of the *Makedones*, a tribe described as predominantly Greek once it had completed its ethnogenesis (Hsch. fr. 5; cf. also Hdt. 5,22), was the plains to the east and north (Macedonian Plain) of Olympus, with the main urban centres of Aegae (7th-5th cents.) and Pella (from the 5th cent.). Argead princes (the name originally belonged to one of the many Macedonian tribes, then referred to the royal dynasty of the Argeadae; they were recognized as Greek by other Greeks and were admitted to the Olympic Games) had subjugated other Macedonian tribes since the 7th cent.: successively Pieria in the south-west, south of the lower reaches of the Haliacmon, Bottiaea in the east, between the Haliacmon and Axius, and Almopia in the north.

The Macedonians extended their rule into the hinterland of the Chalcidice in the 6th cent., so that their domains bordered directly on Thracian territories that were conquered and incorporated into the Persian Empire by Darius I in 512 BC. The kingdom of Macedonia thus emerges into history in the late 6th/5th cents.: the Macedonian kings made agreements with the Persian Empire and probably paid tribute over the ensuing decades. After the Persian defeats at Greek hands in 479 BC at Plataeae and Mycale, the territories of Mygdonia, Crestonia and Anthemus to the east of Bottiaea seem to have come under Macedonian rule, and the beginnings of Upper Macedonian influence on the largely independent princedoms of Elimea, Eordaea, Lyncus, Orestis and Pelagonia can be observed (Thuc. 2,99,2). Overall, Macedonia was not a unitary kingdom; moreover, succession disputes within the Argead dynasty constantly led to political weakness.

From the mid 5th cent. to the accession of Philip II in 359 BC, Macedonia experienced substantial economic growth thanks to increased exports of timber for shipbuilding to the new Aegean naval power, Athens. At the same time, however, it became embroiled in the conflicts among the southern Greeks. Particular difficulties arose from Athenian campaigns on the Chalcidice in connection with the Delian League (in 432 BC, Chalcidian cities had defected from the league, joining the so-called 'Chalcidian League' headed by Olynthus), and from Macedonian attempts to conquer or otherwise incorporate the Upper Macedonian princedoms. This strat egy changed under King Archelaus (*c.* 413–399 BC), who made improvements to the infrastructure that particularly benefited the army. For example, roads were built, fortifications and bases were constructed and Pella was developed into a royal residence. After Archelaus' murder in 399 BC, the very existence of the enfeebled Macedonia was threatened by the Illyrians (who seemed on the verge of permanent occupation of large parts of the kingdom), by other neighbours (such as the Paeones and Thracians) and by Athens, Sparta, Thebes and the Chalcidian League – until Philip II succeeded in inflicting devastating defeats on the Illyrians and Paeones in 359 BC, and in subsequently keeping them at bay.

Philip, initially as regent for his nephew Amyntas IV (360/59 BC) – who was still a minor – and soon as king, consolidated the Macedonian kingdom externally through negotiation, tribute and military action, and internally by his success in achieving the union of the individual princedoms (esp. Lyncus and Elimea) and permanently incorporating them into the kingdom while preserving the influence of their various elites. He also reformed the army and made it effective, and made donations of land to Macedonians and others wishing to settle in the newly-acquired territories, creating stable, personal ties between these people and the Macedonian royal house. New cities were founded in various contexts (Philippi, Apollonia, Heraclea Lyncestis, all on the future *via Egnatia*, and Beroea).

A priority of the king in the years that followed was to secure the frontiers of his expanding kingdom:

In the east, the frontier was first pushed back to the Nestus (Crenides 356 BC, including the silver mines of the Pangaeum; Edones) and hence to the territory of the Thracian Odrysae. In the process, the Thracian princedoms were either conquered or incorporated as client states (by 340 BC).

The Chalcidice in the south-east was also integrated into the kingdom following the conquest of Potidaea (356 BC) and the destruction of Olynthus (348 BC), as following their own defeat were the Paeones of the Axius valley to the north-west in 359 BC; the latter was one of the most important transport corridors.

As part of his alliance policy, Philip II forged dynastic links with the Epirote Molossi in the west (Olympias), and (according to Satyrus of Callatis) with the Illyrians (Audata, relative of Bardyllis I), Geti (Meda) and Thessalians (Nicesipolis of Pherae; Philinna of Laris(s)a).

After they were conquered, cities of the new Macedonian territories that had been founded by Athens or were Athenian allies, such as Amphipolis (357), Pydna (356) and Methone (355), as well as other Greek cities, esp. those of economic importance, were usually given a garrison but allowed to retain administrative autonomy.

In the south, the territory of the Thessalian League, of which Philip was probably *archon* for some time, at first became a buffer zone against Greece, until most Greek cities joined the Corinthian League newly founded and headed by Philip (338/7 BC; treaty of alliance cf. StV III 403 I) after the battle of Chaeronea in 338 BC; this merely set the seal on the *de facto* Macedonian hegemony, a situation unaffected by the longed-for 'General Peace' (*koine eirene*) which was now declared.

Sources

Overall, for the time before Philip II, source material from which we can begin to reconstruct Macedonian history is only fragmentary. Herodotus gives the kings' list (from *c.* mid 7th cent.: Hdt. 8,137) and describes the Macedonians' participation in the great Persian War (480/79 BC). Thucydides recounts participation in the Peloponnesian War (431–404 BC), while Xenophon's *Hellenica* is notable as a source for the early 4th cent. There are also the references in the work of the Alexander historian Arrian. Otherwise, there are incidental references in later authors, and some incomplete document finds. The situation improves somewhat for the reign of Philip II (fragments of Anaximenes, Theopompus and Ephorus of Cumae; Diodorus Siculus; Justin; orations by Demosthenes, Isocrates, Aeschines).

The rise of Macedonia, esp. under Philip II, has won increasing attention from ancient historians, particular consideration being given to the causes and aims (and implementation) of the Macedonian conquests from the reign of Philip II, which are perhaps to be seen as part of a planned conquest of the entire Persian Empire (357 BC: Corinthian League's assent to a Persian campaign).

Linguistic studies of Macedonian names have sought to give an affirmative response to the question (already controversial in Antiquity) of the Macedonians' Greek ethnicity. However, Macedonian is preserved only as a vestigial language. Over the centuries, until its final displacement by Greek, it was ever more strongly Hellenized, so that this issue continues to be disputed.

We learn of cultural and urban developments from archaeological evidence, albeit mostly from necropoleis and mostly from Central and Upper Macedonia (Aegae/Vergina, Pella, Dion: city and major sanctuary, Aeane), and from inscriptions and coins.

→ Maps pp. 89, 95, 97, 99

Literature

M. Andronikos, Vergina. The Royal Tombs and the Ancient City, 1984; E.N. Borza, In the Shadow of Olympus. The Emergence of Macedon, 1990; A. Demandt, Antike Staatsformen, 1995; J. Engels, Philipp II. und Alexander der Große, 2006; M. Errington, Geschichte Makedoniens, 1986; N.G.L. Hammond, Philip of Macedon, 1994; Id., G.T. Griffith, A History of Macedonia, vol. 2 (550–336 BC), 1979; M.B. Hatzopoulos, Macedonian Institutions under the Kings, 2 vols., 1996; F. Papazoglou, Les villes de Macédoine, 1988, P. Siewert, L. Aigner-Foresti, Föderalismus in der griechischen und römischen Antike, Stuttgart 2005, 53ff.; I. Vokotopoulou (ed.), Makedonen. Die Griechen des Nordens, 1994; G. Wirth, Geschichte Makedoniens, vol. 1: Philipp II., 1985; M. Zahrnt, Olynth und die Chalkidier, 1971.

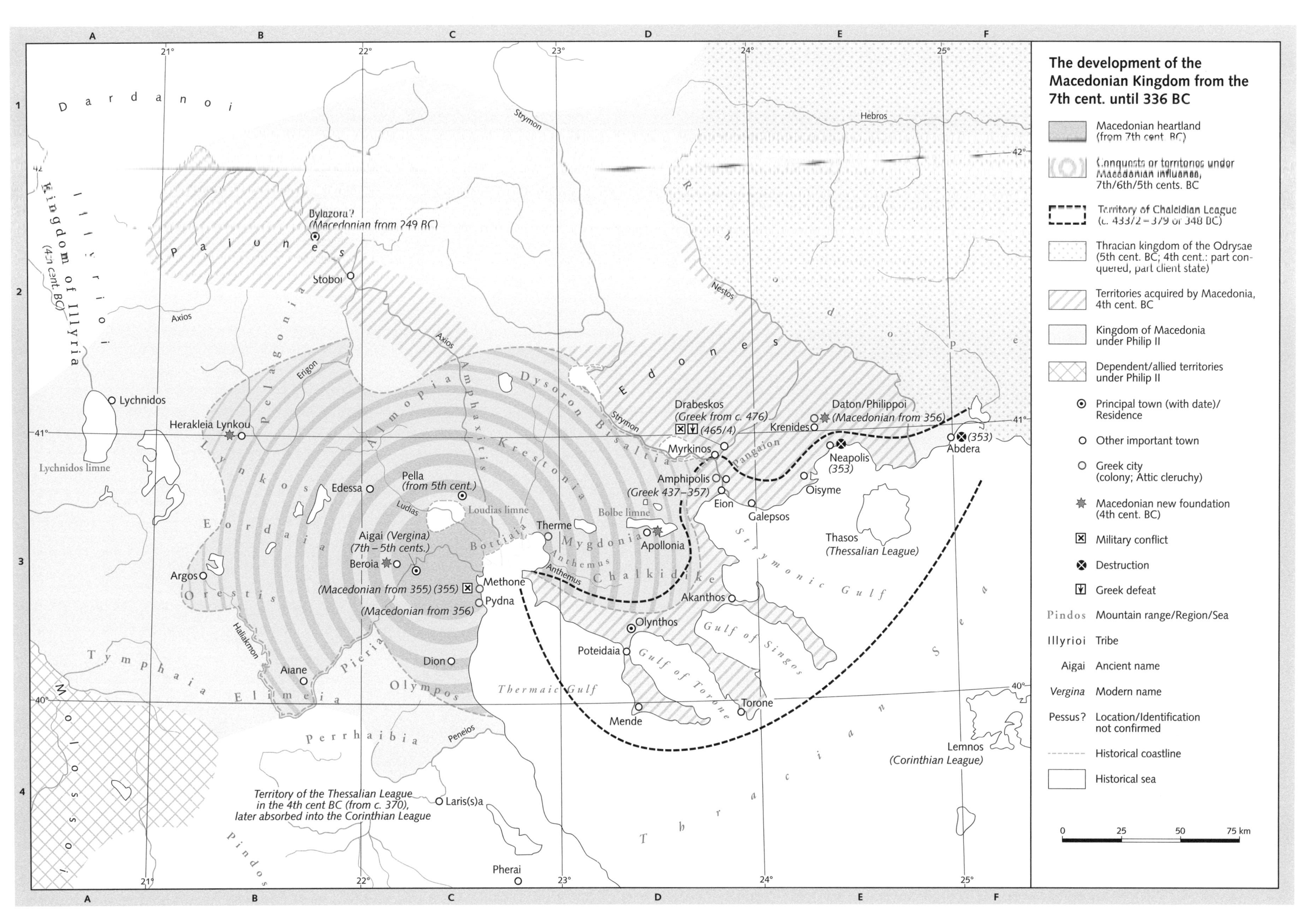
The development of the Macedonian Kingdom from the 7th cent. until 336 BC
Macedonian heartland (from 7th cent. BC)
Conquests or territories under Macedonian influence, 7th/6th/5th cents. BC
Territory of Chalcidian League (c. 433/2 – 379 or 348 BC)
Thracian kingdom of the Odrysae (5th cent. BC; 4th cent.: part conquered, part client state)
Territories acquired by Macedonia, 4th cent. BC
Kingdom of Macedonia under Philip II
Dependent/allied territories under Philip II
Principal town (with date)/Residence
Other important town
Greek city (colony; Attic cleruchy)
Macedonian new foundation (4th cent. BC)
Military conflict
Destruction
Greek defeat
Pindos Mountain range/Region/Sea
Illyrioi Tribe
Aigai Ancient name
Vergina Modern name
Pessus? Location/Identification not confirmed
Historical coastline
Historical sea
0 25 50 75 km
Dardanoi
Illyrioi
Kingdom of Illyria (4th cent. BC)
Paiones
Bylazora? (Macedonian from 249 BC)
Stoboi
Lychnidos
Lychnidos limne
Herakleia Lynkou
Edessa
Pella (from 5th cent.)
Loudias limne
Aigai (Vergina) (7th – 5th cents.)
Beroia
Argos
Orestis
Eordaia
Lynkos
Pelagonia
Almopia
Amphaxitis
Dysoron
Krestonia
Bisaltia
Bottiaia
Mygdonia
Anthemus
Chalkidike
Therme
Methone
(Macedonian from 355) (355)
Pydna
(Macedonian from 356)
Dion
Aiane
Pieria
Olympos
Elimeia
Tymphaia
Perrhaibia
Molossoi
Pindos
Territory of the Thessalian League in the 4th cent BC (from c. 370), later absorbed into the Corinthian League
Laris(s)a
Pherai
Thermaic Gulf
Poteidaia
Olynthos
Mende
Torone
Gulf of Torone
Gulf of Singos
Akanthos
Apollonia
Bolbe limne
Amphipolis
(Greek 437 – 357)
Eion
Galepsos
Myrkinos
Drabeskos (Greek from c. 476) (465/4)
Krenides
Daton/Philippoi (Macedonian from 356)
Neapolis (353)
Oisyme
Abdera (353)
Pangaion
Edones
Rhodope
Thasos (Thessalian League)
Strymonic Gulf
Lemnos (Corinthian League)
Thracian Sea
Axios
Erigon
Ludias
Haliakmon
Peneios
Strymon
Nestos
Hebros

Populus Romanus: the 4 urban and 31 rural *tribus* in Italy (*c.* 500–241 BC)

The term *tribus* denoted divisions, initially exclusively according to ethnic group, of the Roman people. Since the 6th cent. it had also always referred to their place of residence. Varro (Ling. 5,55) derives the term from the three tribes of the Tities, Ramnes and Luceres, but the etymology is uncertain. The *tribus* were named geographically, like the four urban *tribus* (*tribus urbanae – Palatina, Esquilina, Collina, Suburana*, cf. map C), or by *gens* ('clan'), like the rural *tribus Claudia*, named after the Sabine Claudii, who migrated in in 504 BC. After the conquest of Crustumerium and the creation of the *tribus Clustumina*, the new and exclusively rural *tribus* (*tribus rusticae*) were given local names. In 495 BC, 17 rural *tribus* on the newly-annexed *ager publicus* had been added to the original four urban *tribus* (Liv. 2,21,7; cf. map B). In 387 BC, four further rural *tribus* were set up on the territory of the conquered city of Veii (*tribus Stellatina, Sabatina, Arnensis, Tromentina*). The last of the final total of 35 *tribus* to be created were the *tribus Quirina* and *tribus Velina* in 241 BC. Citizens and citizen communities joining after this were assigned to one of the existing *tribus*. Belonging to one of the 35 *tribus* was a distinguishing privilege of Roman citizens; the *tribus* were legal entities with reserved seats at the theatre and their own funerary customs, and they were entitled to sustenance from the public grain supply.

It is striking that from the 4th cent., *tribus* were only created in pairs. We can infer from this the intention of maintaining the odd number of *tribus* in the *comitia tributa* and the *concilium plebis*, useful for technical reasons in elections. These popular assemblies, not divided according to wealth like the *comitia centuriata*, but by the locally-defined *tribus*, could not provide any as it were 'fairer' distribution and weighting of votes, because the individual *tribus* differed greatly according to territory and composition of (richer and poorer) populace, yet each individual *tribus* still only had one vote at the ballot. The voting sequence (cf. Cic. Leg. agr. 2,79), which was retained into the Imperial Period, also prevented a 'fairer' weighting of votes, because the voting results were announced successively. During the Social War (91-87 BC), the creation of new *tribus* was probably considered for the last time. Quite apart from traditionalist and *clientela*-related arguments, to do so would also have hampered ever more seriously the practicalities of elections. Proposals for legislation which ought, by virtue of their importance, to have been voted on by the 193 voting bodies of the *comitia centuriata*, were being brought ever more frequently to this electoral assembly instead, by very reason of its comparatively simple voting system, divided into only 35 *tribus*.

Sources

Most of the information relevant to this area of constitutional history comes from Livy's *History of Rome*. Other references have come to us indirectly, from the lost encyclopaedia of the Augustan grammarian M. Verrius Flaccus, *De verborum significatione*. In the 2nd cent. AD, Sex. Pompeius Festus made an epitome of it, which at least survives in fragments. This epitome in turn was reworked into a digest in the 8th cent. by the Lombard scholar Paulus Diaconus. There are also many inscriptions bearing the names of Roman citizens and giving the designation of the 35 *tribus*.

The maps

The adjacent maps show the locations and sizes of the four *tribus urbanae* from the 6th cent. BC (map C), the 17 oldest *tribus rusticae* (map B) and the 14 *tribus rusticae* created in the 4th and 3rd cents. (map A). The geographical terms from which *tribus* names are probably or certainly derived are also shown where they are localized or where their location is conjectured.

Literature

J. Bleicken, Die Verfassung der römischen Republik, [8]2000; T. J. Cornell, The Beginnings of Rome, 1995; M.H. Crawford, Tribus, tessères et régions, in: CRAI 4, 2002, 1125–1136; G. Forni, Le tribù romane, 1, 1996; 3,1, 1985; 4, 2006; H. Galsterer, Herrschaft und Verwaltung im republikanischen Italien, 1976; W. Kubitschek, Imperium Romanum tributim discriptum, 1889; B. Linke, Von der Verwandtschaft zum Staat, 1995, 117–120; Cl. Nicolet, The World of the Roman Citizen, 1980; M. Rieger, Tribus und Stadt. Die Entstehung der römischen Wahlbezirke im urbanen und mediterranen Kontext (ca. 750- 450 v. Chr.), 2006; L. Ross Taylor, The Voting Districts of the Roman Republic, 1960.

Tribus	Derivation	Date of establishment
1. Suburana/Sucusana	Subura, an area of Rome; Varro, Ling. 5,48f.; Fest. 506	6th cent. BC
2. Palatina	*mons Palatinus* in Rome; Varro, Ling. 5,56; Fest. 506	6th cent. BC
3. Esquilina	Esquiliae (*mons Cispius* and *mons Oppius*), an area of Rome; Varro, Ling. 5,51; 56; Fest. 506	6th cent. BC
4. Collina	area of Rome around the *collis Viminalis* and *collis Quirinalis*; Varro, Ling. 5,51; 56; Fest. 506	6th cent. BC
5. Aemilia	*gens Aemilia*	before 495 (Liv. 2,21,7)
6. Camilia	*gens Camilia*	before 495
7. Claudia	*gens Claudia*; Liv. 2,16	before 495
8. Clustumina	*oppidum Crustumerium*; Fest. 48	before 495
9. Cornelia	*gens Cornelia*	before 495
10. Fabia	*gens Fabia*	before 495
11. Galeria	*gens Galeria*	before 495
12. Horatia	*gens Horatia*	before 495
13. Lemonia	*gens Lemonia*; Fest. 102	before 495
14. Menenia	*gens Menenia*	before 495
15. Papiria	*gens Papiria*; Paul. Fest. 263	before 495
16. Pollia	*gens Pollia*	before 495
17. Pupinia	*gens Pupinia*; Paul. Fest. 265	before 495
18. Romilia	*gens Romilia*; Paul. Fest. 331: *Romulia tribus*	before 495
19. Sergia	*gens Sergia*	before 495
20. Voltinia	*gens Voltinia*	before 495
21. Volturia/Veturia	*gens Volturia/Veturia*	before 495
22. Arniensis/Arnensis	River Aro, an outflow of the *lacus Sabatinus* into the *mare inferum* in Etruria	387 (Liv. 6,5,8)
23. Sabatina	*lacus Sabatinus* in Etruria; Fest. 464; Paul. Fest. 465	387 (Liv. 6,5,8)
24. Stellatina	*campus Stellatinus* in Etruria; Fest. 464	387 (Liv. 6,5,8)
25. Tromentina	*campus Tromentus* in Etruria; Paul. Fest. 505	387 (Liv. 6,5,8)
26. Pomptina	*urbs Pomptia/ager Pomptinus* in Latium; Paul. Fest. 263	358 (Liv. 7,15,11)
27. Poplilia/Poblilia/ Publilia	ager Poplilius in the territory of the Volsci in Latium; Paul. Fest. 265	358 (Liv. 7,15,11)
28. Maecia	settlement Ad Maecium near Lanuvium; Paul. Fest. 121	332 (Liv. 8,17,11)
29. Scaptia	*urbs Scaptia* in Latium; Fest. 464	332 (Liv. 8,17,11)
30. Falerna	*ager Falernus* in Campania	318 (Liv. 9,20,6)
31. Oufentina	after the river Ufens in Latium; Fest. 212	318 (Liv. 9,20,6)
32. Aniensis	after the river Anio in Latium	299 (Liv. 10,9,14)
33. Teretina	after the river Teres/Trerus; Fest. 498	299 (Liv. 10,9,14)
34. Quirina	the Romans' self-designation as 'Quirites'? But cf. Fest. 304: after the Sabine town of Cures	241 (Liv. Per. 19)
35. Velina	*lacus Velinus* at Reate	241. (Liv. Per. 19)

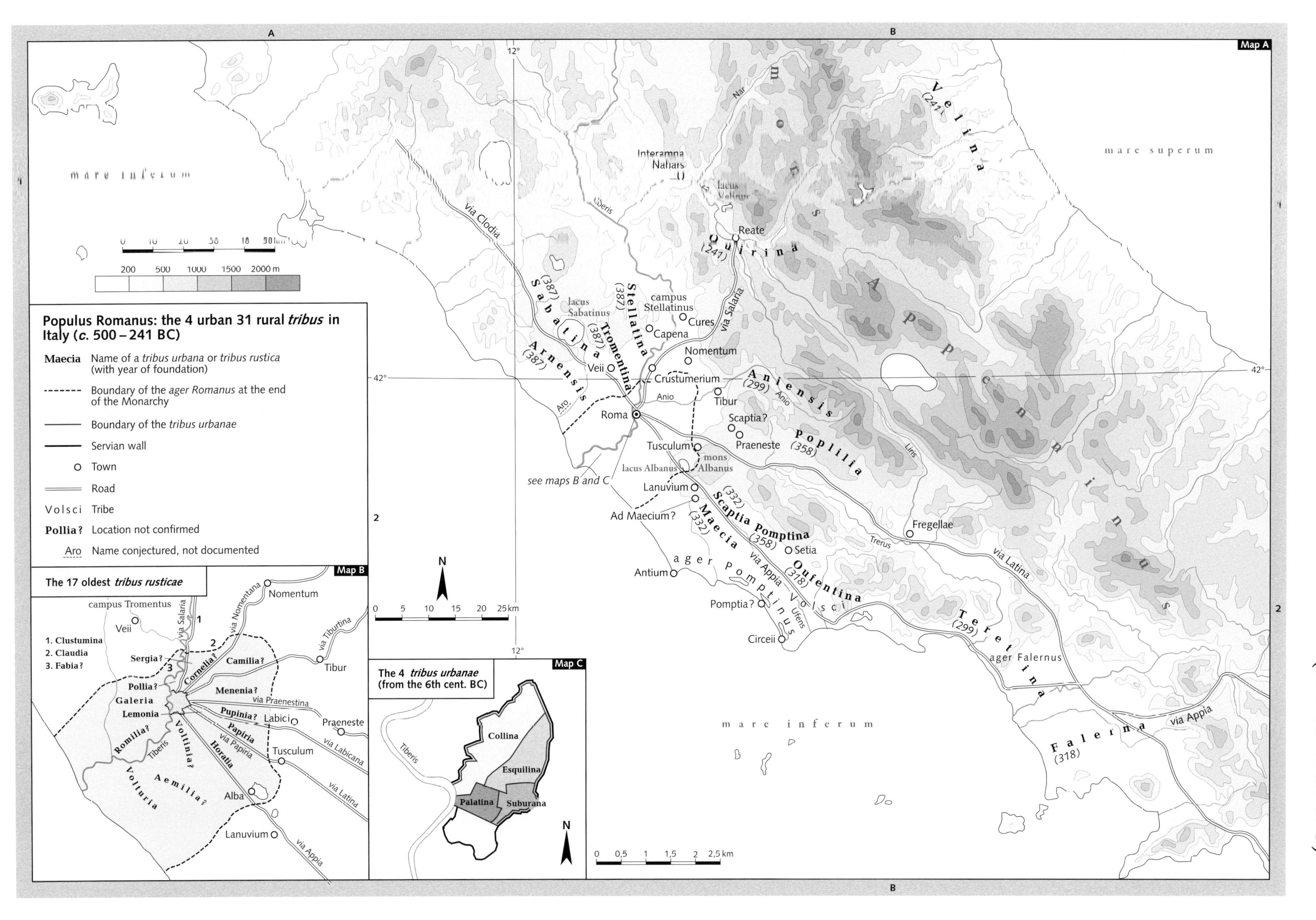
Map A
Populus Romanus: the 4 urban 31 rural tribus in Italy (c. 500 – 241 BC)
Maecia Name of a tribus urbana or tribus rustica (with year of foundation)
Boundary of the ager Romanus at the end of the Monarchy
Boundary of the tribus urbanae
Servian wall
Town
Road
Volsci Tribe
Pollia ? Location not confirmed
Aro Name conjectured, not documented
200 500 1000 1500 2000 m
0 5 10 15 20 25 km
mare inferum
mare superum
mons Appenninus
Velina (241)
Quirina (241)
Sabatina (387)
Stellatina (387)
Tromentina (387)
Arnensis (387)
Aniensis (299)
Poplilia (358)
Scaptia (332)
Pomptina (358)
Maecia (332)
Oufentina (318)
Teretina (299)
Falerna (318)
Volsci
ager Pomptinus
ager Falernus
Interamna Nahars
lacus Velinus
Reate
Nar
Tiberis
via Clodia
via Salaria
via Latina
via Appia
lacus Sabatinus
campus Stellatinus
Cures
Capena
Nomentum
Veii
Crustumerium
Anio
Tibur
Aro
Roma
Scaptia?
Praeneste
Tusculum
mons Albanus
lacus Albanus
see maps B and C
Lanuvium
Ad Maecium?
Antium
Setia
Pomptia?
Ufens
Circeii
Fregellae
Trerus
Liris
Map B
The 17 oldest tribus rusticae
campus Tromentus
1. Clustumina
2. Claudia
3. Fabia?
Sergia?
Pollia?
Galeria
Lemonia
Romilia?
Voltinia?
Horatia
Papiria
Pupinia?
Menenia?
Camilia?
Cornelia?
Aemilia?
Volturia
via Nomentana
via Tiburtina
via Praenestina
via Labicana
via Papiria
Labici
Alba
Map C
The 4 tribus urbanae (from the 6th cent. BC)
Collina
Esquilina
Palatina
Suburana
0 0,5 1 1,5 2 2,5 km

Roman colonial foundations in Italy before the Gracchi

The text to the main map is found alongside the complementary map 'Colonial foundations in Italy after 134/33–121 BC' (p. 155).

Latin city leagues (11th-4th cents. BC) (supplementary map)

I. Prior history

The map shows the territory of Latium, the region settled by the Latins, in the time preceding its political reorganization by Rome following the Latin War (340–338 BC). The politically independent communities (*populi*) in Latium vetus, which were closely linked by the Latin language and cults, started to form their own federal organizations at an early stage, based on religion (amphictyonies) and/or politics and military relations (symmachies), with their own officials. Rome was sometimes a member.

The ethnogenesis of the Latins was a drawn-out process that played out in Latium (continuity of settlement and culture from the preceding Bronze and Early Iron Ages; → maps pp. 31 and 65). There is evidence for the Latial culture from the 10th cent. The region of earliest settlement stretched south from the lower reaches of the Tiber to the cape of Circaeum (*Latium vetus*), forested country with good soil for arable and livestock farming (transhumance) and salt extraction (Tiber Delta). It was crossed by the north-south route between Etruria and Campania. The early period (so-called pre-urban period) was characterized by autonomous settlements in the Alban Hills and on the Tiber Plain, centred on Alba Longa. The towns in the Alban Hills declined in importance from the late 8th cent., while the communities in the plain and along the coast began to flourish. The final transition to an urban society took place between *c.* 650 and 570 BC.

II. City leagues

Associations in the early period of pre-urban Latium (to *c.* mid 7th cent. BC) centred on Alba Longa were probably above all sacral, organized around central sanctuaries. A list of thirty peoples/*populi* said to have taken part in the *feriae Latinae* is found in Plin. HN 3,69; some of these names survive in place-names (e.g. Bubetani, Querquetulani, Tolerienses). One of the cults important to the Latin identity was that of Iuppiter Latiaris on the summit of the *mons Cabus*. The *Ferentinum* (spring sanctuary) and the cult of Diana Ferentina near Aricia flourished from the 6th cent. BC. There was another sanctuary at Lavinium, near which was said to be the tomb of Aeneas, of great importance to Latium.

As far as can be judged from fortified settlements (*oppida*), defensive ramparts and so-called weapon graves, the formation of the *nomen Latinum/koinon* of the Latins probably served communal defence against neighbours hostile to the tribe (Etruscans, Umbrians, Oscans) in the 6th-4th cents., but also defence against Rome. To determine membership of the Latin League from the 6th (?)/5th cents. to 389/85, we rely on references to particular cities in Cato Orig. 4,129 H and Dion. Hal. Ant. Rom. 5,61,3, not all of which have been located. Tusculum and Aricia emerged as the dominant centres in the 5th cent.

The relationship between the Latin cities and Rome cannot always exactly be clarified. Rome was at least a sometime member (from 470 BC) of the League. The League already became to some extent dependent on Rome in the Monarchical Period; Rome was able to extend its domain by annexing Latin peoples. According to the literary tradition, the Roman king Tullus Hostilius destroyed Alba Longa around the mid 7th cent. and settled Albanes on the *mons Caelius* in Rome, to legitimize his dominion over Latium (Enn. Ann. 120–126; Cic. Rep. 2,31f.; Liv. 1,22–31; Dion. Hal. Ant. Rom. 3,1–35). Other Latin cities were said to have suffered a similar fate, and the cult of Diana was transferred to Rome. Rome's dominion over Latium under the latter Tarquin kings is confirmed in the first treaty between Rome and Carthage (509). There were also bilateral agreements/treaties (*foedera*). After the battle at *lacus Regillus* (*c.* 496 BC), in which Rome did not win a clear-cut victory, the historically attested *foedus Cassianum* (Dion. Hal. Ant. Rom. 6,95,2; Liv. 2,33,9; Cic. Balb. 53) was concluded in 493 BC; it regulated the rights of the Roman and Latin members of the League. In the face of constant external threats (e.g. advancing Oscan and Sabellian tribes), a strategy of shared defence was agreed (establishment of the *coloniae* of Ardea, Velitrae, Cora, Norba, Circeii and Setia). In the teeth of the persistent threat of the Celts after 387/6, the federal structure of the Latins was revived in 359 BC, and a joint army of the cities was set up, marked by the hegemony of Rome. The League broke apart in 341 when the Latins joined the Campanians to oppose Rome and the Samnites. Around 340, an attempted reconciliation of all the Latins and the attempted union (*unum populum, unam rem publicam fieri*, Liv. 8,5,4–6) with Rome failed. After a victorious campaign (340–338) against the Latin League, Rome finally dissolved it in 338/7, but it survived as a sacral association into the Roman Imperial Period.

Depending on their behaviour, the cities were annexed by Rome (*ager publicus*) to belong to the Roman state as *municipia* (Tusculum as early as 381, Aricia, Nomentum, Lanuvium, Pedum, Lavinium, Velitrae, Antium). *Coloniae civium Romanorum* of Roman citizens (Antium) were newly established. The other cities of Latium, on the one hand the *coloniae Latinae* of the Latin League (established colonies, e.g. Signia, Ardea, Norba) and autonomous cities (Tibur, Praeneste, Cora), on the other hand the newly-founded *coloniae Latinae* under Latin law, were made allies (*socii nominis Latini*) and compelled to provide troops. In 90 BC, they acquired Roman citizenship.

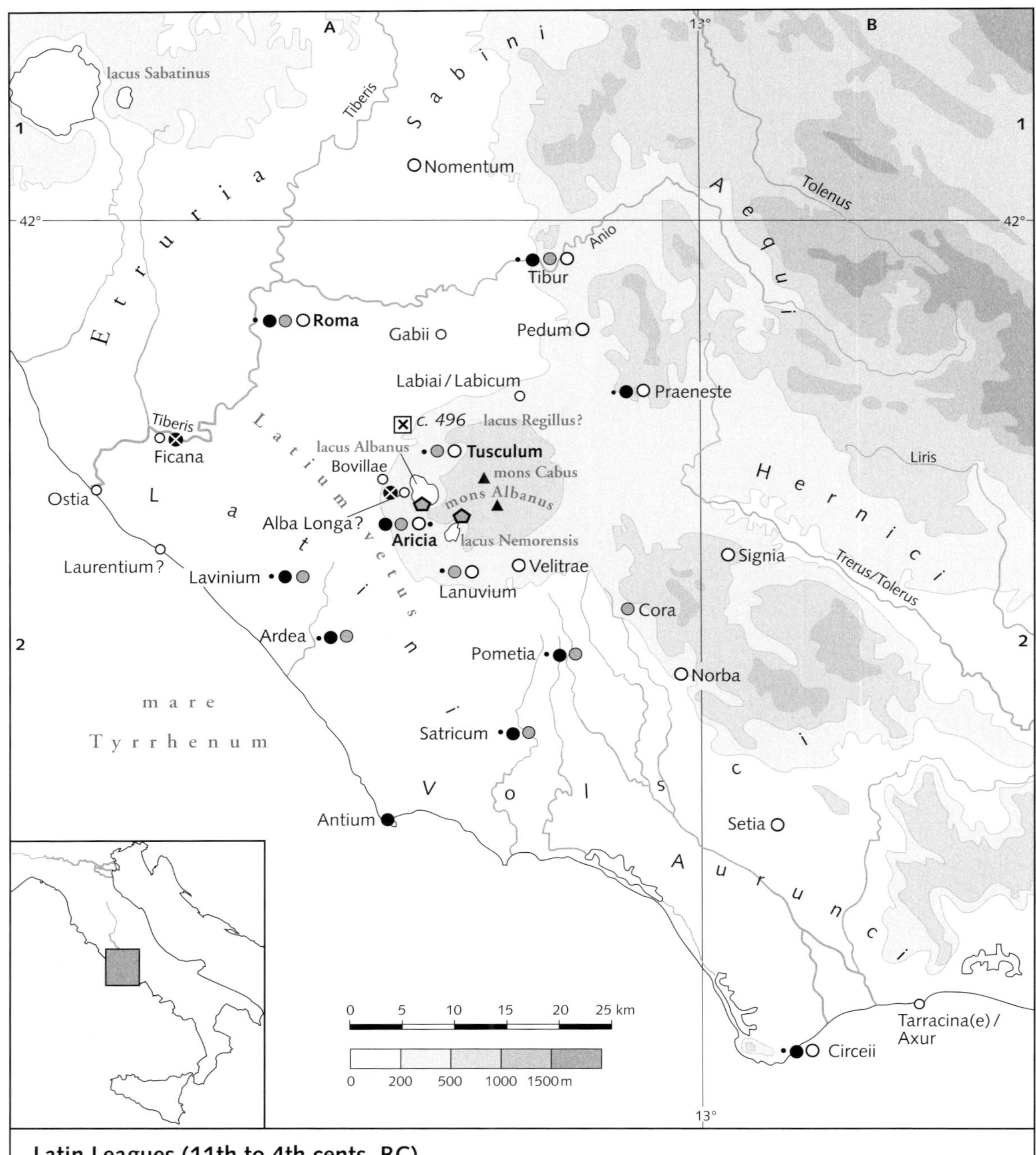

Sources

For the early period, the 'princely graves' of Praeneste (*c.* 700 BC) are an important source. Alongside inscriptions transmitted in secondary sources, the most important literary sources are Livy, Dionysius of Halicarnassus, Diodorus Siculus, Pliny (*Naturalis historia*), Festus, Cato, Plutarch (*Romulus*) and Cicero.

→ Maps pp. 65, 67, 111, 137, 155

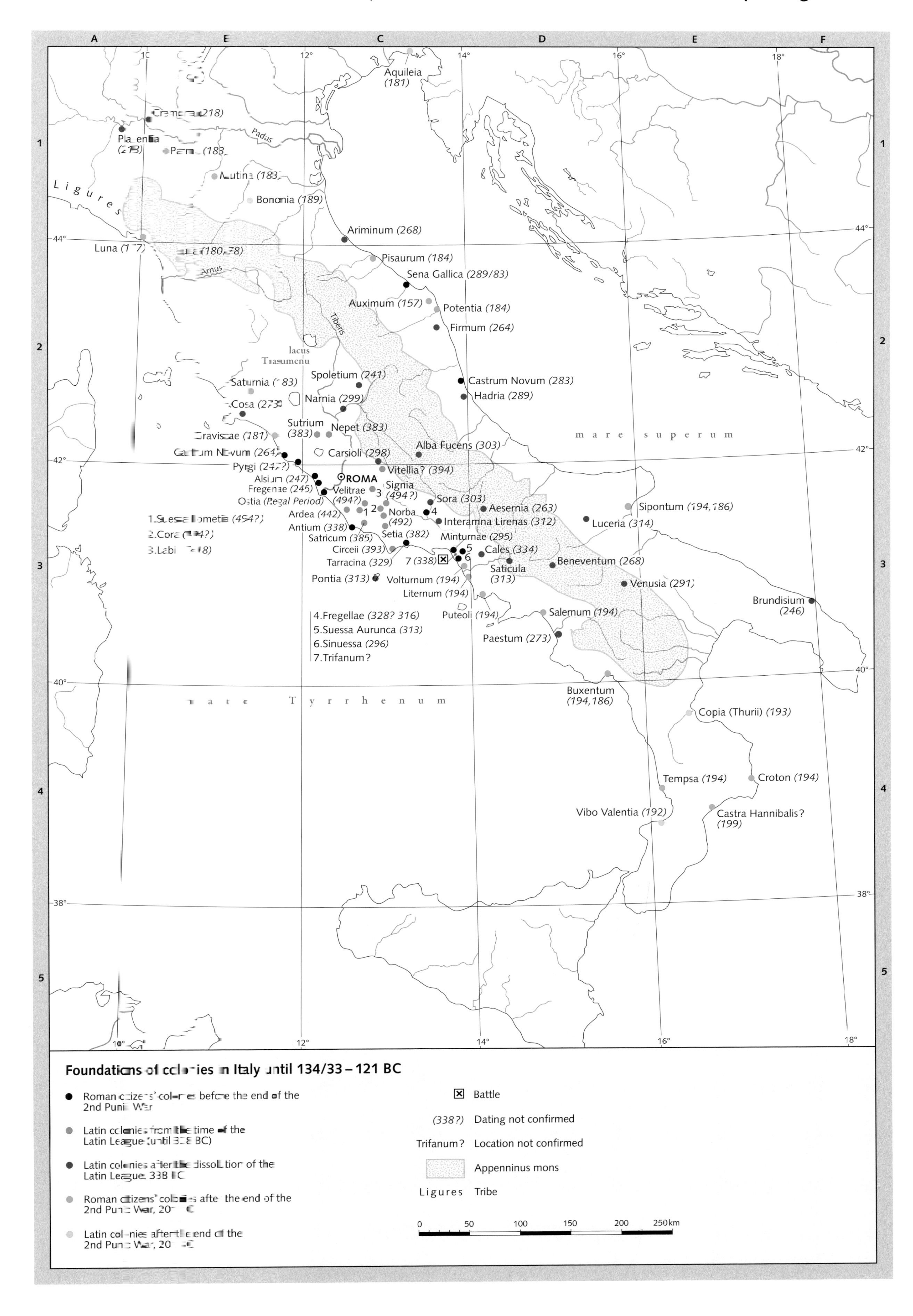
Aquileia (181)
Ariminum (268)
Pisaurum (184)
Sena Gallica (289/83)
Auximum (157)
Potentia (184)
Firmum (264)
Bononia (189)
Padus
Arnus
Tiberis
lacus Trasumenu
Spoletium (241)
Castrum Novum (283)
Hadria (289)
Narnia (299)
Nepet (383)
Sutrium (383)
Alba Fucens (303)
Carsioli (298)
Vitellia? (394)
ROMA
Signia (494?)
Velitrae (494?)
Sora (303)
Aesernia (263)
Norba (492)
Interamna Lirenas (312)
Ardea (442)
Antium (338)
Satricum (385)
Setia (382)
Minturnae (295)
Circeii (393)
Cales (334)
Tarracina (329)
Saticula (313)
Pontia (313)
Volturnum (194)
Liternum (194)
Puteoli (194)
Beneventum (268)
Luceria (314)
Sipontum (194,186)
Venusia (291)
Brundisium (246)
Salernum (194)
Paestum (273)
Buxentum (194,186)
Copia (Thurii) (193)
Tempsa (194)
Croton (194)
Vibo Valentia (192)
Castra Hannibalis? (199)
mare superum
Tyrrhenum
4.Fregellae (328? 316)
5.Suessa Aurunca (313)
6.Sinuessa (296)
7.Trifanum?
Battle
(338?) Dating not confirmed
Trifanum? Location not confirmed
Appenninus mons
Ligures Tribe
0 50 100 150 200 250km

The expansion of Rome into Etruria and Umbria up to the *lex Iulia* (4th cent. to 90/88 BC)

The Roman expansion into Etruria and Umbria up to the *lex Iulia* (4th cent.–90/88 BC)

The map shows significant aspects of Rome's expansion into the region of the Etruscan and Faliscan cities and the region of Umbrian settlement (which has only recently been the object of closer scholarly attention) in the period from the 4th to the early 1st cents. The beginnings of this process of expansion in the territories of Etruria and Umbria can be traced to the military conflicts (434–396 BC) between Latin Fidenae, Veii's bridgehead on the Tiber (conquered 426 BC), and Veii itself (conquered 396 BC) on the one hand and Rome on the other. It came to an end *de iure* with the grant of Roman citizenship under the *lex Iulia de civitate danda* in 90 BC, the introduction of Latin as the official language and the incorporation of the territories concerned into the Augustan *regiones* VI and VII in 7 BC.

I. Sources and material basis

The expansion of Roman territory by successive conquest, colonization and integration of the annexed cities and tribes, or by the conclusion of *foedera* with Etruscan and Umbrian cities, has literary attestation in Livy and Polybius in particular. There are also references in Diodorus Siculus, Appian and Cassius Dio.

Particularly worthy of note are significant epigraphic sources, which can be used to demonstrate more than just the spread of the Latin language before the introduction of Latin as the official language (90/88 BC). Finds of Latin inscriptions in Etruscan and Umbrian cities firstly suggest at least a certain degree of Roman influence, and secondly offer information to complement the literary sources, with important information on the integration of the Etruscan and Umbrian cities into the Roman city-state system. They show how treaties (*foedera*) linked the cities (*socii* or *civitates foederatae*) to Rome and how, sometimes in conjunction with such treaties, they relinquished parts of their *territorium* to Rome, for *coloniae Latinae* or *coloniae civium Romanorum* to be founded there. No attempt has been made to map Latin, Etruscan or Umbrian language monuments or bilingual inscriptions where the content is of no historical relevance.

In many cities and some areas of Etruria and Umbria, indications of Roman expansion have been discovered in archaeological surveys, and have continued to emerge in the very recent past.

II. The course of Roman expansion in central Italy

Roman expansion, or the expansion of the *ager Romanus*, in central Italy is shown in three phases. In some cases, the 'frontiers' shown follow the Etruscan territorial boundaries (mostly known), which are orientated towards features of the natural landscape.

1. Between c. 400 and 350 BC
In the first half of the 4th cent., the territory of Vei(s)/Veii was annexed after a ten-year war (405–396 BC: Liv. 4,60,9; Liv. 5,1ff.; transfer of the goddess Juno to Rome: Liv. 5,22,5), as were some Faliscan lands. The resistance of Capena and Falerii, which had supported Veii, was broken. In their immediate vicinity, Rome founded Nepet (383/73 BC) and Sutrium (373 BC) as colonies under Latin law. The natural Tiber frontier between Latins and Etruscans was now dissolved. The conquest of the large and fertile city-state of Veii is regarded as a turning-point in Roman history. Not only had Rome at a stroke enlarged its territory from approx. 250 km² to approx. 800 km², but it had asserted itself against the Etruscans in general, and begun to pave the way towards hegemony over central Italy. Moreover, the allocation of land to Roman citizens formed a precedent for future conquests.

But Rome did not only rely on military success. The unparalleled development of the Roman Empire was based on the ability to proceed flexibly and pragmatically. This approach encompassed peaceful co-existence by the conclusion of treaties (*foedera*) and the readiness to integrate allies and even former enemies into the commonwealth by the award of a limited form of Roman citizenship (*cives/civitas sine suffragio*). In this way, the conflicts with Tarch(u)na/Tarquinii (358-351 BC; Liv. 7,12,5ff.) were settled by a truce, regulated by treaty and lasting for forty years (Liv. 7,22,5); on expiry, it was renewed. Firm evidence also survives of treaties with the Umbrian cities of Camerinum, Iguvium and Ocriculum. According to Livy, there were many other treaties between Roman and Umbrian and Etruscan cities, although these are not explicitly described or preserved. Sources reflect a lasting amicable relationship between Rome and Caisri/Caere. It is disputed whether the city received the status of *civitas sine suffragio* as early as 353 BC or only in 293 BC (Sordi).

2. Mid 4th cent. to mid 3rd cent. BC
The period from 350 to 250 BC, which brought Rome enormous territorial gains, was characterized by many engagements ending in Roman victory, including against Etruscan cities: Sutrium was besieged from 311–309 BC (Liv. 10,32ff.). In 310 BC, the Romans plundered the Etruscan frontier region around the *mons Ciminus* (Liv. 9,36,12), and the same year saw the battle of Perusna/Perusia. Conquests in the region of Velsna/Volsinii followed in 308 BC.

In 302, there was a revolt in Arretium against the Cilnii, an Etruscan noble clan. This led to a Roman intervention (Liv. 10,3,2). A campaign was conducted against Rusellae in the same year, and in 295 BC, the Romans moved against Clevsin/Clusium and once more against Perusna/Perusia (Liv. 10,3,6ff.). The war against the Falisci and the Samnites (343-275 BC) was resumed from 298–293 BC, and ultimately led to the dissolution of the Samnite League. The Third Samnite War ended in the decisive battle of Sentinum in Umbria in 295 (Pol. 2,19,6; Liv. 10,27ff.; Diod. Sic. 21,6) between the victorious Romans and an alliance of Senones, Umbri, Samnites and Etruscans. A truce with Velsna/Volsinii and Perusia was declared in 294 BC; Perusia was later conquered, as was Arretium (Liv. 10,37,1ff.). The conquest of Rusellae followed in 294/3 or 292 BC, and peace treaties were signed with the Falisci in 293 BC (Liv. 10,46,10ff.).

In 284 BC, Velsna/Volsinii made an alliance with the Gauls, with whom Rome had been in constant conflict since *c.* 390 BC, esp. since the traumatic defeat of 387 BC at the Allia and the conquest and sacking of the city of Rome (Roman victory over the Gauls only in 225 BC at Telamon: Pol. 2,27). Arretium was besieged. There was a battle, probably *c.* 283/81 BC, at the *lacus Vadimonis*, with a Roman victory over the Etruscans (Pol. 2,19ff.). Victories over Tarch(u)na/Tarquinii, Velsna/Volsinii and Velch/Volci followed in 281 and 280 BC (Liv. Per. 12,13) and over Caisri/Caere in 273 (Cass. Dio fr. 33). In 264 BC, Velsna/Volsinii veteres was destroyed, the population moved to Volsinii novi and the deity Vertumna/Vertumnus evoked to Rome. In 241 BC, Falerii veteres suffered the same fate – it was destroyed and its population moved to Falerii novi (Pol. 1,65). Through these conflicts, Rome succeeded in extending its territory and sphere of influence, esp. in southern Etruria, in the contact zone between the Etrusci and the Umbri, and in Umbria itself as far as the Adriatic. The cities and regions were tied to Rome by *foedera* and integrated into the Roman alliance system, but they remained largely autonomous in administrative terms (Galsterer).

3. Mid 3rd cent. to 1st cent. BC
The period from the mid 3rd to the 1st cent. BC was mostly one of consolidation of these conquests, and of expansion of the Roman/Romanized territory in the interior and esp. in northern Etruria. This was achieved by the construction of important consular roads (some on older existing routes, as archaeology has demonstrated), achieving a better infrastructure for military and commercial purposes. They ran through areas in which it has been shown that most of the (new) Roman settlements were also located, but generally paid no attention to the old Etruscan and Umbrian centres.

The conclusion of a *foedus* (also *aequum/iniquum*), i.e. a treaty of alliance making the partner a *socius populi Romani*, was always associated with the relinquishment of at least part of the *territorium* to Rome (cf. archaeological surveys in the *territorium* of the Latin colony of Cosa); *coloniae Latinae* or *coloniae civium Romanorum* were set up and Roman villas and so-called *fattorie* constructed here as enclaves (Dyson, Enei). The foundation of fortified *coloniae* is attested in literature, and most have been archeologically identified. They were mostly created in coastal areas in the context of the Punic Wars.

The exact extent of the Etruscan and Umbrian *territoria* around their urban centres remains unknown and can only be reconstructed to a limited extent, the courses of boundaries mostly being hypothetical. However, recent research involving surveys and excavations in some *territoria* of the Tyrrhenian littoral and the Etrurian interior has at least improved our understanding of the ongoing Romanization process. Comparative studies of Etruscan settlement development and Roman settlement by means of estates and *fattorie* within the *territoria* permit insight into the Roman presence in general and the differences in approach in the settlement of coastal and inland areas respectively. The coastal area was occupied earlier and more intensively by Roman settlers, and Etruscan settlements were destroyed or abandoned, while a less aggressive picture emerges inland: a smaller number of Roman estates developed alongside Etruscan settlements (Hemphill, Terrenato).

Overall, it is clear, although difficult to map, that Rome followed no strict plan for the integration of central Italy (or, later, of other regions) in its acts of conquest and measures of Romanization. Rather, it took many discrete steps, each of which must likewise be evaluated separately. These steps included in particular the granting of limited Roman citizenship, the inclusion, by force and/or treaty, of the Etruscans and Umbrians in the Roman alliance system as *socii* in so-called *civitates foederatae*, the associated recognition of Roman supremacy, the relinquishment of *territorium*, which became *ager Romanus*, the foundation of *coloniae* under Latin (*colonia Latina*) and Roman law (*colonia civium Romanorum*) as enclaves, the construction of consular roads and the spread of the Latin language.

→Maps pp. 75, 77, 109, 137, 139, 175

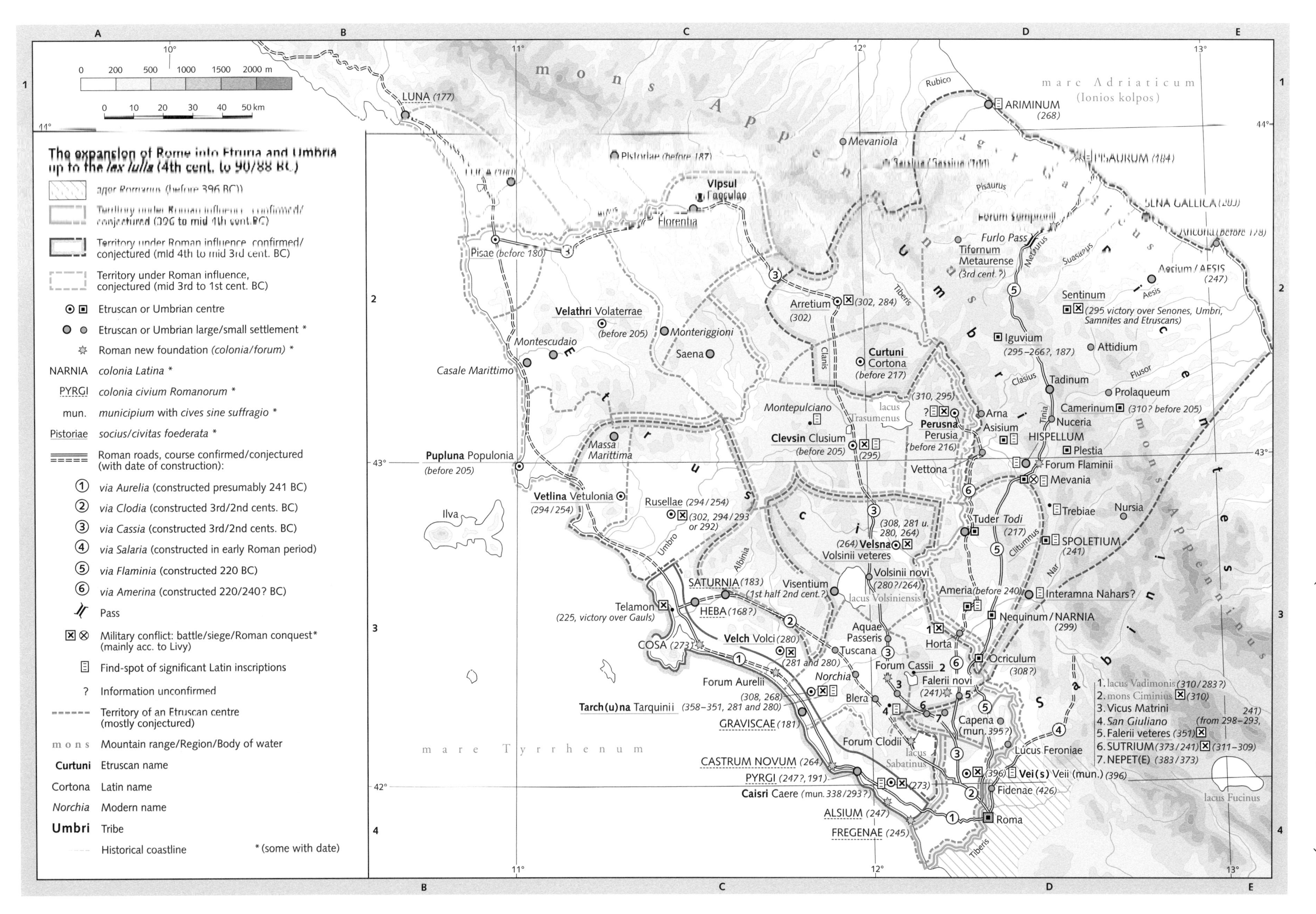
The expansion of Rome into Etruria and Umbria up to the *lex Iulia* (4th cent. to 90/88 BC)
0 200 500 1000 1500 2000 m
0 10 20 30 40 50 km
ager Romanus (before 396 BC)
Territory under Roman influence, confirmed/conjectured (396 to mid 4th cent. BC)
Territory under Roman influence, confirmed/conjectured (mid 4th to mid 3rd cent. BC)
Territory under Roman influence, conjectured (mid 3rd to 1st cent. BC)
Etruscan or Umbrian centre
Etruscan or Umbrian large/small settlement *
Roman new foundation (colonia/forum) *
NARNIA colonia Latina *
PYRGI colonia civium Romanorum *
mun. municipium with cives sine suffragio *
Pistoriae socius/civitas foederata *
Roman roads, course confirmed/conjectured (with date of construction):
1 via Aurelia (constructed presumably 241 BC)
2 via Clodia (constructed 3rd/2nd cents. BC)
3 via Cassia (constructed 3rd/2nd cents. BC)
4 via Salaria (constructed in early Roman period)
5 via Flaminia (constructed 220 BC)
6 via Amerina (constructed 220/240? BC)
Pass
Military conflict: battle/siege/Roman conquest* (mainly acc. to Livy)
Find-spot of significant Latin inscriptions
? Information unconfirmed
Territory of an Etruscan centre (mostly conjectured)
mons Mountain range/Region/Body of water
Curtuni Etruscan name
Cortona Latin name
Norchia Modern name
Umbri Tribe
Historical coastline
* (some with date)
1. lacus Vadimonis (310/283?)
2. mons Ciminius (310)
3. Vicus Matrini
4. San Giuliano (from 298–293, 241)
5. Falerii veteres (351)
6. SUTRIUM (373/241) (311–309)
7. NEPET(E) (383/373)
mare Adriaticum (Ionios kolpos)
mare Tyrrhenum
mons Appenninus
LUNA (177)
ARIMINUM (268)
Velathri Volaterrae (before 205)
Pupluna Populonia (before 205)
Vetlina Vetulonia (294/254)
Arretium (302)
Curtuni Cortona (before 217)
Clevsin Clusium (before 205)
Perusna Perusia (before 216)
Velsna Volsinii veteres
Velch Volci (280)
Tarch(u)na Tarquinii (358–351, 281 and 280)
Caisri Caere (mun. 338/293?)
Vei(s) Veii (mun.) (396)
Roma

The campaigns of Alexander the Great (336–323 BC)

Having secured the northern (Thrace and up to the Istrus/Danube) and western (southern Illyria) frontiers of Macedonia and put down a revolt in Thebes in 335 BC, Alexander the Great crossed to Asia Minor in 334 as *hegemon* of the Panhellenic League (Corinthian League) to pursue the war of revenge against Persia which Philip II had declared in the spring of 337 (Diod. Sic. 17,17). No-one, ancient or modern, has been able to determine whether it was his intention from the outset to conquer the Persian Empire.

Course of events

In 334, after defeating a small satrap army at the Granicus, Alexander occupied Lydia (taking Sardis) almost unopposed, and the west coast (resistance at Miletus, Halicarnassus), where freedom was declared for all the Greeks of Asia Minor and the fleet of the Corinthian League was released. After conquering Caria, moving on through Lycia (Sagalassus), Pamphylia and Phrygia (Celaenae), wintering at Gordium (334/33) and crossing the Taurus, he took the field in early November 333 at Issus, defeating Darius III (336–330 BC).

While most Phoenician coastal cities joined Alexander, Tyre only fell after a lengthy siege (332) Tyre, like Gaza farther south, an important trans-shipment centre on the Egyptian frontier, was severely punished for its resistance. Darius had initiated peace negotiations during the siege of Tyre, offering a partition of the empire with (initially) the Euphrates as the frontier. Alexander refused: from now if not before, it must have been his intention to bring the whole Persian Empire, indeed the entire *oikoumene*, under his rule.

Egypt was occupied without resistance by the 'Son of Zeus-Ammon', liberator from the Persian yoke, and Alexandria was founded in early 331 (cf. supplementary map), the first of many foundations and refoundations made by Alexander.

The army crossed the Euphrates and Tigris in the summer of 331, and clashed with Darius again at Gaugamela on 1 October. Darius fled to Ecbatana, thus abandoning Babylon and Susa and making Alexander 'King of Asia'. Alexander occupied Persepolis (plunder) in January 330, remaining there until May because of unrest in Greece (Agis III of Sparta) before moving on to Ecbatana.

Pursuing Darius, who had withdrawn eastwards (only to be murdered by Bessus, probably for wanting to give up the fight against Alexander), Alexander reached Areia, where he put down a rebellion, and moved on to Phrada. Through Phrada, Alexander took his army to Alexandria (Qandahār), then over the mountains to Alexandria (Ghazni). Although his satraps had constantly to put down new revolts in his wake, he himself encountered no resistance from the southern Iranian tribes. With colossal exertion, the army crossed the Paropamisus (Hindu Kush) in winter, continuing the advance from there through the Kabul Valley to Bactra (329). Everywhere, Alexander founded veteran colonies as support bases, as well as larger 'colonies' such as Alexandria Eschate ('the Uttermost') on the Iaxartes. At the Oxus the Achaemenid Bessus, who had ruled as Artaxerxes V in eastern Iran since the murder of Darius and who had also found support in India, was handed over to him and executed. Alexander faced three years of small but punishing wars in Bactria and Sogdiana (the region north of Bactria, capital Maracanda) against the fortresses of Iranian rulers, the Iranian tribes and their allies, the Scythian-Massagetan nomad cavalry. He was only able to prevail after allying with the north-eastern Iranian nobility by marrying Roxane, the daughter of the local Sogdian prince Oxyartes, early in 327.

Persian suzerainty was to an extent recognized in India (present-day Punjab, northern Pakistan), although there were no satrapies there. The northwest frontier was conquered after Persia in 327, while the Indus extended a friendly welcome. The region up to the Indus and Hydaspes was handed over to Macedonian satraps, the Hydaspes was crossed east of Taxila and in 326, after skirmishes, the realm of the Indian King Porus was set up as a client kingdom. The army's refusal to go further signalled the easternmost extent of the campaign by the Hyphasis, an event marked by the erection of twelve altars. Alexander turned south, using his fleet on the Indus (November 326), and reached the Ocean after many skirmishes with the tribes. Nonetheless, two more satrapies were set up in southern India and colonies were founded, although these did not last.

Preparations were made for the homeward march in the summer of 325 at Patala (port and naval base) in the Indus Delta. The wounded and elderly were to take the easier northern route with Craterus, while Nearchus sailed along the coast, subjecting it to study. In the autumn of 325, Alexander himself led the rest of the army through the hostile Gedrosian Desert to Pura, suffering enormous losses. The great campaign ended in March 324 after a visit to Pasargadae: a mass wedding took place of Macedonian soldiers to Persian women at Susa, as the symbolic inauguration of a new empire, the 'Macedonian-Iranian Empire', entirely focussed on the person of Alexander. Further subsequent landmarks along the way were Opis, where the Macedonians who had completed their service were released, and Ecbatana (death of Hephaestion); then, after a brief campaign against a mountain tribe, came the return to Babylon.

The exploratory expeditions and naval construction work attested in the sources suggest that Alexander's next great ambition was probably the conquest of the Arabian peninsula. However, while preparing for this campaign, Alexander fell ill in 323 BC, and he died on 10 June, as King of the Macedons, *hegemon* of the Corinthian League, Pharaoh of Egypt and Great King of Persia.

Sources

The state of the sources presents problems, as almost the only writings giving the history and biography of Alexander the Great are late, secondary literature of the so-called 'Alexander historians'. They drew on two traditions: Callisthenes was a contemporary of the events, and took part in the Alexander campaign as court historian. Ptolemy and the Alexander historian Aristoboulus described all the campaigns. Although they had been eye witnesses, they only recorded their experiences in writing much later, and in the process they made use of extant works. Other contemporaries were Nearchus, Onesicritus, Chares and Ephippus. In Rome, Cleitarchus was the most widely read source. The most that survives of these first-generation works and of most of the subsequent literature from the 1st cent. BC to the 2nd cent. AD (so-called 'vulgate' tradition: Justin, Plutarch, Arrian (the most important source), Diodorus Siculus, Curtius Rufus; these are mostly based on Cleitarchus, Ptolemy and Aristoboulus) is fragments cited by other writers. The authenticity of the correspondence between Alexander and Darius is today unquestioned. Epigraphic sources are sparse, coins almost impossible to assess historically. The secondary literature is vast beyond control; there are some very recent monographs.

→ Maps pp. 8f., 87, 105, 115, 133; Alexander [4] 'the Great', BNP 1, 2002

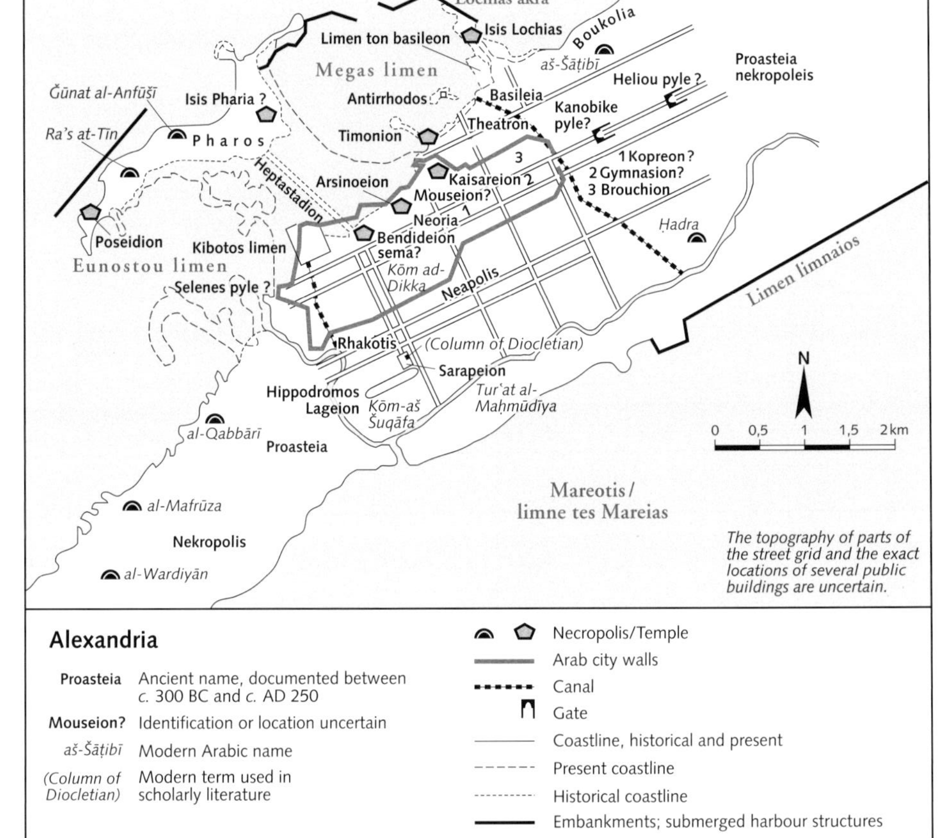

Literature

R.S. Bagnall, P. Derow (eds.), The Hellenistic Period. Historical Sources in Translation, 2004; A.B. Bosworth, A Historical Commentary on Arrian's History of Alexander, 2 vols., 1980, 1995; Id., Alexander and the East, 1996; Id., Conquest and Empire. The Reign of Alexander the Great, 1996; M. Clauss, Alexandria, 2003; H.-J. Gehrke, Geschichte des Hellenismus, [3]2003; Id., Alexander der Große, [4]2005; G. Grimm, Alexandria. Die erste Königsstadt der hellenistischen Zeit, 1998; R. Lane Fox, Alexander the Great, 1974; M. Pfrommer, Alexandria. Im Schatten der Pyramiden, 1999; Id., Alexander der Große. Auf den Spuren eines Mythos, 2001; J. Wiesehöfer, Das frühe Persien, 1999; J. Seibert, Alexander der Große, 1972; P. Siewert, L. Aigner-Foresti, Föderalismus in der griechischen und römischen Antike, 2005; H.-U. Wiemer, Alexander der Große, 2005.

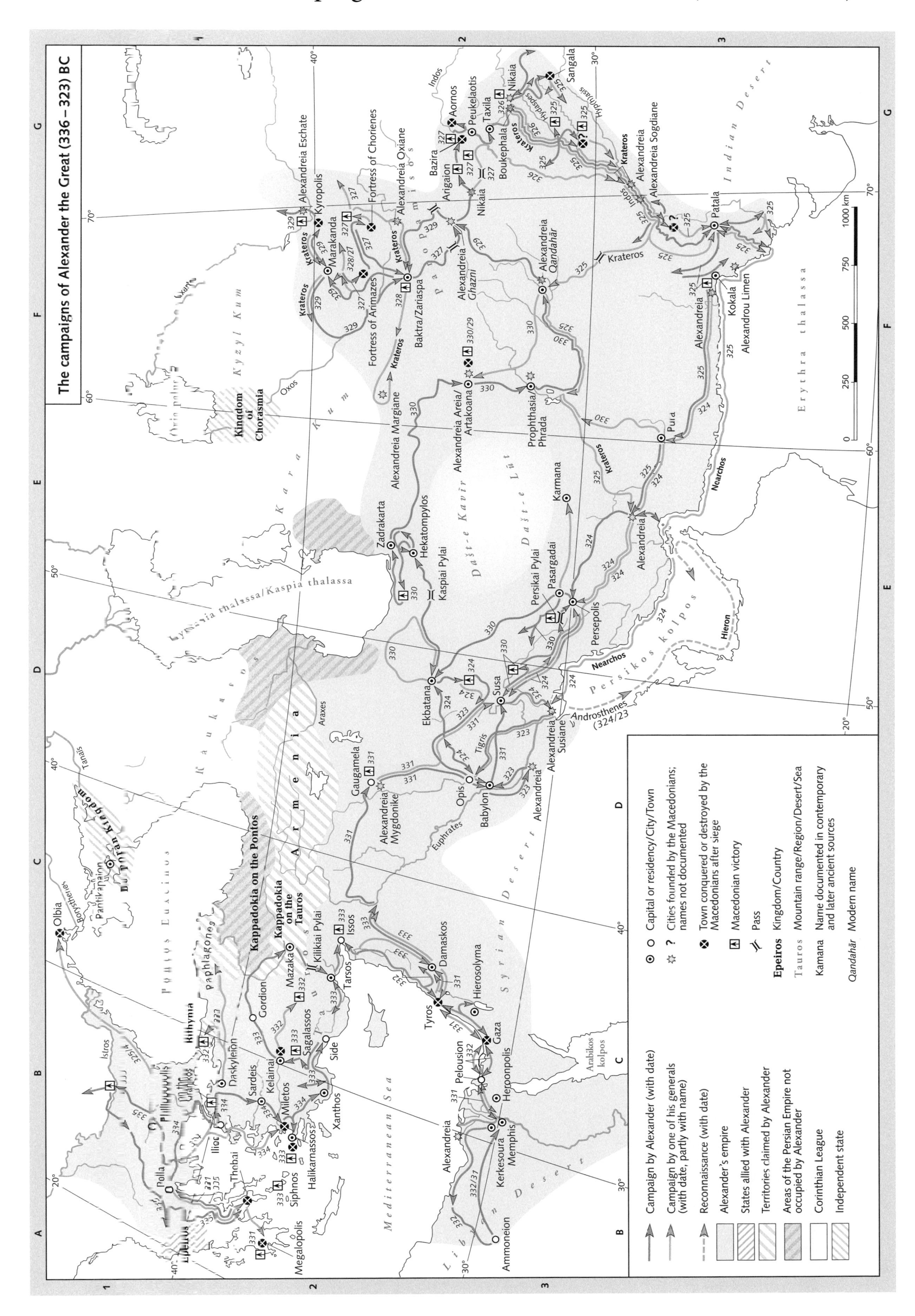
The campaigns of Alexander the Great (336 – 323) BC
Campaign by Alexander (with date)
Campaign by one of his generals (with date, partly with name)
Reconnaissance (with date)
Alexander's empire
States allied with Alexander
Territories claimed by Alexander
Areas of the Persian Empire not occupied by Alexander
Corinthian League
Independent state
Capital or residency/City/Town
Cities founded by the Macedonians; names not documented
Town conquered or destroyed by the Macedonians after siege
Macedonian victory
Pass
Epeiros Kingdom/Country
Tauros Mountain range/Region/Desert/Sea
Kamana Name documented in contemporary and later ancient sources
Qandahār Modern name

The kingdoms of the Diadochi (*c.* 303 BC)

At his sudden death in 323 BC, Alexander the Great left his empire with no provision for succession. The period leading up to the emergence of the Hellenistic states in the 3rd cent. (272) was thus characterized by many disputes among his 'successors' (*diadochoi*), his closest companions and officers, who held important offices or received or already held satrapies.

The emergence of the kingdoms of the Diadochi can be roughly divided into two phases: a) the four Wars of the Diadochi up to the death of Antigonus Monophthalmus (who had primarily fought to preserve the unity of the empire) at the battle of Ipsus in 301 BC, and b) the phase beginning as early as 306/305 with the assumption of royal titles by the various Diadochi, in which the successor states to the Alexandrian empire (of whom three remained by 281: those of the Ptolemies, Seleucids and Antigonids) gradually acquired the shape of states in their own right, albeit without establishing what we might call stable political conditions.

As a whole, this was a historical period of great complexity, not readily susceptible to scrutiny. There were many theatres of activity, often simultaneous, and in a very wide range of locations. Only the main strands of the events can be described and mapped here.

I. The initial situation: the struggle to keep the empire intact versus the personal interests of the Diadochi

The first arrangements, agreed at Babylon by the Diadochi after bitter disputes, still show attempts to preserve the unity of the empire. Alexander's mentally incompetent brother was to be installed as Philip III with the former's (as yet unborn) son by Roxane, the future Alexander IV, as kings with equal powers, under the guardianship of the absent Craterus and the regency of Perdiccas. This construction, however, neither proved practicable, nor, as the sources attest (Diod. Sic. 20, 37, 4) did it suit the ambitions of the potential direct successors, namely Antipater, still *strategos* of Europe, Lysimachus, now satrap of Thrace, Leonnatus, satrap of Hellespontic Phrygia, Antigonus, still satrap of Greater Phrygia, Ptolemy, satrap of Egypt, or Perdiccas.

Moreover, the state of the empire in general was unstable. Immediately after Alexander's death, many Greek states, including Athens and the Aetolian League, had risen against Macedonian rule (Lamian War 323/2), and there were mutinies in the east. The satrapy of Cappadocia, allocated to the Greek Eumenes, had first to be reconquered (322), while Leonnatus fell in battle against the Greeks.

II. First to Third Wars of the Diadochi (cf. also supplementary map)

Perdiccas, who as regent had the entire military array of the empire at his disposal, claimed the whole empire for himself, in the process uniting Antipater, Craterus, Antigonus, Lysimachus and Ptolemy in a coalition against him and in 321/320 occasioning the First War of the Diadochi, the theatres of which were in Asia Minor (Eumenes against the coalition) and Egypt (Antipater against Ptolemy). Provisions of a compromise agreed at Triparadisus in north-eastern Syria (east of Tripolis) after the death of Craterus and the murder of Perdiccas in the summer of 320 included the bestowal of the satrapy of Babylonia on Seleucus and the appointment of Antipater as guardian of the kings and regent. However, this compromise failed as early as the autumn of 319, with Antipater's death.

The Second War of the Diadochi was instigated by Cassander, the son of Antipater, who saw himself as the legitimate successor following his marriage to the last daughter of Philip II. He fought it in Europe against Polyperchon, Antipater's appointed successor, and Olympias, whom he eliminated along with various other members of the Macedonian royal house (late 317).

At the same time, Antigonus, joined by Ptolemy and Lysimachus, attacked Eumenes, who was loyal to Polyperchon and in 318/317 succeeded in extending and consolidating his dominion beyond Mesopotamia and the Persis until his murder in 315. The theatre of war here was Mesopotamia (Zagros Mountains). Antigonus purportedly fought in the name of the unity of empire, but it rapidly became clear that his aim was total dominion. He occupied the eastern satrapies, including that of Seleucus, and demanded the release of Alexander IV (proclamation of Tyre), whom Cassander had taken prisoner. He also won Greek cities to his side, demanding their freedom and autonomy.

Cassander, Lysimachus, Ptolemy and Seleucus joined forces against him for the Third War of the Diadochi in 314. Ptolemy undertook operations in southern Asia Minor, and Seleucus fought a naval campaign in the eastern Aegean. In 312, Demetrius, the son of Antigonus, was defeated at Gaza, and Seleucus won back his satrapy, although Antigonus managed to bring Syria back under his control. In 311, he was compelled to accept terms, or rather a truce, confirming the *status quo* until King Alexander IV reached maturity, whereupon Cassander had Alexander murdered and re-ignited the fight for the succession.

Antigonus moved against Seleucus in the east, while Ptolemy began a major offensive against Cassander in southern Asia Minor, the Aegean and Greece – although he failed to gain the support of the Greek cities. He concluded a peace with Cassander in 308 and withdrew to Egypt. Towards the end of 308, there was also an agreement between Antigonus and Seleucus, whom the former had failed to weaken significantly in Mesopotamia.

III. The assumption of royal titles by the Diadochi

After the spectacular successes of his son Demetrius in Greece (liberation of Athens in 307) and the Aegean and the expulsion of Ptolemy from Cyprus (naval battle off Salamis) in 306, Antigonus had himself proclaimed king alongside his son at the newly-founded capital of Antigonea on the Orontes. Ptolemy, Cassander, Lysimachus and Seleucus soon followed his example (305, 'Year of the Kings'). Late in 306, the Antigonid kings tried in vain to conquer Egypt directly, and from 306 the Ptolemaic base of Rhodes was besieged (cf. also supplementary map). A treaty brought an end to the siege (304). In 303, Demetrius started a new offensive against Greece, and in 302, an alliance system, the League of Corinth, was founded under Antigonid leadership; most of the Greek cities joined.

IV. The Fourth War of the Diadochi

The year 302 then saw the outbreak of the Fourth War of the Diadochi, between Antigonus and the allies Ptolemy, Cassander, Lysimachus and Seleucus.

Seleucus had brought the eastern Iranian regions under his control after 308, and had advanced deep into India. There he signed a treaty with the most prominent of the local princes, Sandracottus/Chandragupta, formally subjecting the latter to Seleucus' sovereignty. Thereafter, in 302, Seleucus had to return for more campaigning in Asia Minor, where Lysimachus was confronting Antigonus. In 301, the war ended with the battle of Ipsus in Phrygia, at which the aged Antigonus Monophthalmus fell and Demetrius fled.

With that, the possibility of preserving Alexander's empire as a political unit was definitively extinguished. The ensuing years instead saw the establishment of several major successor kingdoms.

Sources

Although the Hellenistic period saw the most profuse writing activity in Greek Antiquity, hardly any historical or philosophical works have survived. There are fragments of the contemporary accounts of Timaeus of Tauromenium, Hieronymus of Cardia and Poseidonius of Apamea. Important secondary sources from the Roman period are Diodorus Siculus, Pompeius Trogus and Appian (survey of the Seleucids), as well as the *Vitae* of Plutarch (Eumenes, Demetrius and Pyrrhus). There are also Jewish texts in Greek and Aramaic (Flavius Josephus; Old Testament, Book of Daniel). As well as inscriptions, Egyptian papyri and Mesopotamian cuneiform texts are important for the period of the Diadochi. As for archaeological finds, there are only sparse remains of the capitals of the great Diadochi kingdoms, but more substantial finds have been made in Miletus, Ephesus and Pergamum. The titles and portraits of the Diadochi are known primarily from coin images and marble busts.

Overall, the ambivalent state of the sources and the complex, convoluted sequences of events mean that scholars have been, and continue to be, faced by many problems of detail, although at the same time the main historical narratives can be accepted as factual knowledge.

→Maps pp. 100–103, 113, 121, 125, 133; Diadochi, Wars of the, BNP 4, 2004; Diadochi and Epigoni, BNP 4, 2004

Literature

R.S. Bagnall, P. Derow (eds.), The Hellenistic Period. Historical Sources in Translation, 2004; R.A. Billows, Antigonos the One-Eyed and the Creation of the Hellenistic State, 1990; G.R. Bugh, The Cambridge Companion to the Hellenistic world, 2006; W.M. Ellis, Ptolemy of Egypt, 1994; H.-J. Gehrke, Geschichte des Hellenismus, [3]2003; H.S. Lund, Lysimachus, 1992; A. Mehl, Seleukos Nikator und sein Reich. 1. Teil, 1986; O. Müller, Antigonos Monophthalmos und »Das Jahr der Könige«. Untersuchungen zur Begründung der hellenistischen Monarchien 306–304 v. Chr., 1972; W. Orth, Die Diadochenzeit im Spiegel der historischen Geographie, 1993; M. Rathmann, Perdikkas zwischen 323 und 320. Nachlassverwalter des Alexanderreiches oder Autokrat?, 2005; C. Schäfer, Eumenes von Cardia und der Kampf um die Macht im Alexanderreich, 2002; L. Schober, Untersuchungen zur Geschichte Babyloniens und der oberen Satrapien von 323–303 v. Chr., 1981; F.W. Walbank (ed.), The Hellenistic World, [2]1984.

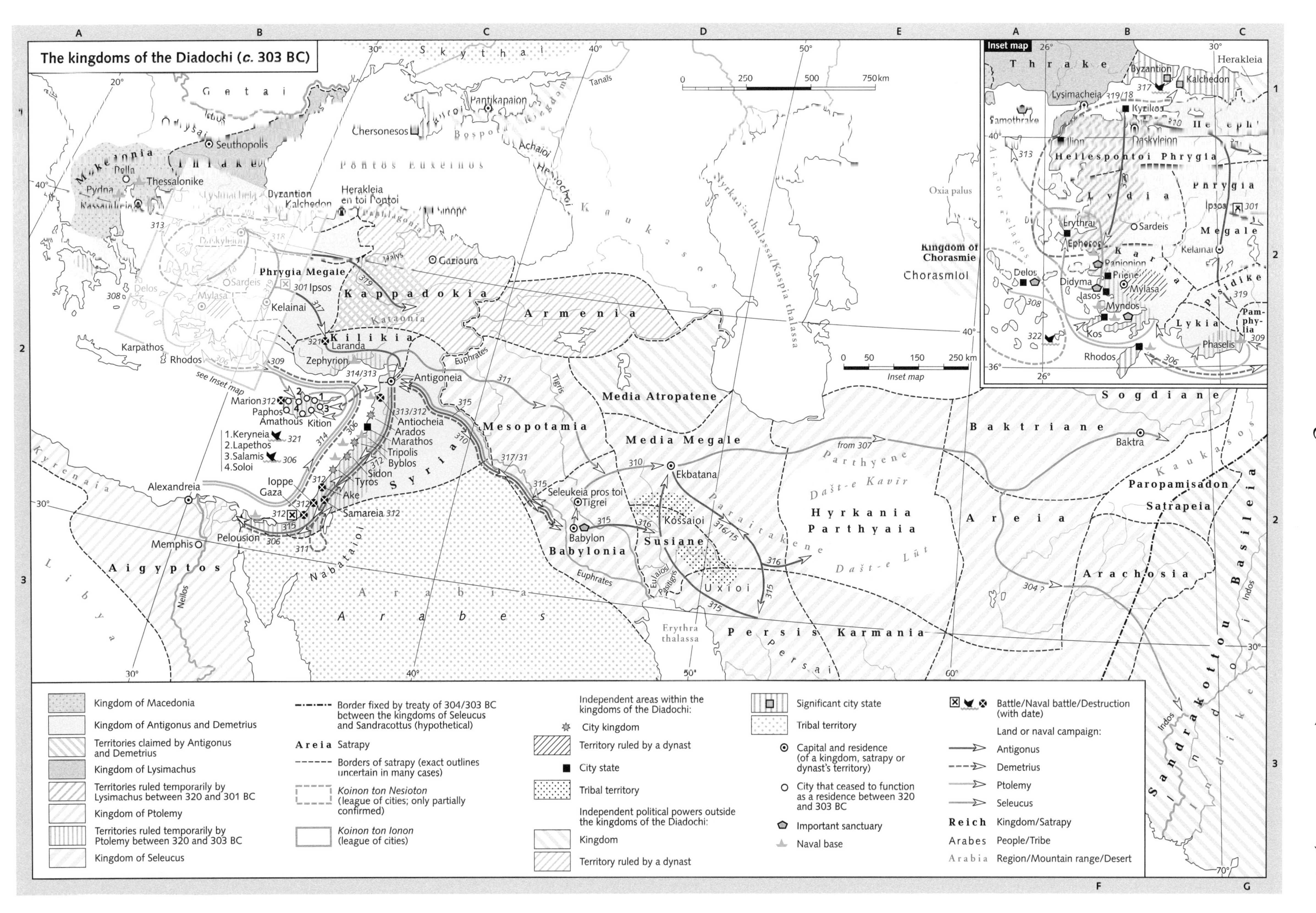
The kingdoms of the Diadochi (c. 303 BC)
Inset map
see Inset map
Makedonia
Thrake
Getai
Skythai
Pontos Euxeinos
Kaukasos
Phrygia Megale
Kappadokia
Armenia
Kilikia
Mesopotamia
Media Atropatene
Media Megale
Syria
Babylonia
Susiane
Persis
Karmania
Hyrkania
Parthyaia
Baktriane
Sogdiane
Areia
Arachosia
Paropamisadon Satrapeia
Sandrakottou Basileia
Aigyptos
Libya
Arabes
Kyrenaia
Kingdom of Chorasmie
Chorasmioi
Alexandreia
Memphis
Pelousion
Gaza
Ioppe
Antigoneia
Ekbatana
Babylon
Seleukeia pros toi Tigrei
Baktra
Kingdom of Macedonia
Kingdom of Antigonus and Demetrius
Territories claimed by Antigonus and Demetrius
Kingdom of Lysimachus
Territories ruled temporarily by Lysimachus between 320 and 301 BC
Kingdom of Ptolemy
Territories ruled temporarily by Ptolemy between 320 and 303 BC
Kingdom of Seleucus
Border fixed by treaty of 304/303 BC between the kingdoms of Seleucus and Sandracottus (hypothetical)
Areia Satrapy
Borders of satrapy (exact outlines uncertain in many cases)
Koinon ton Nesioton (league of cities; only partially confirmed)
Koinon ton Ionon (league of cities)
Independent areas within the kingdoms of the Diadochi:
City kingdom
Territory ruled by a dynast
City state
Tribal territory
Independent political powers outside the kingdoms of the Diadochi:
Kingdom
Territory ruled by a dynast
Significant city state
Tribal territory
Capital and residence (of a kingdom, satrapy or dynast's territory)
City that ceased to function as a residence between 320 and 303 BC
Important sanctuary
Naval base
Battle/Naval battle/Destruction (with date)
Land or naval campaign:
Antigonus
Demetrius
Ptolemy
Seleucus
Reich Kingdom/Satrapy
Arabes People/Tribe
Arabia Region/Mountain range/Desert
1. Keryneia
2. Lapethos
3. Salamis
4. Soloi

The Hellenistic world in the 3rd cent. BC

I. Sources and the current state of research

The source material on the development and formation of the Hellenistic states of the 3rd century is particularly sketchy for the period from the death of Pyrrhus of Epirus (319/18–272 BC), whose life is included in Plutarch's collection of biographies, and 220 BC, the starting-point of Polybius' *Histories*. All that has survived is some disconnected remarks by various ancient writers, such as the Latin excerpts made by Justin in the 4th cent. AD from a history written in the first cent. AD. There are also inscriptions offering a wealth of individual pieces of information, papyri and some Egyptian and Babylonian documents dating from that period. If new finds were to be made, they could force scholars to thoroughly revise the existing reconstructions. In spite of this situation it is the *communis opinio* among scholars – at least for the time being – that the Ptolemies in particular were the driving force within the world of the nascent Hellenic states; this does not imply, however, that the Seleucid and Antigonid kingdoms remained passive. There is still controversy as to whether the Ptolemies aimed at shoring up their rule and their zones of economic influence or if their primary aim was expansion to achieve 'world domination'.

II. The establishment of the great kingdoms

After 305 BC, independent royal dynasties (Antigonids, Ptolemies, Seleucids among others) emerged, whose position was confirmed after the battle of Ipsus in 301 BC and was further strengthened in 281 BC after Lysmachus was defeated by Seleucus I at Curupedium in western Asia Minor – despite the murder of Seleucus I, who was in control of the east and all of Asia Minor and had been about to conquer the European territories. After defeating the Galatian invaders (who were forced off into Anatolia) in 277 BC and after the death of Pyrrhus of Epirus, Antigonus II Gonatas, son of Demetrius, was able to set himself up as king of Macedonia in 276 or 272. This completed the partition of Alexander's Empire into the three great kingdoms of the Antigonids, the Seleucids and the Ptolemies respectively. This state of affairs remained largely unchanged for about two hundred years, even though at least the decades up to the end of the 3rd cent. were characterized by unstable power relationships.

1. The Antigonid Kingdom

The Antigonid Kingdom as ruled over by Antigonus II Gonatas (*c.* 277–239 BC), Demetrius II (239–229 BC), Antigonus III (229–222 BC) and Philip V (221–179 BC) was centred on the core territory of Macedonia with its capital Pella and the important residences of Aegae and Demetrias. The kingdom also comprised (temporarily) Thrace, Thessaly and parts of Greece (with Euboea, Piraeus and Corinth under direct rule), where right from the start the Antigonids had to fend off incursions by the Ptolemies and later the Seleucids. Thanks to a combination of political restraint toward the independent-minded Greek city states and military pressure from Demetrias, Chalcis and Corinth (either Macedonian settlements or military bases under direct Macedonian control) and despite attempts at expansion by the Spartan king Areus and the Achaean and Aetolian Leagues, the Antigonids managed to hold on to their kingdom. After three Macedonian wars, the Romans dissolved the Macedonian kingdom in 168 BC and changed its status to the province of Macedonia in 148 BC.

2. The Seleucid Kingdom

The Seleucid Kingdom under Seleucus I's son Antiochus I (293/81–261 BC) Antiochus II (268/61–246 BC), Seleucus II (246–226 BC), Seleucus III (226–222 BC) and Antiochus III (223/2–187 BC) had its heartland in northern Syria and Babylonia. Its size was subject to considerable fluctuations: around 303 BC, it extended from Asia Minor (see supplementary map on the distribution of languages around 270 BC) and the eastern edge of the Mediterranean to India. From the middle of the 3rd cent., conflicts within the dynasty (Antiochus Hierax ruled over Asia Minor 242–238(?), d. 226 BC) and the growing strength of the Parthians led to a shrinking process; Bactria seceded from the kingdom. The kingdom's centres included Antiochia (the former Antigonia) on the Orontes, Seleucia Pieria, Laodicea and Apamea in northern Syria. Other important cities were Seleucia on the Tigris near ancient Babylon and Seleucia/Susa as well as Apamea Kibotos and Sardis in Asia Minor.

3. The Ptolemaic Kingdom

The Lagids' Ptolemaic Kingdom (→ map p. 121) under the rule of Ptolemy II (285/83–246 BC), Ptolemy III (246–221 BC) and Ptolemy IV (221–204 BC) comprised the core territory of Egypt with Alexandria as its capital and large, but fluctuating external tracts of land ('non-contiguous territory') in the Cyrenaica, the Levant (including the Jewish temple state), Cyprus, southern and western Asia Minor as well as the city states of the Nesiotic League and the northern Aegean.

4. Further state and tribal territories

There was a great variety of 'territorial states' of differing degrees of dependence or autonomy as the case might be: alongside 'core kingdoms' which preserved a measure of traditional indigenous rule there are examples of autonomous Greek *polis* statehood as well as relics of tribal organization juxtaposed with new federations.

Another feature, – particularly salient in Asia Minor and the Middle East, but also a factor in Europe –, was the competition between Greek and indigenous political concepts and traditions. Mirroring the great kingdoms, smaller units had come into existence such as the kingdom of Bithynia in Asia Minor (297/96, Nicomedes), the so-called Paphlagonian principalities and the kingdoms of Cappadocia and Pontus (Mithradates I) respectively.

Pergamum (Attalus I, 241–197 BC) benefited more than most from the Seleucids' internal strife; it set up a government that was independent of the Seleucids and developed into a medium-sized power (defeating the Galatians in 238 BC, acquiring regal status).

In addition, autonomous temple-states (e.g. Pessinus) in Asia Minor and elsewhere managed to hold their ground. Alongside them, as well as in mainland Greece and on the shores of the Black Sea, existed – at least temporarily – autonomous Greek cities and federations of *poleis* (see → maps pp. 100–103). Cases in point are the Aetolian and Achaean Leagues (*koina*). Although the latter were busy fighting off the expanding Macedonian kingdom, they still achieved considerable, if temporary territorial gains. On the Peloponnese, they made war against Sparta (Cleomenean War, 229/8–222 BC) and later against the Aetolians (Social War, 220–217 BC).

In the same period, nomads on horseback from Iran made incursions into the Seleucid kingdom. From the 230's onward, they began to establish the 'Parthian' empire on Seleucid territory (under Arsaces I and Arsaces II). In Bactria, the satrap Diodotus had himself proclaimed king. During the reign of Aśoka (269/8–233/2 BC), the Indian empire of the Mauryans began to flourish again; it extended from Arachosia in the north almost to the southern tip of the subcontinent. There probably were political and trade relations with the Hellenistic kingdoms, but the source material is rather fragmentary. After Aśoka's death, the empire rapidly disintegrated. In the reign of his successor Sophagasenu, the Seleucids reconquered the territory and restored their suzerainty, as they did from 212 to 205 in Armenia, Atropatene, Parthia and Bactria.

Asia Minor: approximate areas of language distribution, *c.* 270 BC

Anatolian languages: Luwian languages; Lydian. Languages introduced through later migrations: Phrygian and Mysian; Galatian. Greek spoken along the coasts

→ Maps pp. 100–103, 113, 123, 125, 129, 151ff., 215; Hellenistic States, BNP 6, 2005; Asia Minor V. Languages, BNP 2, 2003

Literature

R.S. Bagnall, P. Derow (eds.), The Hellenistic Period. Historical Sources in Translation, 2004; G.R. Bugh, The Cambridge Companion to the Hellenistic World, 2006; P. Cartledge et al. (eds.), Hellenistic Constructs. Essays in Culture, History, and Historiography, 1997; A. Chaniotis, War in the Hellenistic World, 2005; B. Funck (ed.), Hellenismus. Beiträge zur Erforschung von Akkulturation und politischer Ordnung in den Staaten des hellenistischen Zeitalters, 1996; H.-J. Gehrke, Geschichte des Hellenismus, [3]2003; G. Hölbl, Geschichte des Ptolemäerreiches, 1994; J. Kobes, 'Kleine Könige'. Untersuchungen zu den Lokaldynasten im hellenistischen Kleinasien (323–188 v. Chr.), 1996; A. Kuhrt, S. Sherwin-White (eds.), Hellenism in the East. The Interaction of Greek and Non-Greek Civilizations from Syria to Central Asia after Alexander, 1987; O. Mørkholm, Early Hellenistic Coinage from the Accession of Alexander the Great to the Peace of Apamea (336–188 B.C.), 1991; G. Shipley, The Greek World after Alexander. 323–30 B.C., 2000; F.W. Walbank (ed.), The Hellenistic World, [2]1984.

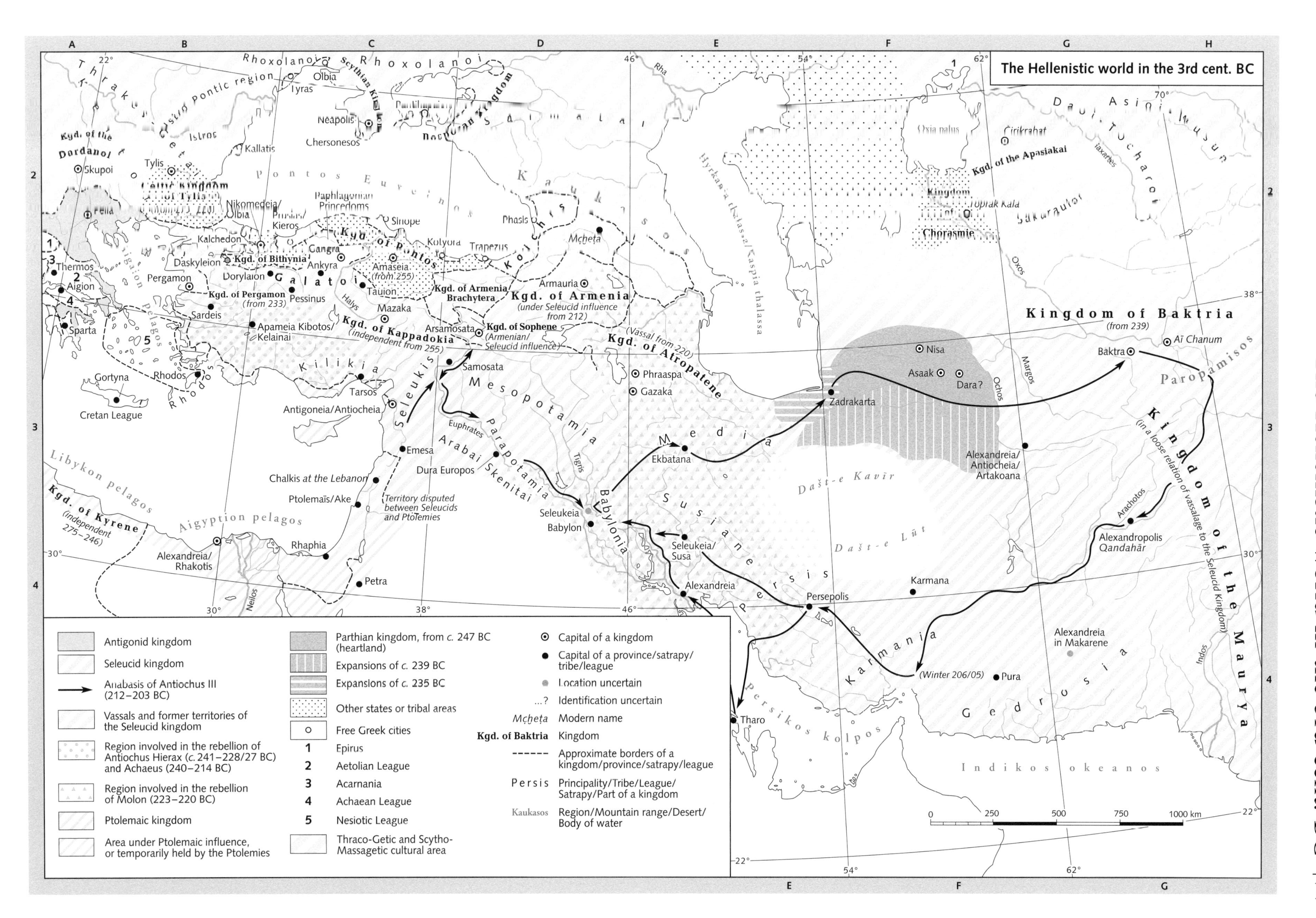
The Hellenistic world in the 3rd cent. BC
Kingdom of Baktria
(from 239)
Kingdom of the Maurya
(in a loose relation of vassalage to the Seleucid Kingdom)
Kgd. of Armenia
(under Seleucid influence from 212)
Kgd. of Atropatene
(Vassal from 220)
Kgd. of Kappadokia
(independent from 255)
Kgd. of Sophene
(Armenian/Seleucid influence)
Kgd. of Armenia Brachytera
Kgd. of Pontos
Kgd. of Bithynia
Kgd. of Pergamon
(from 233)
Kgd. of Kyrene
(independent 275–246)
Galatoi
Mesopotamia
Parapotamia
Arabai Skenitai
Babylonia
Media
Susiane
Persis
Karmania
Gedrosia
Paropamisos
Kilikia
Seleukis
Pontos Euxeinos
Hyrkania thalassa/Kaspia thalassa
Aigyption pelagos
Libykon pelagos
Persikos kolpos
Indikos okeanos
Dašt-e Kavīr
Dašt-e Lūt
Baktra
Aï Chanum
Nisa
Asaak
Dara?
Zadrakarta
Alexandreia/Antiocheia/Artakoana
Alexandropolis Qandahār
Alexandreia in Makarene
Pura
Karmana
(Winter 206/05)
Persepolis
Tharo
Alexandreia
Seleukeia/Susa
Ekbatana
Phraaspa
Gazaka
Seleukeia
Babylon
Dura Europos
Emesa
Samosata
Arsamosata
Armauria
Mazaka
Tauion
Amaseia (from 255)
Sinope
Trapezus
Phasis
Mcheta
Ankyra
Pessinus
Apameia Kibotos/Kelainai
Sardeis
Pergamon
Daskyleion
Dorylaion
Kalchedon
Nikomedeia/Olbia
Kieros
Tarsos
Antigoneia/Antiocheia
Chalkis at the Lebanon
Ptolemaïs/Ake
(Territory disputed between Seleucids and Ptolemies)
Rhaphia
Petra
Alexandreia/Rhakotis
Rhodos
Gortyna
Cretan League
Sparta
Aigion
Thermos
Pella
Skupoi
Tylis
Kallatis
Chersonesos
Neapolis
Olbia
Tyras
Rhoxolanoi
Antigonid kingdom
Seleucid kingdom
Anabasis of Antiochus III (212–203 BC)
Vassals and former territories of the Seleucid kingdom
Region involved in the rebellion of Antiochus Hierax (c. 241–228/27 BC) and Achaeus (240–214 BC)
Region involved in the rebellion of Molon (223–220 BC)
Ptolemaic kingdom
Area under Ptolemaic influence, or temporarily held by the Ptolemies
Parthian kingdom, from c. 247 BC (heartland)
Expansions of c. 239 BC
Expansions of c. 235 BC
Other states or tribal areas
Free Greek cities
1 Epirus
2 Aetolian League
3 Acarnania
4 Achaean League
5 Nesiotic League
Thraco-Getic and Scytho-Massagetic cultural area
Capital of a kingdom
Capital of a province/satrapy/tribe/league
Location uncertain
...? Identification uncertain
Mcheta Modern name
Kgd. of Baktria Kingdom
Approximate borders of a kingdom/province/satrapy/league
Persis Principality/Tribe/League/Satrapy/Part of a kingdom
Kaukasos Region/Mountain range/Desert/Body of water
0 250 500 750 1000 km

City foundations and educational establishments in the Hellenistic period (4th–2nd cents. BC)

City foundations and places of education in the Hellenistic Period (4th-2nd cents. BC)

'The matter of whether one was a Greek or a "barbarian" was decided on the basis not of origin or race, but of education' (H. GEHRKE). The Greek term for education, *paideia*, denoted education as a process of 'upbringing' and, at the same time, as 'cultivation', as an attribute and result of the process of schooling. The Greek educational ideal in the Hellenistic Period was characterized by free personal development. The concept of schooling encompassed gymnastics, artistic and literary education, imparted through private tuition or at public institutions. There was no place, in the Hellenistic Period, for the cycle of the seven *artes liberales* as a foundation for a 'general education' of young boys. It must be stressed that there was no unified system of schooling or education in the *oikoumene*.

I. Hellenistic city foundations (330–133 BC) (map B)

For Alexander the Great and the Hellenistic rulers who succeeded him (Antigonids, Ptolemies, Seleucids and, esp. in the 2nd cent. BC, Attalids), the literary and/or archaeological record attests to many foundations and refoundations of cities that were then named after the rulers themselves or their wives or parents. In their respective heartlands, these cities would be residences/metropoleis that were generally also important crossroads or trading centres, either by the sea with good harbours (Alexandria in Egypt, Seleucia Pieria and Antigonea/Antioch on the Orontes, Demetrias in Thessaly) or inland on important land or river routes (Seleucia/Opis, Dura Europus). City foundations in the hinterland, where autochthonous cultures still had a strong presence, primarily served the process of Hellenization (in the sense of promoting Greek culture as a means of legitimizing the Hellenistic rulers) or, as from Rome's intervention in the Balkans, Asia and the Levant, of Romanization. Like all centres inhabited by Greeks, they had communal educational institutions providing general as well as specialist schooling (e.g. architecture, geography, medicine; scholar as specialist: Aristot. Part. an. 639a 1–13; Aristot. Pol. 1282a 1–7; Aristot. Eth. Nic. 1094b 23–27): gymnasia, theatres and libraries, as well as (private) schools of philosophy, grammar and rhetoric, out of which particular projects led to the development of new fields of scholarship that were then nurtured at institutions such as the Mouseion in Alexandria.

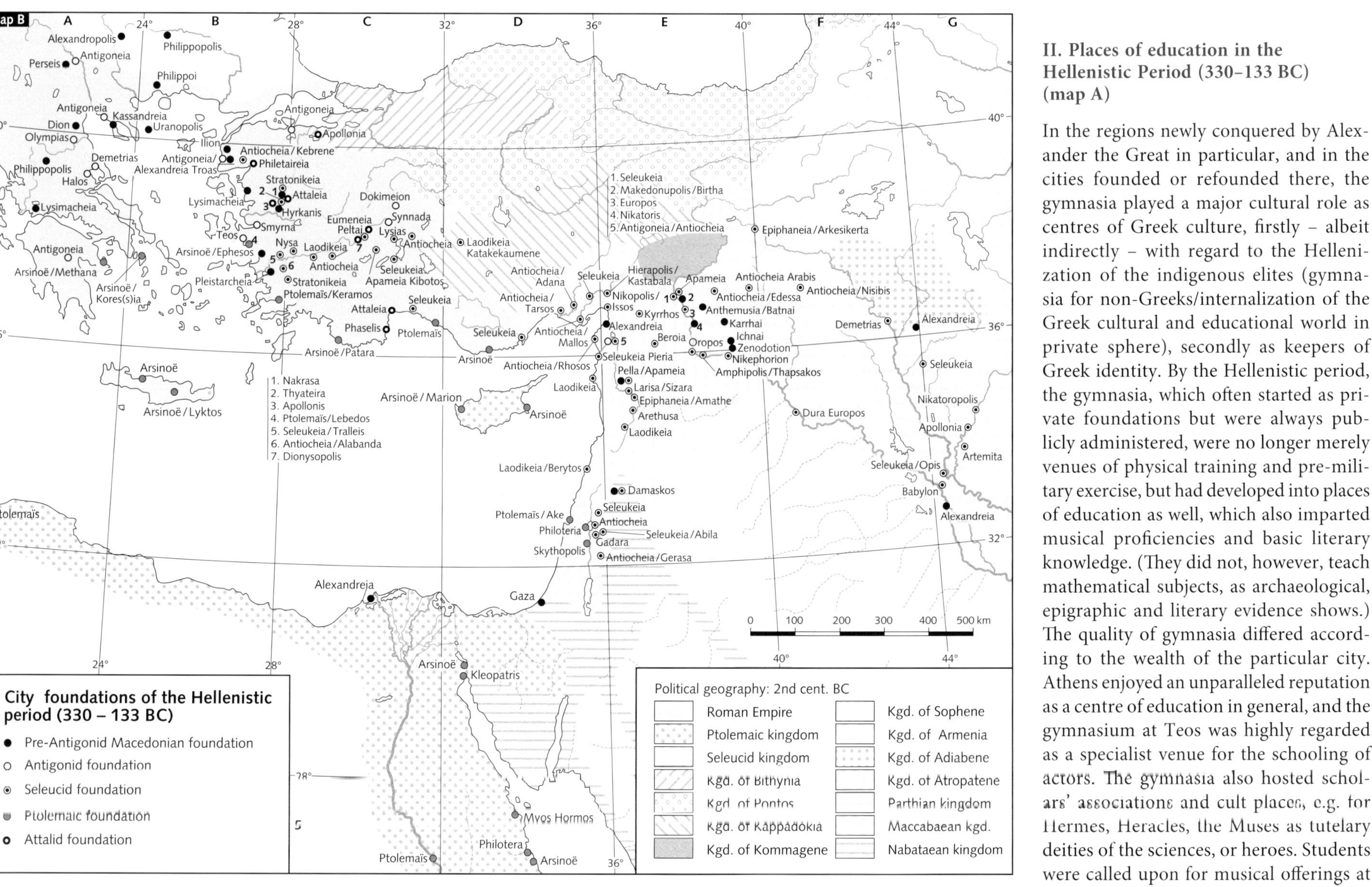

II. Places of education in the Hellenistic Period (330–133 BC) (map A)

In the regions newly conquered by Alexander the Great in particular, and in the cities founded or refounded there, the gymnasia played a major cultural role as centres of Greek culture, firstly – albeit indirectly – with regard to the Hellenization of the indigenous elites (gymnasia for non-Greeks/internalization of the Greek cultural and educational world in private sphere), secondly as keepers of Greek identity. By the Hellenistic period, the gymnasia, which often started as private foundations but were always publicly administered, were no longer merely venues of physical training and pre-military exercise, but had developed into places of education as well, which also imparted musical proficiencies and basic literary knowledge. (They did not, however, teach mathematical subjects, as archaeological, epigraphic and literary evidence shows.) The quality of gymnasia differed according to the wealth of the particular city. Athens enjoyed an unparalleled reputation as a centre of education in general, and the gymnasium at Teos was highly regarded as a specialist venue for the schooling of actors. The gymnasia also hosted scholars' associations and cult places, e.g. for Hermes, Heracles, the Muses as tutelary deities of the sciences, or heroes. Students were called upon for musical offerings at city festivals for the gods and for the ruler cult. Rather than examinations, there were contests and competitions (cf. victor lists).

Theatra and *odeia*, here in the sense of places of assembly for festive, cultic or sporting presentations, are among the most numerous, most widespread and archaeologically best-attested buildings of Antiquity. For some others, there is only epigraphic or literary testimony. Many new theatres were built in the Hellenistic period, esp. in the 2nd and 1st cents. BC and particularly in Asia Minor.

Other than libraries in private houses, including private schools of philosophy and associations, there is generally only epigraphic/literary evidence of collections of books and the associated requisite storage, study and refectory space, mostly within Hellenistic royal palaces (Alexandria, Pergamum, Pella, Sinope) and public and sacral urban buildings (e.g. the Lyceum at Athens) or facilities within such buildings (e.g. sometimes in gymnasia or baths). Only a few are attested through archaeology, since the furniture generally used for storage was mostly wooden, and has therefore not survived.

→ Maps pp. 113, 115, 117, 125, 129; Theatre II., BNP 14, 2009

Literature

G.M. COHEN, The Hellenistic Settlements in Europe, the Islands, and Asia Minor, 1995; Id., The Seleucid Colonies, 1978; Id., The Hellenistic Settlements in Syria, the Red Sea Basin, and North Africa, 2006; J.D. GRAINGER, The Cities of Seleucid Syria, 1990; P. GREEN (ed.), Hellenistic History and Culture, 1993; D. KAH, P. SCHOLZ (eds.), Das hellenistische Gymnasion, 2004; M.P. NILSSON, Die hellenistische Schule, 1955; W. ORTH, Königlicher Machtanspruch und städtische Freiheit. Untersuchungen zu den politischen Beziehungen zwischen den ersten Seleukidenherrschern (...) und den Städten des westlichen Kleinasiens, 1977; C. SCHNEIDER, Kulturgeschichte des Hellenismus, 2 vols., 1967, 1969; K.-W. WEEBER, Panem et circenses, 1994.

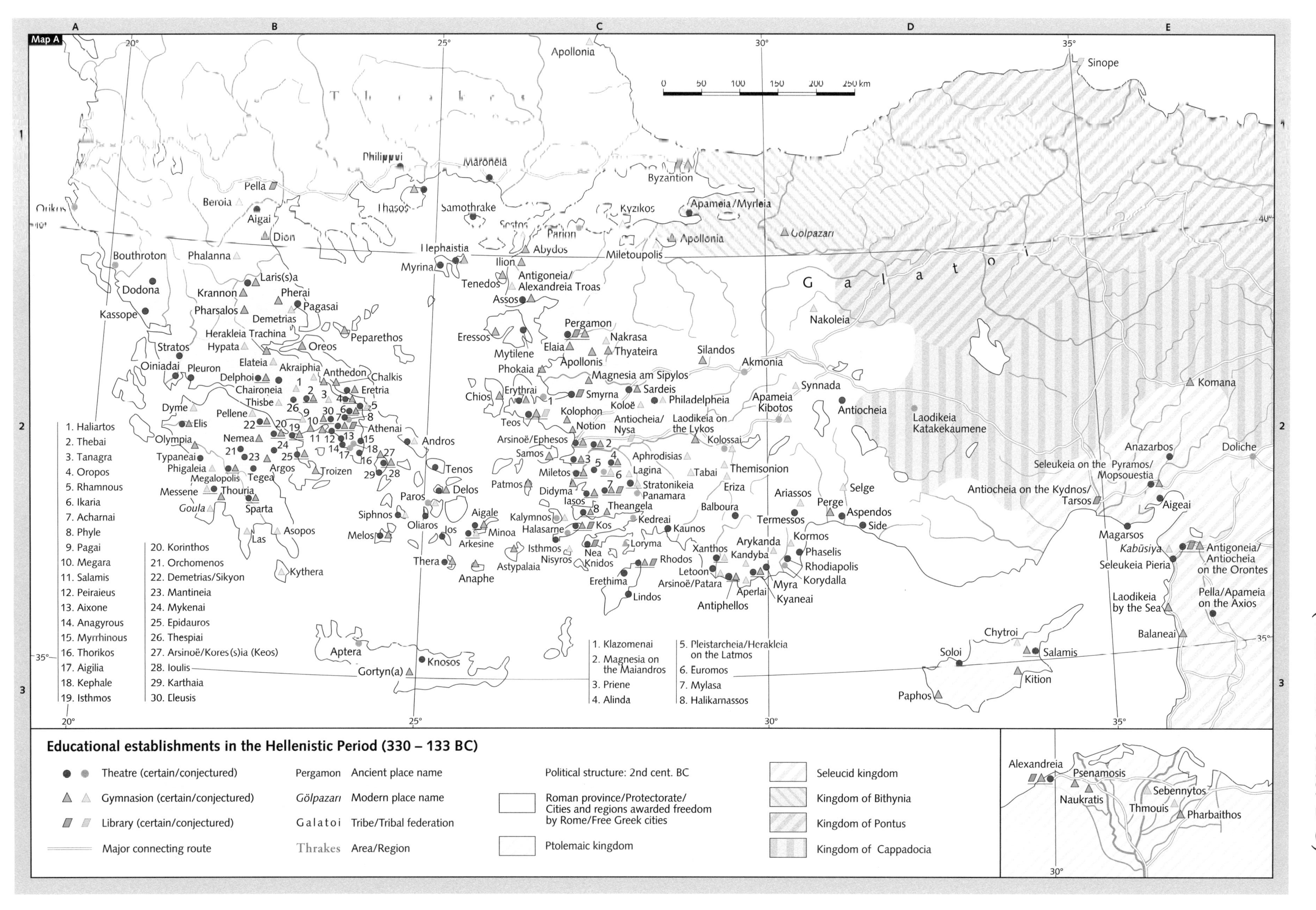
Map A
Educational establishments in the Hellenistic Period (330 – 133 BC)
Theatre (certain/conjectured)
Gymnasion (certain/conjectured)
Library (certain/conjectured)
Major connecting route
Pergamon Ancient place name
Gölpazarı Modern place name
Galatoi Tribe/Tribal federation
Thrakes Area/Region
Political structure: 2nd cent. BC
Roman province/Protectorate/Cities and regions awarded freedom by Rome/Free Greek cities
Ptolemaic kingdom
Seleucid kingdom
Kingdom of Bithynia
Kingdom of Pontus
Kingdom of Cappadocia
1. Haliartos
2. Thebai
3. Tanagra
4. Oropos
5. Rhamnous
6. Ikaria
7. Acharnai
8. Phyle
9. Pagai
10. Megara
11. Salamis
12. Peiraieus
13. Aixone
14. Anagyrous
15. Myrrhinous
16. Thorikos
17. Aigilia
18. Kephale
19. Isthmos
20. Korinthos
21. Orchomenos
22. Demetrias/Sikyon
23. Mantineia
24. Mykenai
25. Epidauros
26. Thespiai
27. Arsinoë/Kores(s)ia (Keos)
28. Ioulis
29. Karthaia
30. Eleusis
1. Klazomenai
2. Magnesia on the Maiandros
3. Priene
4. Alinda
5. Pleistarcheia/Herakleia on the Latmos
6. Euromos
7. Mylasa
8. Halikarnassos
0 50 100 150 200 250 km
Galatoi
Sinope
Komana
Doliche
Antigoneia/Antiocheia on the Orontes
Pella/Apameia on the Axios
Anazarbos
Aigeai
Seleukeia on the Pyramos/Mopsouestia
Antiocheia on the Kydnos/Tarsos
Magarsos
Kabūsiya
Seleukeia Pieria
Laodikeia by the Sea
Balaneai
Salamis
Kition
Chytroi
Soloi
Paphos
Alexandreia
Psenamosis
Naukratis
Sebennytos
Thmouis
Pharbaithos
Laodikeia Katakekaumene
Antiocheia
Selge
Aspendos
Side
Perge
Phaselis
Rhodiapolis
Korydalla
Kormos
Termessos
Ariassos
Myra
Kyaneai
Arykanda
Kandyba
Aperlai
Antiphellos
Xanthos
Letoon
Arsinoë/Patara
Balboura
Eriza
Themisonion
Synnada
Apameia Kibotos
Akmonia
Nakoleia
Gölpazarı
Apameia/Myrleia
Apollonia
Byzantion
Kyzikos
Miletoupolis
Parion
Abydos
Ilion
Antigoneia/Alexandreia Troas
Tenedos
Assos
Pergamon
Nakrasa
Thyateira
Elaia
Apollonis
Magnesia am Sipylos
Sardeis
Philadelpheia
Silandos
Smyrna
Koloë
Kolophon
Notion
Antiocheia/Nysa
Laodikeia on the Lykos
Kolossai
Aphrodisias
Tabai
Lagina
Stratonikeia
Panamara
Theangela
Kedreai
Kaunos
Loryma
Rhodos
Lindos
Erethima
Knidos
Nisyros
Kos
Iasos
Didyma
Miletos
Samos
Arsinoë/Ephesos
Teos
Erythrai
Chios
Phokaia
Mytilene
Eressos
Patmos
Kalymnos
Halasarne
Isthmos
Astypalaia
Minoa
Aigiale
Arkesine
Anaphe
Thera
Ios
Oliaros
Paros
Delos
Tenos
Andros
Siphnos
Melos
Knosos
Gortyn(a)
Aptera
Kythera
Maroneia
Samothrake
Hephaistia
Myrina
Thasos
Philippoi
Apollonia
Pella
Beroia
Aigai
Dion
Phalanna
Laris(s)a
Pherai
Pagasai
Demetrias
Krannon
Pharsalos
Herakleia Trachina
Hypata
Elateia
Akraiphia
Anthedon
Chalkis
Eretria
Oreos
Peparethos
Delphoi
Chaironeia
Thisbe
Athenai
Troizen
Argos
Tegea
Asopos
Las
Sparta
Thouria
Messene
Megalopolis
Goula
Phigaleia
Typaneai
Olympia
Elis
Dyme
Pellene
Nemea
Pleuron
Oiniadai
Stratos
Kassope
Dodona
Bouthroton
Orikos

Egypt from the 4th to 1st cents. BC

I. The Ptolemaic kingdom (323 BC–12 August 30 BC) (map A)

The Ptolemies established themselves in Egypt, which Ptolemy I initially (323 BC) held as a satrapy, after the death of Alexander the Great (Ptolemy's assumption of the royal title: 305 BC). They ruled with the help of the immigrant Graeco-Macedonian upper class (functionaries, colonists, esp. in the region of the drained Lake Moiris/Faiyum, and *klerouchoi*), who were socially and politically dominant, but they always had to fulfil the expectations of the indigenous elite (priests, later Egyptian *klerouchoi* and functionaries) and the Egyptian populace. Thus, an Egyptian byname and/or regnal name is recorded for almost all the Ptolemaic kings. Their main focus was on the Mediterranean world: Nubia, the indigenous Kushite (Kuš) kingdom of Napata (9th-4th/3rd cents. BC) and esp. Meroë (4th/3rd cents. BC–AD 350, cf. Diod. Sic. 1,60; 3,6), were largely left to their own devices. However, Egyptian and Ptolemaic cultural influence is well attested in these regions, too, and there were close trading relations (caravan trade after the introduction of the camel), just as there were with the kingdom of the Nabataeans, who controlled the intermediary trade with Arabia.

The Ptolemaic heartland, and hence the region of direct rule in Egypt, was the lands surrounding the capital and residence, Alexandria, as well as the entire Nile Delta and Nile Oasis. Lower and Upper Egypt were each divided into 20 *nomoi*. Cities such as Ptolemaïs (map B) in Upper Egypt were founded as Greek *poleis*, but centres like Memphis, and esp. Thebes, did not wane in importance alongside them. The temples had an important role as guardians of the old traditions. They won more and more privileges from the king, not least for their economic interests (temple economy), and in the process gained considerable autonomy. From 322/321, there were also the foreign possessions: large sections of the Levantine coast, disputed between the Ptolemies and the Seleucids (cf. The Six Syrian Wars, → map p. 123), Cyprus, territories in southern and western Asia Minor, the northern Aegean and Cyrenaica, all of these initially with differing administrative relationships to the crown, but tending towards unification. These possessions also served the security of Egypt. Most were lost in the 2nd cent., apart from Cyprus and (the kingdom of) Cyrene, and dependence on Rome steadily grew from 168 BC ('Day of Eleusis', intervention of the Roman army on the side of Ptolemy VI against Antiochus IV). The 2nd cent. BC was characterized by internal unrest and resistance from indigenous populations. On all frontiers, it became necessary to introduce security measures. In addition to difficulties abroad, the country was shaken by constant succession disputes within the royal dynasty. The kingdom was finally divided under Ptolemy VIII (163 BC), who named Rome as heir to Cyrene in his will as early as 162/61. Ptolemy X made the same arrangement for all of Egypt in 88 BC (Cic. Leg. agr. 1,1,1), but Rome at first refrained from acquisition (PP III/IX 5253; VI 14555 /=14556). Only later did Cyrene (74 BC) and Cyprus (58/30 BC) pass to Rome.

II. The kingdom of Cleopatra VII (47–30 BC) (map C)

Ptolemy XII (*c.* 59/51 BC) had named his daughter Cleopatra as his heir by testament, and Rome was to guarantee the independence of Egypt. Egypt thus became *de facto* a client state of Rome. With the reign of Cleopatra VII (52/47 BC–12 August 30 BC), whom Caesar confirmed as sole ruler in 47, Ptolemaic rule in Egypt came to an end. During her volatile reign, Egypt once more achieved great territorial expansion. From mid-43, it again controlled Cyprus. In 41, Mark Antony made substantial territorial concessions, affecting what was already Roman provincial territory, because he needed the financial support of Egypt against the Parthians. These territories comprised, by 38 at the latest, parts of Cilicia, in 37/36 the principality of Chalcis, then Phoenicia (except Sidon, Tyre, Ascalon), parts of Judaea and Arabia, land on Crete (except Gortyna; until 34?, province of Creta) and Cyrene (province of Cyrenae), with civil administration being Ptolemaic, but military administration Roman. Ptolemy XV Caesar (b. 23 June 47 BC), the son of Cleopatra and Caesar, became Cleopatra's co-regent, and in 34 BC acquired, like her, the title of 'King of Kings'. He nominally shared the expansion of her power, and Mark Antony recognized him as the son of Caesar. He attempted to flee after his mother's death in 30 BC, but Octavian had him killed. In the same year, Octavian also had Ptolemy Philadelphus (b. 36 BC), the son of Cleopatra and Mark Antony, who in 34 BC had been crowned King of Phoenicia, Syria, Cilicia (Plut. Antonius 54,7) and all lands between the Euphrates and the Hellespont (Cass. Dio 49,41,3), brought to Rome. With that, Egypt ceased to exist as an independent kingdom, and became a Roman province.

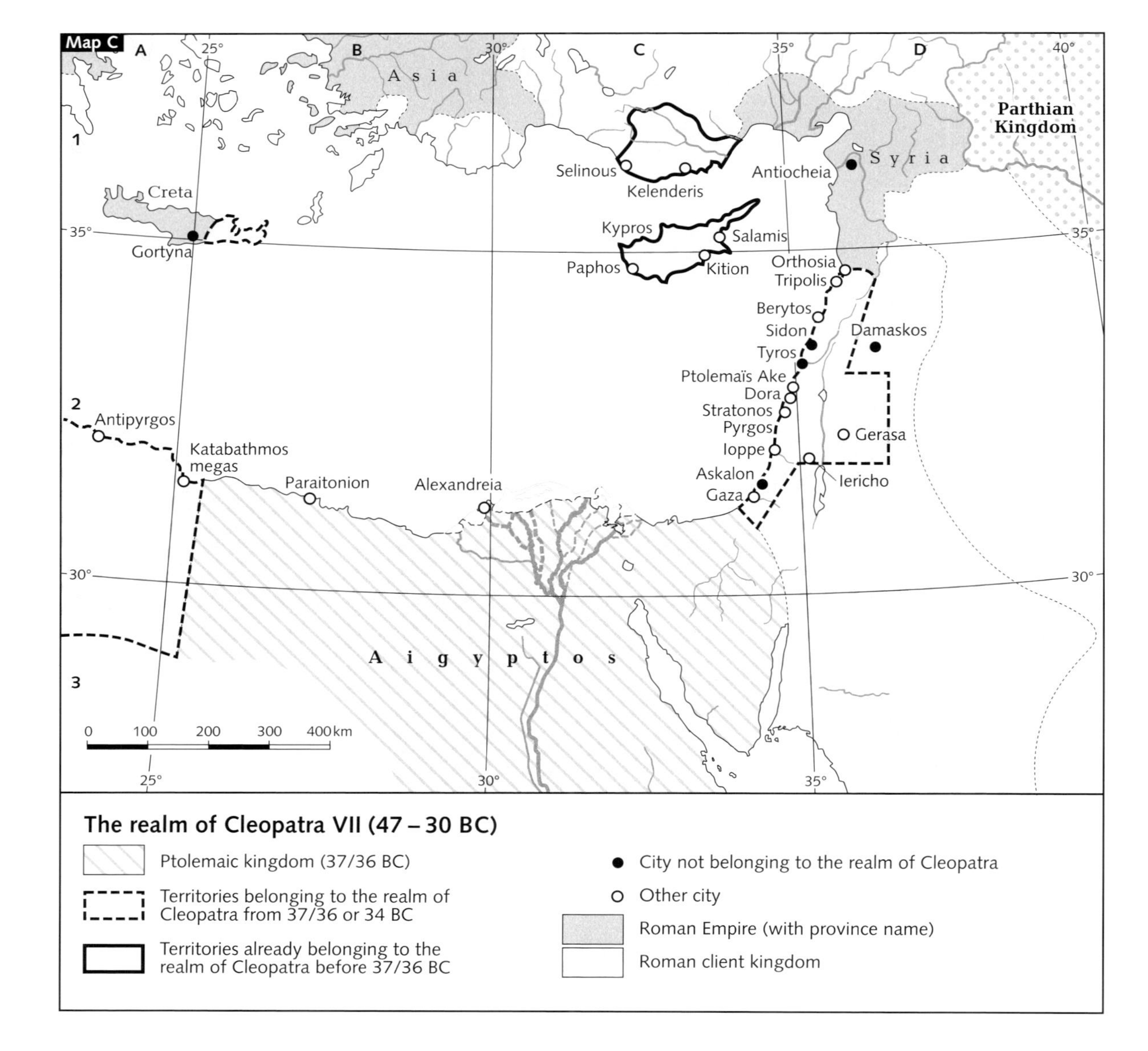

The realm of Cleopatra VII (47 – 30 BC)

III. Egypt: economy (4th-2nd. cents. BC; map B)

In the pharaonic tradition, the king assumed responsibility for the Nile Oasis. The construction of dykes and canals and the sowing of seed and harvesting of crops were therefore organized and registered centrally by the state, in accordance with ancient administrative structures. Control of economic production ('royal farmers', *basilikoi georgoi*), e.g. of grain, wine and oil-producing plants, was in the hands of the king, though he delegated the collection of taxes and duties in a regulated system to private tax farmers. Under Ptolemy II, Lake Moiris (Faiyum) was drained, winning new cultivable land. The same king also had a Nile-Suez Canal built, and founded cities on the Red Sea and in Palestine at important trading centres. As power was briefly extended over Arabia, including the Nabataean region, plans included the rerouting of the Arabian-Syrian caravan trade from Petra to Alexandria, *the* trans-shipment centre for Inner Egypt, to the west and east and for maritime trade. Bases were secured in Meroitic Kuš (gold and ivory) and expeditions in quest of gold were undertaken as far as Lower Nubia. Rich mineral resources and abundant revenues, an elaborate financial authority with its own (internal) currency and wide-ranging trading links brought substantial incomes, which financed major construction projects but also the expansionist foreign policy (e.g. the Syrian Wars). Ptolemy III pursued this strategy further, by founding cities in the Cyrenaica and securing the southern frontier and the connection to the Red Sea. The reign of Ptolemy IV saw revolts in the Nile Delta and the Thebaid, while the Dodekaschoinos came under Meroitic rule and inflation weakened the silver currency. All of this created socio-economic problems, the effects of which continued in subsequent years and coincided with defeat abroad. In Cleopatra's reign, moreover, there was a series of poor harvests and epidemics.

Sources

Thanks to the good state of the sources based on a wealth of papyrus finds in the Faiyum (albeit with qualifications because of the single, unrepresentative find site), the political and economic history of Ptolemaic Egypt is among the best-researched fields of the Hellenistic Period. Sources (among others): Diodorus Siculus (based on Hieronymus of Cardia); Arrian; Ophellas [2]; Plutarch; Cassius Dio; also numerous inscriptions, coins, archaeological finds

→ Maps pp. 115, 123, 135, 163, 171, 179; Cleopatra (II 12) VII., BNP 3, 2003; Ptolemaeus I (1. 3–24) BNP 12, 2008; Egypt E. Late Period, BNP 4, 2004 Ptolemaeus I (2), BNP 12, 2008

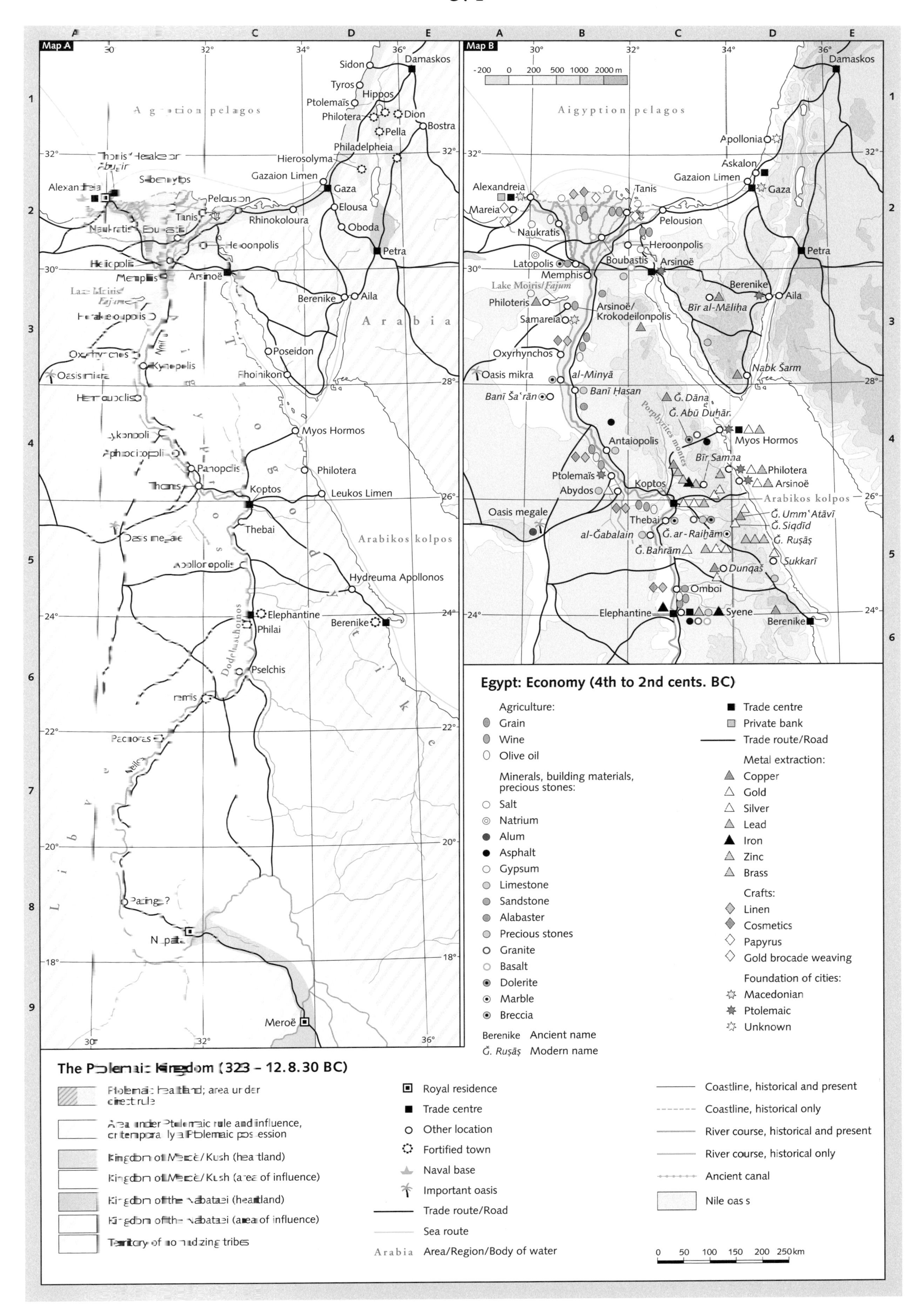

Map A
Map B
Sidon
Damaskos
Tyros
Hippos
Ptolemaïs
Philotera
Dion
Pella
Bostra
Philadelpheia
Hierosolyma
Gazaion Limen
Gaza
Pelousion
Rhinokoloura
Elousa
Oboda
Heroonpolis
Petra
Arsinoë
Berenike
Aila
Arabia
Poseidon
Phoinikon
Myos Hormos
Philotera
Koptos
Leukos Limen
Thebai
Arabikos kolpos
Hydreuma Apollonos
Elephantine
Philai
Berenike
Pselchis
Meroë
Aigyption pelagos
Apollonia
Askalon
Alexandreia
Mareia
Tanis
Naukratis
Latopolis
Memphis
Boubastis
Lake Moiris/Fajum
Philoteris
Samareia
Arsinoë/ Krokodeilonpolis
Bīr al-Mālịha
Oxyrhynchos
Oasis mikra
al-Minyā
Nabk Šarm
Banī Ša'rān
Banī Ḥasan
Ǧ. Dāna
Ǧ. Abū Duḫār
Porphyrites montes
Antaiopolis
Bīr Samna
Ptolemaïs
Abydos
Oasis megale
al-Ǧabalain
Ǧ. ar-Raiḫām
Ǧ. Bahrām
Ǧ. Umm'Atāvī
Ǧ. Siqdīd
Ǧ. Ruṣāṣ
Sukkarī
Dunqāš
Omboi
Syene
Egypt: Economy (4th to 2nd cents. BC)
Agriculture:
Grain
Wine
Olive oil
Minerals, building materials, precious stones:
Salt
Natrium
Alum
Asphalt
Gypsum
Limestone
Sandstone
Alabaster
Precious stones
Granite
Basalt
Dolerite
Marble
Breccia
Berenike Ancient name
Ǧ. Ruṣāṣ Modern name
Trade centre
Private bank
Trade route/Road
Metal extraction:
Copper
Gold
Silver
Lead
Iron
Zinc
Brass
Crafts:
Linen
Cosmetics
Papyrus
Gold brocade weaving
Foundation of cities:
Macedonian
Ptolemaic
Unknown
The Ptolemaic Kingdom (323 – 12.8.30 BC)
Kingdom of Meroë/Kush (heartland)
Kingdom of Meroë/Kush (area of influence)
Kingdom of the Nabataei (heartland)
Kingdom of the Nabataei (area of influence)
Royal residence
Trade centre
Other location
Fortified town
Naval base
Important oasis
Trade route/Road
Sea route
Arabia Area/Region/Body of water
Coastline, historical and present
Coastline, historical only
River course, historical and present
River course, historical only
Ancient canal
Nile oasis
0 50 100 150 200 250 km

The six Syrian Wars (275/74 –168 BC)

Modern scholarship uses the term 'Syrian Wars' to denote the group of six conflicts that took place from 275/274 to 168 BC between the Ptolemies and Seleucids, over possession of southern Syria. The initial cause was the state of occupation following the partition of the kingdom of Antigonus in 301 BC. Seleucus I gave southern Syria, which had been allocated to him, to his ally Ptolemy I, who had recently (302/01) conquered the territory; however, Seleucus did not give up his claim to it (Diod. Sic. 21,1,5). The region was of economic importance (cedar of Lebanon, harbours, caravan routes), but was more important to the Ptolemies, because, like Cyrene and Cyprus, it provided a military outpost zone for Egypt and could act as a starting-point for expansionist expeditions (as it did throughout Antiquity, as evidence shows from the 2nd millennium BC). The 'Syrian Wars' are thus primarily to be understood as a single element within the whole field of Ptolemaic and Seleucid military operations throughout the Mediterranean region.

Depicting the campaign routes of each of the six wars would go beyond the scope of the map, particularly since the main routes by water and on land were used repeatedly. H. Waldmann presents a detailed cartographic reconstruction.

I. The First Syrian War (275/74–271 BC)

The First Syrian War began as a serious threat to Ptolemy II, with a pincer attack by Antiochus I and his son-in-law Magas of Cyrene to be expected (Paus. 1,7,3; → Map p. 117). Magas was the stepson of Ptolemy I, but his relations with his half-brother Ptolemy II were strained. However, a revolt forced Magas to return to Cyrene (275?), before Antiochus, who was still busy in Asia Minor, was able to intervene in 274. Ptolemy II then counterattacked in the same year, making raids in various regions, advancing into northern Syria and even sending naval expeditions as far as the Black Sea (Steph. Byz.). His attempt to reduce Seleucid pressure on Coele Syria with a military expedition through the Persian Gulf into Seleucid Babylonia was thwarted by the resistance of the Seleucid satraps. Antiochus took Damascus in his own counterattack (by water and land). The remaining course of the war, which lasted until around 271 BC, is unknown. Essentially, it brought about the continuation of the *status quo*.

II. The Second Syrian War (260–253? BC)

The Second Syrian War changed nothing in this *status quo*. As a whole, the course of hostilities in the south is rather uncertain (Porph. FGrH 260 F 43), and the main theatre was rather in or near Asia Minor (Miletus; 260 BC Ptolemaic defeat outside Ephesus, 255 BC another off Cos, with the participation of Antigonus II Gonatas), which implies that the initiative lay with Ptolemy II. The war cost the aggressor, Ptolemy II, who was trying to exploit the Seleucid successional transition (Antiochus II from 261 BC), influence in the Aegean and Asia Minor, while Antiochus gained territory in Ionia, Cilicia and Pamphylia. The war was over by 253 BC, when the first dynastic connection was agreed, between Berenice, daughter of Ptolemy, and Antiochus (wedding 252 BC; one son), who consequently repudiated his wife Laodice, with whom he had two sons. It is unclear whether Antiochus II later became reconciled with Laodice. He died in mysterious circumstances in 246 BC, perhaps even murdered by Laodice. Ptolemy II died at almost the same time.

III. The Third Syrian War (246–241 BC)

The Third Syrian War, also the 'War of Laodice', was probably (the sources are contradictory and confused) the consequence of dynastic intrigues, and began immediately after the death of Antiochus II (246 BC). Laodice wanted to secure the Seleucid throne for her son, the future Seleucus II, and Berenice wanted it for hers. Berenice called on the aid of her just-crowned brother, Ptolemy III, who (scholars suppose) nursed the ambition of taking control of the Seleucid kingdom, at least through his nephew. He rapidly took Seleucia Pieria at the mouth of the Orontes (Ptolemaic until 219), was received with joy in Antioch (where he probably found Berenice and certainly her son murdered by Laodice), and from there advanced, apparently without resistance, to Mesopotamia (Babylon). This led to his recognition in the entire east of the Seleucid kingdom (stela of Adulis, OGIS 54). A revolt forced him to return to Egypt. He did not thereafter resume his Asian campaign, which casts doubt on his intention of conquering the Seleucid kingdom. Instead, from 245 to 241 BC, he considerably expanded the foreign Ptolemaic possessions, from southern Asia Minor to Thrace, a process which drew the third Epigoni ruler, Antigonus II Gonatas of Macedonia, into this war. Although Ptolemy III had not achieved his goals in Asia (Seleucus II held on to his kingdom and peace was agreed between the two kingdoms in 241 BC), the Ptolemaic kingdom remained the more powerful of the two in the ensuing decades.

IV. The Fourth Syrian War (221/19–217 BC)

Coele Syria itself became the theatre of the Fourth Syrian War. Antiochus III, on the throne since 223/22 BC, was attempting with great energy to countervail the collapse that threatened his kingdom (reorganization of the kingdom from satrapies to *strategiai*; separation of fiscal administration; introduction of a ruler cult across the whole kingdom forging a national identity), and in 221 BC he decided to invade Coele Syria, provoked by the aid provided by the Ptolemies to Attalus I of Pergamum. Only in 219 did he take Seleucia Pieria. When the Ptolemaic governor, Theodotus, defected to his side, the Ptolemaic court (under Ptolemy IV from 221 BC) adopted delaying tactics, and used the time to build an army, which then defeated the Seleucid forces at Rhaphia in southern Coele Syria in 217 BC. Ptolemy refrained from taking Seleucia, and confined himself to re-incorporating Coele Syria. Antiochus III at first (216-213 BC) contented himself with conquering large areas of Asia Minor, then beginning his eastern campaign (212–205 BC), which may have won him the byname 'the Great', but, lacking permanent conquests, did not suffice to restore stability to the Seleucid kingdom. Nonetheless, Antigonus III had won much prestige among his contemporaries, and was in a recognized position of power.

V. The Fifth Syrian War (202–198/94 BC)

Only in the Fifth Syrian War (202–198/94) did Antiochus III succeed in winning the disputed territory. The war took place in the context of the death of Ptolemy IV in 204 BC, and the so-called 'secret agreement' between Antiochus and Philip V of Macedonia, which provided for the partition of the Ptolemaic foreign possessions (and perhaps even the entire kingdom) after the accession of Ptolemy V (204 BC), who was just six years old. In 202 BC, Antiochus undertook the conquest of Syria and Judaea, while Philip prevailed against the Ptolemies in the northern Aegean as far as the Hellespont. From 201 BC, the medium-sized powers of Pergamum and Rhodes were also involved, which subsequently led to Roman intervention, and hence to the beginning of Roman expansion throughout the Eastern Mediterranean region (e.g. Second Macedonian War, → Map p. 151). After battles varying in result, esp. in southern Syria and Egypt (200 BC: victory at Panium (localization uncertain) in the Thebaid/Egypt, not on map section), Antiochus completed his annexation of the region in 198 BC. The territory remained Seleucid even after the marriage of his daughter Cleopatra to Ptolemy V (winter 194/93).

VI. The Sixth Syrian War (170–168 BC)

After the death of Cleopatra (176 BC), who had been regent for Ptolemy VI since 180 BC, an anti-Seleucid atmosphere immediately arose, and this ultimately led to the Sixth Syrian War. Antiochus IV, reigning since 175 BC, had pre-emptively stationed forces in Coele Syria, and in 170 he met the approaching Egyptian army even before it crossed the border, defeating it at Pelusium, occupying Lower Egypt and besieging Alexandria, posing as protector of his nephew, Ptolemy VI. As in the Fifth Syrian War, the Ptolemies called on Rome for aid and mediation (169 BC), and once more their pleas at first went unheard, as Rome was engaged in its war with Perseus (171–168 BC). Only after the peace (168) did the Roman emissary, C. Popilius Laenas, in the Alexandrian suburb of Eleusis, finally order Antiochus (who had now occupied not only Alexandria, but all of Egypt) out of Egypt and force him to return Cyprus, but not Coele Syria. This decision saved the Ptolemaic kingdom, but at the cost of increasing dependence on Rome.

The sources

As is the case in respect of the history of the three great kingdoms (Ptolemaic, Seleucid and Antigonid) that had formed *c.* 272 BC after the Wars of the Diadochi, and of their ensuing attempts to establish an empire comparable to that of Alexander, the sources for the 'Syrian Wars' are also highly fragmentary and sporadic, at least until the beginning of the *Histories* of Polybius (*c.* 220). There are occasional references in various ancient authors (e.g. Strabo, Flavius) Josephus, Zeno) and in Egyptian and Babylonian inscriptions (e.g. stele of Heroonpolis/Patumus; stele of Adulis; cuneiform tablets) and papyri. The state of the sources thereafter improves considerably, not only because of the first books of Polybius' history, which survive intact, but thanks to instructive documentary sources. The Syrian Wars have received little focused attention from scholars, but have mostly been treated in the context of Hellenistic history as a whole. However, it has been determined that it was mostly the Ptolemies who for long periods were the driving force, and who had farther-reaching goals in view.

→Maps pp. 115, 117, 121, 125, 129, 131; Syrian Wars, BNP 14, 2009

Literature

R.S. Bagnall, P. Derow (eds.), The Hellenistic Period. Historical Sources in Translation, 2004; H.-J. Gehrke, Geschichte des Hellenismus, [3]2003 (with sources); H. Heinen, The Syrian-Egyptian Wars and the New Kingdoms of Asia Minor, in: CAH 7.1, 21984, 412–445; G. Hölbl, Geschichte des Ptolemäerreiches. Politik, Ideologie und religiöse Kultur von Alexander dem Großen bis zur römischen Eroberung, 1994; W. Huss, Ägypten in hellenistischer Zeit, 2001; H. Waldmann with R. Rademacher, Syrien und Palästina in hellenistischer Zeit. Die Syrischen Kriege (280–145 BC), TAVO B V 16.1, 1987 (detailed depiction of the campaign routes).

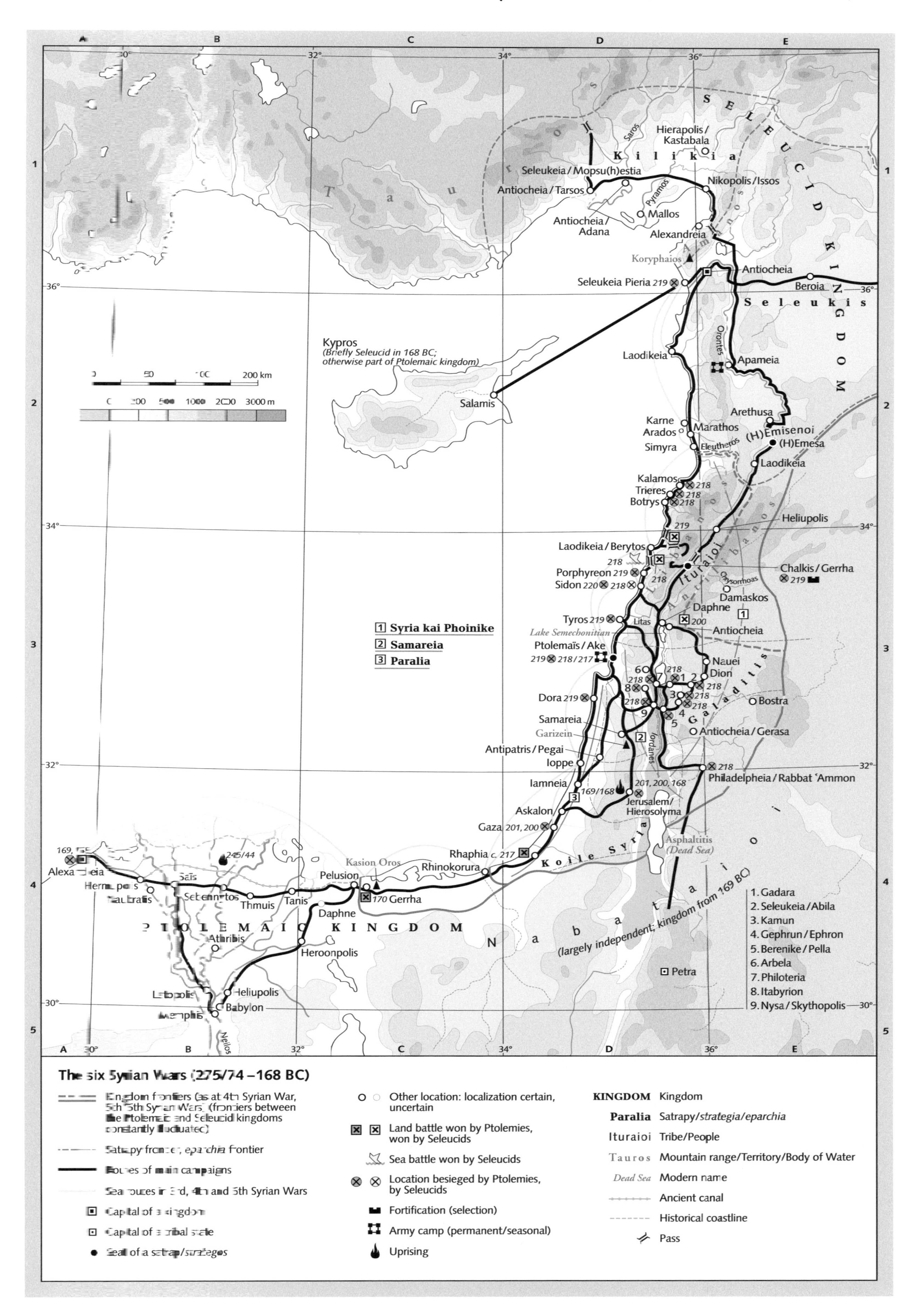

Seleukeia / Mopsu(h)estia
Antiocheia / Tarsos
Hierapolis / Kastabala
Kilikia
Nikopolis / Issos
Mallos
Antiocheia / Adana
Alexandreia
Koryphaios
Antiocheia
Beroia
Seleukeia Pieria 219
Seleukis
SELEUCID KINGDOM
Kypros
(Briefly Seleucid in 168 BC; otherwise part of Ptolemaic kingdom)
Salamis
Laodikeia
Apameia
Orontes
Arethusa
Karne
Arados
Marathos
Simyra
Eleutheros
(H)Emisenoi
(H)Emesa
Laodikeia
Kalamos
Trieres
Botrys
Heliupolis
Laodikeia / Berytos
Porphyreon 219
Sidon 220 218
Ituraioi
Chalkis / Gerrha
Damaskos
Daphne
Tyros 219
Litas
Lake Semechonitian
Antiocheia
Ptolemaïs / Ake
219 218/217
Naueі
Dion
Bostra
Dora 219
Samareia
Garizein
Galaditis
Antiocheia / Gerasa
Antipatris / Pegai
Ioppe
Iordanes
Iamneia
Philadelpheia / Rabbat ʿAmmon
Jerusalem / Hierosolyma
201, 200, 168
169/168
Askalon
Gaza 201, 200
Asphaltitis (Dead Sea)
Rhaphia c. 217
Rhinokorura
Koile Syria
Kasion Oros
Pelusion
170 Gerrha
Tanis
Thmuis
Daphne
Saïs
Alexandreia
PTOLEMAIC KINGDOM
Athribis
Heroonpolis
Heliupolis
Babylon
Memphis
Nabataioi
(largely independent; kingdom from 169 BC)
Petra
1 Syria kai Phoinike
2 Samareia
3 Paralia
1. Gadara
2. Seleukeia / Abila
3. Kamun
4. Gephrun / Ephron
5. Berenike / Pella
6. Arbela
7. Philoteria
8. Itabyrion
9. Nysa / Skythopolis
0 50 100 200 km
0 200 500 1000 2000 3000 m
The six Syrian Wars (275/74 –168 BC)
Routes of main campaigns
Sea routes in 3rd, 4th and 5th Syrian Wars
Capital of a kingdom
Capital of a tribal state
Other location: localization certain, uncertain
Land battle won by Ptolemies, won by Seleucids
Sea battle won by Seleucids
Location besieged by Ptolemies, by Seleucids
Fortification (selection)
Army camp (permanent/seasonal)
Uprising
KINGDOM Kingdom
Paralia Satrapy/strategia/eparchia
Ituraioi Tribe/People
Tauros Mountain range/Territory/Body of Water
Dead Sea Modern name
Ancient canal
Historical coastline
Pass

The Pergamene kingdom of the Attalids (241 to *c.* 185 BC)

The terms 'Pergamene Kingdom' and 'Attalid Kingdom' denote a region of Asia Minor centred upon the fortified city of Pergamum. Growing steadily from 281 BC, its territory fluctuated considerably in the course of the 2nd cent. Initially a dynastic structure (Philetaerus, Eumenes I) tolerated by the Seleucids, it became a kingdom proper (maps A and B) around 238 BC when Attalus I accepted the title of king (regnal period 241-197). At its period of greatest extent under Eumenes II (197-159 BC, maps C and D) it reached from Thrace to the Taurus Mountains. Its early support of Rome allowed the kingdom to survive until the extinction of the dynasty. In 133 BC, Attalus III bequeathed it to Rome, which thereupon founded the province of Asia (129, Liv. Per. 58f.).

The citadel of Pergamum is the best-researched of all the Hellenistic royal palaces, and shows how the representation of the kings also found expression in architecture (Temple of Athena Nikephoros/'Dying Gaul', monument to victory over the 'Gauls' (Galatians), Altar of Zeus with gigantomachy). Important sources for this period are Livy, Strabo and Pausanias, along with inscriptions, coins and the excavations throughout the kingdom.

I. Development of the kingdom up to Attalus I (maps A and B)

The emergence of the kingdom was closely associated with the problems to which the Seleucid kingdom was exposed. The latter's intractable land mass, the chronic Seleucid rivalry with the Ptolemies (Syrian Wars) and numerous internal dynastic conflicts facilitated the development of regions under the independent control of local dynasts (e.g. the kingdoms of Bithynia, Cappadocia, Pontus).

Philetaerus (281–263), to whom Lysimachus' command of the fortress and the treasury stored there had been transferred, restricted his rule to the middle Caicus valley after Lysimachus' death in 281, observing loyalty to Seleucus I and Antiochus I. However, he systematically expanded Pergamum to a residence in the style of a Greek *polis*. Philetaerus' successor, Eumenes I (263–241), extended the territory northward to the foot of Mount Ida and eastward into the upper Caicus valley, marking it out with garrison towns (e.g. Philetaerea). Access to the sea was provided by the harbour of Elea, south of the mouth of the Caicus. The region to the south, up to Myrina, was probably annexed during the 2nd Syrian War. However, substantial tributes had to be paid to the Galatians in exchange for this expansion.

The dispute within the Seleucid house between Seleucus II and his brother Antiochus Hierax enabled Attalus I, after the victories over the Galatians (238) and that over the coalition of the Galatians with Antiochus Horax (230–227?), temporarily to extend the kingdom as far as the Taurus Mountains. It was reduced to the immediate environs of Pergamum by 223, but by around 218 – now with the aid of Celtic mercenaries – it presided stably over the region from Colophon in the west to the northern shore of the Aegean, and eastwards at least as far as the Macestus. An alliance with the Aetolian League against Philip V brought Attalus into contact with the Romans for the first time, and he supported them in the First Macedonian War (gaining Aegina). For their part, they maintained diplomatic relations with Pergamum. Still, in spite of his friendship with Rome, Attalus was unable to expand his kingdom any further by the time of his death in 197. On the contrary: it shrank continually under Seleucid pressure.

II. The Pergamene Kingdom under Eumenes II (maps C and D)

Only under Attalus' son, Eumenes II, did the kingdom rise to become the prime political power of Asia Minor until the shift in Rome's eastern policy after 168. It became the regional centre of Greek culture and scholarship (second-largest library of the ancient world at Pergamum). Eumenes' expansion plans depended entirely on Roman support, and he succeeded in coaxing Rome to war with Antiochus III (191–188). Not only did he play a prominent role in the Roman victory at Magnesia on the Sipylus, but he also used the Roman forces against the Galatians (189). He was also the one to profit most from the terms of the treaty of Apamea (188), which granted him the Seleucid territory as far as the Taurus, assured him of rule over numerous Greek cities of Asia Minor, including Ephesus and Telmessus, and enabled him, with Roman support, to win further territorial gains from war with Prusias I of Bithynia and Pharnaces I of Pontus. After this zenith, the political power of the kingdom waned rapidly.

III. The Galatian tribal states of Asia Minor until their dissolution into the Roman province of Galatia (3rd cent.–25 BC)

The establishment of the Galatians in Asia Minor and the assertion of their statehood until 25 BC were consequences of the policies of the Hellenistic powers and later of Rome. In 278/7, the nomadic tribes, i.e. the three major tribes of the Tolistobogii, Trocmi and Tectosages, were brought to Asia Minor as allies of Nicomedes I of Bithynia and other members of the anti-Seleucid alliance, and were then sent into battle (277–275). Around 275/4, they were given the eastern part of Phrygia to settle, and in 274/73 also part of western Cappadocia, which, in spite of its prior 'Phrygian' (→ map p. 117) population they shaped in terms of ethnic identity and language as a newly-emergent historical territory of Galatia, which endured into the Byzantine period. In turn, the major tribes were each subdivided into four politically independent tetrarchies, which were assembled in a federation, with a federal sanctuary.

→ Maps pp. 114f., 119, 123, 159, 183; Pergamum I.–IV., BNP 10, 2007

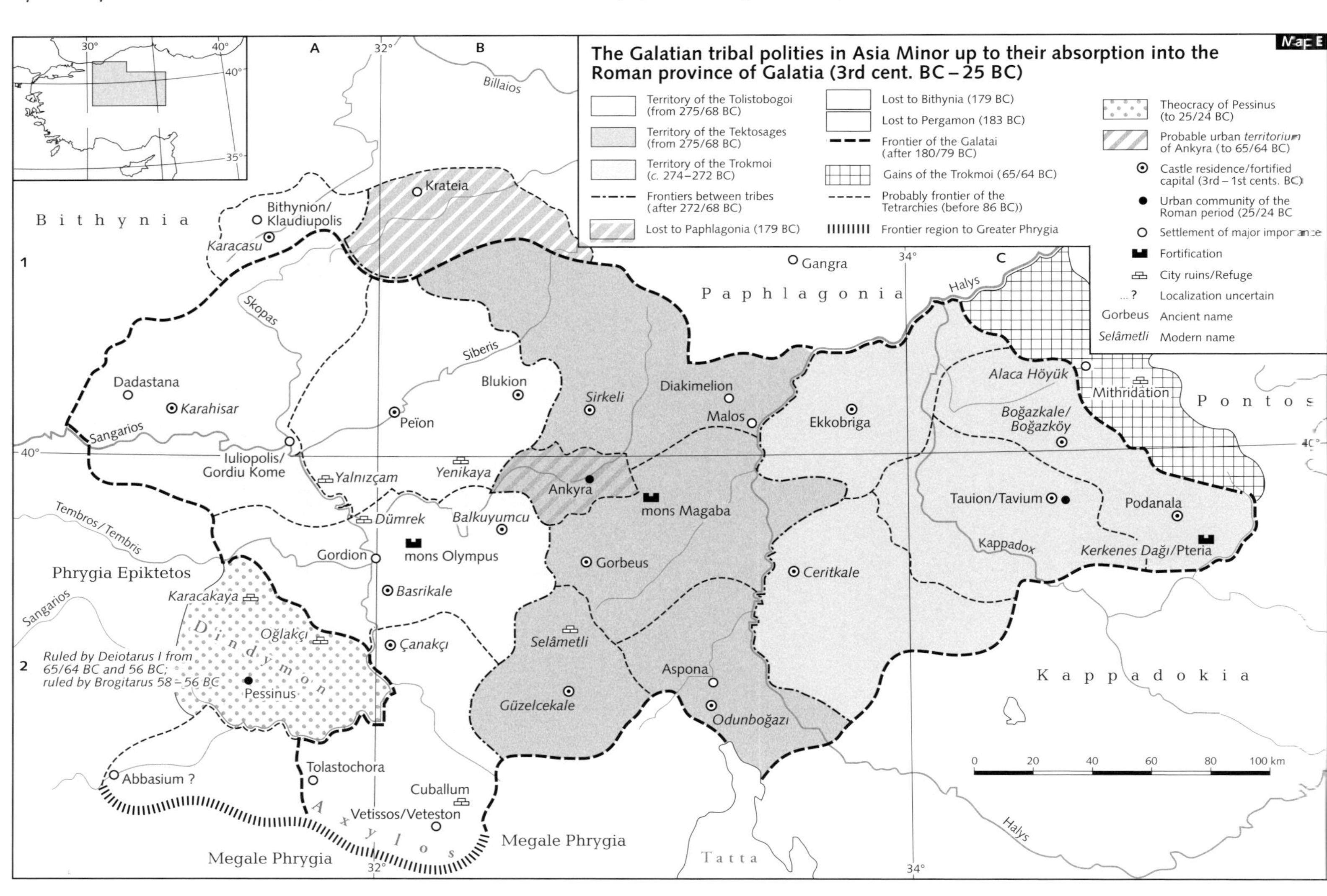

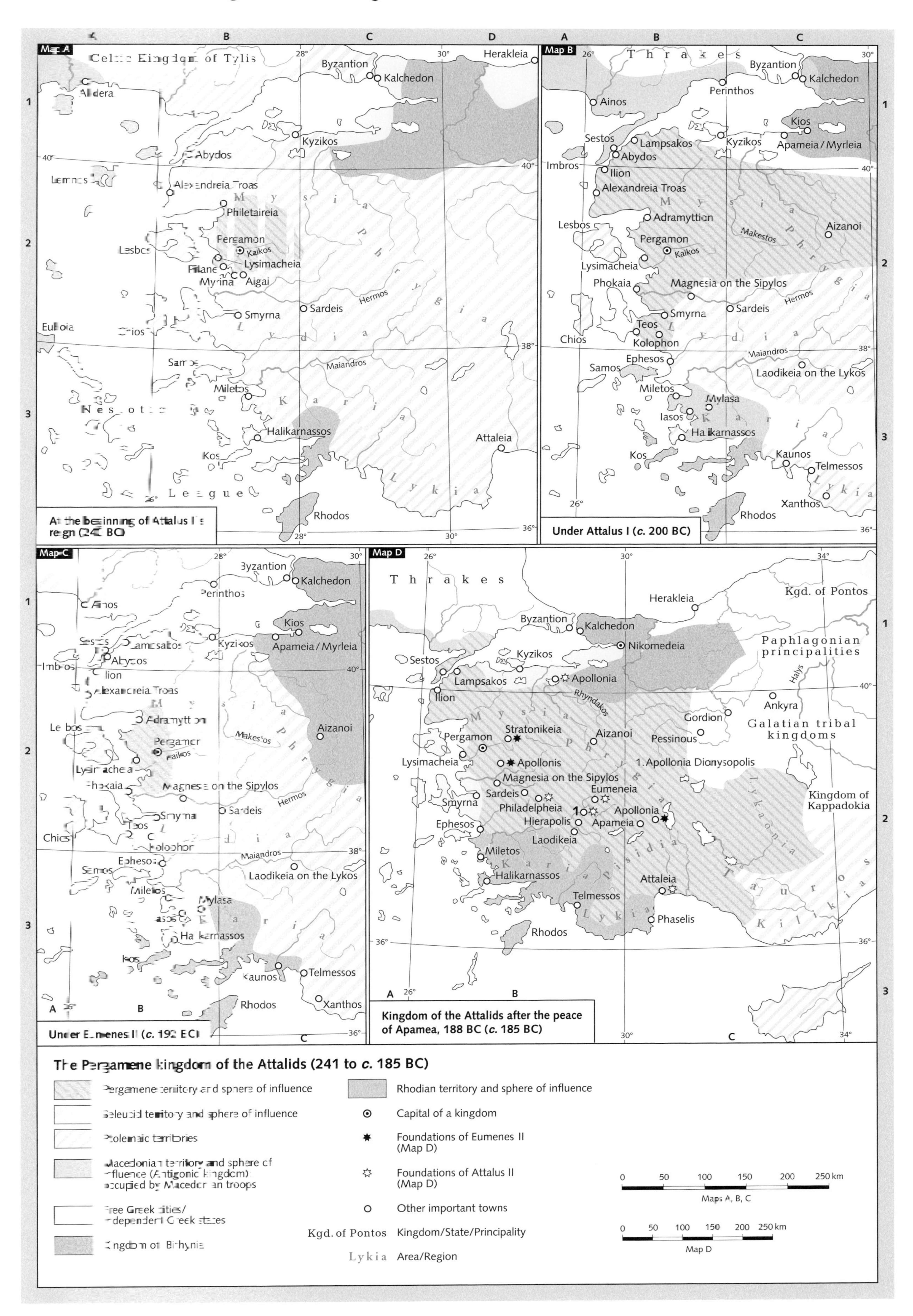
Map A
Byzantion
Kalchedon
Herakleia
Kyzikos
Abydos
Philetaireia
Pergamon
Kaikos
Lysimacheia
Aigai
Smyrna
Sardeis
Hermos
Maiandros
Miletos
Halikarnassos
Attaleia
Rhodos
At the beginning of Attalus I's reign (241 BC)
Map B
Byzantion
Kalchedon
Perinthos
Ainos
Kios
Sestos
Lampsakos
Kyzikos
Apameia / Myrleia
Abydos
Imbros
Ilion
Alexandreia Troas
Adramyttion
Lesbos
Aizanoi
Makestos
Pergamon
Kaikos
Lysimacheia
Phokaia
Magnesia on the Sipylos
Smyrna
Sardeis
Hermos
Teos
Kolophon
Chios
Ephesos
Maiandros
Samos
Laodikeia on the Lykos
Miletos
Mylasa
Iasos
Kos
Kaunos
Telmessos
Xanthos
Rhodos
Under Attalus I (c. 200 BC)
Map C
Byzantion
Kalchedon
Perinthos
Kios
Kyzikos
Apameia / Myrleia
Aizanoi
Makestos
Hermos
Maiandros
Laodikeia on the Lykos
Mylasa
Telmessos
Rhodos
Xanthos
Map D
Thrakes
Heraklei a
Kgd. of Pontos
Byzantion
Kalchedon
Nikomedeia
Paphlagonian principalities
Sestos
Kyzikos
Lampsakos
Apollonia
Halys
Ilion
Rhyndakos
Ankyra
Gordion
Galatian tribal kingdoms
Stratonikeia
Pergamon
Aizanoi
Pessinous
Lysimacheia
Apollonis
1. Apollonia Dionysopolis
Magnesia on the Sipylos
Sardeis
Eumeneia
Kingdom of Kappadokia
Smyrna
Philadelpheia
Apollonia
Hierapolis
Apameia
Ephesos
Laodikeia
Miletos
Halikarnassos
Attaleia
Telmessos
Phaselis
Rhodos
Kingdom of the Attalids after the peace of Apamea, 188 BC (c. 185 BC)
The Pergamene kingdom of the Attalids (241 to c. 185 BC)
Rhodian territory and sphere of influence
Capital of a kingdom
Foundations of Eumenes II (Map D)
Foundations of Attalus II (Map D)
Other important towns
Kgd. of Pontos Kingdom/State/Principality
Lykia Area/Region
0 50 100 150 200 250 km
Maps A, B, C
0 50 100 150 200 250 km
Map D

The Bosporan Kingdom from the 5th cent. BC to the 1st cent. AD

The so-called Bosporan Kingdom established itself in the mostly Scythian-settled region on the north-eastern shore of the Black Sea (cf. also map C) after the phase of colonization, mostly by Greeks of Asia Minor (perhaps under pressure from the Lydians, later the Persians), from the 7th/6th cents. BC. The kingdom was a union of Greek cities to either side of the Cimmerian Bosporus (Straits of Kerč), on the Chersonesus Taurica (Crimea) and its eastward counterpart the Taman Peninsula, south of the Maeotis (Sea of Azov). The union was under the leadership of Panticapaeum.

At first (*c.* 480 BC), this alliance was probably meant to provide protection against indigenous tribes. Many of the cities were extended and powerfully fortified in the 2nd half of the 5th cent., and a rampart 25 km long was built from Tyritace to the Maeotis. Over time, the Greeks and indigenous peoples (some of whom were nomadic, some settled) developed closer political, economic and cultural ties. This co-existence was the basis of the region's lasting economic florescence and the consequent expansion of its trading links with the Greek world.

Map B

The development of the Bosporan Kingdom

- Approximate extent of the Bosporan Kingdom under the Archaeanactids in the 5th cent. BC
- Under the Spartocids:
- Conquests of Leucon I (389–349 BC)
- Conquests of Paerisades I (349–311/10 BC)
- Conquests of Pharnaces (63–47 BC)
- Conquests of Aspurgus (8? BC – AD 38)
- Conquests of Cotys I (AD 45–71)

I. The importance of the Bosporan Kingdom

The Bosporan Kingdom was among the most important suppliers of grain and raw materials, and was also a handling centre for the slave trade. It controlled the coastal regions, the entrance into the Maeotis and the venues of contact with the indigenous population (see below and map C), and hence also trade with the hinterland, grain cultivation and salted fish production.

Apart from grain and fish, it also exported pelts, wool, textile fibre plants, timber, iron ore, precious metals (from the Urals and Siberia), limestone and clay. Polybius (4,38) also mentions cattle, slaves, honey and wax. It imported wine, though this was also produced locally, as well as olives, artisanal products, luxury goods and clothing, as the archaeological record to some extent attests. The kingdom had several trading centres and harbours, and was an important trading partner of Athens (esp. from the 5th cent.), and later of Delos, Rhodes and the southern Pontic cities (Sinope).

II. The development of the Bosporan Kingdom

The league of cities and subsequent kingdom were ruled by the Anatolian Greek (?) Archaeanactids, and thereafter by the Hellenized Scytho-Thracian (?) Spartocid Dynasty. However, the royal title is only attested from the 4th cent.

The Spartocids expanded the territory of their kingdom (cf. map B) by the incorporation (not always voluntary) of free Greek cities (e.g. Nymphaeum, Theodosia) and by conquests in Scythian territory on the Crimea and in the Sindo-Maeotian region from the 4th cent. BC under Leucon I and Paerisades I. These expansions are often deducible only from the altered forms of address for the Bosporan rulers. It is to be assumed that the cities were able to retain a degree of *polis* autonomy and that the individual tribes were governed by their own princes.

An economic and cultural decline is apparent in the first half of the 3rd cent., probably because of the Celtic attacks in Asia Minor, the weakness of Athens and Egyptian competition in grain supply. After a period of recovery in the 2nd half of the 3rd cent., there were political and economic crises in the late 3rd and 2nd cents., as well as military conflicts with the Sarmatians and Scythians. The Scythian kingdom on the Crimea, with its capital at Neapolis Scythica, was expanding (deployment of defensive ramparts on the Bosporan western frontier). Southward realignments of trade (towards Egypt), esp. by the cities of Athens, Thasus and Heraclea Pontica, also exacerbated difficulties. Around 109 BC, the situation culminated in the fall of the Spartocids, who handed the kingdom, initially of their own free will, to Mithridates VI of Pontus. He drew the kingdom into his disputes with Rome, which led to its reduction to the status of a Roman vassal state (cities as Roman bases) under his son Pharnaces. Only with Aspurgus was a royal dynasty founded once more, *c.* 8 BC. It ruled until around the mid 4th cent. AD, albeit dependent upon Rome to a greater or lesser degree.

III. Extent of Scythian settlements and kurgans (7th-4th/3rd cents.; supplementary map C)

Recent and spectacular finds have brought the origins, homelands and major developmental phases of the primarily nomadic Scythian tribes and culture (tumulus burials in so-called kurgans; individual finds) into the focus of research once more, and not only for Russian scholars. But ancient tradition already paid attention to the Scythians (Scythae; ancient Near Eastern sources from 8th/7th cents.; Greek sources: esp. Herodotus, but also e.g. Thucydides, Aristophanes, Strabo, Aelianus). Around the mid 6th cent. BC, the centre of Scythian rule shifted away from North Caucasia and the Hypanis (Kuban) region, westwards into the steppes and forests of the lower Borysthenes (Dnieper), between the Thracians (Thraci) in the south-west and the Sarmatians (Sarmatae) in the east. These territories had good pasture land and bordered on the Greek coastal cities of the northern Black Sea shore – later also the Bosporan Kingdom (sometimes a military ally) where goods were exchanged at particular trading points as part of a wider trade into and out of the Mediterranean. Hellenization is discernible in the Scythian elite from the 5th cent. (mixed population in the cities; Greek luxury items in finds, esp. fine pottery). Classical Scythian culture disappeared around 300 BC (reasons including pressure from the Sarmatae, climate change, political and military causes), but a Late Scythian kingdom was founded in the 3rd/2nd cents. on the Crimea (capital: Neapolis Scythica; until 1st cent. BC; cultural legacy until 3rd cent. AD).

Map C

Distribution of Scythian settlements and kurgans (7th to 4th/3rd cents. BC)

- Settlement
- Earthworks
- Kurgan(s)
- Necropolis
- Neapolis — Ancient name
- Sachama — Modern name
- Greek colony

1. Arcyz
2. Adžigol
3. Širokaja Balka
4. Rožnovskij
5. Širokoe
6. Ševčenko
7. Novaja Rozanovka
8. Želtokamenka
9. Apostolovo
10. Ordžonikidze
11. Tolstaja Mogila
12. Krasnokutsk
13. Kičkas
14. Tomakovka (Ostraja Mogila)
15. Čertomlyk
16. Geremesov
17. Nikopol'
18. Kirovvo
19. Kapulovka
20. Kut
21. Kozël
22. Kamenskoe gorodišče
23. Cimbalka
24. Čmyrev
25. Mordinov
26. Deev
27. Oguz
28. Solocha
29. Gajmanova Mogila
30. Melitopol'
31. Šul'govka
32. Konstantinovka
33. Dvugorbaja Mogila
34. Berdjansk
35. Ždanov
36. Elisavetovskoe gorodišče
37. Inkerman
38. Al'ma
39. Uč-Baš
40. Krestovaja Gora
41. Ašlama-Dere
42. Simferopol'
43. Nejzackoe
44. Kizil-Koba
45. Kostel'
46. Aj-Todor
47. Koška
48. Karaul-Oba
49. Ajvazovsk
50. Frontovoe
51. Kul'-Oba
52. Il'ičevo
53. Temir-Gora
54. Nymphaion
55. Cukur-Liman
56. Bol'šaja Bliznica
57. Merdžany-Kurgan
58. Seven Brothers Kurgan
59. Karagodeuaš
60. Kurgans near Elizavetinskaja Stanica
61. Kurgan Cetuk
62. Voronežskaja Stanica, Kurgan 17
63. Razdol'naja Stanica
64. Ust'-Labinskaja Stanica
65. Ladožskaja Stanica, Kurgan 22
66. Necropolis of Ul' and Uljap (Ul'skaja Stanica)
67. Necropolis of Kelermes
68. Goverdovskaja Stanica
69. Kurdžips
70. Tuľskaja Stanica
71. Necropolis on the river Fars
72. Machoševskaja Stanica
73. Kostromskaja Stanica
74. Gubskaja Stanica
75. Besleneevskij-Kurgan 27
76. Podgornaja Stanica
77. Zolotoj

Sources

Information on the material culture is provided esp. by old and recent archaeological surveys (mostly Russian) of the Greek settlements (Panticapaeum, Olbia, Berezan, Tanaïs), and their finds, esp. coin finds, but clay stamps as well; also by archaeological processing of autochthonous features and finds (cf. supplementary map C). Literary and epigraphic sources fluctuate with the condition of the kingdom (approx. 1,500 known Greek and Latin inscriptions: amphora stamps, graffiti, coins).

→ Maps pp. 85, 113, 115, 117, 129, 135, 141, 159; The Bosporan Kingdom, 5th cent. BC – 1st cent. AD. The northern Black Sea region as an economic sphere in the Hellenistic period, BNP 12, s.v. Regnum Bosporanum; Regnum Bosporanum, BNP 12, 2008; Scythae I., BNP 13, 2008 (with map)

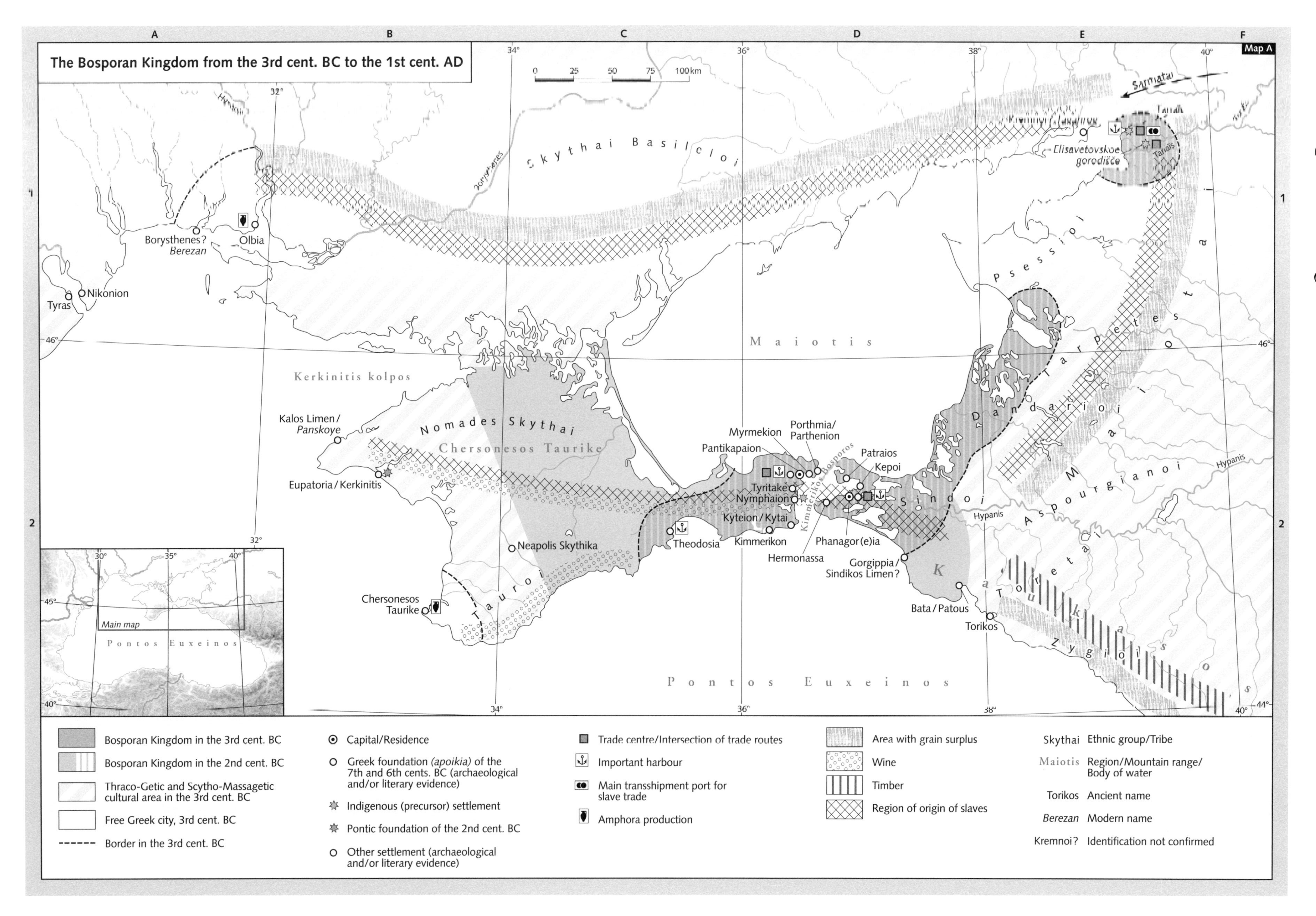
The Bosporan Kingdom from the 3rd cent. BC to the 1st cent. AD
Map A
0 25 50 75 100 km
Skythai Basilcloi
Sarmatai
Tanais
Elisavetovskoe
gorodisce
Borysthenes?
Berezan
Olbia
Tyras
Nikonion
Psessoi
Maiotis
Tarpetes
Dandarioi
Maiotai
Mourgianoi
Aspourgianoi
Hypanis
Kerkinitis kolpos
Kalos Limen/
Panskoye
Nomades Skythai
Chersonesos Taurike
Eupatoria / Kerkinitis
Myrmekion
Porthmia/
Parthenion
Pantikapaion
Patraios
Kepoi
Kimmerikos Bosporos
Tyritake
Nymphaion
Kyteion / Kytai
Sindoi
Theodosia
Kimmerikon
Phanagor(e)ia
Hermonassa
Gorgippia /
Sindikos Limen?
Neapolis Skythika
Chersonesos
Taurike
Tauroi
Bata / Patous
Torikos
Toretai
Kerketai
Zygioi
Main map
Pontos Euxeinos
Pontos Euxeinos
Bosporan Kingdom in the 3rd cent. BC
Bosporan Kingdom in the 2nd cent. BC
Thraco-Getic and Scytho-Massagetic cultural area in the 3rd cent. BC
Free Greek city, 3rd cent. BC
Border in the 3rd cent. BC
Capital/Residence
Greek foundation (apoikia) of the 7th and 6th cents. BC (archaeological and/or literary evidence)
Indigenous (precursor) settlement
Pontic foundation of the 2nd cent. BC
Other settlement (archaeological and/or literary evidence)
Trade centre/Intersection of trade routes
Important harbour
Main transshipment port for slave trade
Amphora production
Area with grain surplus
Wine
Timber
Region of origin of slaves
Skythai Ethnic group/Tribe
Maiotis Region/Mountain range/Body of water
Torikos Ancient name
Berezan Modern name
Kremnoi? Identification not confirmed

The Hellenistic world in the 2nd cent. BC

The 2nd cent. BC was the period during which the Hellenistic world of the eastern Mediterranean came under Roman rule. This came about not so much through a policy of aggression on Rome's part (in fact, the Romans long sought to avoid involvement in the east, not least because of their commitments in the west, e.g. Punic Wars) as in consequence of the self-seeking prestige politics typically pursued, as throughout the preceding Hellenistic period, by the Hellenistic monarchs and the Greek cities and leagues.

Ultimately, Rome established a presence in the Balkans and Asia Minor because of the appeals for Roman help against the great Hellenistic kingdoms which the medium-sized powers of Pergamum, Rhodes and the Greek states in particular issued ever more frequently from 201 BC. These roused Rome to diplomatic and/or military intervention, which would lead at first to indirect rule, then increasing dominance and finally the direct exercise of power (province of Macedonia; province of Asia; regions under Roman control, client kingdoms/spheres of influence).

I. Phases and consequences of Roman intervention in the Balkans, Greece and Asia Minor

1. Rome's wars against Philip V and Antiochus III

On the death of Ptolemy IV in 204 BC, the Ptolemaic kingdom was now represented by a mere six-year-old child. The Seleucid Antiochus III and Philip V of Macedonia now called a truce in their struggle for power in Asia Minor and concluded a probably secret treaty, this Egyptian weakness seemingly having presented both rulers with the opportunity to secure their respective spheres of interest, Antiochus in the Levant (5th Syrian War 202–200, → map p. 123), Philip V in the northern Aegean. This recourse to arms also drew in the kingdom of Pergamum and the republic of Rhodes. The balance of power in the complex of Hellenistic states, fragile at the best of times, now threatened to come entirely out of kilter, and the two medium-sized powers therefore asked Rome for help in 201. The Romans, here deploying their own troops in Greece for the first time, together with their Greek allies (Aetolian League, later Achaean League) inflicted defeat on Philip V in 197 at Cynoscephalae (Second Macedonian War, →map p. 151). The following year, Philip concluded a peace with Rome restricting him to his territorial heartland. He was entirely driven out of Greece. In 196, the Romans proclaimed the freedom of the Greeks. For reasons of political calculation, however, they refrained at first from imposing direct rule, and withdrew their troops from Greece again until 194.

Rome's next intervention was necessitated by the expansionist behaviour of Antiochus III, who was trying to restore the Seleucid kingdom and had advanced into Europe in 196 after conquering Abydus on the Hellespont. Not only the Pergamenes felt this to be a threat. Rome at first used diplomacy. But when the Aetolian League asked for Antiochus' help against Rome in 192, full-scale war broke out in Greece. Antiochus was forced to withdraw to Asia Minor, where he was defeated early in 189 at Magnesia on the Sipylus by a Roman-Pergamene alliance. The war ended in 188 with the treaty of Apamea. The Seleucids now lost all their territory in Asia Minor north of the Taurus, Pergamum and Rhodes in particular reaping the benefits.

2. Rome's war against Perseus of Macedonia and the beginnings of direct rule over Macedonia, Greece and western Asia Minor

Friendly Greek states increasingly appealed to Rome for help in the ensuing period, but Rome's involvement increasingly came to be seen as interference, until the formerly valued friend was gradually transformed into a detested overlord.

The plea from Eumenes II of Pergamum in 172, who felt the east of his territory to be threatened by the expansionist aspirations of Perseus of Macedonia, led to a new Roman intervention, to the Third Macedonian War (171–168,→ map p. 151) and, after the battle of Pydna in 168, to the liquidation of the Macedonian Antigonid state. After a revolt in 148, Macedonia was placed under direct rule and became a Roman province. Rhodes suffered indirect, economic punishment from Rome for its attempts to mediate: a free port was set up on Delos, a blow at Rhodes' greatest vulnerability. In exchange for the *pax Romana*, Greece was in future expected to demonstrate absolute and unconditional loyalty to Rome. Resistance, e.g. at Epirus or in Achaia, led to drastic punishments. In 146, Corinth was destroyed. The Pergamene kingdom was bequeathed to Rome in 133. The province of Asia was set up on its territory and local rulers were set up under Roman control. The Roman sphere of influence now reached deep into Asia Minor already. Even the (albeit 1st-cent. BC) wars with Mithridates VI of Pontus (→ map p. 159), who succeeded in much extending his kingdom – briefly as far as the Chersonesus Taurica (modern Crimea, cf. → map p. 127) – could no longer imperil Roman dominance, which was associated with considerable economic exploitation.

The two remaining major Hellenistic kingdoms, those of the Seleucids and Ptolemies, had little to offer to counter this development, which they recognized too late if at all. Internal dynastic difficulties, increasing resistance on various levels from indigenous populations (cf. e.g. Hasmonaeans, Maccabean Revolts, → map p. 131), renewed military conflicts and other factors permitted no new internal consolidation of these kingdoms. In spite of their wealth and opportunities, they failed to co-ordinate their strengths to regenerate their kingdoms and stand up to the expansion of Rome.

3. The death throes of 'political' Hellenism

Territorial losses to the Parthians had confined the Seleucid kingdom to northern Syria, and with further persistent dynastic disputes it sank into obscurity. Northern Syria was also lost in 63 BC, becoming Roman when Pompey incorporated it as the province of Syria.

Ptolemy VI (180–145) tried again to expand at the expense of the Seleucids, but after his death the Ptolemaic kingdom became paralyzed by constant dynastic disputes. Although it remained independent even after losing its foreign possessions, it was now firmly under Roman influence.

II. The emergence of the Parthian kingdom

The Seleucid kingdom lost more territory, faced above all with the strengthening Parns/Parthians. They had entered Seleucid territory from the mid 3rd cent., one of the many nomadic tribes moving down from the north. They succeeded in establishing themselves in what would become Parthia. After a short period of expedient recognition of the suzerainty of Antiochus III, they expanded their kingdom (probably after 188) far beyond Parthia, southwards, eastwards and westwards. Under Mithridates I (171–139/38), they conquered parts of the Graeco-Bactrian kingdoms and western Iran, as well as one of the Seleucid cities of Mesopotamia, Seleucea on the Tigris, in whose immediate vicinity they founded Ctesiphon as their capital, in the Hellenistic manner.

Sources

The condition of the sources for this period is relatively favourable, as Polybius gives a thorough treatment of Rome's conquest of the great kingdoms and the establishment of Roman 'world rule' in his *Histories*. The work begins around 220 BC; its first books are preserved complete, along with extensive excerpts. His work was used by the Roman annalists, and hence also by Livy; later historians (Appian) also dealt with the period. There are also papyri from Egypt, cuneiform texts from Mesopotamia, written testimonies (e.g. of Jewish history), epigraphic and numismatic materials and archaeological finds. If these conflicts have hitherto been viewed mostly from the perspective of Rome, recently attention has also been increasingly devoted to the Greek and Hellenistic side.

There are also substantial recent studies of the aspiring kingdom of the Parthians, and its amalgamation esp. with the disintegrating Seleucid kingdom. These are based on a substantial inventory of contemporary written sources, mostly in Parthian, Greek and Aramaic (ostraka, documents, inscriptions), on the literary tradition (e.g. Strabo, Tacitus), some of it from the territory of the Parthian kingdom itself (e.g. Isidorus of Charax), and on Chinese historiography. Archaeological studies and a wealth of numismatic materials complete the picture.

→ Maps pp. 115, 117, 121, 127, 131, 133, 153, 159, 161, 215; Hellenistic States, BNP 6, 2005

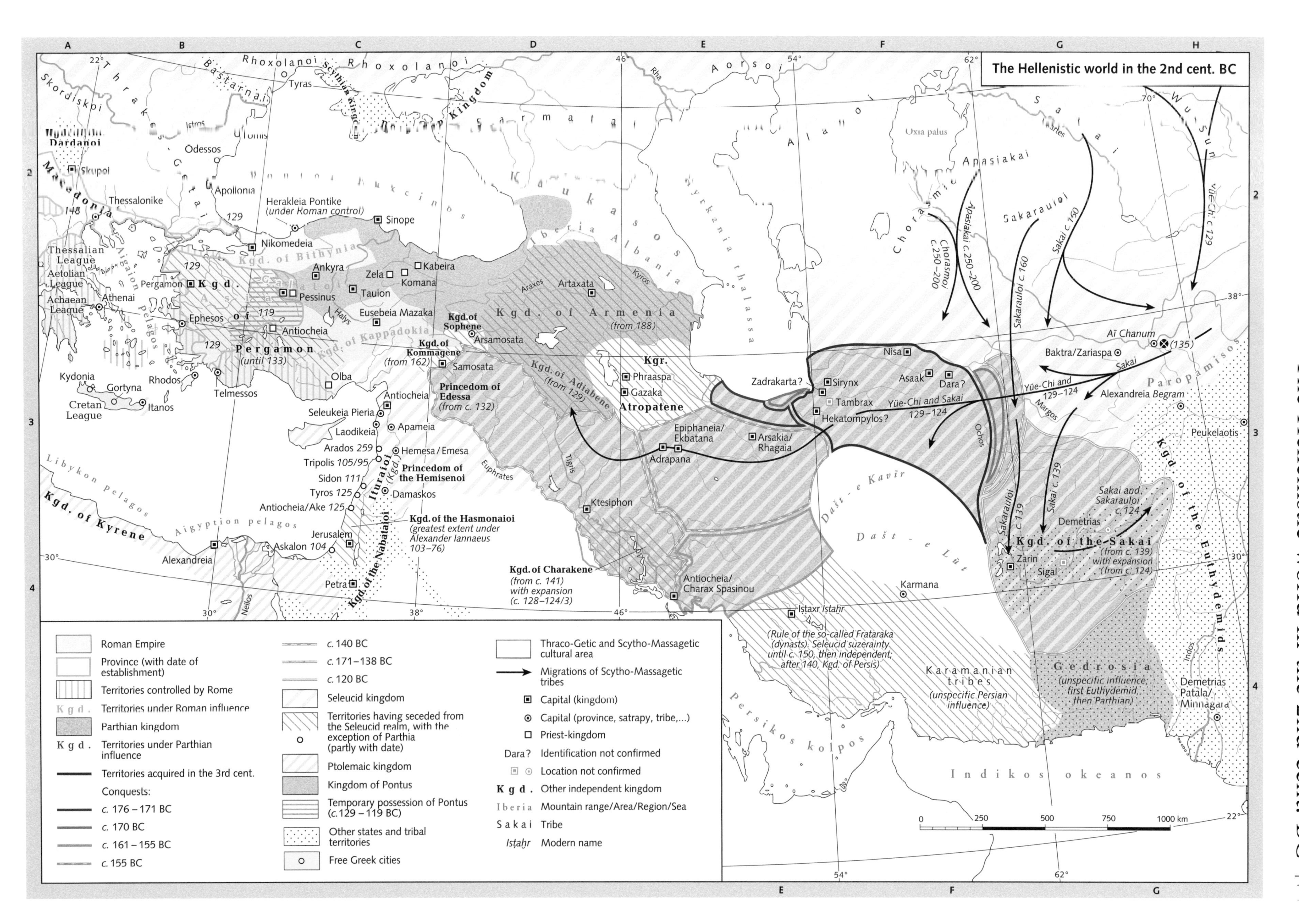
The Hellenistic world in the 2nd cent. BC
Roman Empire
Province (with date of establishment)
Territories controlled by Rome
Territories under Roman influence
Parthian kingdom
Territories under Parthian influence
Territories acquired in the 3rd cent.
Conquests:
c. 176 – 171 BC
c. 170 BC
c. 161 – 155 BC
c. 155 BC
c. 140 BC
c. 171 – 138 BC
c. 120 BC
Seleucid kingdom
Territories having seceded from the Seleucid realm, with the exception of Parthia (partly with date)
Ptolemaic kingdom
Kingdom of Pontus
Temporary possession of Pontus (c. 129 – 119 BC)
Other states and tribal territories
Free Greek cities
Thraco-Getic and Scytho-Massagetic cultural area
Migrations of Scytho-Massagetic tribes
Capital (kingdom)
Capital (province, satrapy, tribe,...)
Priest-kingdom
Dara? Identification not confirmed
Location not confirmed
Kgd. Other independent kingdom
Iberia Mountain range/Area/Region/Sea
Sakai Tribe
Iṣṭaḫr Modern name

Syria and Palestine, 2nd and 1st cents. BC

The Syrian-Palestinian region was an important link between Asia Minor and the south (Arabia and Egypt) throughout Antiquity. This was where overland routes from north, south and east met, and there were important trading centres and harbour cities here through which trade was conducted into the Mediterranean region.

Accordingly, the region was constantly fought over, e.g. in the Hellenistic period by the Ptolemies and Seleucids (Syrian Wars). Although the overlords changed (e.g. Assyrians, Achaemenids), certain local entities or independent quasi-states survived the centuries from the early Iron Age (from *c.* 12th/11th cents. BC) and others developed, e.g. the coastal Phoenician cities, the Jewish temple state in the Canaanite region, the Nabataean kingdom to the south and the Ituraean to the north.

I. Political developments in the 2nd and 1st cents. BC

After the 5th Syrian War (202–198/94 BC), the Jewish temple state centred at Jerusalem came under Seleucid rule from *c.* 200 BC. It had been part of the Ptolemaic kingdom in the 3rd cent. and had been integrated into its economic administration (tax farming, etc.), but its autonomy and statehood were respected, as previously by the Persians. Antiochus III also guaranteed all its privileges, but in the reign of Antiochus IV, Judaea was so exposed to gradual Hellenizing tendencies (e.g. renaming of Jerusalem to Antioch and its organization as a Greek *polis*) that a popular revolt ensued. The result of this so-called Maccabean Revolt (167–157 BC) was the emergence of an independent Jewish state, a theocracy under the Hasmonaean Dynasty, so called after the Maccabean progenitor Ḥašmōn. However, over time this state, too, developed characteristics typical of Hellenistic monarchies, including the adoption of the royal title, despite the fact that its rulers played the role of champions of Judaism. Armed conflicts with the Seleucids and the tense relationship between radical advocates of Hellenism, Graecized Jews and groups 'true to the faith' (Chassidim: 'true to the faith', Essenes: Dead Sea scrolls, Pharisees: establishment of synagogues and schools) who rebelled against politicization and 'secularization' typified the period up to the absorption of the entire region into the Roman province of Syria (64/63 BC), and beyond into the long reign of Herod I (40–4 BC). He ruled over the Jews from 37 BC with his conquest of Jerusalem, while at the same time being a Roman client prince and Hellenistic king.

II. The Maccabean Revolt (167–157 BC) and Hasmonaean rule

A guerrilla-style uprising began in 167 BC against the profaning of the Temple of Jerusalem, the ban on practising the Jewish religion and the imposition of taxes by Antiochus IV (175-164 BC). It was led by Mattathias and his son Judas Maccabaeus; the latter succeeded in 165 or 164 BC in retaking and reconsecrating the Jerusalem temple. The rescuing and resettling of Jewish minorities led to the expansion of the war. The battle of Beth Zacharia in the summer, in which Judas Maccabaeus was defeated by Seleucid troops led by Lysias, led to a peace imposed by the Seleucids. The Jewish fortress of Beth Sura was taken and the pro-Seleucid Alcimus was installed as High Priest at Jerusalem against the will of the Hasmonaeans. Antiochus V (164-162 BC) thus restored the Jewish theocracy.

Judas Maccabaeus led a successful underground campaign under Demetrius I (162–151/50 BC), inflicting a serious defeat on Seleucid troops under Nicasar at Adasa (location not certain) in 161. He also made an alliance with Rome. However, Demetrius pre-empted the Roman intervention. In April 160, Judas Maccabaeus was defeated by the Seleucids under Bacchides, and died in the battle.

Nonetheless, the Hasmonaeans gradually gained control of Judaea, exploiting dynastic conflicts and the Seleucid overlords' external difficulties to annex large tracts of the periphery of the Jewish area of settlement. Their obligation to pay tribute to the disintegrating Seleucid kingdom ended in 143/42 BC, the Seleucid citadel in Jerusalem was taken and from 141 or 140 BC, the hereditary High Priesthood was assigned by the Jewish people. The royal title in Hellenistic style is attested with some likelihood from Aristobulus I (104 BC), at all events certainly from Alexander Jannaeus (103–76 BC), under whom the kingdom reached its greatest extent. It led to open civil war with the Pharisees.

After the death of Alexander Jannaeus in 76 BC and that of his wife Alexandra Salome in 67, a dynastic conflict broke out between Aristobulus II and Hyrcanus II. This was a decisive factor in the subjection of the diminished Jewish state to the Roman governor of Syria from 63 BC, the year of Pompey's conquest of Jerusalem. Hyrcanus II was stripped of the royal title and the real power went to Antipater, the father of Herod I. Hasmonaean rule was abolished in the course of the Parthian invasion and the Roman reconquest of the east (40–37 BC); Herod I ruled until 4 BC. He had the Temple of Jerusalem restored, but also founded cities in the manner of a Hellenistic ruler (e.g. Caesarea, Samaria-Sebaste).

III. The Nabataean kingdom

The Nabataeans, an Arabic nomadic people (main urban centre: Petra) attested in inscriptions and literary sources, gradually became settled and culturally highly Hellenized in the course of the Hellenistic Period, forming a kingdom (ruler list from 169 BC to AD 106). The greatest extent of this kingdom dates from the 1st cent. BC, under Arethas III, a contemporary of Pompey's, with whom he signed an alliance. The kingdom probably extended northwards as far as Damascus, and southwards to Teima. The Nabataeans traditionally controlled the trading routes to and from Arabia ('Incense Road').

IV. The Ituraean kingdom

The Arabic tribe of the Ituraeans (from the eponym Jeṭūr, supposedly son of Ishmael: Gn 25: 15; 1 Chr 1: 31) is attested by biblical and other sources east of the Jordan in the early Hellenistic Period (1 Chr 5: 19; Eupolemus in Euseb. Praep. evang. 9, 30). The tribe then settled in the region of the Antilebanon, in the plain of Massyas-Biqac, and in the Lebanon. They were regarded as raiders; there were attacks on Byblus, Berytus and even Damascus. Ancient sources (Str., Jos.) report that in the 1st cent. BC, Ptolemy, son of Mennaeus (*c.* 85-40 BC) founded a kingdom in Chalcis which subsequently expanded eastwards and southwards. Pompey compelled Ituraea to pay tribute to Rome. Later, Cleopatra VII received the region as a gift of thanks for her support for the exploits of Mark Antony, who had had Ptolemy's son executed in 36 or 34 BC.

In the Roman period, to stabilize the region further, Heliopolis (Baalbek), the former cultic centre of the Ituraean Tetrarchs of Chalcis, was founded as a veterans' colony (16 BC). The remainder of the region was controlled by local dynasts loyal to Rome. For instance, Augustus gave the south to Herod I around 20 BC.

Sources

The condition of the sources is particularly good, especially for Jewish history, Hellenistic-Jewish culture, the dichotomy between the traditional-minded rural population and lower social strata, who tended to dissociate themselves from all things foreign and Gentile, and the Hellenized upper strata, who were much more open to Greek influences (language, philosophy), and for their relationship with non-Jewish surroundings and powers (Ptolemies, Seleucids). Firstly, there are the surviving religious texts, such as the Books of the Maccabees, secondly the (often hostile) references in Greek literature (e.g. in Hecataeus of Abdera). Recent research has paid more attention to Hellenistic Jewish culture and the religious changes of this period (Apocalypticism, Messianism, Chassidim).

The books of Flavius Josephus are important sources for the entire region, as are those of Strabo, Cassius Dio and Diodorus Siculus and the Old and New Testaments. There are also material sources from many important excavation sites (Jerusalem, Qumran, Gadara, Gerasa, Pella, etc.).

→ Maps pp. 121, 123, 135, 161, 215; Ituraea, BNP 6, 2005; Hasmonaeans, BNP 6, 2005

Literature

B. Bar-Kochva, Judas Maccabaeus, 1989; E.J. Bickerman, The Jews in the Greek Age, 1988; K. Bringmann, Hellenistische Reform und Religionsverfolgung in Judäa, 1983; M. Goodman (ed.), Jews in a Graeco-Roman World, 1998; J.D. Grainger, Hellenistic Phoenicia, 1991; E.S. Gruen, Heritage and Hellenism. The Reinvention of the Jewish Tradition, 1998; M. Hengel, Judentum und Hellenismus, [3]1988; A. Kasher, Jews and Hellenistic Cities in Eretz-Israel (332 BCE–70 CE), 1990; A. Schalit, König Herodes, 2001; E. Schürer, G. Vermès, The History of the Jewish People in the Age of Jesus Christ (175 B.C.–A.D. 135), 3 vols., 1973–1987; M. van Ess, T. Weber (eds.), Baalbek, 1999.

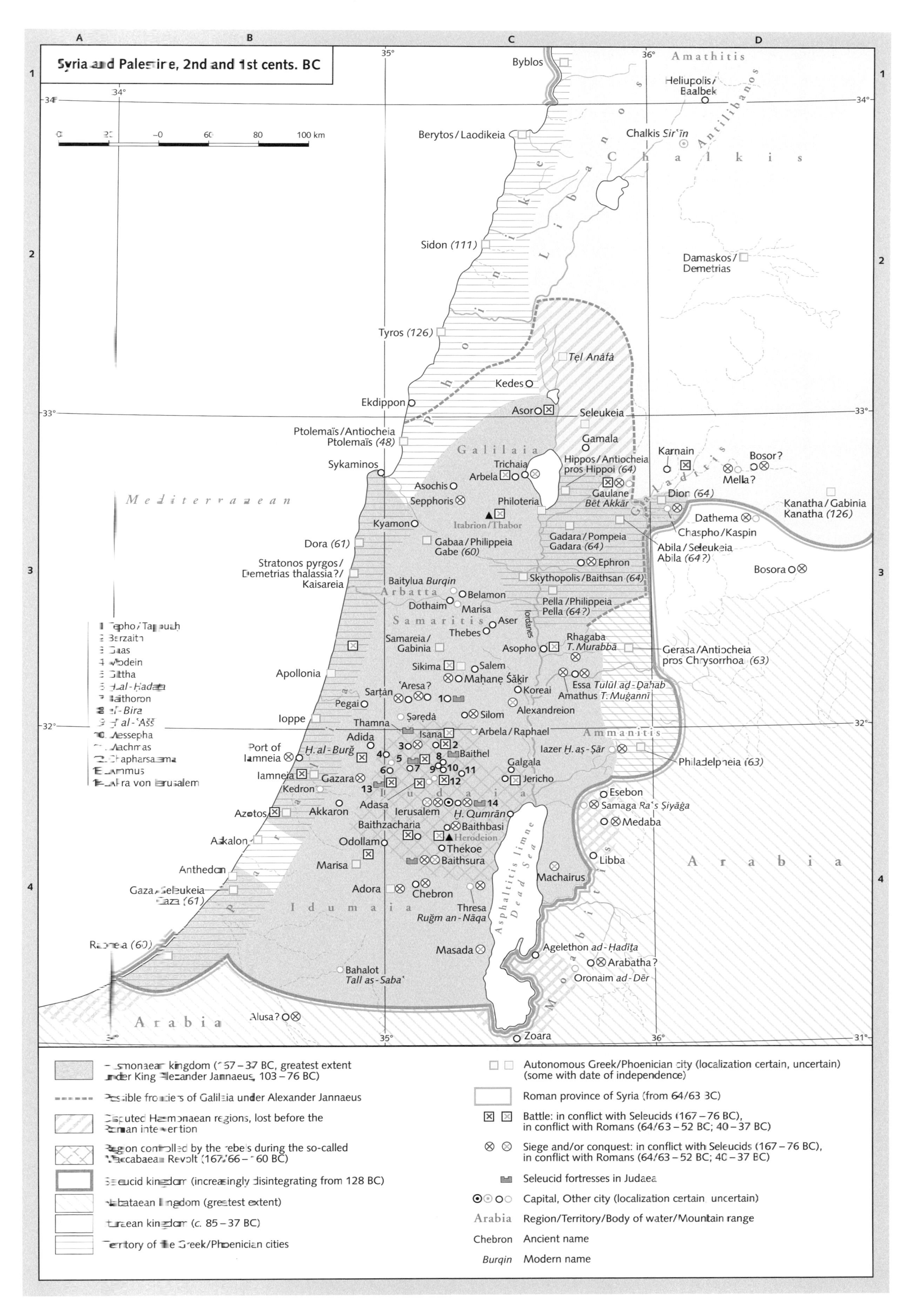
100 km
Byblos
Amathitis
Heliupolis / Baalbek
Antilibanos
Berytos / Laodikeia
Chalkis Sirʿīn
Chalkis
Libanos
Sidon (111)
Damaskos / Demetrias
Tyros (126)
Phoinike
Tel Anáfá
Kedes
Ekdippon
Asor
Seleukeia
Ptolemaïs / Antiocheia
Ptolemaïs (48)
Gamala
Galilaia
Karnain
Bosor?
Mella?
Sykaminos
Trichaia
Arbela
Hippos / Antiocheia pros Hippoi (64)
Asochis
Gaulane
Bet Akkār
Sepphoris
Philoteria
Dion (64)
Kanatha / Gabinia
Kanatha (126)
Mediterranean
Itabrion / Thabor
Dathema
Kyamon
Chaspho / Kaspin
Gabaa / Philippeia
Gabe (60)
Gadara / Pompeia
Gadara (64)
Abila / Seleukeia
Abila (64?)
Dora (61)
Ephron
Stratonos pyrgos / Demetrias thalassia? / Kaisareia
Baitylua Burqin
Skythopolis / Baithsan (64)
Bosora
Arbatta
Belamon
Dothaim
Marisa
Pella / Philippeia
Pella (64?)
Samaritis
Aser
Iordanes
Thebes
Rhagaba
T. Murabbā
Samareia / Gabinia
Asopho
Gerasa / Antiocheia pros Chrysorrhoa (63)
Apollonia
Sikima
Salem
Maḥanē Šākir
ʿAresa?
Sartān
Koreai
Essa Tulūl ad-Dahab
Amathus T. Muġannī
Pegai
Alexandreion
Ioppe
Ṣarēdā
Silom
Thamna
Isana
Arbela / Raphael
Ammanitis
Adida
Baithel
Iazer Ḫ. aṣ-Ṣār
Port of Iamneia
Philadelpheia (63)
Galgala
Iamneia
Gazara
Jericho
Kedron
Esebon
Adasa
Samaga Raʾs Šiyāġa
Azotos
Akkaron
Ierusalem
Ḫ. Qumrān
Medaba
Baithzacharia
Baithbasi
Askalon
Herodeion
Odollam
Thekoe
Libba
Arabia
Anthedon
Marisa
Baithsura
Machairus
Asphaltitis limne
Dead Sea
Adora
Chebron
Gaza / Seleukeia
Idumaia
Thresa
Ruğm an-Nāqa
Masada
Agelethon ad-Ḥadīṯa
Arabatha?
Moabitis
Oronaim ad-Dēr
Bahalot
Tall as-Saba'
Alusa?
Arabia
Zoara
Seleucid fortresses in Judaea
Capital, Other city (localization certain, uncertain)
Region/Territory/Body of water/Mountain range
Chebron
Ancient name
Burqin
Modern name
Autonomous Greek/Phoenician city (localization certain, uncertain) (some with date of independence)
Roman province of Syria (from 64/63 BC)
Battle: in conflict with Seleucids (167 – 76 BC), in conflict with Romans (64/63 – 52 BC; 40 – 37 BC)
Siege and/or conquest: in conflict with Seleucids (167 – 76 BC), in conflict with Romans (64/63 – 52 BC; 40 – 37 BC)

The Hellenistic kingdoms of Indo-Bactria in the 2nd and 1st cents. BC

The region of Bactria in north-eastern Iran, an old cultural centre of Iranian population with an urban culture that developed at an early stage, was richly endowed by nature (grain, livestock, resources, e.g. gold). Its capital, Bactra, lay on what later became the Silk Road. Another long-distance trade route followed the course of the Oxus and its tributary the Bactrus west to the Caspian Sea. To the south, yet another route — which, despite the difficult passes, had been in use since the 4th millennium BC — linked the region with the Indian subcontinent, following the river Cophen through the territory of Gandaritis and crossing the strategically important Khyber Pass. A period of two centuries of Greek rule and cultural influence (the so-called 'Graeco-Bactrian' style in art) began — sporadically at first — under the Achaemenids, consolidating with Alexander the Great. This development was not confined to Bactria, but also took hold in the neighbouring Indian regions of Alexander's empire.

Process of Hellenization and political history

After the two-year-long campaign to subject the territory in the 4th cent. BC, a number of settlements (e.g. Alexandria Oxiana and Alexandria of the Caucasus) and fortresses were founded, some for Alexander's veterans, and these established themselves alongside the old urban cultures. The Iranian population mixed with the Greek, and more Macedonian and Thracian settlers arrived in waves. Hellenization gained momentum under the Seleucids, when the territory became a satrapy and more Greek settlements sprang up. Seleucus I secured (around 300 BC), and Antiochus I reasserted and expanded, Seleucid possession of Bactria by treaties with Indian princes, e.g. with the Maurya king Androcottus/Chandragupta (318–294).

Around the mid 3rd cent., the satraps Diodotus I (250–239/8) and II (239/8–230), who were pressing for independence from the Seleucid kingdom, established an independent Graeco-Bactrian kingdom. An attempted reconquest by the Seleucids in 206 was ended by a treaty between Antiochus III and King Euthydemus I (230–200).

He and his son Demetrius (200–190), and later Menander I (155–130), conquered wide tracts of north-western India, creating the 'Graeco-Bactrian kingdom' (so-called Indo-Greeks), which expanded into the Indian subcontinent after 180 BC. There, in honour of his father, Demetrius founded the city of Euthydemia. The realm is said to have extended as far as the territory of the Seres ('Silk People', Str. 11,11.1) around 170 BC; their location has not been securely established.

The kingdom passed to Eucratides I (170–145 BC), who had incited an uprising in Bactria. After his murder by his son (? Eucratides II, 145–140), two short-lived dynasties ensued: that of the Eucratids in the north-west (kings attested until 130 BC), Bactria proper, and that of the Euthydemids in the south-east of the Paropamisus/Indian Caucasus /Hindu Kush (kings attested until 130 BC).

Probably around the mid 2nd cent., shortly before (around 160) or, more probably, after (around 140) — the various dates are still hotly discussed among scholars — the region was conquered by the nomadic Sacae, who had been driven from the Aral Sea region and the Iaxartes/Syr Darya valley by the Central Asian Indo-Scythians (called Yüe-chi in Chinese sources). In the north east, though, the fragmentary kingdoms of what became the Indo-Greeks were able to survive until the 1st cent. BC, having conquered south-eastern Afghanistan (Paropamisus and Arachosia) and north-western India (modern Pakistan) in the early 2nd cent. Thus, kings are attested for Paropamisus and Arachosia from 155–70 BC, for Gandhara/Gandaritis (Menander around 100 BC) and the Punjab from 155–110 BC. Around the same time (after 160), the west was also subjected to Parthian conquests under Mithridates I. The Yüe-chi/Indo-Scythians migrated farther west in the 2nd cent. BC, conquering parts of Graeco-Bactria and later moving on to India ('Indo-Scythia' in Ptolemy *et al.*).

The Iranians displaced the Indo-Greeks from Indo-Bactria in the 1st cent. BC. Sources indicate Sacaean or Indo-Scythian/Yüe-chi rulers between *c.* 85 BC and AD 20, and probably a Parthian dynasty from 50 BC.

The sources

The reconstruction of the history of the Hellenistic kings after Alexander and their Iranian successors presents great difficulties, as the few, sparse sources only cover the first hundred years and most kings are attested only by coin finds, esp. their (Middle) Indian names.

Thus the dates are mostly hypothetical, and there is uncertainty as to the extent of territories and the succession and numbers of homonymous rulers. Other than the testimonies from Chinese historiography (the most important primary source here, esp. for the end of Graeco-Bactrian rule, is the Chinese dynastic history (Chapter 123), the Shiji of Sima Qian), only secondary Greek sources are available for Bactria in the period from Alexander the Great, firstly Arrian's account of Alexander the Great (Arr. Anab.) and that of Curtius Rufus, which derive from Aristobulus, Ptolemy and Cleitarchus.

For coherent accounts, we depend on Strabo (11,11) and Ptolemy (6,11 N), and there are sporadic reports in Aelian, Aeschylus, Aristotle, Diodorus Siculus, Pliny (Plin. HN), Polybius, Theophrastus and Xenophon (Xen. Cyr.).

The literary record declines further until the reign of Eucratides I, and for the subsequent period we rely entirely on archaeological sources, e.g. the excavations at Ai Khanum (perhaps Alexandria Oxiana) on the Oxus in the north of present-day Afghanistan. They revealed a Greek *polis* with acropolis, temples, gymnasium and theatre. The inscriptions include Greek monumental and funerary inscriptions, and many *ostraka* (economic content), the remains of two literary papyri and Hellenistic, Iranian, Indian and Indo-Greek coins were also unearthed. The town was destroyed in the mid 2nd cent. There are only sparse remains of Bactra, the capital. Remains of a temple of the Dioscuri were found at Dilberjin. Find sites in the north (Uzbekistan and Tajikistan), such as Termez (Demetrias) and Dalverzin Tepe, mostly date from the Indo-Scythian period.

Scholars have supposed that the famous Oxus Treasure came from the temple at Taḫt-i Sangīn (some way to the east of Demetrias/Termez, on a northern tributary of the Oxus) — the merchants who acquired it were less specific (from the Oxus region). The Treasure is a hoard of approx. 1,500 coins (Greek imports of the Achaemenid period and copies, coins from the Hellenistic period), gold and silver work (e.g. statuettes, bracelets, parts of vessels, ornamented gold plaques, animal figures, jewellery in Assyrian, Greek and esp. Sacaean style) and some cylinder seals and gems.

There are Hellenistic wall remains at Afrasiab/Marakanda (modern Samarkand) in Sogdiana. Greek inscriptions have been found esp. in Ai Khanum, elsewhere only occasionally (e.g. Juga Tepe near Dilberjin). The numismatic sources have been considerably enlarged by substantial hoard finds (e.g. Qunduz).

All results and finds shed light on the art, some Greek (coin minting with purely Hellenistic traits) and some indigenous Iranian (temples) in character, but often mixed (Oxus river cult). Most examples come from Ai Khanum. The Gandaritis, a flourishing Buddhist cultural landscape under the Graeco-Bactrians and the Kushans (1st–4th cents. AD), was probably the cradle of the so-called 'Gandhara art' of the 2nd and 3rd cents. AD, in which Graeco-Roman and nomadic art blended with Indian content and traditions, forming the famous hybrid Hellenistic-Buddhist style.

→ Maps pp. 113, 117, 129, 205, 215; Graeco-Bactria, BNP 5, 2004; Bactria, BNP 2, 2003

Literature

M. Alram, Die Geschichte Ostirans von den Griechenkönigen in Baktrien und Indien bis zu den iranischen Hunnen (250 v. Chr.–700 n. Chr.), in: W. Seipel (ed.), Weihrauch und Seide. Alte Kulturen an der Seidenstraße, 1996, 119–140; P. Bernard, Fouilles d'Ai Khanoum 1–8, 1973–1992; W. Eder, J. Renger (ed.), Herrscherchronologien der antiken Welt, 2004, 127–132; A.K. Narain, The Greeks of Bactria and India, in: CAH 8, 1989, 388–421; I.R. Picikjan, Oxos-Schatz und Oxos-Tempel. Achämenidische Kunst in Mittelasien, 1992; W. Posch, Baktrien zwischen Griechen und Kuschan. Untersuchungen zu kulturellen und historischen Problemen einer Übergangsphase, 1995; L. Renou (ed.), La géographie de Ptolémée. L'Inde (vol. 7, 1–4), 1925; S. Sherwin-White, A. Kuhrt, From Samarkand to Sardis, 1993; W.W. Tarn, The Greeks in Bactria and India, [3]1997.

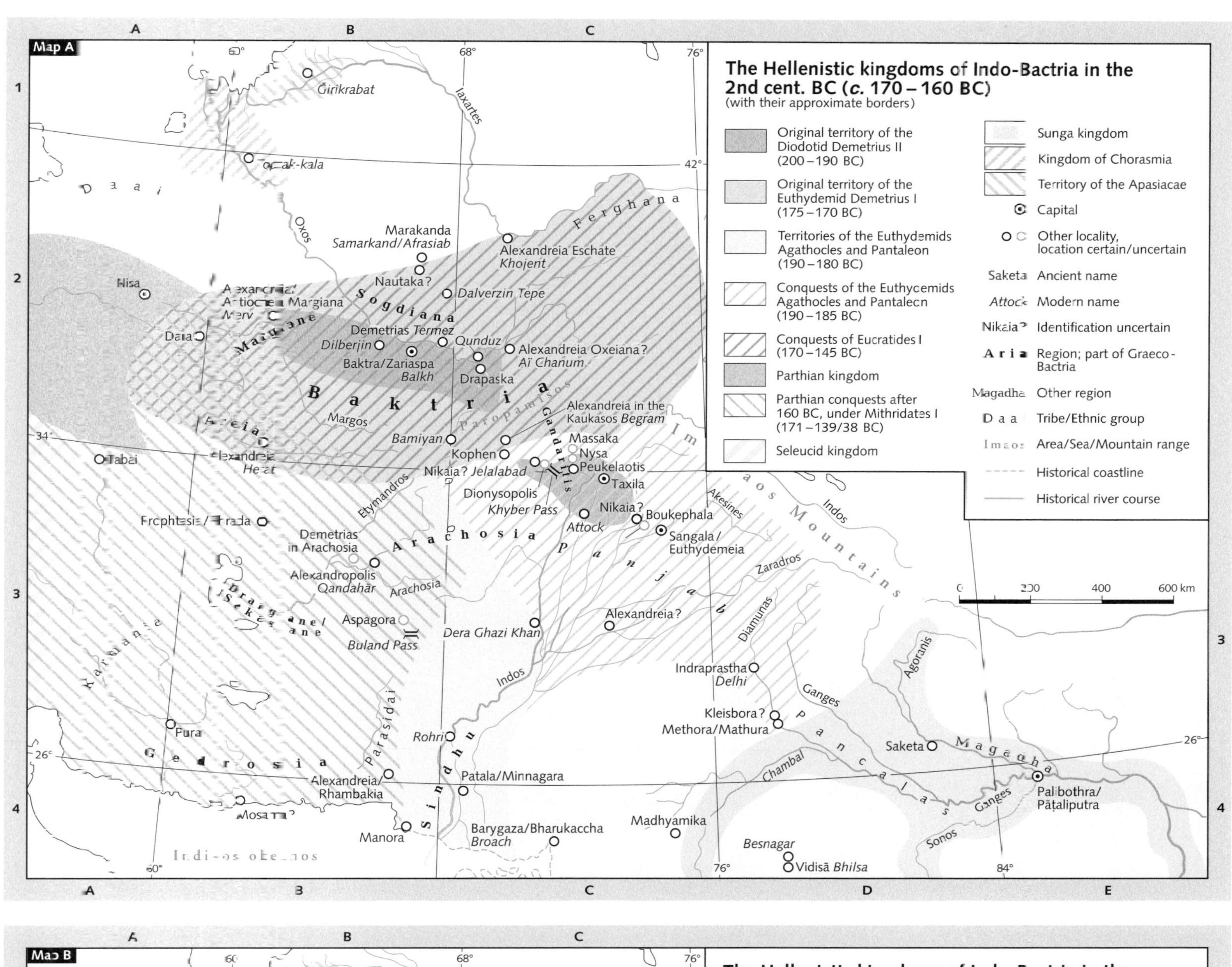
Map A
The Hellenistic kingdoms of Indo-Bactria in the 2nd cent. BC (c. 170 – 160 BC)
(with their approximate borders)
Original territory of the Diodotid Demetrius II (200 – 190 BC)
Original territory of the Euthydemid Demetrius I (175 – 170 BC)
Territories of the Euthydemids Agathocles and Pantaleon (190 – 180 BC)
Conquests of the Euthydemids Agathocles and Pantaleon (190 – 185 BC)
Conquests of Eucratides I (170 – 145 BC)
Parthian kingdom
Parthian conquests after 160 BC, under Mithridates I (171 – 139/38 BC)
Seleucid kingdom
Sunga kingdom
Kingdom of Chorasmia
Territory of the Apasiacae
Capital
Other locality, location certain/uncertain
Saketa Ancient name
Attock Modern name
Nikaia? Identification uncertain
Aria Region; part of Graeco-Bactria
Magadha Other region
Daai Tribe/Ethnic group
Imaos Area/Sea/Mountain range
Historical coastline
Historical river course
0 200 400 600 km
Girikrabat
Iaxartes
Marakanda
Samarkand/Afrasiab
Alexandreia Eschate
Khojent
Ferghana
Oxos
Nisa
Nautaka?
Dalverzin Tepe
Sogdiana
Margiana
Demetrias Termez
Dilberjin
Qunduz
Alexandreia Oxeiana?
Aï Chanum
Baktra/Zariaspa
Balkh
Drapaska
Baktria
Margos
Paropamisos
Alexandreia in the Kaukasos Begram
Bamiyan
Massaka
Nysa
Kophen
Peukelaotis
Taxila
Nikaia? Jelalabad
Dionysopolis
Khyber Pass
Nikaia?
Attock
Boukephala
Sangala/Euthydemeia
Imaos Mountains
Indos
Akesines
Zaradros
Tabai
Herat
Etymandros
Demetrias in Arachosia
Arachosia
Alexandropolis
Qandahar
Aspagora
Buland Pass
Dera Ghazi Khan
Alexandreia?
Diamunas
Indraprastha
Delhi
Agoranis
Ganges
Kleisbora?
Methora/Mathura
Pura
Gedrosia
Parasidai
Rohri
Sindhu
Saketa
Magadha
Chambal
Alexandreia/Rhambakia
Patala/Minnagara
Palibothra/Pāṭaliputra
Manora
Barygaza/Bharukaccha
Broach
Madhyamika
Besnagar
Vidisā Bhilsa
Sonos
Pañcāla

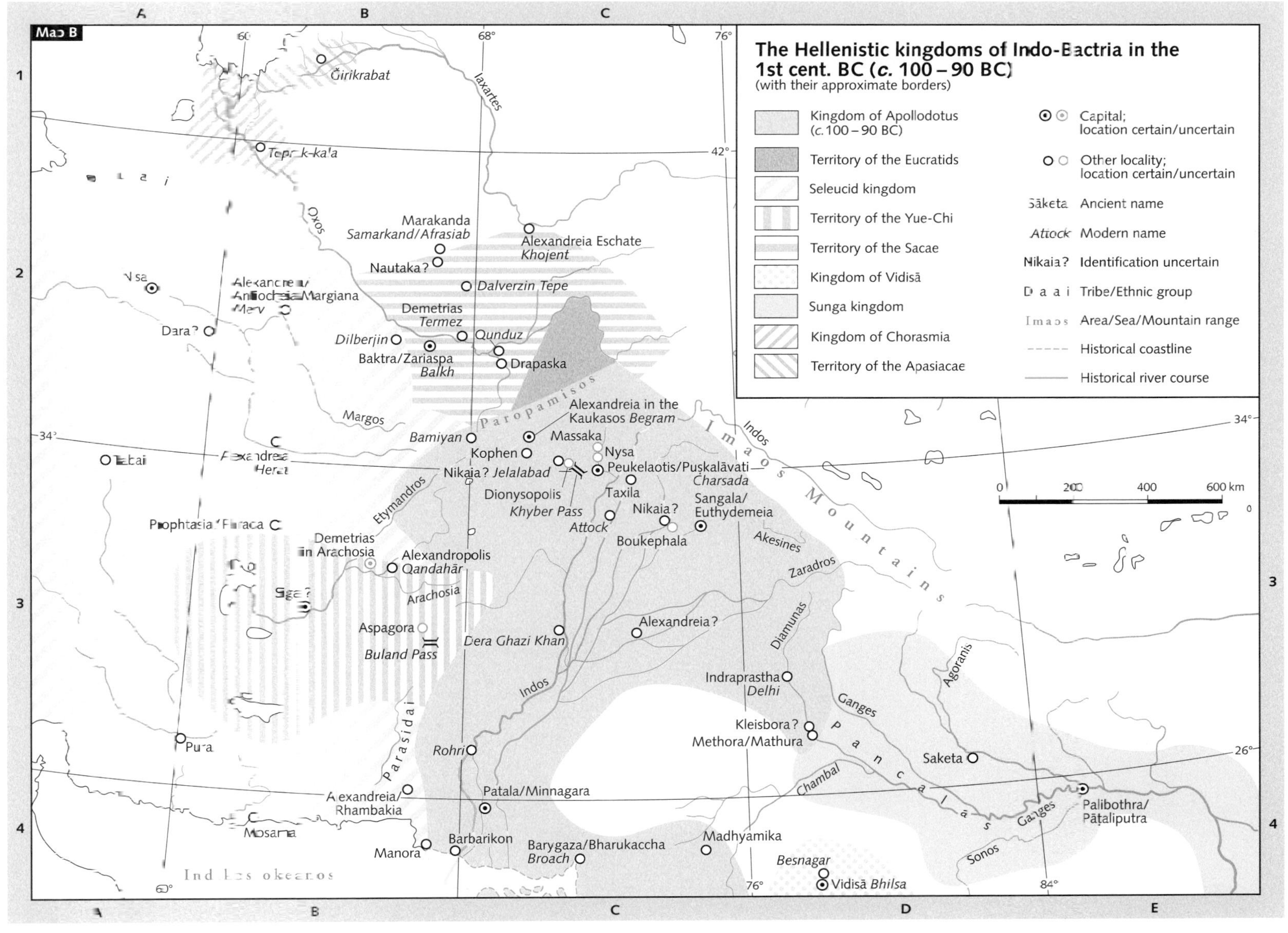
Map B
The Hellenistic kingdoms of Indo-Bactria in the 1st cent. BC (c. 100 – 90 BC)
(with their approximate borders)
Kingdom of Apollodotus (c. 100 – 90 BC)
Territory of the Eucratids
Seleucid kingdom
Territory of the Yue-Chi
Territory of the Sacae
Kingdom of Vidisā
Sunga kingdom
Kingdom of Chorasmia
Territory of the Apasiacae
Capital; location certain/uncertain
Other locality; location certain/uncertain
Sāketa Ancient name
Attock Modern name
Nikaia? Identification uncertain
Daai Tribe/Ethnic group
Imaos Area/Sea/Mountain range
Historical coastline
Historical river course
0 200 400 600 km
Girikrabat
Iaxartes
Marakanda
Samarkand/Afrasiab
Alexandreia Eschate
Khojent
Nautaka?
Dalverzin Tepe
Demetrias Termez
Dilberjin
Qunduz
Baktra/Zariaspa
Balkh
Drapaska
Margos
Paropamisos
Alexandreia in the Kaukasos Begram
Bamiyan
Massaka
Nysa
Kophen
Peukelaotis/Puskalāvati
Charsada
Nikaia? Jelalabad
Taxila
Dionysopolis
Khyber Pass
Nikaia?
Attock
Sangala/Euthydemeia
Boukephala
Imaos Mountains
Indos
Akesines
Zaradros
Etymandros
Demetrias in Arachosia
Alexandropolis
Qandahār
Arachosia
Aspagora
Buland Pass
Dera Ghazi Khan
Alexandreia?
Diamunas
Indraprastha
Delhi
Agoranis
Ganges
Kleisbora?
Methora/Mathura
Pura
Parasidai
Rohri
Saketa
Chambal
Alexandreia/Rhambakia
Patala/Minnagara
Palibothra/Pāṭaliputra
Manora
Barbarikon
Barygaza/Bharukaccha
Broach
Madhyamika
Besnagar
Vidisā Bhilsa
Sonos

Trading routes in the Hellenistic period (4th–1st cents. BC)

In spite of the many military conflicts, the Near East and the Mediterranean region formed an increasingly close-knit zone of economic and commercial activity in the Hellenistic period. Participants were the Hellenistic kingdoms of the eastern Mediterranean, the Carthaginian thalassocracy, the Greek cities and leagues and, from at least the 3rd cent., the ascendant power of Rome. This process of integration was assisted by the increasing establishment of the money economy. The most important commercial centres and harbours were Alexandria in Egypt, Seleucia on the Tigris, Athens, Miletus, Delos (a free port from 166 BC), Rhodes, Carthage, Cirta, Rome and Aquileia.

I. Foundations of the economy

The basis of the Hellenistic economy was agriculture. That we can speak of 'the Hellenistic economy' in spite of great regional disparities depends upon certain recognizable basic overall tendencies: a) interventions in the agricultural economy and land tenure on the basis of land ownership, b) the introduction of coinage and c) urbanization (Scheidel, Von Reden). The efficient administration of infrastructure conditions (irrigation, transport and canal systems, political unification, etc.), esp. in the Hellenistic kingdoms, facilitated improvements in productivity.

As well as 'internal trade' (including by navigable rivers), long-distance trade also intensified, as a result of 4th- and 3rd-cent. voyages of exploration (by water and land) and the deliberate creation of trading centres at domestic points of intersection and on coasts. Some of this trade followed routes already established, while some created new ones (Red Sea) to Arabia (Incense Road), India (sea routes), China (Silk Road), the African heartland and via central Europe as far as the British Isles.

II. Eastern Mediterranean: Hellenistic kingdoms and Greece

The kingdoms of the Diadochi and Epigoni implemented planned economic policies. This is particularly evident in Egypt, where the targeted elimination of corruption, economic waste and often chaotic private initiatives generated enormous wealth. Alexandria remained the greatest commercial centre of the 'known world' until the time of the Roman *princeps* Augustus.

The production of staples such as oil, salt, fish, beer, honey and dates, the manufacture of papyrus, textiles, glass and luxury articles, the transport system, banking and foreign trade were all the preserve of the Egyptian state. It protected its own economy with tolls of up to 50%, and achieved substantial foreign trade surpluses, not least by its expansion of eastward trade.

The introduction of modern cultivation methods made Egypt the bread-basket of the eastern Mediterranean, and the king received around a third of the proceeds. The minting of coins and banking as a whole were also in the hands of the state. State incomes comprised the revenues from the royal warehouses, revenues from crown lands, tolls and taxes collected by tax farmers.

The most important items in the state budget were court expenses, army and civil service pay and foreign expenses such as tributes. Tax fraud was punished with imprisonment or sale into slavery.

Private entrepreneurs had more scope in the s phere of trade. Commodities were agricultural products such as grain, oil, wine and salt, as well as slaves and raw materials of all kinds: coral, pearls, pigments, glass, incense, myrrh, timber, papyrus, rice (India), beer, silk, wool, cotton, linen, hemp, horses, elephants and ivory. Persistent price increases and social unrest are documented from the 2nd cent. BC.

The Ptolemaic kingdom was not alone in experiencing backlashes against Hellenization from its indigenous population, with demonstrative entrenchments of the native culture and language. The increasing power of the 'peripheral powers' (Rome, Parthia) contributed to a general economic crisis, which most seriously affected the Seleucids, who controlled the long-distance trade between Greece (good contacts with Miletus and Athens attested) and Asia Minor, south-west and central Asia (e.g. with Androcottus in India/Silk Road). Its sheer extent rendered the Seleucid kingdom vulnerable: its political stability was badly shaken by many military conflicts and its population was extremely diverse. As well as the Levantine harbour cities, its trading centres were Antioch on the Orontes and Seleucia on the Tigris. Among the factors promoting commerce was the profusion of Macedonian and Greek settlements in new and revived city foundations. River (Euphrates) and sea (Gulf of Persia – Dilmun/Bahrain – Indian Ocean) routes were also used. In Syrian territory, connections followed traditional routes (Palmyra).

Control of the Incense Road, and hence of the trade with Arabia, was sometimes disputed between the Ptolemies and the Nabataeans, but in the end it came to be dominated by Petra.

The Greek cities (not only of the so-called motherland) and Greek traders were integrated into the various trading networks of the Mediterranean region (including Italy and what would become the Roman territories of the western Mediterranean), and they had connections with those in central Europe (Celtic regions) and Asia (finds of stamped Greek transport amphorae; coin finds). They had a well-developed banking and monetary system (credit, loans, bottomries). The Aegean remained an important trans-shipment centre for the grain and slave trades (Black Sea region, eastern Mediterranean).

III. Western Mediterranean: Carthage and Rome

After the Italian wars of the 4th cent., the western Mediterranean region was characterized by the struggle for primacy and the definition of economic spheres of influence between the land power of Rome and the sea power of Carthage (Punic Wars). The purposes of these struggles were to acquire or conquer land and labour, to gain access to raw materials and to obtain markets for trade.

The Western Phoenicians pursued productive agriculture in Africa, controlling territories on the Iberian Peninsula (where they also had ore deposits at their disposal), Corsica (?), Sardinia and Sicily. Their zone of production yielded wheat, barley, figs, pomegranates, nuts, dates, olives, wine, cabbage, garlic, leeks, peas, sheep, goats, cattle and pigs, and manufactured pottery, textile and leather products, carpets, gypsum, pigments and jewellery. But they were not only producers, indeed their primary role was as traders (trading colonies) who used shrewdly-designed treaties (e.g. between Carthage and Rome) to secure themselves economic influence. The trade empire of the Western Phoenicians (sc. Carthaginians) stretched from the African heartland through the entire Mediterranean to distant Britain ('Tin Route'). The Romans, too, political opponents though they were, imported Western Phoenician commodities and adopted technological achievements, e.g. 'threshing-machines' and so-called 'Punic windows'.

Agriculture was the linchpin of the Italian economy until the 3rd cent. BC. Only then did coinage begin to take hold, and contacts with the Hellenistic kingdoms furthered the commercial transformation of economic structures. Firstly, large-scale land ownership developed, with the use of *ager publicus* acquired during the war with Hannibal. This triggered the development of a market-orientated estate economy in central Italy in particular, helped by the renovation/creation of an Italian road network. Secondly, the period of the Punic Wars saw Rome become a maritime power as well. Traders followed in the wake of the legions. They first went to Africa, but from the 3rd cent. there is also evidence of Roman and Italic merchants among the camp followers on Roman campaigns in the Adriatic (Illyria) and Aegean (Macedonia) regions. In the latter, from 166 BC Romans made Delos the most important commercial centre of the eastern Mediterranean. As well as agricultural produce, luxury goods were increasingly imported, while exports included wine, of which there is especially good evidence in Gaul. The slave trade was of great economic importance in the 2nd and 1st cents. Slaves came to Italy from Gaul and the eastern Mediterranean (Str. 14,5,2). With the destruction of Carthage in 146 BC, Rome became the leading commercial power in the western Mediterranean. The agrarian crisis (Gracchi) of the last third of the 2nd cent., which was triggered by increasing large-scale land ownership, the limitless availability of slave labour, a stagnation of colonization activity, the expansion of the urban working class and heavy losses from many wars, did nothing to hamper this development.

Sources

The state of the sources varies from region to region, and generalizations or inferences for elsewhere based on regional sources are impossible. In places, we have information in the minutest detail (for 3rd-cent. Faiyum from the 'archive' of Zeno), and in others, the sources are most imprecise and fragmentary. The economy of the Ptolemaic kingdom is relatively well documented thanks to relevant papyrus finds, and it has been much studied by scholars. The situation is far less favourable for the Seleucid kingdom – scholars even dispute whether the *Oeconomica* of Pseudo-Aristotle can be used at all as a basis for interpretation of the Seleucid economy, not least because of problems concerning its date.

There is no contemporary history of early Rome before the Punic Wars, and as a whole the state of the sources is poor, in spite of the accumulating epigraphic and numismatic evidence. The Elder Cato's (234-149 BC) treatise *De agricultura* (On Agriculture) affords a glimpse of agrarian conditions and some background for assessment of the Gracchan agrarian legislation. There are references in Diodorus Siculus, Appian and Plutarch to slave revolts in the context of the *latifundium* economy. For Carthage, there is only indirect transmission, and for the Western Phoenicians across their entire range of settlement there is an almost complete lack of written sources, and hardly a plethora of material ones (excavations, pottery finds).

→ Maps pp. 8f., 81, 83, 85, 115, 119, 121, 127, 139, 201, 203, 205

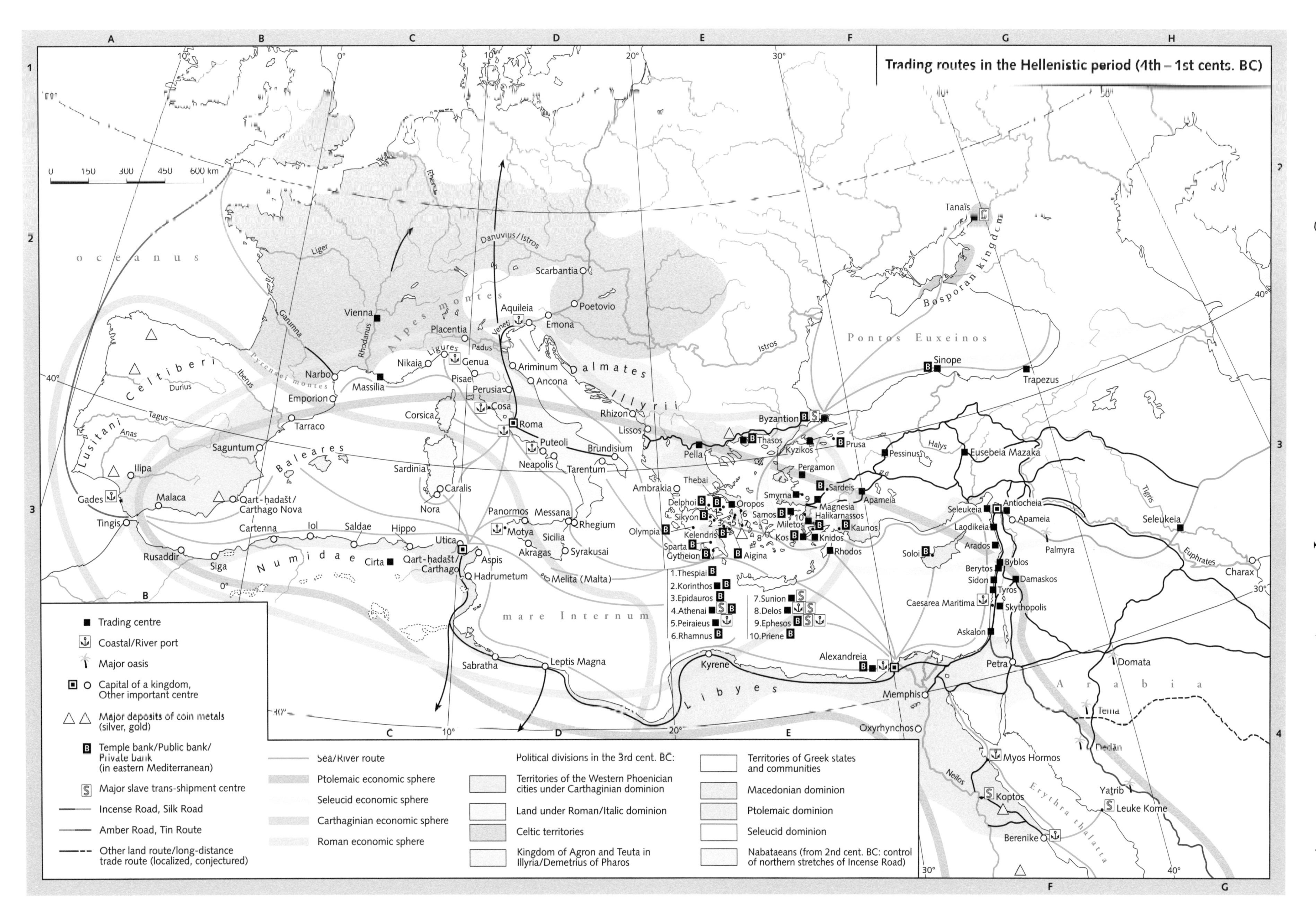
Trading routes in the Hellenistic period (4th – 1st cents. BC)
Trading centre
Coastal/River port
Major oasis
Capital of a kingdom, Other important centre
Major deposits of coin metals (silver, gold)
Temple bank/Public bank/Private bank (in eastern Mediterranean)
Major slave trans-shipment centre
Incense Road, Silk Road
Amber Road, Tin Route
Other land route/long-distance trade route (localized, conjectured)
Sea/River route
Ptolemaic economic sphere
Seleucid economic sphere
Carthaginian economic sphere
Roman economic sphere
Political divisions in the 3rd cent. BC:
Territories of the Western Phoenician cities under Carthaginian dominion
Land under Roman/Italic dominion
Celtic territories
Kingdom of Agron and Teuta in Illyria/Demetrius of Pharos
Territories of Greek states and communities
Macedonian dominion
Ptolemaic dominion
Seleucid dominion
Nabataeans (from 2nd cent. BC: control of northern stretches of Incense Road)
1.Thespiai
2.Korinthos
3.Epidauros
4.Athenai
5.Peiraieus
6.Rhamnus
7.Sunion
8.Delos
9.Ephesos
10.Priene
0 150 300 450 600 km

Italy under Roman rule: the Social system (338–89/82 BC)

I.

The Roman Republic's system of organizing its Italian *socii* ('allies') from 338 BC (with the end of the Latin War) until the so-called 'Social War' (also *bellum sociale*, *bellum Marsicum*, waged by the *socii* against Rome from 91-89 BC to win Roman citizenship) is generally referred to by scholars by use of the term 'confederation'. The ancients had no term for it, using only circumlocutions (e.g. *cives Romani, socii et nomen Latinum*). This 'system' is to be understood in the context of Rome's gradual subjection of Italy. The main source for the conflicts associated with this process is Livy, who describes the course of events and battles in great detail, though not always reliably, from the Roman perspective.

The geographical range of this arrangement, which was formerly also called the 'Italic League' (Beloch), covered all territories of the *socii* in Italy, including those to the north of the Arnus-Rubicon frontier, i.e. also Ligurian and Gaulish tribes (after the 2nd Punic War), until this region was added to the province of Gallia cisalpina. The islands of Sardinia and Corsica did not belong to it; neither did Sicily, with the exception of the city of Messana. These became Roman provinces in 227. The main purpose of the system of allies was to entitle Rome to access to the allies' military resources – a facility which greatly increased military capabilities – and to control their foreign policies. The sole joint institution was the pan-Italian army under Roman supreme command. The alliance system ended when the Roman Republic came to consist only of the city of Rome, the *ager Romanus* and Roman *municipiae* and *coloniae* (in the 80s BC extent up to the Padus, from 49 to the Alps).

II. Scholarly explanatory models

J. Beloch (1880) imputed to Rome a policy of conscious Romanization, a policy T. Hantos (1983) argued to be nowhere verifiable. Rather, Hantos' model is based on the assumption that the Romanization of Italy was actually an unwelcome development as far as Rome was concerned.

Both models attribute to the Romans deliberate action and an understanding of what was taking place in terms of the conquest of Italy. As more recent research, e.g. on the Etruscan cities, has shown (→ Map p. 111), there is no evidence of systematic 'Romanization' in the sense of a concept that Italy was being somehow incorporated into the Roman regime, irrespective of whether such a concept might have been seen in a positive or negative light. Apart from anything else, the institutional conditions were absent. The process by which Rome became the dominant power in Italy, with the consequent fusion of Italy into one entity, was characterized by quite different Roman actions in individual situations, which were invariably dealt with pragmatically and situatively (cf. the 'patchwork quilt' on the map). At most, certain recurrent factors can be gleaned. The arrangements in the wake of the Latin War of 338 already show the full spectrum of ways in which conquered or self-affiliating communities might be integrated into the *civitas Romana* or tied to Rome:

- Persons, communities, regions were conquered and annexed, land and populace simply added to the *civitas*.
- Cities (e.g. Capua) were incorporated into the zone of Roman citizenship as *municipiae* while retaining their urban autonomy.
- From 338, Rome used the old institution of the 'colonies of the Latin League' (*colonia civium Latinorum*): this created legally independent, but politically clearly Roman fortresses, mostly on the coast or at strategically important inland locations, which by virtue of their sovereignty (citizenship, administration) were in a position to act instantly when needed.
- Some old treaty stipulations were renewed and updated (cession of territory for the settlement of Roman citizens).
- Union, coerced or (mostly) peaceful, by means of bilateral *foedera* of a defensive nature, with cities, tribes and tribal polities as *socii*, who were not connected with one another.

For the period in question, Italy can be roughly divided into the territory of Roman citizenship and the territory of the *socii*. The former consisted of the city of Rome; the territory of the full Roman citizens (*ager Romanus*) and the territories added to the *ager Romanus* after the 2nd Punic War, i.e. the territories of the rural *tribus* (*tribus rustica*) under Roman administration; so-called 'citizens' towns' (*oppida*, later *municipia civium Romanorum*, e.g. Tusculum); then the Latin cities and towns incorporated in 338 BC; communities which had formerly possessed limited citizenship; citizens' colonies (*coloniae civium Romanorum*), and communities of 'half-citizens' of foreign language and culture, who kept wide-ranging internal autonomy and self-rule, but were obliged to contribute militarily (e.g. Caere, Capua).

The territory of the *socii* was composed of the *oppida* counted among the *nomen Latinum* (later *municipia Latinorum*), e.g. Tibur and the Latin colonies, and the other *socii*. They retained their sovereignty (citizenship, administration; social arrangements) and a treaty of confederation (*foedus*) was signed with them (e.g. Tibur, Praeneste), obliging them to provide troops, and leading to dependence on Rome in foreign policy. This arrangement held sway over by far the greater part of Italy.

III. History

The territory of the city of Rome was greatly increased by the conquest and destruction of Veii in 396 BC and the settlement of Roman citizens on the territory of that formerly Etruscan city-state. Subsequently, allies and even former enemies were also integrated at an early stage into the Roman community, e.g. by the award of a limited (half-) Roman citizenship (*civitates sine suffragio*, e.g. Tusculum 383, Caere 353 or 351?), and there were many subsequent military confrontations, such as the Latin War (340-338). At the end of that war, Rome gave some Latin cities Roman citizenship and bound the communities that remained independent with bilateral treaties as *socii*. Not only these, but also cities of Campania, e.g. the Greek Cyme/Cumae, were incorporated as *civitates sine suffragio* (Liv. 8,14). With hindsight, it can be seen that all these steps led to structural change at Rome: where before the Latin War, Rome had been a city-state with a surrounding territory and few other associated communities, it now reached as far as the Bay of Naples, and its sphere of influence was extensive. Rome had become the leading political power in Italy.

All wars of the subsequent period (326–272) directly or indirectly served the sustenance and extension of the hegemony of the Roman Republic, a process underpinned by the accompanying foundation of new colonies. This phase included the Samnite Wars of 326–290, to which scholars also add the ensuing conflicts with the Etruscans and Gauls in 285-280 and with Taras/Tarentum, southern Italian tribes and Pyrrhus from 282–272, because Samnites and other Italian tribes also took part. Rome won more territory through military action (battles; capture of Bovianum, one of the main Samnite centres), changes to battle tactics and armament, the exhaustion of diplomatic means (conclusion of treaties e.g. with Apulian tribes behind the Samnite League, peace treaties e.g. with Etruscan cities), road-building (e.g. via Appia: 264 to Brundisium) and the systematic construction of advance fortifications, the so-called *propugnacula*, as Latin colonies (e.g. Cales 334, Fregellae 328/316, Suessa Aurunca 313, Luceria 315/14?, Venusia 291, Beneventum 268), but esp. by the use of its allies: it reinforced its position in Campania, annexed the *ager Gallicus*, incorporated southern Italy (colony at Paestum 273). The Samnite League was dissolved and the remaining tribes concluded alliances with Rome. The Lucani (around Paestum), likewise Tarentum and all other cities of southern Italy signed treaties, the last being Rhegium in 270.

In these war years, Italy under Roman hegemony grew together into a political unity which brought social, economic, legal and cultural unity in its wake. External enemies were successfully held at bay. Exceptions were the intervention of Pyrrhus of Epirus and the 2nd Punic War, the Carthaginians under the command of Hannibal. And with the exception of the Social War (91–89 BC), general internal peace reigned in Italy.

The next period is characterized by Rome's expansion outside Italy (264–241: 1st Punic War, 237 annexation of Sardinia and Corsica, 227 Sicily, Sardinia and Corsica become Roman provinces; 226 Ebro Treaty: delineation of spheres of interest between Carthage and Rome; 229–228: 1st Illyrian War, etc.). In spite of some efforts by allies to secede, the system attained a particularly high effectiveness in the 2nd Punic War (218–201, → Map p. 139), and new territory was won and added to the *ager Romanus*. When that war ended, the Roman takeover of Italy resumed and continued, esp. through the accelerated expansion of the network of *viae publicae* – for military actions, but also for trade and communication, and esp. in northern Italy.

Thereafter, calls for the general granting of Roman citizenship became ever more vociferous. The Social War (91–89) united the Italian allies in a struggle to win the full citizenship Rome denied them. In the end, in 90 BC, Rome awarded Roman citizenship to its loyal allies by the *lex Iulia*. In 89, the *lex Plautia Papiria* extended Roman citizenship to all the Italic confederates. In the same year, the *lex Pompeia de Transpadanis* gave Latin citizenship to inhabitants north of the Padus.

The sources

Although there is no contemporary historiography for the 4th and 3rd cents. until the Punic Wars, and the account of this period, relying as it must on the annalists and later authors such as Livy (the main source), is not always reliable in detail, important laws, treaties and senatorial resolutions were recorded. Evidently the number of inscriptions on public buildings, monuments, sarcophagi, etc., had increased compared to earlier periods, so that secondary sources could base their work on such materials. This period also saw the beginnings of Roman coin minting. Polybius' History begins around 168 BC, and gives an account of the period 220-144 BC.

→ Maps pp. 107, 108, 109, 111, 141, 143, 151, 155, 157, 195

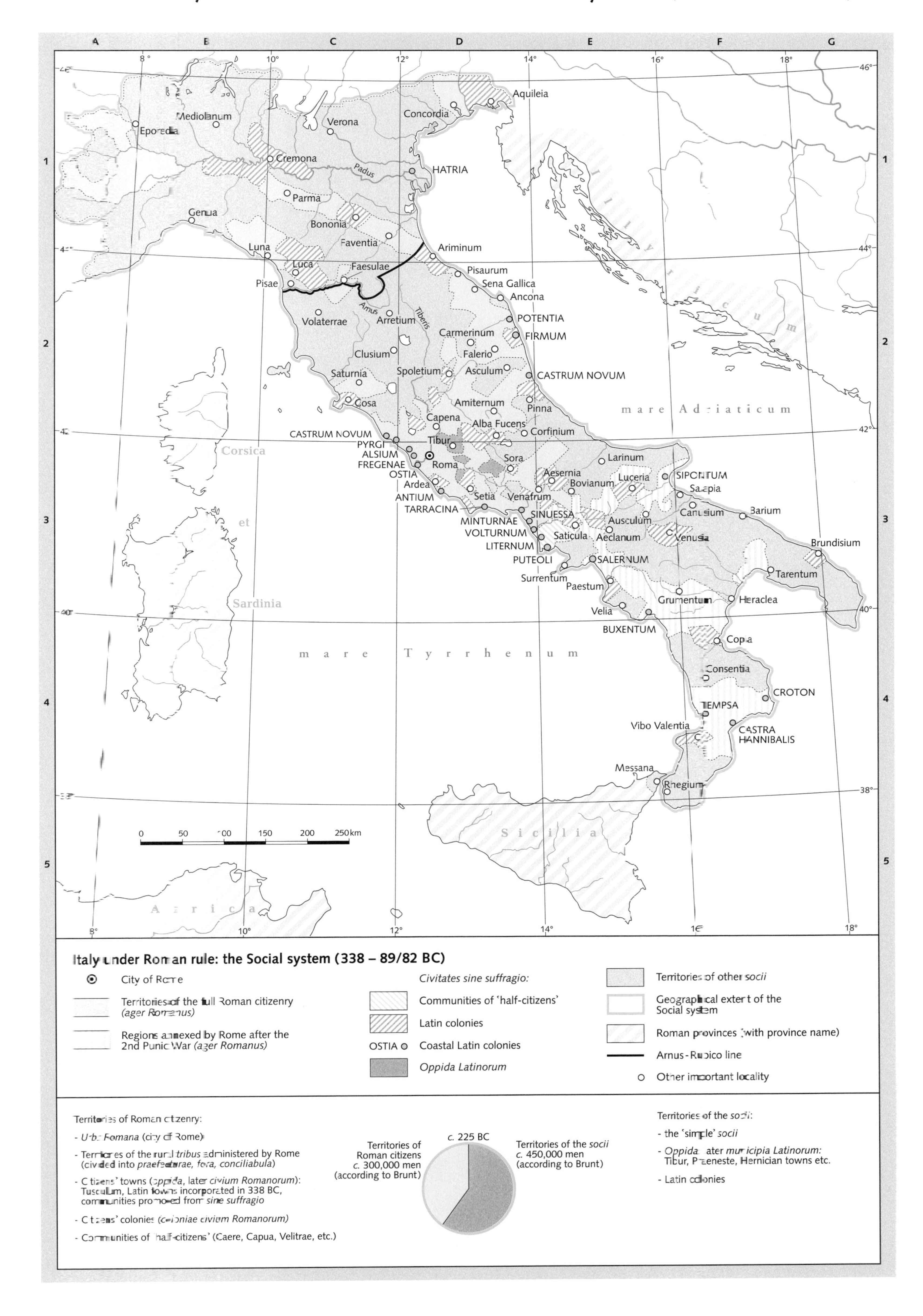
Aquileia
Concordia
Mediolanum
Eporedia
Verona
Cremona
Padus
HATRIA
Parma
Genua
Bononia
Faventia
Luna
Ariminum
Luca
Faesulae
Pisaurum
Pisae
Sena Gallica
Ancona
Arnus
Tiberis
Volaterrae
Arretium
POTENTIA
Carmerinum
FIRMUM
Falerio
Clusium
Saturnia
Spoletium
Asculum
CASTRUM NOVUM
Cosa
Amiternum
Pinna
mare Adriaticum
Capena
Alba Fucens
Corfinium
CASTRUM NOVUM
PYRGI
ALSIUM
FREGENAE
OSTIA
Tibur
Roma
Sora
Larinum
Corsica
Aesernia
Bovianum
Luceria
SIPONTUM
Salapia
Ardea
ANTIUM
Setia
Venafrum
TARRACINA
MINTURNAE
SINUESSA
Ausculum
Barium
VOLTURNUM
Saticula
Aeclanum
Venusia
Brundisium
LITERNUM
PUTEOLI
SALERNUM
Surrentum
Tarentum
Paestum
Sardinia
Heraclea
Velia
BUXENTUM
Copia
mare Tyrrhenum
Consentia
CROTON
TEMPSA
Vibo Valentia
CASTRA HANNIBALIS
Messana
Rhegium
Sicilia
0 50 100 150 200 250 km
8° 10° 12° 14° 16° 18°
46° 44° 42° 40° 38°
A B C D E F G
1 2 3 4 5
Italy under Roman rule: the Social system (338 – 89/82 BC)
City of Rome
Territories of the full Roman citizenry (ager Romanus)
Regions annexed by Rome after the 2nd Punic War (ager Romanus)
Civitates sine suffragio:
Communities of 'half-citizens'
Latin colonies
OSTIA Coastal Latin colonies
Oppida Latinorum
Territories of other socii
Geographical extent of the Social system
Roman provinces (with province name)
Arnus-Rubico line
Other important locality
Territories of Roman citizenry:
- Urbs Romana (city of Rome)
- Territories of the rural tribus administered by Rome (divided into praefecturae, fora, conciliabula)
- Citizens' towns (oppida, later civium Romanorum): Tusculum, Latin towns incorporated in 338 BC, communities promoted from sine suffragio
- Citizens' colonies (coloniae civium Romanorum)
- Communities of 'half-citizens' (Caere, Capua, Velitrae, etc.)
c. 225 BC
Territories of Roman citizens c. 300,000 men (according to Brunt)
Territories of the socii c. 450,000 men (according to Brunt)
Territories of the socii:
- the 'simple' socii
- Oppida, later municipia Latinorum: Tibur, Praeneste, Hernician towns etc.
- Latin colonies

The First and Second Punic Wars

These maps deal with the 1st and 2nd Punic Wars, military conflicts in pursuit of primacy in the western Mediterranean fought between Rome and Carthage. Before 264 BC, the relationship between the two powers was one of peaceful coexistence, their respective sphere of influence precisely defined by several treaties (509 BC, several in the 4th cent. BC). However, the situation changed radically with a dispute over the city of Messana on Sicily, a situation which arose above all from Roman expansion within Italy. For the first time, Romans and Carthaginians faced each other across the straits and felt their respective territories to be threatened by the other.

Because of the destruction of Carthage in the 3rd Punic War (146 BC), there are no historic sources describing the wars and their background from a Carthaginian perspective, only works by Greek and Roman authors (e.g. Philinus of Acragas, Fabius Pictor, Polybius and Livy) and other source genres.

I. 1st Punic War (264–241 BC) (map inset)

The local conflict between Messana (Roman ally) and Hieron I of Syracuse led to a Roman intervention in Sicily. The victory over Hieron, the signing of a treaty (263) and the taking of Tauromenium (263) and Acragas (262) provoked Carthage to immediate hostilities on Sicily, accompanied by raids on the Italian coast (e.g. 261). Rome, hitherto a land power, defeated the maritime power of Carthage with the help of its Southern Italian allies and its construction of a naval fleet of its own: the first sea victory was at Mylae (260). The Romans took Corsica in 259, leading to Carthage's abandonment of other parts of Sardinia (258: Sulcis). There were simultaneous hostilities on and off the coast of Sicily (258: conquest of Camarina, Enna; 257: Tyndaris; 256: Cape Ecnomus). Rome rejected Carthaginian peace overtures.

Rome's attempt in 256/55 to take the war to Africa failed (sea and land battles in 256 at Aspis, Adys, Tunes; 255 at Kerkouane (and Cape Bon?) and Cossura). From 254, the Romans succeeded in conquering several cities of northern and southern Sicily while Carthage was occupied fighting the Numidians. Only returning to Sicily in 251, the Carthaginians' attack on Panormus was repelled by Rome in 250. The war thereafter focused on western Sicily, where Hamilcar Barca was able to repel all attacks (including on Lilybaeum, Drepanum and Eryx (249)) with the help of mercenaries. The war was decided in 241 with the victory over the Carthaginian fleet off the island of Aegusa (Battle of the Aegates Islands). If the Romans' original ambition had merely been to keep Carthage out of eastern Sicily (cf. Pol.), their war aim changed later, as the peace treaty (StV III 493) shows: Carthage had to relinquish all its Sicilian territory. Power relations in the western Mediterranean had fundamentally shifted. Rome now had supremacy, not least because she had at last understood the strategic importance of the sea.

After a mercenary revolt (241–238) involving Sardinian mercenaries, which the Carthaginians were able to put down only with great effort, Carthage lost Sardinia (StV III 497) and Corsica to Rome for good (237). Like Sicily (except for the Roman ally Syracuse, which remained independent), they now became Roman provinces (227).

Carthage tried to compensate for its losses by extending its rule on the Iberian Peninsula (238/37–219: conquest of Iberian tribal regions; foundation of Qart-ḥadašt/Carthago Nova in 227). In 216, Carthage and Rome, preoccupied with conflicts with Ligurians and Celts and with the 2nd Illyrian War, signed the so-called Ebro Treaty (StV III 503), defining the River Iberus/Ebro as the frontier of their respective spheres of influence. However, the Romans also signed a (probably) formal alliance with the coastal city of Saguntum, which was south of this frontier. After Hannibal had conquered regions of inner Iberia, he besieged Saguntum (in 219) and took it.

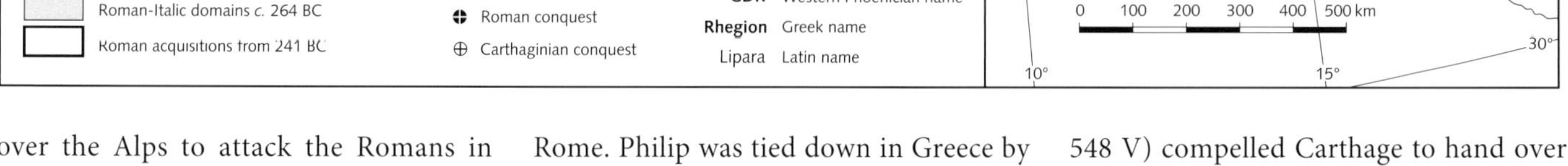

II. 2nd Punic War (218–201 BC) (main map)

It is not entirely clear whether the Romans declared war in the spring of 218 because of Hannibal's conquest of Saguntum or because of his breach of the Iberus frontier. Thereafter, Hannibal took his army over the Alps to attack the Romans in Italy, the Romans sent an army to Spain and gathered troops on Sicily to invade North Africa.

Late in 218 came the first Roman defeats, on the Ticinus and on the Trebia. Hannibal moved south in 217, and attempted to recruit Italic tribes to his cause. He defeated C. Flaminius in the Battle of Lake Trasimene (*lacus Trasumenus*). On 2 August 216, the Romans suffered a cataclysmic defeat at the Battle of Cannae. Hannibal's war aim was to reduce Rome to the status of a medium-sized power, as his treaty (StV III 528) of 215 with the Macedonian King Philip V shows. He also needed to destroy the powerful alliance system of Rome with its *socii*. Taras/Tarentum (212, reconquered 209), Capua (lost 211) and some other southern Italian cities defected to Carthage, as did Hieronymus of Syracus and the eastern Numidian King Massinissa. Nonetheless, Hannibal did not succeed in establishing a cohesive power bloc in southern Italy or in persuading enough *socii* to desert Rome. Philip was tied down in Greece by an alliance between Rome and the Aetolian League. Carthage's attempts to reconquer its former possessions in Sardinia and Sicily failed (212: Roman conquest of Syracuse). The Romans won military victories on the Iberian Peninsula, extending their influence southwards by 211. Although they were then defeated, they were not entirely driven out of Spain. In 209, Carthago Nova was conquered.

In spite of many skirmishes and sieges, Hannibal spent the ensuing years fruitlessly criss-crossing southern Italy. Rome won a victory on the Metaurus in 207 which prevented the joining of Hannibal's army and that of Hasdrubal approaching from the north. In 206, P. Cornelius Scipio won a decisive victory at Ilipa (the Iberian Peninsula was lost in 205). Massinissa seceded from Carthage. Scipio crossed to North Africa, forcing Hannibal to return there, and the Carthaginians were defeated in 202 at the decisive Battle of Zama, which ended the war. The peace treaty of 202/01 (StV III 548 V) compelled Carthage to hand over of its fleet and imposed upon it the loss of all its possessions outside North Africa, swingeing reparation payments, a ban on making war without Roman consent and the recognition of the kingdom of Massinissa. However, Carthage remained a trading centre with very substantial agricultural resources.

→ Maps pp. 71, 73, 93, 111, 137

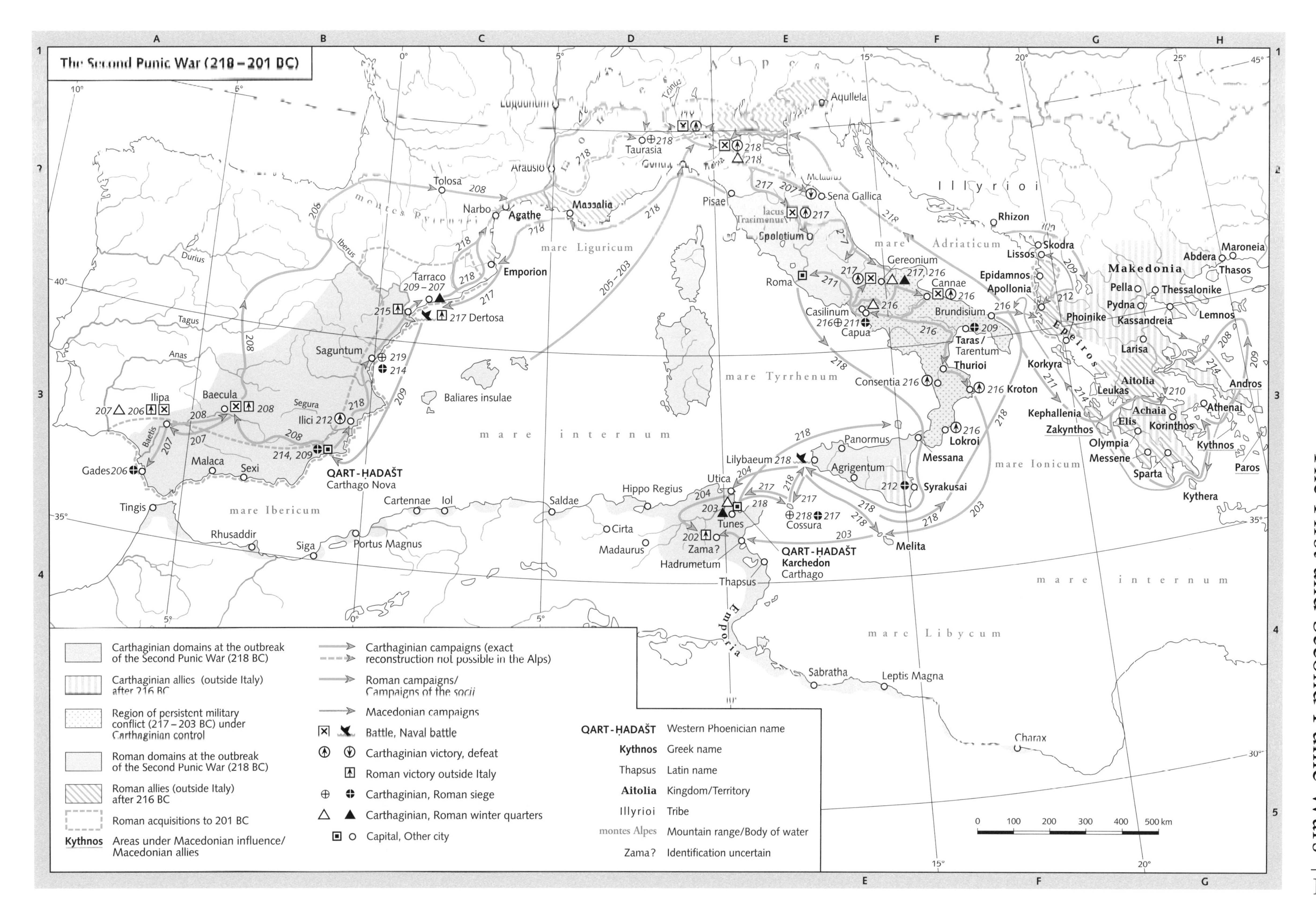

The Second Punic War (218 – 201 BC)
Carthaginian domains at the outbreak of the Second Punic War (218 BC)
Carthaginian allies (outside Italy) after 216 BC
Region of persistent military conflict (217 – 203 BC) under Carthaginian control
Roman domains at the outbreak of the Second Punic War (218 BC)
Roman allies (outside Italy) after 216 BC
Roman acquisitions to 201 BC
Kythnos Areas under Macedonian influence/ Macedonian allies
Carthaginian campaigns (exact reconstruction not possible in the Alps)
Roman campaigns/ Campaigns of the socii
Macedonian campaigns
Battle, Naval battle
Carthaginian victory, defeat
Roman victory outside Italy
Carthaginian, Roman siege
Carthaginian, Roman winter quarters
Capital, Other city
QART-ḤADAŠT Western Phoenician name
Kythnos Greek name
Thapsus Latin name
Aitolia Kingdom/Territory
Illyrioi Tribe
montes Alpes Mountain range/Body of water
Zama? Identification uncertain
0 100 200 300 400 500 km
Tolosa
Narbo
Agathe
Arausio
Massalia
Emporion
Taurasia
Tarraco 209 – 207
217 Dertosa
Saguntum
Iberus
Durius
Tagus
Anas
Baetis
Segura
Ilipa
Baecula
Ilici
QART-ḤADAŠT
Carthago Nova
Gades
Malaca
Sexi
Tingis
Rhusaddir
Siga
Portus Magnus
Cartennae
Iol
Saldae
Cirta
Madaurus
Hippo Regius
Utica
Tunes
Zama?
Hadrumetum
Thapsus
QART-ḤADAŠT
Karchedon
Carthago
Emporia
Sabratha
Leptis Magna
Charax
Baliares insulae
mare Liguricum
mare internum
mare Ibericum
mare Tyrrhenum
mare Adriaticum
mare Ionicum
mare Libycum
montes Pyrenaei
Aquileia
Pisae
Sena Gallica
lacus Trasimenus
Spoletium
Roma
Gereonium
Cannae
Casilinum
Capua
Brundisium
Taras / Tarentum
Thurioi
Consentia
Kroton
Lokroi
Lilybaeum
Panormus
Agrigentum
Messana
Syrakusai
Cossura
Melita
Illyrioi
Rhizon
Skodra
Lissos
Epidamnos
Apollonia
Phoinike
Epeiros
Korkyra
Kephallenia
Zakynthos
Leukas
Makedonia
Pella
Pydna
Thessalonike
Kassandreia
Larisa
Abdera
Maroneia
Thasos
Lemnos
Aitolia
Achaia
Elis
Olympia
Messene
Sparta
Korinthos
Athenai
Andros
Kythnos
Paros
Kythera

The territorial development of the Imperium Romanum in the Republican Period

The map shows the development of the Roman Empire in 19 chronological steps, marked in the legend and commentary with the years in which the expansions took place (→ table p. 265).

264: The Roman allies in Italy
A web of interrelations ('confederacy system') defined as the 'allies and all those called Latins' (*socii nominisque Latini*) and centred on Rome formed in Italy after the Latin War (340–338). It extended from Ligurian and Celtic tribes north of the Rubicon, south as far as Messana on Sicily. It was based solely on the Romans' power of disposal over military resources and their authority in matters of foreign policy, and it formed the basis for Roman rule in Italy. This 'confederation' had no federal institutions and its only external manifestation was in the form of the army under the command of Roman magistrates. The 'confederation' was dissolved when the Italian allies were granted equal citizenship status to Roman citizens following the Social War (91–87, → maps pp. 137, 157).

237: 1st Punic War (264–241 BC)
This war extended Roman rule outside Italy for the first time, to Sicily, Sardinia, Corsica and Africa. The peace concluded by Rome with Carthage in 241 expanded Rome's territory to incorporate two geographical entities, the administration of which necessitated the creation of two extra praetorships in 227. Thus emerged the provinces of Sicilia (which excluded the Kingdom of Syracuse) and Sardinia et Corsica in 241 and 238/37 respectively (→ map p. 138).

219: 1st and 2nd Illyrian Wars (232–229 and 219 BC)
The Illyrian Princess Teuta was held responsible for the piracy which increasingly interfered with Adriatic trade in the 3rd cent. She did not even shrink from attacking Illyrian coastal cities. The Senate resolved upon war, in which Teuta capitulated and abdicated. When her successor Demetrius of Pharus attempted in 220/19 to expand his territory beyond what was acceptable to Rome, the Romans intervened again, conquering Pharus in 219. The Senate now took over the cities of Corcyra, Issa, Pharos, Apollonia and Dyrrhachium as permanent protectorates.

206: 2nd Punic War (218–201 BC)
The treaties that ended this war described the sovereign territory of Carthage. The territories Carthage was compelled to cede were apparently self-explanatory, i.e. the lands formerly under Carthaginian rule in Spain. Two spheres of activity (*provinciae* – later Hispania citerior and Hispania ulterior) were annexed and administered here from 206 BC (→ map p. 139). The territory of the Kingdom of Syracuse also fell to Rome in 210, and was incorporated into the province of Sicilia.

168: 3rd Macedonian War (171–168 BC)
The Senate had claimed no territorial annexations after the 2nd Macedonian War (200–196). Its aim seems to have been solely the permanent exclusion of Macedonian influence from the Greek world. After the 3rd Macedonian War, the Macedonian monarchy was abolished, and four autonomous regions were created from the ancestral royal lands. The kingdom of Genthius, the Illyrian king allied to the Macedonians, was however incorporated into the Roman protectorate of Illyricum (→ map p. 151).

148: The Macedonian Revolt under Philip VI/Andriscus (150–148 BC)
Andriscus, supposedly the son of the last Macedonian king, Perseus, had himself crowned Philip VI at Pella in 150, and quickly won support. Only when a Roman propraetor was defeated and killed in battle against Philip VI did Rome deploy a substantial force against the king, whom it defeated at Pydna. The Senate now ordered the establishment of Macedonia as a province, with its administrative capital at Thessalonica (→ map p. 151).

146 (1): 3rd Punic War (149–146 BC)
In 150, when Carthage took action to protect its interests without permission from the Roman Senate following various frontier disputes with the Numidian king Massinissa, Rome declared war on the city. After a three-year siege, Carthage fell and was razed to the ground. The territory was now organized into the Roman province of Africa. The frontier with the Numidian kingdom formed the *fossa regia*. It is preserved in archaeological and epigraphic remains.

146 (2): The Achaean War (146 BC)
The usual frontier squabbles were the justification in 147 BC for a senatorial decree according to which the Achaean League was to be reduced to comprise essentially only the cities of the territory of Achaea. Achaean rejection of this imposition led to war. The Achaeans were defeated at Corinth in the autumn of 146. The peace imposed by the Senate incorporated Greece (Achaia) into the province of Macedonia.

129: The Roman war with Eumenes III/Aristonicus (133–129 BC)
The Attalid Aristonicus, whose regnal name was Eumenes III, claimed the throne of Pergamum contrary to the testament of his half-brother Attalus III (d. 133), who had bequeathed it to Rome. A Roman army arrived on the Ionian coast in 131. Eumenes did succeed in defeating the Romans at Leucae in 130, where the consul P. Licinius Crassus fell. But in that same year, the king was besieged and forced to capitulate at Stratonicea by the consul M. Perperna. Rome then set up the former Pergamene kingdom as the province of Asia (→ map p. 153).

123 (1): The war against the Salluvii, Vocontii, Allobroges and Arverni (123 BC)
In 125, the city of Massalia called for Roman help against its neighbouring Celtic tribes, the Salluvii and Vocontii, who were devastating the region. The Arverni and Allobroges joined the tribal forces. The Gaulish allies were defeated several times in the ensuing conflict. The region north of the *territorium* of Massalia as far as the *lacus Lemanus* (Lake Geneva), won in these campaigns, was organized into the province of Gallia transalpina.

123 (2): Conquest of the Balearic Islands (123/22 BC)
One of the consuls for 123, Q. Caecilius Metellus, conquered the Baleares at the Senate's request: most of the islands' inhabitants lived from piracy. The islands were added to the province of Hispania citerior.

75: Elimination of the scourge of piracy (102–75 BC)
With the elimination of the Seleucid kingdom and Rhodes as policing powers in the Aegean in 188 and 168 respectively, there was no longer any authority able to act against piracy in the region. The Senate gradually became aware of the problem, appointing a propraetor to counter piracy on the Cilician coast in 102. It took until 78–74 for this region to be fully pacified and for the piracy in the eastern Mediterranean, which was based here, to be stemmed. From this time on, Roman magistrates were sent yearly into this permanently secured province (→ map p. 183).

74/67: Establishment of the province of Creta et Cyrenae (74/67 BC)
Q. Caecilius Metellus conquered the island of Crete in 69–67, an island hitherto dominated by pirates. The Cyrenaica, bequeathed to Rome in 96 BC by the will of Ptolemy Apion, a son of Ptolemy VIII, and annexed as the province of Cyrenaica by the Senate in 74, was now united with Crete into a single province, Creta et Cyrenae (→ map p. 163).

64/63: 3rd Mithridatic War (74–63 BC)
With his victory over Mithridates, Pompey placed the entire east of the Empire on a more secure footing, both domestically and in terms of foreign relations. He created the new provinces of Bithynia et Pontus and Syria, and considerably expanded those of Asia and Cilicia. His creation of a cordon of client states in the region completed the consolidation of the Empire's position (→ maps pp. 159, 161).

58: Incorporation of Cyprus into the province of Cilicia (58 BC)
In 58 BC, M. Porcius Cato was charged with the annexation of Cyprus, ruled by Ptolemy, a son of Ptolemy IX. Ptolemy offered no resistance, but chose to take his own life, and Cato undertook the incorporation of the island as part of the province of Cilicia.

51: The conquest of Gaul (58–51 BC)
As proconsul of Gallia transalpina, Caesar conquered the entire mainland territory of Gaul in 58–51 BC. However, the political crisis at Rome, which culminated in the Civil War, prevented him from reorganising Gallia transalpina, a task only undertaken by Agrippa under Augustus (→ map p. 165).

46: Creation of the province of Africa nova and the military zone around Cirta (46 BC)
After his victory at Thapsus, Caesar organized the region of the emporia in the east and west of the *fossa regia* as the province of Africa nova. The mercenary leader P. Sittius, who had put his band of mercenaries at Caesar's disposal and had conquered Cirta, the residence of the Numidian king Juba, was rewarded by Caesar with the region around Cirta (→ map p. 147).

In 49 BC, Caesar besieged the city of Massalia, which had taken the Senate's side against Caesar at the outbreak of the Civil War. It lost substantial territory, but kept its autonomy.

33: Conquest of further territory in the Illyricum by the younger Caesar (35–33 BC)
The younger Caesar and future Augustus undertook three campaigns in Illyricum between late 35 and 33 BC. From the sea, he secured the upper Adriatic, fought the Iapodes and other Illyrian tribes and conquered many of their cities. He celebrated a triumph for these victories only in 29 BC.

30: Egyptian campaign of the younger Caesar (30 BC)
After his naval victory at Actium on 2 September 31, the younger Caesar took his army to Egypt. Alexandria fell on 1 August, Mark Antony and Cleopatra VII taking their own lives. The former Ptolemaic kingdom was now organized into the Roman province of Aegyptus, but was placed under the rule of a *praefectus* as a 'crown domain': he was directly answerable to the *princeps*.

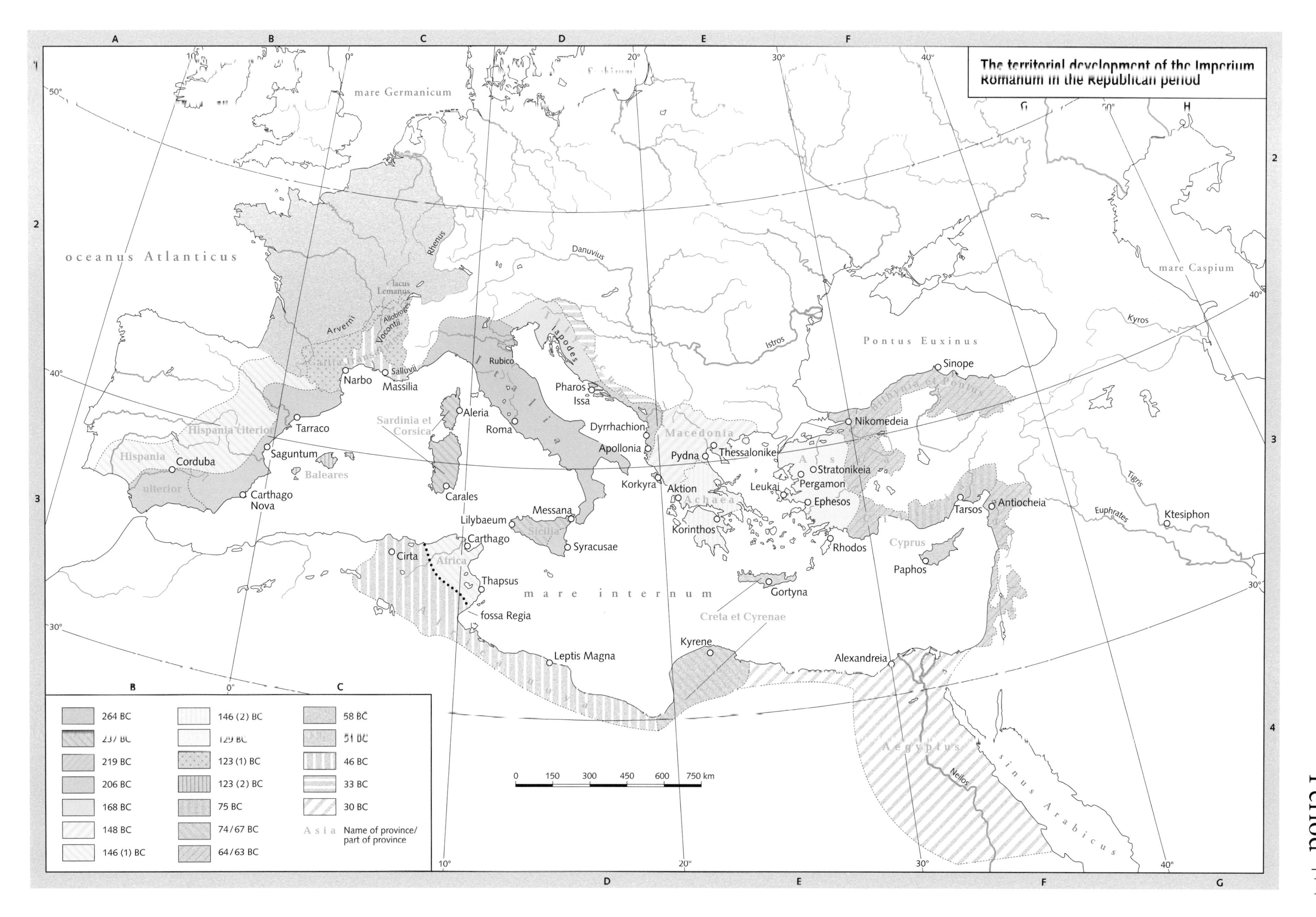
The territorial development of the Imperium Romanum in the Republican period
oceanus Atlanticus
mare Germanicum
mare Caspium
Pontus Euxinus
mare internum
sinus Arabicus
Hispania citerior
Hispania
ulterior
Baleares
Sardinia et Corsica
Italia
Sicilia
Macedonia
Achaea
Cyprus
Creta et Cyrenae
Africa
Aegyptus
Iapodes
Arverni
Vocontii
Allobroges
Salluvii
Lacus Lemanus
Rhenus
Danuvius
Istros
Kyros
Tigris
Euphrates
Nellos
Corduba
Carthago Nova
Saguntum
Tarraco
Narbo
Massilia
Aleria
Carales
Roma
Rubico
Pharos
Issa
Messana
Syracusae
Lilybaeum
Carthago
Thapsus
fossa Regia
Cirta
Leptis Magna
Dyrrhachion
Apollonia
Korkyra
Pydna
Thessalonike
Aktion
Korinthos
Leukai
Pergamon
Ephesos
Stratonikeia
Nikomedeia
Rhodos
Gortyna
Kyrene
Sinope
Tarsos
Antiocheia
Paphos
Alexandreia
Ktesiphon
0 150 300 450 600 750 km
264 BC
237 BC
219 BC
206 BC
168 BC
148 BC
146 (1) BC
146 (2) BC
129 BC
123 (1) BC
123 (2) BC
75 BC
74/67 BC
64/63 BC
58 BC
51 BC
46 BC
33 BC
30 BC
Asia Name of province/part of province

The development of the Roman provinces of Sardinia, Corsica and Sicilia (3rd cent. BC to 5th/6th cents. AD)

I. Development of the Roman provinces of Sardinia and Corsica (237 BC – AD 534)

Records from Sardinia are far more plentiful and informative. The smaller island of Corsica was always in the shadow of its neighbour. Sardinia was known for its fertility. With Africa, Sicily and Egypt, the island delivered the grain so important to the urban population of Rome. Sardinia was also of economic importance for its ore resources. Corsica, on the other hand, could offer only the products of its forests (timber, resin) and the livestock on its good pastureland (sheep, cattle, goats), especially in the wetter west of the island. Apiculture (honey, beeswax) also flourished on Corsica. In terms of infrastructure, both islands were of interest to the Etruscans, Carthaginians, Greeks and Romans in equal measure for their numerous natural harbours. We discover from (albeit uncertain) sources that the Romans had made an ultimately fruitless attempt in the early days of the Republic to establish a settlement on Corsica (Theophr. Hist. pl. 5,8,2; Diod. Sic. 15,25 for the year 386). This explains why, in the first treaty they signed with the Romans, the Carthaginians (508/07 BC) insisted on guaranteeing themselves access to both islands.

The Roman Senate seems only gradually to have understood the strategic importance of the islands after the 1st Punic War, laying claim to Sardinia and Corsica only in an addendum to the peace treaty of 241 BC. The islands were occupied after initially hard-fought battles with the remnants of the Carthaginian mercenary forces, and in 227 BC they were organized into the province of Sardinia et Corsica. The table on p. 266 makes clear the difficulty of the Roman position in the new province even after the departure of the Carthaginians. There were constant revolts, to which the lengthy catalogue of Roman generals' triumphs over the Sardinians and Corsicans bears eloquent testimony. The history of the provincial administration also attests to the presence on Rome's very doorstep of a permanent problem of public order. Augustus at first left the administration of the province in the hands of the Senate, then took the islands under his own direct rule as two separate provinces in AD 6. Subsequently, both islands were repeatedly transferred between senatorial and imperial administration.

At the time of the elder Pliny (1st cent. AD), the populace thus constantly rising against Roman rule (and it is open to question to what extent this was politically-motivated unrest and to what extent mere banditry) was living on Sardinia in various tribes (*populi*), 18 cities (*oppida*) and one Roman colony (Turris Libisonis/Porto Torres), and on Corsica in 32 communities (*civitates*) and two Roman colonies (Mariana/Golo and Aleria).

II. Development of the Roman province of Sicilia (241 BC – AD 474)

This fertile, volcanic island with plentiful water was one of Rome's most important suppliers of grain (wheat). But its abundant pastures were also significant (breeding of cattle and horses; skins, wool), as were viticulture and apiculture (honey, beeswax).

Before the 1st Punic War (264 241 BC), the Carthaginians held the west of the island (the former Phoenician colonies) as a protectorate, while Hiero II of Syracuse ruled the east. After 241, having expelled the Carthaginians, the Romans installed their first provincial administration, leaving the kingdom of Hiero untouched. But when Hiero's grandson and successor Hieronymus defected to the Carthaginian side in 215, the Romans no longer saw themselves bound by their treaty obligations with the kingdom of Syracuse, and after conquering the capital in 212, they added the kingdom to their province.

M. Antonius gave Roman citizenship to all freeborn Siculi, supposedly on the basis of a regulation planned by Caesar. However, the Senate cancelled this regulation; Sex. Pompeius then restored its legal force. The communities of the late Republic, as described by Cicero in his Verrine Orations, fell into four categories:

1. three *civitates foederatae*; these had the sole obligation of military service to Rome (Messana, Tauromenium, Netum);
2. five *civitates liberae atque immunes*; these had substantial autonomy and were not obliged to pay tributes (Centuripae, Halaesa, Segesta, Panormus, Halicyae);
3. 34 *civitates decumanae*; these were obliged to pay tributes to Rome (e.g. Catina and Leontini) as they had previously done to Hieron II;
4. 26 *civitates censoriae*; their land was *ager publicus* (e.g. Syracusae and Lilybaeum).

Under Augustus, Sicilia became a senatorial province under a proconsul with two quaestors. The elder Pliny (HN 3,88–93) describes (not entirely without contradiction) the political structure of the communities in this phase:

The province had 63 communities, including:

1. five Roman colonies (Syracusae, Catina, Thermae, Tyndaris, Tauromenium);
2. one *oppidum civium Romanorum* (Messana);
3. three communities under Latin law (Centuripae, Netum, Segesta);
4. 46 *civitates stipendiariae.*

The legal status of 13 *oppida* remains unknown.

Strabo (6,2 f.) was aware of changes already made to the Republican system by Augustus, but he is of no help in describing these without ambiguity.

Cf. also the table → p. 266

The sources

The six speeches of Cicero against Verres are a vast treasure-trove for the history of Sicily. Cicero's activities as quaestor at Lilybaeum in 75 BC and his fifty days of preparatory research for the trial of Verres provided him with a wealth of information. Strabo and Pliny the Elder also provide a wide range of information about the island.

Literature

Sardinia et Corsica: J. Jehasse, L. Jehasse, La Corse antique, 1993; A. Mastino, Rustica plebs id est pagi in provincia Sardinia. Il santuario rurale dei »pagani Uneritani« della Marmilla, in: S. Bianchetti (ed.), Poikilma. FS M.R. Cataudella, 2001, 781–807; P. Meloni, La Sardegna romana. I centri abitati, in: ANRW II 11.1, 1988, 491–551; Id., La Sardegna romana, 21990; E. Pais, Storia della Sardegna e della Corsica durante il dominio Romano (ed. A. Mastino), [2]1999; C. Urso, Storia, società ed economia in Sardegna e Corsica, 1997; C. Vismara, Funzionari civili e militari nella Corsica romana, in: Studi per L. Breglia 3, 1987, 57–68; R. Zucca, La Corsica romana, 1996.

Sicilia: G. Bejor, Gli insediamenti della Sicilia romana, in: A. Giardina (ed.), Società romana e impero tardoantico, vol. 3, 1986, 463–519; J. Dubouloz, S. Pittia (eds.), La Sicile de Cicéron, 2007; W. Eck, Ein Quästor oder zwei Quästoren im kaiserzeitlichen Sizilien?, in: ZPE 86, 1991, 107–114; M.A.S. Goldsberry, Sicily and its Cities in Hellenistic and Roman Times, 1982; D. Kienast, Die Anfänge der römischen Provinzialordnung in Sizilien, in: V. Giuffrè (ed.), Sodalitas. FS A. Guarino, 1984, 105–123; P. Lévêque, La Sicilie, 1967; G. Manganaro, La Sicilia da Sesto Pompeo a Diocleziano, in: ANRW II 11.1, 1988, 3–89; A. Pinzone, Provincia Sicilia, 1999; A. Di Vita, Un milliarium del 252 a.C. e l' antica via Agrigento – Panormo, in: Kokalos 1, 1955, 10–21; R.J.A. Wilson, Towns of Sicily during the Roman Empire, in: ANRW II 11.1, 1988, 90–206; Id., Sicily under the Roman Empire, 1990

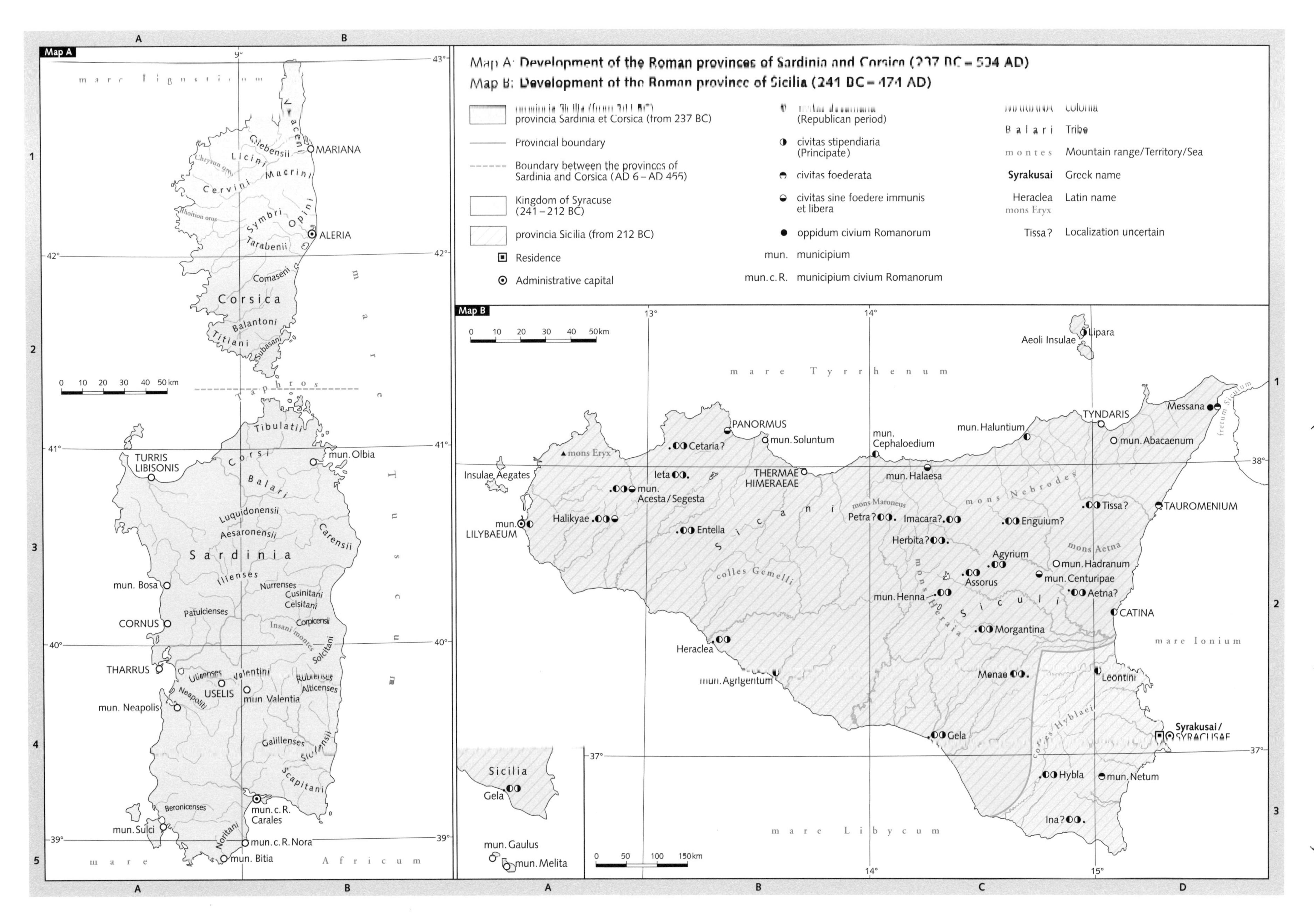
Map A: Development of the Roman provinces of Sardinia and Corsica
Map B: Development of the Roman province of Sicilia (241 BC – 171 AD)
provincia Sardinia et Corsica (from 237 BC)
Provincial boundary
Boundary between the provinces of Sardinia and Corsica (AD 6 – AD 455)
Kingdom of Syracuse (241 – 212 BC)
provincia Sicilia (from 212 BC)
Residence
Administrative capital
(Republican period)
civitas stipendiaria (Principate)
civitas foederata
civitas sine foedere immunis et libera
oppidum civium Romanorum
mun. municipium
mun. c. R. municipium civium Romanorum
Balari Tribe
montes Mountain range/Territory/Sea
Syrakusai Greek name
Heraclea Latin name
mons Eryx
Tissa? Localization uncertain
Map A
mare Ligusticum
Corsica
MARIANA
ALERIA
Sardinia
TURRIS LIBISONIS
mun. Olbia
mun. Bosa
CORNUS
THARRUS
USELIS
mun. Valentia
mun. Neapolis
mun. Sulci
mun. c. R. Carales
mun. c. R. Nora
mun. Bitia
mare Africum
mare Tuscum
Taphros
Map B
mare Tyrrhenum
mare Libycum
mare Ionium
Aeoli Insulae
Lipara
Insulae Aegates
mun. LILYBAEUM
PANORMUS
mun. Soluntum
mun. Cephaloedium
THERMAE HIMERAEAE
mun. Halaesa
mun. Haluntium
TYNDARIS
mun. Abacaenum
Messana
TAUROMENIUM
Tissa?
Enguium?
Imacara?
Petra?
Herbita?
Agyrium
Assorus
mun. Hadranum
mun. Centuripae
Aetna?
CATINA
mun. Henna
Morgantina
Menae
Leontini
Syrakusai / SYRACUSAE
Gela
Hybla
mun. Netum
Ina?
Heraclea
mun. Agrigentum
Halikyae
Entella
Ieta
mun. Acesta / Segesta
Cetaria?
mons Eryx
mons Aetna
mun. Gaulus
mun. Melita
Sicilia

The development of the Roman provinces on the Iberian Peninsula (2nd cent. BC – 5th cent. AD)

The development of the Roman provinces on the Iberian Peninsula (2nd cent. BC – 5th cent. AD)

Their acquisition of the Iberian Peninsula brought the Carthaginians territories of enormous economic importance. Quite apart from the products of the fertility of the land, mineral wealth was particularly striking, especially in the south, the future Hispania Baetica. Here were precious gems, gold, silver, iron, tin and lead: the makings of an expansive overseas trade.

After the Roman victory at Ilipa in 207 BC and the expulsion of the Carthaginians from the Iberian Peninsula (→ map p. 139), the Senate announced its intention of sending officials on an annual basis to the two Spanish provinces, whose shared frontier was only officially set in 197 BC to the south-west of Carthago Nova. After many conflicts with the indigenous population, though, Roman rule in Spain was only finally secured under Augustus. In the end, partly thanks to the foundation of numerous Roman colonies, the Spanish provinces established themselves as particularly loyal areas of the Roman Empire. For this reason, hardly any administrative reforms were required during the Roman Imperial period, as cartographic history and the table below show.

The Balearic group (Baliares insulae), consisting of Insula minor (Menorca), Insula Maior (Majorca) and the Pityussae (Ibiza and Formentera), were known for their good wine and wheat, as well as for the mules bred and kept here. The inhabitants of the Balearics were renowned as mercenaries throughout the Mediterranean world. Their expertise with slingshots made them elite troops. They often undertook raids on shipping, finally endangering shipping in the western Mediterranean to such a degree that the Romans gave one of the consuls for 123 BC, Q. Caecilius Metellus, the task of conquering the islands. He succeeded within a few months, earning himself the by-name *Balearicus*. On Insula maior he founded the city of Palma, and he took a Latin colony to Pollentia. This was one of the first colonies the Romans set up outside Italy. The archipelago was added to the province of Hispania citerior.

During the Age of Migrations, the Iberian Peninsula was to some extent a mere corridor (Vandals; → map p. 233), but some Germanic tribes such as the Suebi and the Visigoths made new homelands here (→ map p. 235). In AD 551, the Byzantine Emperor Justinian I (527–565) took advantage of the conflict between the Visigoth Athanagild and the Visigothic King Agila I (549–555) to occupy the province of Carthaginiensis, which the Byzantines were able to hold until 625 (→ map p. 237). In AD 711, the tensions between the Visigothic King Roderic and followers of his predecessor Witiza provided an opportunity for the mainly Berber troops crossing from Africa to Spain in the context of the great Arab expansion. On 23 June of that year, the Visigoths were defeated in the first battle, where Roderic fell.

The sources

Early histories of the Iberian provinces are told by the historians Livy, Appian and Cassius Dio (book 53). The various administrative registers, such as the Laterculus Veronensis (*c.* AD 313), are important for the period after Diocletian.

The map

The most important administrative changes are portrayed up to the age of the Germanic successor states. The many Roman colonies and titular colonies that made such a vital contribution to the Romanization of the Iberian Peninsula are also shown.

Sources

Str. 3,1,1–3,5,11; Ptol. 2,4–6; Mela 2,85–96; 3,3–3,15; Plin. HN 3,6–30; 4,110–118; *on the islands:* Str. 3,5,1 f.; Plin. HN 14,71; 18,67; Mela 2,124–126; Flor. Epit. 1,43 (*bellum Balearicum*).

Literature

J. Arce, El último siglo de la España romana, 1982; J.M. Blázquez, Hispanien unter den Antoninen und Severern, in: ANRW II 3 (1975), 452–522; P. Bosch-Gimpera, Katalonien in der römischen Kaiserzeit, in: ANRW II 3 (1975), 572–600; R. Collins, Early Medieval Spain. Unity and Diversity, 400–1000, 1983; L.A. Curchin, The Local Magistrates of Roman Spain, 1990; F. Diego Santos, Die Integration Nord- und Nordwestspaniens als römische Provinz in der Reichspolitik des Augustus, in: ANRW II 3 (1975), 523–571; A.T. Fear, Rome and Baetica. Urbanization in Southern Spain c. 50 BC–AD 150, 1996; A. Ferreiro (ed.), The Visigoths in Gaul and Spain. A Bibliography, 1988; R. Haensch, Capita provinciarum. Statthaltersitze und Provinzialverwaltung in der römischen Kaiserzeit, 1997; S. Hamann, Vorgeschichte und Geschichte der Sueben in Spanien, 1971; E. James (ed.), Visigothic Spain. New Approaches, 1980; S.J. Keay, Roman Spain, 1988; A. Lintott, Imperium Romanum, 1993; N. Mackie, Local Administration in Roman Spain A.D. 14–212, 1983; M. Ponsich, Pérennité des relations dans le circuit du Détroit de Gibraltar, in: ANRW II 3 (1975), 655–684; J.S. Richardson, The Romans in Spain, 1996; E.A. Thompson, The End of Roman Spain IV, in: Nottingham Mediaeval Studies 23, 1979, 1–21; Id., The Goths in Spain, 1969; A. Tovar, J.M. Blázquez Martínez, Forschungsbericht zur Geschichte des römischen Spanien, in: ANRW II 3 (1975), 428–451; J.B. Tsirkin, Romanisation of Spain: Socio-Political Aspect III. Romanisation during the Early Empire, in: Gerión 12, 1994, 217–253. *Maps:* Tabula Imperii Romani, Porto, Madrid, Lisbon, Tarraco/Balearics, Valencia, 5 vols., 1993–2001.

Date	Province	Event	Sources
206 BC	Hispania/Hispaniae	Administrative organization based at Tarraco	App. Hisp. 152; Tarraco: Cass. Dio 53,12,5
197 BC		Establishment of frontier between Hispania citerior, capital Tarraco, and Hispania, capital Corduba	Liv. 32,28,11; cf. 29,3,5; 40,41,10. Corduba: Bell. Alex. 49,1–55,3; Cic. Fam. 10,31,6; 10,32,5
123 BC		Conquest of Baliares insulae, incorporation into province of Hispania citerior	Liv. Per. 60; Str. 3,5,1
27 BC		H. citerior renamed Tarraconensis, H. ulterior Baetica	Cass. Dio 53,12,5; ILS 103
	Lusitania	Province created after conquest of mid-west, capital Emerita	Cass. Dio 53,12,5; CIL 2 Suppl. 87; Emerita: AE 1990, 514
AD 211/217	Hispania superior, western part of future Callaecia	NW of Tarraconensis split off as separate province; capital Bracara Augusta	G. Alföldy, Provincia Hispania Superior, 2000 (inscription); ILS 1157 (on division of province of H. citerior)
Diocletian (284–305) and Constantine (306–337)	Carthaginiensis	SE of Tarraconensis split off as separate province; capital Carthago Nova	Laterculus Veronensis 11; Polemius Silvius 4,3; Breviarium 5
AD 409		Invasion of Suebi, Vandals and Alani	Chron. min. 2,17,42
c. AD 425		Baliares Vandal	Hydatius, Chronica 86
AD 456		Invasion of Visigoths under Theoderic II	Iord. Get. 231; Chron. min. 2,28,173
AD 534		Baliares Byzantine	Procop. Vand. 4,5,7

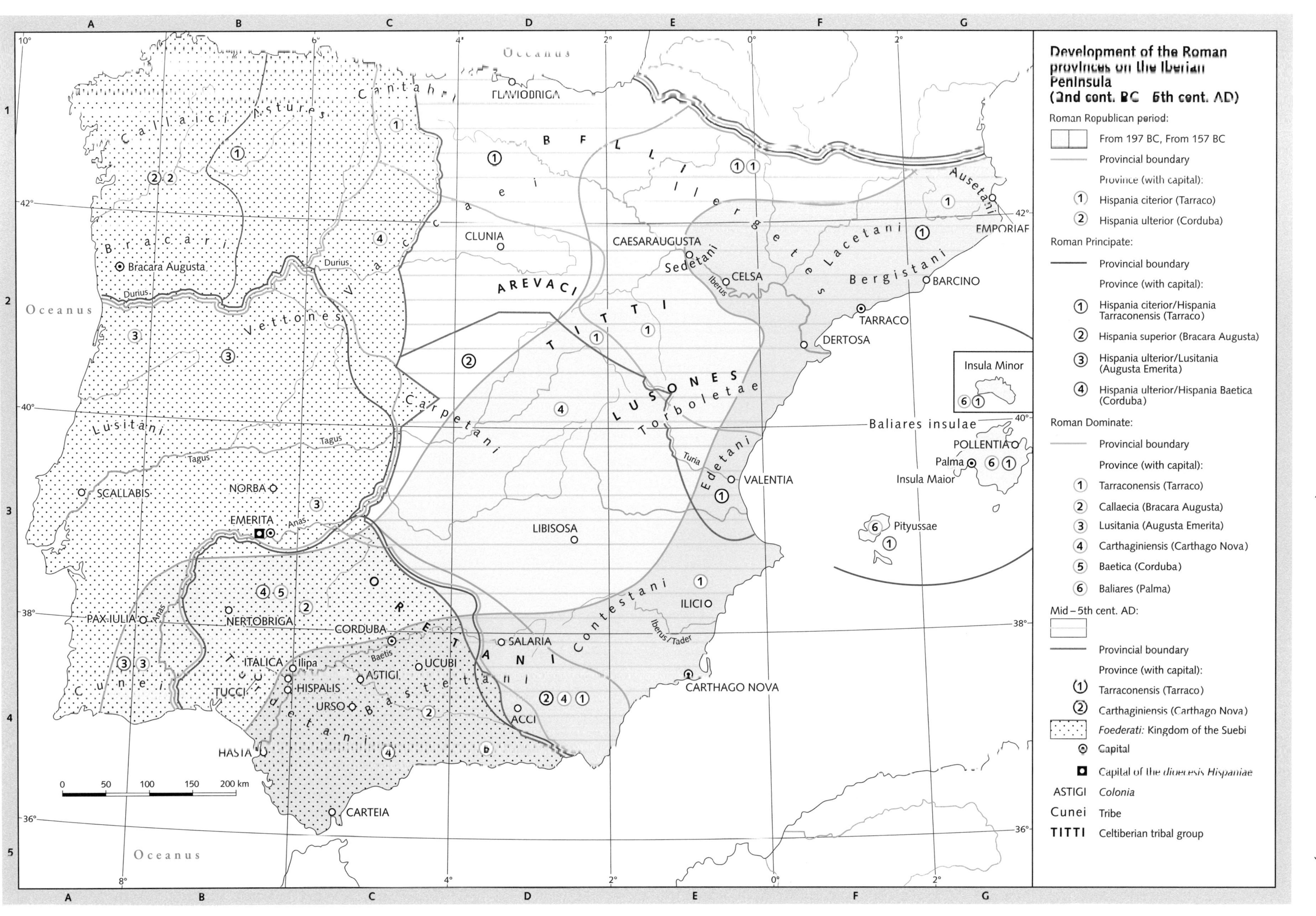
Development of the Roman provinces on the Iberian Peninsula (2nd cent. BC – 5th cent. AD)
Roman Republican period:
From 197 BC, From 157 BC
Provincial boundary
Province (with capital):
① Hispania citerior (Tarraco)
② Hispania ulterior (Corduba)
Roman Principate:
Provincial boundary
Province (with capital):
① Hispania citerior/Hispania Tarraconensis (Tarraco)
② Hispania superior (Bracara Augusta)
③ Hispania ulterior/Lusitania (Augusta Emerita)
④ Hispania ulterior/Hispania Baetica (Corduba)
Roman Dominate:
Provincial boundary
Province (with capital):
① Tarraconensis (Tarraco)
② Callaecia (Bracara Augusta)
③ Lusitania (Augusta Emerita)
④ Carthaginiensis (Carthago Nova)
⑤ Baetica (Corduba)
⑥ Baliares (Palma)
Mid – 5th cent. AD:
Provincial boundary
Province (with capital):
① Tarraconensis (Tarraco)
② Carthaginiensis (Carthago Nova)
Foederati: Kingdom of the Suebi
Capital
Capital of the dioecesis Hispaniae
ASTIGI Colonia
Cunei Tribe
TITTI Celtiberian tribal group
Oceanus
Callaici
Astures
Cantabri
Bracari
Bracara Augusta
Vaccaei
Vettones
Lusitani
Carpetani
Cunei
Turdetani
Bastetani
Oretani
Contestani
Edetani
Torboletae
LUSONES
TITTI
AREVACI
BELLI
Ilergetes
Sedetani
Lacetani
Bergistani
Ausetani
FLAVIOBRIGA
CLUNIA
CAESARAUGUSTA
CELSA
EMPORIAE
BARCINO
TARRACO
DERTOSA
VALENTIA
LIBISOSA
ILICIO
CARTHAGO NOVA
SALARIA
ACCI
CORDUBA
UCUBI
ASTIGI
URSO
HISPALIS
Ilipa
ITALICA
TUCCI
HASTA
CARTEIA
NERTOBRIGA
PAX IULIA
EMERITA
NORBA
SCALLABIS
Insula Minor
Baliares insulae
POLLENTIA
Palma
Insula Maior
Pityussae
Durius
Tagus
Anas
Baetis
Iberus
Turia
Iberus/Tader
0 50 100 150 200 km

The development of the Roman provinces in North Africa (146 BC – AD 395)/Rome's war against Jugurtha (112–105 BC)

146–46 BC
After the destruction of the city of Carthage in the spring of 146 BC, P. Cornelius Scipio had the seventh Roman province defined by laying down a demarcation line to the Numidian kingdom. This *Fossa regia*, of which evidence survives in the form of boundary stones, ran from Thabraca in the north-west to Thenae in the south-east. The province was named Africa, probably after the Libyan Afri tribe from the lower reaches of the Muthul (cf. smaller map). The province's administrative seat was at Utica.

46–40/39 BC
After his victory at Thapsus in 46 BC, Caesar created a second African province, Africa nova. This adjoined the existing province (now renamed Africa vetus) behind the *Fossa regia* to the south-west. A boundary stone placed during the reign of Vespasian, which refers to the renovation of the *Fossa*, mentions the 'boundary between the new province and the old, where the *Fossa regia* was'. Caesar assigned Cirta and the surroundings of this former royal capital of the Numidians to P. Sittius for the settlement of his veterans. Sittius had provided Caesar with support during the Civil War with privately-recruited troops. In 44, this region was incorporated into Africa nova, but it kept its special status as a military district (*dioecesis Numidia*).

40/39 BC – AD 284
The two African provinces of Africa vetus and Africa nova were united as Africa proconsularis under the *triumvir* M. Aemilius Lepidus. The administrative capital of Africa proconsularis was Carthage. In the Senate session of 13.1.27 BC, Africa proconsularis was placed under the control of the Senate. The governor commanded the *legio III Augusta*. Under Caligula, command of the legion passed to a legate who also ruled the military district of Numidia. After the suppression of several uprisings by Suetonius Paullinus and Hosidius Geta, the provinces of Mauretania Caesariensis and Mauretania Tingitana were established to the west of Africa proconsularis in AD 42. Under Septimius Severus, the military district of Numidia was established as an imperial province in its own right.

From AD 284
Diocletian's imperial reforms divided Africa proconsularis into the three provinces of Africa proconsularis, Africa Byzacena and Africa Tripolitana. Numidia was split into two provinces, Numidia Cirtensis and Numidia Militiana. Mauretania Caesariensis was also divided in two as Mauretania Sitifensis and Mauretania Caesariensis. These seven provinces were constituted as the *dioecesis* Africa. The remaining, westerly part of Mauretania, Mauretania Tingitana, was added to the *dioecesis* Hispaniae, and the two eastern provinces of Libya superior and Libya inferior were incorporated in the *dioecesis* Oriens.

A word on the numerous Roman colonies established in the African provinces. It is probably impossible in many cases to distinguish between genuine and titular colonies. But even if only a fraction of the African colonies were to be attributable to the settlement of new colonists, their number would be considerable in comparison to the number of colonies in other parts of the Roman Empire. In this we may recognize the enormous – and not only economic – importance that North Africa had, and was to have, for Rome.

Rome's war against Jugurtha (112–105 BC)

During his 55-year reign, the Numidian king Massinissa had built up a respectable realm with his unconditional championship of the interests of the Roman Senate following the collapse of Carthage in the 2nd Punic War and the subsequent mercenary revolt. His kingdom extended from the River Muthul in the west to Philainon Bomoi (Arae Philaenorum under Roman rule) in the east. Its southern frontier cannot be securely located. The court at his royal capital of Cirta was profoundly influenced by Carthaginian and, through Punic agency, Hellenistic traditions. Massinissa himself had grown up and been educated in Carthage, and had contacts with various rulers in the east (e.g. the Bithynian king Nicomedes II). But Massinissa's primary diplomatic relations were with the Roman Senate, which had granted him full recognition as King of the Numidians.

On Massinissa's death in 148 BC, his Numidian kingdom was bequeathed to his three sons to rule jointly. Micipsa, who had two sons of his own, gave particular esteem and support to his nephew Jugurtha. He probably also facilitated Jugurtha's posting at the head of a force of elite troops to support the Romans in their war with the Celtiberians in Numantia. It was there that Jugurtha won the friendship of the younger Scipio (P. Cornelius Scipio Aemilianus Africanus), who recommended Micipsa to adopt him. When Micipsa died in 118, a period of joint rule by Micipsa's two sons and Jugurtha began. In view of the ambition displayed by Jugurtha (with the support of Rome), there was a series of disputes over the Numidian throne.

Jugurtha had Micipsa's son Hiempsal eliminated, and banished his brother Adherbal, who turned to the Senate for help. The Senate negotiated a division of the kingdom, but Jugurtha did not recognize this. At Cirta in 112, he attacked and killed Adherbal along with the Italian merchants of the city. The Senate now declared war on Jugurtha.

The conduct of this war by various Roman commanders was a scandal in all its aspects, which Sallust (1st cent. BC), whose historical monograph is, for all the problems posed by its chronology, the central source for the Jugurthan War, found reason enough to use as an example proving the decline of Rome. The remark Sallust attributes to Jugurtha, *'urbem venalem et mature perituram, si emptorem invenerit'* ('a venal city, ripe to perish should it ever find a buyer'), is indicative.

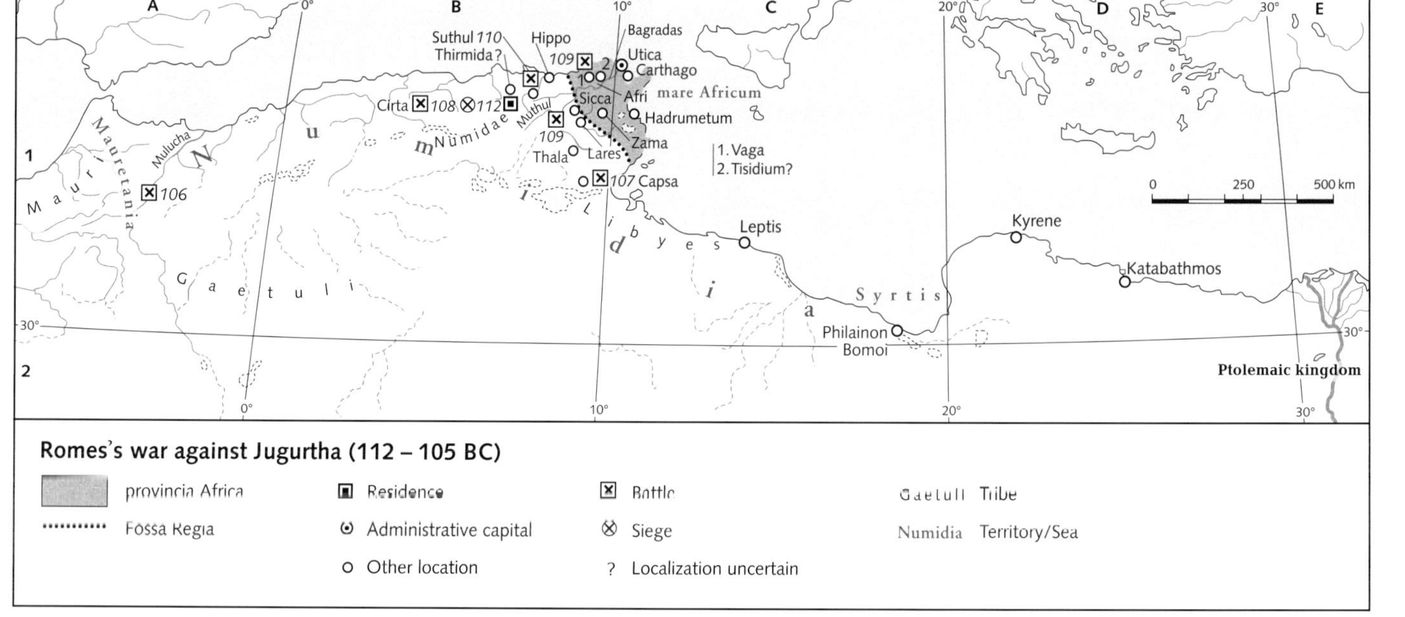

Year	Event
112	Jugurtha conquers Cirta
111	Senate declares war on Jugurtha. The consul L. Calpurnius Bestia, assigned the campaign, follows up initial successes (Leptis Magna allying with Rome) by concluding a barely achieved peace. Jugurtha obeys the call of the Senate (compelled by the people's tribune C. Memmius) to present himself in person at Rome as a witness regarding the bribery of Roman politicians. He is expelled for improper behaviour (bribery, assassinations).
110	Continued prosecution of the war by the consul Sp. Postumius Albinus, in whose absence his brother, the legate Aulus, is heavily defeated by Jugurtha at Suthul and is forced to capitulate.
109	The consul Q. Caecilius Metellus restores army discipline, defeats Iugurtha at the Muthul, attacks Zama without success, takes Vaga.
108	Metellus, now proconsul, takes Thala. However, he interrupts his storming of the forces of Jugurtha and his son-in-law, the Mauretanian king Bocchus, stationed at Cirta, when he finds that his legate, C. Marius, has been elected consul and has been charged with further prosecution of the war against Jugurtha.
107	Lengthy rearmament prevents the consul, C. Marius, from resuming battle to any great extent. He takes Capsa and neighbouring localities.
106	Now proconsul, Marius marches west as far as the Mulucha. He succeeds in fighting off attacks by Bocchus at the Mulucha and again on the return march, and he reaches winter quarters at Cirta. Bocchus begins negotiations with Marius.
105	L. Cornelius Sulla, Marius' legate, succeeds in inducing Bocchus to betray Jugurtha and to deliver him into Roman hands.
104	Marius celebrates a triumph over Jugurtha on January 1.

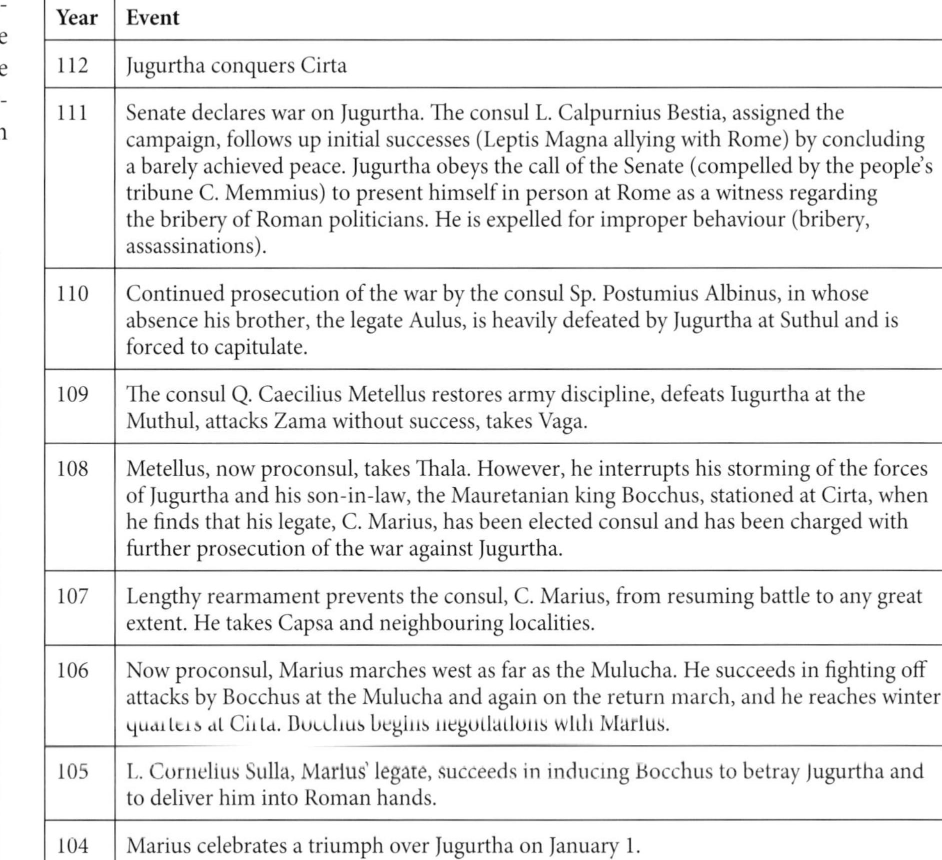

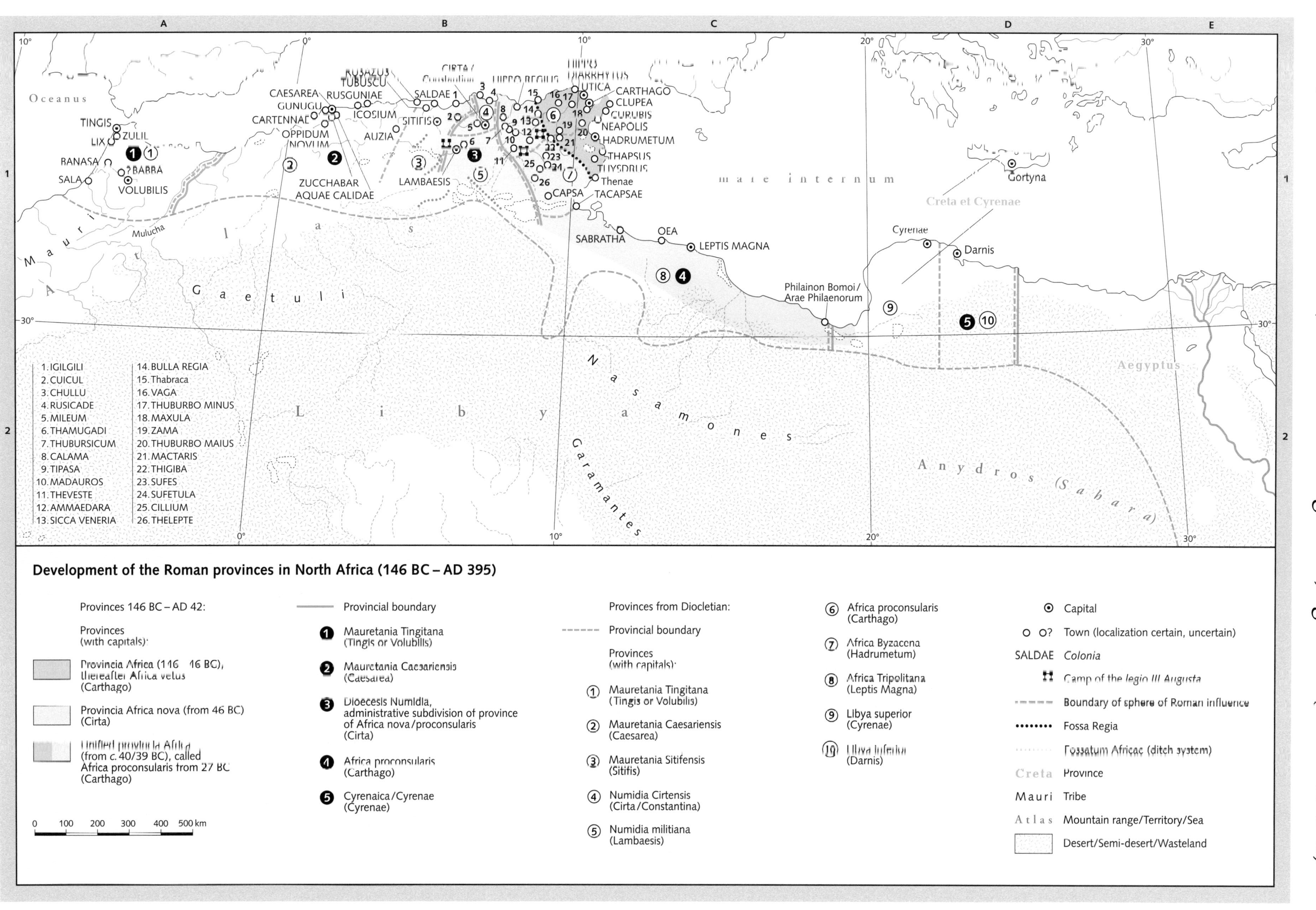
Oceanus
TINGIS
ZILIL
LIX
BANASA
BABBA
SALA
VOLUBILIS
Mulucha
Mauri
Gaetuli
Libyes
CAESAREA
GUNUGU
CARTENNAE
OPPIDUM NOVUM
RUSGUNIAE
ICOSIUM
AUZIA
SALDAE
SITIFIS
ZUCCHABAR
AQUAE CALIDAE
LAMBAESIS
CAPSA
TACAPSAE
Thenae
THAPSUS
HADRUMETUM
NEAPOLIS
CURUBIS
CLUPEA
CARTHAGO
UTICA
SABRATHA
OEA
LEPTIS MAGNA
mare internum
Philainon Bomoi/
Arae Philaenorum
Cyrenae
Darnis
Creta et Cyrenae
Gortyna
Nasamones
Garamantes
Anydros (Sahara)
Aegyptus
1. IGILGILI
2. CUICUL
3. CHULLU
4. RUSICADE
5. MILEUM
6. THAMUGADI
7. THUBURSICUM
8. CALAMA
9. TIPASA
10. MADAUROS
11. THEVESTE
12. AMMAEDARA
13. SICCA VENERIA
14. BULLA REGIA
15. Thabraca
16. VAGA
17. THUBURBO MINUS
18. MAXULA
19. ZAMA
20. THUBURBO MAIUS
21. MACTARIS
22. THIGIBA
23. SUFES
24. SUFETULA
25. CILLIUM
26. THELEPTE
Development of the Roman provinces in North Africa (146 BC – AD 395)
Provinces 146 BC – AD 42:
Provinces (with capitals):
Provincia Africa (146–46 BC), thereafter Africa vetus (Carthago)
Provincia Africa nova (from 46 BC) (Cirta)
Unified province of Africa (from c. 40/39 BC), called Africa proconsularis from 27 BC (Carthago)
0 100 200 300 400 500 km
Provincial boundary
Mauretania Tingitana (Tingis or Volubilis)
Mauretania Caesariensis (Caesarea)
Dioecesis Numidia, administrative subdivision of province of Africa nova/proconsularis (Cirta)
Africa proconsularis (Carthago)
Cyrenaica/Cyrenae (Cyrenae)
Provinces from Diocletian:
Provincial boundary
Provinces (with capitals):
Mauretania Tingitana (Tingis or Volubilis)
Mauretania Caesariensis (Caesarea)
Mauretania Sitifensis (Sitifis)
Numidia Cirtensis (Cirta/Constantina)
Numidia militiana (Lambaesis)
Africa proconsularis (Carthago)
Africa Byzacena (Hadrumetum)
Africa Tripolitana (Leptis Magna)
Libya superior (Cyrenae)
Libya inferior (Darnis)
Capital
Town (localization certain, uncertain)
SALDAE Colonia
Camp of the legio III Augusta
Boundary of sphere of Roman influence
Fossa Regia
Fossatum Africae (ditch system)
Creta Province
Mauri Tribe
Atlas Mountain range/Territory/Sea
Desert/Semi-desert/Wasteland

Rome's wars in the west (206 –101 BC)

I. Conflicts on the Iberian Peninsula, 206–154 BC

In the atmosphere of tension between Rome and Carthage, the interests of both powers clashed after the 1st Punic War (264–241) on the resource-rich Iberian Peninsula. The Carthaginians, driven by economic need, were led by the Barcids, and the Romans were urged on by their ally, the trading metropolis of Massalia. With the delineation of spheres of interest at the Iberus (Ebro, Júcar or Segura) in 226/25, the conflict over Saguntum in 221/19 and the 2nd Punic War (218–201), Spain was for two hundred years constantly recalling itself to the attention of the Senate.

Having entirely driven the Carthaginians out of the peninsula during the 2nd Punic War in 206, the Romans gradually claimed dominion over all of Spain. The domestic political pressure to which Roman provincial governors were subjected allows the inference that in many instances of conflict, the provocation or aggression came from the Roman side.

From 206, the Senate organized the territory it claimed into two provinces, Hispania citerior in the east and Hispania ulterior in the south-west. The table (→ p. 267) lists the well-attested conflicts of the first half of the 2nd cent. BC. The level of deployment by the Romans and their allies here is astonishing, given that Roman troops were also simultaneously fighting in the eastern Mediterranean, e.g. in the 2nd (214–205) and 3rd Macedonian Wars (171–168) and the war against Antiochus III (191–188). At the end of this phase of permanent war, the territory of the two Roman provinces covered the entire eastern half of the peninsula (→ map p. 145).

II. The Celtiberian War (154–133 BC)

Two phases of military conflict are referred to jointly according to Polybius' example as the 'Celtiberian War' (πύρινος πόλεμος/ *pýrinos pólemos*, a 'fiery war', Pol. 35,1). The first phase describes the conflicts of 154–151, the second those of 143–133, which ended with the fall of Numantia and which we term the Numantine War (cf. Liv. Per. 56).

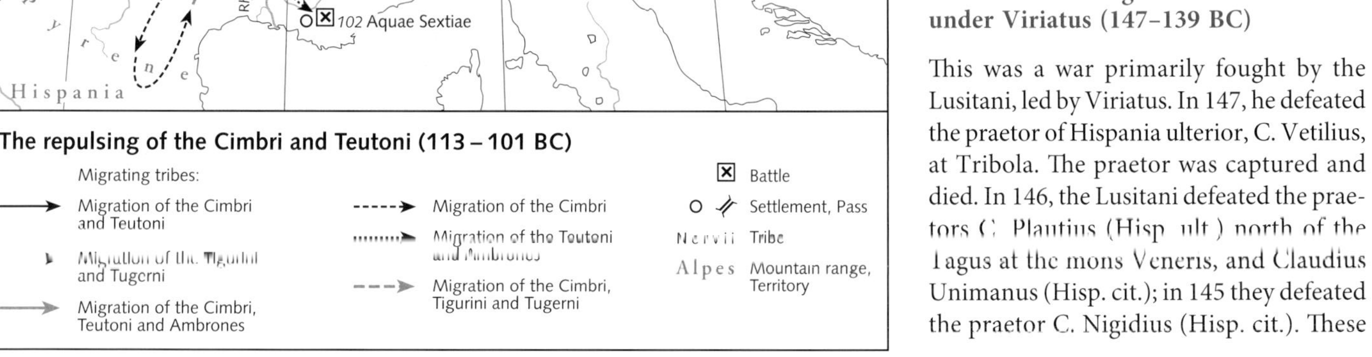

The repulsing of the Cimbri and Teutoni (113 – 101 BC)

The conflicts began in 154 with the invasion of Hispania citerior by the Lusitani, inflicting a severe defeat on a praetorian army. In 153, the Lusitani crossed to Mauretania, probably only to plunder, soon returning to the province. The successes of the Lusitani emboldened tribes such as the Arevaci and Belli to make their own attempts to break free of dependence on Rome. Only the proconsul for 151, M. Claudius Marcellus, was able to bring a tentative end to the insurgency by military and diplomatic action.

A second phase of conflict began in 143, when the spark of rebellion probably spread from the Lusitani to the Celtiberian Arevaci, Titti and Belli. After battles of varied outcome (capture of Contrebia by Q. Caecilius Metellus in 142), the proconsul of Hispania citerior, Q. Pompeius, floated a peace treaty with the tribes in 140. A senatorial commission was even dispatched to rearrange conditions in the province. Negotiations, however, foundered upon the shadowy role of Pompeius, and the Senate decided to continue the war. In 137, it sent the consul C. Hostilius Mancinus, but he and his army were encircled near Numantia and forced to capitulate. The Numantines stipulated Mancinus' quaestor, Ti. Sempronius Gracchus, as the guarantor of the act of capitulation. In 178, Gracchus' father had reached a lasting peace settlement with the Celtiberians, and they hoped for something similar through the agency of the son. However, the treaty was swallowed up into the mire of internal Roman politics, and was repudiated by the Senate in 136. Numantia was the focal point of the ensuing battles. The city was besieged for nine months in 134 by a force under P. Cornelius Scipio Aemilianus Africanus. It was conquered in 133 and destroyed.

III. Rome's war against the Lusitani under Viriatus (147–139 BC)

This was a war primarily fought by the Lusitani, led by Viriatus. In 147, he defeated the praetor of Hispania ulterior, C. Vetilius, at Tribola. The praetor was captured and died. In 146, the Lusitani defeated the praetors C. Plautius (Hisp. ult.) north of the Tagus at the mons Veneris, and Claudius Unimanus (Hisp. cit.); in 145 they defeated the praetor C. Nigidius (Hisp. cit.). These battles, difficult to locate, probably mainly took place in the territory of the Carpetani and Vaccaei and the cities of Segovia and Segobriga in Hispania citerior. The Lusitani at this time were probably operating with several armies and not only under Viriatus' command. After these successes, the Lusitani conquered large expanses of both provinces. In 145, they were defeated by C. Laelius (Hisp. cit.), and in 144 by Q. Fabius Maximus (Hisp. ult.; capture of Tucci). However, there was no decisive success for either side. After a fruitless siege of Numantia by Q. Pompeius (Hisp. ult.) and a defeat of Q. Fabius Maximus Servilianus (Hisp. ult.) at Erisane, a peace treaty was made and approved by the popular assembly at Rome. By it, Viriatus was recognized as *socius atque amicus* of the Roman people. But in the same year, the Senate withdrew its consent to the peace. In the course of renewed negotiations, Viriatus was murdered at the instigation of the proconsul, Q. Servilius Caepio (Hisp. ult.). After his death, the Lusitani gave up the war and surrendered unconditionally.

Sources and the map

Most of our information comes from Livy (Liv. 28–45; Liv. Per. 46–59) and Appian (2nd cent. AD) in his *Iberica*. Coins and inscriptions make a significant contribution to our knowledge of the Roman governors in particular.

In view of the sketchy reports of the Spanish wars in the sources, no complete campaign routes can be reconstructed with good conscience. Only those locations are therefore marked, ordered by period of conflict, which the sources specifically assign to particular military actions.

IV. The repulsing of the Cimbri and Teutoni (113–101 BC; map inset)

The migration that brought the Cimbri and Teutoni into conflict with Rome is to be seen in context with various westward migrations of Germanic peoples in the 2nd/1st cents. BC. Both tribes moved from the Danube to Noricum, where the consul Cn. Papirius Carbo was defeated in 113. The tribes probably crossed the Rhine at Mainz, where they were joined by Celts, the Ambrones and the Helvetian Tigurini and Tugerni. Turned away by the Belgae, they turned southwards, defeating a Roman army under the consul M. Iunius Silanus on the Liger (Loire) in 109. The Helvetian tribes now separated from the Germans and moved southwards into the territory of the Nitiobroges. Here, they defeated a Roman army under the consul L. Cassius Longinus on the Garumna (Garonne) in 107. The consul fell in this battle. They joined up with the Germans again on the lower reaches of the Rhodanus (Rhone) to defeat two armies, led by the proconsul Q. Servilius Caepio and the consul Cn. Mallius Maximus respectively, at Arausio on 5 and 6 October 105 BC. The Cimbri then broke away, migrating over the Pyrenees. Turned away by the Celtiberians, they retraced their steps and moved up the Rhone with the Tigurini and Tugerni, intending to break into Italy over the 'Tridentina iuga' (Fern/Resia Passes or the Brenner). Meanwhile, in 104, C. Marius had taken over the *provincia Gallia* and reorganized and reformed the army. By 102, he was thus in a position utterly to annihilate the Ambrones and Teutoni in two battles at Aquae Sextiae. On 30 July 101 BC, the Roman army under Marius and the proconsul Q. Lutatius Catulus met the Cimbri, Tigurini and Tugerni, who had crossed the Alps into the valley of the Padus (Po), at Vercellae. This Roman victory, too, was catastrophic for the defeated adversaries. The danger which had raised the *furor Teutonicus* (Lucan. 1,255 f.) to the status of a political rallying cry was warded off for now.

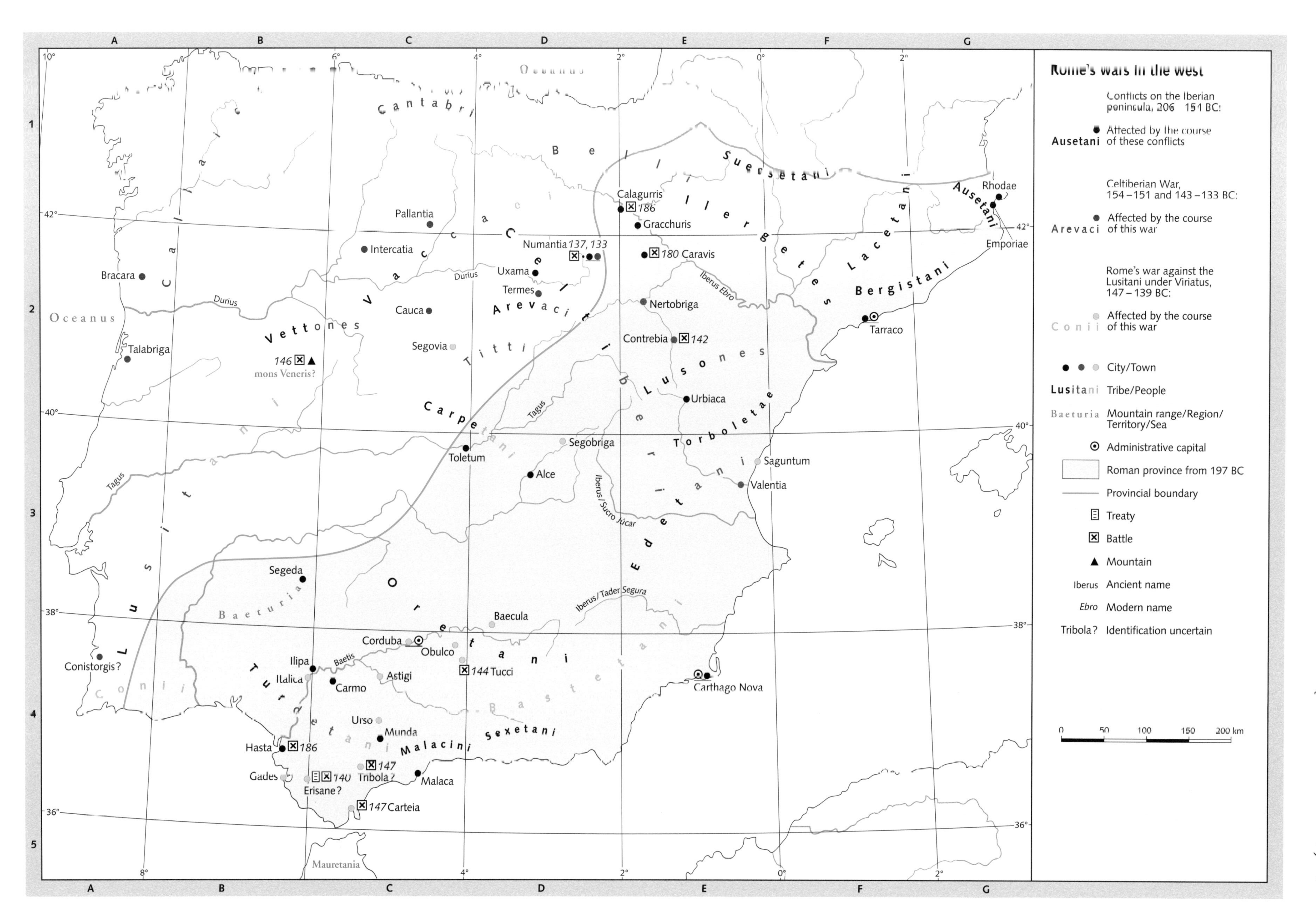
Rome's wars in the west
Conflicts on the Iberian peninsula, 206 151 BC:
Ausetani Affected by the course of these conflicts
Celtiberian War, 154–151 and 143–133 BC:
Arevaci Affected by the course of this war
Rome's war against the Lusitani under Viriatus, 147–139 BC:
Conii Affected by the course of this war
City/Town
Lusitani Tribe/People
Baeturia Mountain range/Region/Territory/Sea
Administrative capital
Roman province from 197 BC
Provincial boundary
Treaty
Battle
Mountain
Iberus Ancient name
Ebro Modern name
Tribola ? Identification uncertain
0 50 100 150 200 km
Oceanus
Cantabri
Callaici
Bracara
Talabriga
Conistorgis ?
Pallantia
Intercatia
Cauca
Segovia
Vaccaei
Uxama
Termes
Numantia 137, 133
Arevaci
Vettones
146 mons Veneris?
Durius
Tagus
Toletum
Carpetani
Titti
Belli
Celtiberi
Calagurris 186
Gracchuris
180 Caravis
Nertobriga
Contrebia 142
Urbiaca
Segobriga
Alce
Iberus Ebro
Iberus / Sucro Júcar
Iberus / Tader Segura
Saguntum
Valentia
Carthago Nova
Tarraco
Emporiae
Rhodae
Ausetani
Bergistani
Lacetani
Suessetani
Ilergetes
Lusones
Torboletae
Edetani
Baecula
Corduba
Obulco
144 Tucci
Astigi
Carmo
Urso
Munda
147 Tribola ?
147 Carteia
Malaca
Malacini
Sexetani
Bastetani
Oretani
Segeda
Ilipa
Italica
Baetis
Hasta 186
Gades
140 Erisane ?
Turdetani
Baeturia
Lusitani
Conii
Mauretania

Rome's wars in the east I (214–129 BC)

Through the 2nd cent. BC, Rome fought wars in Italy, in the west and then also in the east of the Mediterranean world. The year 168, in which the Romans finally defeated the Macedonians at Pydna, is generally seen as the culmination of Rome's development into a world power – this was how Polybius saw it at the time. But the map continues up to the establishment of the province of Asia in 129. From that year on, more than forty years passed before a Roman legion saw battle in the eastern Mediterranean.

I. The 1st Macedonian War (214–205 BC)

In 229, the Roman Senate pressed home its interests in the Adriatic by military means. Trade was constantly being disrupted by the quasi-piratical foreign policy of the Illyrian Queen Teuta. The Romans intervened here in 229/28, setting up a protectorate in Illyria over various tribes (e.g. the Parthini) and cities (e.g. Corcyra, Apollonia) and securing this position in the 2nd Illyrian War in 219.

Roman and Macedonian interests overlapped on the Illyrian coast. It seemed to the Macedonian King Philip V that his opportunity to cement his position permanently had come when Hannibal annihilated the Roman armies at Cannae in 216. The alliance between Philip and Hannibal (spring 215) was the provocation for the 1st Macedonian War.

At first, the Roman fleet sought to keep Philip away from the Illyrian coast. But when Philip succeeded in conquering Lissus in 212, the Romans began to seek allies in Greece, finding them in the Aetolian League, Sparta, Elis and Messene. But so preoccupied was Rome with its struggle against Hannibal in Italy that it was unable to provide support to its Greek allies. The Aetolians therefore concluded a separate peace with Philip in 206. The Senate followed their example in 205, giving up the Illyrian Atintani (Treaty of Phoenice).

II. The 2nd Macedonian War (200–196 BC)

No sooner had the Senate ended its war with Carthage in the spring of 201 than it made plans for war with Philip V. Scholars dispute the motives behind this decision, which were probably also less than plain to the Roman popular assembly, which voted against the war in a first vote, and the legionaries, who mutinied at Apollonia.

The Hellenistic east was in a politically unstable conditon. Ptolemy IV had died in 204, and his successor was just six years old. Various alternating court factions conducted government at Alexandria, a situation which Philip and Antiochus III exploited, concluding a treaty in 203/02 to divide between them the Ptolemaic overseas possessions (Aegean, western Asia Minor, Thrace, Coele Syria, Phoenice, Cyprus, Cilicia, Lycia). It was obvious that this would involve them not only with the Ptolemaic kingdom but also with other powers, for instance with Attalus I of Pergamum and with Rhodes. These promptly petitioned the Senate to intervene.

Following diplomatic soundings, the Senate in 200 BC stepped in with an attack on Epirus, an area the Macedonian king claimed for himself. After a victory in 198 at the gorge formed by the River Aous at Antigonea, the the interior of Epirus was at the Romans' mercy. The Achaean League emulated the action of the Aetolian League the previous year in joining the Roman side. The Boeotians joined them in 197. After fruitless peace negotiations at Nicaea (Locris), the proconsul T. Quinctius Flamininus won the decisive victory over the Macedonians in June 197 at Cynoscephalae. Under the terms of the peace, Philip was compelled to give up all possessions in Greece and withdraw his garrisons from Chalcis, Demetrias and Corinth, the 'three fetters of Greece'.

III. The 3rd Macedonian War (171–168 BC)

Perseus, the son of Philip V, re-armed his kingdom purposefully. He activated old contacts and forged new ones, courting the Achaean, Aetolian and Boeotian Leagues and the Delphic Amphictyony. He secured his authority in the north by repelling the Sapaeans, who had occupied the ore-rich Pangaeum, and by intensifying diplomatic contacts with the Odrysae, Illyrians and Bastarnae and the Thracian prince Teres, who married a sister of Perseus. He forged dynastic links with the Bithynian royal house, giving his sister Apame in marriage to Prusias II, and he also allied with the Seleucid house, marrying a daughter of Seleucus IV.

The Senate and people of Rome decided on war with Perseus in 172. The praetor Cn. Sicinius crossed from Brundisium to Apollonia. Roman emissaries sought military support from Ariarathes IV of Cappadocia, Eumenes II at Pergamum, Antiochus IV at Antioch, Prusias II at Nicomedia, the Numidian King Massinissa, Ptolemy VI at Alexandria, the Illyrian King Genthius and Greek cities on the mainland and the islands. From Corcyra they visited Elis, Messene and Argos, as well approaching the Aetolians, Thessalians (Larisa), Boeotians (Thisbe, Haliartus, Corone), Euboea (Chalcis), the Epirotes and Acarnanians, Byzantium and Rhodes. A truce obtained from Perseus provided breathing space for these activities.

In 171, a battle was fought at the eastern entrance to the Vale of Tempe, a gorge formed by the lower Peneius between Olympus and Ossa. The battlefield was on an elevation, called 'Callinicus', at Atrax. Perseus had the better of the battle. Had he deployed his phalanx with determination, he may well have won a complete victory. But he hoped that his leniency might induce the Romans to relent. This was a mistake.

The year 170 BC passed with actions by both sides in Boeotia and southern Illyria. There was no conclusive moment. The consul for 169, Q. Marcius Philippus, likewise did not succeed in landing a decisive blow. He did evade the Macedonian position at the western end of the Vale of Tempe by taking a route to the north of Lake Ascuris – a risky manoeuvre in which, incidentally, Polybius took part in person (Pol. 28,13,3) – to reach the coast by way of the mountains. However impressive this move may have been, it was in vain, as the relieving fleet did not appear and the Macedonians blocked off the coastal plain at Dion.

169 brought the decisive actions. While in the west, the praetor L. Anicius Gallus defeated the fleet of Genthius, who had by now allied with Perseus, and compelled Genthius to surrender at Scodra, the consul, L. Aemilius Paullus bypassed the Macedonian position at Dion by crossing the pass at Petra, forcing the Macedonian king to withdraw to Pydna to prevent access to the Macedonian heartland. Thus it was that the Romans' final and decisive victory over the Macedonians took place at Pydna on 22 June. The king fled to Samothrace, where he surrendered to the Romans. The Senate abolished the Macedonian kingdom and divided Macedonia into four autonomous commonwealths (*res publicae*) with administrative capitals at Amphipolis, Thessalonica, Pella and Heraclea. The Romans refrained from outright annexations. Genthius' kingdom was also abolished and its territory added to the Roman protectorate of Illyricum, in existence since 219.

IV. The Macedonian Revolt under Philip VI/Andriscus (150–148 BC)

Several crises shook the Balkan peninsula between 156 and 146. They were centred on Dalmatia, Macedonia and the Peloponnese. The Dalmatae caused problems with attacks on states directly or indirectly allied to Rome (Daorsi, Tragurium and Epetium, colonies of Issa, a city allied to Rome), on whose behalf the Senate felt compelled to intervene militarily. In 156, the consul, C. Marcius Figulus, conquered various Dalmatian cities and besieged Delminium, the capital of the Dalmatae, which was conquered in 155 by the consul P. Cornelius Scipio Nasica Corculum.

A crisis in the Macedonian commonwealths proved more momentous. Here, one Andriscus, an adventurer (or so our sources portray him, Polybius calling him Pseudophilippus), had himself enthroned at Pella as Philip VI, son of Perseus. The Macedonians took up arms against him in two skirmishes at the River Strymon, but could not defeat his Thracian troops. In 150/49, a Roman military tribune at the head of an Achaean contingent attempted to shield Thessaly against him. By 149/48, actions against Andriscus were already under the command of a praetor (or propraetor), P. Iuventius Thalna. There was a battle in Thessaly at which Iuventius was defeated and killed. Andriscus now made contact with Carthage. The conflict threatened to erupt on a worldwide scale. The praetor Q. Caecilius Metellus then (in 148) succeeded in defeating Andriscus with two legions. Andriscus fled to the Thracian dynast Byzes, who, however, handed him over to Metellus. This marked the end of the autonomy of the Macedonian republics, which were now organized as a Roman province. Even in 142, a man called Philip who called himself a son of Perseus could still attract much support. He was killed by L. Tremelius Scrofa, the quaestor of the governor of Macedonia.

V. The Achaean War, 146 BC

When the Romans left the Balkans, they left the seeds of unrest everywhere in Greece, and these duly germinated. There was almost universal dissatisfaction with the arrangements imposed by the Senate. Pro-Roman circles used the sympathy of the Senate to excessively brutal effect. Those Achaeans who, suspected of pro-Macedonian sympathies, had been taken like Polybius as hostages to Italy in 168, fewer than 300 of whom remained alive, now also returned to Greece, rekindling memories of the good old days.

The trigger for the Achaean uprising against Rome was a conflict between Sparta and Megale Polis over frontier issues. The Senate wanted this dispute settled as an internal affair of the Achaean League. The synhedrion ruled in favour of Megale Polis, but the disagreement persisted. The Senate distrusted the policies of the Achaean League, which was trying to prevent Sparta from breaking away. In the summer of 147, an embassy under L. Aurelius Orestes arrived at Corinth, where the synhedrion was in session. The senatorial ruling proclaimed by the emissaries dictated the separation of Sparta, Corinth, Argos, Orchomenus in Arcadia and Heraclea Trachinia from the Achaean League. The Achaeans were not responsive to the Roman agenda, and a *casus belli* was thus established. The Boeotians, Euboeans, Phocians and Locrians joined the Achaean side.

The propraetor, Q. Caecilius Metellus, soon had the situation under control in Thessaly and Boeotia, before the consul, L. Mummius, supported by Pergamene contingents, defeated the Achaeans at Corinth. The city was taken and largely destroyed. Men were killed, women and children enslaved. Greece (Achaea) was incorporated in the administration of the province of Macedonia.

→ Maps pp. 117, 141, 152, 187

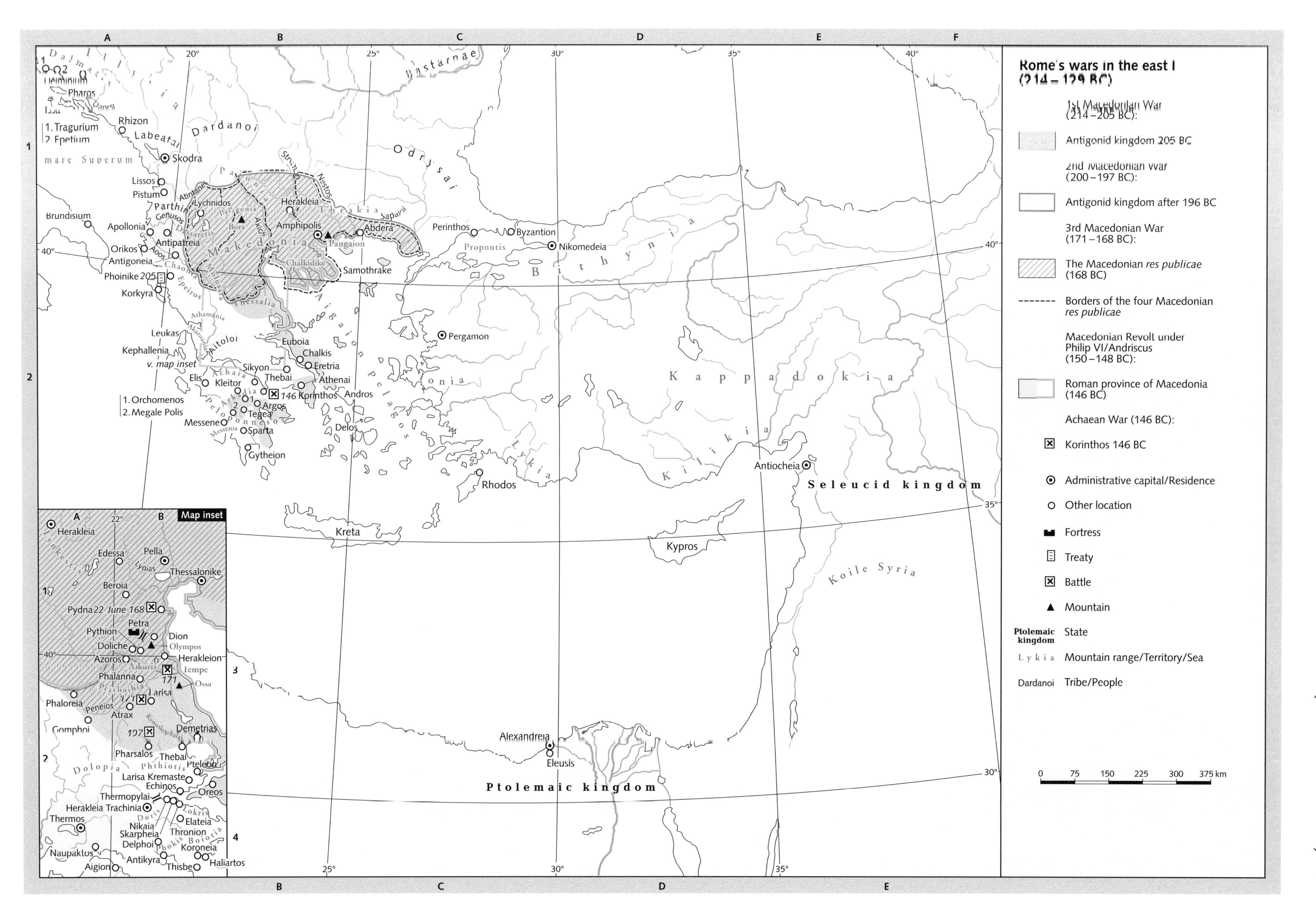
Rome's wars in the east I (214 – 129 BC)
1st Macedonian War (214 – 205 BC):
Antigonid kingdom 205 BC
2nd Macedonian War (200 – 197 BC):
Antigonid kingdom after 196 BC
3rd Macedonian War (171 – 168 BC):
The Macedonian res publicae (168 BC)
Borders of the four Macedonian res publicae
Macedonian Revolt under Philip VI/Andriscus (150 – 148 BC):
Roman province of Macedonia (146 BC)
Achaean War (146 BC):
Korinthos 146 BC
Administrative capital/Residence
Other location
Fortress
Treaty
Battle
Mountain
Ptolemaic kingdom
State
Lykia
Mountain range/Territory/Sea
Dardanoi
Tribe/People
0
75
150
225
300
375 km
Map inset
Herakleia
Edessa
Pella
Thessalonike
Beroia
Pydna 22 June 168
Petra
Pythion
Dion
Olympos
Doliche
Azoros
Herakleion
Tempe
Phalanna
171
Ossa
Larisa
Phaloreia
Peneios
Atrax
Gomphoi
197
Demetrias
Pharsalos
Thebai
Dolopia
Phthiotis
Pteleon
Larisa Kremaste
Echinos
Oreos
Thermopylai
Herakleia Trachinia
Thermos
Doris
Lokris
Elateia
Nikaia
Skarpheia
Thronion
Delphoi
Phokis
Boiotia
Koroneia
Naupaktos
Antikyra
Haliartos
Aigion
Thisbe
Dalmatia
Illyria
Pharos
1. Tragurium
2. Epetium
Rhizon
Labeatai
Dardanoi
mare Superum
Skodra
Lissos
Pistum
Brundisium
Apollonia
Parthini
Atintanes
Lychnidos
Genusos
Antipatreia
Orikos
Aoos
Antigoneia
Chaonia
Phoinike 205
Epeiros
Korkyra
Makedonia
Bora
Axios
Herakleia
Strymon
Nestos
Thrakia
Sapaioi
Amphipolis
Pangaion
Abdera
Chalkidike
Samothrake
Thessalia
Athamania
Aitoloi
Leukas
Kephallenia
v. map inset
Euboia
Chalkis
Eretria
Sikyon
Thebai
Athenai
Elis
Kleitor
Achaia
146 Korinthos
Andros
1. Orchomenos
2. Megale Polis
Argos
Tegea
Messene
Peloponnesos
Messenia
Sparta
Delos
Gytheion
Aigaion Pelagos
Bastarnae
Odrysai
Perinthos
Byzantion
Propontis
Nikomedeia
Bithynia
Pergamon
Ionia
Lykia
Rhodos
Kappadokia
Kilikia
Antiocheia
Seleucid kingdom
Kreta
Kypros
Koile Syria
Alexandreia
Eleusis
Ptolemaic kingdom

Rome's wars in the east II (214–129 BC)

VI. The Syrian War (191–188 BC)

As part of the partition of the Ptolemaic overseas possessions planned with Philip, Antiochus III gained Coele Syria, Phoenice and Judaea. Advised by Hannibal, who had taken refuge with him, he concentrated his ambitions on Asia Minor and Thrace, taking Lysimachea and expanding the city as a residence for his son. He also sought a presence in Greece. Expecting to be welcomed as a liberator, he landed at Demetrias in the October of 192 BC. The Aetolians there proclaimed him commander of the federation. While some Greeks took his side (Chalcis, Elis), though, Philip, the Achaeans, Attalus II and Rhodes sided with Rome. At first, Antiochus contented himself with taking a few Thessalian cities, but the most important of them, Larisa, did not open its gates to him. The actions of Philip and the propraetor M. Baebius Tamphilus had already lost Antiochus Thessaly again by the time the consul, M' . Acilius Glabrio, arrived in April 191 with new troops. The coalition forces under the consul's command were far superior in numbers to the royal army when battle came at Thermopylae. The battle was a catastrophe for Antiochus. He fled with 500 men to Ephesus by way of Chalcis (→ map inset).

In 191, the consul L. Cornelius Scipio decided to take the war to Asia Minor. Although the praetor C. Livius Salinator had won a victory over the royal fleet at Mount Corycus (Ionia) in the autumn of 191, transferring large contingents of troops over the Aegean by ship was still not advisable, so Scipio chose instead to march the army through Thrace to the Hellespont. Antiochus lost maritime supremacy in the summer of 190 BC when his fleet was defeated off Myonnesos by the praetor L. Aemilius Regillus. From Lysimachea, where the Roman army had arrived, Scipio could now cross the Hellespont without risk. Following negotiations, which the king abandoned in view of the Romans' extreme demands, battle was joined in the winter of 190/89 at Magnesia on the Sipylus. Antiochus III suffered a heavy defeat. A truce was followed in the summer of 188 by the final peace. The king gave up all his possessions in Asia Minor, the demarcation line being the Taurus. Antiochus was thus left with only eastern Cilicia (Cilicia Pedias).

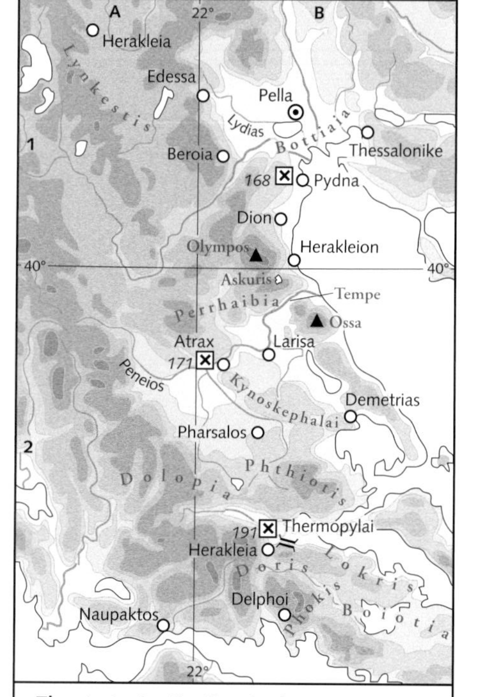

The strategic situation in the region between the Vale of Tempe and Thermopylae

VII. The Roman consul Cn. Manlius Vulso's raid on the Galatians, 189 BC

Following the truce between Rome and Antiochus, the consul Cn. Manlius Vulso undertook a raid on the Galatians of Asia Minor, who had supported the Seleucid king in the recent war. He passed through Caria, Lycia and Pisidia on his way to Galatia, where the Galatians suffered two defeats, at Mt. Olympus (Gordium) and Mt. Magaba. The consul pulled back to Ephesus with his booty-laden army in the autumn of 189. On the march back through Thrace, bound for Rome with the spoils, his army was ambushed by Thracians and the booty thoroughly plundered.

VIII. Rome's war with Eumenes III/ Aristonicus (133–129 BC)

Aristonicus, regnal name Eumenes III, was a genuine son of Eumenes II (197–159) and half-brother of King Attalus III of Pergamum, who had died in the spring of 133 BC. He sought to override Attalus' will, which bequeathed Pergamum to Rome and freed the cities of the kingdom, and to usurp the throne. Eumenes III gained support not so much from the cities (except among the non-Greek population) as from the rural population, including the poor and those in bondage. He gathered them all under a programme of social revolution as the *Heliopolitai* (Citizens of the Sun). The philosopher Blossius, formerly a supporter of Tib. Sempronius Gracchus at Rome, joined him.

Eumenes' bid for power began with a broad-based offensive which even extended beyond the frontiers of the kingdom itself (Hellespont, Mysia, Lydia, Ionia, coast and hinterland) before the Romans intervened to protect their interests in Asia Minor. Many cities formerly subject to the king quickly joined him (Leucae, Thyatira, Apollonis, Stratoniceia), while others formerly autonomous generally opposed him (Pergamum, Bargylia, Ephesus, Colophon, Myndus, Sestus, Cyzicus, Byzantium, but not Phocaea). As early as 132, however, Eumenes already had to accept a defeat, at Cyme in a battle against the fleet of Ephesus. Four Anatolian dynasts – Nicomedes II, Ariarathes V, Mithridates V and Pylaemenes (Paphlagonia) – took part in the struggle against Eumenes III, probably at the urging of the Roman Senate.

In 131, the Romans joined the hostilities with an army under the consul P. Licinius Crassus. Our sources do not permit a clear view of the course of the war. Crassus operated on the Carian and Ionian coasts, where the cities of Mylasa and Halicarnassus took his side.

Early in 130, Crassus besieged the coastal city of Leucae at the northern entrance to the Gulf of Smyrna. There, he was ambushed, defeated, captured and finally killed. M. Perperna, one of the consuls for that year, finally succeeded in defeating Eumenes III in the Caicus Valley at Stratoniceia. He surrounded Eumenes in the city until he capitulated.

M'. Aquillius, a consul for 129, succeeded with some effort in overcoming the last resistance in the Mysian hinterland as far as Caria, and organized the Roman inheritance as the province of Asia.

The maps

The maps on pp. 151 and 153 show eight military campaigns in the Greek-Anatolian region. To prevent the abundance of sometimes highly fragmentary information from obscuring the whole, only the raid of Cn. Manlius Vulso in Asia Minor is marked with arrows.

The sources

Documentation of the period covered by these maps (214–129 BC) varies considerably in terms of sources. For 216-144 BC, the history of Polybius (2nd cent. BC) would be a solid source foundation were it not for its fragmentary preservation for this period. The excerpts compiled according to perspectives of content under the Byzantine Emperor Constantine VII (945-959) are of value here. From Livy, a contemporary of Augustus', only books 21 – 45, on the events of 218-167 BC, survive for the period of relevance to these maps. For the subsequent period, Livy's history is available only in extracts. We have another such précis from P. Annius Florus (1st/2nd cents. AD). The matter of the sources becomes more problematical for the second half of the second century. The histories of various authors – from Diodorus Siculus, a contemporary of Cicero's, to Zonaras, a Byzantine courtier of the 11th/12th cents. – are listed in the table on p. 268. The table indicates which of these authors are relevant for which events marked on the maps.

The strategic situation in the Tempe Valley/Thermopylae region (map inset)

It would probably have been possible for the Romans to achieve their aim of breaking through to the Macedonian east coast by sea. But the risks of such an invasion were very high. Instead, they tried to reach their goal from the west and south by land. The usual approach led from the south, through the pass of Thermopylae and the Tempe gorge formed by the Peneius, then northwards along the coast past Heracleum, Dion and Pydna. The Macedonians therefore had to secure the switch-lines at Thermopylae and Tempe. But both could be circumvented, which Cato the Elder achieved at Thermopylae in 191 and the consul L. Aemilius Paullus at Tempe in 169.

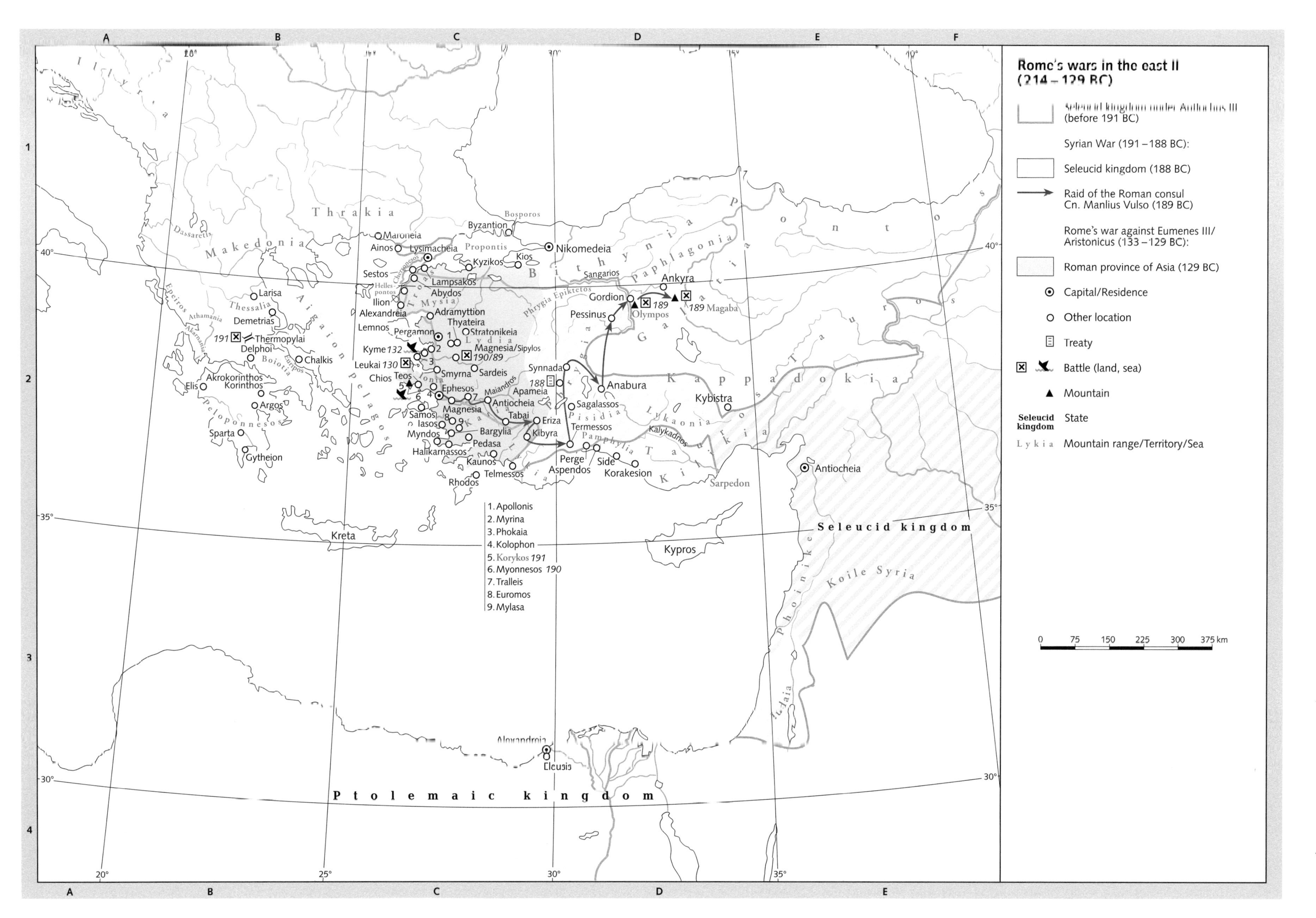

Rome's wars in the east II (214 – 129 BC)
Seleucid kingdom under Antiochus III (before 191 BC)
Syrian War (191 – 188 BC):
Seleucid kingdom (188 BC)
Raid of the Roman consul Cn. Manlius Vulso (189 BC)
Rome's war against Eumenes III/ Aristonicus (133 – 129 BC):
Roman province of Asia (129 BC)
Capital/Residence
Other location
Treaty
Battle (land, sea)
Mountain
Seleucid kingdom
State
Lykia
Mountain range/Territory/Sea
0 75 150 225 300 375 km
1. Apollonis
2. Myrina
3. Phokaia
4. Kolophon
5. Korykos 191
6. Myonnesos 190
7. Tralleis
8. Euromos
9. Mylasa
Illyria
Thrakia
Makedonia
Dassaretis
Epeiros
Athamania
Akarnania
Thessalia
Boiotia
Euripos
Peloponnesos
Aigaion pelagos
Larisa
Demetrias
191
Thermopylai
Delphoi
Chalkis
Akrokorinthos
Korinthos
Elis
Argos
Sparta
Gytheion
Maroneia
Ainos
Lysimacheia
Propontis
Byzantion
Bosporos
Nikomedeia
Kios
Kyzikos
Sestos
Hellespontos
Lampsakos
Abydos
Ilion
Alexandreia
Adramyttion
Lemnos
Mysia
Thyateira
Pergamon
Stratonikeia
Lydia
Kyme 132
Magnesia/Sipylos
190/89
Leukai 130
Smyrna
Sardeis
Chios
Teos
Ionia
Ephesos
Maiandros
Apameia
188
Synnada
Samos
Magnesia
Antiocheia
Iasos
Tabai
Eriza
Myndos
Bargylia
Kibyra
Pedasa
Halikarnassos
Kaunos
Telmessos
Rhodos
Bithynia
Sangarios
Paphlagonia
Phrygia Epiktetos
Gordion
Ankyra
189
Olympos
189 Magaba
Pessinus
Galatia
Anabura
Sagalassos
Pisidia
Termessos
Pamphylia
Lykaonia
Kalykadnos
Perge
Aspendos
Side
Korakesion
Kilikia
Kybistra
Kappadokia
Pontos
Tauros
Sarpedon
Antiocheia
Seleucid kingdom
Kreta
Kypros
Koile Syria
Phoinike
Judaia
Alexandreia
Eleusis
Ptolemaic kingdom

Roman colonization

I. Introduction

It was certainly not a purpose of ancient 'colonization', as in the modern sense, to establish dominion over wide expanses of territory. Rather, the newly-established settlements were largely or entirely independent of the settlers' home city in both economic and political terms. Roman colonization was an exception among colonization movements of the ancient world insofar as it served the primary purpose of militarily and politically securing Roman rule. Only in a secondary and later phase did it acquire the purpose, divorced from this political aim, of supplying veterans and the urban Roman proletariat with land and a new means of existence.

As a rule, the Romans set up their colonies on the territories of conquered cities. Originally, it was the people who took the initiative to dispatch a colony. From the end of the 2nd cent. BC, holders of the *imperium* also did so, and in the last phase the order came from the emperor. After the territory had been measured, the colony was founded in accordance with traditional rituals. The *tresviri coloniae deducendae* issued the settlement's constitution. New settlers (*supplementa*) were from time to time sent into existing colonies, e.g. to Cosa, Aquileia, Minturnae, Bononia and Placentia. The Romans had two forms of colony:

1. Colonies of Roman citizens (*coloniae civium Romanorum*) were set up by around 300 colonists (*coloni*) with their families. Each family was allotted two *iugera*, i.e. 0.5 ha, of land. Later, the citizens' colonies founded between 184 and 157 BC had up to 2,000 settlers (recorded for Mutina, Parma and Luna), each allotted between 5 and 10 *iugera*. The colonists retained their Roman citizenship and their community remained part of the Roman commonwealth, so that they enjoyed only a limited degree of self-rule under the leadership of *praetores* or *duoviri*.

2. The Romans were already taking part in various colonial foundations during the Monarchical Period, as members of the Latin League. Following the dissolution of the League after the Battle of Trifanum in 338 BC, in which the Romans decisively defeated the Latins, Rome continued to establish colonies in Italy in line with Latin tradition. The Romans did meanwhile keep to their custom of sending out citizens' colonies, too: most of these *coloniae civium* were coastal cities, which the sources call *coloniae maritimae*.

The Latin colonies set up after 338 BC served unambiguously military purposes, i.e. the protection of the region of Roman rule (cf. Aquileia at the invader's gateway into Italy after crossing the Julian Alps, Placentia and Cremona against the ever-unruly Ligurians). They were much larger in terms of territory and population than the early citizens' colonies, and a local census gave them their own military units who took the field with the Roman allies. They ruled themselves, and their citizens did not enjoy Roman citizenship.

As a result of the Social War (91–87 BC), communities of Gallia transpadana were promoted into the ranks of Latin colonies without receiving new colonists. In the wake of this development, most of the colonial foundations undertaken through the 1st cent. BC under Caesar and Augustus in Gallia transalpina and on the Iberian Peninsula also took place with no dispatch of colonists. In the end, the establishment of 'true' colonies became ever more the exception. In most cases, existing cities were simply accorded the title of *colonia*. These are referred to generally as 'titular' or 'honorary' colonies. Distinguishing between these and actual colonies can be difficult if not by now impossible in individual cases.

Latin colonies ceased to be established in the 1st cent. BC, Roman citizens' colonies early in the 2nd cent. AD. In total, there is epigraphic or literary evidence of around 400 Roman and Latin colonies, their legal statuses highly diverse. They were particularly highly regarded in the Imperial period, and across the Empire they functioned as a significant element of Romanization. – In clarification of the *coloniae maritimae*: this appears not to have been a *terminus technicus*, but merely a term used by the sources denoting colonies founded by the sea (e.g. Antium, Croton, Brundisium, Puteoli, Minturnae). These are therefore not specifically highlighted on the map.

II. Colonial foundations in Italy to 134/33–121 BC (→ map p. 109)

13 colonies of the Latin League are attested which were founded with the sometimes considerable participation of the Romans prior to 338 BC: Velitrae (494?), Suessa Pometia (494?), Signia (494?), Cora (494?), Norba (492), Ardea (442), Labici (418), Vitellia (394), Circeii (393), Satricum (385), Sutrium (383), Nepet (383) and Setia (382).

It is striking that only four Latin colonies were established after the catastrophic Gaulish invasion of 387/86 BC. The other 21 colonies were founded between 338 and 201 BC (i.e. after the end of the Latin War) under Roman leadership.

In the period after the 2nd Punic War (218-201), the Romans established many colonies, but after 157 apparently not a single one. There was probably no more land available. For this period, the map shows 22 colonial foundations, including five under Latin law. *Citizens' colonies* were: Castra Hannibalis? (199), Volturnum (194), Tempsa (194), Sipontum (194, 186), Salernum (194), Puteoli (194), Liternum (194), Croton (194), Buxentum (194, 186), Pisaurum (184), Potentia (184), Saturnia (183), Parma (183), Mutina (183), Graviscae (181), Luna (177) and Auximum (157). *Latin colonies* were: Copia (193), Vibo Valentia (192), Bononia (189), Aquileia (181) and Luca (180/178).

III. Colonial foundations in Italy after 134/33–121 BC (→ map p. 155)

The lack of arable land in Italy for further settlements is explained by the generally tolerated habit of occupying more *ager publicus* than the law permitted. It is therefore understandable that the calls for new legal measures to limit this practice grew ever more strident. Such a law was implemented by the people's tribune for 133 BC, Tib. Sempronius Gracchus. A commission was constituted, charged with distributing the available parcels of land. This agricultural commission continued its work even after the violent death of Tib. Gracchus. Its efficacy was severely compromised by the withdrawal of its adjudicatory competence in 129. C. Gracchus, Tiberius' younger brother, revived his brother's plans for agrarian reform. But when he, too, was murdered in 121, and the sale of the parcels of land already allotted was declared legal, the agrarian commission's activities fell entirely into abeyance. Some colonies established in the course of the Gracchan agrarian reforms are attested, e.g. Fabrateria Nova (124) to the south-east of Fregellae, a city destroyed in the previous year, and the *colonia maritima* Neptunia near Tarentum.

IV. Colonial foundations outside Italy

Before Caesar, only a few colonies were set up outside Italy. The first was Carteia in Hispania Baetica in 171, a colony under Latin law. The second colony on Iberian soil was Valentia in the territory of the Edetani on the lower reaches of the Turia. It was set up early in the 1st cent. BC. The first citizens' colony outside Italy was the Colonia Iunonia Carthago set up on a legislative initiative of C. Gracchus in 122. It was named after Tinnit, the tutelary goddess of the city of Carthage in her *interpretatio Romana* guise as Juno. After the death of C. Gracchus in 121, the settlement was stripped of its status as a citizens' colony. Caesar revived the plan in 45 BC and brought it to fruition (Colonia Iulia Carthago). – As part of the organization of the province of Gallia transalpina, the citizens' colony of Narbo Martius was established in 118 on the road from the mouth of the Rhône to Spain. It became the administrative capital of the new province of Gallia Narbonensis. Marius (Colonia Mariana, *c.* 100 BC) and Sulla (Colonia Veneria Aleria, *c.* 80 BC) successively set up colonies on Corsica. – Other colonies and titulary colonies set up outside Italy at various times are shown on other maps: Spain (p. 145), Gaul (p. 167), Britain (p. 193), the upper and lower Danubian provinces (p. 189), the northern Balkans (p. 184 f.), the Balkan peninsula (p. 187), Asia Minor (p. 183), the Levant (p. 181), North Africa (p. 177) and the islands of Crete (p. 163), Sicily, Sardinia and Corsica (p. 143). Egypt was without colonies, a fact explained by the special status of the province of Aegyptus. Pliny (Plin. HN 5,128) does report a colony on the island of Pharos, set up by Caesar while dictator, but it appears that Augustus withdrew its colonial status.

The sources

Inscriptions are our most reliable sources for knowledge of Roman colonial history, and there are also coins and archaeological evidence in individual cases. The literary transmission in the works of Livy and Pliny ('Natural History') and the writings of the Roman land surveyors are far less revealing by comparison.

The maps

The commentary deals with the situations documented in the maps on pp. 109 and 155. The map on p. 109 shows, insofar as it is deducible from the sources, Roman colonization from its beginnings under the Roman kings, through the Gaulish invasion of 387/86 BC, the dissolution of the Latin League (which had hitherto undertaken colonization) and the Second Punic War up to the ebb of colonization activities after 157 BC. The map on p. 155 deals with Roman colonization activities in connection with the Gracchan agricultural reforms, which by their nature promised a considerable increase in foundations. However, after the failure of the Gracchi, colonial foundations were suspended again, until more were again gradually established to provide for the flood of veterans from the numerous campaigns of the warlords Marius, Sulla, Lucullus and Pompey. Marius was said to have settled around 80,000 veterans in newly-founded colonies, and around 350,000 veterans were supposedly accommodated in colonies between 59 BC and AD 14. No more actual new foundations were made after Trajan, the granting of the title of *colonia* mutating into an effective imperial political tool in relation to the cities of the Empire.

→ Map p. 161

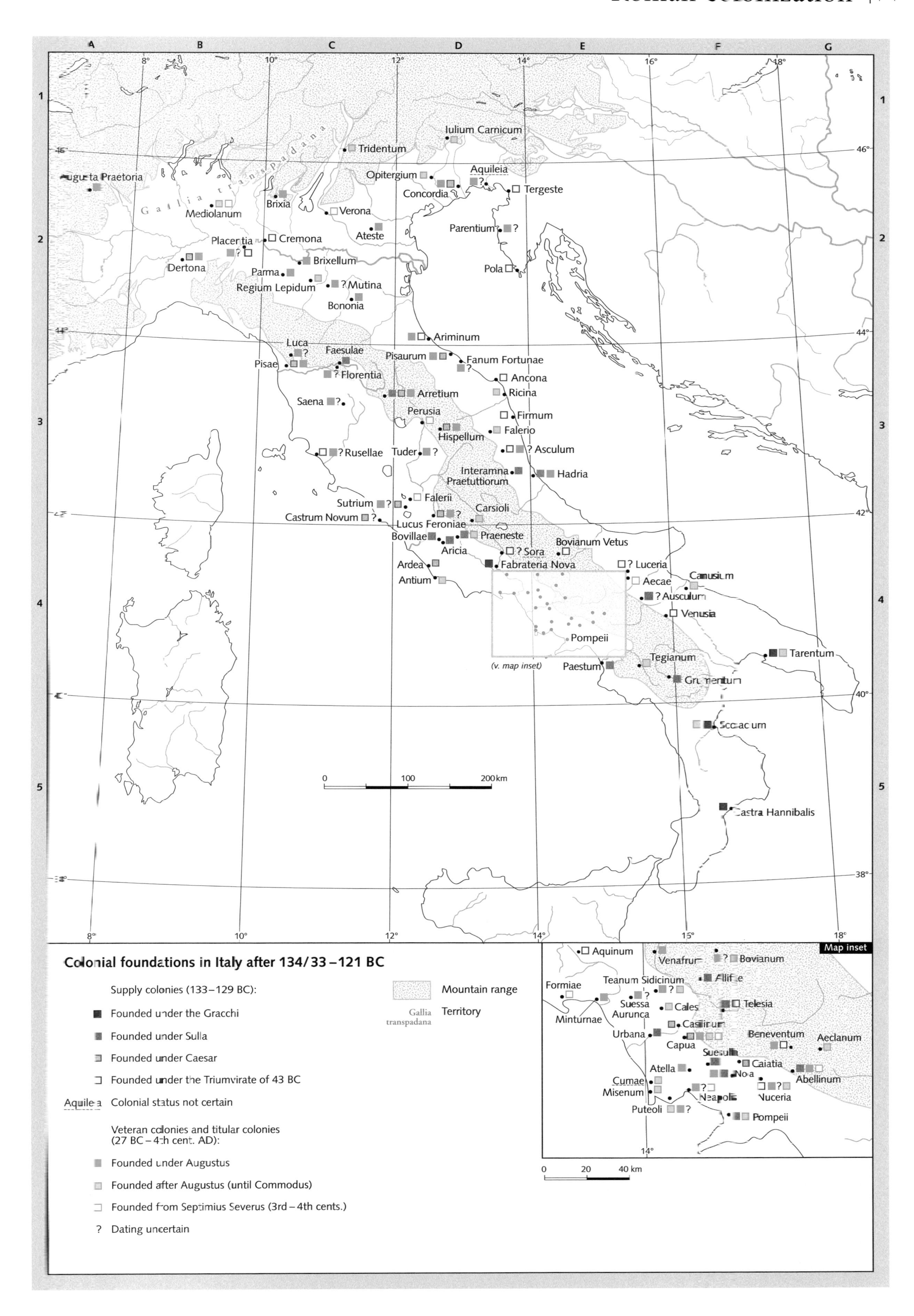
Colonial foundations in Italy after 134/33–121 BC
Supply colonies (133–129 BC):
Founded under the Gracchi
Founded under Sulla
Founded under Caesar
Founded under the Triumvirate of 43 BC
Aquileia Colonial status not certain
Veteran colonies and titular colonies (27 BC – 4th cent. AD):
Founded under Augustus
Founded after Augustus (until Commodus)
Founded from Septimius Severus (3rd – 4th cents.)
? Dating uncertain
Mountain range
Gallia transpadana Territory
Map inset
Augusta Praetoria
Mediolanum
Brixia
Verona
Tridentum
Iulium Carnicum
Opitergium
Concordia
Aquileia
Tergeste
Parentium
Pola
Ateste
Cremona
Placentia
Dertona
Brixellum
Parma
Regium Lepidum
Mutina
Bononia
Ariminum
Luca
Pisae
Faesulae
Florentia
Pisaurum
Fanum Fortunae
Ancona
Ricina
Arretium
Saena
Perusia
Hispellum
Firmum
Falerio
Asculum
Rusellae
Tuder
Interamna Praetuttiorum
Hadria
Sutrium
Falerii
Castrum Novum
Lucus Feroniae
Carsioli
Bovillae
Aricia
Praeneste
Sora
Bovianum Vetus
Ardea
Fabrateria Nova
Antium
Luceria
Aecae
Canusium
Venusia
Pompeii
(v. map inset)
Paestum
Tegianum
Tarentum
Scolacium
Castra Hannibalis
Aquinum
Venafrum
Bovianum
Formiae
Teanum Sidicinum
Allifae
Suessa Aurunca
Cales
Telesia
Minturnae
Casilinum
Urbana
Capua
Beneventum
Aeclanum
Suessula
Atella
Caiatia
Nola
Abellinum
Cumae
Misenum
Neapolis
Nuceria
Puteoli
Gallia transpadana
0 100 200 km
0 20 40 km

The Roman Social War (91– 87 BC)

This war grew out of a revolt within the Roman alliance system (*bellum Marsicum, bellum Italicum, bellum sociale*). In the main, it was the Italic *socii* of central and southern Italy who fought the Romans, and their real purpose was to obtain Roman citizenship for all the Italic peoples. It is possible that some Italic groups were also striving for complete independence from Rome (according to MOURITSEN, it was their intention to win more participation in the process of political governance). The conflict had long been brewing, fuelled by the Italic peoples' growing discontent at their situation. War was triggered by the assassination of the M. Livius Drusus, who as people's tribune at Rome had endeavoured to achieve the legal equivalence of the Italic *socii* with Roman citizens. The result of this bloody war was at all events impressive: the inhabitants of all of Italy south of the Padus now became Roman citizens. The Roman allies who had fought so many battles shoulder to shoulder with the Romans were now essentially equal to them in political terms.

The battles, bitterly fought on both sides and initially posing a serious threat to the very survival of Rome, mainly took place in central and southern Italy, from Picenum to Campania to Calabria and Bruttium. Rome was opposed by the Picentestini, Marsi, Paeligni, Vestini, Marrucini, Frentani, Hirpini, Samnites, Lucani and Poediculi, with Celts from Gallia cisalpina also joining in. Osci also joined the rebels in the Campanian cities, such as Pompeii. Overall, the Italians may have had 100,000 men of all military branches in the field, but over the course of time it would become clear that Rome had by far the greater resources to deploy. The Italians' political centre was at Corfinium (which they now, with great symbolic resonance, renamed Italica), an important crossroads in the territory of the Paeligni. Apart from Venusia, all the Latin colonies remained loyal to Rome, as did the Greek coastal cities. Moreover, the tribes and communities of the Umbri and Etrusci defected only at a late stage to the Italic side, and they were also the first to avail themselves of the opportunity to take Roman citizenship offered them by the Roman people in late 90 BC with the *lex Iulia de civitate* (*sociis danda*).

Military conflict began in the autumn of 91 BC with the murder of the Roman praetor Q. Servilius in the theatre at Asculum. The Italic strategy consisted in the first place of isolating and eliminating the Roman bases (i.e. the Latin colonies and prefectures) scattered throughout Italy. Thus, within a few days, all Roman colonies were thrown back on their own resources, their access roads blocked by Italic forces. On the Roman side, the first priority was to reinstate the connection between Rome and the Adriatic, thereby to split the rebel forces in two.

The Italic side enjoyed some success at first, especially in Campania, Picenum and Apulia. For instance, led by Vettius Scato, they defeated the troops of the consul L. Iulius Caesar at Aesernia. But reverses then began to mount, and through the course of the war they were twice compelled to relocate their political and military centre, in the spring of 89 from Italica (formerly Corfinium) to Bovianum, and the summer of the same year to Aesernia. In the northern battle zone, a battle was fought on the Tolenus west of Alba Fucens on 11 June 90 BC; in it the Romans, led by Marius, defeated the Paeligni under Vettius Scato. Shortly afterwards, Sulla defeated the Marsi south of lacus Fucinus. The Romans thus succeeded, step by step, in splitting the armies of the rebels along the Rome-Adriatic line. North of this line, the Romans under Pompeius Strabo won a victory at Asculum, probably in October 90, even conquering Corfinium after fierce fighting, so that they now controlled this theatre of war. The Italic stronghold was surrounded. It was probably in this precarious situation that the Italians attempted in vain to forge an alliance with the Pontic King Mithridates VI. However, the Senate viewed the chances of a satisfactory conclusion to the war as so remote that – still in 90 BC – they relented to an extent, on the basis of a law introduced by the consul L. Iulius Caesar (*lex Iulia de civitate*), granting Roman citizenship to those Italic groups which had remained loyal to Rome, and allowing commanders to grant citizenship to Italians (ILS 8888). Thus began a development which, through two further pieces of legislation, would finally lead to the settlement of the conflict. A few months later, two people's tribunes sponsored a more far-reaching law (*lex Plautia Papiria*), according to which all Italian rebels south of the Padus would receive Roman citizenship provided that they laid down their arms within 60 days and presented themselves to the praetor at Rome. For the moment, it is true that the battles raged on undiminished – Strabo defeated the Marsi at Asculum, Sulla the Samnites under L. Cluentius at Nola and under Papius Mutilus at Aesernia. Later, he also took Bovianum after fierce fighting. A law introduced by the consul Cn. Pompeius Strabo, the father of the *triumvir*, awarded Roman citizenship to the Gaulish cities south of the Padus, including those in Gallia cisalpina, and Latin citizenship to all *socii* north of the Padus. One of the last Roman successes in the field was the victory of the propraetor Q. Caecilius Metellus Pius over the Marsian Q. Poppaedius Silo at Aesernia early in 88.

The map

We limit ourselves in the map to showing those localities, rivers, lakes and mountain ranges which were of some importance in the context of the Roman Social War. The most decisive successes of either party to the war are highlighted. The roads of the Appennine peninsula played an important part, determining troop movements, and so the most important are shown here. The map reveals that the centre of conflict was clearly to the east and south-east of Rome in the Appenninus and by the Adriatic, and that Italy to the north of Rome was much less affected by the war.

The sources

The contemporary accounts of the Social War are lost to us: the history of Posidonius, the histories of Sisenna, the memoirs of Sulla and the Greek monograph on the Social War by Lucullus. Our main source is Appian, a contemporary of the emperors Trajan and Hadrian, and the 13th book of his *Roman History*. Other reports, meanwhile, e.g. of the Augustan historians Diodorus Siculus and Livy, are preserved only in fragments. Among the few inscriptions to date from these years, two edicts of the consul Pompeius Strabo stand out, made while he was with his army outside Asculum in Picenum (ILS 8888). In these, he grants Roman citizenship to an entire division of Spanish cavalry. The annex contains the names of officers in the consul's *consilium* – including those of the future Pompeius Magnus, Cicero and Catilina. – The Italians' coins present a particular language. On their obverse, these coins show the central theme of the entire movement, ITALIA, and the reverse shows various scenes of symbolic importance to the Italians' main concern (SYDENHAM no. 617 ff.).

→ Map p. 137

Sources

Motives of the rebels: Rhet. Her. 13; *Celts of Gallia cisalpina:* App. Civ. 1,219 f.; *Troops from Gallia cisalpina:* Str. 5,4,2; *Italica:* Vell. Pat. 2,16,4; Diod. Sic. 37,2,6 f.; *Troop strengths of the Italic tribes:* App. Civ. 1,177; *Murder of the praetor at Asculum:* Liv. Per. 72; App. Civ. 1,171–174; *Campania:* Liv. Per. 72 f.; App. Civ. 1,183; 199; *Picenum:* App. Civ. 1,204; *Apulia:* App. Civ. 1,190; *Scato's victory at Aesernia:* App. Civ. 1,182; *Battle of the Tolenus:* App. Civ. 1,191–195; Ov. Fast. 6,563; *Battle of lacus Fucinus:* App. Civ. 1,21–203; *Battle of Asculum:* App. Civ. 1,205 f.; *Mithridates VI:* Posidonius FGrH 87 F 36; Diod. Sic. 37,2,11; *lex Iulia:* Cic. Balb. 21; Gell. NA 4,4,3; App. Civ. 1,212–214; *lex Plautia Papiria:* Cic. Arch. 7; *lex Pompeia:* Ascon. in Pis. 2; Plin. HN 3,138; *Strabo's victory at Asculum:* App. Civ. 1,216; *Battle of Corfinium:* App. Civ. 1,127; Diod. Sic. 37,2,9; *Sulla's victory at Nola:* App. Civ. 1,217–221; Cic. Div. 1,72; *Sulla's victory at Aesernia:* App. Civ. 1,223 f.; *Battle of Bovianum:* App. Civ. 1,223–225.

Literature

P.A. BRUNT, Italian aims at the time of the Social War, in: JRS 55, 1965, 90–109; C. CICHORIUS, Römische Studien, 1922, 130–185; W. DAHLHEIM, Der Staatsstreich des Konsuls Sulla und die römische Italienpolitik der achtziger Jahre, in: J. BLEICKEN (ed.), Colloquium aus Anlaß des 80. Geburtstages von A. Heuß (Frankfurter Althistorische Studien 13), 1993, 97–116; E. GABBA, Rome and Italy. The Social War, in: CAH 9, ²1994, 104–128; H. GALSTERER, Rom und Italien vom Bundesgenossenkrieg bis zu Augustus, in: M. JEHNE, R. PFEILSCHIFTER (eds.), Herrschaft ohne Integration? Rom und Italien in republikanischer Zeit (Studien zur Alten Geschichte 4), 2006, 293–308; A. KEAVENEY, Sulla, the Marsi and the Hirpini, in: CPh 76, 1981, 292–296; H.D. MEYER, Die Organisation der Italiker im Bundesgenossenkrieg, in: Historia 7, 1959, 74–7[illegible]; H. MOURITSEN, Italian Unification. A study in ancient and modern historiography, 1998; S. PEAKE, A note on the dating of the Social War, in: G&R 44, 1997, 161–164; E.T. SALMON, Notes on the Social War, in: TAPhA 89, 1959, 184; E.A. SYDENHAM, The Coinage of the Roman Republic, 1952, no. 617 ff.

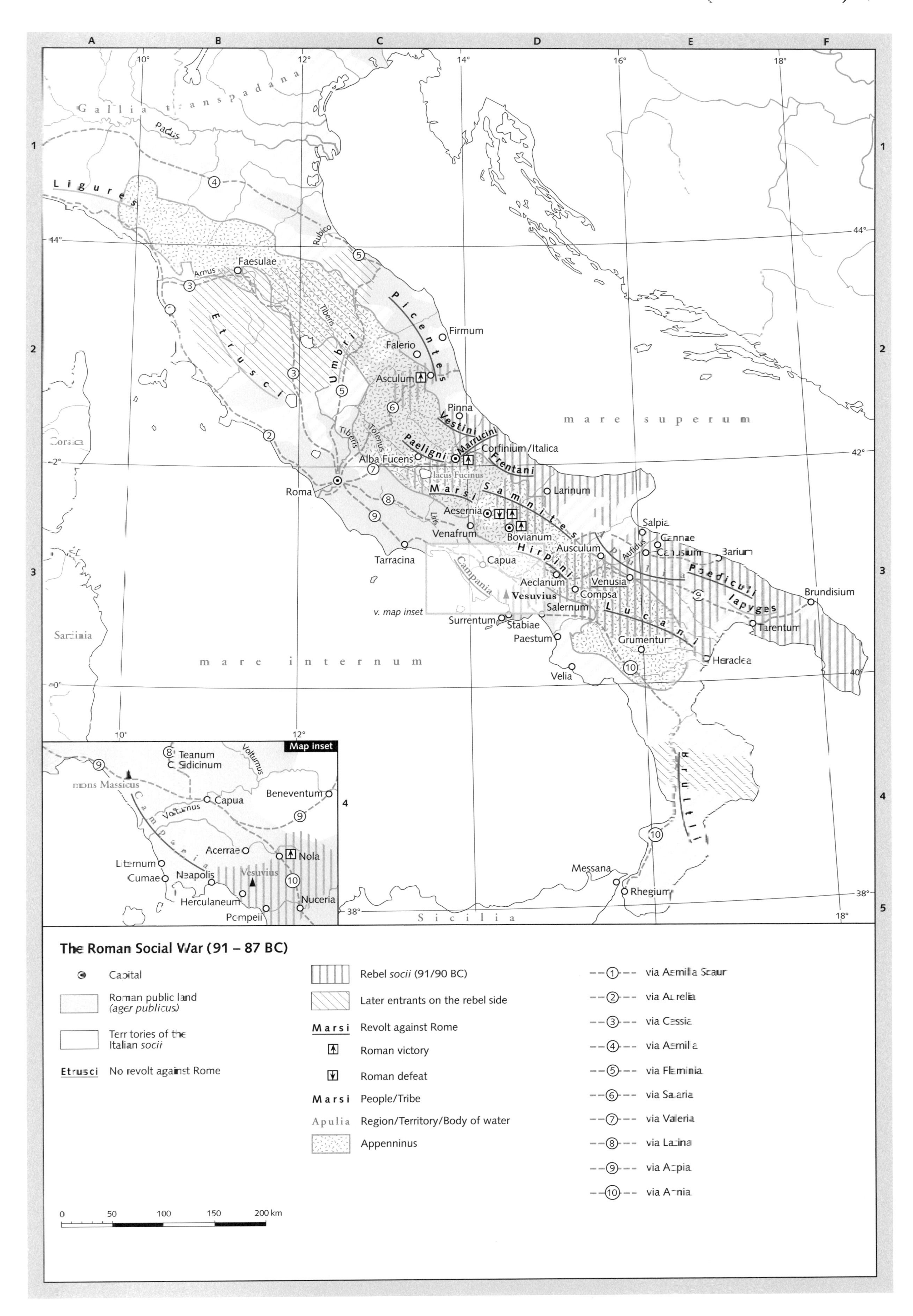
Gallia transpadana
Padus
Ligures
Rubico
Faesulae
Arnus
Etrusci
Tiberis
Umbri
Picentes
Firmum
Falerio
Asculum
Pinna
Vestini
Marrucini
Paeligni
Corfinium/Italica
Frentani
Tolenus
Alba Fucens
lacus Fucinus
Roma
Marsi
Samnites
Larinum
Aesernia
Bovianum
Liris
Venafrum
Hirpini
Tarracina
Capua
Campania
Aeclanum
Ausculum
Aufidus
Salpia
Cannae
Canusium
Barium
Venusia
Compsa
Vesuvius
Salernum
Poediculi
Iapyges
Brundisium
Tarentum
Lucani
Surrentum
Stabiae
Paestum
Grumentum
Heraclea
Velia
v. map inset
Corsica
Sardinia
mare superum
mare internum
Bruttii
Messana
Rhegium
Sicilia
Map inset
Teanum Sidicinum
Volturnus
mons Massicus
Capua
Beneventum
Acerrae
Nola
Liternum
Cumae
Neapolis
Herculaneum
Pompeii
Nuceria
The Roman Social War (91 – 87 BC)
Capital
Roman public land (ager publicus)
Territories of the Italian socii
Etrusci No revolt against Rome
Rebel socii (91/90 BC)
Later entrants on the rebel side
Marsi Revolt against Rome
Roman victory
Roman defeat
Marsi People/Tribe
Apulia Region/Territory/Body of water
Appenninus
via Aemilia Scaur
via Aurelia
via Cassia
via Aemilia
via Flaminia
via Salaria
via Valeria
via Latina
via Appia
via Annia
0 50 100 150 200 km

The three Roman wars against Mithridates VI (89–85, 83/82, 74–63 BC)

I. The 1st Mithridatic War (89–85 BC)

From the outset, Mithridates VI pursued the same path as his grandfather, Pharnaces I, and his father, Mithridates V, in conducting a policy characterized by aggressive diplomacy and military intervention. This won him dominion of the Bosporan kingdom in 115/14, then protectorship of the trading cities on the north-western coast, from Olbia to Apollonia. In Asia Minor, he won Colchis and Armenia on the right bank of the Euphrates and eastern Paphlagonia in 104/03. Arrogating a role policing conflicts here and in Galatia and Cappadocia, he came into contact with the Roman Senate for the first time. His attempt to subject the Bithynian kingdom and so fashion the Black Sea into a *mare Mithradaticum* after the death of Nicomedes III in 94 BC led to the first military conflict with Rome.

In 90 BC, when the Senate regulated the details of the succession in Bithynia (Nicomedes IV) and Cappadocia (Ariobarzanes I), and when in both cases Mithridates failed to pursue his claims, the Roman emissary, M'. Aquillius, interpreted this compliance as weakness and incited Nicomedes to attack Pontus. This was a foolhardy provocation of the highly-armed Pontians, and Mithridates now in turn invaded Bithynia and the Roman province of Asia, encountering sporadic resistance (Laodicea on the Lycus, Magnesia on the Sipylus, Rhodes). The 'Asiatic Vespers', in which 80,000 Italians are said to have been massacred in a single day in Mithridates' new territories following a proclamation at Ephesus, compromised most Greek cities' relations with Rome, and bound them more closely to the king. In the autumn of 88, Mithridates sent two armies to Greece, where they were welcomed almost everywhere (except at Sparta and Thespia). Mithridates may have seen Athens, under Aristion, as the administrative capital of a European satrapy.

When L. Cornelius Sulla crossed to Epirus with a Roman army in the spring of 87, Mithridates was at the height of his powers. The rapidity with which Sulla took control of the situation is therefore astonishing. He moved against Athens, besieging the city and taking it on 1 March 86 BC. He then went on to annihilate Pontic armies at Chaeronea and, in the autumn of 86, at Orchomenus.

Sulla now took the war to Asia Minor. In the meantime, though, his *populares* enemies at Rome had declared him an enemy of the people (*hostis publicus*). For this reason, the efforts of L. Licinius Lucullus on Sulla's behalf to persuade Rome's allies to assemble a fleet to fight Mithridates did not meet with universal cooperation. From 86, a *popularis* army under C. Flavius Fimbria was operating against Mithridates in competition with Sulla. Fimbria had accompanied the consul and *popularis* L. Valerius Flaccus to the east as legate, but had soon fallen out with him and, following an army mutiny in which Flaccus was killed, he took command of the army himself. He destroyed Ilium, conquered Bithynia and trapped Mithridates in Pitane. The king would have fallen into Roman hands at this early stage, had Lucullus, in spite of the ships he had by now succeeded in recruiting, not let him escape to sea.

Hence, it was domestic political constraints that compelled Sulla in 85 BC to impose a hasty peace on the king at Dardanus. This restricted his rule to Pontus once more, but the treaty was only a verbal one, not fixed in writing.

II. The 2nd Mithridatic War (83/82 BC)

Re-armament, which Mithridates VI set about immediately after the Treaty of Dardanus in 85, provided the propraetor of Asia, L. Licinius Murena, with a pretext for an attack on the Pontic monarch. Murena's incursions from Cappadocia into Pontus in 83/82 were no more than raids, not a declared war. But in 82, Mithridates succeeded in bringing Murena to battle and defeating him. A command from Sulla, now dictator, to Murena to desist, prevented Mithridates from exploiting this victory. No form of peace settlement was attempted. Murena founded the city of Licinea in the borderlands between Cappadocia and Pontus.

III. The 3rd Mithridatic War (74–63 BC)

Late in 74, on the death of King Nicomedes IV of Bithynia, Mithridates was armed to the teeth and diplomatically well-equipped with contacts, secured by marriage or treaty, with the Armenian King Tigranes II, the Ptolemies, Q. Sertorius in Spain and the pirates who operated worldwide. He felt strong enough to take the provocative step of denouncing as a forgery Nicomedes' will, which named Rome as heir to the kingdom. He then enthroned his own pretender in Nicomedia in 73. The Senate could not but see this as a declaration of war. For the sake of clarity, the war is divided here into five phases:

First phase of war (73/72): Mithridates secured Pontus' borders to the south by sending his general Diophantus into Cappadocia. The king himself commanded an army of invasion through Bithynia, accompanied by the fleet, and he defeated the land army of the consul M. Aurelius Cotta at Calchedon, sinking his ships in the harbour. Without taking the city, the king proceeded along the Propontic coast to Cyzicus. He spent the whole winter of 73/72 besieging this city, whose defences were organized by L. Licinius Lucullus, until in the following spring, stalemated by Lucullus, he fled with his fleet, first to Nicomedia and then, in the summer of 72, to Pontus. His land army was largely destroyed in flight from the Romans.

Second phase of war (71/70): Lucullus pursued the king to Pontus and besieged Amisus with the royal residence of Eupatoria. The city was finally conquered, in dramatic circumstances. However, Mithridates shut himself away in the fortress of Cabira in 71 to assemble a new army. Within a short time, he had gathered infantry and cavalry in formidable numbers. There were constant skirmishes with Roman troops in the plain of Phanaroea, but neither side achieved a decisive success. The king then relinquished his safe fortress, for reasons that are not clear, and sought refuge with his son-in-law Tigranes, the Armenian king. His army drifted away, and after a number of defeats at the hands of Roman units and a catastrophic storm on the voyage from Nicomedia into the Black Sea, his fleet was no longer serviceable. Meanwhile, Lucullus conquered Heraclea Pontica, Tium, Amastris, and finally also the two residences of Sinope and Amasea. The entire Pontic heartland was thus in Roman hands by the summer of 70 BC. Tigranes interned Mithridates for twenty months in one of his own fortresses, until it seemed advantageous to him to free him and send him into battle against Lucullus in Pontus with an army of 10,000 men.

Third phase of war (69–67): In the end, Lucullus sent a demand for Mithridates' extradition through his brother-in-law P. Clodius Pulcher, whom he sent to Tigranes as an emissary. The demand was not met. Lucullus therefore invaded Armenia. Tigranes was now unable to spare Mithridates and the army he had provided for him, so he kept him back in Armenia. In spite of this, after fierce fighting, Tigranes lost his residences of Tigranocerta and Artaxata. Nisibis also fell into Roman hands in spring 67, after months of siege.

Fourth phase of war (68/67): Mithridates finally entered Pontus in the autumn of 68 with 8,000 cavalry. After several successes, especially a victory over Lucullus' legate, C. Valerius Triarius, in the spring of 67 at Zela, he was once more lord of his own domain, and was even able to win back parts of Bithynia and Cappadocia. The Roman army mutinied against Lucullus, whose *imperium* the Senate had withdrawn, and it consequently no longer posed an acute danger to the king.

Fifth phase of war (66–63): The Pontic king's position was not as strong as it seemed. He had by now lost important allies. Sertorius had been killed in 72, the pirates had been eliminated by Cn. Pompeius (Pompey) in 67, Tigranes had been put out of action, and the Romans had forced the Parthian king Phraates III to declare neutrality. Now, in the spring of 66 BC, Pompey took command of the war against Mithridates. In Lesser Armenia on the right bank of the Euphrates, the armies faced off in a war of position, in the course of which Mithridates withdrew ever farther to the north east. Pompey inflicted a final, annihilating defeat on the king at the place where he would subsequently found Nicopolis.

Mithridates himself escaped. In the winter of 66/65, he withdrew by way of Dioscurias to Phanagoria. There, in 63 BC, his son Pharnaces II finally compelled him to take his own life. At the news of the death of Mithridates VI, Pompey, who was in Judaea at the time, returned to Amisus to direct affairs in Pontus and Bithynia.

The sources

The story of Mithridates as shown in this map was told in the literature of the time and of later periods. Plutarch, however, does not seem to have written his biography. The historians made up for this with their avid interest. Appian, for instance, (2nd cent. AD), devoted an entire book (12: *Mithridateios*) of his *Roman History* to the king. Memnon of Heraclea Pontica (probably 2nd cent. AD) transmitted much important information on Mithridates VI in his chronicle of his home city, which is preserved for us only in fragments. Understandably, the historically-proficient geographer Strabo, being related to Mithridates VI on his mother's side and having been born in the Pontic residence of Amasea, refers often to the king in his geographical work (his historical treatise is lost). Inscriptions attest, among other things, to the reception granted to the king by various communities. The coins issued in the Pontic kingdom give eloquent testimony – the large mercenary armies in Mithridates' service evidently generated a considerable demand for them.

→ Map p. 161

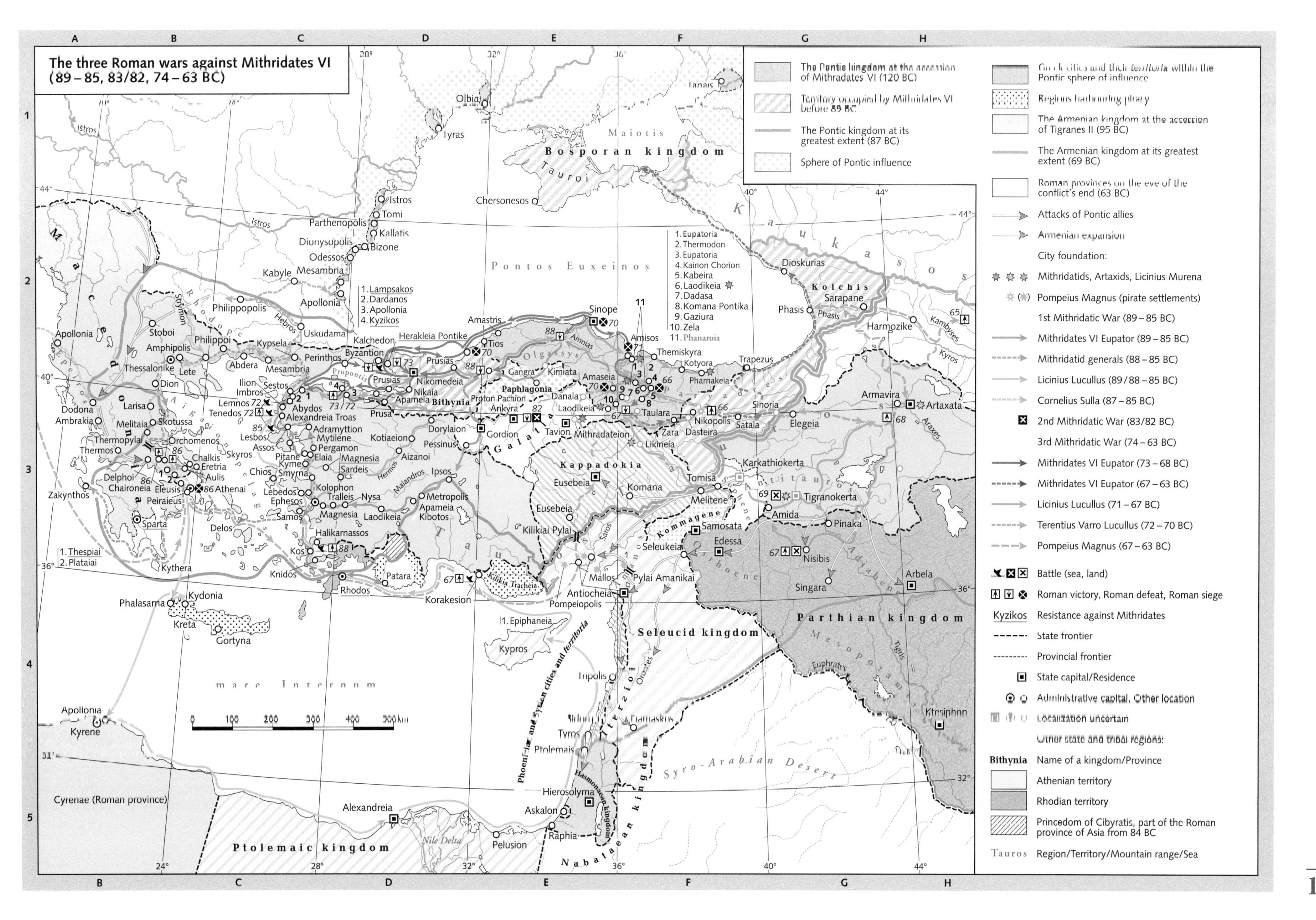

The three Roman wars against Mithridates VI (89 – 85, 83/82, 74 – 63 BC)
The Pontic kingdom at the accession of Mithradates VI (120 BC)
Territory occupied by Mithridates VI before 89 BC
The Pontic kingdom at its greatest extent (87 BC)
Sphere of Pontic influence
Greek cities and their territoria within the Pontic sphere of influence
Regions harbouring piracy
The Armenian kingdom at the accession of Tigranes II (95 BC)
The Armenian kingdom at its greatest extent (69 BC)
Roman provinces on the eve of the conflict's end (63 BC)
Attacks of Pontic allies
Armenian expansion
City foundation:
Mithridatids, Artaxids, Licinius Murena
Pompeius Magnus (pirate settlements)
1st Mithridatic War (89 – 85 BC)
Mithridates VI Eupator (89 – 85 BC)
Mithridatid generals (88 – 85 BC)
Licinius Lucullus (89/88 – 85 BC)
Cornelius Sulla (87 – 85 BC)
2nd Mithridatic War (83/82 BC)
3rd Mithridatic War (74 – 63 BC)
Mithridates VI Eupator (73 – 68 BC)
Mithridates VI Eupator (67 – 63 BC)
Licinius Lucullus (71 – 67 BC)
Terentius Varro Lucullus (72 – 70 BC)
Pompeius Magnus (67 – 63 BC)
Battle (sea, land)
Roman victory, Roman defeat, Roman siege
Kyzikos Resistance against Mithridates
State frontier
Provincial frontier
State capital/Residence
Administrative capital, Other location
Localization uncertain
Other state and tribal regions:
Bithynia Name of a kingdom/Province
Athenian territory
Rhodian territory
Princedom of Cibyratis, part of the Roman province of Asia from 84 BC
Tauros Region/Territory/Mountain range/Sea
1. Lampsakos
2. Dardanos
3. Apollonia
4. Kyzikos
1. Eupatoria
2. Thermodon
3. Eupatoria
4. Kainon Chorion
5. Kabeira
6. Laodikeia
7. Dadasa
8. Komana Pontika
9. Gaziura
10. Zela
11. Phanaroia
1. Thespiai
2. Plataiai
1. Epiphaneia
Bosporan kingdom
Pontos Euxeinos
Maiotis
Tauroi
Kolchis
Paphlagonia
Bithynia
Galatia
Kappadokia
Kommagene
Seleucid kingdom
Parthian kingdom
Ptolemaic kingdom
Mesopotamia
Syro-Arabian Desert
mare Internum
Phoenician and Syrian cities and territoria
Hasmonaean kingdom
Nabataean kingdom
Cyrenae (Roman province)
Olbia
Tyras
Chersonesos
Istros
Tomi
Kallatis
Bizone
Parthenopolis
Dionysopolis
Odessos
Mesambria
Apollonia
Kabyle
Philippopolis
Uskudama
Kypsela
Perinthos
Byzantion
Kalchedon
Herakleia Pontike
Amastris
Tios
Sinope
Amisos
Themiskyra
Kotyora
Pharnakeia
Trapezus
Dioskurias
Phasis
Sarapane
Harmozike
Armavira
Artaxata
Nikomedeia
Prusias
Nikaia
Apameia
Prusa
Gangra
Kimiata
Amaseia
Danala
Laodikeia
Ankyra
Gordion
Tavion
Mithradateion
Taulara
Zara
Dasteira
Nikopolis
Satala
Sinoria
Elegeia
Likineia
Eusebeia
Komana
Tomisa
Melitene
Karkathiokerta
Tigranokerta
Amida
Pinaka
Samosata
Edessa
Nisibis
Singara
Arbela
Ktesiphon
Seleukeia
Pylai Amanikai
Antiocheia
Pompeiopolis
Mallos
Kilikiai Pylai
Korakesion
Kypros
Tripolis
Damaskos
Tyros
Ptolemais
Hierosolyma
Askalon
Raphia
Pelusion
Alexandreia
Kyrene
Apollonia
Phalasarna
Kydonia
Kreta
Gortyna
Kythera
Sparta
Peiraieus
Athenai
Eleusis
Chaironeia
Delphoi
Thermopylai
Thermos
Orchomenos
Chalkis
Eretria
Aulis
Skyros
Larisa
Melitaia
Skotussa
Dion
Thessalonike
Lete
Amphipolis
Philippoi
Stoboi
Abdera
Apollonia
Dodona
Ambrakia
Zakynthos
Ilion
Sestos
Imbros
Lemnos
Tenedos
Abydos
Alexandreia Troas
Adramyttion
Mytilene
Lesbos
Assos
Pitane
Elaia
Kyme
Pergamon
Magnesia
Sardeis
Chios
Smyrna
Lebedos
Kolophon
Ephesos
Tralleis
Nysa
Samos
Magnesia
Laodikeia
Halikarnassos
Kos
Knidos
Rhodos
Patara
Delos
Kotiaeion
Dorylaion
Pessinus
Aizanoi
Ipsos
Metropolis
Apameia
Kibotos
Proton Pachion
Istros
Strymon
Hebros
Rhodope
Propontis
Hermos
Maiandros
Amnias
Olgassys
Phasis
Kambyses
Kyros
Araxes
Tigris
Euphrates
Orontes
Saros
Nile Delta
Kilikia Tracheia
Tauros
Antitauros
Kaukasos
0 100 200 300 400 500 km

Pompey's reorganization of the Near East (67– 59 BC)

I. Pompey's career

During the years portrayed on this map, Cn. Pompeius Magnus had a level of personal power unprecedented in Roman history, the like of which would not be seen again until Augustus. The foundation of this power was his family's vast clientele, which his father Pompeius Strabo, among others, had accumulated during the Social War of 91-89.

Pompeius Strabo served the consul Rutilius Lupus as legate in 90 BC at Picenum, where his family had estates. As consul in 89, he then led his troops against the Italic tribes, conquering Asculum and fighting the Marsi, then winning a triumph over the Picentes in December 88 – all of this, incidentally, without transferring his booty to the public treasury. He granted Roman citizenship to Gaulish tribes as far as the Padus and Latin law to communities beyond it. When the consul Pompeius Rufus was about to remove his proconsular army command in 88, he incited his troops to mutiny. In the course of the uprising, the consul was killed. At approximately the same time, the other consul, L. Cornelius Sulla, accepted his army's urging to march on Rome. These are the first examples of the political involvement of the army clienteles which Marius' reforms had enabled to come into being.

Cn. Pompeius ('Pompey') provided three legions of his father's veterans and other clients in 83 BC to augment the forces of Sulla on his return from the east. In doing so, he laid the foundations of his own unequalled career.

That career culminated in the two *imperia extraordinaria* awarded to him, the first in 67 by the *lex Gabinia*, to fight piracy throughout the Mediterranean, the second in 66 by the *lex Manilia*, to fight King Mithridates VI of Pontus. On these were founded the competences which empowered Pompey to reorganize the empire's eastern territories. Consequently, it is with these *imperia* that the map's portrayal of political developments begins.

II. The war against the pirates, 67 BC

It took Pompey three months to accomplish the task, for which he was granted *imperium* for three years over the entire region to 400 *stadia* (approx. 75 km) inland, of freeing the entire Mediterranean from the scourge of piracy. He completed it by annihilating a pirate fleet at Coracesium in Cilicia. To root out the malady, Pompey provided the pirates with homes and land, at places including Mallus, Adana, Epiphaneia and Soli (later Pompeiopolis) in Cilicia Pedias (between Seleucia and the Syriae Pylae), Dyme on the western coast of Achaea, and in southern Italy. At the time, he did not seek senatorial approval for these measures.

III. The3rd Mithridatic War (73–63 BC)

In 68 BC, when Mithridates returned home to Pontus from house arrest with Tigranes II in Armenia, a fourth phase in Rome's third war against the Pontic king began, partly overlapping the war against the pirates. In a few months, Mithridates won back the heartland of his kingdom. The final chapter was the defeat inflicted on the legate Valerius Triarius at Zela in 67 BC.

Shortly before that, the proconsul Licinius Lucullus had been forced to relinquish his *imperium* for the war against Mithridates. In the spring of 66, Pompey replaced him, taking up his new duties by pursuing Mithridates from Cilicia (fifth phase of war, 67-63 BC). Mithridates fled into the Bosporan region of his kingdom, to Phanagoria, where he died in 63 during an uprising led by his son, Pharnaces.

The settlements reached by Pompey in the course of and following his campaigns against the pirates and Mithridates still had to be ratified by the Senate. Pompey's impending return triggered grave concern in the Senate recalling Sulla's reign of terror, which had come about with Pompey's support. Unlike Sulla, though, Pompey discharged his legions and entered Rome in January 61 without military backing. The Senate breathed again. Misreading the political situation, the Optimates felt it safe to fob Pompey off by granting him a triumph while rejecting his reorganization in the east and his plans for settling his veterans. This lack of compliance was a crucial factor in driving Pompey to join forces with Caesar and Crassus in what became known as the 'First Triumvirate'. Early in 59 BC, during his consulate, Caesar made provision for Pompey's veterans and confirmed Pompey's reorganization of the east by a *lex Iulia de actis Cn. Pompei in transmarinis provinciis confirmandis* (Vell. Pat. 2,44,2).

The map

The primary provisions of Pompey's reorganization of the Near East were the establishment of the provinces of Bithynia et Pontus and Syria and an expansion of the province of Cilicia. The map shows the eastern territories of the Roman Empire as they appeared in 59 BC. Worthy of note is how much territory the empire gained by the conquests and organizational measures of Pompey (solid boundary lines) compared to the prior position (dashed boundary lines). The designation of various communities in the province of Syria as *civitates liberae et immunes* is sometimes problematic. The sources do not always allow clear understanding of this state of affairs. The inland boundaries of provinces are also sometimes only approximately known, though this is hardly surprising given the extent of the regions concerned and the sparseness of our sources. The allocation of the various Aegean islands to the provinces of Macedonia or Asia is not the central task of this map. It is not individually known how far the Roman governor at Thessalonica, say, might have regarded Melos as part of the empire without this being set down by treaty. The Roman view of the world tended sometimes to make arbitrary presumptions of competences for which there was no legal support.

Bithynia et Pontus

On the basis of the testament of the Bithynian King Nicomedes IV, who died in 74 BC, the proconsul of Asia, Iunius Iuncus, was commissioned by the Senate to redesignate the kingdom as a province, and in the autumn of 74, the consul Aurelius Cotta was entrusted with the administration of the new province of Bithynia. A single customs region was established for Asia and Bithynia. The third war against Mithridates VI, who raised a claim to the Bithynian throne in 73, was prosecuted at first by the proconsul Licinius Lucullus (73–66), then by Pompey, armed with his extraordinary *imperium* (66–63). In the winter of 65/64, Pompey at Amisus drew up the new double province of Bithynia et Pontus. Pontus was divided into eleven urban *territoria*: Amastris, Amisus, Eupatoria (renamed Magnopolis), Cabira (renamed Diospolis), Phazemon (renamed Neapolis), Sinope, Zela, the new foundations of Megalopolis, Nicopolis, Pompeiopolis and one other city, possible candidates for which are Abonutichus, Amasia, Heraclea Pontica and Tium. After the death of Mithridates VI in the winter of 63/62, Pompey issued an edict, according to Pliny (Plin. Epist. 10,79 f.; 112; 114 f.) a *lex Pompeia*, for the province, setting up two assemblies. Amastris, Calchedon, Prusa, Sinope and Amisus became *civitates liberae et immunes*. The foundation of numerous cities, some of whose names pay tribute to their founder, secured the administrative functionality of the Pontic part of the province in particular.

Syria

Pompey set up the new province in the winter of 64/63. It essentially consisted of the heartland of the Seleucid kingdom up to, i.e. south east of, the Taurus. Antiochus XIII Philadelphus was deposed. His kingdom was conquered by Tigranes II (95 – c. 55) before being taken from the latter by Pompey. The provincial capital was set up at Antioch, and Pompey based this province, too, on various urban *territoria*.

Cilicia

M. Antonius, who won great renown as an orator, was entrusted with the *provincia Cilicia*, at first as a praetor, later with *prorogatio imperii*. His job was to combat the piracy which threatened trade and traffic on Mediterranean routes, and which chiefly originated in the Cilician mountain country. Only a narrow coastal strip was under Roman control. The foundation for the organization of the province was laid with the conquest of Cilicia Tracheia (from Attaleia to Seleucia) in 78-74 BC by Servilius Vatia, at first a praetor, then with *prorogatio imperii*. In 67, Pompey extended the province northwards.

Pompey designed the borderlands of the empire's eastern fringes in the pattern which Augustus would subsequently systematically consolidate. The provinces of the empire bordered throughout on client states, never on enemy territory. Thus, on the eastern and southern frontiers of Bithynia et Pontus, the neighbours were the Tolistobogii, Trocmi and Paphlagonians. Asia bordered on the Tolistobogii and the Tectosages to the east. Cilicia's northern frontier was with the temple state of Olba and the kingdom of Tarcondimotus, the territory of the Tectosages and the kingdoms of Antipater of Derbe and the Cappadocian Ariobarzanes Philorhomaios. The eastern frontier of Syria was with the Ituraeans under Ptolemy, son of Mennaeus, and the Nabataeans under Aretas III Philhellen.

The sources

The map is mainly based on information from ancient literary sources, esp. the contemporary reports of Cicero and Caesar. It also relies on information provided by Augustan writers such as Pompeius Trogus, Diodorus Siculus and Strabo. Velleius Paterculus served as an officer under Tiberius (AD 14–37). Plutarch and Appian (1st/2nd cents. AD) are rich sources for details of the life of Pompey. Two authors from the reign of Vespasian (69–79) and somewhat later are the historian Flavius Josephus of Jerusalem and Memnon, a local historian from Bithynia. The poet Lucan (d. AD 65) left an epic in which he sides with Pompey. Both Plinies (d. AD 79 and 112/13 respectively) had personal professional experience of the empire's eastern half which Pompey shaped. From the 3rd to the 5th cents., we have the historian Cassius Dio, the writer Porphyry, the historian Ammianus Marcellinus, his contemporary Eutropius and the Christian Orosius (early 4th cent.).

→ Maps pp. 157, 159, 163, 181, 183

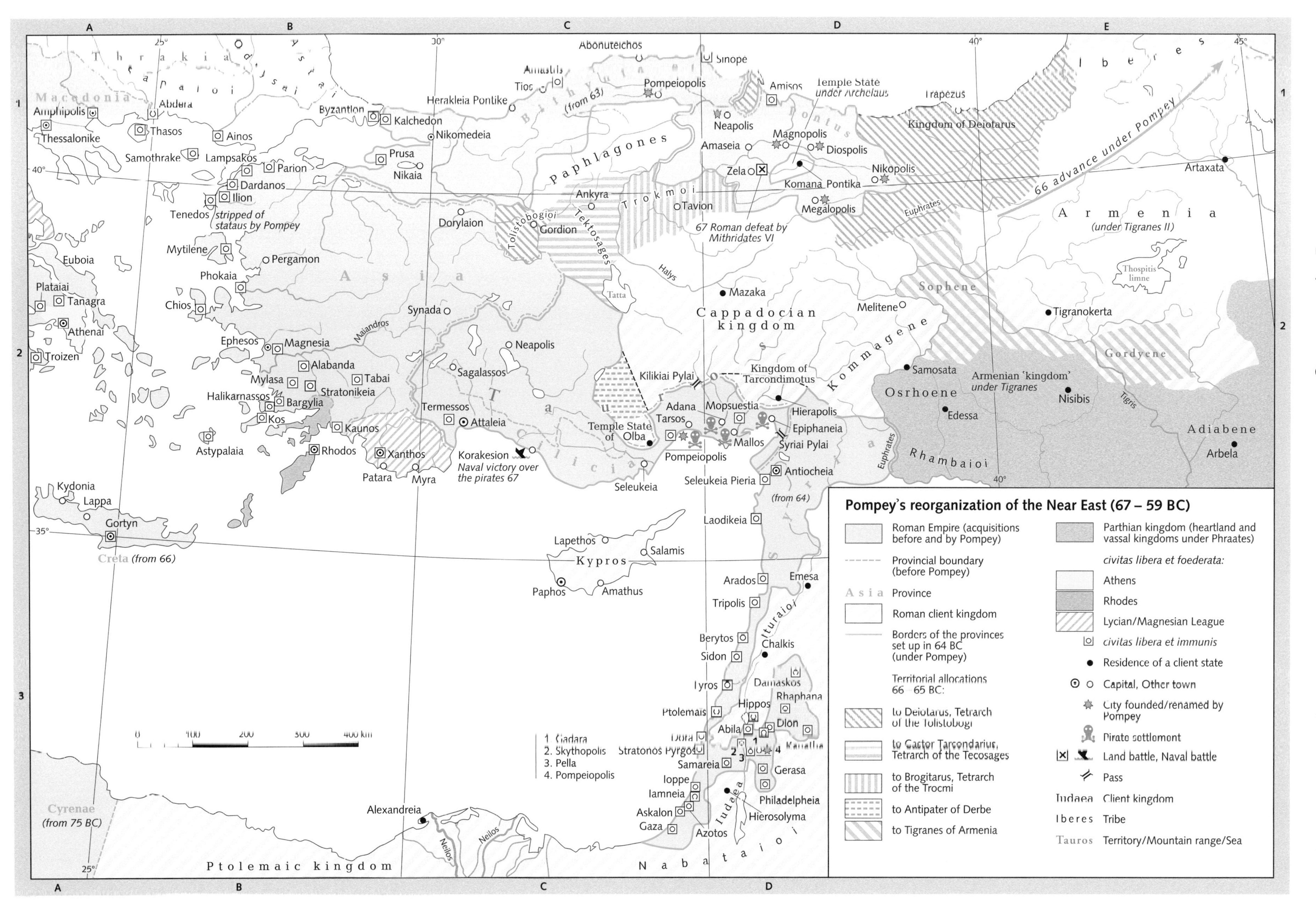
Pompey's reorganization of the Near East (67 – 59 BC)
Roman Empire (acquisitions before and by Pompey)
Provincial boundary (before Pompey)
Asia Province
Roman client kingdom
Borders of the provinces set up in 64 BC (under Pompey)
Territorial allocations 66–65 BC:
to Deiotarus, Tetrarch of the Tolistobogi
to Castor Tarcondarius, Tetrarch of the Tecosages
to Brogitarus, Tetrarch of the Trocmi
to Antipater of Derbe
to Tigranes of Armenia
Parthian kingdom (heartland and vassal kingdoms under Phraates)
civitas libera et foederata:
Athens
Rhodes
Lycian/Magnesian League
civitas libera et immunis
Residence of a client state
Capital, Other town
City founded/renamed by Pompey
Pirate settlement
Land battle, Naval battle
Pass
Iudaea Client kingdom
Iberes Tribe
Tauros Territory/Mountain range/Sea
66 advance under Pompey
Kingdom of Deiotarus
Armenia (under Tigranes II)
Cappadocian kingdom
Kingdom of Tarcondimotus
Armenian 'kingdom' under Tigranes
Temple State under Archelaus
Temple State of Olba
67 Roman defeat by Mithridates VI
Korakesion Naval victory over the pirates 67
Tenedos stripped of status by Pompey
Creta (from 66)
Cyrenae (from 75 BC)
Ptolemaic kingdom
1. Gadara
2. Skythopolis
3. Pella
4. Pompeiopolis

162 | Administrative history of Crete and the Cyrenaica in the Roman period (96 BC – 6th cent. AD)

Administrative history of Crete and the Cyrenaica in the Roman period (96 BC – 6th cent. AD)

I. Cyrenaica

The coastal strip of North Africa from the Great Syrte at the Philainon Bomoi (Arae Philaenorum) in the west to Catabathmus Megas in the east took its name of Cyrenaica from the Theran colony of Cyrene. Together with Ptolemaïs, Berenice, Arsinoë and Apollonia, this city formed the so-called Libyan Pentapolis. The highlands of the Cyrenaica (up to 900 m a.s.l.), which still have plentiful springs today, were particularly fertile in antiquity and thus relatively densely populated. Wheat thrived here, along with barley, legumes and vines. There was also silphium, a plant that yielded a juice which was used as a drug: this highly lucrative worldwide export trade from the cities of the Cyrenaica lasted into the Hellenistic period, but silphium was gradually over-exploited to extinction in the Roman Imperial period. The distance to Crete from the promontory of Phycus, west of Apollonia, is less than 300 km.

A Ptolemaic secundogeniture had ruled over the Cyrenaica, but this died out with Ptolemy Apion, a son of Ptolemy VIII, in 96 BC. This last king of the territory bequeathed his kingdom to the Romans. At first, the Romans refrained from incorporating the Cyrenaica as a province. Rather, they simply adopted the royal territories and used them via leasing arrangements. Only in 74 BC did they organize the Cyrenaica into the *provincia Cyrenaica*.

II. Crete

After Sicily, Sardinia and Cyprus, Crete is the fourth-largest Mediterranean island, stretching some 260 km from west to east and an average of 50 km from north to south. Four massifs (Leuka Ore, Ida, Lassithi and Sitia) determine the island's landscape. These range from over 2,400 m to 1,400 m in diminishing altitude from west to east. In the south, these mountain ranges fall directly to the shoreline, hardly allowing room for settlements or even harbours, but in the north and east they slope out more gently towards the sea. Numerous bays have formed here, which trading ships used as convenient natural harbours on the important route from the west to the Levant (cf. Cydonia, Knossos with its four harbours, Itanus). The north and east coasts were thus particularly attractive for settlement. The mountain country has only a few plateaux, but these are quite fertile. The largest among them is the Mesara in south central Crete, in which was located the city of Gortyn, the administrative centre in the Roman period.

On this mountainous island, agriculture was never a source of wealth, and from the earliest times Cretans had sought their fortune far afield. They were particularly valued in the eastern Mediterranean as mercenaries, and Ptolemies and Antigonids alike for this reason courted the islanders in the Hellenistic period. Opportunity arose for friendship-fostering interventions amid the constant squabbles between the many Cretan cities. The Roman Senate, too, repeatedly exerted itself as a mediator, and Roman emissaries visited the island several times, some attested dates being 184, 180 and 174 BC.

Apart from mercenary activities, piracy was a gainful line of business for the Cretans. Like the highlands of Cilicia and the Balearics, the mountainous interior of Crete offered pirates opportune refuge. Neither Athens nor any of the Hellenistic monarchies was strong enough to operate as a maritime police force. Rhodes, too, in league with other Greek states, had fought a debilitating war (155–153) on Crete, but had failed to get to grips with Cretan piracy. Only when maritime insecurity began to have a grave impact on Roman interests did the Romans bestir themselves to remedy the situation once and for all. They made a beginning in Cilicia late in the 2nd cent. BC, but their permanent presence in what became the province of Cilicia failed to bring a real halt to piracy in the eastern Mediterranean.

While consul in 69 BC, and thereafter as proconsul, Q. Caecilius Metellus, who later took the *cognomen* Creticus, fought a protracted war against the pirates on Crete. His *imperium* collided with that granted to Cn. Pompeius (Pompey) in 67 BC by the *lex Gabinia* to fight piracy in the entire Mediterranean. According to the provisions of this law, Pompey's *imperium* overrode the jurisdictions of all governors whose provinces bounded the Mediterranean to approx. 75 km inland – in other words, it covered the whole of Crete. The battles Metellus was forced to fight from fortress to fortress on the island were hard and unrelenting. When the Cretans heard how conciliatory Pompey's dealings with his opponents were, they themselves asked him to intervene in the conflict on Crete. Pompey did indeed intervene, in 67 BC, with his legate, L. Octavius, and military conflict would probably have ensued between the two *imperatores* if Pompey had not been reassigned to the war against Mithridates VI in early 66 by the *lex Manilia*, which compelled him to leave matters on Crete in the hands of Metellus. Metellus thereupon pursued the conquest of the island with all severity, and subdued it to form a province in the same year, 66 BC.

III. Creta and/et Cyrenaica

The two provinces of Creta and Cyrenaica were governed as separate administrative units until they were affected by the donations of the *triumvir* M. Antonius (Mark Antony) – Antony presented the whole of Cyrenaica, but only part of Crete, to the Ptolemaic kingdom. The status of the two provinces in the period following the death of Cleopatra on 12 August 30 BC is unknown. The act of state of 13 January 27 BC united Creta et Cyrenaica as one province, under senatorial rule. The imperial reforms of Diocletian split the province into its constituent parts once more as Creta and Cyrenaica, the latter now itself divided into Libya superior and inferior. Allocated to different dioceses, these provinces were also retained after 395.

In the 5th cent., the Cyrenaica was claimed by the Vandal kingdom (457), and in 643 it was conquered by Muslim Arabs on behalf of the Caliph Umar ibn al-Ḫaṭṭāb. Crete, on the other hand, as part of the Byzantine Empire, fell into the hands of the Venetians during the Fourth Crusade in 1204.

The map

The map is split in two on different scales, as it would not be possible to give a fitting depiction of both parts on a single map because of their difference in size (Crete: 8,331 km²; Cyrenaica: over 800,000 km²). While the coastline naturally gives a clear delineation of the island, the southern border between the Roman province of Cyrenaica and the Sahara can only be suggested.

→ Maps pp. 147, 235, 237, 243

Date	Province	Event	Sources
96 BC		Romans inherit Cyrenaica from Ptolemy Apion; cities declared free	Liv. Per. 70; Obseq. 49; App. Mithr. 600; Iust. 39,5,2; Jer. Chron. for the year 96; Cities: SEG 9, 3; Liv. Per. 70; Amm. Marc. 22,16,24
		Rome takes over administration of the royal lands but does not found a province; cities remain free	Cic. Leg. agr. 2,51; Tac. Ann. 14,18
74	Cyrenaica	Foundation of province	Sall. Hist. 2,43
66	Creta	Metellus conquers Crete and organizes it as province; capital Gortyn	Cic. Flac. 30; 100; Liv. Per. 100; Plut. Pompeius 29; Iust. 39,5,3; Eutr. 6,11; Rufius Festus, Breviarium 7,1
44	Creta and Cyrenaica	Senate allots provinces to Brutus (Creta) and Cassius (Cyrenaica)	Cic. Phil. 2,97; 11,27; App. Civ. 3,8; 12; 36; Plut. Brutus 19,5; Cass. Dio 45,32,4; 46,23,3
36–30		Triumvir M. Antonius annexes Cyrenaica and part of Crete to the Ptolemaic kingdom	Cass. Dio 49,32,5; 49,41,3; cf. Plut. Antonius 54
30–27		Status of both provinces unknown	
27 BC	Creta et Cyrenaica	Provinces amalgamated and placed under senatorial rule; capital Gortyna	Cass. Dio 53,12,4; Gortyn: AE 1933, 100; 1979, 636
AD 284–395	Creta in dioecesis Moesiae; Libya superior (Pentapolis) and Libya inferior (Sicca, Marmarica) in dioecesis Oriens	Whole of Creta one province: capital Gortyna; Cyrenae divided: Libya superior, capital Cyrene; Libya inferior, capital Darnis	Laterculus Veronensis 5,17; 1,1 f.; Darnis: Hierocles, Synekdemos 734,3
after AD 395	Creta in dioecesis Macedonia, Libya superior and Libya inferior in dioecesis Aegyptus	Diocesan reform	Not. Dign. Or. 3,10; 1,81 f.

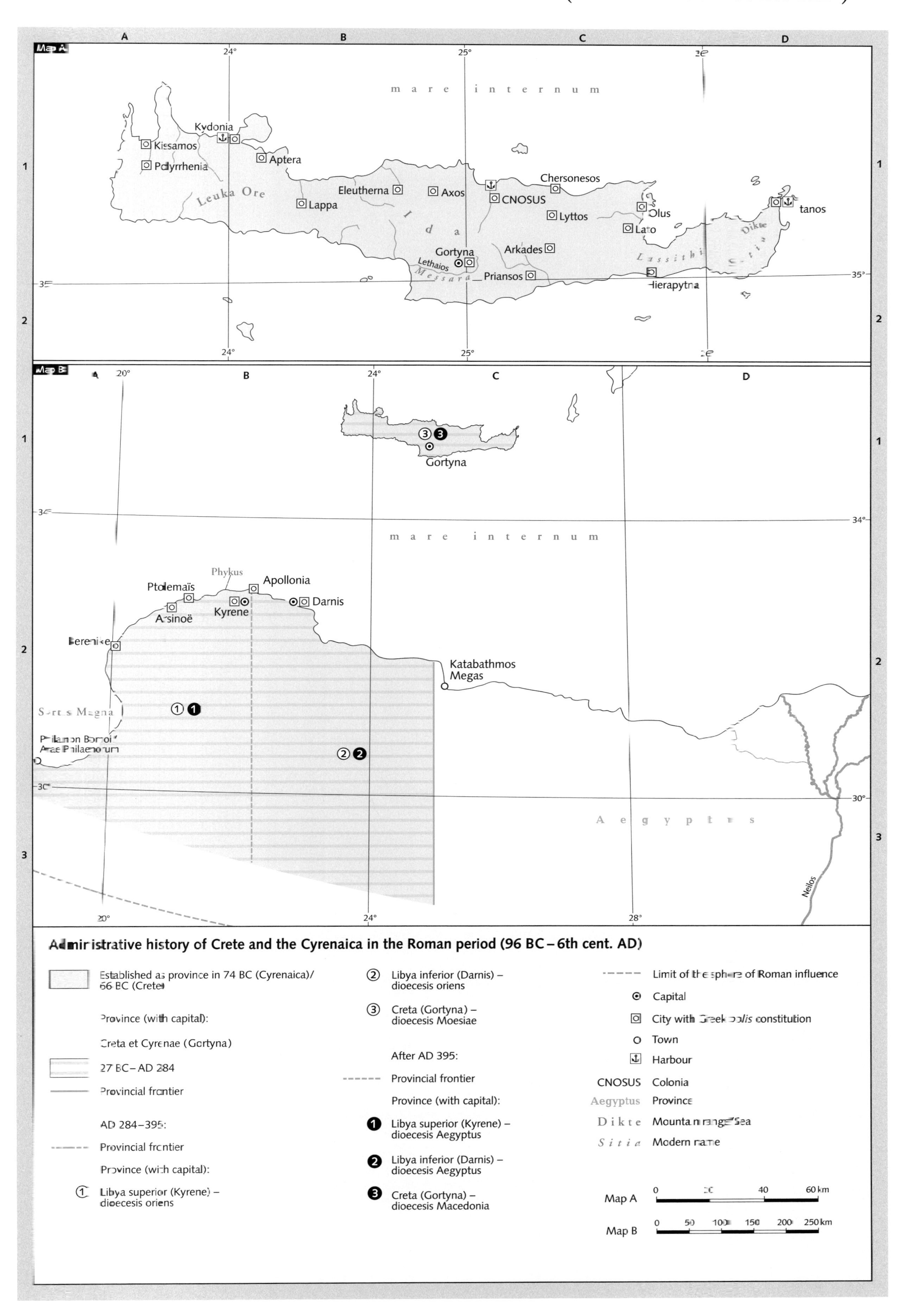

Map A
m a r e i n t e r n u m
Kydonia
Kissamos
Polyrrhenia
Aptera
Leuka Ore
Lappa
Eleutherna
Axos
Chersonesos
CNOSUS
Lyttos
Olus
Lato
tanos
Dikte
I d a
Gortyna
Lethaios
Messara
Arkades
Priansos
Lassithi
Sitia
Hierapytna
Map B
Gortyna
m a r e i n t e r n u m
Phykus
Apollonia
Ptolemaïs
Kyrene
Darnis
Arsinoë
Berenike
Katabathmos Megas
Syrtis Magna
A e g y p t u s
Neilos
Administrative history of Crete and the Cyrenaica in the Roman period (96 BC – 6th cent. AD)
Established as province in 74 BC (Cyrenaica)/ 66 BC (Crete)
Province (with capital):
Creta et Cyrenae (Gortyna)
27 BC – AD 284
Provincial frontier
AD 284–395:
Provincial frontier
Province (with capital):
① Libya superior (Kyrene) – dioecesis oriens
② Libya inferior (Darnis) – dioecesis oriens
③ Creta (Gortyna) – dioecesis Moesiae
After AD 395:
Provincial frontier
Province (with capital):
❶ Libya superior (Kyrene) – dioecesis Aegyptus
❷ Libya inferior (Darnis) – dioecesis Aegyptus
❸ Creta (Gortyna) – dioecesis Macedonia
Limit of the sphere of Roman influence
Capital
City with Greek polis constitution
Town
Harbour
CNOSUS Colonia
Aegyptus Province
Dikte Mountain range/Sea
Sitia Modern name
Map A 0 20 40 60 km
Map B 0 50 100 150 200 250 km

Caesar's proconsulship in Gaul (58–50 BC)

Caesar held the office of proconsul for the provinces of Gallia citerior (Gallia cisalpina), Illyricum and Gallia ulterior (Gallia transalpina) from March 58 BC until 10/11 January 49 BC. This was the period through the course of which he laid the practical foundations for the further ascent of his career towards the position of *dictator perpetuus*. The numerous festivals of thanksgiving (*supplicationes*) by which the Senate recognized his services to the Roman commonwealth could only add to his *dignitas*, drowning out the memory of the legal breaches he had committed during his earlier consulship. It may be that he perceived an alternative venue for a similar course of action in Illyricum, in a confrontation with the Dacian king Burebista. At all events, he opted for Gaul.

Just as ancient warfare was strictly bound to the shifting of the seasons, winter campaigns taking place only in exceptional cases (cf. Caesar's campaign against the Bituriges in the winter of 52/51 BC), so ancient historiography likewise consistently saw the winter as a caesura between two periods of war. Caesar's work on the Gallic Wars is thus divided into eight books, corresponding to his eight years in office as proconsul. He did not write the last book himself: it was written in Caesar's spirit by one of his most trusted legates, A. Hirtius.

Caesar justified his military actions in Gaul consistently in terms of the interests of Rome and of his Gaulish allies. The fact that Caesar was primarily pursuing quite personal, career-related goals in Gaul is quite another matter. Thus, during the first year of his proconsulship (58 BC), he secured the Roman province beyond the Alps in a few months, in battles with the Celtic Helvetii and the Germanic Suebi (under Ariovistus). The winter camp in which his legate T. Labienus quartered the Roman troops was at Arar/Saône, outside the province of Gallia Narbonensis.

In the second year (57 BC), a pact among the Belgic tribes gave Caesar a motive to intervene. He saw the parties to this pact as "conspiring against the Roman people" (*contra populum Romanum coniurare*, Caes. B Gall. 2,1,1). The Romans had to fight hard here to gain just one victory (over the Nervii on the Sabis). In the end, however, Caesar was able to win himself the desired respect in the territory of the Belgae, and it was in this territory that the Roman troops wintered on this occasion, by the Liger/Loire, as well as among the Veragri on the upper Rhodanus/Rhône at Octodurus.

In the third year (56 BC), Caesar fought the Veneti (in Brittany) on land and at sea, as well as the Belgic Morini and Menapii, while his legate P. Licinius Crassus conquered the Sotiates in Aquitania. The main Roman winter camp for the year was with the Lexovii (in Normandy).

In the fourth year (55 BC), Caesar confronted the Germanic Usipetes and Tencteri, who had crossed the Rhine under pressure from the Suebi, entering the territory of the Menapii. Matters oscillated between negotiation and battle, and Caesar prevailed, though with such a lack of scruple that M. Porcius Cato in the Senate moved (albeit without success) that Caesar be handed over to the Germans. In the same year, Caesar demonstrated his technical and military excellence by building a pile bridge across the Rhine and spending eighteen days on the eastern bank of the river. He then ferried an expeditionary force over to Britain for a short time. This time, the Romans' central winter camp was in the territory of the Belgae.

In the fifth year (54 BC), Roman troops again landed on the island of Britain, but now with the intent (which in the end they achieved only nominally) of permanent occupation. Apart from a few actions against Belgic tribes, this operation occupied Caesar for the whole year. For this season, he quartered his troops in five winter camps: with the Morini, Nervii, Essuvii, Remi, Eburones and in the westerly region of the Belgica.

The sixth year (53 BC) was mostly devoted to a second Rhine crossing to prevent the right-bank tribes from again entering Gaul. The Belgic Eburones were a constant problem, but so were the Treveri, which whom Caesar had also had dealings in 54. Strong units therefore made their winter camp among the Treveri, the Lingones and the Senones. The Romans seemed to be masters of the country.

However, in the seventh year (52 BC) Caesar's control of all the regions he had up to now won for Rome threatened to slip away. The Arvernian chieftain Vercingetorix succeeded in uniting almost all the Gaulish tribes (except the Lingones, Remi and Treveri) under his leadership, and this skilful strategist repeatedly embarrassed Caesar. He successfully defended Gergovia, the headquarters of his tribe, against the Romans, but was finally forced to capitulate at Alesia. For this winter too, the Romans split into various camps – among the Sequani, Remi, Ambivareti, Bituriges, Ruteni and Haedui.

Caesar and his legates only had a few trouble spots left to concern them in the eighth year (51 BC): the Bituriges, Carnutes, Belgic Bellovaci and the Cadurci. The Roman winter quarters were again distributed across the whole of Gaul to secure the peace throughout the province.

The final year of Caesar's governorship in Gaul (50 BC) presents to us a proconsul visibly striving to win the favour of the Gauls, imposing no new tributes upon them and leaving their ways of government and life untouched. The impending civil war was beginning to cast its shadow.

Caesar's expeditions to Britannia

The tin trade had brought the British Isles into contact with the Mediterranean world as early as the 10th cent. BC, through the agency of the Phoenicians, with Tartessus on the lower Guadalquivir as an entrepot. Pytheas of Massalia (4th cent. BC) learned much about the British tin trade. One important entrepot harbour was in the island group of the Cassiterides (generally identified with the Isles of Scilly), and there was another on the Belerion akron (Land' s End). Tin was obtained on the nearby island of Ictis (possibly St. Michael's Mount in Mount's Bay, Cornwall).

Caesar also tells of the metallic wealth of the British island in his work on the Gallic Wars. However, if we are to believe his report, it was strategic considerations that led him to make two short expeditions to Britain in 55 and 54 BC. Britannia was a constant threat to Caesar's Gaulish conquests. The Britons supported rebellious Gauls with troops and materials, and offered asylum to fleeing Gauls. Lastly, the Druids on the island constituted a particular potential risk to the extent that they were able to unite the Gaulish tribes in resistance to Rome.

An expedition in the autumn of 55 BC, for which a coastal inspection under C. Volusenus prepared the way, served to gather information on geographical, social and political conditions. The following summer, Caesar intended to establish a permanent presence on the island, but in spite of some military successes he did not succeed. After agreeing treaties with the tribes of the Trinovantes, Cenimagni, Segontiaci, Ancalites, Bibroci and Cassi and with the British king Cassivellaunus, he and his troops left the island.

The sources

The central source for the subject matter of this map are the *Commentarii de bello Gallico*. They are probably to a large extent based on the accounts of the individual legates active on Caesar's behalf in Gaul. Brimming with information though these *Commentarii* are, the difficulties they present to historians seeking to use them are just as great. Apart from the few notes by Cicero, there are no parallel contemporary sources available with whose help we might be able to verify the reliability of Caesar's report. Subsequent accounts of Caesar's work in Gaul are largely dependent on the information of Caesar himself. Of great importance are the archaeological excavations which have made it possible in various locations to pinpoint such features as Roman camps.

The map

The map does not show individual military campaigns. This could be done on a year-by-year basis, though even then no claim to completeness could be made, but the eye of the observer would be confused rather than meaningfully informed by the totality. All geographic and ethnic details included in the commentary are shown, and together with the commentary and the table (→ p. 268) these enable reliable orientation in respect of Caesar's activities in Gaul.

→ Maps pp. 167, 193

Sources

British expedition: Caes. B Gall. 4,20–36; 5,1–23; *Tin:* Hdt. 3,115; Pol. 3,57,3; Caes. B Gall. 5,12,5; Str. 2,5,15; 2,5,30; 3,2,9; 3,5,11; Diod. Sic. 5,22; 5,38,4 f.; Plin. HN 7,119; 7,197; 34,156; Ptol. 2,3,3.

Literature

Gaul in general:
F. FISCHER, Caesars strategische Planung für Gallien, in: H. HEFTNER, K. TOMASCHITZ (eds.), Ad fontes! FS G. Dobesch, 2004, 305–315; H. GESCHE, Caesar (Erträge der Forschung 51), 1976; T.R. HOLMES, Caesar's Conquest of Gaul, 1911; G. SCHULTE-HOLTEY, Untersuchungen zum gallischen Widerstand gegen Caesar, thesis, Münster 1968.
Helvetii: G. WALSER, Bellum Helveticum. Studien zum Beginn der caesarischen Eroberung von Gallien (Historia Einzelschriften 118), 1998; Id., Caesar und die Germanen. Studien zur politischen Tendenz römischer Feldzugsberichte (Historia Einzelschriften 1), 1956; W. WIMMEL, Caesar und die Helvetier, in: RhM 123, 1980, 126–137.
Bibracte: L. FLUTSCH, A. FURGER-GUNTI, Recherches archéologiques sur le site présumé de la bataille de Bibracte, in: Revue suisse d' art et d' archéologie 44, 1987, 241–251; L. FLUTSCH, La localisation de la bataille de Bibracte: historique et bilan des recherches récentes, in: Revue suisse d' art et d' archéologie 48, 1991, 38–48; D. LOHMANN, Bibracte. Lesermanipulation im Bellum Helveticum, in: AU 36, 1993, 37–52.
Vesontio: K. CHRIST, Caesar und Ariovist, in: Chiron 4, 1974, 251–292; G. SCHEDA, Caesars Marsch nach Vesontio, in: AU 14, 1971, 70–74.
Rhine crossing: R.C. GILLES, How Caesar Bridged the Rhine, in: CJ 64, 1969, 359–365.
Vercingetorix: G. BORDONOVE, Vercingétorix, 1959.
Gergovia: P. EYCHART, Gergovie. Légende et réalité, 1969.
Alesia: J. HARMAND, Une campagne césarienne: Alésia, 1967; J. LE GALL, La bataille d' Alésia, 1999; D. PORTE, L' imposture Alésia, 2004; M. REDDÉ, A. MIRON, Alesia. Vom nationalen Mythos zur Archäologie, 2006.
British expedition: R. DION, Les campagnes de César en l' année 55, in: REL 41, 1963 (publ. 1964), 186–209; C. HAWKES, Britain and Julius Caesar, in: Proceedings of the British Academy 63, 1977, 125–192; C.E. STEVENS, 55 B.C. and 54 B.C., in: Antiquity 1947, 3–9; Id., Britain and the Lex Pompeia Licinia, in: Latomus 12, 1953, 14–21; G. URSO, Cesare e l' ideologia della conquista: la Britannia, in: Acta Classica Universitatis Scientiarum Debreceniensis 38/39, 2002/03, 225–235

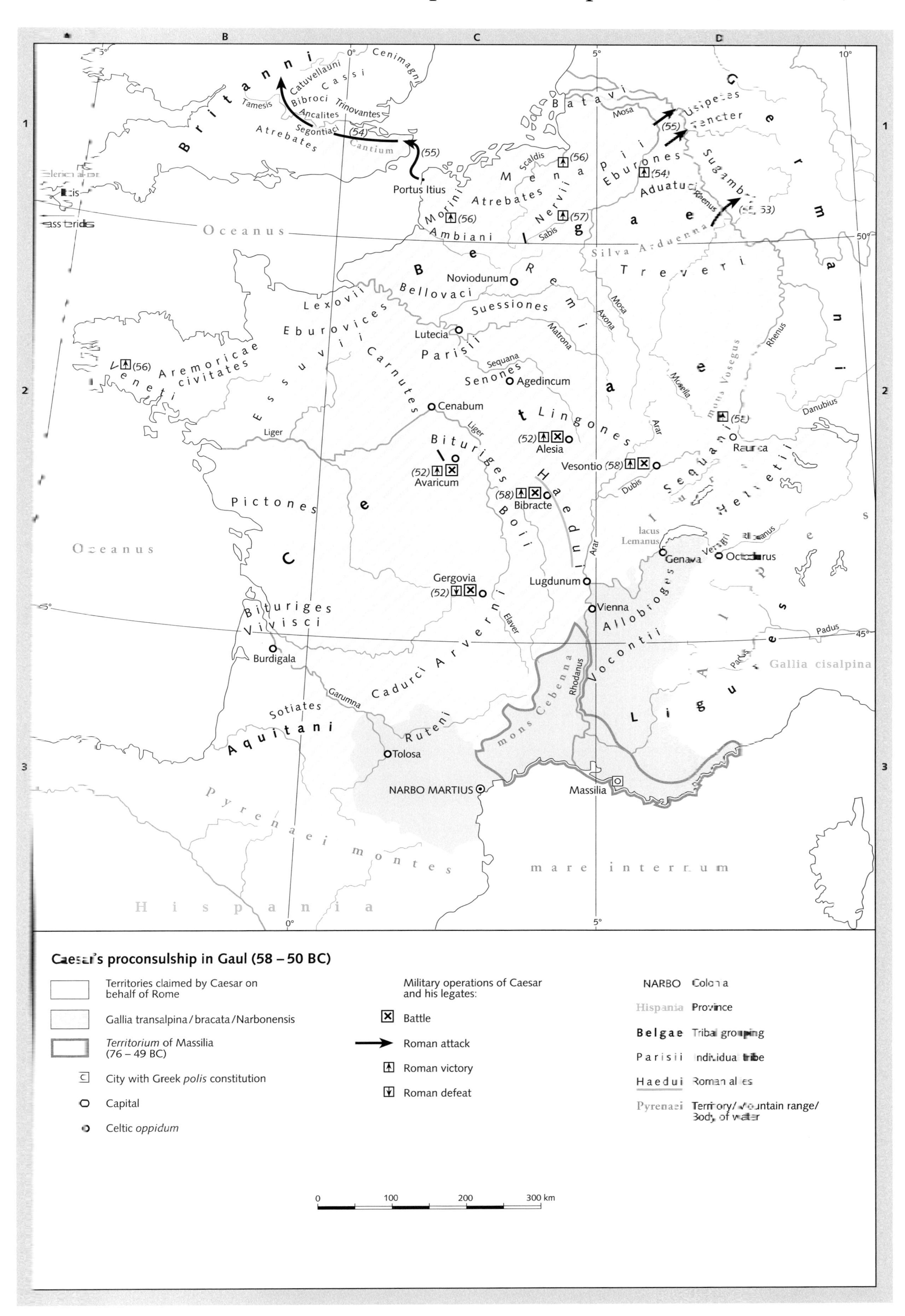

Britanni
Catuvellauni
Cassi
Cenimagni
Bibroci
Trinovantes
Ancalites
Segontiaci
Atrebates
Tamesis
Cantium
(54)
(55)
Portus Itius
Oceanus
Batavi
Mosa
Usipetes
Tencter
Germani
Sugambri
Menapii
Scaldis
(56)
Eburones
(54)
Aduatuci
Rhenus
(55, 53)
Morini
(56)
Atrebates
Nervii
(57)
Ambiani
Sabis
Belgae
Silva Arduenna
Treveri
Remi
Noviodunum
Bellovaci
Suessiones
Lexovii
Eburovices
Lutecia
Matrona
Axona
Mosa
Aremoricae civitates
Veneti
(56)
Esuvii
Carnutes
Parisii
Sequana
Senones
Agedincum
Mosella
mons Vosegus
Rhenus
Cenabum
Lingones
(58)
Danubius
Liger
Bituriges
(52)
Alesia
Arar
Avaricum
Vesontio (58)
Dubis
Sequani
Helvetii
Pictones
Haedui
(58)
Bibracte
Celtae
Boii
lacus Lemanus
Genava
Veragri
Octodurus
Oceanus
Gergovia
(52)
Lugdunum
Elaver
Arverni
Bituriges Vivisci
Vienna
Allobroges
Alpes
Burdigala
Vocontii
Padus
Gallia cisalpina
Cadurci
Rhodanus
mons Cebenna
Garumna
Sotiates
Aquitani
Ruteni
Ligures
Tolosa
NARBO MARTIUS
Massilia
Pyrenaei montes
mare internum
Hispania
Caesar's proconsulship in Gaul (58 – 50 BC)
Territories claimed by Caesar on behalf of Rome
Gallia transalpina / bracata / Narbonensis
Territorium of Massilia (76 – 49 BC)
City with Greek polis constitution
Capital
Celtic oppidum
Military operations of Caesar and his legates:
Battle
Roman attack
Roman victory
Roman defeat
NARBO Colonia
Hispania Province
Belgae Tribal grouping
Parisii Individual tribe
Haedui Roman allies
Pyrenaei Territory/ Mountain range/ Body of water
0
100
200
300 km

The development of the Roman provinces in Gaul (1st cent. BC – 4th cent. AD)

The development of the Roman provinces in Gaul (1st cent. BC – 4th cent. AD)

The map deals with Gaul bounded by the Pyrenaei montes, the Mare internum, the Alps, the Jura, the Rhenus and the Oceanus (cf. table p. 269). The overall impression of the territory is conditioned by a plethora of individual landscapes including highlands (Massif Central, Vosegus, Cebenna, Arduenna), plains (basin around Lutecia, plateau of Aquitania) and fracture zones (Rhenus, Rhodanus and Arar, Liger). Agriculture, the basis of ancient settlement, thrives throughout the entire region thanks to good irrigation and moderate contours. Five major rivers cross the territory – the Rhodanus/Rhône flows into the Mediterranean, the Garumna/Garonne, Liger/Loire, Sequana/Seine and Mosa/Maas into the Atlantic. All were navigable, but countless smaller rivers and waterways also provided inland navigation with the necessary momentum. Land routes open up the whole of Gaul from north to south and from west to east. Connections between the Atlantic and Mediterranean were available through the valleys of the Sequana and Rhodanus and the Garumna and Atax. All in all, then, optimum conditions were in place for the formation of an infrastructure in support of a rich diversity of commerce.

The land link between Italy and the Iberian Peninsula and its two Roman provinces, which had been particularly important since the 2nd Punic War (218–201 BC), could only be accomplished through the coastal region of southern Gaul. This route necessarily entered the territory of Massilia (Greek Massalia), a powerful trading city with a large *territorium* (cf. map inset), which stood loyally with Rome from the outbreak of the war. The Romans recognized this loyalty by repeatedly coming to the Massaliots' aid when Massilia was pressured by the Celtic tribes living around it (e.g. the Ligures, Oxybii, Deciates). From 125 BC, inconsiderable beginnings developed into outright war between Massilia and a confederation of Salluvii, Vocontii, Arverni and Allobroges. The two proconsuls, M. Fulvius Flaccus (123) and Cn. Domitius Ahenobarbus (121), and the consul Q. Fabius Maximus (later 'Allobrogicus', 121) helped and were victorious. Cn. Domitius Ahenobarbus, who was involved with conditions in Gaul on a number of occasions from his consulship in 122 to around 118, also had the connecting road (named after him) built between Italy and Spain. To guard it, the proconsul C. Sextius Calvinus in 122 established the military station (named after him) of Aquae Sextiae/Aix-en-Provence, in the territory of the Salluvii. The name of the road station of Forum Domitii, south-west of Nemausus, refers to the builder of the road. The territory won in this war outside the Massaliot *territorium* as far up as the lacus Lemanus/Lake Geneva was organized into the province of Gallia transalpina, probably in 121. It was also known as Gallia togata, as distinct from Gallia comata, non-Roman Gaul. The very name of Narbo Martius, west of Massilia, the first Roman citizens' colony permanently established (in 118 BC) outside Italy, betrays its primary purpose of giving military protection to the new province, which it also served as administrative capital.

C. Julius Caesar, while proconsul of Gallia transalpina (58-51 BC), occupied the whole of Gallia comata for Rome, and governed it, initially as a single entity including Gallia Narbonensis (→ map p. 165). Shortly before his death, Caesar separated Gallia Narbonensis from Gallia comata. Gaul was divided into four provinces under Augustus, probably by the act of state of 13 January 27 BC. At first, Augustus held Gallia Narbonensis, so important to domestic politics because of its proximity to Rome, under his own personal rule. In 22 BC, however, he passed it to senatorial administration. The three Imperial provinces of Aquitania, Lugdunensis and Belgica organized themselves around the cultic focus of the Ara Romae et Augusti at Lugdunum.

In AD 69, in the so-called Batavian Revolt, the Treveri and Lingones threatened to secede from the Roman Empire and form an *imperium Galliarum*. The project, however, soon failed (→ map p. 191). Domitian (AD 71-96) then organized the two Germanic military districts of Germania inferior and Germania superior as civilian administrative entities, i.e. provinces. Parts of Gallia Belgica were added to the new province of Germania. The history of the Gaulish provinces is more clearly interrupted by the so-called *imperium Galliarum* (AD 260-274) set up in opposition to the emperors at Rome. Like the other splinter-realm of Palmyra, this breakaway 'empire', too, was won back for Rome by Aurelian (270–275) (→ map p. 223).

Germans invaded Gaul for the first time in the reign of Marcus Aurelius (AD 161–180). Not only this, but also social and economic problems among the rural Gaulish populace led to severe unrest ('*bagaudae* uprisings') in the late 3rd cent.

In a programme of constitutional reform spanning the whole empire, Diocletian (284–305) and Constantine I (306–337) divided the land mass of Gaul into two *dioeceses*, comprising at least eight (dioecesis Galliarum) and seven (dioecesis Viennensis) provinces respectively. Clearly though he distanced himself from Diocletian in other aspects of social and political policy, in terms of the mechanisms of administration Constantine I continued the reform plans designed by his predecessor, at least with his radical redesign of the praetorian prefecture (→ map p. 225) – which shows itself on this map in the foundation of the praefectura Galliarum with four *dioeceses* and its capital at Augusta Treverorum/Trier (from c. AD 400 Arelate/Arles).

Various kingdoms formed on Gaulish soil in the 5th cent. in the course of the great Migration Period (Franks, Alamanni, Burgundians, Visigoths; → map p. 235).

The sources

Literary and epigraphic attestations support the map's depiction of the administrative structure of the Roman Empire. Particularly imporant for Late Antiquity is the *Notitia provinciarum et civitatum Galliae,* a register of the 17 Gaulish provinces and their communities, compiled in the 4th/5th cents. AD probably for purposes of ecclesiastical organization and expanded and corrected accordingly. The respective provincial capitals are highlighted. The so-called *Laterculus Veronensis*, dated to AD 313 under Constantine I (306-337) can be used as a parallel source to the *Notitia*: it is a register, named after the place of origin of its earliest copy, of the Roman provinces arranged by *dioecesis*, reproducing the Diocletianic imperial reform decreed a few years before. Also worthy of mention here is Iohannes Laurentius Lydus (6th cent.), who held the office of *praefectus praetorio* for forty years and composed a treatise 'On the offices of the Roman state' (Latin *De Magistratibus*) based on the experience he thereby gained.

The map

In spite of intensive research on Gaulish provincial history, points of uncertainty constantly arise, which are always discussed but never resolved. For instance, there is the issue of the administrative capitals of the particular Gaulish provinces. These problems are not explicitly highlighted in the map. Exhaustive information on the subject is found in R. Haensch (cf. literature).

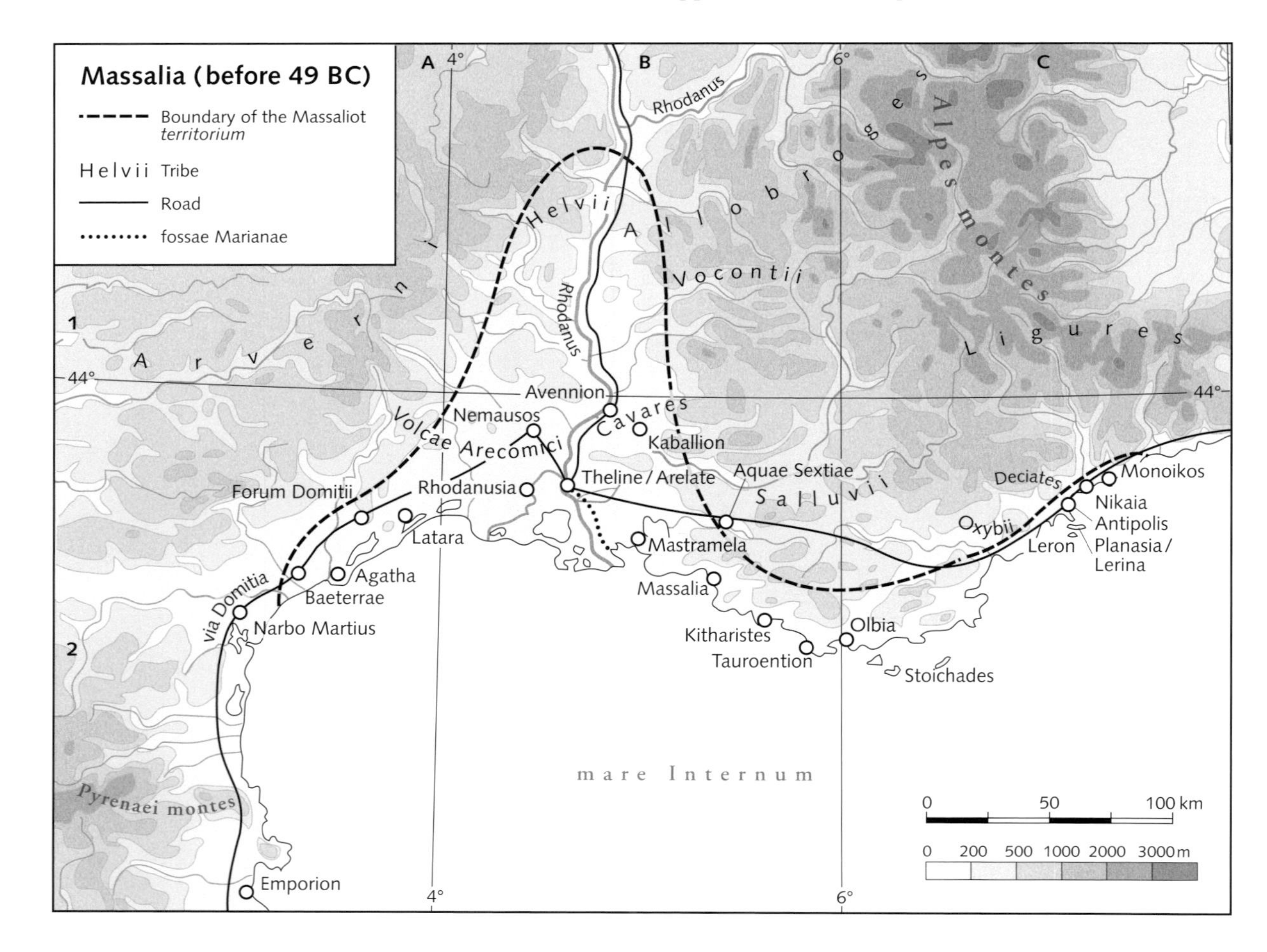

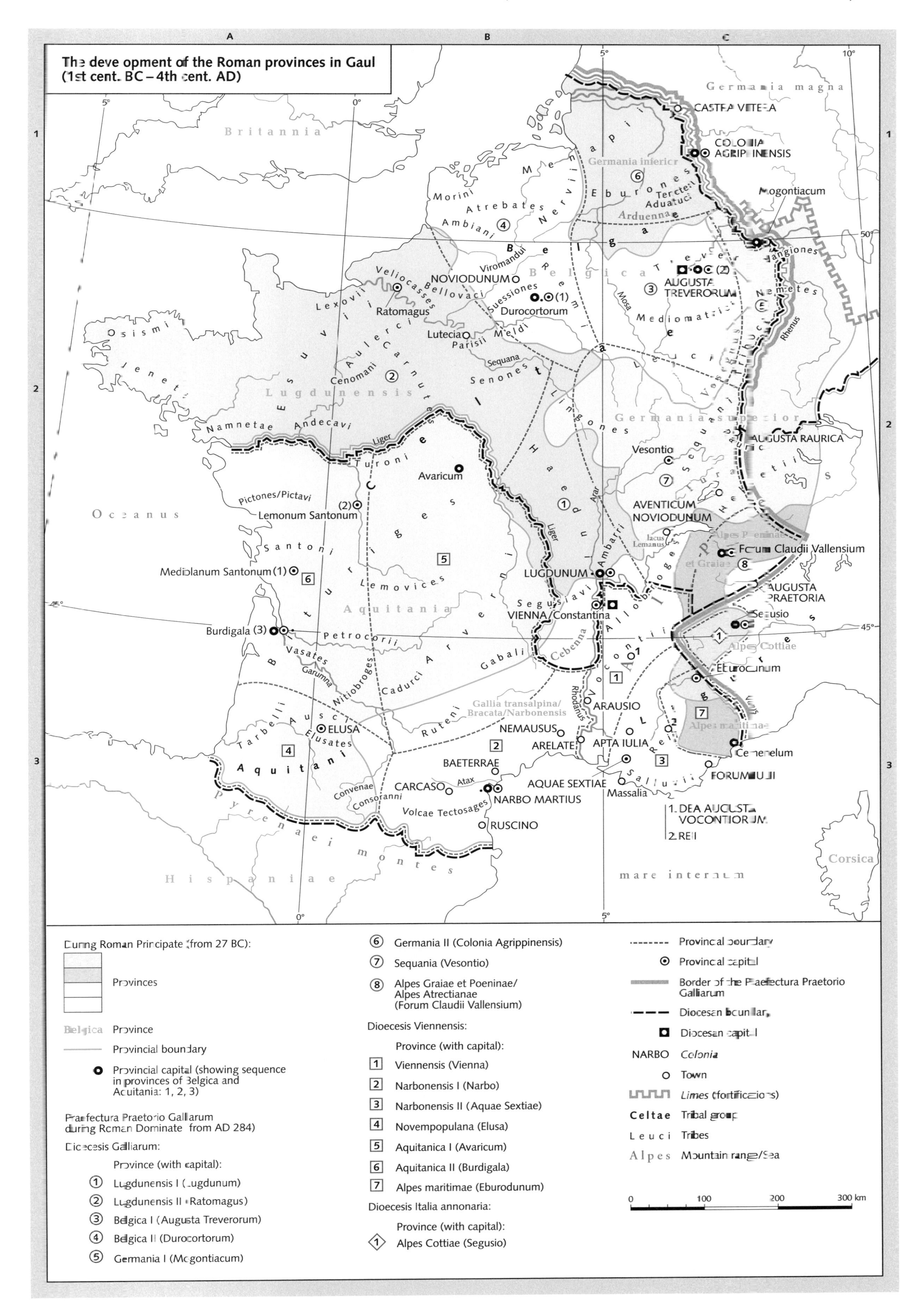
The development of the Roman provinces in Gaul (1st cent. BC – 4th cent. AD)
Britannia
Germania magna
Germania inferior
Germania superior
Belgica
Lugdunensis
Aquitania
Gallia transalpina/Bracata/Narbonensis
Alpes Cottiae
Alpes maritimae
Oceanus
mare internum
Hispaniae
Pyrenaei montes
Corsica
Cebenna
Arduenna
Rhenus
Mosa
Sequana
Liger
Arar
Rhodanus
Garumna
Atax
CASTRA VETERA
COLONIA AGRIPPINENSIS
Mogontiacum
AUGUSTA TREVERORUM
NOVIODUNUM
Ratomagus
Durocortorum
Lutecia
Avaricum
Lemonum Santonum
Mediolanum Santonum
Burdigala
ELUSA
LUGDUNUM
VIENNA/Constantina
Vesontio
AVENTICUM
NOVIODUNUM
Forum Claudii Vallensium
AUGUSTA RAURICA
AUGUSTA PRAETORIA
Segusio
Eburodunum
ARAUSIO
NEMAUSUS
ARELATE
APTA IULIA
AQUAE SEXTIAE
Massalia
Cemenelum
BAETERRAE
CARCASO
NARBO MARTIUS
RUSCINO
1. DEA AUGUSTA VOCONTIORUM
2. REII
Morini
Atrebates
Ambiani
Nervii
Menapii
Eburones
Tencteri
Aduatuci
Treveri
Mediomatrici
Remi
Suessiones
Bellovaci
Viromandui
Veliocasses
Lexovii
Osismi
Veneti
Namnetae
Andecavi
Aulerci
Cenomani
Carnutes
Parisii
Meldi
Senones
Lingones
Haedui
Sequani
Helvetii
Nemetes
Vangiones
Turoni
Pictones/Pictavi
Santoni
Lemovices
Bituriges
Arverni
Petrocorii
Vasates
Nitiobroges
Cadurci
Gabali
Ruteni
Tarbelli
Ausci
Elusates
Convenae
Consoranni
Volcae Tectosages
Segusiavi
Ambarri
Allobroges
Vocontii
Salluvii
Aquitani
Celtae
During Roman Principate (from 27 BC):
Provinces
Belgica Province
Provincial boundary
Provincial capital (showing sequence in provinces of Belgica and Aquitania: 1, 2, 3)
Praefectura Praetorio Galliarum during Roman Dominate (from AD 284)
Dioecesis Galliarum:
Province (with capital):
1 Lugdunensis I (Lugdunum)
2 Lugdunensis II (Ratomagus)
3 Belgica I (Augusta Treverorum)
4 Belgica II (Durocortorum)
5 Germania I (Mogontiacum)
6 Germania II (Colonia Agrippinensis)
7 Sequania (Vesontio)
8 Alpes Graiae et Poeninae/Alpes Atrectianae (Forum Claudii Vallensium)
Dioecesis Viennensis:
Province (with capital):
1 Viennensis (Vienna)
2 Narbonensis I (Narbo)
3 Narbonensis II (Aquae Sextiae)
4 Novempopulana (Elusa)
5 Aquitanica I (Avaricum)
6 Aquitanica II (Burdigala)
7 Alpes maritimae (Eburodunum)
Dioecesis Italia annonaria:
Province (with capital):
1 Alpes Cottiae (Segusio)
Provincial boundary
Provincial capital
Border of the Praefectura Praetorio Galliarum
Diocesan boundary
Diocesan capital
NARBO Colonia
Town
Limes (fortifications)
Celtae Tribal group
Leuci Tribes
Alpes Mountain range/Sea
0 100 200 300 km

The Roman Empire in the Civil War 49–45 BC

The causes and provocations of wars have been a constant subject of lively discussion since Thucydides, and the Civil War brought about by Caesar is no exception. One crucial cause of this war was probably the circumstance that Caesar ran the risk, in his confrontation with opponents in the Senate and with Pompey, of losing his political identity and with it his *dignitas*. However, the war was provoked by the *senatus consultum ultimum* issued by the Senate on 7 January 49 BC, which aimed at forcing the proconsul to relinquish his army. Caesar reacted to this as close to instantaneously as the speed of communications permitted: in the night of the 10-11 January, he and an armed force crossed the Rubicon, the river that marked the southern boundary of his province of Gallia citerior, thereby entering Italy and triggering war by this breach of the constitution. Caesar's awareness of the gravity of the step he had just taken is attested by his quotation of Menander, "Let the die be cast!" (*anerríphtho kýbos*). Peace negotiations were initiated several times by both sides before and during the hostilities. However, on every occasion they failed, often foundering on the resistance of the senators, who could not countenance the prospect of another accommodation between Caesar and Pompey.

Caesar broke into Italy with a rapidity that astonished everyone. He kept up this tempo in all his subsequent military actions in this war, as the calendar at the end of the book shows (→ table p. 269). The dates follow the pre-Julian calendar, which at this time was between six and eleven weeks ahead of the solar season. After one year of fifteen months – twelve as usual followed by three intercalary months (*menses intercalares*) – the reformed calendar came into force on 1 January 45 BC, synchronized with the seasons once more.

The senators, under the leadership of Pompey, to whose protection against the insubordinate proconsul they had entrusted themselves, had not counted on the decisiveness with which Caesar pursued the power of the state. They rushed to vacate Rome, intending to recapture the capital of the empire later from the east. Caesar, who wanted to prevent them from leaving Italy, reached Brundisium too late.

He had thus won Italy. Now, instead of pursuing Pompey and the Senate as most probably expected, he moved to keep his rear secure, turning his attention west to Spain. First, a brief stay at Rome was devoted to an attempt (which would prove fruitless) to justify his position before a rump Senate. Only by using the threat of violence to override the veto of the people's tribune, L. Caecilius Metellus, did he succeed in laying hands on the state treasury.

He reached Spain not without complications. Massilia had taken the side of the Senate and Pompey, and would only be taken after some five months of siege. There were also exceedingly difficult battles against L. Afranius and M. Petreius, the two legates who administered Hispania citerior for Pompey, after Caesar had bypassed the eastern Pyrenees, in the region where the Cinga and the Sicoris, left tributaries of the lower Hiberus/Ebro, flow into the mainstem. Caesar's victory at Ilerda finally opened the way for him to enter Hispania ulterior, which he took from M. Terentius Varro, the legate there, with the capitulation of Corduba. After a short visit to Gades, Caesar went by sea to Tarraco and on to Massilia. Only now did the latter surrender.

After a short stay at Rome, Caesar crossed the Adriatic to the Illyrian coast at the turn of the year. A lengthy and debilitating (to both sides) war of position ensued in the region of the cities of Dyrrhachium, Petra, Apollonia, Amantia and Oricus. From this, Caesar finally withdrew east into Thessalia, and Pompey followed him. Battle was joined near Pharsalus, where Caesar's forces were victorious. Pompey broke away, fleeing to Mytilene by way of Amphipolis, then southwards to Cyprus, from where he hoped to find refuge in Ptolemaic Egypt. However, the crown council of the young Ptolemy XIII advised his murder as he landed at Pelusium.

Caesar had followed a short distance behind Pompey, reaching the Ptolemaic capital of Alexandria by way of Ephesus. It was at Alexandria that he received word of Pompey's death. While there, he intervened in succession disputes, in the course of which he put his own life at risk, finally assisting Cleopatra VII and her younger brother Ptolemy XIV to the throne by winning a battle whose precise location by the Nile has not been established.

Caesar had spent roughly nine months in Egypt when he left in June 47 BC to go to Syria. There, Pharnaces II, the son of Mithridates VI was planning to regain his father's kingdom under the cloak of the civil war. Caesar crushed him in a short but brutal battle at Zela. This was the setting for Caesar's famous message of victory to his friend C. Matius at Rome: "I came, I saw, I conquered" (*veni, vidi, vici*) – words which adorned a large placard the following year at his Pontic triumph.

The remains of the defeated army from Pharsalus united with troops sent by the Numidian King Juba in Africa, forming a substantial battle force under the leadership of Q. Caecilius Metellus Pius Scipio, to continue the fight against Caesar. Caesar moved against it in December 47 BC, in a winter campaign. After various operations near the cities of Ruspina, Uzita and Leptis Minor, the final battle came at Thapsus: Caesar won.

The civil war that had developed from Caesar's constitutional breach at the Rubicon only came to a clear-cut end in Spain. Trouble was brewing there in Hispania ulterior around two legions which had formerly served under M. Terentius Varro, Pompey's legate, and which had since served Caesar. Augmented by others fleeing the defeat at Thapsus and led by Pompey's elder son, Gnaeus, these now threatened to become a real danger to Caesar. Caesar now again ventured to undertake a winter campaign. Fighting came to a head near the cities of Ategua and Ucubis, culminating in the decisive victory on the plain at Munda – a victory Caesar personally played a part in securing.

The sources

The calendar is mainly based on Caesar's own *Commentario de bello civile*. We thus have a source that is very comprehensive, but probably also biased. It is also based on three reports that carry Caesar's name, and are at least contemporary, on the wars in Alexandria, Africa and Hispania ulterior. The letters of Cicero also contain important information from a contemporary. Later authors who sometimes provide information these contemporary sources do not are the historian Appian (in book 2 of his *Civil Wars*, 2nd cent. AD), Cassius Dio (books 41–43 of his *Roman History*; 2nd/3rd cents.) and the biographers Suetonius (biography of Caesar; 1st/2nd cents.) and Plutarch (biographies of Caesar and Pompey; 1st/2nd cents.).

→ Map p. 171

Sources

anerriphtho kybos: Ath. 13,8, cf. Suet. Iul. 32; Plut. Caesar 32,8; Plut. Pompeius 60,2; *clementia Corfiniensis*: Cic. Att. 9,16; *veni, vidi, vici:* Suet. Iul. 37,2; Plut. Caesar 50,3f.; *Letters of Cicero:* still important for chronology is O.E. Schmidt, Der Briefwechsel des M. Tullius Cicero von seinem Prokonsulat in Cilicien bis zu Caesars Ermordung, 1893, 103–260; *to Pharsalus*: Caes. B Civ. 1,10,3; 1,23,5; 1,25,1f.; 1,27f.; 1,37–87; 2,1–16; 3,2–5; 3,22; 3,41–49; Cic. Att. 7,14,1; 9,1,1; Cic. Fam. 16,12,3; Cic. Phil. 8,18f.; 13,32; Cic. Lig. 19; Cic. Deiot. 33f.; Liv. Per. 109–112; Suet. Caes. 31f.; Plut. Caesar 32; 35–39; 48; Plut. Pompeius 60; 62; 64–73; Plut. Cato Minor 53; App. Civ. 2,35; 2,38; 2,40–43; 2,47–49; 2,52–57; 2,60–82; Cass. Dio 41,12; 41,20–26; 41,35f.; 41,39; 41,44; 41,46; 41,50–63; 42,8; *bellum Alexandrinum*: Caes. B Civ. 3,111,3–6; Bell. Alex. 12; 17–33; Liv. Per. 112; App. Civ. 2,90; 2,150; Plut. Caesar 49; Cass. Dio 42,38,2; 42,40–44; *Pharnaces II*: Bell. Alex. 65–78; Liv. Per. 113; Plut. Caesar 50; App. Civ. 2,91; App. Mithr. 120f.; Cass. Dio 42,45–49; *bellum Africum*: Bell. Afr. 1–7; 37–61; 79–86; Liv. Per. 113f.; Plut. Caesar 52f.; Plut. Cato Minor 58; App. Civ. 2, 95–97; Cass. Dio 42,56–58; 43,4–8; *bellum Hispaniense*: Bell. Hisp. 20–39; Liv. Per, 115; Plut. Caesar 56; App. Civ. 2,104; Cass. Dio 43,35–38.

Literature

H. Bruhns, Caesar und die römische Oberschicht in den Jahren 49–44 v. Chr. Untersuchungen zur Herrschaftsetablierung im Bürgerkrieg, 1978; P. Jal, La guerre civile à Rome, 1963; E. Konik, *Clementia Caesaris* als System der Unterwerfung, in: [illegible] (eds.), Forms of Control and Subordination in Antiquity, 1988, 226–238; O. Leggewie, *Clementia Caesaris*, in: Gymnasium 65, 1958, 17–36; K. Raaflaub, *Dignitatis contentio*. Studien zur Motivation und politischen Taktik im Bürgerkrieg zwischen Caesar und Pompeius, 1974; M. Rambaud, Les marches des Césariens vers l' Espagne au début de la guerre civile, in: Mélanges Heurgon, 1976, 845–861; S. Rochlitz, Das Bild Caesars in Ciceros Orationes Caesarianae. Untersuchungen zur clementia und sapientia Caesaris, 1993; P. Simelon, Aspects de la situation socio-économique en Italie entre 49 et 45 avant J.C., in: Acta Classica Universitatis Scientiarum Debreceniensis 21, 1985, 73–100.

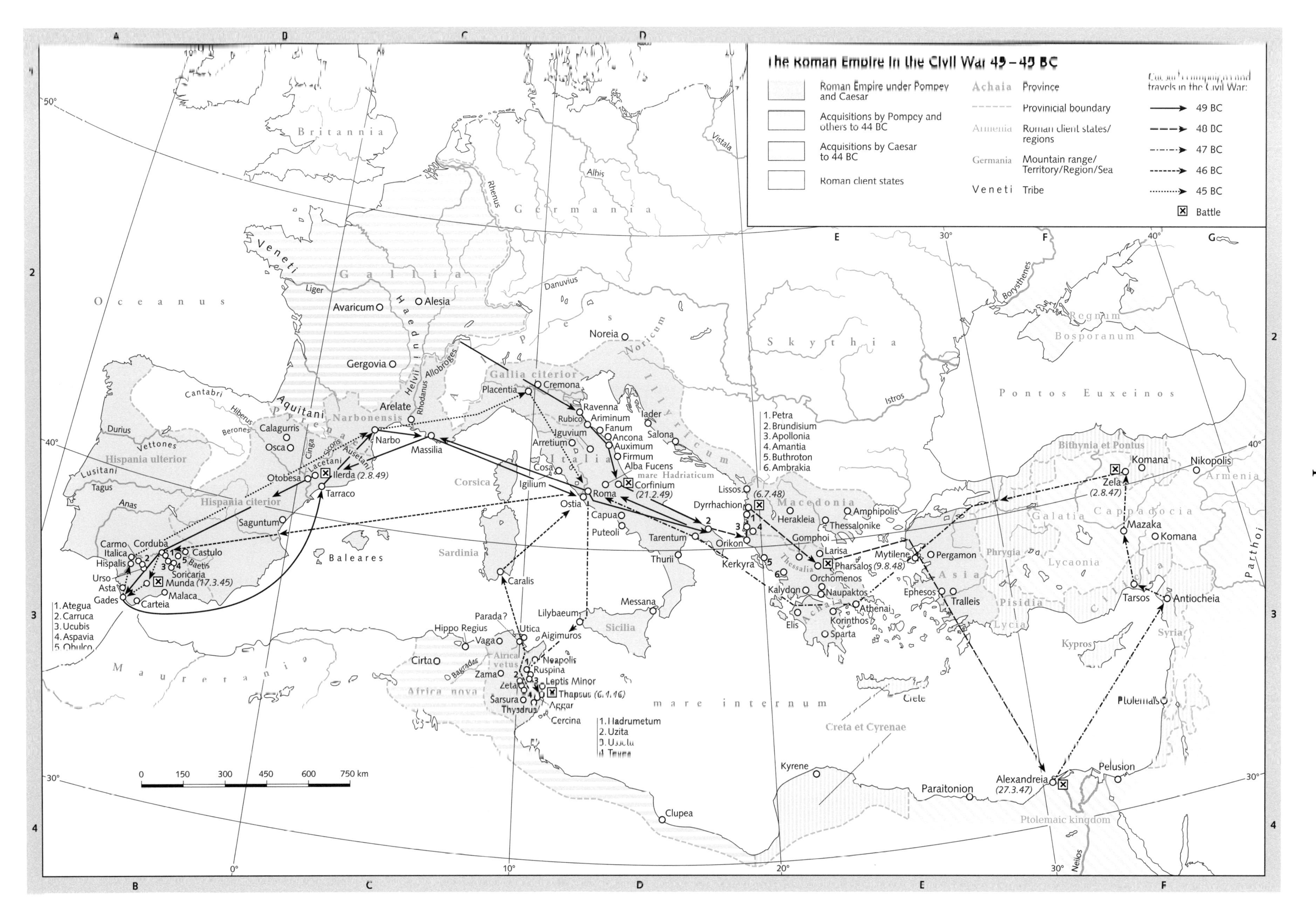
The Roman Empire in the Civil War 49 – 45 BC
Roman Empire under Pompey and Caesar
Acquisitions by Pompey and others to 44 BC
Acquisitions by Caesar to 44 BC
Roman client states
Achaia Province
Provincial boundary
Armenia Roman client states/regions
Germania Mountain range/Territory/Region/Sea
Veneti Tribe
travels in the Civil War:
49 BC
48 BC
47 BC
46 BC
45 BC
Battle
Oceanus
Britannia
Germania
Gallia
Veneti
Aquitani
Haedui
Helvii
Allobroges
Alesia
Avaricum
Gergovia
Gallia citerior
Narbonensis
Hispania ulterior
Hispania citerior
Lusitani
Vettones
Cantabri
Berones
Lacetani
Ausetani
Calagurris
Osca
Otobesa
Ilerda (2.8.49)
Tarraco
Saguntum
Narbo
Arelate
Massilia
Carmo
Corduba
Italica
Hispalis
Urso
Asta
Gades
Castulo
Soricaria
Munda (17.3.45)
Malaca
Carteia
1. Ategua
2. Carruca
3. Ucubis
4. Aspavia
5. Obulco
Baleares
Corsica
Sardinia
Caralis
Sicilia
Lilybaeum
Messana
Placentia
Cremona
Ravenna
Rubico
Ariminum
Fanum
Ancona
Auximum
Firmum
Iguvium
Arretium
Alba Fucens
Corfinium (21.2.49)
Cosa
Igilium
Roma
Ostia
Capua
Puteoli
Tarentum
Thurii
Italia
mare Hadriaticum
Illyricum
Iader
Salona
Noricum
Noreia
Danuvius
Rhenus
Albis
Vistala
Skythia
Istros
Borysthenes
Pontos Euxeinos
Bosporanum
Regnum
1. Petra
2. Brundisium
3. Apollonia
4. Amantia
5. Buthroton
6. Ambrakia
Lissos
Dyrrhachion (6.7.48)
Orikon
Kerkyra
Macedonia
Amphipolis
Herakleia
Thessalonike
Gomphoi
Larisa
Pharsalos (9.8.48)
Orchomenos
Kalydon
Naupaktos
Athenai
Korinthos
Elis
Sparta
Achaia
Mytilene
Pergamon
Ephesos
Tralleis
Asia
Phrygia
Lycaonia
Pisidia
Lycia
Galatia
Cappadocia
Bithynia et Pontus
Komana
Zela (2.8.47)
Nikopolis
Armenia
Mazaka
Komana
Tarsos
Antiocheia
Syria
Kypros
Ptolemais
Parthoi
Pelusion
Alexandreia (27.3.47)
Paraitonion
Ptolemaic kingdom
Neilos
Kyrene
Creta et Cyrenae
Crete
mare internum
Mauretania
Cirta
Hippo Regius
Vaga
Parada?
Utica
Aigimuros
Africa vetus
Africa nova
Bagradas
Zama
Neapolis
Ruspina
Leptis Minor
Zeta
Sarsura
Thapsus (6.4.46)
Aggar
Thysdrus
Cercina
1. Hadrumetum
2. Uzita
3. Uscta
4. Tegea
Clupea
0 150 300 450 600 750 km

The Roman Empire in the Civil War, 44-30 BC

I. On 27 November 43 BC, on the basis of the *lex Titia*, the proconsuls M. Aemilius Lepidus and M. Antonius (Mark Antony), with C. Julius Caesar Octavianus ('Caesar the Younger'), who had been elected *consul suffectus* on 19 August of that year, were equipped with consular *imperium* for five years, i.e. until the end of 38 BC, and charged with the task of reconstituting the commonwealth (*triumviri rei publicae constituendae*). In this function, they decreed sweeping proscriptions (to which Cicero was among those who fell victim, on 7 December 43), chose 18 Italian towns (including Capua, Rhegium, Beneventum, Nuceria, Ariminum and Vibo Valentia) which were to provide land in their *territoria* for the maintenance of veterans, and divided the provinces among themselves, as follows:

- Lepidus claimed Gallia Narbonensis and Hispania citerior,
- Antony took Gallia transalpina (except Gallia Narbonensis) and Gallia cisalpina, and
- Caesar the Younger took Africa vetus and nova, Sicilia, Sardinia and Corsica

Sex. Pompeius Magnus, who like Caesar the Younger had been commissioned by the Senate for its struggle against Antony and named *praefectus classis et orae maritimae*, and who was himself proscribed, established himself in Sicily with his substantial fleet, denied Caesar the Younger access to his province and created a reservoir of all the opponents of the *triumviri* there.

II. Following the two battles between the Caesarians and the Republicans at Philippi, the first of which took place early in October, the second on 23 October 42 BC, the armies and provinces were newly assigned:

- Lepidus was allocated Africa vetus and Africa nova, which Caesar the Younger ceded to him in the spring of 41,
- Antony receives Gallia transalpina and Gallia Narbonensis; Gallia cisalpina lost its provincial status and was added to Italy,
- Caesar the Younger initially took Africa vetus and nova, as well as Hispania citerior, Sicilia and Sardinia et Corsica, which he kept. The distribution of land to the veterans was also transferred to him.

Sex. Pompeius, who had been excluded from these arrangements, received still more adherents following the deaths of M. Iunius Brutus and C. Cassius Longinus, and implemented an ever more effective blockade on Rome's grain supply in an attempt to force the *triumviri* to recognize him. At all events, Sicily was firmly in his hands, and was not available to Caesar the Younger.

L. Antonius, younger brother of the *triumvir* and consul for the year 41, came into conflict with Caesar the Younger over the accommodation of the veterans. Supported, or egged on, by Fulvia, the wife of the *triumvir* M. Antonius (Antony), he intervened on behalf of his brother against the preferential treatment given to the veterans of Caesar the Younger, and also for the towns deeply affected by these measures to be spared. But a fundamental point of conflict was the rivalry between the consular and the triumviral *imperium*. In the end, civil war was the result, with Sex. Pompeius taking the side of L. Antonius and Fulvia. Starting at Rome, the fighting spread north into Etruria, where, however, Caesar the Younger trapped and besieged L. Antonius at Perusia in the autumn of 41. Reports of the ritual killing of 300 senators and equestrians from Perusia, ordered by Caesar the Younger on 15 (the Ides of) March 40 BC in revenge for the killing of his father, are probably not fictions for propaganda purposes.

III. In September of 40 BC, the invasion of Italy envisaged by Mark Antony threatened to set off another civil war. This time, officers and soldiers from both sides successfully pushed Antony and Caesar the Younger to reach an understanding. Now, at Brundisium, a treaty was agreed (and secured by the marriage of Antony to Octavia, the sister of Caesar the Younger) which regulated the distribution of the provinces among the *triumviri* as follows:

- Lepidus received Africa vetus and nova,
- Antony received Macedonia, Bithynia et Pontus, Asia, Cilicia, Syria and Creta et Cyrenae,
- Caesar the Younger received both Spanish provinces, both Gaulish provinces, Illyricum, Sicilia, Sardinia et Corsica and provision for Italy, where, however, Antony as well as Caesar was allowed to raise troops.

Scodra, the most southerly town of the province of Illyricum, was chosen as the border point when the provinces were shared out between Antony and Caesar the Younger. The historian Appian (App. Civ. 5,65; 2nd cent. AD) describes this as follows: "Octavianus [Caesar the Younger] and Antonius now repartitioned the Roman Empire, setting the boundary between them at the Illyrian town of Scodra, which was thought to be the place closest to the midpoint of the Ionian gulf. All provinces to the east of this point as far as the River Euphrates were to fall to Antonius, while Octavianus was to have everything to the west as far as the Ocean. Lepidus was to have Africa, as Octavianus had granted it to him."

Once more, Sex. Pompeius was excluded from negotiations. Antony even declared himself ready to support Caesar the Younger in his fight with Pompey. But Pompey already ruled Sicily, Sardinia and Corsica, and had troops march into Gaul and Africa. In the end, the supply of the city of Rome became so precarious that the *triumviri* had no choice but to come to an understanding with Pompey.

In 40 and 39 BC, the Parthians invaded the province of Syria (Sidon, Aradus, Apamea, Laodicea and Antioch conquered) under the crown prince Pacorus and Q. Labienus, the son of a former legate of Caesar's in Gaul. They also invaded Cilicia and Asia (Miletus possibly conquered, also Mylasa, Alabanda, Aphrodisias; Stratoniceia besieged without success).

IV. In August 39, an accord was reached with Sex. Pompeius at Misenum, by which the state of the possessions of the *triumviri* remained unaltered, but Sex. Pompeius was granted Sicilia, Sardinia, Corsica and the Peloponnese for five years. Caesar the Younger had prepared this agreement well in advance: in 40 BC, he had married Scribonia, the sister of the influential senator L. Scribonius Libo, whose daughter was married to Pompey. By this connection, Caesar the Younger hoped to create better contacts with Pompey. But the peace was called into question at once when Caesar proceeded to divorce Scribonia to marry Livia, and it disintegrated completely in the winter of 39/8 when Menodorus, a fleet admiral in the service of Pompey, defected and presented Sardinia and Corsica to Caesar the Younger. In the spring of 38, Pompey won two naval battles against Caesar, at Cumae and in the Gulf of Messana, but he failed to exploit these victories.

V. Late in the summer of 37, Caesar the Younger again concluded an agreement with Antony, who promised to aid him in his fight with Pompey. Furthermore, the term of the *triumviri* had run out in the autumn of 38, and had to be renewed as a matter of urgency if the triumvirate was not to lose whatever mask of legality it had and with it the acceptance of the army. After a well-equipped fleet was built, the proconsul M. Vipsanius Agrippa succeeded twice (first at Mylae and then at Naulochus on 3 September 36) in defeating Pompey on Caesar's behalf, then landing on Sicily and driving Pompey out. Pompey's troops capitulated. When Lepidus, who had supported Caesar the Younger in his conflict with Pompey, now sought to capitalize on the victory and claimed Sicily, Caesar relieved him of his powers, permitting him to retain his life, his fortune and his position as *pontifex maximus*.

VI. Caesar the Younger (in Illyricum) and Antony (in the Armenian-Parthian borderlands, Armenia occupied 34/33) undertook various campaigns in 36 BC after the death of Sex. Pompeius.

During his operations in the east (40-33 BC; in that year Antony also planned a second Parthian campaign), Antony made various regulations which were comparable in dimension to the reorganization of the Near East by Pompey in 67-59 BC:

- He enacted measures to reinforce various priestly states, such as Comana Pontica and Zela in Pontus.
- Countering the urbanization efforts of Pompey the Great, he breathed new life into the kingdom of Pontus, enthroning Darius, son of Pharnaces II and grandson of Mithridates VI in 39 BC. On the death of the Galatian prince Deiotarus, his possessions in Pontus were passed to Darius.
- He installed Castor, probably a grandson of Deiotarus, as King of Galatia, granting him Paphlagonia. After his death in 38, he was succeeded by Amyntas, a Galatian aristocrat, who had held the position of a 'scribe' under Deiotarus and had already received from Antony the eastern part of Pisidia with parts of Lycaonia and Pamphylia as his kingdom. His son Deiotarus was enfeoffed with Paphlagonia in 38.
- Archelaus Sisines of the priestly dynasty of Comana Pontica was nominated king of Cappadocia by Antony in 41, but only in 36 did he succeed in overcoming a pretender to the throne.
- Polemon, a wealthy citizen of Laodicea on the Lycus, received the western part of Cilicia and parts of Lycaonia as his kingdom. He was also assigned Lesser Armenia in 33 BC.

The most substantial reforms on the soil of the Roman Empire and in the territories of the client states were made by Antony in favour of Cleopatra and her children. In 39, she was given Cyprus and part of Cilicia Tracheia. Coele Syria and regions of the Phoenician coast, Nabataean Arabia, Jericho in Judaea, Ituraea with its adjacent *territorium* and territories on Crete and in the Cyrenaica were also signed over to her, as were countries yet to be conquered, such as Armenia, Media and Parthia.

VII. After the victory of Caesar the Younger at Actium on 2 September 31 BC, and the capture of Alexandria on 1 August 30 BC, the continental Ptolemaic kingdom was established as the province of Aegyptus, with the special status of a 'crown domain'.

The sources

The state of the sources is good. The summaries (*Periochae*) of the History of the contemporary scholar Livy (Liv. Per. 116–113) give year-by-year reports, and we have the pertinent passages of the historians Velleius Paterculus (2,82–88; 1st cent. AD), Appian (*Civil Wars* 4 f.; 2nd cent.) and Cassius Dio (Books 48–51; 2nd/3rd cents.). There are also the biographies of Caesar and Augustus (Caesar the Younger) by Suetonius (1st/2nd cents.) and biographies by Plutarch (Brutus and Antony; 1st/2nd cents.). Finally, Augustus' record of his accomplishments, the *Res gestae divi Augusti*, reveals his perspective on these matters.

→ Maps pp. 155, 169

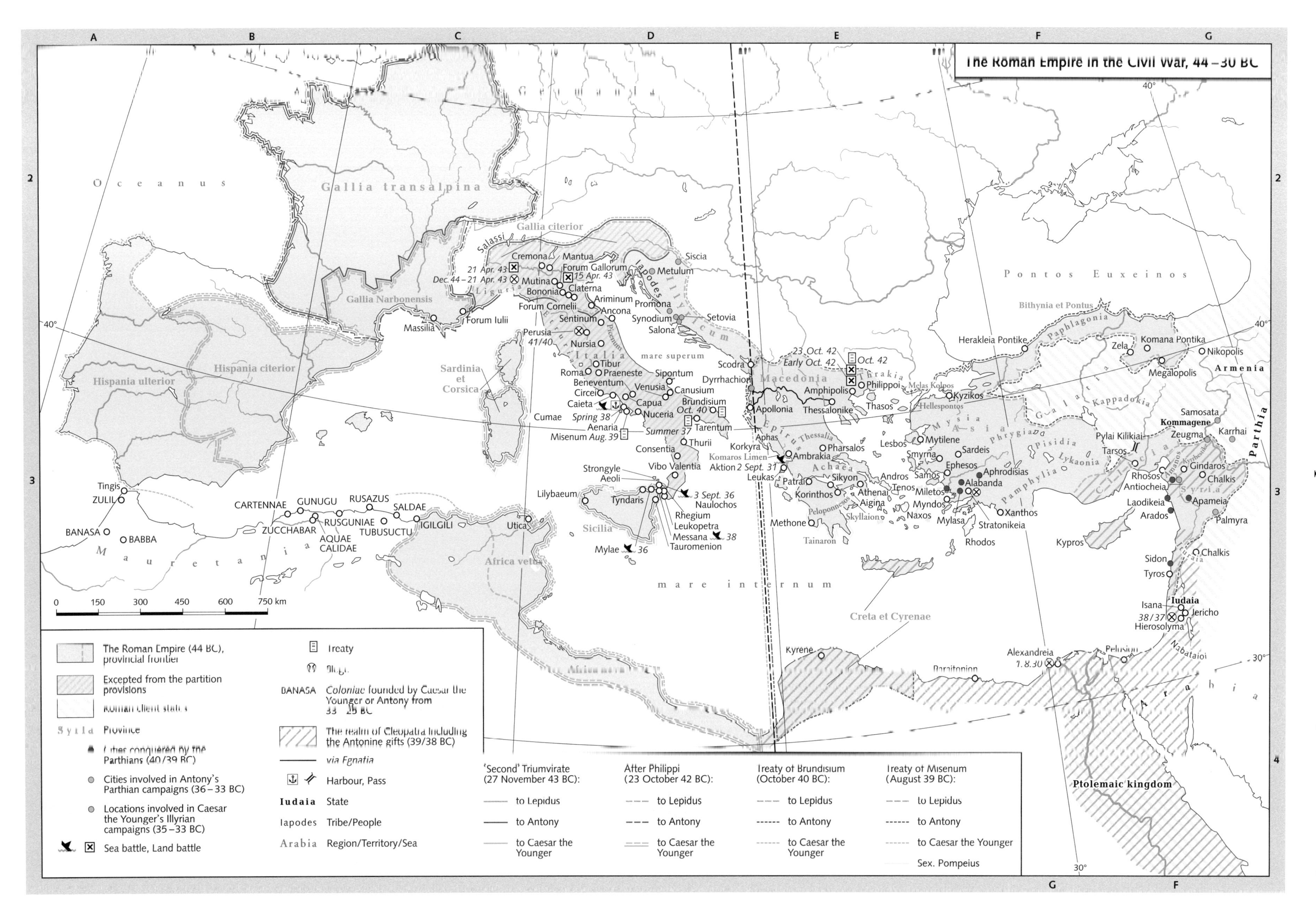
The Roman Empire in the Civil War, 44 – 30 BC
The Roman Empire (44 BC), provincial frontier
Excepted from the partition provisions
Province
Cities involved in Antony's Parthian campaigns (36 – 33 BC)
Locations involved in Caesar the Younger's Illyrian campaigns (35 – 33 BC)
Sea battle, Land battle
Treaty
The realm of Cleopatra including the Antonine gifts (39/38 BC)
via Egnatia
Harbour, Pass
Iudaia State
Iapodes Tribe/People
Arabia Region/Territory/Sea
'Second' Triumvirate (27 November 43 BC):
After Philippi (23 October 42 BC):
Treaty of Brundisium (October 40 BC):
Treaty of Misenum (August 39 BC):
to Lepidus
to Antony
to Caesar the Younger
Sex. Pompeius
0 150 300 450 600 750 km
Oceanus
Pontos Euxeinos
mare internum
mare superum
Gallia transalpina
Gallia Narbonensis
Gallia citerior
Hispania citerior
Hispania ulterior
Mauretania
Africa vetus
Sardinia et Corsica
Sicilia
Italia
Macedonia
Achaea
Creta et Cyrenae
Bithynia et Pontus
Armenia
Parthia
Kommagene
Ptolemaic kingdom
Iudaia
Kappadokia
Galatia
Thrakia
Illyricum
Iapodes
Massilia
Forum Iulii
Mutina
Bononia
Cremona
Mantua
Forum Gallorum
Claterna
Ariminum
Ancona
Perusia 41/40
Roma
Tibur
Praeneste
Capua
Nuceria
Brundisium
Tarentum
Rhegium
Messana
Tauromenion
Mylae
Naulochos
Lilybaeum
Utica
Tingis
Dyrrhachion
Apollonia
Thessalonike
Amphipolis
Philippoi
Aktion 2 Sept. 31
Korinthos
Athenai
Kyzikos
Ephesos
Miletos
Tarsos
Antiocheia
Apameia
Palmyra
Tyros
Sidon
Hierosolyma
Jericho
Alexandreia 1.8.30
Kyrene
Rhodos
Kypros

The city of Rome

The most striking topographical feature of Rome at first glance at this map is its riparian location. This guarantees water in abundance – an advantage which might rapidly have become a disadvantage had not nearby heights (spurs of a tufa plateau) provided refuge. And indeed, these were the first sites of settlement. However, the particular advantage of the location was neither water nor protection from it. It was the position on an important trade route leading from the salt flats at the Tiber mouth, through fords and over the Tiber Island, and on through the northern sector of what became the urban area into the territory of the Sabini. The Romans, accordingly, named this road the *via Salaria*. The benefits of this location, however, were vitiated by the fact that all the low-lying ground between the hills, crossed by streams, ran out into flat flood-plains which were swampy and constantly threatened by the Tiber floods. Archaeology provides evidence of substantial reconstruction and drainage work from the mid 8th cent. BC which made possible the creation of open venues (Velabrum, Forum, Comitium).

Such restructuring operations to assist communications between the various settled hills, all of which betrays the initiative of a central power, indicate that we should date the foundation of the city to this phase of development – just as the Romans' own foundation myth did (753 BC).

The maps

The map showing the entire urban area of Rome and the inset of the city centre show the whole structural extent of Rome as it is known today. At no point did the city ever look like this. However, this albeit unhistorical portrayal, assembling all demonstrated historical layers, is intended to provide insight into all phases of urban Roman history. Many details remain (or repeatedly become) controversial, especially, for instance, the drawing of the various boundaries within and around the city. The basis of all urban research is the findings provided by archaeology. Inscriptions are also important, especially a city plan, the so-called *Forma urbis Romae*, which dates from the period between AD 203 and 208. Literary records, e.g. the *Notitia regionum urbis XIV*, the essence of which dates to the reign of Constantine I (306–337), but also many other literary sources, contribute to the reconstruction of the ancient urban profile of Rome.

→ Maps pp. 107, 175

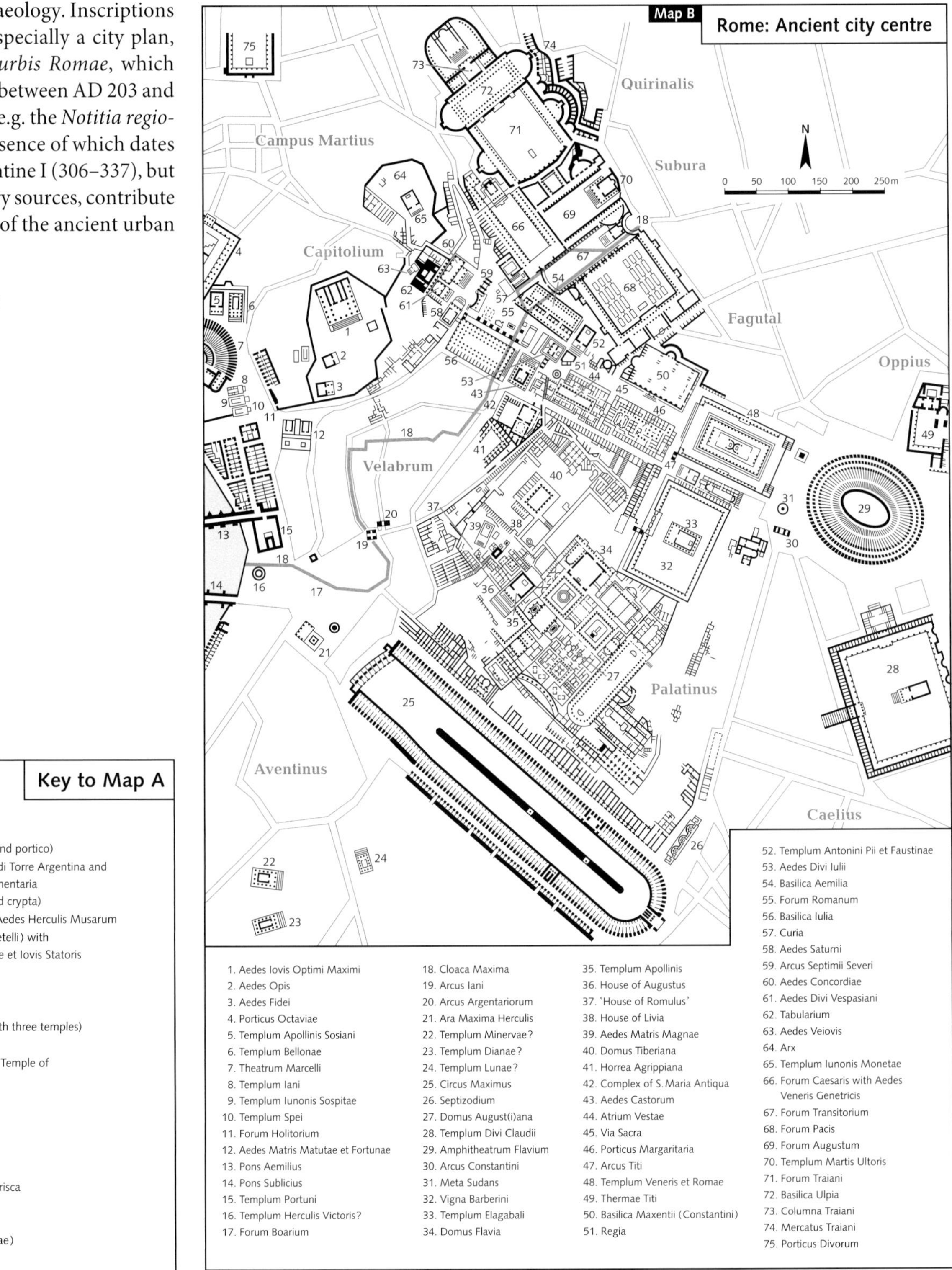

Key to Map A

Pontes (bridges):
A. Aelius
B. Neronis
C. Agrippae
D. Aurelius
E. Cestius
F. Fabricius
G. Aemilius
H. Sublicius
I. Probi

Portae (gates):
a. Aurelia
b. Portuensis
c. Ostiensis
d. Ardeatina
e. Appia
f. Latina
g. Metrovia
h. Asinaria
i. Labicana and Praenestina
j. Tiburtina
k. 'Chiusa'
l. Principalis dextra
m. Praetoria
n. Principalis sinistra
o. Nomentana
p. Salaria
q. Pinciana
r. Flaminia
s. Cornelia
t. Triumphalis?
u. Septimiana
v. Lavernalis
w. Raudusculana
x. Naevia
y. Capena
z. Caelimontana
aa. Querquetulana
bb. Esquilina
cc. Viminalis
dd. Collina
ee. Quirinalis
ff. Salutaris
gg. Sanqualis
hh. Fontinalis
ii. Carmentalis
jj. Flumentana
kk. Trigemina

Important monuments:
1. Capitolium (with Templum Iovis Optimi Maximi Capitolini, Tabularium and Aerarium)
2. Arx (with Templum Iunonis Monetae)
3. Carcer, Temple of Concordia
4. Arch of Septimius Severus
5. Forum Romanum/Imperial forums
6. Atrium Vestae
7. Temple of Venus and Rome
8. Temple of Apollo
9. Vestibulum Domus Palatinae
10. Domus Tiberiana
11. Domus Flavia and Domus Augustana
12. Domus Severiana
13. Aedes Caesarum
14. Temple of Elagabalus
15. Arch of Constantine
16. Amphitheatrum Flavium (Colosseum)
17. Ludus Magnus et Ludus Matutinus
18. Portico and Temple of Claudius
19. Macellum Magnum?
20. Castra Nova Equitum Singularium (Lateran)
21. Amphitheatrum Castrense (Ludus)
22. Domus Sessoriana
23. Baths of Helena
24. Circus Varianus
25. Castra Priora Equitum Singularium
26. Nymphaeum (in the Horti Liciniani)
27. Temple of Minerva Medica (so-called)
28. Mithraeum under S. Clemente
29. Domus Aurea
30. Baths of Titus
31. Baths of Trajan
32. 'Sette sale', Piscina
33. Porticus Liviae
34. Templum Iunonis Lucinae
35. Macellum Liviae
36. Baths of Diocletian
37. Castra Praetoria
38. Templum Veneris Erycinae
39. Aedes Trium Fortunarum
40. Porticus Miliarensis
41. Templum Quirini
42. Templum Salutis
43. Baths of Constantine
44. Horrea
45. Temple of Serapis
46. Graves
47. Temple of Sol
48. Insulae
49. Mausoleum of Augustus and Ustrinum Domus Augustae
50. Horologium Augusti
51. Ara Pacis Augusti
52. Ustrinum Divi Marci Aurelii
53. Ustrinum et Columna Divi Antonini Pii
54. Column of Marcus Aurelius
55. Templum Matidiae and Templum Divi Hadriani
56. Baths of Nero
57. Stadium of Domitian
58. Odeon of Domitian
59. Pantheon
60. Baths of Agrippa
61. Saepta Iulia and Diribitorium
62. Sanctuary of Isis and Serapis
63. Porticus Divorum (Templum Divorum)
64. Ustrinum Hadriani
65. Theatre of Pompey (and portico)
66. Temple on the Largo di Torre Argentina and Porticus Minucia Frumentaria
67. Theatre of Balbus (and crypta)
68. Porticus Philippi and Aedes Herculis Musarum
69. Porticus Octaviae (Metelli) with Aedes Iunonis Reginae et Iovis Statoris
70. Temple of Neptune
71. Circus Flaminius
72. Theatre of Marcellus
73. Forum Holitorium (with three temples)
74. Templum Aesculapii
75. Forum Boarium (with Temple of Fortuna Virilis?)
76. Circus Maximus
77. Templum Lunae?
78. Temple of Minerva
79. Temple of Diana?
80. Thermae Suranae
81. Mithraeum under S. Prisca
82. Baths of Decius
83. Baths of Caracalla (Thermae Antoninianae)
84. Tomb of Scipio
85. Emporium
86. Porticus Aemilia (Navalia?)
87. Horrea Galbana
88. Horrea Lolliana
89. Pyramid of Cestius
90. Sanctuary of Iuppiter Heliopolitanus
91. Naumachia Augusti
92. Domus Clodiae?
93. Circus Gai et Neronis
94. Meta Romuli
95. Mausoleum Hadriani
96. Naumachia Vaticana or Naumachia Traiani

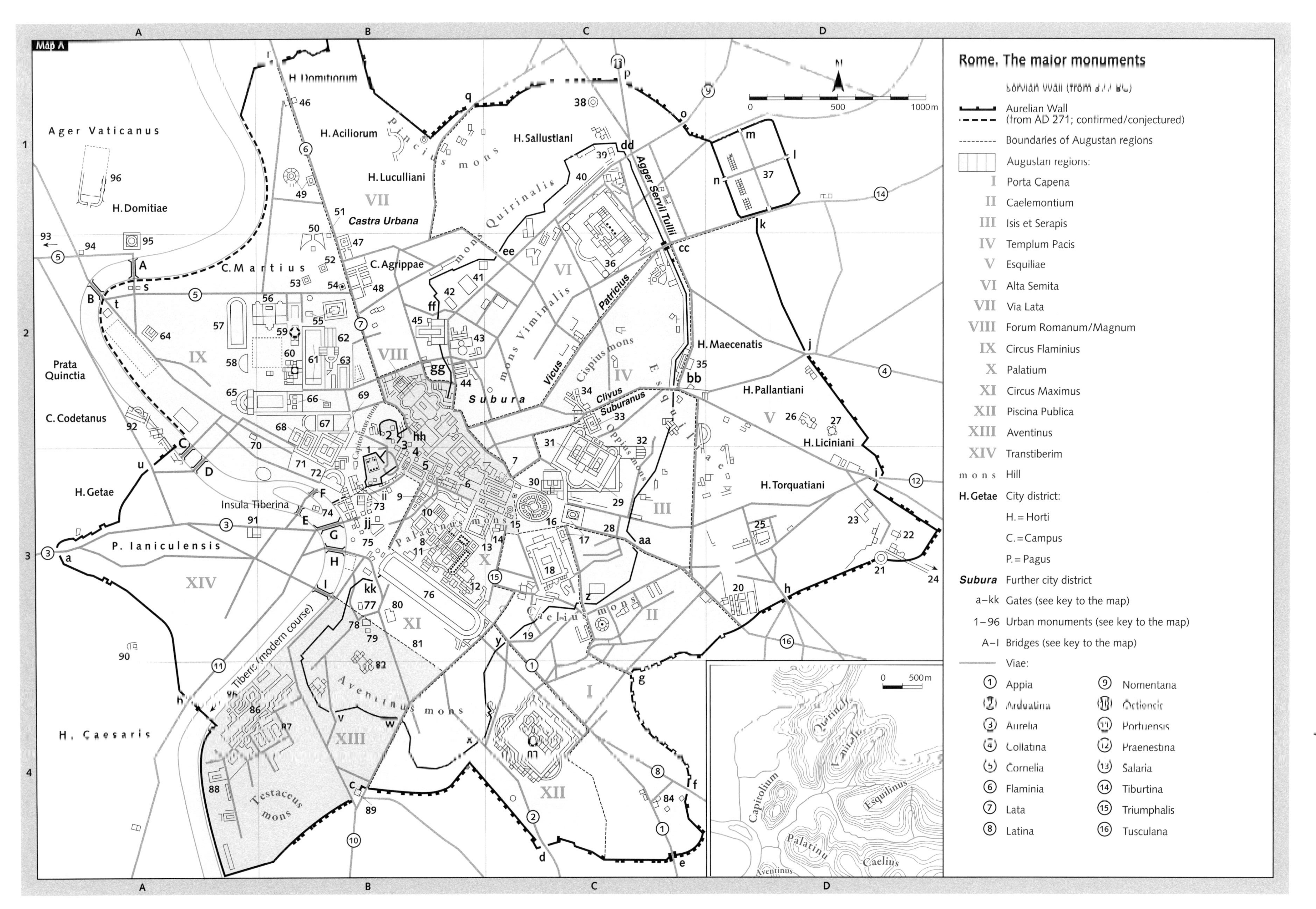
Map A
Rome. The major monuments
Servian Wall (from 378 BC)
Aurelian Wall (from AD 271; confirmed/conjectured)
Boundaries of Augustan regions
Augustan regions:
I Porta Capena
II Caelemontium
III Isis et Serapis
IV Templum Pacis
V Esquiliae
VI Alta Semita
VII Via Lata
VIII Forum Romanum/Magnum
IX Circus Flaminius
X Palatium
XI Circus Maximus
XII Piscina Publica
XIII Aventinus
XIV Transtiberim
mons Hill
H. Getae City district:
H. = Horti
C. = Campus
P. = Pagus
Subura Further city district
a–kk Gates (see key to the map)
1–96 Urban monuments (see key to the map)
A–I Bridges (see key to the map)
Viae:
1 Appia
2 Ardeatina
3 Aurelia
4 Collatina
5 Cornelia
6 Flaminia
7 Lata
8 Latina
9 Nomentana
10 Ostiensis
11 Portuensis
12 Praenestina
13 Salaria
14 Tiburtina
15 Triumphalis
16 Tusculana
0 500 1000m
N
Ager Vaticanus
H. Domitiae
H. Domitiorum
H. Aciliorum
H. Luculliani
H. Sallustiani
Castra Urbana
C. Martius
C. Agrippae
Prata Quinctia
C. Codetanus
H. Getae
Insula Tiberina
P. Ianiculensis
H. Caesaris
Tiberis (modern course)
Testaceus mons
Aventinus mons
Caelius mons
Palatinus mons
Capitolinus mons
Pincius mons
Quirinalis
mons Viminalis
Cispius mons
Oppius mons
Esquiliae
Subura
Vicus Patricius
Clivus Suburanus
Agger Servii Tullii
H. Maecenatis
H. Pallantiani
H. Liciniani
H. Torquatiani
Capitolium
Quirinalis
Viminalis
Esquilinus
Palatinus
Caelius
Aventinus
0 500m

The Augustan division of Rome and Italy into regions

The term *regio*, which in augural terminology denoted a section of sky whose dimensions were ascertained by arcane principles, also referred (synonymously with the term *tribus*) to one of the four urban regions into which the sixth of the Roman kings, Servius Tullius, was said to have divided the area of the city of Rome within the Pomerium. The four original urban *regiones* were 1. the *regio Palatina,* 2. the *regio Collina*, 3. the *regio Esquilina* and 4. the *regio Suburana* (→ map p. 107, map C).

I. Rome

In 7 BC, Augustus enacted an administrative reform under which the urban area of Rome was divided into 14 *regiones*. Applying an apparently random numbering sequence, he divided the inner city into ten *regiones* (*regio I, II, III, IV, VI, VIII, X, XI, XII, XIII*) and the urban area outside the Servian Walls into four (*regio V, VII, IX, XIV*). These *regiones* were then in turn divided into a total of 265 *vici*, i.e. groups of houses grouped into administrative units; these sometimes comprised entire streets of houses. The names by which the individual *regiones* are listed in the registers of Late Antiquity must have come about only in the late 1st cent. AD. Such a reform was necessitated by the sheer extent to which Rome had grown since the Monarchical period. The four old *regiones* had become too large. Regulative measures accompanied or followed these reforms. Thus, in AD 6, seven *cohortes vigilum* of 500 men apiece were set up to fight fires. Each of these cohorts was responsible for fire-watching in two of the fourteen *regiones*. The *vici* developed a function of particular importance in the ruler cult: Augustus associated the cults of the Lares of each *vicus* with the *genius Augusti*.

The 14 *regiones* of the city of Rome (cf. also Map p. 173)

I. Porta Capena, named after the gate in the Servian Wall by which the *via Appia* made its southward exit from the city. However, the Porta Capena was not in *regio I*, but in *regio XII*, which indicates that the naming of the *regiones* was not done officially, but by the inhabitants themselves. The case of the *regio III Isis et Serapis* is similar.
II. Caelemontium, mainly took in the *mons Caelius*.
III. Isis et Serapis, in the valley between the *mons Caelius* and the *mons Oppius*, which this *regio* also included.
IV. Templum Pacis, included the Temple of Juno Lucina and the *mons Cispius*.
V. Esquiliae, the large expanse between the Servian and Aurelian Walls from the Porta Viminalis and Porta 'Chiusa' in the north to the Porta Asinaria in the south.
VI. Alta Semita, covered the *montes Quirinalis* and *Viminalis* and the valley of the Horti Sallustiani, thus later also the Baths of Diocletian and the Castra Praetoria.
VII. Via Lata, covered the eastern part of the Campus Martius and the western part of the *mons Pincius*.
VIII. Forum Romanum or *Forum Magnum*, also included the imperial fora and part of the *mons Capitolinus* with the Temple of Iuppiter Optimus Maximus.
IX. Circus Flaminius, the western portion of the Campus Martius with the Circus Flaminius and the Theatre of Marcellus. This *regio* also had the Mausoleum of Augustus.
X. Palatium, the *mons Palatinus* with the imperial residences.
XI. Circus Maximus, with the Circus itself, the Forum Boarium and the north-eastern slopes of the *mons Aventinus*.
XII. Piscina Publica, named after an old reservoir which was also used as a swimming-pool. In a later period, the Baths of Caracalla were in this *regio*.
XIII. Aventinus, covered the *mons Aventinus*, on the south-western slopes of which were markets and warehouses, and the *mons Testaceus*.
XIV. Transtiberim, the entire section of the city on the right bank of the Tiber, including the Tiber Island.

II. Italy

Augustus also divided Italy into *regiones*. Their arrangement mostly followed the geographical dictates of the landscape. Ignoring the traditional geographical names, these *regiones*, too, were at first merely numbered, from I to XI: southwards from Rome (I–III) and northwards from Rome (IV–XI). The names, mostly of ethnic origin, probably date back to the late 1st cent. AD. The function of these *regiones* is unclear. They may have played a part in the conduct of the census. Pliny bases the arrangement of his description of Italy on this system.

The 11 *regiones* of Italy

I. Latium et Campania, from the Tiberis/Tiber in the north to the Silarus/Sele in the south. Latium was divided in two: *Latium vetus* from Rome to the *mons Circeius*/Monte Circeo and *Latium adiectum* from here to the *mons Massicus*/Monte Massico. The harbours of Ostia and Puteoli, important to the supply of Rome, and the military harbour of Misenum were on the coast of this *regio*. Rome naturally outshone all cities of the region in importance, but Campania, with its lovely landscape and many villas belonging to influential Romans, was, as it were, the 'second capital' of the Empire (Plin. HN 3,56–70).
II. Apulia et Calabria, almost identical to the modern region of Apulia (Puglia), from Tifernus/Biferno to the Bradanus/Brádano with the *mons Garganus*/Promontorio del Gargano. The port of Tarentum/Taranto was on the south-western coast of (ancient) Calabria, and the port of Brundisium/Brindisi on the east coast, the destination of the *via Appia* from Rome via Capua and Tarentum (Plin. HN 3,99–105).
III. Lucania et Bruttium. The *regio* covered the present-day regions of Basilicata and Calabria from the Silarus by way of Rhegium/Reggio di Calabria to the Bradanus. As in *regiones* I and II, there were various Greek cities along the coast, and these had a significant impact on the cultural landscape of the region (Plin. HN 3,71–75; 3,95–97).
IV. Samnium. This *regio,* dominated by the Appenninus, stretched from the Nar/Nera and Tiberis to the upper reaches of the Volturnus/Volturno and from the Aternus/Aterno-Pescara to the Tifernus, and included the lands of the Samnites, Sabini and Marsi, among other more minor tribes (Plin. HN 3,106–109).
V. Picenum. This region comprised the coastal strip from the mouth of the Aesis/Esino to the mouth of the Aternus. In the interior, it stretched as far as the foot of the Appenninus (Plin. HN 3,110 f.). The *via Salaria* reached the sea at Truentum in Picenum.
VI. Umbria. This region comprised the coastal zone from Ariminus/Marecchia to the Aesis, and in the mountains stretched from the Arnus/Arno and the Tiberis to the Nar. Sena Gallica and Fanum Fortunae were important ports here (Plin. HN 3,112–114).
VII. Etruria, the territory from the Macra/Magra to the Tiberis and inland to the Appenninus (Plin. HN 3,50–55).
VIII. Aemilia, from the Ariminus to the Padus/Po, bounded to the south by the Appenninus. The eponymous *via Aemilia* runs its whole length (Plin. HN 115–122). Ravenna, Italy's most important military port after Misenum and, from AD 403, also the imperial residence, was here.
IX. Liguria, from the Var/Varus to the Macra, in the interior as far as the Padus. The main port here was Genua/Genoa (Plin. HN 47–49).
X. Venetia et Histria. This *regio* bordered the Adriatic, with the port of Aquileia, and stretched from the Padus to the Alps and from the Ollius/Oglio to the Arsia/Raša (Plin. HN 126–138).
XI. Transpadana, from the Padus to the Alps, eastwards as far as the Ollius (Plin. HN 123–125). Among the many important cities of this *regio* was Mediolanum, the imperial residence from AD 285-402.

The sources

The most comprehensive source for Rome's division into *regiones* is the *Notitia regionum urbis XIV* (the so-called *Libellus de regionibus urbis Romae*), the main content of which dates back to the reign of Constantine I (306–337). The marble city plan dating from AD 203-208, which has survived in part (the so-called *Forma urbis Romae*), also gives important indications.

The boundaries between the various *regiones* both of Rome and of Italy are shown schematically. Their details are the subject of constant debate.

The main source for the regional divisions of Italy is Pliny the Elder (Plin. HN 3,47–125).

Literature

G. Carettoni *et al.*, La pianta marmorea di Roma antica, 1960; W. Eck, Die staatliche Organisation Italiens, 1979; A. von Gerkan, Grenzen und Größen der vierzehn Regionen Roms, in: BJ 149, 1949, 5–65; J.P. Heisel, Antike Bauzeichnungen, 1993, 193–197; D. Manacorda, Un nuovo frammento della »Forma Urbis« e le calcare romane del cinquecento nell' area della »Crypta Balbi«, in: MEFRA 114, 2002, 693–715; R. Mancini, Le mura aureliane di Roma: atlante di un palinsesto murario, 2001; J. Martínez-Pinna, Reflexiones en torno a los orígenes de Roma: a propósito de recientes interpretaciones, in: Orizzonti 2, 2001, 75–83; C. Nicolet, L' origine des regiones Italiae augustéennes, in: Cahiers du Centre Gustave Glotz 2, 1991, 73–97; P. Ørsted, Regiones Italiae, Ehreninschriften und Imperialpolitik, in: Studies in Ancient History and Numismatics, presented to R. Thomsen, 1988, 124–138; L. Polverini, Le regioni nell' Italia romana, in: Geographia Antica 7, 1998, 23–33; L. Richardson (jr.), A New Topographical Dictionary of Ancient Rome, 1992, 330–332; E. Rodríguez-Almeida, Forma Urbis Marmorea, 1981; R. Thomsen, The Italic Regions, 1947.

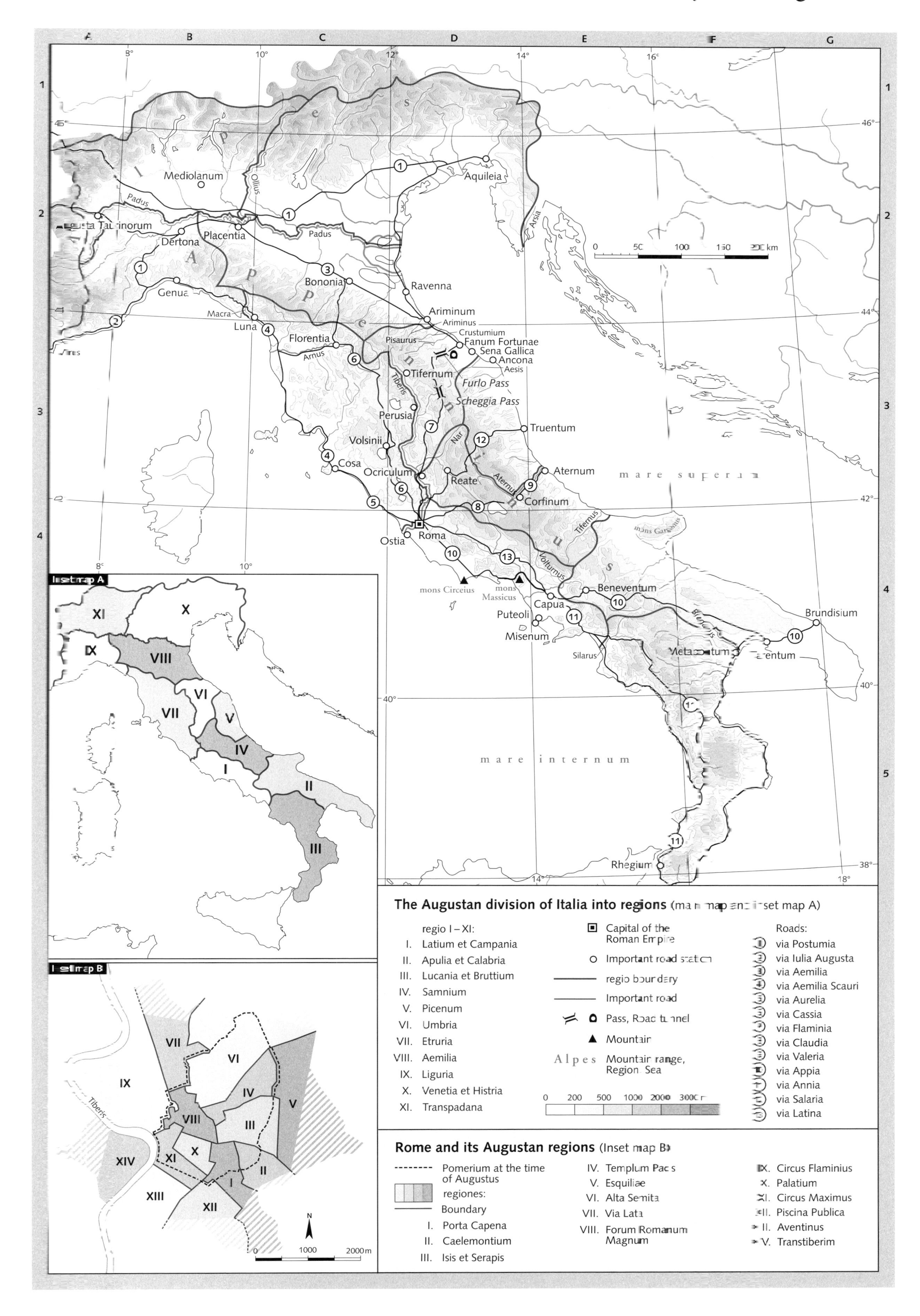
Inset map A
XI
X
IX
VIII
VII
VI
V
IV
I
II
III
Inset map B
VII
VI
IX
IV
V
VIII
III
XI
X
XIV
II
I
XIII
XII
Tiberis
0 1000 2000 m
A l p e s
Mediolanum
Ollius
Aquileia
Padus
Placentia
Dertona
Genua
Macra
Luna
Florentia
Arnus
Bononia
Ravenna
Ariminum
Ariminus
Crustumium
Pisaurus
Fanum Fortunae
Sena Gallica
Ancona
Aesis
Tifernum
Tiberis
Furlo Pass
Scheggia Pass
Perusia
Truentum
Volsinii
Nar
Cosa
Ocriculum
Reate
Aternum
Aternus
Corfinum
Roma
Ostia
Tifernus
mons Circeius
mons Massicus
Volturnus
Beneventum
Capua
Puteoli
Misenum
Silarus
Brundisium
Rhegium
Arsia
mare superum
mare internum
0 50 100 150 200 km
The Augustan division of Italia into regions (main map and inset map A)
regio I – XI:
I. Latium et Campania
II. Apulia et Calabria
III. Lucania et Bruttium
IV. Samnium
V. Picenum
VI. Umbria
VII. Etruria
VIII. Aemilia
IX. Liguria
X. Venetia et Histria
XI. Transpadana
Capital of the Roman Empire
Important road station
regio boundary
Important road
Pass, Road tunnel
Mountain
Alpes Mountain range, Region Sea
0 200 500 1000 2000 3000 m
Roads:
via Postumia
via Iulia Augusta
via Aemilia
via Aemilia Scauri
via Aurelia
via Cassia
via Flaminia
via Claudia
via Valeria
via Appia
via Annia
via Salaria
via Latina
Rome and its Augustan regions (Inset map B)
Pomerium at the time of Augustus
regiones:
Boundary
I. Porta Capena
II. Caelemontium
III. Isis et Serapis
IV. Templum Pacis
V. Esquiliae
VI. Alta Semita
VII. Via Lata
VIII. Forum Romanum Magnum
IX. Circus Flaminius
X. Palatium
XI. Circus Maximus
XII. Piscina Publica
XIII. Aventinus
XIV. Transtiberim

The provinces of the Imperium Romanum from Augustus to Septimius Severus (27 BC to AD 211)

The provinces of the Imperium Romanum from Augustus to Septimius Severus (27 BC to AD 211)

At the Senate sitting of 13 January 27 BC, the future Augustus (he received this honorific title at the Senate sitting three days later) returned to the Senate and people of Rome all the extraordinary powers he had arrogated or been assigned during the preceding period of civil war. However, the Senate pressed him to reassume at least some of the imperial administration, which he finally consented to do (Cass. Dio 53,12,5–7; 53,13,1; Str. 17,3,25; Suet. Aug. 28; 47). On the basis of an *imperium proconsulare*, then, he took over the administration of the following nine provinces:

1. Cyprus, administrative capital Paphus
2. Syria with Cilicia, administrative capital Antioch
3. Aegyptus, administrative capital Alexandria
4. Lusitania, administrative capital Augusta Emerita
5. Tarraconensis, administrative capital Tarraco
6. Narbonensis, administrative capital Narbo
7. Aquitania, administrative capital Burdigala
8. Lugdunensis, administrative capital Lugdunum
9. Belgica, administrative capital Durocortorum

In addition, for the protection of the empire against the Parthians and Germans, Augustus took over command of the Roman forces on the Euphrates (provinces 1–3) and Rhine (provinces 6–9) and the areas respectively bordering them to the west, and to put down the Cantabrians he also took command of the legions in Spain (provinces 4 and 5).

The Senate retained the following ten provinces:

1. Illyricum, administrative capital Salona
2. Achaia, administrative capital Corinthus
3. Macedonia, administrative capital Thessalonica
4. Asia, administrative capital Ephesus
5. Bithynia et Pontus, administrative capital Nicomedia
6. Creta et Cyrenae, administrative capital Gortyn
7. Sicilia, administrative capital Syracusae
8. Sardinia et Corsica, administrative capital Carales
9. Africa proconsularis, administrative capital Carthago
10. Baetica, administrative capital Corduba

With three exceptions (provinces 1, 2 and 9), these were all provinces that were regarded as pacified, and in which no troops were therefore stationed.

Augustus later added more provinces, as did the emperors who succeeded him. Trajan took this development to its zenith: under him, the empire bounded the oceans to both west and east (Atlantic and Persian Gulf).

The table (contd. p. 270) shows this development of the Roman imperial provinces and their administrative capitals from Augustus to Septimius Severus, along with indications of the relevant sources.

The map

The map shows the allocation of provinces to Augustus and the Senate, as well as the provinces subsequently added up to the reign of Septimius Severus, with their capitals. The historical sequence of the major changes to the Roman provincial system is shown in the detailed individual maps.

→ Maps pp. 141, 143, 145, 147, 163, 167, 179, 181, 183–185, 187, 189, 193, 207

Literature

M.G. Angeli Bertinelli, I Romani oltre l' Eufrate nel II secolo d.C. (le provincie di Assiria, di Mesopotamia e di Osroene), in: ANRW II 9,1, 1976, 3–69; T. Bechert, Die Provinzen des römischen Reiches. Einführung und Überblick, 1999; M.-L. Chaumont, L' Arménie entre Rome et l' Iran I. De l' avènement d' Auguste à l' avènement de Dioclétien, in: ANRW II 9,1, 1976, 71–194; R. Haensch, Capita provinciarum. Statthaltersitze und Provinzialverwaltung in der römischen Kaiserzeit (Kölner Forschungen 17), 1997; A. Lintott, Imperium Romanum

Date	Province	Event	Sources
27 BC	Lusitania	province established; capital Augusta Emerita	Cass. Dio 53,12,5; CIL 2, Suppl. LXVII
25/24 BC	Galatia	province established; capital Ancyra	Cass. Dio 53,26,3; Str. 12,5,1; *Ancyra*: AE 1969/70, 601; 1971, 462; IGR 3,189
15 BC	Noricum	administrative prefecture (*praefectus civitatium*) established; administrative capital Virunum	Vell. Pat. 2,39; Tac. Ann. 2,63; *Virunum*: discussion of evidence in Haensch, 353–359
14 BC	Alpes Maritimae	province established; administrative capital Cemenelum	Cass. Dio 54,24,3; Str. 4,6,3; Diod. Sic. 29,28; *Cemenelum*: discussion of evidence in Haensch, 142
AD 6	Iudaea	incorporated in province of Syria; administrative capital Antioch	Jos. Bl. 2,117; 2,167; Jos. Ant. Iud. 17,355; *Antioch*: Str. 16,2,5; Tac. Hist. 2,78; Cass. Dio 69,2,1
AD 9	Province of Illyricum superius (later Dalmatia) and military district of Illyricum inferius (later Pannonia)	division of province of Illyricum	ILS 938
AD 17	Cappadocia	province established; administrative capital Caesarea	Tac. Ann. 2,42,4; Str. 12,1,4; Cass. Dio 57,17,7; Suet. Tiberius 37,4; Vell. Pat. 2,39,3; *Caesarea*: Str. 12,2,7; others in Haensch, 272–276
probably under Tiberius (AD 14–37)	Rhaetia et Vindelicia	province established ('... et Vallis Poenina' until Claudius); administrative capital at first Cambodunum (?), then Augusta Vindelicum	Vell. Pat. 2,39,3; Suet. Aug. 21,1; Cambodunum: Tac. Germ. 41,1; *Augusta Vindelicum*: ILS 2386; 3203; others in Haensch, 146–149
AD 40	Mauretania	province established; adminsitrative capital Iol (later Caesarea)	Cass. Dio 59,25
under Claudius (AD 41–54)	Alpes Graiae/ Atrectianae	province established; administrative capital Forum Claudii Ceutronum (previously and later Axima) – mostly governed together with province of Alpes Poeninae (Alpes Graiae et Poeninae); administrative capital Forum Claudii Vallensium (Octodurus)	ILS 1348; 1368; *Forum Claudii Ceutronum*: ILS 3528; *Forum Claudii Vallensium (Octodurus)*: evidence in Haensch, 144–146
AD 42	Mauretania	province divided into Mauretania Caesariensis (adm. capital Caesarea, formerly Iol) and Mauretania Tingitana (previously ulterior); adm. capital Tingis or Volubilis	*Caesarea and Tingis*: Cass. Dio 60,9,5; discussion in Haensch, 186–192
AD 43	Britannia	province established; adm. capital at first Camulodunum, from Flavian period Londinium	Suet. Claud. 17,1; Cass. Dio 62,3,2; Tac. Agr. 13,5; 14,1; discussion of capitals in Haensch, 120–130
	Lycia	province established; capital Patara	Suet. Claud. 25,3; *Patara*: CIL 3, 14181 from AD 68/69
AD 45	Thracia	province established; capital Perinthus	Cassiod. Chronica 659; Jer. Chron. 2064; Sync. p. 630 Bonn; *Perinthus*: ILS 1093; CIL 3, 731
AD 45/46	Moesia	province (including ripae Thraciae) established; capital Tomi	ILS 986; *Tomi*: SEG 19, 1963, 1109,51–58
AD 63	Alpes Cottiae	province established; capital Segusio	Suet. Nero 18,1; Segusio: CIL 5, 7254
mid 1st cent. AD	Pannonia	province established (formerly Illyricum inferius); capital Carnuntum or Poetovio or Savaria	Vell. Pat. 2,116,2; discussion of capital in Haensch, 349 note 155 with 693 f.
	Dalmatia	province established (formerly Illyricum superius); capital Salona	Vell. Pat. 2,116,2
AD 72	Iudaea	province established; capital Caesarea	Jos. Bl. 2,111; 2,117; 2,167; Jos. Ant. Iud. 17,344; 17,355; *Caesarea*: Tac. Hist. 2,78,4
	Cilicia	province re-established; capital Tarsus	ILS 8971; IGR 3, 840; Imhoof-Blumer, Kleinasiatische Münzen, 1901/02, 445; *Tarsus*: Dion Chrys. 33,17; 33,46;

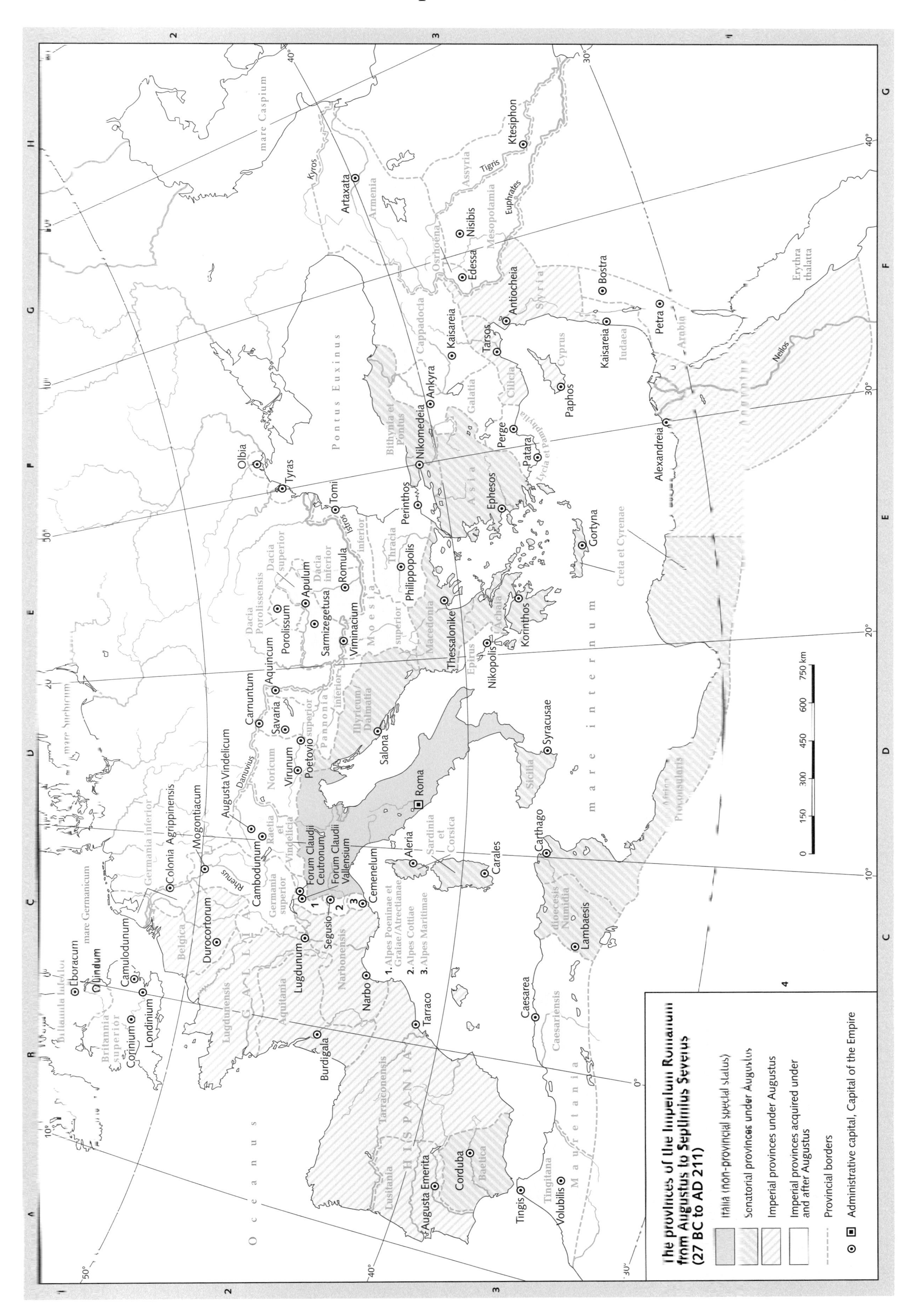
The provinces of the Imperium Romanum from Augustus to Septimius Severus (27 BC to AD 211)
Italia (non-provincial special status)
Senatorial provinces under Augustus
Imperial provinces under Augustus
Imperial provinces acquired under and after Augustus
Provincial borders
Administrative capital, Capital of the Empire

The development of the Roman provinces in Egypt and Arabia (1st cent. BC – 6th cent. AD)

I. Aegyptus

The annexation of the Ptolemaic heartland as a Roman province took place after Roman forces occupied Alexandria on 1 August 30 BC. Augustus, at the time still bearing the name C. Iulius Caesar Divi filius, commanded this action in his capacity as consul IV and by authority of the oath obliging him to prosecute the war against Cleopatra on behalf of all Italy and the western provinces. Aegyptus was the first province he established, and the act was still, as it were, infused by the spirit of the triumvirate. After all, when had a consul or a praetor ever arranged the establishment of a province in pursuance of his own personal interests in this way? For example, Augustus used the Ptolemaic state treasury to pay off his vast debts. He had the Egyptians worship him as the new Pharaoh. He immediately issued the order that no senator might visit the province without his personal permission. Nor did Augustus set aside these arbitrary provisions when he subsequently distanced himself from the oppressive machinery of the triumvirate.

He appointed Cornelius Gallus prefect for the administration of the new province. The *praefectus Aegypti* headed a hierarchically-structured civil service and also represented the *princeps* in his capacity as heir to the Pharaohs. His was the highest office available to the equestrian imperial officials after the *praefectus praetorio* at Rome. At the time of Augustus, he had two legions at his disposal stationed at Alexandria (one from the reign of Hadrian), their main task being to secure public order in the city. After Diocletian's (284–305) division of Aegyptus into three provinces, he was of higher rank, at his residence in Alexandria, than the other Egyptian governors. However, Diocletian's imperial reforms removed his military competences, transferring these to *duces*. The new title *praefectus Augustalis*, with the rank of a *vir clarissimus*, is attested from the late Constantinian period, and may be earlier. The residence of the *praefectus Aegypti* was at Alexandria, which was an independent Greek *polis*, not part of the Egyptian kingdom, and thus bore the title *Alexandria ad Aegyptum*.

The geographical focus of the province of Aegyptus, naturally enough, was the Nile. To the north, the province bounded the Mediterranean; to the west, at Catabathumus Megas, the province of Cyrenae (the Ammonium in the south-west was within the province's territory); to the east the Sinai Desert and the province of Arabia, and to the south the northern Nubian Dodekaschoinos between Syene and Hierasycaminus, which served Rome as a buffer zone to the kingdom of Meroë and the Blemmyae from its conquest by the *praefectus Aegypti* P. Petronius in 24/22 BC until the withdrawal of the last garrisons under Diocletian in AD 298.

The economic policies of the Ptolemies were at first left in place without significant alteration. The long-standing dependence of the imperial capital of Rome on grain shipments from Egypt now became more evident. Withholding or releasing the grain ships from Alexandria could be used as an instrument of political influence on the court, society and people of Rome. As Rome depended economically on Egypt, so too would Constantinople, and it was a disaster for the Byzantine Empire when the emperor Heraclius finally lost Egypt to the Arabs shortly before his death in AD 641.

II. Arabia

Originally, in the course of his planning of a new order for the east of the Roman Empire, Pompey may have envisaged more than just a fleeting expedition against the Nabataeans. They had earlier attacked Jerusalem, but had given up their siege when M. Aemilius Scaurus, Pompey's proquaestor, intervened in Jewish-Nabataean relations. Detained by other duties, Pompey left the administration of the new province of Syria – and with it the task of subjecting the Nabataeans – to Scaurus. The latter marched to Petra, but a lavish bribe from the Nabataean king Aretas III persuaded him to conclude an alliance which would last for 168 years, securing the Nabataeans a dignified peace.

In AD 105/06, however, their kingdom was annexed and established as the province of Arabia on the orders of the emperor Trajan. While it is uncertain for this period whether the administration was based at Petra or Bostra, Bostra was certainly the capital from the reign of Septimius Severus. As far as can be told, the brief occupation of the province by Zenobia's Palmyrenes (268-272) left no permanent trace. Diocletian's imperial reform and other subsequent reorganizations sought to bind Arabia more tightly into the empire. The province was conquered by the Muslim Arabs in 640.

Date	Event	Sources
AD 105/06	Province of Arabia established; administrative capital Bostra or Petra	Cass. Dio 68,14,5; Fest. 14,3; Amm. Marc. 14,8,13; Eutr. 8,3; Jer. Chron. 2118; Chron. pasch. 472; Bostra: CIL 3, 93; Petra: P.Mich. VIII 466; discussion of capital in Haensch, 238–241
from Septimius Severus (193–211)	Administrative capital Bostra	CIL 3, 93; IGLS 13, 9075; RIC 2, 250 no. 94; 261 no. 244; 278 nos. 610–614
268–272	Occupied by Zenobia	Ioh. Mal. 299,4
after 284	Province divided: Arabia I Augusta Libanensis (or nova), adm. capital Bostra; Arabia II, adm. capital Petra	Laterculus Veronensis 1,6 f.; Not. dign. or. passim
after 337	Amalgamation of the two provinces as Palaestina II	Hierocles, Synekdemos 721,12

The sources

The material of the map is based in large part on the late Roman historians, such as Cassius Dio (2nd/3rd cents.) and Ammianus Marcellinus (4th cent.), the breviary literature such as the *Breviarium* of Eutropius and that of Rufius Festus (both 4th cent.) and lists such as the compendium for the official use of the Roman authorities, the *Notitia dignitatum* (5th cent.), and the *Synekdemos* ('Travel Companion') of Hierocles (6th cent.), a statistical list of the Roman provinces and cities. Also important are inscriptions revealing to us the activities of various imperial officials.

→ Maps pp. 147, 161, 181, 221

Sources

praefectus Aegypti: P.Oxy 237; Cass. Dio 54,51; Dig. 1,17,1; Tac. Ann. 12,60,1; Str. 17,1,12; 17,1,53; *praefectus Augustalis:* Not. Dign. Or. 1,29; 23,1; 23,8; 23,24; *dux Aegypti Thebaidos utrarumque Libyarum:* AE 1934, 7–8; *Exclusion of senators from the province:* Tac. Ann. 2,59; CIL 3, 74; *Nubia:* Str. 17,1,53 f.; Plin. HN 6,53; *Catabathmus Megas:* Str. 17,1,13; Mela 1,40; *Nabataeans:* Jos. Ant. Iud. 14,80 f.; Jos. Bl. 1,159.

Literature

Aegyptus: A. Bastianini, Il prefetto d' Egitto (30 a.C.–297 d.C.), in: ANRW II 10.1, 1988, 503–517; P. Bureth, Le préfet d' Égypte (30 av. J.C.–297 ap. J.C.). État présent de la documentation en 1973, in: ANRW II 10.1, 1988, 472–502; G. Chalon, L' édit de Tiberius Julius Alexander, 1964; G. Geraci, 'Eparcía dè nûn esti. La concezione augustea del governo d' Egitto, in: ANRW II 10.1, 1988, 383–411; H. Heinen, W. Schlömer, Ägypten in hellenistisch-römischer Zeit (TAVO B V 21), 1989; H. Hübner, Der Praefectus Aegypti von Diokletian bis zum Ende der römischen Herrschaft, 1952; E.G. Huzar, Arabia, Heir of the Ptolemies, in: ANRW II 10.1, 1988, 343–382; M.A. Levi, L' esclusione dei Senatori romani dall' Egitto Augustea, in: Aegyptus 5, 1914, 231–235; W. Reinmuth, The Prefect of Egypt from Augustus to Diocletian, 1935; M.P. Speidel, Nubia' s Roman Garrison, in: ANRW II 10,1, 1988, 767–798; A. Stein, Die Präfekten von Ägypten in der römischen Kaiserzeit, 1950; TIR N/G 36, 1958.
Arabia: G.W. Bowersock, Roman Arabia, 1983; Id., The Annexation and Initial Garrison of Arabia, in: ZPE 5, 1970, 37–47; D.F. Graf, The Via Nova Traiana in Arabia Petraea, in: JRA (Suppl. 14), 1995, 241–267; R. Haensch, Capita provinciarum, 1997; B.H. Isaac, The Near East under Roman Rule, 1998; D.L. Kennedy, The Frontier of Settlement in Roman Arabia, in: MediterrAnt 3, 2000, 397–453; E. Kettenhofen, Östlicher Mittelmeerraum und Mesopotamien. Die Neuordnung des Orients in diokletianisch-konstantinischer Zeit (284–337 n. Chr.) (TAVO B VI 1), 1984; Id., Östlicher Mittelmeerraum und Mesopotamien. Spätrömische Zeit (337–527 n. Chr.) (TAVO B VI 4), 1984; P. Weiss, M.P. Speidel, Das erste Militärdiplom für Arabia, in: ZPE 150, 2004, 253–264

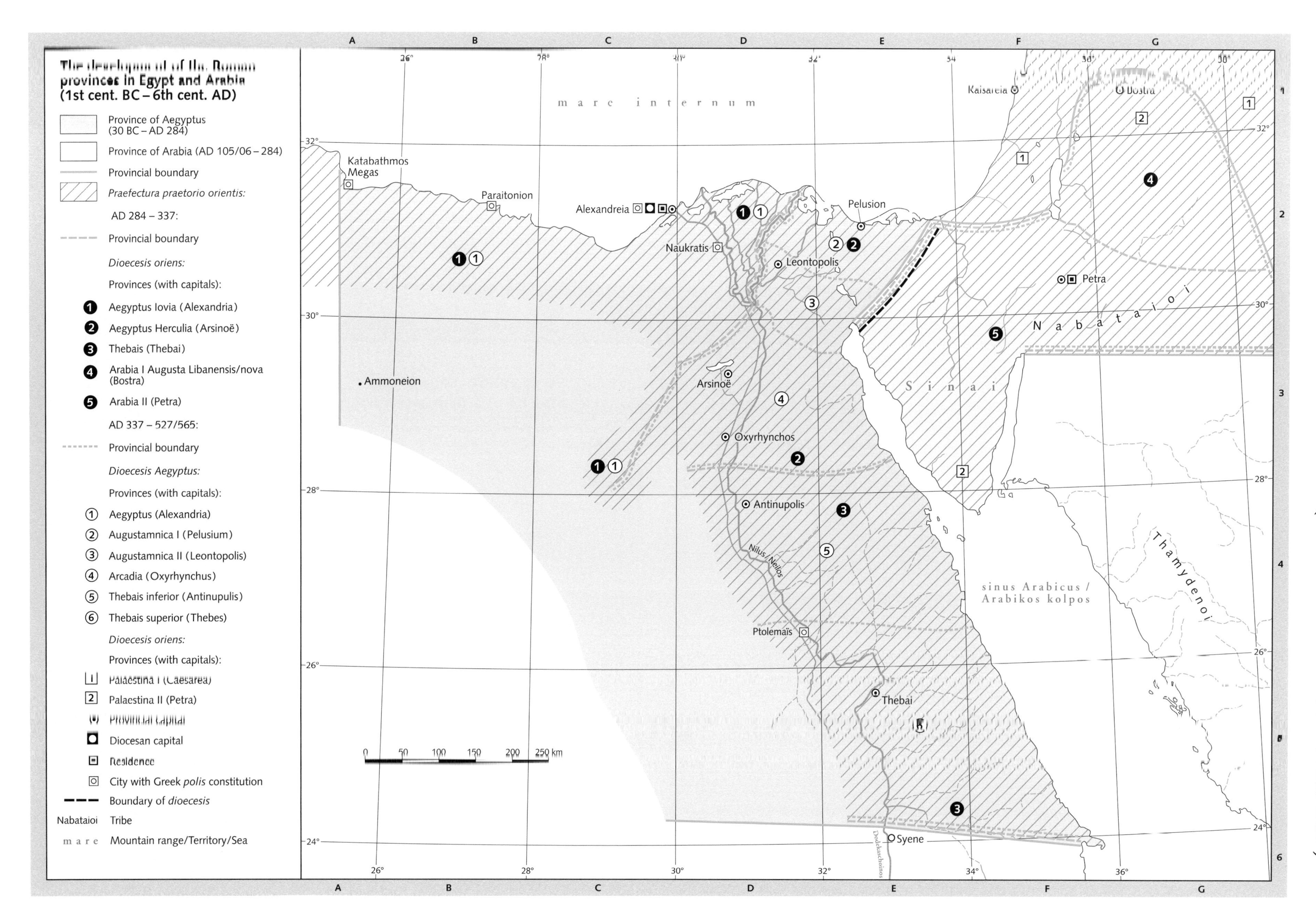
provinces in Egypt and Arabia
(1st cent. BC – 6th cent. AD)
Province of Aegyptus (30 BC – AD 284)
Province of Arabia (AD 105/06 – 284)
Provincial boundary
Praefectura praetorio orientis:
AD 284 – 337:
Provincial boundary
Dioecesis oriens:
Provinces (with capitals):
1 Aegyptus Iovia (Alexandria)
2 Aegyptus Herculia (Arsinoë)
3 Thebais (Thebai)
4 Arabia I Augusta Libanensis/nova (Bostra)
5 Arabia II (Petra)
AD 337 – 527/565:
Provincial boundary
Dioecesis Aegyptus:
Provinces (with capitals):
1 Aegyptus (Alexandria)
2 Augustamnica I (Pelusium)
3 Augustamnica II (Leontopolis)
4 Arcadia (Oxyrhynchus)
5 Thebais inferior (Antinupulis)
6 Thebais superior (Thebes)
Dioecesis oriens:
Provinces (with capitals):
2 Palaestina II (Petra)
Diocesan capital
City with Greek polis constitution
Boundary of dioecesis
Nabataioi Tribe
mare Mountain range/Territory/Sea
mare internum
Katabathmos Megas
Paraitonion
Alexandreia
Naukratis
Pelusion
Leontopolis
Ammoneion
Arsinoë
Oxyrhynchos
Antinupolis
Nilus/Neilos
Ptolemaïs
Thebai
Syene
Kaisareia
Petra
Nabataioi
Sinai
sinus Arabicus / Arabikos kolpos
Thamydenoi
0 50 100 150 200 250 km

The development of the Roman provinces in the Levant (1st cent. BC to 4th cent. AD)

The development of the Roman provinces in the Levant (1st cent. BC to 4th cent. AD)

→ Cf. tables below and on p. 270

I. Cyprus

Cyprus is the third-largest island in the Mediterranean after Sicily and Sardinia. It lies 68 km from the Cilician coast at Cape Anemurium, 95 from the west coast of Syria at Laodicea and 325 km from Egypt. Two chains of mountains determine the relief of the island: the Pentadactylus chain in the north-east, some 160 km long and falling away towards the coast, and the volcanic Trogodus in the south-western interior (1,952 m a.s.l.). The five prominent capes have always been reliable landmarks for seafarers on a coastline that is lacking in natural harbours.

Domestic political disputes influenced Roman foreign policy in the east in 58 BC. In that year, the people's tribune P. Clodius Pulcher instigated a *lex*, probably manoeuvring solely in his own interests but entirely suiting the purposes of Pompey, Crassus and Caesar. By this law, M. Porcius Cato, equipped with an extraordinary *imperium*, was charged with taking and incorporating as a Roman province the island of Cyprus, which at the time was ruled by a son of Ptolemy IX, also a Ptolemy. This device sidelined Cato, a politican who caused the *populares* considerable discomfort, in a way that did him honour, but was no less effective for it. Cato offered Ptolemy the office of Priest of Aphrodite at New Paphus. Ptolemy tried to salvage his fortune, but then, when his attempt failed, chose suicide. We learn of the incorporation of Cyprus into the province of Cilicia through the correspondence of Cicero, who was active there as proconsul in 51/50. During the Civil War, Cyprus was a morning-gift of Caesar's, then Antony's, for Cleopatra VII, so that the island was outside the Roman imperial organization for the period 48-30 BC. We do not know its status between 30 and 27 BC. In 27 BC it became a province in its own right under the personal rule of the *princeps*, before being transferred to senatorial administration in 22 BC. Cyprus retained this status even after the imperial reform of Diocletian, becoming part of the *dioecesis oriens*.

II. Syria

When he deposed Antiochus XIII, Pompey entirely dissolved the kingdom of the Seleucids – or rather, all that remained of it, which was, incidentally, still occupied by the Armenian king Tigranes II (95–55).

The heartland of the former Seleucid kingdom, from the Amanus, a southern offshoot of the Taurus, in the north to the Yarmūk, a left tributary of the Iordanes/Jordan in the south, and from the Mediterranean in the west to the Syrian Desert as far as the Euphrates in the east, became under the aegis of Rome a single political unit for the first time in its history, as the province of Syria. It is crossed from north to south by a high plain which the Jordan Rift divides into the heights of the Libanus and Antilibanus. These fall away to both north and south from their highest altitudes (3,083 and 2,214 m), dropping so far especially in the north at the mouth of the Orontes and between Aradus and Arca that westerly winds bring moist maritime air into the interior, making a productive rain-fed agriculture possible, especially around Beroea and Emesa.

As in the province of Bithynia et Pontus, the basis of the provincial organization Pompey set up in Syria was a network of urban centres with communal self-government. Some of these urban centres were in turn associated in their own organization, the Dekapolis (cities of Abila, Dium, Gadara, Gerasa, Hippus, Canatha, Pella, Philadelphia, Raphana, Scythopolis). There is considerable uncertainty as to which cities belonged to the Dekapolis at which periods, because it subsequently lost and gained various members. Seleucia Pieria, Antioch and Laodicea were set apart as 'free cities' (*civitates liberae*), i.e. they were independent of the provincial administration. That administration had its seat at the former Seleucid residence of Antioch.

The actions of the triumvir M. Antonius, who acquired the province of Syria as one of the provisions of the Treaty of Brundisium in October 40 BC, created a brief caesura in the history of the province. In 36 BC, he presented his wife, Cleopatra VII, with various territorial possessions of the Roman Empire as a gift, including, in Syria, the princedom of Chalcis in the Libanus, all cities from the Eleutherus north of Arca as far south as Egypt, except for Tyre and Sidon, the hinterland (Coele Syria) and the 'Grove of Jericho', where the aromatic balm was harvested.

III. The eastern provinces

The provinces east of Cappadocia and Cilicia were exposed to constant threat from the Parthians, and from AD 224 from the Sassanids. Their divisions and reorganizations are attributable to this problem.

The map

The map shows the eastern portion of the *dioecesis oriens*. Its western provinces of Isauria and Cilicia are shown on the map on p. 183; Aegyptus Iovia, Aegyptus Herculia, Thebais, Arabia and Arabia nova are shown on p. 179 and the map on p. 147 shows Libya I and Libya II.

→ Map p. 161

Date	Province	Event	Sources
58 BC	Part of province of Cilicia	Cyprus organized as province by M. Porcius Cato with propraetorial *imperium*	lex Clodia: Cic. Dom. 52f.; 65; Cic. Sest. 57–60; Liv. Per. 104; Vell. Pat. 2,38,6; 2,45,4f.; App. Civ. 2,23; Cass. Dio 38,30,5; 39,22,2f.; Plut. Cato Minor 34–39; Plut. Brutus 3; Plut. Pompeius 48; Flor. Epit. 1,44; Rufius Festus, Breviarium 13,1; Amm. Marc. 14,8,15; Cilicia: Cic. Fam. 13,48; Cic. Att. 5,21,6
48/47–44 BC	Part of Ptolemaic kingdom	Given by Caesar to a sister and brother of Cleopatra	App. Civ. 5,35; 5,9; Cass. Dio 42,35
44–30 BC		Given by Antony to the children of Cleopatra	Str. 14,6,6; Plut. Antonius 36,2; Cass. Dio 49,32; 49,41
30–27 BC		Status uncertain	
27–22 BC	Province of Cyprus	Under personal rule of *princeps*	Cass. Dio 53,12,7
from 22 BC		Under senatorial rule; administrative capital Nea Paphus	Cass. Dio 54,4,1; Str. 14,6,6; 17,3,25; Paphus: Acts 13,6–13
from Diocletian (AD 284–305) and Constantine I (306–337)		Province of the *dioecesis oriens*; administrative capital Constantia, formerly Salamis	Not. Dign. Or. 1,63; 2,13; 22,5; Pol. Silv. 8,7; Hierocles, Synecdemus 706,3; Constantia: Hierocles, Synecdemus 706,4

Literature

S. Applebaum, Judaea as a Roman Province. The Countryside as a Political and Economic Factor, in: ANRW II 8, 1977, 355–396; E. Badian, M. Porcius Cato and the Annexation and Early Administration of Cyprus, in: JRS 55, 1965, 110–121; T. Bekker-Nielsen, The Roads of Ancient Cyprus, 2004; H. Bietenhard, Die syrische Dekapolis von Pompeius bis Traian, in: ANRW II 8, 1977, 220–261; G.W. Bowersock, Syria under Vespasian, in: JRS 63, 1973, 133–140; K. Buschmann, Östlicher Mittelmeerraum und Mesopotamien. Von Antoninus Pius bis zum Ende des Parthischen Reiches (138–224 n. Chr.), TAVO B V 9, 1992; R. Duncan-Jones, Praefectus Mesopotamiae et Osrhoenae, in: CPh 64, 1969, 229–233; L.-M. Günther, Herodes der Große, 2005; G. Hill, A History of Cyprus, 1940; V. Karageorghis, Early Cyprus Crossroads of the Mediterranean, 2002; E. Kettenhofen, Die römisch-persischen Kriege des 3. Jahrhunderts n. Chr. nach der Inschrift Šāhpuhrs I. an der Ka' be-ye Zartošt (ŠKZ), 1982; Id., Östlicher Mittelmeerraum und Mesopotamien. Die Neuordnung des Orients in diokletianisch-konstantinischer Zeit (284–337 n. Chr.), TAVO B VI 1, 1984; Id., Östlicher Mittelmeerraum und Mesopotamien. Die Zeit der Reichskrise (235–284 n. Chr.), TAVO B V 12, 1983; Id., Vorderer Orient, Römer und Sāsāniden in der Zeit der Reichskrise (224–284 n. Chr.), TAVO B V 11, 1982; M. Konrad, Der spätrömische Limes in Syrien, 2001; F.G. Maier, Zypern, ²1982; F. Millar, The Roman Near East 31 BC–AD 337, 1993; E. Oberhummer, Die Insel Cypern, 1903; I. Pill-Rademacher, Vorderer Orient, Römer und Parther (14–138 n. Chr.), TAVO B V 8, 1988; J.-P. Rey-Coquais, Syrie Romaine de Pompée à Dioclétien, in: JRS 68, 1978, 44–73; M. Sartre, Syria and Arabia, in: CAH 11, 2000, 635–663; J. Wagner, Östlicher Mittelmeerraum und Mesopotamien. Die Neuordnung des Orients von Pompeius bis Augustus, TAVO B V 7, 1983.

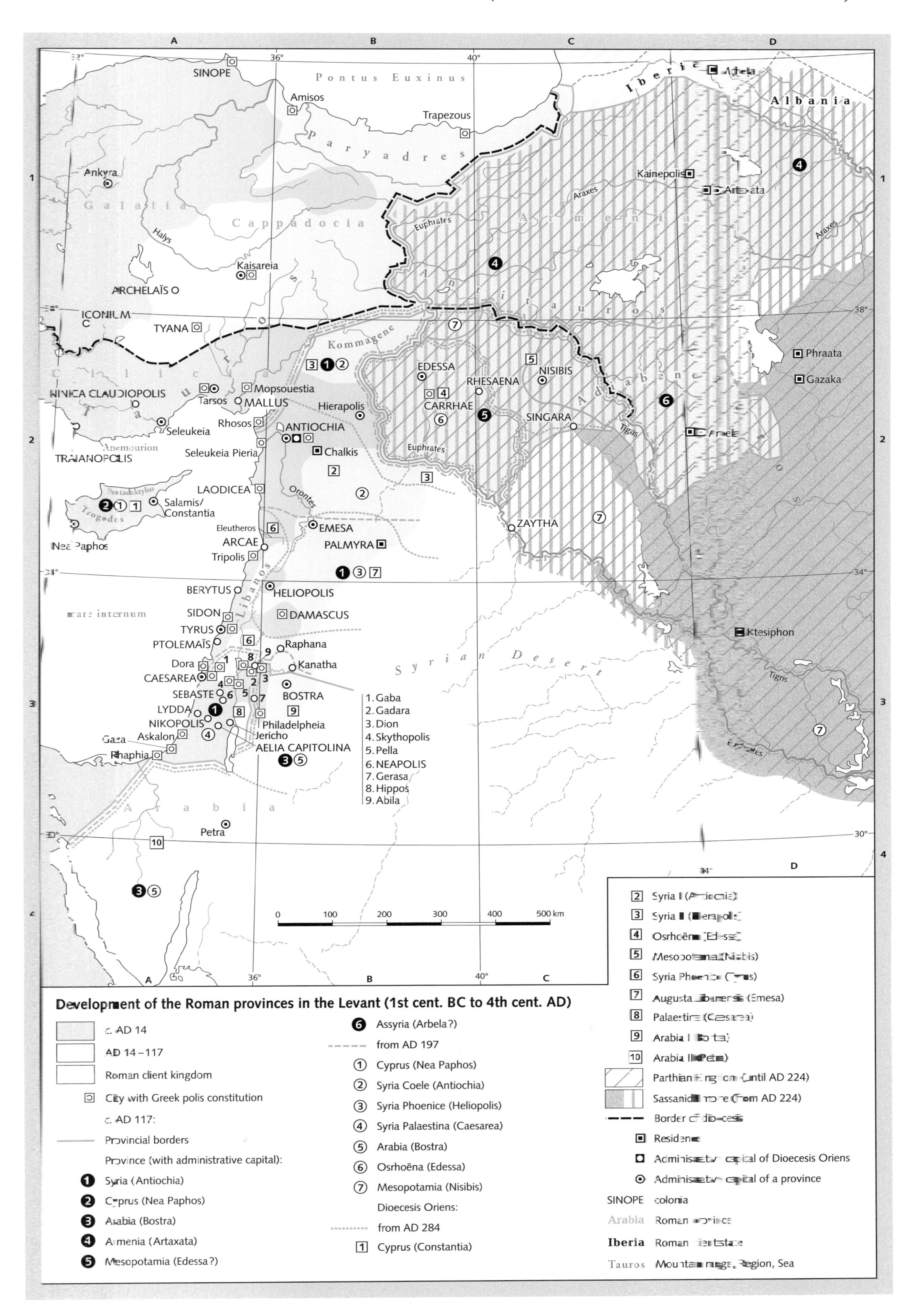
Development of the Roman provinces in the Levant (1st cent. BC to 4th cent. AD)
AD 14–117
Roman client kingdom
City with Greek polis constitution
Provincial borders
Province (with administrative capital):
Syria (Antiochia)
Arabia (Bostra)
Armenia (Artaxata)
Mesopotamia (Edessa?)
Assyria (Arbela?)
from AD 197
Cyprus (Nea Paphos)
Syria Coele (Antiochia)
Syria Phoenice (Heliopolis)
Syria Palaestina (Caesarea)
Arabia (Bostra)
Osrhoëna (Edessa)
Mesopotamia (Nisibis)
Dioecesis Oriens:
from AD 284
Cyprus (Constantia)
Syria Phoenice (Tyrus)
Augusta Libanensis (Emesa)
Administrative capital of Dioecesis Oriens
Administrative capital of a province
SINOPE colonia
Arabia
Iberia
Tauros
Region, Sea
1. Gaba
2. Gadara
3. Dion
4. Skythopolis
5. Pella
6. NEAPOLIS
7. Gerasa
8. Hippos
9. Abila
0 100 200 300 400 500 km
Pontus Euxinus
Paryadres
Albania
Iberia
SINOPE
Amisos
Trapezous
Ankyra
Galatia
Cappadocia
Kaisareia
ARCHELAÏS
ICONIUM
TYANA
Kainepolis
Armenia
Euphrates
Araxes
Halys
Antitauros
Kommagene
EDESSA
RHESAENA
NISIBIS
CARRHAE
SINGARA
Adiabene
Tigris
Phraata
Gazaka
Mopsouestia
Tarsos
MALLUS
Hierapolis
Seleukeia
Rhosos
ANTIOCHIA
Chalkis
Seleukeia Pieria
Anemourion
LAODICEA
Orontes
Salamis/
Constantia
Eleutheros
EMESA
ARCAE
PALMYRA
Tripolis
BERYTUS
HELIOPOLIS
Libanos
SIDON
DAMASCUS
TYRUS
PTOLEMAÏS
Raphana
Dora
Kanatha
CAESAREA
SEBASTE
BOSTRA
LYDDA
NIKOPOLIS
Philadelpheia
Jericho
Askalon
AELIA CAPITOLINA
Rhaphia
Arabia
Petra
ZAYTHA
Syrian Desert
Ktesiphon

The development of the Roman provinces in Asia Minor (2nd cent. BC to 5th cent. AD)

The bewildering frequency with which provincial frontiers in Asia Minor were moved, new provinces founded and provinces merged only to be separated again clearly betrays both the intractability and the importance of this peninsula to the Roman Empire. It was essential to defend the economic potency of the Anatolian provinces continuously against the covetous Parthians (and, from AD 227, Sassanids) at the Euphrates and Tigris. To do this, it was necessary to secure the military deployment route from the Danube to the Euphrates, which led along the southern shore of the Black Sea.

On the administrative changes to the provincial structures in Asia Minor up to the reign of Diocletian (AD 284-305), cf. the table (which continues on p. 271). We are curtly but sometimes revealingly informed about Diocletian's provincial reforms by the so-called *Laterculus Veronensis*, which preserves in a 7th-cent. Veronese manuscript a list containing the 12 *dioeceses*, divided among four praefectures and containing a total of 95 provinces, that comprised the Roman Empire *circa* AD 313. It shows that Diocletian and Constantine created the following provincial order in Asia Minor: the whole of Asia Minor was part of the *praefectura Oriens*, under the command of the *praefectus praetorio per Orientem*. Most of the peninsula was part of two *dioeceses*: the *dioecesis Pontica* with 7 provinces (Laterculus Veronensis 2) and the *dioecesis Asiana* with 9 provinces (Laterculus Veronensis 3). A small part of the far south-eastern corner of the peninsula, comprising the provinces of Cilicia (administrative capital Tarsus), Isauria (administrative capital Seleucia) and Cyprus (administrative capital Constantia), belonged to the *dioecesis Oriens*.

Dioecesis Pontica:

1. Bithynia (administrative capital Nicomedia)
2. Cappadocia (administrative capital Caesarea)
3. Galatia (administrative capital Ancyra)
4. Paphlagonia (administrative capital Gangra Germanicopolis, cf. Nov. 29,1,29 f.)
5. Diospontus (administrative capital Amasea)
6. Pontus Polemoniacus (administrative capital Neocaesarea)
7. Armenia Minor (administrative capital Sebasteia)

Dioecesis Asiana:

1. Lycia et Pamphylia (administrative capital Perge or Patara)
2. Phrygia I (administrative capital Laodicea)
3. Phrygia II (administrative capital Synnada)
4. Asia (administrative capital Ephesus)
5. Lydia (administrative capital Sardeis)
6. Caria (administrative capital Aphrodisias)
7. Pisidia (administrative capital Antioch)
8. Hellespontus (administrative capital Cyzicus)
9. Insulae (53 islands, but 63 in the *Expositio totius mundi et gentium*; administrative capital Rhodos)

Our main sources for the changes to this provincial system after the reign of Diocletian are:

1. the *Laterculus Veronensis*, a register of Roman provinces arranged by *dioecesis* (*c.* 313)
2. the *Breviarium* of Festus Rufius (d. 380)
3. Ammianus Marcellinus (*c.* 330–400; Amm. Marc. 14,8 for Cilicia, Isauria, Cyprus in the *praefectura Oriens*)
4. the *Notitia Dignitatum* (4th/5th cents.)
5. the provincial catalogue of Polemius Silvius (448/49)
6. the 'Travel Companion' (*Synekdemos*) of Hierocles (5th/6th cents.)

→ Map p. 161

Literature

C. Habicht, New Evidence on the Province of Asia, in: JRS 65, 1975, 64–91; R. Haensch, Capita provinciarum. Statthaltersitze und Provinzialverwaltung in der römischen Kaiserzeit, 1997; A. Lintott, Imperium Romanum, 1993; D. Magie, Roman Rule in Asia Minor to the End of the Third Century after Christ, 2 vols., 1950; S. Mitchell, Anatolia, 2 vols., 1993; K.L. Noethlichs, Zur Entstehung der Diözesen als Mittelinstanz des spätrömischen Verwaltungssystems, in: Historia 31, 1982, 70 81; Ch. Roueché, Rome, Asia, and Aphrodisias in the Third Century, in: JRS 71, 1981, 103–120; Maps: J. Wagner, W. Stahl, Die Neuordnung des Orients von Pompeius bis Augustus (67 v. Chr. – 14 n. Chr.),

Date	Province	Event	Sources
AD 116–117	Armenia	Province status	Cass. Dio 68,19 ff.; 68,29; Eutr. 8,3,2; 8,6,2; Rufius Festus, Breviarium 14; 20; RIC 2,289 no. 642
133 BC	Asia	Transfer to Rome by testament of Attalus III	Liv. Per. 58 f.; Plut. Tib. Gracchus 15; Iust. 36,4; Str. 13,4,2; Plin. HN 33,148
129 BC		Province established; administrative capital Ephesus	Str. 14,1,37; *Ephesus*: Cic. Att. 5,13,1
43 BC		Pamphylia incorporated	Cic. Fam. 12,15,5
13.1.27 BC		Senatorial province under a proconsul	Str. 17,3,25; Cass. Dio 53,14,2
AD 44–53; under Domitian		Rhodes part of province of Asia	*Part of province*: 44–53: Cass. Dio 60,24,4; 70–79: Suet. Vesp. 8; *under Domitian*: Plut. Mor. 815 d – *civitas libera*: 53–70: Tac. Ann. 12,58; Suet. Nero 7; 79–81: Dion Chrys. 31; after 81: Plut. Mor. 815d
under Philippus Arabs (244–249)		Caria et Phrygia separated from province	C. Roueché, Rome, Asia and Aphrodisias in the Third Century, in: JRS 71, 1981, 103–120
AD 117	Assyria	Province status	Eutr. 8,3,2; 8,6,2; Rufius Festus, Breviarium 14; 20
74 BC	Bithynia	Transfer to Rome by testament of Nicomedes IV; province established	Cic. Leg. Agr. 2,40; Liv. Per. 93; Vell. Pat. 2,4,1; 39,2; App. Civ. 1,111
64/63 BC	Bithynia et Pontus	Province established; administrative capital Nicomedia	Str. 12,3,1; Plut. Pompeius 38; Liv. Per. 102; *Nicomedia*: Dion Chrys. 38,31; 38,39
1st cent. AD		South-eastern part transferred to Galatia	Str. 12,8,11; 13,1,3
AD 109–114/15, AD 134/35 and under Antoninus Pius (138–161)		Under Imperial administration	Plin. Epist. 10; Cass. Dio 69,14,4; cf. IGR 3,654
c. AD 162		Abonutichus/Ionopolis, Sinope, Amisus added to province of Galatia	Ptol. 5,6,1–3
AD 284–305	Bithynia	Separated from province, in *dioecesis pontica*	Laterculus Veronensis 2
AD 17/18	Cappadocia	Province established	Tac. Ann. 2,42,4
AD 54		United with province of Galatia	Tac. Ann. 13,35,4; 15,6,5; Cass. Dio 62,20,4; 62,22,3 f.
AD 71/72		Armenia Minor, Pontus Galaticus and Pontus Polemoniacus added	W.H. Waddington, E. Babelon, T. Reinach, Recueil général des monnaies grecques d' Asie Mineure 1, [2]1925, 136 nos. 5–8
AD 76		United with province of Galatia	ILS 8904
AD 112		Cappadocia, separated from Galatia, again a province, incorporating Armenia minor and Armenia maior	Cass. Dio 68,19,1; ILS 1041
after AD 116		Reorganization of province and Armenia minor, Pontus Galaticus and Pontus Polemoniacus	IGR 3,100; 111; 4,815; ILS 8801
under Diocletian (284–305)		Pontus Polemoniacus and Armenia Minor separated: provinces in their own right	Laterculus Veronensis 2
AD 379/86		Division into Cappadocia I and II	Cod. Theod. 6,30,2; 13,11,2 – *or earlier, 371/72*: A.H. Jones, The Later Roman Empire 3, 1964, App. 3
AD 386		Armenia Minor: division into Armenia I and II	Not. Dign. Or. 1,109 f.
from *c.* AD 250	Caria et Phrygia	Separated from province of Asia: province in its own right	C. Roueché, Rome, Asia and Aphrodisias in the Third Century, in: JRS 71, 1981, 103–120, here 108 ff.

Cont'd. p. 271

The development of the Roman provinces in Asia Minor (2nd cent. BC to 5th cent. AD)

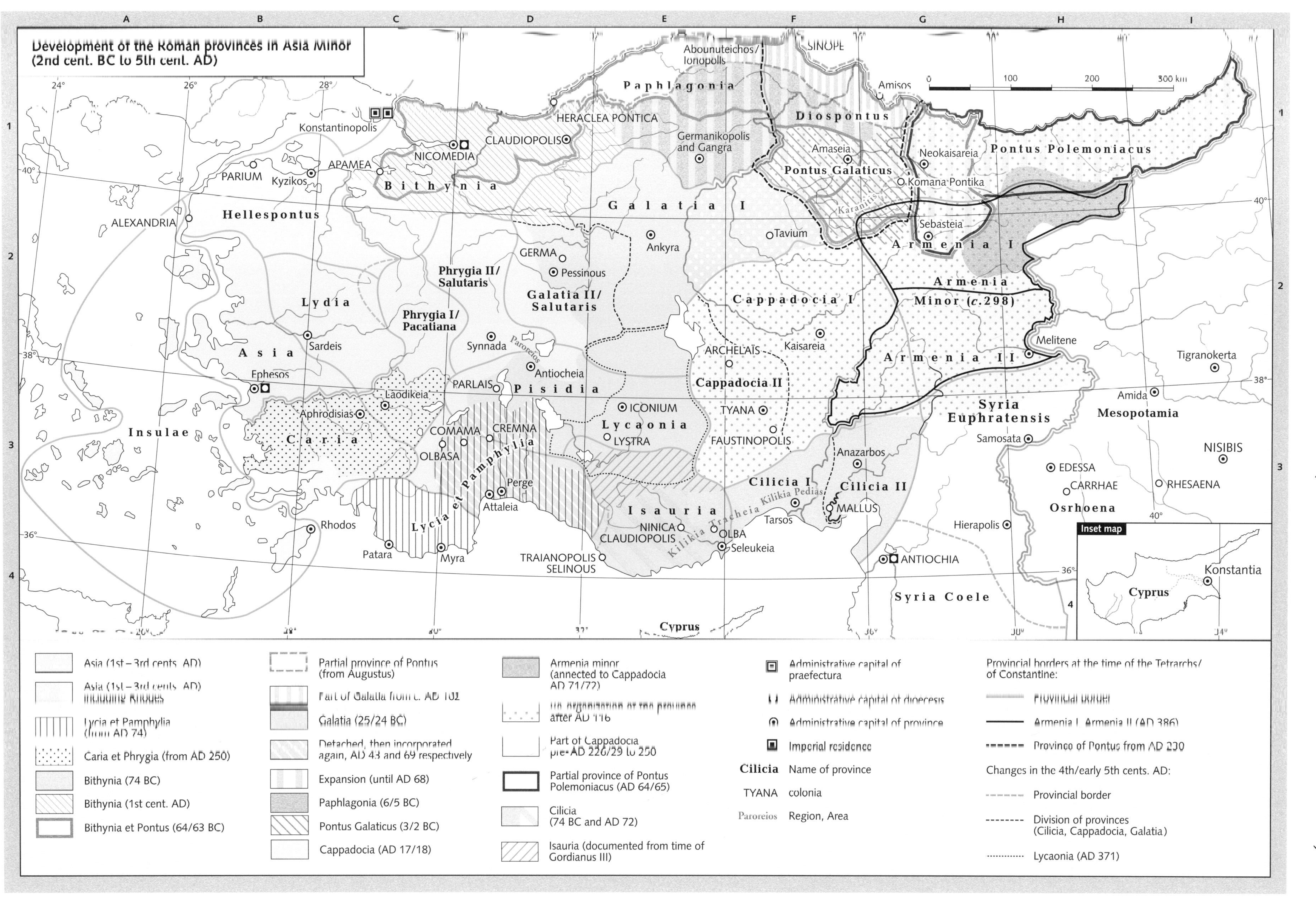

The development of the Roman provinces in the northern Balkan Peninsula (1st cent. BC to 4th cent. AD)

I. The Roman Danube frontier territory from the confluence with the Tisza to the Delta

At Singidunum/Belgrade, the abundant Savus/Sava flows into the Danube's middle reaches from the right and the Pathissus/Tisza from the left. The Danube then leaves the Great Hungarian Plain after the conflux with the Margus/Morava (right tributary) at the *municipium* of the same name. It now breaks through the barrier of the southern Carpathian Alps in a 130 km section of narrower, faster flow which culminates at the Iron Gates (Serbian Đerdap, Romanian Porţile de Fier), where Trajan had Apollodorus of Damascus build a stone bridge more than a kilometre long, a feat commemorated by a memorial stone on the right bank. The river flows on along the southern edge of Wallachia (Romanian Campia Romana), fringed on the left bank by countless marshes and small lakes, until it reaches the western extremity of the hilly country of Dobruja (Romanian Dobrogea), where, after the *municipium* of Troesmis, the vast, swampy Delta (which today covers an area in excess of 4,000 sq. km) begins.

On the provinces' administrative history → tables on p. 272

II. Dalmatia

The Romans named their province on the Illyrian coast after the Illyrian tribe of the Dalmati. Dalmatia, which was separated from the province of Illyricum (formerly Illyricum superius) under Nero (AD 54–68), bordered the Istrian peninsula, considered part of Italy from 18/12 BC, in the north-west (frontier: the small river Arsia), and the province of Macedonia in the south, at Lissus. Its neighbours to the north and east were the provinces of Pannonia and Moesia.

The Imperial reform of Diocletian (284–305) left the province, whose administrative capital was at Salona, intact and assigned it to the *dioecesis Pannoniae*. After Theodosius I (379–395), the north of Dalmatia fell to the western Roman Empire as part of the *dioecesis Pannoniae*, and the south (around Scodra) to the eastern part of the Empire as part of the *dioecesis Moesiae*, called Dalmatia praevalitana.

III. Moesia

This province, named after the Thracian tribe of the Moesi, bordered the Danube to the north and the Black Sea to the east. To the south, it had a frontier with Thracia along the ridges of the Skardon oros/Šar planina and Haemus/Balkans, and to the west with Dalmatia at the Drinus/Drina. But the province of Moesia was inhabited by many other tribes, such as Dardani, Triballi, Scythae and Getae. Roman troops had already pushed as far as the lower Danube on several occasions. Then, in 28 BC, while proconsul of the province of Macedonia, M. Licinius Crassus, a grandson of the *triumvir*, followed up several military successes against various tribes by incorporating this region into his province as a military district. Moesia was organized into a province in AD 46, probably at the same time as Thracia, and in the same year and shortly afterwards, other territorial gains were added to it. Apart from domestic political considerations, it was above all the growing threat posed by the Daci across the Danube that led Domitian to divide the province in AD 86. The frontier between the two provinces of Moesia superior and Moesia inferior ran to the west of the River Ciambrus/Tsibrica to the south. When Trajan established the province of Dacia across the Danube in 106, both Moesian provinces became part of the interior of the Roman Empire, but in view of the constant threat to which Dacia was subject, they were always required as defensive bases. When Dacia was abandoned in AD 271, the Moesian provinces were directly exposed to the Goths, who were driving directly into the Empire. The various divisions of these provinces under Diocletian and his successors in the end proved fruitless, failing to stabilize the situation on this section of the frontier.

IV. Thracia

The Roman province of Thracia reached the sea in several places – the Black Sea from Mesambria to the Bosporus, the Propontis and the northern Aegean. The Aegean islands of Proconnesus, Thasos and Samothrace also belonged to Thracia. Its neighbour provinces were: to the north, Moesia inferior; to the west, Moesia superior, and to the south-west, Macedonia. Thracia was divided into four provinces by the Diocletianic imperial reform, and this region became one of the heartlands of the Byzantine Empire, thanks especially to its proximity to the new capital of Constantinople.

V. Dacia (map inset)

After several wars under Domitian (AD 85/86) and Trajan (101/02 and 105/06), the region enclosed by the arc of the Carpathian Alps, which was mostly settled by Thracian Daci, was occupied by the Romans. In AD 106 it was organized into a province. However, the problems posed by the raids of the Iazyges and Rhoxolani, neighbours of the new province, led to Dacia's being split into two, and soon even into three provinces. The martial character of this province, dictated by its exposed position, is illustrated by the fact that Dacia superior had a resident praetorian legate to whom the senior officials of the other two smaller provinces, both imperial procurators, were subordinate. This arrangement lasted at least until the time of the Marcomannic Wars (AD 167–182). The emperors from Hadrian (117–138) made special efforts to defend the security of this province, whose mineral wealth made it valuable. For example, a contiguous network of bases was set up along the River Alutus (*limes Alutanus*), as was a system of forward positions farther to the east (*limes Transalutanus*). Aurelian then relinquished the province of Dacia in 271, as part of his efforts to secure the eastern and northern frontiers of the Empire. The military was evacuated along with the Roman population to regions south of the Danube.

→ Maps pp. 189, 225

Sources

Tabula Traiana: ILS 5863; *Danube Bridge:* Cass. Dio 68,13; Procop. Aed. 4,6,11–14 – cf. K. Lehmann-Hartleben, Die Trajanssäule, 1926, pl. 45.

Literature

G. Alföldy, Die illyrischen Provinzen Roms: Von der Vielfalt zur Einheit, in: G. Urso (ed.), Dall' Adriatico al Danubio, 2004, 207–220; F. Fodorean, Le strade della Dacia romana, in: Quaderni Catanesi di Studi antichi e medievali 3, 2004, 331–446; B. Gerov, Beiträge zur Geschichte der römischen Provinzen Moesien und Thrakien, 1980, 1997, 1998; Id., Die Grenzen der römischen Provinz Thrakien bis zur Gründung des Aurelianischen Dakien, in: ANRW II 7,1, 1979, 212–240; N. Gudea, T. Lobüscher, Dacia. Eine römische Provinz zwischen Karpaten und Schwarzem Meer, 2006; R. Haensch, Capita provinciarum, 1997; A. Jagenteufel, Die Statthalter der römischen Provinz Dalmatia von Augustus bis Diokletian, 1958; M. Šašel Kos, The Roman Conquest of Dalmatia in the Light of Appian' s "Illyrike", in: G. Urso (ed.), Dall' Adriatico al Danubio, 2004, 141–166; M. Sordi, La Pacificazione dell' Illirico e Tiberio, in: G. Urso (ed.), Dall'Adriatico al Danubio, 2004, 221–228; P. Weiss, Neue Diplome für Soldaten des Exercitus Dacicus, in: ZPE 141, 2002, 241–251; J. Wilkes, Dalmatia, 1969. Also: TIR L 34 Budapest, 1968; TIR L 35 Bucarest, 1969.

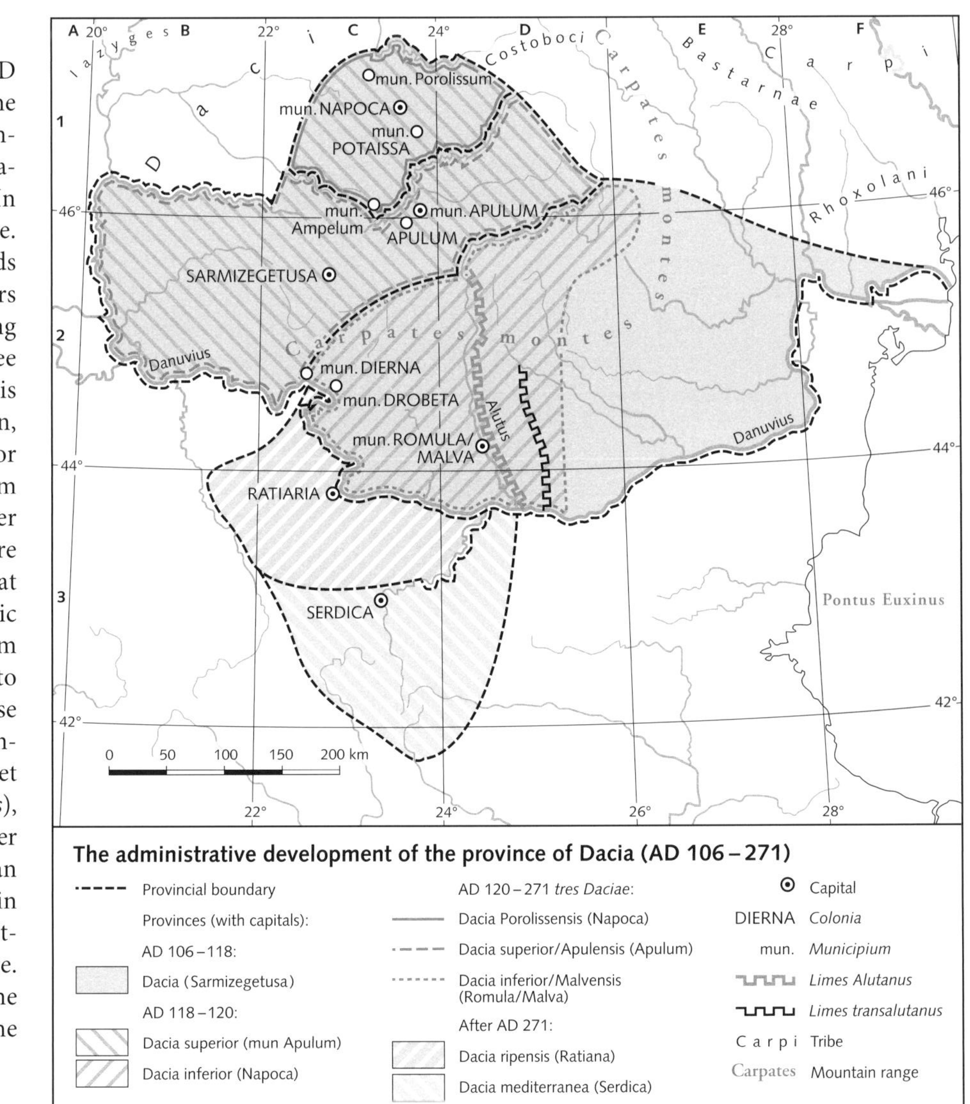

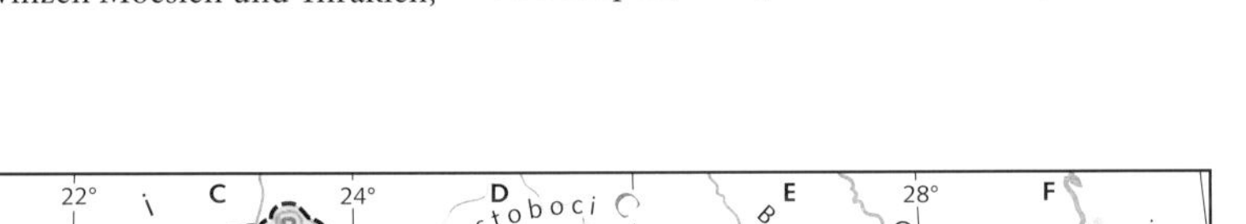

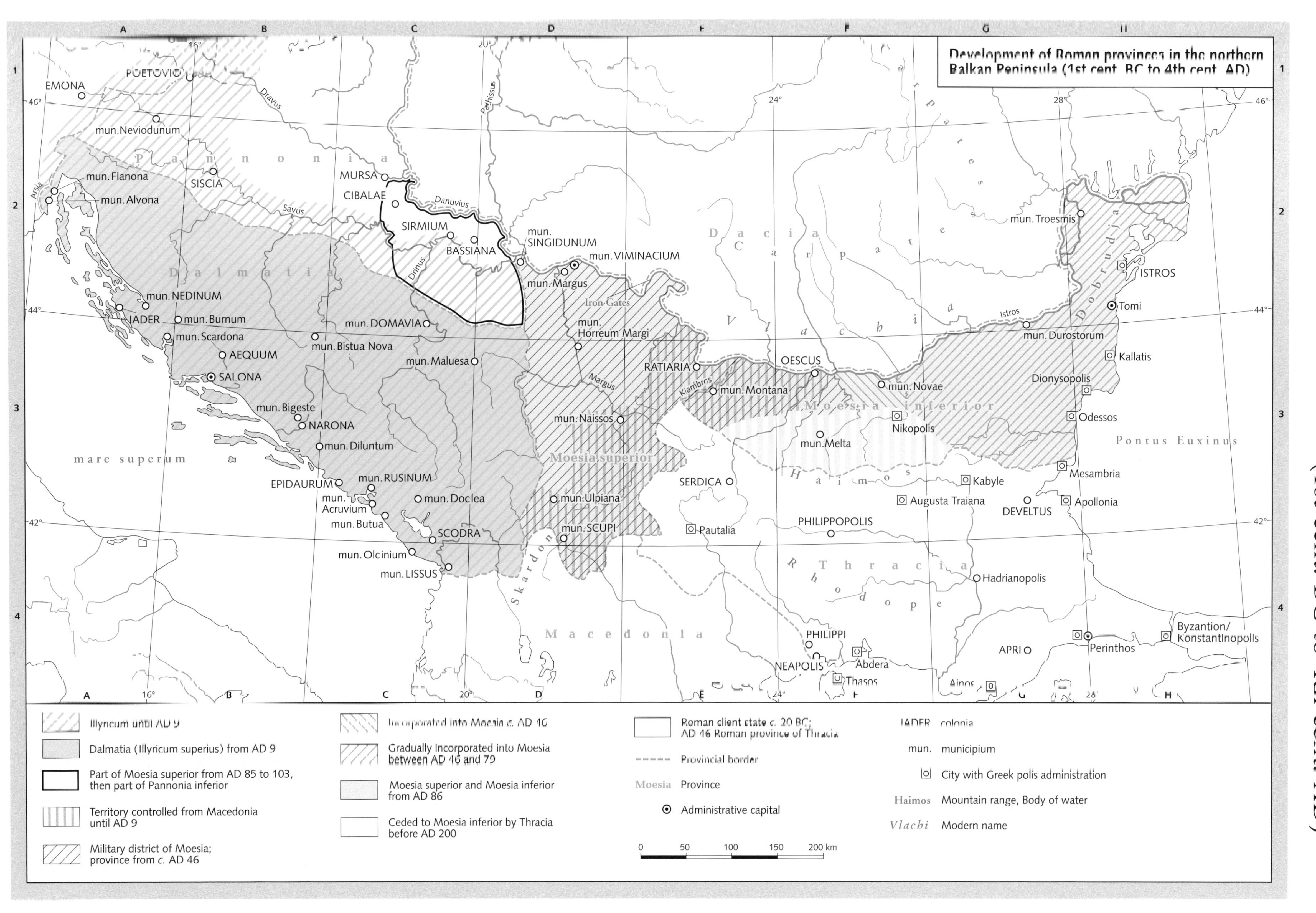
Development of Roman provinces in the northern Balkan Peninsula (1st cent. BC to 4th cent. AD)
Illyricum until AD 9
Dalmatia (Illyricum superius) from AD 9
Part of Moesia superior from AD 85 to 103, then part of Pannonia inferior
Territory controlled from Macedonia until AD 9
Military district of Moesia; province from c. AD 46
Incorporated into Moesia c. AD 16
Gradually incorporated into Moesia between AD 16 and 79
Moesia superior and Moesia inferior from AD 86
Ceded to Moesia inferior by Thracia before AD 200
Roman client state c. 20 BC; AD 16 Roman province of Thracia
Provincial border
Moesia Province
Administrative capital
IADER colonia
mun. municipium
City with Greek polis administration
Haimos Mountain range, Body of water
Vlachi Modern name
0 50 100 150 200 km
EMONA
POETOVIO
mun. Neviodunum
Pannonia
mun. Flanona
mun. Alvona
SISCIA
Arsia
Dravus
Savus
MURSA
CIBALAE
SIRMIUM
BASSIANA
Danuvius
Drinus
Pathissus
mun. SINGIDUNUM
mun. VIMINACIUM
mun. Margus
Iron Gates
mun. Horreum Margi
Margus
Dalmatia
mun. NEDINUM
IADER
mun. Burnum
mun. Scardona
AEQUUM
SALONA
mun. DOMAVIA
mun. Bistua Nova
mun. Maluesa
mun. Bigeste
NARONA
mun. Diluntum
mare superum
EPIDAURUM
mun. RUSINUM
mun. Acruvium
mun. Butua
mun. Doclea
SCODRA
mun. Olcinium
mun. LISSUS
mun. Naissos
Moesia superior
mun. Ulpiana
mun. SCUPI
Skardon
Macedonia
RATIARIA
Kiambros
mun. Montana
OESCUS
SERDICA
Pautalia
Dacia
Carpates
Vlachia
mun. Troesmis
Dobrudja
ISTROS
Tomi
Istros
mun. Durostorum
Kallatis
Dionysopolis
Odessos
mun. Novae
Moesia inferior
Nikopolis
mun. Melta
Pontus Euxinus
Haimos
Mesambria
Kabyle
Augusta Traiana
DEVELTUS
Apollonia
PHILIPPOPOLIS
Thracia
Rhodope
Hadrianopolis
PHILIPPI
NEAPOLIS
Abdera
Thasos
Ainos
APRI
Perinthos
Byzantion/ Konstantinopolis

The development of the Roman provinces in the southern Balkan Peninsula

Some 80 years after establishing their protectorate over parts of Illyricum, the Romans successively established three provinces in the southern Balkans: Macedonia, Achaia and, briefly, Epirus.

After the end of the 3rd Macedonian War (→ Map p. 151), in 167 BC, the senate charged a ten-man commission under the direction of the proconsul L. Aemilius Paullus with organizing the Macedonian heartlands into four *res publicae* (Greek *merides*). Livy (Liv. 45,29,5-9) describes the regional structure of these four *merides* in detail, but does not allow their northern frontiers to be accurately deduced:

Res Publica I: the territory between the Rivers Nessus and Strymon. To the east of the Nessus, all the land that had belonged to Perseus except Aenus, Maronea and Abdera; to the west of the Strymon, Bisaltica with Heraclea Sintica; administrative capital Amphipolis.

Res Publica II: the territory enclosed to the east by the Strymon from its source (except Bisaltica with Heraclea Sintica) and bounded to the west by the River Axius. Also the lands of the Paeones to the east of the Axius; administrative capital Thessalonica.

Res Publica III: the territory between the Rivers Axius and Peneius, bounded by Mount Bora/Vermion, also the part of Paeonia west of the Axius, and Edessa and Beroea; administrative capital Pella.

Res Publica IV: the territory beyond Mount Bora, bordering Illyricum and Epirus; administrative capital Pelagonia (the city is elsewhere called Heraclea Lyncestis).

Following the Macedonian revolt under Philip VI or Andriscus (150–148 BC), the Senate ordered the four Macedonian *res publicae* to be reformed into a province of Macedonia (also including Epirus and Illyricum). The administrative capital was to be Thessalonica. Greece, officially called Achaia, was added to the province of Macedonia in 146 after the Achaean War – the result was a province of enormous territorial extent. It is thus understandable that when the provinces came to be classified at the Senate sitting of 13 January 27 BC, Achaia (with the Greek spelling) was made a province in its own right and put under senatorial administration. The Roman colony of Corinthus became the administrative capital of Achaia (*Laus Iulia Corinthus*).

A typical feature of Hellenistic Greece was the formation of city leagues, so-called *koina*, which gradually lost their political functions under Roman rule, but continued to be of cultic importance. Pausanias (Paus. 7,16,9) was therefore no doubt wrong when he asserted that all the *koina* were dissolved by the peace of 146 BC. Many of them even survived the provincial reforms of Augustus (cf. Syll.3 796A). Rather, developments reflected the transition to the provincial assemblies of the Imperial Period. The map shows most of the *koina* – at least the most important of them – together with their respective capitals.

The sources

The sources for the events reflected in the map are diverse, because it covers some ten centuries. For the late Republic and the Principate, we have the reports of various authors. Later, inscriptions increase in importance, until lists of information like the *Laterculus Veronensis* begin to help us understand the administrative history of these provinces from Late Antiquity onwards.

The map

The map attempts to present complex processes in a simplified way, without losing sight of its primary topic: the administrative history of the provinces of the southern Balkans.

Literature

J.K. Anderson, A Topographical and Historical Study of Achaea, in: ABSA 49, 1954, 72–92; R. Haensch, Capita provinciarum. Statthaltersitze und Provinzialverwaltung in der römischen Kaiserzeit (Kölner Forschungen 17), 1997; N.G.L. Hammond, Epirus. The Geography, the Ancient Remains, the History and the Topography of Epirus and Adjacent Areas, 1967; Id., The Koina of Epirus and Macedonia in: IC 16, 1992, 183–192; M. Heil, Zwei spätantike Statthalter aus Epirus und Achaia, in: ZPE 108, 1995, 159–165; S.F. Johnson, The Dorian States of the Province of Achaea, 1969; B. Levy, When Did Nero Liberate Achaea and Why?, in: A. Rizakis (ed.), Archaia Achaia kai Eleia, 1991, 189–194; A. Lintott, Imperium Romanum, 1993; F. Papazoglu, Quelques aspects de l' histoire de la province de Macédoine, in: ANRW II 7,1 302–369; R.K. Sherk, Roman Imperial Troops in Macedonia and Achaea, in: AJPh 78, 1957, 52–62.

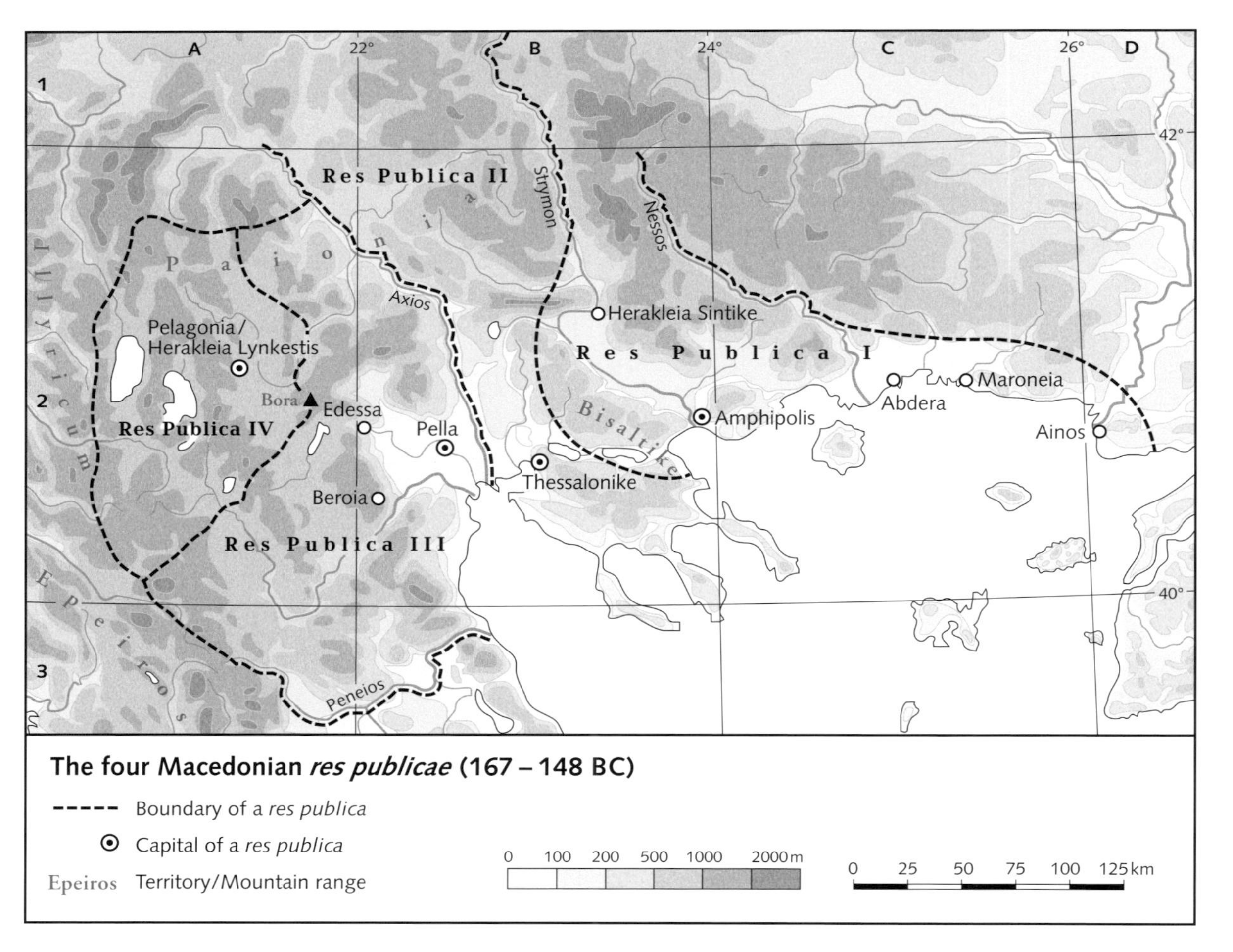

The four Macedonian *res publicae* (167 – 148 BC)

Date	Province	Event	Sources
148 BC	Macedonia	Province established	Liv. Per. 50; Plin. HN 34,64; Val. Max. 7,5,4; Rufius Festus 7
146 BC		Achaea incorporated into province of Macedonia	cf. Syll.³ 684 (proconsul Macedoniae to Dymae); Plut. Cimon 2,1
Mid 2nd cent. AD		Thessalia and Phthiotis incorporated into province of Macedonia; capital Thessalonica	Ptol. 3,12–14; Thessalonica: ILS 1067; 9490
13.1.27 BC	Achaia	Senatorial province established; capital Corinthus	Cass. Dio 53,12; Str. 17,3,25; *Corinthus:* Acts 18,12
AD 15–44		Incorporated into province of Moesia	Tac. Ann. 1,76,4; 80,1; Suet. Claud. 25,3; Cass. Dio 60,24,1
7th–10th cents.		Incorporation of western Greece into *thema* of Nicopolis; central Greece into *thema* of Hellas; Peloponnesus a *thema* in its own right	Konstantinos Porphyrogennetos, De thematibus 2,5 f.; 2,8
Reign of Trajan (98–117)?	Epirus	Province established, including Ionian Islands and parts of Acarnania; capital Nicopolis	Ptol. 3,13,1; 3,14,1; Nicopolis: ILS 8849
under Diocletian (284–305)	Insulae	Province established, excluding Euboea, Scyros, Lemnos and Imbros	Laterculus Veronensis 3

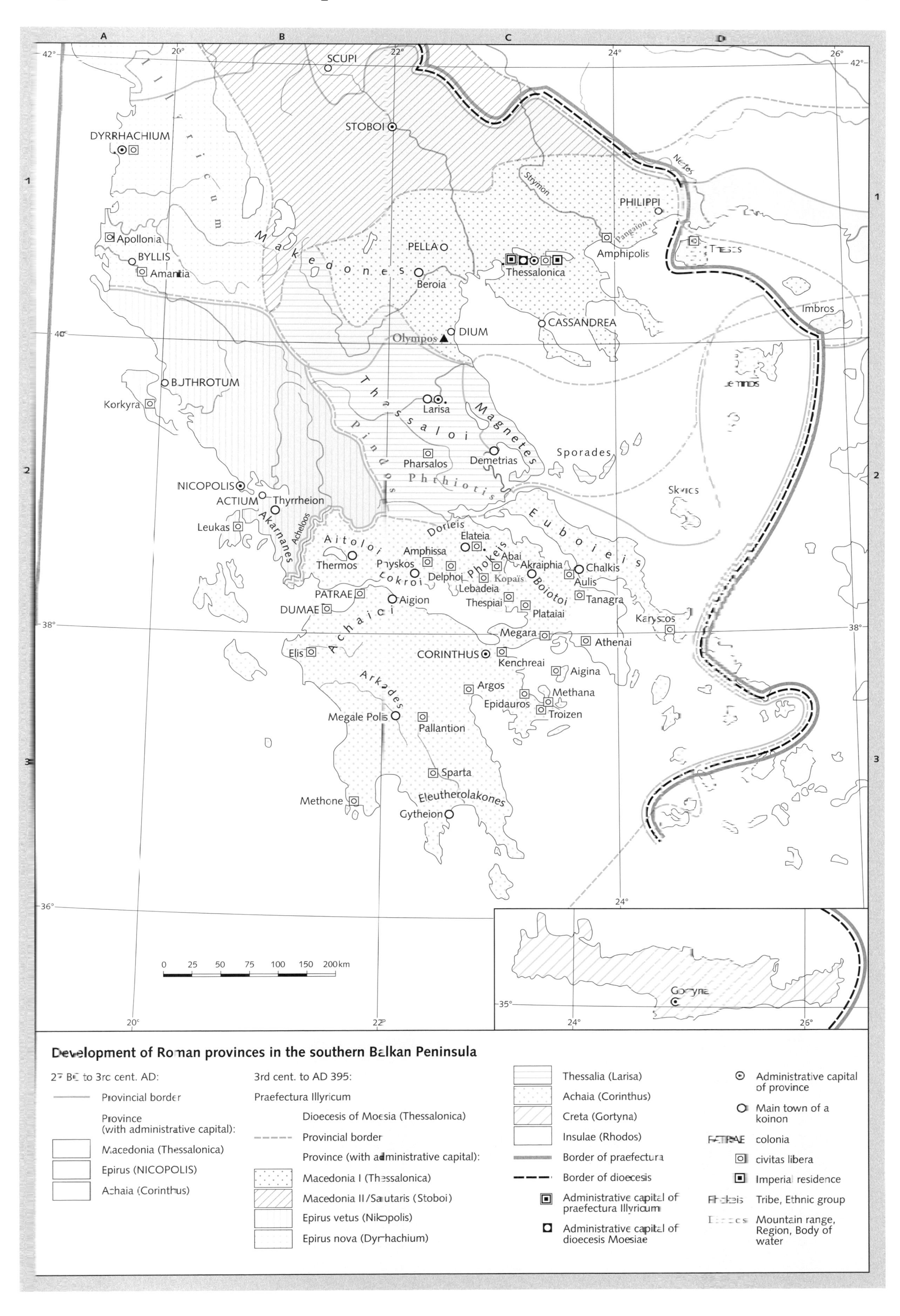
SCUPI
STOBOI
DYRRHACHIUM
Apollonia
BYLLIS
Amantia
Makedones
PELLA
Beroia
Thessalonica
PHILIPPI
Amphipolis
Strymon
Pangaion
CASSANDREA
Imbros
DIUM
Olympos
BUTHROTUM
Korkyra
Thessaloi
Larisa
Pindos
Magnetes
Pharsalos
Demetrias
Phthiotis
Sporades
NICOPOLIS
ACTIUM
Thyrrheion
Leukas
Akarnanes
Acheloos
Aitoloi
Thermos
Dorieis
Elateia
Amphissa
Abai
Akraiphia
Chalkis
Euboieis
Delphoi
Phokeis
Kopaïs
Lebadeia
Boiotoi
Aulis
Tanagra
Thespiai
Plataiai
Lokroi
PATRAE
Aigion
DUMAE
Achaici
Karystos
Megara
Athenai
Elis
CORINTHUS
Kenchreai
Aigina
Arkades
Argos
Methana
Epidauros
Troizen
Megale Polis
Pallantion
Sparta
Eleutherolakones
Methone
Gytheion
0 25 50 75 100 150 200 km
42°
40°
38°
36°
35°
20°
22°
24°
26°
A
B
C
D
1
2
3
Development of Roman provinces in the southern Balkan Peninsula
3rd cent. to AD 395:
Provincial border
Province (with administrative capital):
Macedonia (Thessalonica)
Epirus (NICOPOLIS)
Achaia (Corinthus)
Praefectura Illyricum
Dioecesis of Moesia (Thessalonica)
Provincial border
Province (with administrative capital):
Macedonia I (Thessalonica)
Macedonia II/Salutaris (Stoboi)
Epirus vetus (Nikopolis)
Epirus nova (Dyrrhachium)
Thessalia (Larisa)
Achaia (Corinthus)
Creta (Gortyna)
Insulae (Rhodos)
Border of praefectura
Border of dioecesis
Administrative capital of praefectura Illyricum
Administrative capital of dioecesis Moesiae
Administrative capital of province
Main town of a koinon
colonia
civitas libera
Imperial residence
Tribe, Ethnic group
Mountain range, Region, Body of water

The development of the Roman provinces on the middle Danube (1st cent. BC – 3rd/4th cents. AD)

The development of the Roman provinces on the middle Danube (1st cent. BC – 3rd/4th cents. AD)

I. The Roman Danube frontier from the headwaters to the influx of the Tisza

Two great river frontiers framed the Roman Empire to the north at the time of its greatest extent under Trajan (AD 98–117): the Rhine and the Danube.

The Danube was navigable by the flat-bottomed barges of internal trade all the way from near its headwaters to the Black Sea. Abundant inflowing Alpine rivers, such as the Hilaria/Iller, Licca/Lech, Isara/Isar and Aenus/Inn, swell the river, and by the time it leaves what was the province of Rhaetia at Batavis/Passau and enters the province of Noricum it has expanded to a substantial width. The Danube cuts through upland ridges like that of the Franconian Jura at Weltenburg. Flat, often marshy lowlands (e.g. fenlands of the Donauried and Donaumoos) alternate with narrow valley sections (e.g. the Wachau). Two more Alpine rivers flow in, the Druna/Traun and Anisus/Enns, before the Danube enters the Vienna Basin. The Amber Route, from the Baltic to the Adriatic at Aquileia, crossed the Danube near Petronell-Carnuntum/Bad Deutsch-Altenburg, at the influx of the the Marus/March (left tributary), which flows down from the Sněžník Massif. This is where the Danube left Noricum to enter Pannonia, the third great Danubian province. Breaking through the Transdanubian Uplands (Hungarian: Dunántúli-középhegység), it takes a right-angle turn to the south. Below Aquincum/Budapest, it flows on a clearly diminishing gradient through broad, marshy flood plains. Reaching the south of the Great Hungarian Plain, it receives major tributaries from the Eastern Alps, such as the Dravus/Drau and Savus/Sava (right) and Pathisus/Tisza (left), as it leaves Pannonia and enters Moesia.

On the administrative history of the provinces → Tables p. 273

II. Raetia

In 15 BC, Drusus and Tiberius, the two stepsons of Augustus, undertook a campaign that took them separately over the Alps. The result of this campaign was the creation of a military district under the command of a prefect and incorporating the Raeti, understood to be a group of peoples of the Alps and the northern Alpine foothills. The necessary infrastructure was provided by various road building projects to construct the *via Claudia*, from Augusta Vindelicum/Augsburg over the Alps to Altinum/Altino and Hostilia/Ostiglia, and the road leading west-east from Brigantium/Bregenz via Cambodunum/Kempten to Iuvavum/Salzburg. The Roman province, whose northern frontier was the Danube and which included what is now south-eastern Switzerland, Vorarlberg, the Tirol, the Central Alps and the Alpine foothills as far as Noricum, was organized in the mid 1st cent. AD. The frontier zone was first pushed back northwards across the Danube under the Flavian emperors (AD 69–96). After further advances, it was secured initially by a wall-and-ditch construction, until a stone wall was built late in the 2nd cent. to regulate border controls. This border arrangment lasted until AD 260, after which the Danube again assumed the function of a frontier.

In 260, the governor of Raetia recognized the authority of Postumus in the 'Gallic Empire'. Around 265, the province again came under the rule of Gallienus, the emperor at Rome. Finally, the province was divided in two by the imperial reforms of Diocletian and Constantine I.

III. Noricum

By the time it attracted Roman interest, the territory of the Celtic Norici stretched from the Aenus/Inn in the west to *lacus Pelso*/Lake Balaton in the east, and from the Danube in the north to the northern foothills of the Julian Alps in the south. In the Roman period, Noricum adjoined Raetia to the west and Pannonia to the east. This eastern Alpine region was of great economic and commercial importance. For one thing, Noricum was rich in natural resources, such as valuable ores (iron and gold). Iron from Noricum (*ferrum Noricum*) had a legendary reputation in the ancient world. In spite of thin soils, the high plains were also very agriculturally productive (livestock, wool, woollen materials, leather). Furthermore, the Amber Route passed through Norican territory between the influx of the Marus at Carnuntum and Poetovio/Ptuj on its way to Aquileia.

It is not generally thought that the Noricans were ruled by a single king in this period: Livy (Liv. 43,5,8) writes of 'royal brothers' agreeing an official relationship of 'guest-friendship' with the Romans in 170 BC. This agreement facilitated trade in and through Noricum for Italian traders, who, for instance, founded a trading post at Virunum on the Magdalensberg.

The occupation of Noricum was one consequence of the large-scale Alpine campaign under Drusus, the stepson of Augustus. The region was of particular strategic importance as the link between Italy and the Empire's northern frontier. Noricum was officially organized into a province under Claudius (41–54). The *legio II Italica* was stationed here in connection with the Marcomannic Wars (167–182). The province was divided into Noricum ripense and Noricum mediterraneum under Diocletian (284–305), and both provinces were assigned first to the *dioecesis Pannoniae*, then the *dioecesis Illyrici* under Constantine I (306–337).

Noricum ripense became the theatre of the great migration of the 5th cent., in the course of which Vandals, Huns and Germanic tribes made their way up the Danube, severely affecting the province.

IV. Pannonia

This province was named after a group of diverse Illyrian and Celtic tribes, the Pannonii, whose territories were bounded to the north and east by the Danube, to the south by the Savus/Sava and to the west by the *lacus Pelso*/Lake Balaton and the Julian Alps.

What the Romans called Pannonia was the northern part of Illyricum, whose conquest they heralded with the establishment (229-219 BC) of a protectorate over various tribes and cities of the region between the Rivers Naro/Neretva in the north and Aous/Vjose in the south. That conquest was to all intents and purposes completed by the younger Caesar, the future Augustus, in 35-33 BC. The province of Illyricum could only be called pacified after the suppression of revolts that recurred on an annual basis from 16 BC, especially those of 9 BC and AD 6-9. Illyricum was divided in two in AD 9, forming the province of Illyricum superius, later Dalmatia, and the military district of Illyricum inferius. Under Claudius (41–54), the latter was organized into a province, Pannonia. It included the entire Pannonian region as described above, as far as the Danube (→ Map p. 185).

The growing importance which Trajan (AD 98–117) attached to the Danube region in connection with his plans for Dacia also necessitated tighter organization of the Danubian provinces. In 103/106, Pannonia was therefore divided into the provinces of Pannonia superior (to the west) and Pannonia inferior (to the east). The boundary between the two provinces ran east of the *lacus Pelso*, southwards as far as the border of the province of Dalmatia, then west to the Julian Alps.

Military considerations (distribution of legions, border security) led to the transfer of the districts around Sirmium (AD 85–103), Emona (from 200) and Brigetio (from 214) to the provinces of Moesia superior, Venetia et Histria and Pannonia inferior respectively. Under Diocletian (284–305), Pannonia superior was divided into Pannonia I and Savia, Pannonia inferior into Pannonia II and Valeria. In the 4th cent., the Danube frontier in the Pannonian provinces had to be defended, particularly against the Suebian Quadi. Following the Roman defeat at Adrianople in 378 at which the Emperor Valens fell, the Danubian frontier defences collapsed.

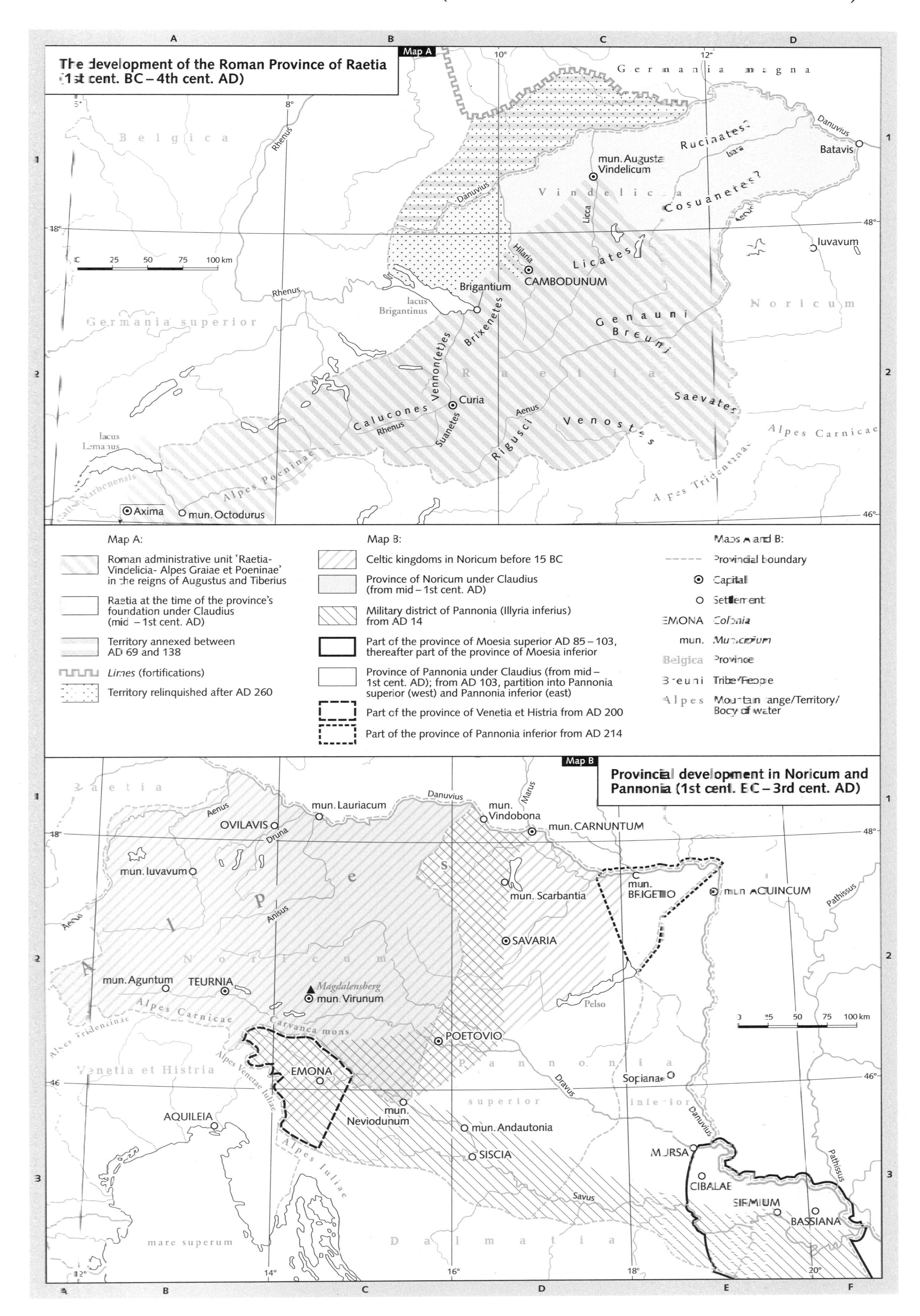
Map A
The development of the Roman Province of Raetia (1st cent. BC – 4th cent. AD)
Germania magna
Belgica
Germania superior
Vindelicia
Raetia
Noricum
mun. Augusta Vindelicum
Batavis
Iuvavum
CAMBODUNUM
Brigantium
lacus Brigantinus
Curia
lacus Lemannus
Axima
mun. Octodurus
Alpes Poeninae
Alpes Carnicae
Alpes Tridentinae
Runicates?
Cosuanetes?
Licates
Genauni
Breuni
Saevates
Venostes
Rigusci
Calucones
Brixenetes
Vennon(et)es
Suanetes
Danuvius
Rhenus
Aenus
Licca
Hilaria
Isara
0 25 50 75 100 km
Map A:
Roman administrative unit 'Raetia-Vindelicia- Alpes Graiae et Poeninae' in the reigns of Augustus and Tiberius
Raetia at the time of the province's foundation under Claudius (mid – 1st cent. AD)
Territory annexed between AD 69 and 138
Limes (fortifications)
Territory relinquished after AD 260
Map B:
Celtic kingdoms in Noricum before 15 BC
Province of Noricum under Claudius (from mid – 1st cent. AD)
Military district of Pannonia (Illyria inferius) from AD 14
Part of the province of Moesia superior AD 85 – 103, thereafter part of the province of Moesia inferior
Province of Pannonia under Claudius (from mid – 1st cent. AD); from AD 103, partition into Pannonia superior (west) and Pannonia inferior (east)
Part of the province of Venetia et Histria from AD 200
Part of the province of Pannonia inferior from AD 214
Maps A and B:
Provincial boundary
Capital
Settlement
EMONA Colonia
mun. Municipium
Belgica Province
Breuni Tribe/People
Alpes Mountain range/Territory/Body of water
Map B
Provincial development in Noricum and Pannonia (1st cent. BC – 3rd cent. AD)
Raetia
Alpes
Noricum
Pannonia superior
Pannonia inferior
Venetia et Histria
Dalmatia
mare superum
mun. Lauriacum
OVILAVIS
mun. Iuvavum
mun. Vindobona
mun. CARNUNTUM
mun. Scarbantia
mun. BRIGETIO
mun. AQUINCUM
SAVARIA
mun. Aguntum
TEURNIA
Magdalensberg
mun. Virunum
POETOVIO
EMONA
mun. Neviodunum
mun. Andautonia
SISCIA
AQUILEIA
Sopianae
MURSA
CIBALAE
SIRMIUM
BASSIANA
Alpes Carnicae
Carvanca mons
Alpes Venetae Iuliae
Alpes Iuliae
Pelso
Danuvius
Aenus
Druna
Anisus
Marus
Dravus
Savus
Pathissus
0 25 50 75 100 km

The so-called Batavian Revolt and the Roman civil war of AD 69/70

The events we generally refer to as the Batavian Revolt probably began as aspects of the civil war which broke out in the wake of the death of the Emperor Nero, only gradually taking on the character of a full-scale revolt of Germanic and Gaulish tribes against Roman rule as matters unfolded.

Nero was forced to take his own life on 9 June AD 68. The brief reign of his successor Galba, who was never equal to his position – particularly in view of his lack of understanding of the political power games surrounding the throne – ended with his murder on 15 January 69. Otho had worked towards Galba's elimination, and now duly succeeded him, but he failed to prevail against Vitellius, the *legatus Augusti pro praetore* who was now acclaimed *imperator* by the army in Germania, and following what was in fact rather an insignificant defeat at Betriacum/Calvatone on 16 April, he chose suicide. Vitellius did not reign unchallenged for long: just two months later, on 1 July, Vespasian, commander of the forces fighting the Jewish rebels, was acclaimed *imperator* at Alexandria. At Betriacum on 24/25 October, Vitellius failed to overcome the forces of the Danubian army, who under their legate Antonius Primus had taken Vespasian's side and marched for Rome. He retreated to Rome, where he died on 20 December of wounds suffered during street-fighting. The Senate acknowledged Vespasian as *imperator* on 21 December.

Germans and Gauls intervened in this civil war: German tribes such as the Batavi, Bructeri, Cannenefates, Cugerni, Frisii, Tencteri, Treveri and Tungri, and the Gaulish Lingones. They were led by the Batavian Iulius Civilis, the Treverans Iulius Classicus, Iulius Tutor and Iulius Valentinus and the Lingonian Iulius Sabinus. These hostilities took place in the two military districts of Germania superior and Germania inferior and in the province of Gallia Belgica.

In August 69, the Batavian Civilis, praefect of a Batavian auxiliary cohort stationed in the Rhine Delta, the Batavian homeland, received a letter from Antonius Primus, legate of the *legio VII*, calling on him to join forces with Vespasian. The implicit incitement to mutiny was intended to keep those legions that were loyal to Vitellius far away from the decisive battle for power in Italy. Whether to take the side of Vespasian or to rise against Roman rule (*bellum internum* or *bellum externum*) was a decision that would only be made once news of Vitellius' death reached the Germanic troops on the Rhine.

First, however, Civilis initiated hostilities against the regular Roman forces in the late summer of 69. Every victory he won against Munius Lupercus, the legate of the *legio XV Primigenia*, won him more adherents on both banks of the Rhine. A cohort of Tungri joined him. Troops of the Bructeri and Tencteri followed their example. In the end, the legion was forced to withdraw to Castra Vetera/Xanten, where Civilis besieged it. Occasionally relieved, the camp garrison succeeded in holding out for seven months or so before capitulating in March 70, whereupon the victorious Germans broke their word and annihilated it. Meanwhile, the Treveri and the Gaulish Lingones had agreed on the foundation of an *Imperium Galliarum*, and sought support for their project from the Gaulish provinces in particular. However, the momentum which the entire movement had thereby developed soon ebbed away as Vespasian, initially represented at Rome by his younger son Domitian and the governor of Syria, C. Licinius Mucianus, took ever firmer control of the Empire. On his orders, the legate Q. Petillius Cerialis assembled a total of nine legions against the rebels, five from Italy, three from Spain and one from Britain. The victories of Sextilius Felix, praesidial procurator of Noricum, at Bingium/Bingen in May and of Cerialis himself at Augusta Treverorum/Trier in June and over the troops of Civilis at Castra Vetera in July were decisive. Civilis withdrew to the difficult terrain of his homeland in the Rhine Delta, but was unable to make any headway against Cerialis. Finally, in October, he negotiated a peace favourable to himself and the Batavi. The Romans suffered heavy losses during these hostilities. Four or five legions were severely depleted and had to be disbanded: the *legio I* at Bonna/Bonn, the *legio IV Macedonica* at Mogontiacum/Mainz, probably the *legio V Alaudae* [?] and certainly the *legio XV Primigenia* at Castra Vetera and the *legio XVI Gallica* at Novaesium/Neuss.

The map

The focus of the map covers the area whipped up by the Batavian Revolt, from the Liger/Loire in the west to the Albis/Elbe in the east, from the North Sea (Oceanus) in the north to Rome in the south. Hostilities were concentrated in the areas on the left and right banks of the lower reaches of the Rhine, from Mogontiacum/Mainz north, i.e. the two Germanic military districts and the province of Gallia Belgica. The map also shows those localities in northern Italy that played a significant part in the actual civil war of AD 69, i.e. the conflict between troops loyal to Otho, Vitellius and Vespasian respectively.

The sources

Source materials for the events of the so-called Batavian Revolt are exceedingly sparse. It is true that we have a very comprehensive account of events from a contemporary, Tacitus. However, his personal viewpoint is so skilfully (i.e. often quite indiscernibly) woven into his work that discussions of Tacitus' credibility as a reporter have never been settled, esp. since the ground-breaking publication of G. WALSER in 1951. The information Tacitus here put to use for his own purposes came partly from the *Bella Germaniae* ('Germanic Wars') and *Historiae* of the elder Pliny. The latter served as an officer in Germania for a long period, and was also a contemporary observer of these events. He may have been involved in the political life of his own day to an even higher degree than Tacitus. It is not possible to identify any other literary sources for Tacitus' portrayal of the Batavian Revolt. There is sporadic information in the works of later authors, but they do not improve on the state of our knowledge obtained from Tacitus. Inscriptions provide important pointers, particularly on the stationing of the legions involved in the conflict.

Chronology of hostilities

9 June 68	Nero's suicide.
1 Jan. 69	The two legions camped at Mogontiacum/Mainz refuse to swear allegiance to Galba.
2 Jan.	Vitellius is acclaimed emperor by the *legio I* from Bonna/Bonn at the camp at Colonia Agrippinensis/Cologne.
3 Jan.	The remainder of the army in Germania inferior swears allegiance to Vitellius.
15 Jan.	Galba murdered.
14 Apr.	Battle of Betriacum/Calvatone: Vitellius victorious over Otho
16 Apr.	Otho's suicide.
Apr./May	Eight Batavian auxiliary cohorts sent back to the Rhine from Italy
1 July	Vespasian acclaimed emperor at Alexandria camp
Aug.	Revolt of the Germanic Cannenefates under Brinno
Aug./Sept.	Batavian rebels win victories in the Rhine Delta
Sept.	The eight *cohortes Batavorum* revolt at the Mogontiacum camp. Rebels defeat Herennius Gallus, legate of the *legio I*, in battle at Bonna. Rebels swear allegiance to Vespasian; besiege camp at Castra Vetera
Oct.	Legate Dillius relieves camp at Castra Vetera, but rebels then again besiege it.
24/25 Oct.	Battle of Betriacum: Vespasian's forces under the legate Antonius Primus defeat Vitellius. Legions at Novaesium/Neuss camp swear allegiance to Vespasian through legate Hordeonius Flaccus.
Mid-Nov.	Battle of Asciburgium/Asberg. Romans under the legate C. Dillius Vocula defeated at Gelduba/Gellep. Camp at Castra Vetera relieved. Civilis wounded.
19 Dec.	Fire at Capitolium in Rome. Battle of Mogontiacum.
20 Dec.	Rome taken by Antonius Primus. Death of Vitellius.
Late Dec.	Cavalry victory for Civilis at Novaesium. Camp of Castra Vetera surrounded again.
21 Dec. 69	Vespasian recognized as Emperor by the Senate.
Early Jan. 70	Death of Hordeonius Flaccus. Camp at Mogontiacum besieged by Civilis.
Mid-Jan.	Vocula frees Mogontiacum camp.
Feb./March	Revolt at Novaesium camp. Death of Vocula. Roman legionaries swear allegiance to *Imperium Galliarum*.
March	Camp at Castra Vetera capitulates, is plundered, garrison (5,000 men) killed, legate Munius Lupercus murdered.
Apr.	Civilis allies with Colonia Agrippinensis, wins territory along the Mosa/Maas. The Lingonian Iulius Sabinus defeated (probably at Andematunnum/Langres) by the Sequani, who are loyal to Rome.
May	Defeat of the Treveran Iulius Tutor at Bingium/Bingen. Legates Herennius Gallus and Numisius Rufus murdered.
May/June	Battle of Rigodulum/Riol: Cerialis victorious over the Treveran Valentinus, who is captured and later executed by Caesar Domitianus. Cerialis at camp of Augusta Treverorum.
June	Battle of Augusta Treverorum/Trier: Cerialis victorious over Civilis, Classicus and Tutor.
July	Rebels relinquish Colonia Agrippinensis. Battle of Castra Vetera: Cerialis defeats Civilis.
Aug.	Battles at Arenacium/Kleve-Rindern, Batavodurum/Nijmegen, Grinnes/Rossum and Vada (?).
Aug./Sept.	Naval battle at the confluence of the Vahalis/Waal and the Mosa.
Sept.	Battles on the Insula Batavorum/Betuwe; negotiations.
Sept./Oct. 70	End of hostilities. Capitulation of the Batavi under Civilis.

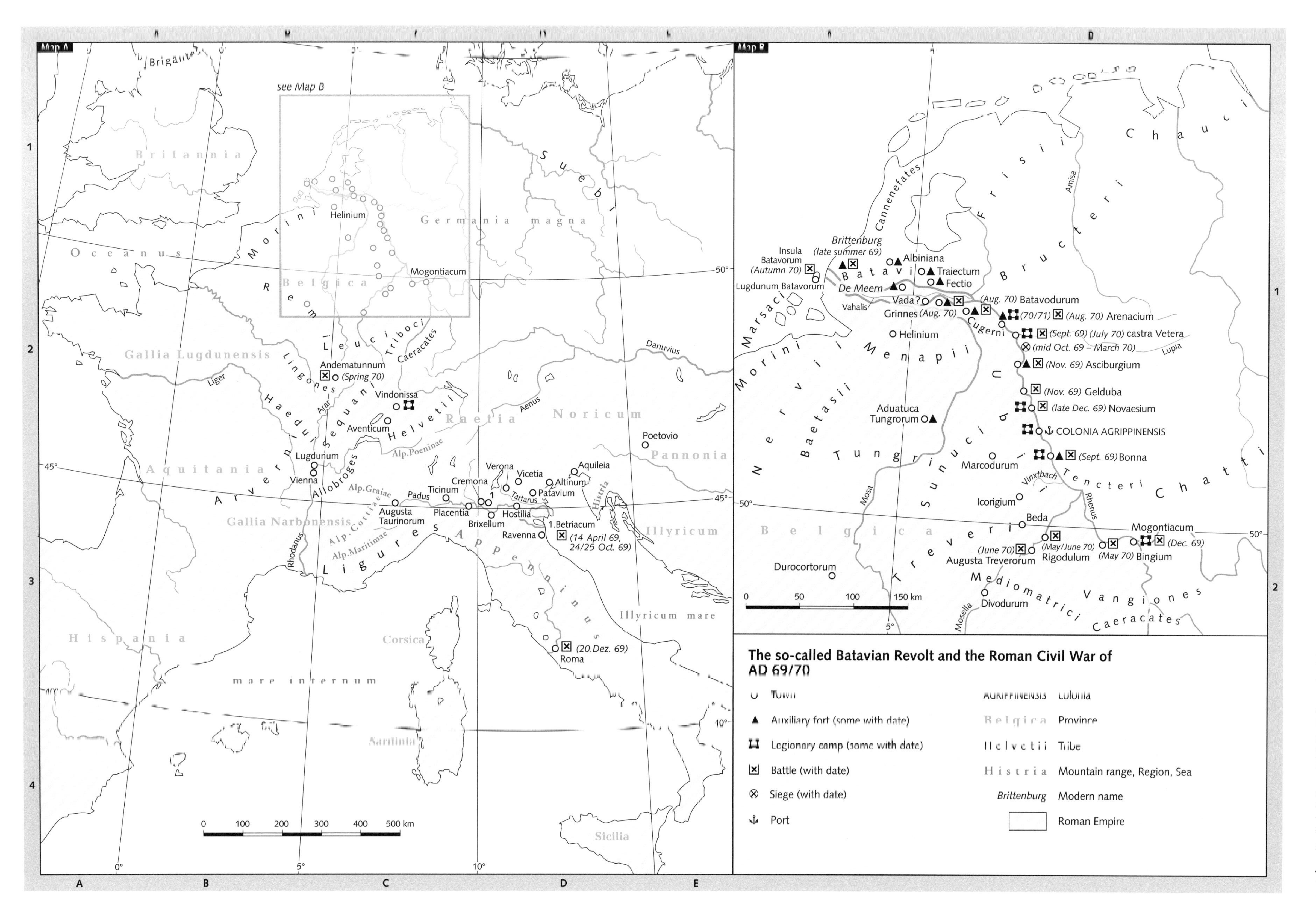
Map A
see Map B
Britannia
Oceanus
Morini
Remi
Belgica
Helinium
Mogontiacum
Germania magna
Suebi
Gallia Lugdunensis
Liger
Leuci
Triboci
Caeracates
Lingones
Andematunnum
(Spring 70)
Vindonissa
Haedui
Arar
Sequani
Aventicum
Helvetii
Raetia
Aenus
Noricum
Danuvius
Poetovio
Pannonia
Lugdunum
Vienna
Allobroges
Aquitania
Arverni
Alp.Poeninae
Alp.Graiae
Padus
Augusta Taurinorum
Ticinum
Cremona
Placentia
Brixellum
Verona
Vicetia
Tartarus
Hostilia
Patavium
Altinum
Aquileia
Histria
1.Betriacum
(14 April 69, 24/25 Oct. 69)
Ravenna
Illyricum
Gallia Narbonensis
Alp.Cottiae
Alp.Maritimae
Rhodanus
Ligures
Appenninus
Illyricum mare
(20.Dez. 69)
Roma
Corsica
Hispania
mare internum
Sicilia
0 100 200 300 400 500 km
Map B
Insula Batavorum
(Autumn 70)
Brittenburg
(late summer 69)
Lugdunum Batavorum
Cannenefates
Albiniana
Batavi
Traiectum
Fectio
De Meern
Frisii
Bructeri
Chauci
Amisa
Marsaci
Vahalis
Vada?
Grinnes (Aug. 70)
(Aug. 70) Batavodurum
(70/71)
(Aug. 70) Arenacium
Cugerni
Helinium
(Sept. 69) (July 70) castra Vetera
(mid Oct. 69 – March 70)
Lupia
Morini
Menapii
(Nov. 69) Asciburgium
(Nov. 69) Gelduba
(late Dec. 69) Novaesium
COLONIA AGRIPPINENSIS
Ubii
Nervii
Baetasii
Aduatuca Tungrorum
Tungri
Sunuci
Marcodurum
(Sept. 69) Bonna
Vinxtbach
Tencteri
Chatti
Icorigium
Mosa
Rhenus
Beda
Mogontiacum
Belgica
Treveri
(June 70)
Augusta Treverorum
(May/June 70)
Rigodulum
(May 70) Bingium
(Dec. 69)
Durocortorum
Mediomatrici
Divodurum
Vangiones
Caeracates
Mosella
0 50 100 150 km
The so-called Batavian Revolt and the Roman Civil War of AD 69/70
Town
Auxiliary fort (some with date)
Legionary camp (some with date)
Battle (with date)
Siege (with date)
Port
AGRIPPINENSIS Colonia
Belgica Province
Helvetii Tribe
Histria Mountain range, Region, Sea
Brittenburg Modern name
Roman Empire

The development of the Roman provinces in Britain (1st cent. AD – AD 410)

The development of the Roman provinces in Britain (1st cent. AD – AD 410)

The Romans first crossed the Oceanus Britannicus in 55 and 54 BC under Caesar when he was proconsul of Gaul (→ Map p. 165). Since then Rome had maintained its claim of dominion over the island, but this claim had never been realized in the form of a Roman provincial administration. The islanders had merely been obliged since to pay import and export duties. Plans to subjugate the island, prepared under Augustus, Tiberius and Caligula, would only be implemented under Claudius (AD 41–54).

A. Plautius, commanding an army of four legions, won the south-east of the island for Rome on Claudius' behalf. He took Camulodunum and celebrated an *ovatio*, or 'mini-triumph', at Rome in AD 47. The propraetor P. Ostorius Scapula expanded Roman influence westwards and northwards in 47–52, fighting the Iceni, Silures, Deceangli, Ordovices and Brigantes.

In 59/60, the propraetor C. Suetonius Paullinus occupied the island of Mona (Ynys Môn/Anglesey), where he annihilated the Druids, but he soon had to withdraw his forces in haste to put down a dangerous rebellion by the Iceni, under their princess Boudicca, and the Trinovantes, in the course of which the *colonia* of Camulodunum and the towns of Verulamium and Londinium were razed to the ground.

After the governors Q. Petillius Cerialis (70/71–73) and Sex. Iulius Frontinus (73–77/78), it was Cn. Iulius Agricola in particular, the father-in-law of the historian Tacitus, who extended the Roman sphere of influence northwards in 77–84, even as far as the Caledonii, on whom he inflicted a decisive defeat at the *mons Graupius* in 83. He considered conquering Hibernia (Ireland) in 81, and on his orders a fleet sailed north round Britannia, subjugating the Orcades (Orkney Islands) on its way and proving that Britannia was an island. It was also thought that the island of Thule had been seen on this voyage (Shetland?).

The far north of the island could not be permanently controlled, so in the reign of Trajan a series of forts was built along a 120-km route from the *mare Germanicum* in the east to the sound separating Britannia from Hibernia in the west. Under Hadrian (from 122), a frontier wall guarded with forts and turrets was constructed. Farther north, between 142 and 145, Antoninus Pius had a wood and turf fortification embankment built on stone foundations along a route just 60 km long in the territory of the Damnonii and Venicones. This, however, was quickly and permanently abandoned, in 168.

Between 208 and 211, Septimius Severus again tried to press forward from both fortifications against the Maeatae and Caledonii. However, after his death, Caracalla and Geta made peace with these tribes, and withdrew again to the line of Hadrian's Wall. Under Caracalla (211–217), the province of Britannia was divided into Britannia superior (in the south) and Britannia inferior (in the north). From 286–296, Carausius and his successor Allectus ruled an *Imperium Britanniarum* as 'Augusti'. After the violent suppression of the breakaway 'empire', the island was divided into four provinces. During the reign of Valentinian I (364–375), Theodosius, the future emperor (379–395), ordered the separation of another province from the four existing ones, and named it Valenti(nian)a. The central government at Rome felt itself increasingly incapable of properly governing the remote island, and finally, in 410, Emperor Honorius (393–423) told the cities of the Britannic provinces to provide for their own security, thereby to all intents and purposes relinquishing control of the island.

The map

The map shows the administrative development of Britain as part of the Imperium Romanum. There is much more that is uncertain here than certain. The localization of most of the tribes of the island is only approximate, the geographical distribution of the four (sometimes five) provinces of Late Antiquity is uncertain and the drawing of the frontiers between the individual provinces is controversial, as are the names of the individual provincial administrative seats. The northern frontier of the province set up under Claudius is doubtful, and it is also open to question whether, in the first years of Roman rule, there was a permanent administrative capital on the island at all, or whether the staff of government travelled around the province in the entourage of the governor.

What is at least certain is that there was to begin with one single province on the island, that this province was divided under Caracalla into a southern Britannia superior and a northern Britannia inferior and that the Diocletianic imperial reform further divided these two provinces into a total of four. It is once more less certain that a fifth province was created in 368 with the Valenti(nian)a. It is possible that one of the four provinces was merely renamed, or that a fifth province whose name is unknown to us was separated from an existing province and was now renamed Valenti(nian)a.

The sources

The geographers Strabo and Pomponius Mela, both writing in the early 1st cent. AD, acquaint us with the geography of Britain, and there is an excellent historical source for this century in the treatise *De vita Iulii Agricolae* by Tacitus, published around AD 98. It deals with his father-in-law, who was governor of the province of Britannia under Domitian (AD 81–96), and who brought the influence of Rome to the far north of the island. Our information about Britain in the period of the High and Late Empire becomes less and less reliable. It comes firstly from the *historia Augusta*, a collection of emperor biographies from Hadrian (117–138) to Carinus (284/85), and secondly from various more or less official lists giving administrative details, e.g. the *Notitia Dignitatum*, an administrative manual from the period around 425/30, and the *Laterculus* of Polemius Silvius, an almanac of history and culture aligned to the calendar of the year, dating from 448/49. The plentiful Latin inscriptions found all over the island are a more generally reliable source for the High and Late Empire in Britain. These include funerary inscriptions, but also honorary inscriptions and milestones. Nothing short of thrilling, and highly revealing of life on the northern limits of the Roman Empire, is a hoard of well over a thousand wax tablets and wooden tablets with ink writing, discovered in 1973 at Vindolanda/Chesterholm, a Roman fort near Hadrian's Wall, some 40 km west of Newcastle.

→ Maps pp. 165, 167, 198

Date	Province	Event	Sources	Capital
43	Britannia	Province established	Suet. Claud. 17; Cass. Dio 60,19–21; ILS 216; CIL 3, 7061	initially Camulodunum, then Londinium from the Flavian period (discussion of evidence in Haensch, 120–123)
197	Britannia superior	Province divided	Cass. Dio 55,23,2; 6	Londinium (CIL 7, 24; AE 1976, 363)
	Britannia inferior			Eboracum (CIL 13, 3162; ILS 2401)
296	Britannia I	Britannia divided into four	ILS 5435; Laterculus Veronensis 7; Not. Dign. Occ. 23,10–15	Corinium (ILS 5435)
	Britannia II			Eboracum
	Britannia Maxima Caesariensis			Londinium
	Britannia Flavia Caesariensis			Lindum
368	Britannia Valenti(nian)a	Valenti(nian)a – a fifth province?	Not. Dign. Occ. 23,11; Polemius Silvius 11,6; cf. Amm. Marc. 27,8	Eboracum or Carlisle

Sources

Ancient writings: Str. 14 f. (S. Radt, vol. 4, 2005); Mela 3,49–54 (K. Brodersen, 1994); Not. Dign. Occ. (O. Seeck, 1876); Polemius Silvius, Laterculus (MGH AA 9,511–551); *Inscriptions:* R.G. Collingwood, R.P. Wright, The Roman Inscriptions of Britain 1. Inscriptions on Stone, 1965 (RIB); *Vindolanda Tablets:* A.K. Bowman, J.D. Thomas, J.N. Adams, The Vindolanda Writing-Tablets (Tabulae Vindolandenses II), 1994; *Britannic tolls:* Str. 4,5,3; *Conquest of Britain:* Tac. Agr. 13,2–5; Suet. Cal. 19,3; Cass. Dio 59,21,4; *A. Plautius*: Tac. Agr. 14,1; Cass. Dio 60,19–21; Suet. Claud. 24,3; *Ostorius Scapula*: Tac. Ann. 12,31–39; *Mona*: Tac. Ann. 14,29 f.; *Boudicca:* Tac. Ann. 14,31–37; Tac. Agr. 15 f.; Cass. Dio 62,1–12; *Hibernia:* Tac. Agr. 24,3; *Thule/Shetland*: Tac. Agr. 10; *Hadrian's Wall:* SHA Hadr. 12,6; *Antonine Wall:* SHA Antoninus Pius 5,4; *Septimius Severus*: Cass. Dio 76,11,1; 76,12,1; Herodian. 3,14,1 f.; *Caracalla and Geta*: Cass. Dio 77,1,1; *Theodosius*: Amm. Marc. 28,3,7; *End of Roman rule*: Zos. 6,10,2.

Literature

General: T. Bechert, Die Provinzen des römischen Reiches. Einführung und Überblick, 1999, 161–166; A.R. Birley, The Fasti of Roman Britain, 1981; R. Haensch, Capita provinciarum. Statthaltersitze und Provinzialverwaltung in der römischen Kaiserzeit (Kölner Forschungen 17), 1997; B. Jones, D. Mattingly, An Atlas of Roman Britain, 1990; A. Lintott, Imperium Romanum, 1993; TIR M 30.31, 1983; TIR N 30.31/O29, 1987. *Britannia under Claudius:* D.C. Braund, Ruling Roman Britain. Kings, queens, governors and emperors from Julius Caesar to Agricola, 1996. *Druids on Mona:* F. Lynch, Prehistoric Anglesey, 1970; F. Le Roux, C.J. Guyonvarc'h, Les Druides, [3]1986. *Boudicca:* G. Webster, Boudica, The British Revolt against Rome A.D. 60, 1978. *Agricola:* G.S. Maxwell, Agricola and Roman Scotland: Some structural evidence, in: J. Bird (ed.), Form and Fabric. Studies in Rome's material past in honour of B.R. Hartley, 1998, 13–20; R.H. Martin, Tacitus on Agricola: Truth and Stereotype, in: Ibid. 9–12. *Hadrian's Wall:* E. Birley, Research on Hadrian's Wall, 1961; C.E. Stevens, The Building of Hadrian's Wall, 1966. *Antonine Wall:* G. MacDonald, The Roman Wall in Scotland, 1934. *Carausius/Allectus:* P.J. Casey, Carausius and Allectus, 1994; N. Shiel, The Episode of Carausius and Allectus, 1977. *End of Roman rule:* S. Esmonde-Cleary, The Ending of Roman Britain, 1989.

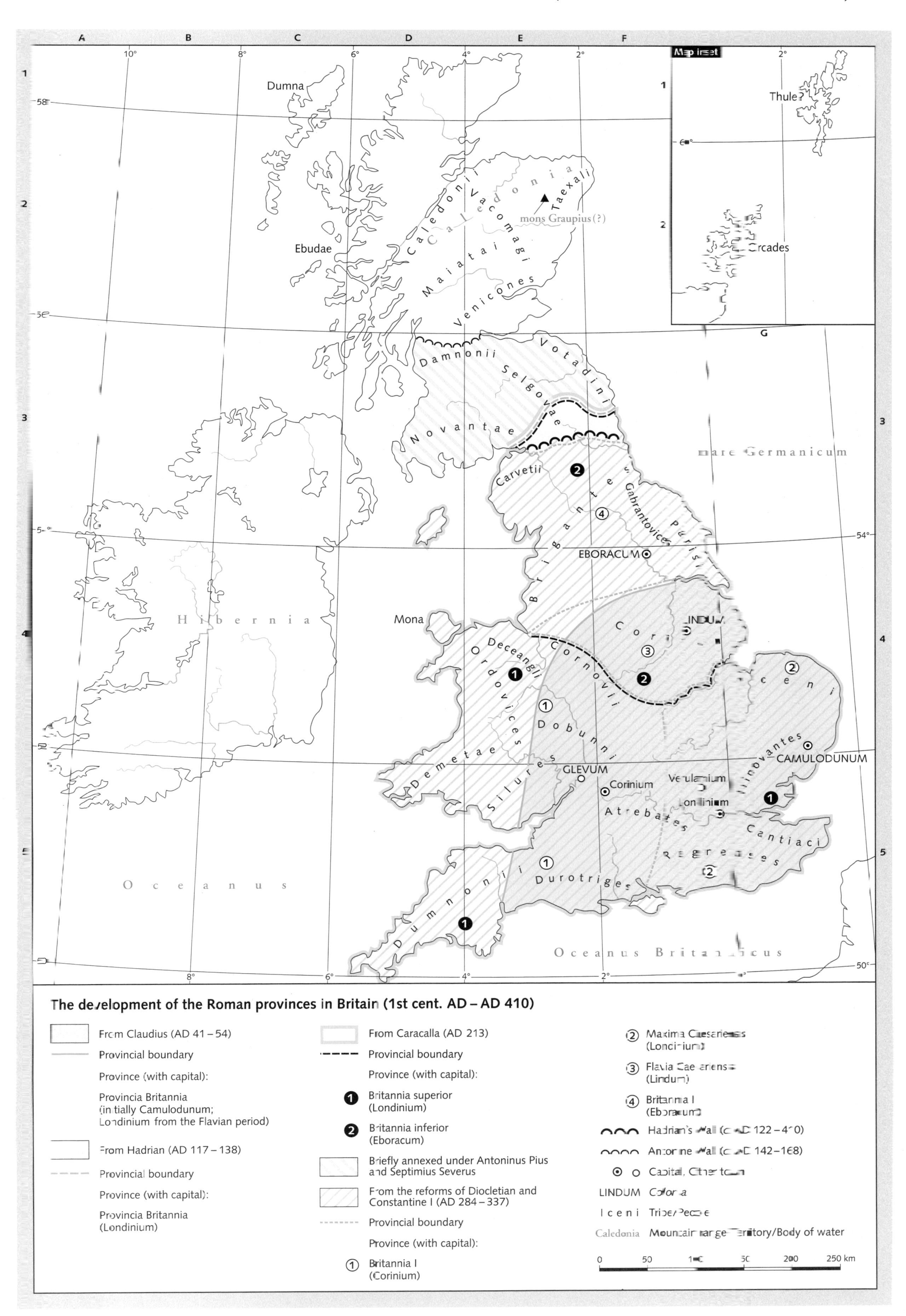
Map inset
Thule?
Dumna
Ebudae
Caledonii
Caledonia
Vacomagi
Taexali
mons Graupius (?)
Maiatai
Venicones
Damnonii
Votadini
Selgovae
Novantae
mare Germanicum
Carvetii
Brigantes
Gabrantovices
Parisi
EBORACUM
Hibernia
Mona
Deceangli
Ordovices
Cornovii
Corieltauvi
Demetae
Dobunni
Silures
GLEVUM
Corinium
Atrebates
Trinovantes
CAMULODUNUM
Cantiaci
Durotriges
Dumnonii
Oceanus
Oceanus Britannicus
The development of the Roman provinces in Britain (1st cent. AD – AD 410)
From Claudius (AD 41 – 54)
Provincial boundary
Province (with capital):
Provincia Britannia (initially Camulodunum; Londinium from the Flavian period)
From Hadrian (AD 117 – 138)
Provincial boundary
Province (with capital):
Provincia Britannia (Londinium)
From Caracalla (AD 213)
Provincial boundary
Province (with capital):
Britannia superior (Londinium)
Britannia inferior (Eboracum)
Briefly annexed under Antoninus Pius and Septimius Severus
From the reforms of Diocletian and Constantine I (AD 284 – 337)
Provincial boundary
Province (with capital):
Britannia I (Corinium)
Capital, Other town
Mountain range Territory/Body of water
0 50 200 250 km

Roads and routes in the Imperium Romanum

The maps show the most imporant roads of the Roman Empire. In particular, they illustrate the tightly-knit nature of the road network. No comparable empire (even the Persian Empire, for all the high priority accorded to communications structures as demonstrated by the organization of the royal roads) had a traffic infrastructure of a density to rival that of the Romans. The portrayal of the roads here pays no heed to dating. It must therefore be remembered that the road network as shown here never existed in this form. Certain roads were neglected or even abandoned and others newly built or refurbished at various times according to need. The routes of the roads are approximate throughout, partly in accordance with the small scale of the maps, but also because of the often incomplete state of research. The state of research into the road and route network of the Roman Empire differs according to region. In some areas, e.g. Gaul and Britain, the study of ancient roads is very advanced. Elsewhere, however, current political conditions or the geographical difficulty of the terrain have hampered its progress, e.g. on the Sinai Peninsula and the Macedonian Highlands.

The Romans distinguished between private roads and so-called *viae publicae*, which were commissioned by senior magistrates and financed from the public purse, ran on public land outside settlements and served public transportation. Roads marked with milestones were undoubtedly such *viae publicae*. It is also assumed that the roads shown on the Tabula Peutingeriana and listed in the Itinerarium Antonini belonged to this class of roads. The roads shown in Italy are all *viae publicae*. For the other parts of the Empire, roads attested as*viae publicae* by milestones are highlighted in colour. Attestation by both milestones and itineraries is shown for the Alpine region by way of example.

The source basis for the portrayal is the archaeological record of road sections, though these can only seldom be dated exactly. Also bridges and roadhouses, though these can only be identified beyond doubt in a few cases. The contribution of epigraphic evidence to our knowledge of Roman roads is considerable – across the Empire, well over 6,000 inscriptions have already been found, dating from (probably) 253 BC (ILS 5801) to the 5th cent. AD. Finally, the literary record must be mentioned. Evidence is dispersed across the whole of ancient literature, e.g. in the letters of Cicero (1st cent. BC), or the works of the historians Ammianus Marcellinus (4th cent. AD) and Procopius (6th cent. AD). More instructive still, however, are the literary itineraries (*itineraria*), such as the Tabula Peutingeriana (original collation 1st-4th cents. AD) and the Itinerarium Antonini (3rd cent. AD).

I. The road network in Italy (Map p. 195)

In Italy itself, the Romans could use the Etruscan route network. New road building (albeit always based on preceding installations) began towards the end of the 4th cent. BC, possibly also stimulated by the examples of the Hellenistic monarchies. The *via Appia*, from Rome to Capua, for instance, was commissioned in 312 BC by the censor Appius Claudius Caecus. Eventually, Rome sat spider-like at the focus of a web of roads across Italy, linking the city with all quarters of the peninsula. Only after some delay were regions outside Italy linked into this system and road networks in the provinces developed. Caesar may already have sketched some plans for what would, under Augustus and over the following two centuries, become the unprecedented development of a pan-imperial road network.

II. The road and track network in the Balkans (Map p. 196)

Most of the northern Balkans is mountainous territory, and many areas are difficult to access, so that the road network here is not particularly dense. North of the Danube, Dacia was served by only a few roads, while a relatively dense strip of roads led along the southern bank of the Danube, motivated above all by military requirements for securing the border.

Farther south, a particularly important road linked the Adriatic with the Aegean. This was the *via Egnatia*, commissioned by Cn. Egnatius, proconsul of the province of Macedonia in 143 BC (?). The road led from Dyrrhachium and Apollonia on the Adriatic, via Heraclea, Edessa and Pella to Thessalonica. From there it continued north of the Chalcidian Peninsula to Amphipolis, bypassing Neapolis to the north on its way to Perinthus on the Propontis and thence to Byzantium. In places, it made use of an existing road route from the time of the Macedonian kingdom.

The map shows only a very few roads in Greece, but appearances here are deceptive. Recent research has already revealed a plethora of connecting roads and tracks in some provinces (Attica, Megaris, Argolid), and the same picture probably applied to the entire Greek peninsula. There is, however, little record of such infrastructure in the form of inscriptions, milestones or itineraries. In some areas, this route network dates back to the Mycenaean period. It included broader roads, sometimes paved and reinforced with supporting embankments and carriage-tracks, and a multitude of narrower tracks, made but not architecturally stabilized, whose routes avoided unfavourable terrain such as extreme ascents and descents, and therefore could do without bridges and levelling. If it is true that fewer roads were built here in the Roman Period than elsewhere in the Empire, this may have been because a well-functioning infrastructure was already in place.

III. The road and route network in the Alps and Asia Minor (Map p. 197)

The Alps, a chain of mountains 1,200 km long on the northern frontiers of Italy, were a source of dread to the Romans before Caesar, but with Caesar's conquest of Gaul they had lost must of their terror. The important passes – the Montgenèvre and the Great and Little St. Bernhard in the western Alps, the Splügen and Julier Passes in the central Alps and the *ad Pirum* ('Pear Tree', now Hrušica Pass) in the Julian Alps – were all gradually made more or less safe, if not exactly effortless, to cross. The Romans took great pains in making the Alps more penetrable with a road system. One of the most important south-north routes led along the *via Claudia Augusta* from Altinum on the Adriatic, and from Hostilia and Verona, via Tridentum, across the Resia and Fern Passes to Augusta Vindelicum/Augsburg, the administrative capital of the province of Raetia. This road was laid out in AD 46/47 under Claudius, on a much older route. From the reign of Septimius Severus (193–211), the road over the Brenner Pass was often preferred instead.

Like in the Levant, Roman road building in Asia Minor could take advantage of the Persian road network and that of the Hellenistic monarchies. Intensive road construction began with the foundation of the province of Asia in 129 BC. It intensified under Augustus (27 BC – AD 14) and the Flavian emperors (AD 69–96). The so-called 'Pilgrims' Route' from Constantinople via Ancyra, Tarsus and Antioch to Hierosolyma/Jerusalem acquired special importance in Late Antiquity.

IV. The road and route network on the Iberian Peninsula and in North Africa (Map p. 198)

The Roman road network on the Iberian Peninsula, which was based on Celtic and Carthaginian predecessors in the south and west, really only developed in its own right under Augustus and his successors and the emperors Trajan (98–117) and Hadrian (117–138). The denser road network evident in Baetica is primarily attributable to the fact that Roman administration took a greater and earlier interest in the infrastructure of this resource-rich region than elsewhere on the peninsula.

In North Africa, Roman road construction could base itself on the Carthaginian roads. The first road building here was probably under Augustus (Carthago – Hadrumetum – Sabratha), but it was only with the reign of Tiberius (AD 14-37) that a real building boom set in. Mapped Roman roads and tracks in the hinterland give an impression of less development than the road network in the coastal strip, but this is deceptive. The African *limes* was occupied by a great number of military outposts, all of which will unquestionably have been linked up by tracks or roads (→ Map p. 212). If these routes, which will often have been covered by desert sand, greatly hampering archaeological detection, are added to the littoral roads, a road network of considerable overall density is perceived.

V. The road and route network in Britain and Gaul (Map p. 199)

The Roman road network in Britain is archaeologically well documented, but it is relatively poorly attested in the written sources. The relevant first page of the Tabula Peutingeriana is lost, and milestone finds on the island are not as numerous as elsewhere in the Roman Empire. One striking road dates from the first period of the Roman Conquest: the Fosse Way. This crosses the entire south of the island from Isca/Exeter in the West Country to Lindum/Lincoln in the East Midlands, by way of Corinium/Cirencester and Ratae/Leicester. This road seems to have marked the western boundary of the Roman province under Claudius (AD 41–54) , and was originally a defensive rampart, as the local name of the road suggests (cf. *fossa*, ditch). Like other Roman roads, long stretches of the Fosse Way also form part of the modern road network in England.

In Gaul, the topographically dominant rivers, the Sequana/Seine, Liger/Loire, Garumna/Garonne and Rhodanus/Rhône, determined the layout of road building. Alongside the many small and likewise navigable tributaries of these rivers, the road network comprised an instrument of infrastructure that was, from economic and military perspectives, uniquely useful. Building on a well-functioning Celtic road system, Roman road building started with the *via Domitia* (118 BC), connecting the Italian and Iberian Peninsulas. Further phases of road construction date from the reigns of Claudius and the Flavians to Antoninus Pius (138–161).

Viae publicae in Italy and the road and route network in Corsica, Sardinia and Sicily

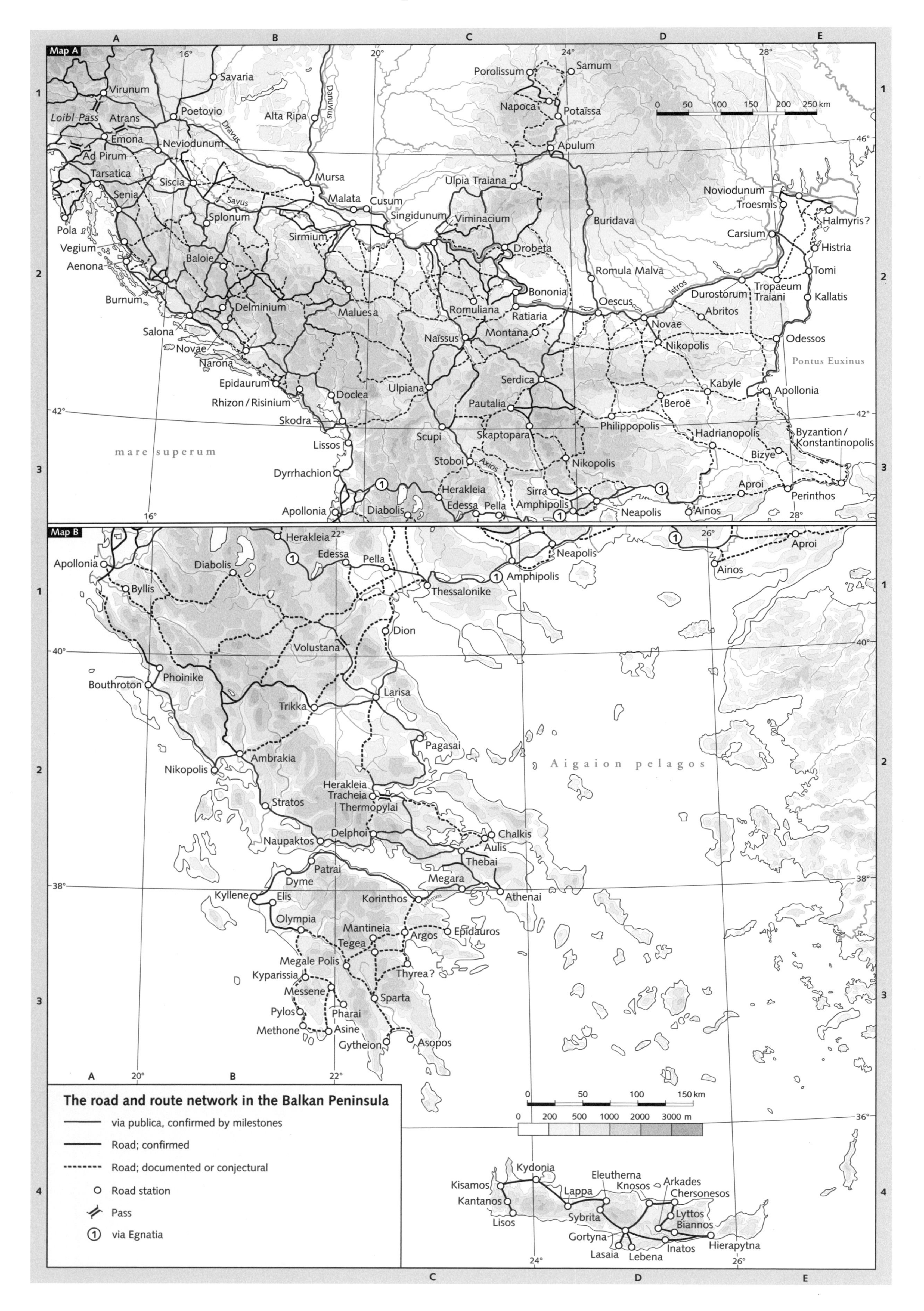

Map A
Virunum
Savaria
Loibl Pass
Atrans
Poetovio
Alta Ripa
Danuvius
Dravus
Emona
Neviodunum
Ad Pirum
Tarsatica
Siscia
Mursa
Senia
Savus
Malata
Cusum
Singidunum
Viminacium
Ulpia Traiana
Pola
Splonum
Sirmium
Vegium
Aenona
Baloie
Burnum
Delminium
Maluesa
Salona
Novae
Narona
Epidaurum
Rhizon / Risinium
Doclea
Skodra
Lissos
mare superum
Dyrrhachion
Apollonia
Diabolis
Porolissum
Samum
Napoca
Potaïssa
Apulum
Buridava
Drobeta
Romula Malva
Bononia
Romuliana
Ratiaria
Montana
Naïssus
Oescus
Istros
Novae
Nikopolis
Ulpiana
Serdica
Pautalia
Scupi
Skaptopara
Stoboi
Axios
Herakleia
Edessa
Pella
Sirra
Amphipolis
Neapolis
Philippopolis
Nikopolis
Noviodunum
Troesmis
Halmyris?
Carsium
Histria
Tomi
Durostorum
Tropaeum Traiani
Kallatis
Abritos
Odessos
Pontus Euxinus
Kabyle
Apollonia
Beroë
Hadrianopolis
Byzantion / Konstantinopolis
Bizye
Aproi
Perinthos
Ainos
0 50 100 150 200 250 km
Map B
Herakleia
Edessa
Pella
Apollonia
Diabolis
Byllis
Neapolis
Amphipolis
Thessalonike
Aproi
Ainos
Dion
Volustana
Phoinike
Bouthroton
Trikka
Larisa
Pagasai
Ambrakia
Nikopolis
Aigaion pelagos
Herakleia Tracheia
Thermopylai
Stratos
Delphoi
Naupaktos
Chalkis
Aulis
Thebai
Patrai
Dyme
Megara
Athenai
Kyllene
Elis
Korinthos
Isthmos
Olympia
Mantineia
Tegea
Argos
Epidauros
Megale Polis
Kyparissia
Thyrea?
Messene
Sparta
Pylos
Pharai
Methone
Asine
Gytheion
Asopos
Kydonia
Kisamos
Kantanos
Lisos
Lappa
Eleutherna
Knosos
Arkades
Chersonesos
Sybrita
Lyttos
Biannos
Gortyna
Lasaia
Lebena
Inatos
Hierapytna
0 50 100 150 km
0 200 500 1000 2000 3000 m
The road and route network in the Balkan Peninsula
via publica, confirmed by milestones
Road; confirmed
Road; documented or conjectural
Road station
Pass
1 via Egnatia

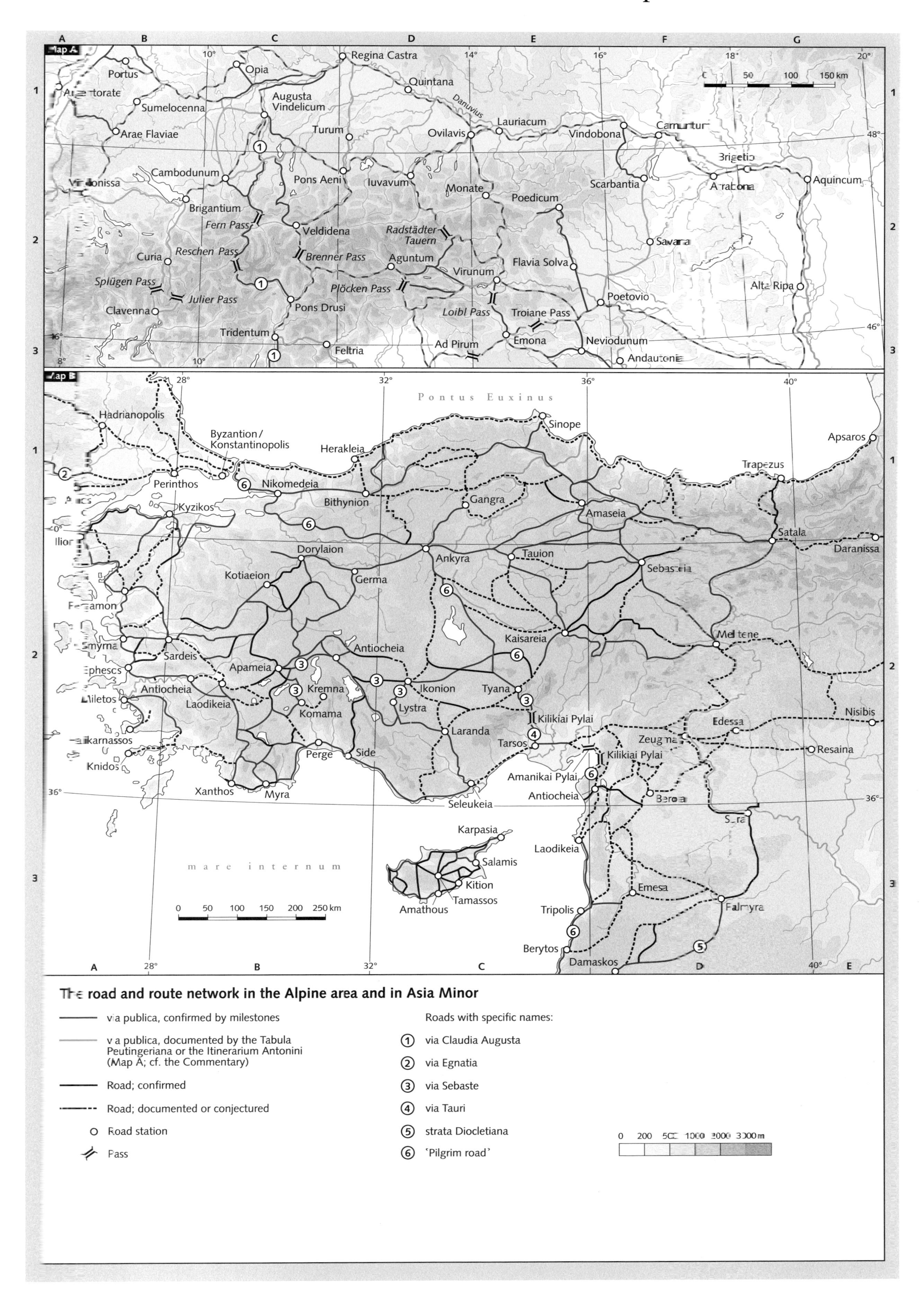

The road and route network in the Alpine area and in Asia Minor

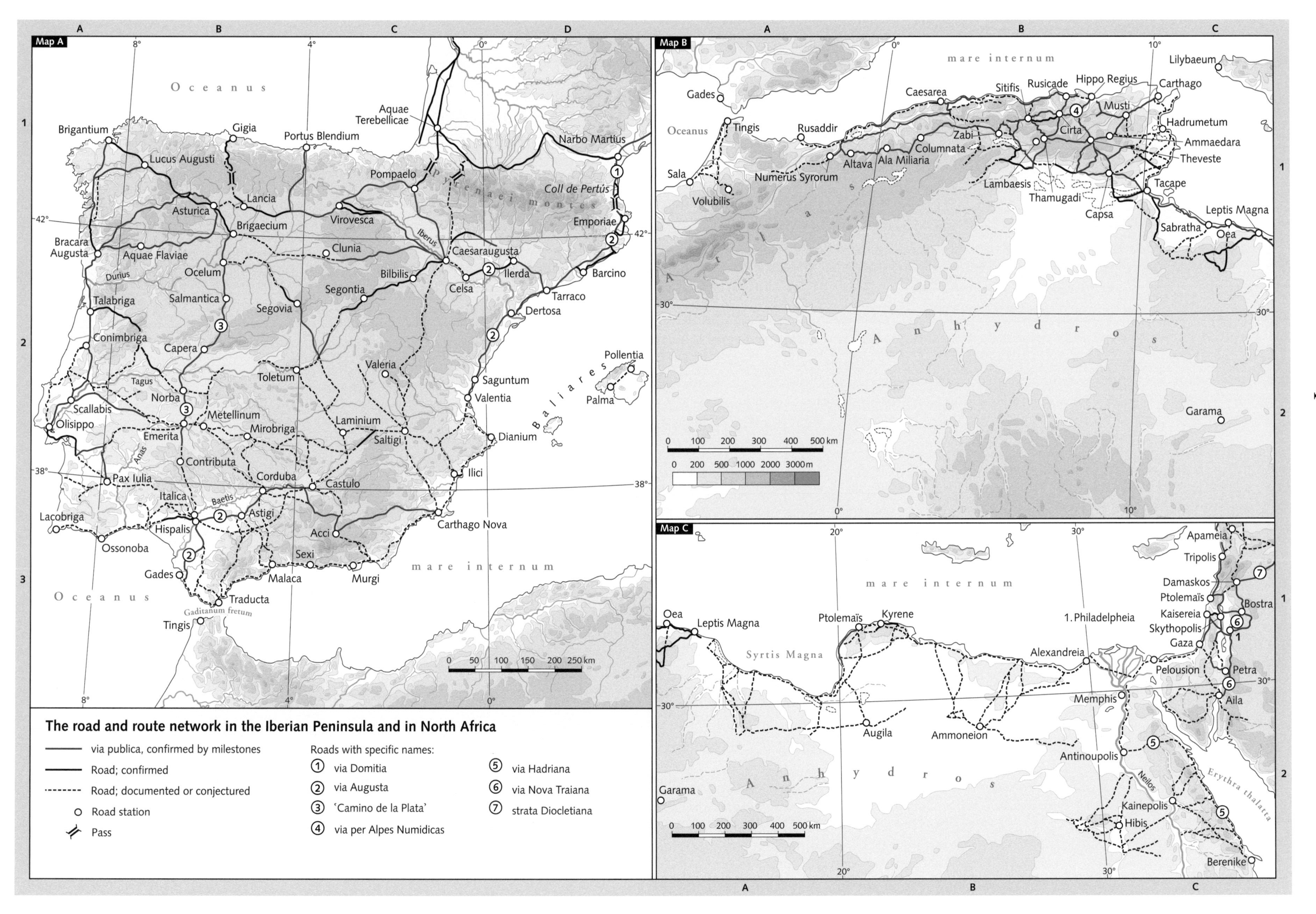
Map A
Oceanus
mare internum
Brigantium
Lucus Augusti
Gigia
Portus Blendium
Aquae Terebellicae
Pompaelo
Pyrenaei montes
Coll de Pertús
Narbo Martius
Emporiae
Barcino
Tarraco
Dertosa
Lancia
Asturica
Virovesca
Brigaecium
Iberus
Caesaraugusta
Ilerda
Celsa
Bilbilis
Clunia
Bracara Augusta
Aquae Flaviae
Durius
Ocelum
Segontia
Salmantica
Segovia
Talabriga
Conimbriga
Capera
Toletum
Valeria
Saguntum
Valentia
Pollentia
Baliares
Palma
Tagus
Norba
Scallabis
Olisippo
Metellinum
Mirobriga
Emerita
Laminium
Saltigi
Dianium
Anas
Contributa
Ilici
Pax Iulia
Corduba
Castulo
Italica
Baetis
Astigi
Hispalis
Carthago Nova
Lacobriga
Acci
Ossonoba
Sexi
Gades
Malaca
Murgi
Traducta
Gaditanum fretum
Tingis
0 50 100 150 200 250 km
The road and route network in the Iberian Peninsula and in North Africa
via publica, confirmed by milestones
Road; confirmed
Road; documented or conjectured
Road station
Pass
Roads with specific names:
1 via Domitia
2 via Augusta
3 'Camino de la Plata'
4 via per Alpes Numidicas
5 via Hadriana
6 via Nova Traiana
7 strata Diocletiana
Map B
mare internum
Lilybaeum
Gades
Caesarea
Sitifis
Rusicade
Hippo Regius
Carthago
Musti
Oceanus
Tingis
Rusaddir
Zabi
Cirta
Hadrumetum
Ammaedara
Columnata
Altava
Ala Miliaria
Theveste
Sala
Numerus Syrorum
Lambaesis
Tacape
Thamugadi
Volubilis
Capsa
Leptis Magna
Sabratha
Oea
Atlas
Anhydros
Garama
0 100 200 300 400 500 km
0 200 500 1000 2000 3000 m
Map C
Apameia
Tripolis
mare internum
Damaskos
Ptolemaïs
Bostra
Oea
Leptis Magna
Ptolemaïs
Kyrene
1. Philadelpheia
Kaisereia
Skythopolis
Gaza
Syrtis Magna
Alexandreia
Pelousion
Petra
Memphis
Aila
Augila
Ammoneion
Antinoupolis
Erythra thalatta
Anhydros
Nellos
Garama
Kainepolis
Hibis
0 100 200 300 400 500 km
Berenike

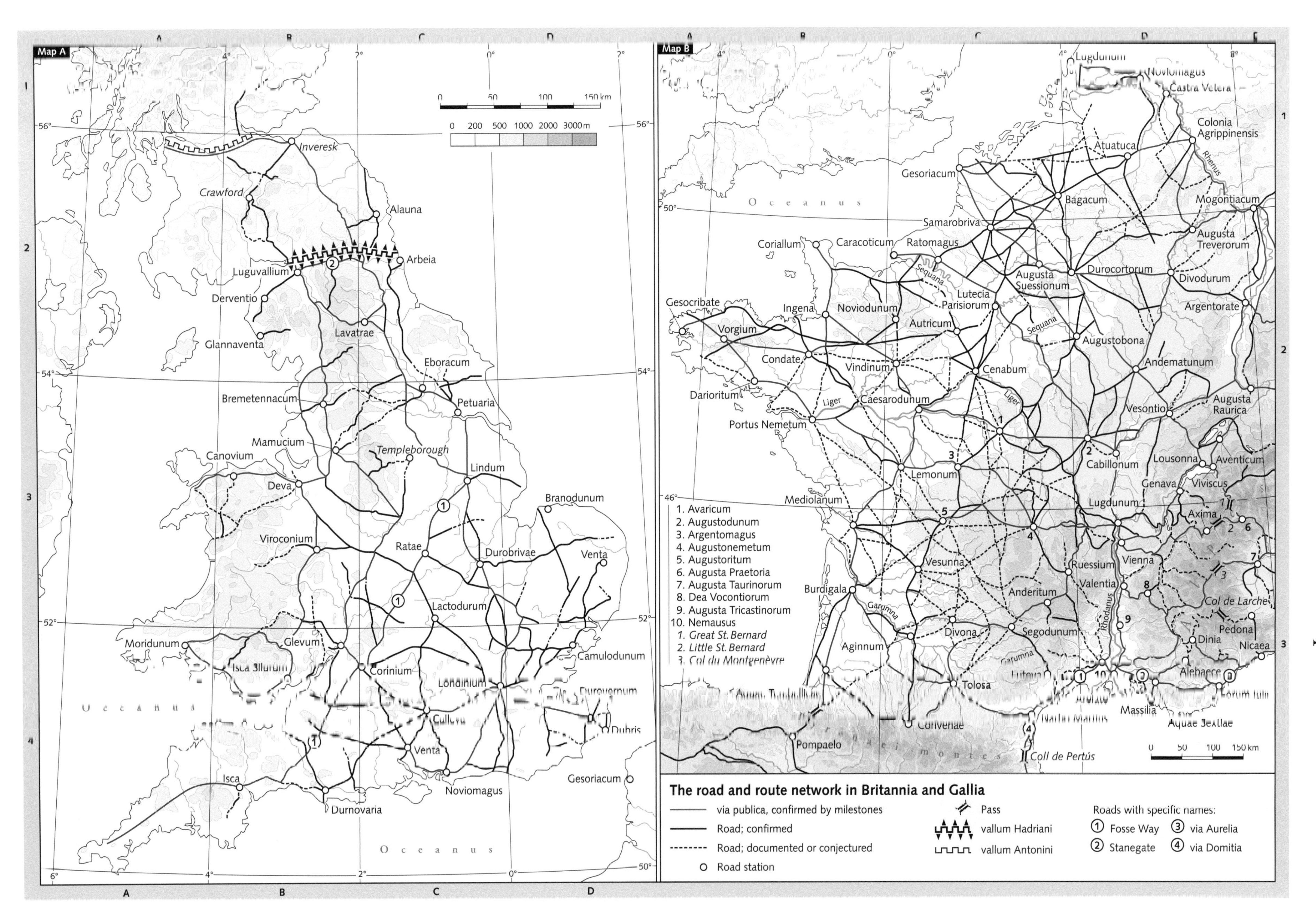
Map A
Map B
The road and route network in Britannia and Gallia
via publica, confirmed by milestones
Road; confirmed
Road; documented or conjectured
Road station
Pass
vallum Hadriani
vallum Antonini
Roads with specific names:
① Fosse Way
② Stanegate
③ via Aurelia
④ via Domitia
1. Avaricum
2. Augustodunum
3. Argentomagus
4. Augustonemetum
5. Augustoritum
6. Augusta Praetoria
7. Augusta Taurinorum
8. Dea Vocontiorum
9. Augusta Tricastinorum
10. Nemausus
1. Great St. Bernard
2. Little St. Bernard
3. Col du Montgenèvre
Col de Larche
Coll de Pertús
Oceanus
Rhenus
Sequana
Liger
Garumna
Rhodanus
Castra Vetera
Noviomagus
Colonia Agrippinensis
Mogontiacum
Augusta Treverorum
Divodurum
Argentorate
Augusta Raurica
Aventicum
Lousonna
Genava
Viviscus
Vesontio
Andematunum
Durocortorum
Augustobona
Cabillonum
Lugdunum
Vienna
Ruessium
Valentia
Segodunum
Anderitum
Tolosa
Divona
Vesunna
Lemonum
Cenabum
Lutecia Parisiorum
Augusta Suessionum
Bagacum
Atuatuca
Samarobriva
Gesoriacum
Ratomagus
Caracoticum
Noviodunum
Autricum
Vindinum
Caesarodunum
Condate
Ingena
Coriallum
Vorgium
Gesocribate
Darioritum
Portus Nemetum
Mediolanum
Burdigala
Aginnum
Pompaelo
Convenae
Massilia
Aquae Sextiae
Nicaea
Pedona
Dinia
Axima
Inveresk
Crawford
Alauna
Arbeia
Luguvallium
Derventio
Glannaventa
Lavatrae
Eboracum
Petuaria
Templeborough
Bremetennacum
Mamucium
Canovium
Deva
Viroconium
Lindum
Ratae
Durobrivae
Branodunum
Venta
Camulodunum
Lactodurum
Londinium
Corinium
Glevum
Calleva
Noviomagus
Durnovaria
Isca
Moridunum
Dubris
Gesoriacum

Important agricultural areas of the Mediterranean (1st and 2nd cents. AD)

I. Grain

The most important areas of grain cultivation were Mesopotamia (although its poor transport links with the Mediterranean negated its importance as a producer for long-distance trade), the entire flood plain of the Nile (shipped down the Nile and loaded onto seagoing ships at Alexandria), North Africa from Numidia to Cyrene (where the main export harbours were Carthage and Leptis Magna), Mauretania, Hispania Baetica (with Gades as the export harbour) and, finally, Sardinia and Sicily.

Because of their proximity, the main traditional sources of supply for Rome from the second half of the 3rd cent. BC were Sardinia and Sicily (import harbours: Puteoli and Ostia). Rome later also began drawing upon North Africa (from mid 2nd cent. BC) and Egypt (from late 1st cent. BC, substantial Egyptian surpluses monopolized for the supply of Italy). Only at a still later date did grain from southern Spain also enter the long-distance trade to a significant degree.

Scholars have long devoted much effort to attempts at assessing the quantitative importance of these regions of grain surplus, and the debates are far from being resolved. The sources allow only an approximate picture of Rome's grain supply. Roughly a quarter of the grain imported into Rome came from Egypt, and a good half from North Africa. The remainder came mostly from Sicily and Sardinia, but some also from Spain (cf. Cic. Verr. 2,3; 2,5,52; Leg. Man. 34; (Ps.-)Aur. Vict. Epit. Caes. 1,6; Jos. Bl. 2,16,4.).

II. Wine

Wine was produced for large-scale export in Syria, on the Aegean Islands, on the west coast of Asia Minor, on Sicily, in Campania and Etruria, on the Garumna/Garonne, in what is now Provence, in Hispania citerior and in Mauretania. The smaller wine-producing region around Augusta Treverorum/Trier, conversely, served only the local region around the *limes*. Italian and especially Aegean wines were certainly much loved and tended to occupy upper price brackets, while Spanish and Gaulish wines tended to be simpler, mass-produced wines. However, by the 2nd cent. AD at the latest, they were succeeding in providing stern competition for Italian wines, even on the Italian market itself.

III. Oil

Centres of oil production were Syria, Asia Minor, south-eastern Greece and the Peloponnese, the Italian regions of Campania, Samnium, Lucania and Apulia, Sicily, North Africa and especially southern Spain. While oil-producing regions of the eastern Mediterranean mostly served their local markets, a similar trend developed here as in the wine trade: from the 1st cent. AD, Spanish and then also North African oil exports (according to Plin. HN 15,8 such provincial oil was not inferior to Italian) began to oust Italian products from the market.

IV. Other products

Dates, figs, and nuts were of lesser importance than the abovementioned staples. Their main areas of cultivation were in the eastern Mediterranean.

Flax and hemp were of some importance in the manufacture of textiles and rope. Centres of production were the Black Sea region, Syria and Spain.

Timber was indispensable to construction, shipbuilding and heating. However, resources had become scarce because of continuous predatory cultivation. Timber was produced and exported on a large scale in northern Greece, northern and southern Asia Minor, Syria, North Africa and on some islands (esp. Cyprus). It is interesting to note that in spite of the widespread over-exploitation of forests, the Roman agrarian writers did also make recommendations for sustainable forestry. It remains open to question, however, to what extent and on what scale these well-intentioned recommendations were followed.

V. Chronology

Conditions regarding important agricultural zones in the Mediterranean may have been constantly and fundamentally changing over the millennium separating Homer from Late Antiquity. For example, the Black Sea region, especially the region around Olbia and Panticapaeum, was a greatly important region of grain cultivation producing large surpluses for the supply of Athens in the 5th and 4th cents. BC. But it appears that the lands around the Cimmerian Bosporus and the Black Sea lost their significance as an export region through the 3rd cent. At all events, reports of famines even in the Pontic cities, become more frequent in this period.

Similarly, Italian grain production (e.g. in the Po Valley, Etruria and the south), not insignificant in early times, gave way as early as the 2nd cent. BC to the cultivation of vines, olives and fruit, where the fields were not extensively given over to pasture. Conversely, such regions as southern Gaul, southern Spain and North Africa only began to develop significant agricultural landscapes as they became Romanized, and in the end they would come to outdo Italy.

The sources

This map is primarily based on references from ancient literature and archaeological finds. Even the Homeric epics have numerous references to agriculture in the 'Dark Ages' (1200–800 BC). Hesiod's *Works and Days* (*Erga kai hemerai*, *c.* 700 BC) gives information about conditions in Archaic Greece, and Xenophon's (*c.* 430 – after 355 BC) *Oikonomikos* provides an overview of agriculture in the Greek Classical *polis* period (5th and 4th cents. BC). Only with the Latin agrarian writers Cato the Elder (234–149 BC), Varro (116–27 BC), Columella (1st cent. AD) and the Elder Pliny (AD 23–79; books 14, 15 and 18 of the Natural History), who based their work on Hellenistic sources, do we learn detailed information on agriculture in Italy and, to some extent, in the Roman provinces. More exact notes on cultivation zones, their extent and their economic importance, however, are only incidentally present in the literature listed here.

Throughout ancient literature, there are also references (albeit widely scattered) to the provisioning of the great metropolitan centres. Information on the origins of these agricultural products, which were sometimes imported over great distances in the context of what was an extensive trade in foodstuffs, enable us to form an approximate picture of the most important (i.e. surplus-producing) Mediterranean agricultural regions.

The ancient geographical treatises add depth to this information, e.g. the Geographies of Strabo (63 BC – *c.* AD 19) and Pomponius Mela (*De chorographia;* 1st cent. AD), Pausanias' 'Description of Greece' (2nd cent. AD) and an ancient economic geography, the anonymous *Expositio totius mundi et gentium* (4th/5th cents. AD). In addition to describing the natural landscapes, these geographical works also provide some information on the agricultural use of the particular regions described.

Otherwise, we have the conclusions drawn from archaeological finds (farm complexes, plantations, agricultural equipment, oil or wine presses, pictorial records, evaluation of residual vegetable matter), as well as some information transmitted in inscriptions and (notably important for Egypt from the 3rd cent. BC) on papyri.

The map

The informative value of the adjacent map is limited for various reasons:

- It shows primarily the conditions as they were at the height of the Roman Empire, i.e. the 1st and 2nd cents. AD.
- Naturally, staples were cultivated for consumption at home throughout the entire Mediterranean region. What the map shows in particular are regions that were producing significant quantities of surplus.
- Vegetables, salad vegetables, herbs, fruit and decorative plants were likewise universally cultivated across the whole Mediterranean world. However, these products are highly perishable, and were always produced close to their place of consumption. Only to a very limited extent were they traded over long distances. These categories of plant production, therefore, cannot be shown on the map.
- In terms of agricultural cultivation, then, only the three most important crops, which were also ancient staples, are shown as areas of distribution: grain, (olive) oil and wine. Symbols indicate significant areas of timber and flax/linen production, as well as date, fig and nut cultivation.

→ Map p. 203

Sources

Space does not permit a listing here of the countless, widely scattered references to cultivation zones in ancient literature. Only those sources are indicated which describe agricultural cultivation zones in context: Columella, praef. 20; Plin. HN 14, 8–76; cf. Varro, Rust. 1,54 (wine); 15,1–9 (olives); 18, 49–156 (field crops); *Expositio totius mundi et gentium*, ed. H.-J. Drexhage, in: MBAH II 1/1983 (however, this source describes conditions in the 4th or 5th cent. AD).

Literature

H.-J. Drexhage, H. Konen, K. Ruffing, Die Wirtschaft des Römischen Reiches (1.–3. Jahrhundert), 2002, 59–100; U. Fellmeth, Eine wohlhabende Stadt sei nahe… Die Standortfaktoren in der römischen Agrarökonomie im Zusammenhang mit den Verkehrs- und Raumordnungsstrukturen im römischen Italien, 2002, 13–50, 107–151; D. Flach, Römische Agrargeschichte, 1990; P. Garnsey, Famine and Food Supply in the Graeco-Roman World, 1993; S. Isager, J.E. Skydsgaard, Ancient Greek Agriculture, 1995; F. De Martino, Wirtschaftsgeschichte des alten Rom, 1991; W. Richter, Die Landwirtschaft im homerischen Zeitalter (Archaeologica Homerica vol. II, ch. H), 1968; G. Rickman, The Corn Supply of Ancient Rome, 1980; M. Schnebel, Die Landwirtschaft im hellenistischen Ägypten, 1925; H. Sonnabend (ed.), Mensch und Landschaft in der Antike. Lexikon der Historischen Geographie, 1999 (in which s. esp.: U. Fellmeth, Ackerbau. 1–6; E. Olshausen, Agrargeographie, 14–17; U. Fellmeth, Agarverfassung, 18–24; M. Nenninger, Forstwirtschaft, 151–153; U. Fellmeth, Getreide, 180–183; Id., Großgrundbesitz, 197–200; Id. Landwirtschaft, 305–308; Id., Nahrungsmittel, 367–372); K.D. White, Roman Farming, 1970.

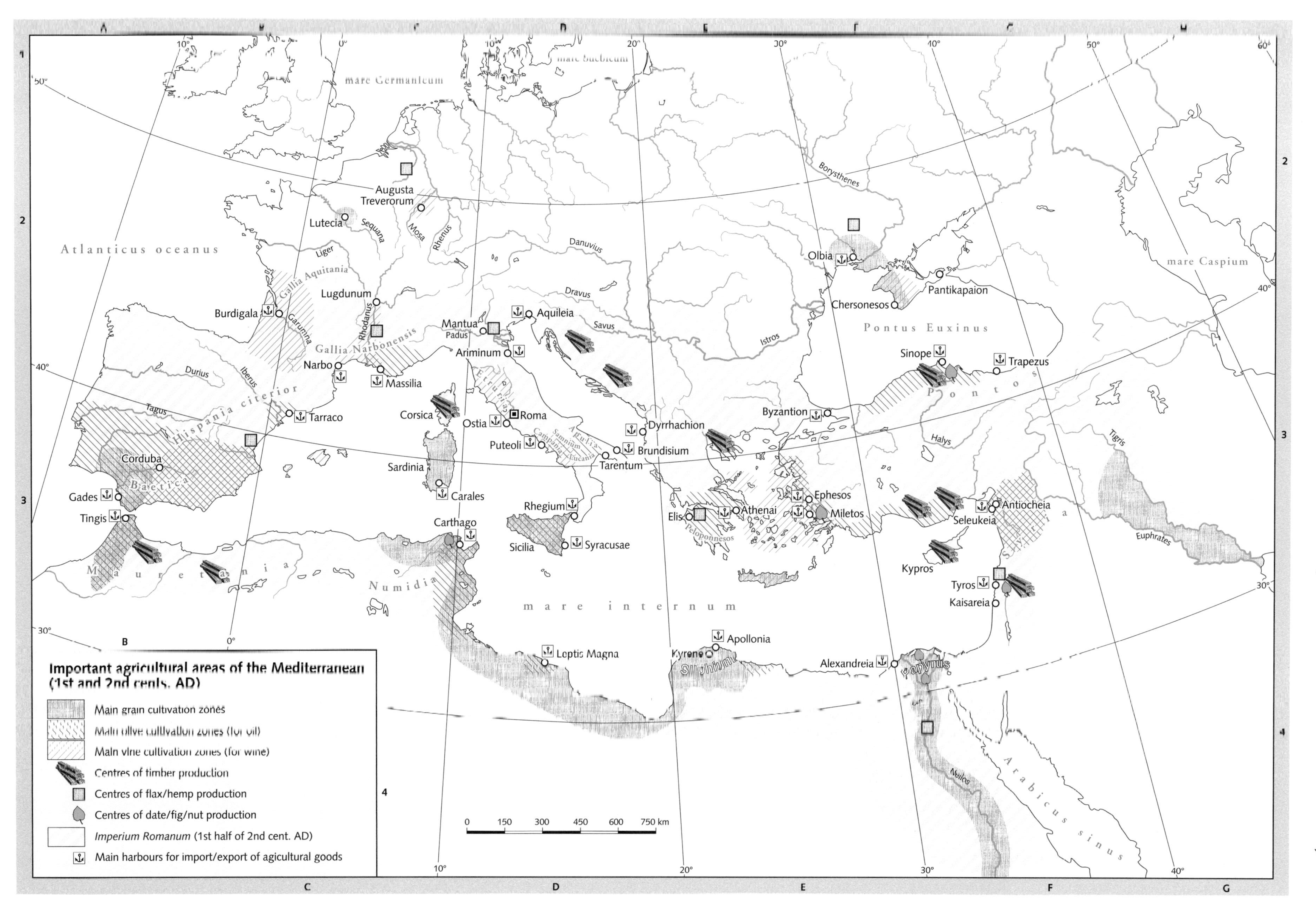
Important agricultural areas of the Mediterranean (1st and 2nd cents. AD)
Main grain cultivation zones
Main olive cultivation zones (for oil)
Main vine cultivation zones (for wine)
Centres of timber production
Centres of flax/hemp production
Centres of date/fig/nut production
Imperium Romanum (1st half of 2nd cent. AD)
Main harbours for import/export of agicultural goods
0 150 300 450 600 750 km
mare Germanicum
Atlanticus oceanus
mare Caspium
Pontus Euxinus
Pontos
mare internum
Arabicus sinus
Augusta Treverorum
Lutecia
Sequana
Mosa
Rhenus
Liger
Gallia Aquitania
Lugdunum
Burdigala
Garumna
Rhodanus
Gallia Narbonensis
Narbo
Massilia
Durius
Iberus
Tagus
Hispania citerior
Tarraco
Corduba
Baetica
Gades
Tingis
Mauretania
Numidia
Mantua
Padus
Ariminum
Aquileia
Danuvius
Dravus
Savus
Istros
Corsica
Etruria
Ostia
Roma
Puteoli
Campania
Samnium
Apulia
Lucania
Sardinia
Carales
Carthago
Rhegium
Sicilia
Syracusae
Tarentum
Brundisium
Dyrrhachion
Leptis Magna
Byzantion
Elis
Peloponnesos
Athenai
Ephesos
Miletos
Kyrene
Apollonia
Alexandreia
Olbia
Borysthenes
Chersonesos
Pantikapaion
Sinope
Trapezus
Halys
Antiocheia
Seleukeia
Kypros
Tyros
Kaisareia
Tigris
Euphrates
Neilos

Trade routes in the Roman Empire (1st–3rd cents. AD)

The map shows the most important connections of transregional and overseas trade in the Roman Imperial period. The patterns of long-distance trade shifted several times, and so accordingly did trade routes. This map shows the situation during the Principate (1st – 3rd cents. AD).

I. Maritime trading routes

The map shows only the most important maritime trading routes, and does not include coastal shipping. The maritime routes shown on the map are not exact descriptions of ancient shipping routes. For one thing, we simply do not know exactly what routes were taken over the high seas. All that we have are destinations and landmarks for some shipping routes. For another, the north-westerly winds that prevail in the Mediterranean in summer often meant that outward and homeward routes had to be different. The square rigging of ancient ships coped well sailing with the wind or in a crosswind, but had difficulty sailing close to the wind or against a headwind. Consequently, although they could sail, say, south from the Straits of Messina (*Fretum Siculum*) past Crete and reach Alexandria with ease in seven to nine days, the route back from the east followed a quite different course: it went via Cyprus, and from there against the wind via Crete to the Straits of Messina. This passage might well take as much as two months. Similarly, from Gades to Ostia took just seven days or so, and from Narbo to Ostia only three days. But the return voyages against the wind had to use different routes here, too, and took much longer. Only voyages across the prevailing wind would usually have taken place along the same route in both directions and over approximately the same time (Rhodes – Alexandria: 4 days – Diod. Sic. 3,34; Africa – Ostia: 2 days – Plin. HN 19,3). It is not possible to show more than one possible passage between two major ports in a map without sacrificing clarity. The necessary compromise is to show the trade flows between individual harbours in a single direct line, even though the actual passage followed quite different routes under certain circumstances.

II. Land trading routes

Trade on land made use of a plethora of tracks and roads, as well as navigable inland waterways and canals.

The relative cost-effectiveness of transportation by ship using inland waterways compared to the (in principle) much more expensive land transport is the subject of lively scholarly debate. The sources (cf. e.g. Str. 4,1,2; Tac. Ann. 13,53; Plin. Ep. 10,41) suggest that inland waterways were preferred for transportation. However, not every region of the Empire was so blessed with navigable inland waterways as Gaul, Germania or the Iberian Peninsula. The most important rivers for shipping were the Anas/Guadiana, Tagus/Tajo (Tejo), Durius/Duero (Douro), Iberus/Ebro, Baetis/Guadalquivir, Garumna/Garonne, Liger/Loire, Sequana/Seine, Rhodanus/Rhône, Arar/Saône, Dubis/Doubs, Mosella/Moselle, Rhenus/Rhine, Danuvius/Istrus/Danube, Padus/Po, Euphrates/al-Furāt, Tigris/Diğla, Neilos/Nile. A combination of waterways and land routes forms the basis for the depiction of land trade routes on this map. The routes shown often follow ancient roads, but the map does not depict trading routes exactly. The routes often also follow navigable inland waterways, leaving them where they cease to be navigable or jumping to the nearest nearby navigable inland waterway along a land route. It must be expected, at least, that where a trading route here depicted followed a river, the waterway was always preferred.

III. Goods traded

The Empire's great cities, especially Rome, depended on the mass importation of grain, the staple foodstuff. Accordingly, the trade routes from the agricultural surplus-producing regions and their ports (Alexandria, Carthage, Gades, Tarraco, Narbo) to Puteoli and Ostia, and from there to Rome, were of supreme importance. It must be assumed that in terms of volume, maritime trade, too, was overwhelmingly dominated by the trade in foodstuffs.

Another important trading category was the metal trade. The raw materials for metal-processing industries came from Spain (iron, gold, silver, lead), Gaul (iron, tin, silver), Britain (copper, gold, iron, lead, tin), Germania (copper), Noricum (iron), Illyria (silver, gold, iron), Macedonia (gold), the Black Sea region (gold, copper), Cyprus (copper), southern Egypt (gold) and Mauretania (copper). As a whole, then, most important metal resources tended to lie on the peripheries of the ancient Mediterranean. But the centres of metal-processing industry were (at least in the Hellenistic and Roman periods) in Italy, Greece and Egypt. The smelted crude metals had to be transported from their places of extraction to the areas where metals were worked. In two particular cases, the raw materials for bronze (copper and tin) and brass (copper and zinc) sometimes had to be combined from different source locations. It is striking in this regard that the much sought-after metals from Britain, northern Spain and Gaul were less often transported over the capricious and much-feared Atlantic and into the Mediterranean through the Straits of Gibraltar, but were usually sent down the north-south routes along the navigable rivers of Gaul and Spain.

Other refined mass products show similar patterns of trade to metals. For example, glass and papyrus, the most important writing medium of antiquity, came from Egypt, purple textiles from Phoenicia (Syria/Judaea), wool and flax/linen from Cilicia (Tarsus), Phrygia and Miletus, and fine pottery at first from northern Italy, later also from Gaul and the Rhineland. These items were produced in one location, but were in demand everywhere. Luxury goods had to an even greater extent been traded since the Hellenistic period throughout the Mediterranean, and were sometimes imported from far afield. Among these goods were jewellery, objects carved from rare woods or ivory, intoxicating substances, unguents, perfumes, cosmetics, gems and silk; all were from the Hellenistic period mostly obtained through an increasing Orient trade, which reached southern India, Siberia and China using caravan routes starting from Egypt and Syria (Palmyra). Other routes went from North Africa and Egypt, deeper into Africa and to Abyssinia.

The origins of fine textiles and luxury goods were mostly concentrated in the eastern Mediterranean. Seleucia, near Antioch, was the most important export harbour for these products. The route from Seleucia over the Aegean to the Straits of Messina was therefore another particularly important maritime link.

The sources

The sources for a map of trans-regional trade in Roman Antiquity are extraordinarily diverse. First worthy of mention are the writings of the ancient geographers, e.g. the 'Geography' (*Geographika*) of Strabo (63 BC – *c.* AD 19), the geography (*De chorographia*) of Pomponius Mela (1st cent. AD), the 'Description of Greece' (*Periegesis hellados*) of Pausanias (2nd cent. AD) and the anonymous *Expositio totius mundi et gentium* (4th/5th cents. AD), an ancient economic geography. As well as describing landscapes, these works also give information on trade and long-distance trade in the regions they describe. There are also the *itineraria* (itineraries), route descriptions for travellers, with some commentary, giving details of roads, distances, stations, etc. The most substantial itinerary is the *Itinerarium provinciarum Antonini Augusti*, a road register of the 3rd cent. AD: it details not only the roads of the entire Empire, but also, in an appendix, the most important maritime routes. Also informative is the *Itinerarium Burdigalense*, in which a pilgrim describes his journey from Bordeaux to Jerusalem, undertaken in the early 4th cent. AD. Another important source for our purpose is the *Tabula Peutingeriana*, a medieval copy of an ancient road map probably completed early in the 5th cent. The *periploi* and *stadiasmoi*, in a manner of speaking 'Itineraries for Seafarers', conversely mostly served coastal shipping. Other ancient sources not directly dealing with geography or trade can also provide much help. For example, when the Emperor Diocletian attempted in AD 301 to rein in the Empire's runaway inflation by setting compulsory maximum prices, the edict included freight tariffs on the most important maritime trading routes. Of course, there are countless scattered references to long-distance trade throughout the whole of ancient literature, fictional and non-fictional. For example, Tacitus in his *Annales* and *Germania* offers numerous references to trade in Gaul, Germania and Britain. Archaeological scholarship can help in many ways to narrow down the locations of trading routes often only roughly sketched in ancient literature. For instance, shipwrecks, harbours, lighthouses and storage facilities, etc., can reveal shipping routes, and the remains inland of harbour facilities, shipwrecks, storage buildings and canals can also give indication of the navigability of rivers. Finds of road sections, milestones and trading or storage buildings can show the routes and significance of roads. Finally, inscriptions have survived in comparative plenty. Funerary inscriptions mostly name traders or shipping agents, and often even give traders' commercial speciality and location, or an agent's area of operation. In some cases, inscriptions indicate traders' associations, their specialities, importance, area of operations and regional scope, or carriers' *collegia* and their transport routes. All these pieces of epigraphic evidence can be of great help in rounding off our picture of trade in Roman antiquity.

→ Map p. 201

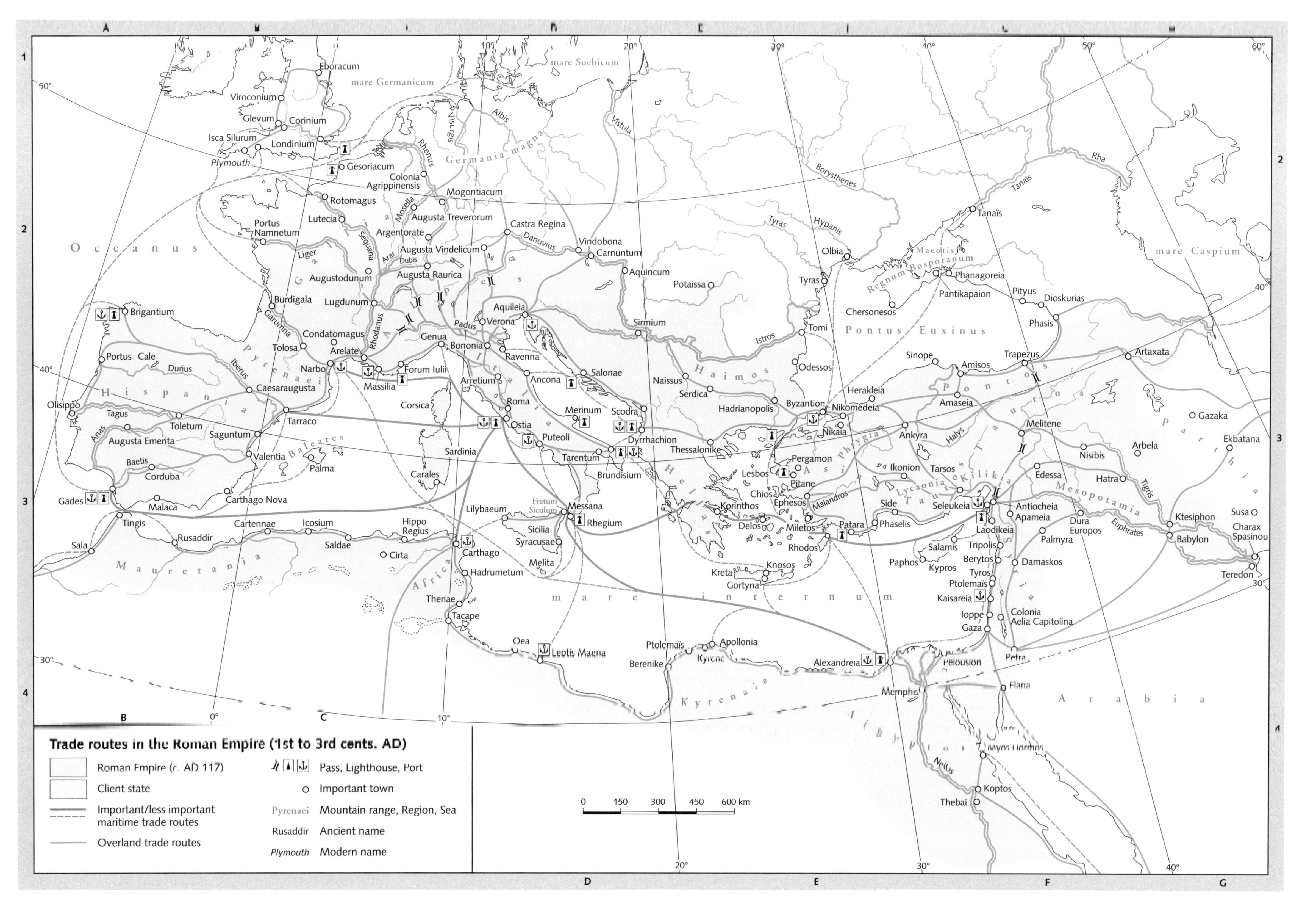
Trade routes in the Roman Empire (1st to 3rd cents. AD)
Roman Empire (c. AD 117)
Client state
Important/less important maritime trade routes
Overland trade routes
Pass, Lighthouse, Port
Important town
Pyrenaei Mountain range, Region, Sea
Rusaddir Ancient name
Plymouth Modern name
0 150 300 450 600 km
Oceanus
mare Germanicum
mare Suebicum
mare Caspium
Pontus Euxinus
mare internum
Germania magna
Hispania
Mauretania
Africa
Kyrenaica
Arabia
Parthia
Mesopotamia
Eboracum
Viroconium
Glevum
Corinium
Isca Silurum
Londinium
Plymouth
Gesoriacum
Colonia Agrippinensis
Mogontiacum
Rotomagus
Lutecia
Portus Namnetum
Augusta Treverorum
Argentorate
Augusta Vindelicum
Augusta Raurica
Augustodunum
Castra Regina
Vindobona
Carnuntum
Aquincum
Potaissa
Burdigala
Lugdunum
Condatomagus
Tolosa
Arelate
Narbo
Massilia
Forum Iulii
Genua
Bononia
Verona
Aquileia
Ravenna
Arretium
Ancona
Salonae
Sirmium
Naissus
Serdica
Brigantium
Portus Cale
Olisippo
Tagus
Toletum
Augusta Emerita
Corduba
Gades
Malaca
Carthago Nova
Caesaraugusta
Tarraco
Saguntum
Valentia
Palma
Corsica
Sardinia
Carales
Roma
Ostia
Puteoli
Merinum
Scodra
Dyrrhachion
Tarentum
Brundisium
Thessalonike
Hadrianopolis
Byzantion
Nikomedeia
Nikaia
Herakleia
Tingis
Sala
Rusaddir
Cartennae
Icosium
Saldae
Cirta
Hippo Regius
Carthago
Hadrumetum
Thenae
Tacape
Oea
Leptis Magna
Lilybaeum
Sicilia
Syracusae
Messana
Rhegium
Melita
Korinthos
Delos
Lesbos
Chios
Ephesos
Pergamon
Pitane
Miletos
Rhodos
Knosos
Kreta
Gortyna
Ptolemaïs
Apollonia
Berenike
Kyrene
Alexandreia
Memphis
Pelousion
Petra
Flana
Myos Hormos
Koptos
Thebai
Tyras
Olbia
Tomi
Odessos
Chersonesos
Pantikapaion
Phanagoreia
Tanaïs
Pityus
Dioskurias
Phasis
Sinope
Amisos
Trapezus
Amaseia
Artaxata
Ankyra
Ikonion
Tarsos
Side
Phaselis
Patara
Melitene
Edessa
Nisibis
Arbela
Hatra
Gazaka
Ekbatana
Seleukeia
Antiocheia
Apameia
Laodikeia
Salamis
Paphos
Kypros
Tripolis
Berytos
Tyros
Ptolemaïs
Kaisareia
Ioppe
Gaza
Damaskos
Palmyra
Colonia Aelia Capitolina
Dura Europos
Ktesiphon
Babylon
Susa
Charax Spasinou
Teredon
Rhenus
Rhodanus
Danuvius
Padus
Istros
Borysthenes
Tanaïs
Rha
Tigris
Euphrates
Nelus
Halys

From the Mediterranean to India. Overland and maritime routes in the 1st/2nd cents. AD

From the Mediterranean to India. Overland and maritime routes in the 1st/2nd cents. AD

It has seldom been mere curiosity that has stimulated people to seek out and explore new regions. The journeys of exploration shown on the maps on pp. 8-9 almost all bear the hallmarks of economic expediency. This is particularly true of the recorded visits to India. It was often monarchs of Egypt who commissioned such voyages. From Egypt above all came the decisive impulse to convey knowledge and goods between the Mediterranean and Indian worlds. Both land journeys and sea voyages had their own particular risks. But the journey by sea was quicker and, in the last analysis, less hazardous, whether it went from the Mediterranean through the Red Sea and across the Indian Ocean, or from Charax Spasinou in the Euphrates-Tigris delta region through the Persian Gulf and so on to the west coast of India.

The earliest information conveyed to the Greeks about India came from the travel account of Scylax of Caryanda, who made a voyage of discovery to the west of India between 519 and 512 BC on behalf of the Persian King Darius I (522–486 BC). He travelled by land to the upper reaches of the Cophen/Kabul, from there by boat to the confluence with the Indus, down the Indus to the Erythra thalatta/Indian Ocean, on to the southern coast of Arabia and south-west into the Red Sea to Egypt. His contemporary Hecateus of Miletus and, slightly later, Herodotus (5th cent. BC) took the remarks on India in their respective works from the account of Scylax (cf. → map pp. 8 f.). The state of knowledge of India in the west hardly then developed further until the time of Alexander the Great. Even Ctesias, the personal physician of the Persian King Artaxerxes (405/04–359 BC), contributed nothing of importance in this regard in his history.

With Alexander (356–323 BC) and the Diadochi, however, new experiences of India came to the west. Alexander's campaign and the Seleucid embassy traffic with Palimbothra, the Maurya residence on the Ganges, via such as Megasthenes (*c.* 350–290 BC) and Daimachus (3rd cent. BC), sparked new interest and new involvement with India and matters Indian. The subcontinent was now reached either by land in the footsteps of Scylax to the upper reaches of the Cophen and the Indus, or by sea from the mouth of the Euphrates and Tigris through the Persian Gulf, then across the Indian Ocean to the west coast of India. The increasing insecurity and difficulty of the land route to India were in the main attributable to the problematic political conditions in the wake of the dissolution of Seleucid rule and the Seleucids' displacement by various dynasts in the course of the 3rd cent., and especially to the rise of the Parthians and the development of tensions between the Parthians and Rome. The land route to China along the so-called 'Silk Road' (this map shows the western section, Seleucia – Nisibis – Ecbatana – Hecatompylus – Antioch – Baktra) was even worse affected.

The safety to some extent provided for the land route to India by the Achaemenid and Seleucid kingdoms had vanished, and routes to India moved seaward by force of necessity. But it was only when the regularity of the monsoon winds was understood and the risks of the sea passage to India had become calculable that the land route actually lost its importance. Voyages now took advantage of the summer south-west monsoon to cross the Indian Ocean from the southern coast of Arabia to India, returning with the winter north-east monsoon. This discovery was attributed to Hippalus, and the wind that blew towards India was named after him. The dating of the discovery is disputed, but there are good arguments for placing it in the late 2nd cent. BC.

Egypt and the Arabs at Bal el-Mandeb were in a position to control sea traffic through the Red Sea and Indian Ocean. In practice, though, this was uncomplicated, because the Ptolemies of Egypt were tied to Rome in a close client relationship from 168 BC at the latest. Egypt was a Roman province from 30 BC, and the Romans had a keen interest in the smooth running of commercial traffic to India. The embassy traffic between Indian dynasts and the imperial court at Rome from the reigns of Augustus and Claudius is well documented in the sources. The Arabs of the Ḥimyaritic and Sabaean kingdoms, meanwhile, earned money from the intermediary trade they operated in their ports, and from harbour dues: it could only be in their interests for such India voyages to proceed smoothly.

Trading volumes

The high intensity of the India trade even in the reign of Augustus is illustrated by Strabo (2,5,12), according to whom 120 ships a year left Myos Hormos on the Red Sea to voyage to India. Only a few, wealthy buyers, however, were able to afford goods from India, which long transport routes and all manner of duties made painfully expensive. A remark in Pliny (HN 12,84) may allow an idea of trade volumes in the mid 1st cent. AD: 100 million *sesterces*, he says, were spent every year on goods from India, China and Arabia. This trade balance was certainly not equalized.

Goods traded

Trade between the two civilizations can be demonstrated by archaeology, with finds in India and Sri Lanka (pottery, lamps, glassware and Roman coins) and in the west (ivory statuette in Pompeii). Literary references to the India trade are especially numerous. Papyri, too, attest to this traffic. One particularly good piece of evidence is a Vienna papyrus of the 2nd cent. AD, recording a contract of sale drawn up in Muziris in India and Alexandria in Egypt. This exchange of goods did not only involve entrepreneurs from the Mediterranean, but also Indian ship owners. India traders mostly imported spices (aloe, amomum, pepper), drugs such as sweet flag (*Acorus calamus*), medicinal plants such as *anchousa*, precious stones and pearls, ivory, ebony, obsidian, onyx and mother-of-pearl. To India they exported pottery, lamps, glassware, textiles, wine and corals.

Trading routes

The ports used by trading ships in India were Patala, and, to the south-east, Barbaricum at the mouth of the Indus, both presumably in the delta region, also Barygaza/Bharuch, Calliena/Kalyan, Zigerus/Jaygarh and the towns of Muziris (near Cranganur?) and Nelcynda in the littoral territory of Limyrice/Kerala (these last two no longer localizable because of major coastal changes). All, then, were on the west coast of India. Only from the 2nd cent. AD did Greek India traders also find their way to Taprobane/Sri Lanka and the Indian east coast. They began their voyages to India at the ports on the west coast of the Red Sea, e.g. Berenice, but mostly Myos Hormos, from where goods were transported by camel to Coptus and thence down the Nile by ship to Alexandria. The extensive land and sea network of routes that linked the Mediterranean with India in the 1st and 2nd cents. AD can be shown with the help of the 'Geography' of Strabo (*c.* 63 BC – AD 24) and the 'Natural History' of Pliny (AD 23–79). Also very helpful are the list-like descriptions of the *Periplus Maris Rubri* (1st cent. AD), the 'Geography' of Ptolemy (2nd cent. AD) and the *Cosmographia* of an anonymous monk of Ravenna (Geographus Ravennas, 7th/8th cents.).

Pliny (Plin. HN 6,100–106), for example, knew of three sea routes from Alexandria to India:

- The first route went from the cape of Syagrum (Ras Fartak) on the south coast of Arabia, and ended at Barbaricum at the mouth of the Indus.
- The second also went from Syagrum, ending in Zigerus on the west coast of India, south of Barygaza.
- The third route went from Ocelis in Arabia (probably Khor Ghurayrah on the Bal el-Mandeb) to Muziris in the far south-west of India.

Pliny describes the third route in detail. It started at Juliopolis (not localized) east of Alexandria, where one boarded a ship which could reach Coptus, known as a storage place for goods from India, in twelve days if the (Etesian) wind was favourable. It continued (mostly by night because of the heat of the day) in four stages by camel through the desert to Berenice, a landing-place on the Red Sea, which was reached on the twelfth day. Another ship was boarded, and this put to sea in summer, before the heliacal rising of Sirius (18 July), or immediately after it. After some thirty days, the ship reached the Arabian coast either at Ocelis, Cane/Bir Ali or Muza (near Salalah). The last of these ports was only used by merchants trading in incense and Arabian perfumes. India traders preferred to use Ocelis, from where they could reach Muziris, the nearest emporium on the south-west coast of India, in forty days with the Hippalus (the summer monsoon). However, according to Pliny, this port was best avoided because of the pirates active in the area at Nitrias (near Mangalore?). Moreover, Muziris was not directly on the sea, and the anchorage was on an island off the coast. Goods had therefore to be taken ashore by barge. Nor were many goods available in exchange at Muziris. A better port in the same region, he says, was Barace (Bakare, not localized). Pepper was carried there in boats hollowed out of a single tree.

Even though the sea route to India was increasingly preferred from the 1st cent. BC, there were still contacts to the Far East that went by land with the caravan trade, depending on political conditions in the intervening Iranian territories. Palmyra, for instance, systematically expanded its westward and eastward commercial contacts from the early Imperial period. The city supported trading posts on the downstream Euphrates at Vologesias, Seleucia, Babylon and Charax Spasinou. There, goods could be transported on to – and in from – India by sea. This overland link to India was cut, however, when the 'Palmyrene Empire', which had split from the Roman Empire under Zenobia and her son Vaballathus, was restored to Rome by the Emperor Aurelian (270–275) in 272: Palmyra was destroyed, and the Palmyrene outposts on the Euphrates were taken over by the Sassanids.

→ Map p. 203

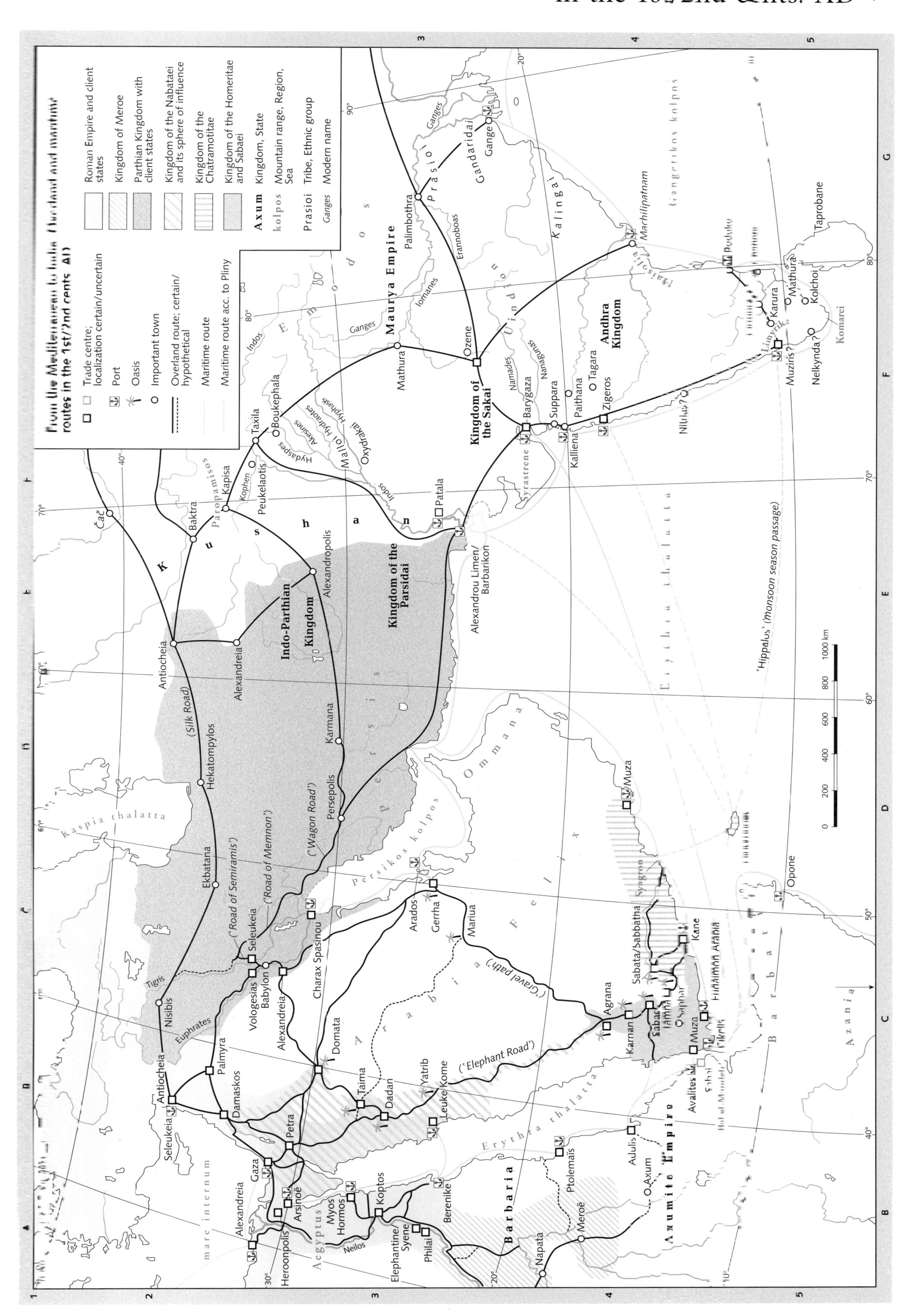
From the Mediterranean to India. Overland and maritime routes in the 1st/2nd cents. AD
Trade centre; localization certain/uncertain
Port
Oasis
Important town
Overland route; certain/hypothetical
Maritime route
Maritime route acc. to Pliny
Roman Empire and client states
Kingdom of Meroe
Parthian Kingdom with client states
Kingdom of the Nabataei and its sphere of influence
Kingdom of the Chatramotitae
Kingdom of the Homeritae and Sabaei
Axum Kingdom, State
kolpos Mountain range, Region, Sea
Prasioi Tribe, Ethnic group
Ganges Modern name

The *Imperium Romanum* in the reign of Septimius Severus (AD 193–211)

As a ruler, Septimius Severus was profoundly shaped by his long struggle to gain the throne. Almost four years elapsed from his acclamation as emperor by the Pannonian legions at Carnuntum (where he had resided as *legatus Augusti pro praetore* of Pannonia superior since 191) in April 193 until his acceptance across the whole empire following his victory over Clodius Albinus at Lugdunum on 19 February 197. The lessons learned during this period made themselves felt in many of his governmental measures. We should always be on our guard against monocausal explanations, but there are undoubtedly motives of greater and lesser weight for human actions, and in the case of Septimius Severus, the trauma of these four years was indisputably a driving force behind his actions as emperor.

Measures that were clearly dictated by the circumstances of a civil war and are thus to be seen only as reactive rather than constructive actions are not considered here – for instance, none of his rivals in the struggle for the throne, neither Didius Iulianus (d. June 193), Pescennius Niger (d. April 194) nor Clodius Albinus and his followers (d. February 197), survived that struggle, but were all in the end killed on the orders of Septimius Severus. It cannot be denied, though, that a certain biographical relevance shines through other matters ostensibly motivated by the simple urge to retaliate.

Antioch and Byzantium, for example, suffered ferocious retribution (long recalled by ancient historians) at Severus' hands for having sided with Pescennius Niger.

Similarly, Niger's Arabian clientele was the real target of his first campaign against the Parthians, which followed his victory over Pescennius Niger in 195 and led to the foundation of the province of Osrhoene (→ map p. 181). The second Parthian campaign, just a few years later, was in no way a political necessity. Septimius Severus needed only small provocation to move against the Arabs again – it may be that they had not yet accepted the fact of his dominion. Such shadows of the past did certainly still exist: Barsemius, the ruler of Hatra, a former confidant of Pescennius Niger, survived two very determined attempts by the emperor to subjugate him – but Hatra by now no longer posed much of a threat.

The division of the provinces of Syria (→ map p. 181) and Britannia (→ map p. 193) also struck at the clienteles of his former foes, Pescennius Niger and Clodius Albinus respectively. These two reforms are generally seen as measures intended to prevent an excessive concentration of power in the hands of individual governors. It would be proved, although Septimius Severus would not live to see it for himself, that such flash points could still flare up again, when fanned by Zenobia (AD 270–272; → map p. 221) in Syria and Carausius (AD 286–293) in Britain. Septimius Severus continued to devote much attention to these two provincial regions. The second Parthian campaign (197–199) led to the creation of the province of Mesopotamia (→ map p. 181), while the major action in Britain against the Maeatae and Caledonii, north of the Antonine Wall (briefly reoccupied in this period) (→ map p. 210), was the last campaign of his life – he died on 4 February 211 at Eboracum (York).

His chief concern and the chief object of his solicitousness was the army. He replaced the entire Praetorian Guard at Rome with his own legionaries of the *legio XIV Gemina* from Carnuntum, personally devoted to him. By doing so, he secured the crucial theatre, so important to the legitimacy of an emperor: the imperial court at Rome. Another drastic measure served to protect it. In preparation for his second Parthian campaign, Septimius Severus had levied three new legions (*legiones I, II, III Parthica*). The *legiones I* and *III* remained stationed at Singara or (probably) Resaena after the end of the war, but against all custom, the *legio II Parthica* took up permanent quarters in Italy, indeed at the very brink of Rome, at the foot of the *mons Albanus*.

The advancement Septimius Severus gave to the law, as embodied by Papinian, Ulpian and Paulus, is probably to be seen as a constructive contribution to state-building on his part, and not as a biographically-conditioned reaction to political events. All three men were renowned jurists. Papinian (Aemilius Papinianus) was head of the chancellery *a libellis*, and Praetorian prefect from 205 to 211. Ulpian (Domitius Ulpianus), was a pupil of Papinian's, and himself rose to become Praetorian prefect in 222. Like Ulpian, the jurist Iulius Paulus was an assessor in the auditorium of Papinian while the latter was Praetorian prefect. These three individuals, through their substantial bodies of writing and their legal activities in positions of high responsibility, made crucial contributions to the development of Roman law.

Another field in which this emperor set new standards was that of the military. Septimius Severus had learned and now knew this profession from the bottom up, and he exploited to the full the opportunities available to him as ruler to implement reforms that were long overdue. His strategy of frontier definition was strictly defensive in intent. He sought to bring the defensive system designed by Hadrian up to date, as attested by the fortifications built under his aegis in Africa and the east, and on the Rhine and Danube. Even the legion stationed in the Alban Hills fitted into a system that would lay the foundations of the army of Late Antiquity, with its *comitatenses* and *limitanei*, i.e. mobile reaction forces and stationary provincial and frontier troops.

Date	Event	Sources
June 193	Abolition of the Praetorian Guard	Cass. Dio 74,1,1 f.; Herodian. 2,13,1–12; 2,14,3; Zon. 12,8
193–195	Siege, conquest of Byzantium; retribution	Herodian. 3,6,9; Cass. Dio 74,12,1–14,6
194	Retribution against Antioch	Herodian. 3,6,9; SHA Sept. Sev. 9,4
	Division of province of Syria: Syria Phoenice and Syria Coele	SHA Hadr. 14,1; Dig. 50,15,1
195	First Parthian War	Cass. Dio 75,1,2 f.; SHA Sept. Sev. 9,9 f.
	Province of Osrhoene established	Cass. Dio 75,3,2; ILS 1353
before 197	Levy of new legions	Cass. Dio 55,24,4
197	Division of province of Britannia: Britannia superior (south) and Britannia inferior (north)	Herodian. 3,8,2
197–199	Second Parthian War	Herodian. 3,8 f.; Cass. Dio 75,9–12; SHA Sept. Sev. 14–16
197	Conquest of Ctesiphon; re-establishment of province of Mesopotamia; capital perhaps Nisibis	SHA Sev. 16,1 f.; Cass. Dio 75,9,3 f.; Mesopotamia already province during 1st Parthian War in 195: Cass. Dio 75,3,2; Nisibis: ILS 1331; 1388; 8847; 9148
198	Unsuccessful siege of Hatra	Cass. Dio 75,10,1; Herodian. 3,9,3–7
between 198 and 203	Numidia separated from Africa proconsularis and made province in its own right; capital Lambaesis	AE 1959, 181; Lambaesis: ILS 2376; CIL 8, 2739; 18083
199	Unsuccessful siege of Hatra	Cass. Dio 75,11,1–75,13,1
208–211	British campaign	Herodian. 3,14 f.; Cass. Dio 76,11–15

Map

Throughout, the map shows geographical information of significance to the biography of Septimius Severus, including matters not mentioned in its commentary.

Literature

A.R. Birley, The African Emperor Septimius Severus, [2]1988; E. Birley, Septimius Severus and the Roman Army, in: Epigraphische Studien 8, 1969, 63–82; J. Hasebroek, Untersuchungen zur Geschichte des Kaisers Septimius Severus, 1921; T. Honoré, Ulpian, 1982; E. Kettenhofen, Die syrischen Augustae in der historischen Überlieferung, 1979; J.D. Leach, J.-J. Wilkes, The Roman Military Base at Carpow, Perthshire, Scotland: Summary of Recent Investigations (1964–70, 1975), in: J. Fitz (ed.), Akten des 11. Internationalen Limeskongresses, 1977, 47–62; D. Liebs, Römische Jurisprudenz in Africa, 1993; Id., Mein Ulpian, in: M. Schauer (ed.), FS W. Suerbaum, 2003, 74–81; V. Marotta, Ulpiano e l' Impero 1, 2000; G.J. Murphy, The Reign of the Emperor L. Septimius Severus from the Evidence of the Inscriptions, 1945; E. Ritterling, legio, in: RE 12 (1924), 1211–1829; R.E. Smith, The Army Reform of Septimius Severus, in: Historia 21, 1972, 481–500; J. Spielvogel, Septimius Severus, 2006.

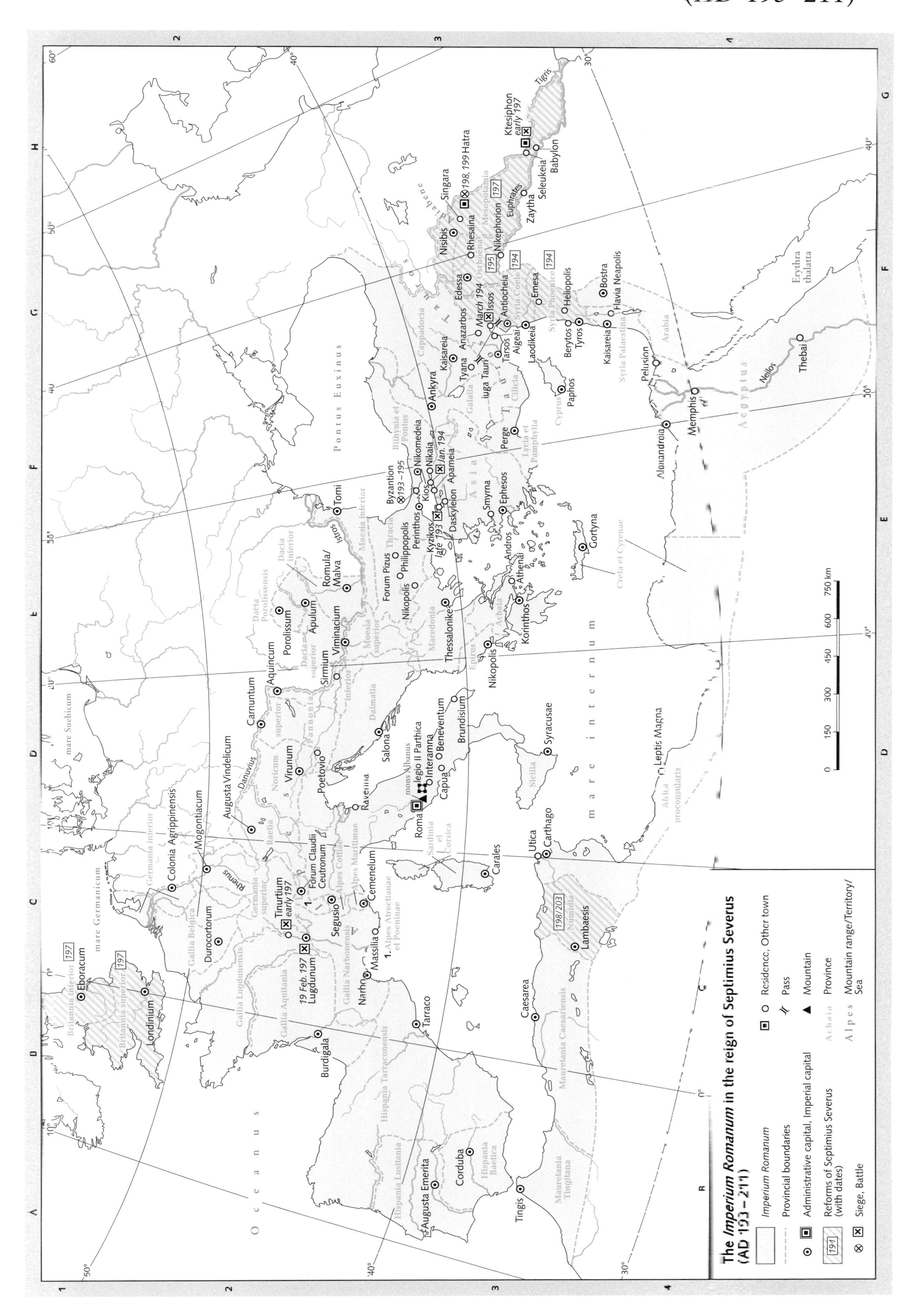
The *Imperium Romanum* in the reign of Septimius Severus (AD 193–211)
Imperium Romanum
Provincial boundaries
Administrative capital, Imperial capital
Reforms of Septimius Severus (with dates)
Siege, Battle
Residence, Other town
Pass
Mountain
Province
Mountain range/Territory/Sea

Distribution of the legions and the frontiers of the Roman Empire

I. Organization of the army from the time of Augustus

In the final days of the Roman Republic, there were probably 68 legions under arms. Including auxiliaries, this is estimated as a force of 350,000 men – more than were required for the maintenance of the peace and internal order of the Empire and its protection at external frontiers. After his victory over Cleopatra and Mark Antony, Augustus, then still C. Iulius Caesar (Octavianus), released a large part of this force, compensating veterans with confiscated wealth and the Egyptian war booty and settling them in various colonies. He then reorganized the whole of the Roman armed forces root and branch. The professional army into which the legions had developed since the Marian reforms now became a standing professional army. Soldiers served a fixed term of 20 years, and at honourable discharge they received a payment from a fund set up for this specific purpose, the *aerarium militare*. Some troops were stationed in and around Rome, but most were in the frontier provinces.

In 27 BC, Augustus assembled nine *cohortes praetoriae* of 500 to 1,000 men apiece, each under the command of its own *tribunus militum*, from 2 BC also under the overall command of two *praefecti praetorio*, and placed them at his own personal disposal, functioning as a kind of body guard. These troops were of elite status, as indicated by their higher pay and shorter term of service. Three of these cohorts were stationed in Rome itself, the other six for the time being in neighbouring towns. Tiberius (AD 14–37) then had a large barracks built east of the *mons Viminalis*, and others were added under Trajan (98–117) and Septimius Severus (193–211). In addition, three (later four) *cohortes urbanae* were founded for policing duties, and in AD 6 seven *cohortes vigiles* were set up for firefighting. Each consisted of 1,000 men.

Approximately 28 legions were stationed in the frontier provinces (25 after the Battle of the Teutoburg Forest in AD 9). Under Septimius Severus the figure was 33 legions. They had a name – to give them a sense of shared identity – as well as an identification number, and bore a legion emblem. Their normal strength was 5,000 men, who had to be Roman citizens and were commanded by a senatorial legionary legate of praetorian rank. These army forces were supplemented by auxiliary contingents of the allies (*auxilia*), whose numbers were roughly equivalent.

Augustus also reorganized the naval forces, establishing two fleets based at Misenum and Ravenna respectively. At first, they were commanded by *liberti Augusti*, later by equestrian *praefecti classis*. Other provincial fleets were created, even in Augustus' time, at Alexandria and on the Rhine and Danube. The following provincial naval units are attested for the reign of Septimius Severus: the *classis Britannica*, the *classis Germanica* on the Rhine, the *classis Moesiaca* on the Danube, the *classis Pontica* in the Black Sea, the *classis Syriaca* off the Levantine coast and the *classis Alexandrina* at Alexandria.

Our knowledge of the Roman military forces established by Augustus, arrangements which survived with only minor changes (e.g. by Septimius Severus) until the reign of Diocletian, comes from inscriptions, including a register of the legions under Septimius Severus (ILS 2288), and from the literature (Jos. Bl. 3,70–109; 115–126; Cass. Dio 55,23).

II. The distribution of the legions under Augustus and under Septimius Severus

Augustus distributed the legions on the basis of the level of threat to the various provinces. Three legions were stationed in the north of the Iberian peninsula, an 'unpacified' region in which the Asturians and Cantabrians were a constant source of trouble. No fewer than eight were stationed on the Rhine and in the north-western foothills of the Alps. This was a region where Augustus initially planned further expansion, but these plans were shelved after the defeat in the Teutoburg Forest in AD 9. That defeat, however, had only increased the intractability of this frontier region, so that withdrawing legions from it was unthinkable. In Illyricum (Dalmatia and military district), where Augustus himself had fought as *triumvir*, gathering relevant experience in the process, five legions were quartered. Two more legions were positioned down the Danube, where only recently M. Licinius Crassus had made striking territorial gains as far as the Danube while proconsul in Macedonia in 28 and 27 BC. The situation in Syria was difficult. This was where Augustus' enemies during the Civil War – Brutus and Cassius and then Mark Antony – had recruited their clienteles. The Parthians also posed a constant threat. Four legions were kept on here, albeit reduced from the original seven. Egypt's economic significance made it an extremely important province, the security of which Augustus had to guarantee, not least in view of domestic political considerations. Conflicts, some even provoked, necessitated various campaigns here against the Arabs and Ethiopians. Two legions were stationed here. Africa, Rome's 'bread basket', which was under senatorial administration, was also secured with three legions (later with only one) in consideration of the disorder stoked in the region by the Gaetuli and Garamantes.

Septimius Severus levied three new legions in preparation for his second Parthian campaign. This constituted a ten percent increase in the size of the army. One of these legions was stationed near Rome – an absolute departure from tradition – while the other two remained in Mesopotamia and Osrhoene to secure the territories recaptured. One effect of provincial reorganization was a measure primarily motivated by domestic political considerations: there were now no more than two legions in any of the frontier provinces. This is probably to be seen as a decision of principle to withhold from provincial governors the means to conduct an insurrection. To achieve it, not a single legion had to vacate its camp – Septimius Severus simply divided the provinces of Britannia and Syria to achieve the reduction of troops. Caracalla then separated the Danube legions in 214 by moving the provincial frontier between Pannonia superior and Pannonia inferior.

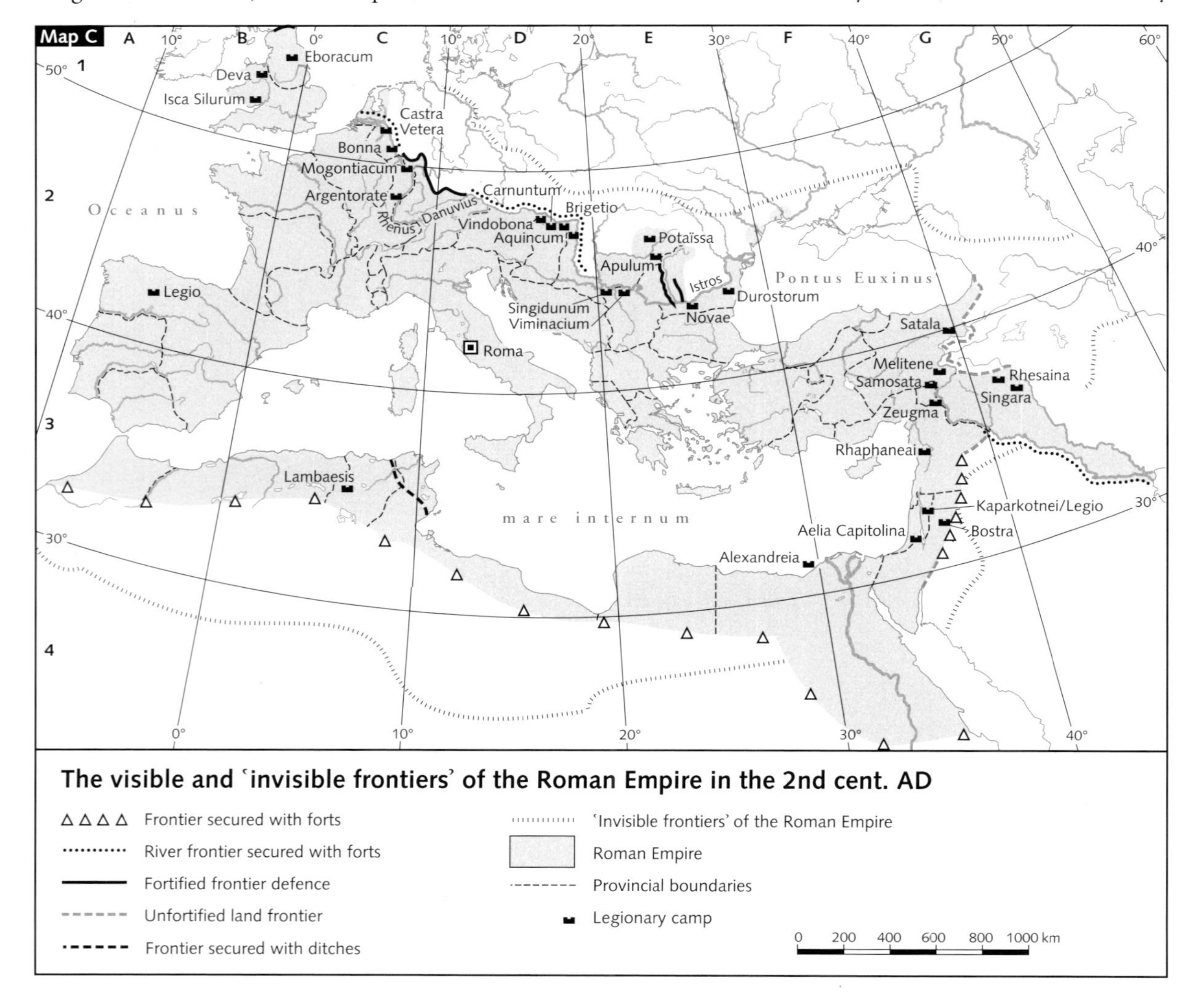

The visible and 'invisible frontiers' of the Roman Empire in the 2nd cent. AD

III. The visible and 'invisible frontiers' of the Roman Empire in the 2nd cent. AD

Ernst Kornemann, in a still-remarkable essay, describes the practice of securing the frontiers by creating a ribbon of client states, a practice dating back to the days of the Elder Scipio. To Tacitus (1st/2nd cents. AD), "the ocean and remote rivers were the boundaries of the Empire" (Tac. Ann. 1,9), the rivers concerned being the Rhine, Danube and Euphrates. The pseudonymous author of the biography of Hadrian (4th cent. AD) also writes of frontier "regions where the barbarians are held back not by rivers but by artificial barriers" (SHA Hadr. 12,6), referring to the artificial defensive ramparts such as the two walls in Britain, the Upper Germanic and Rhaetian *limes*, the Dacian *limes* and the North African *fossatum*. Finally, what would in the time of Augustus become the 'ribbon of client states' characterized by Kornemann had begun in rudimentary form as early as the 2nd Punic War (218-201 BC), and underwent various phases of expansion (Pompey, Caesar) to surround the Empire as a third method of securing its frontiers.

If Kornemann's directions for mapping the 'invisible frontiers' of the Roman Empire are followed, the extent of the Empire expands to a greater or lesser degree at various periods. Overall, though, the further extent is considerable – it includes Britain, Germany and the lands beyond the Danube, the northern shore of the Black Sea, where the Bosporan kingdom was a permanent presence among the Roman clientele, and Armenia to the east of the Euphrates, in Arabian territory. Such client relationships can also be found in Egypt and western parts of North Africa.

→ Maps pp. 210–213

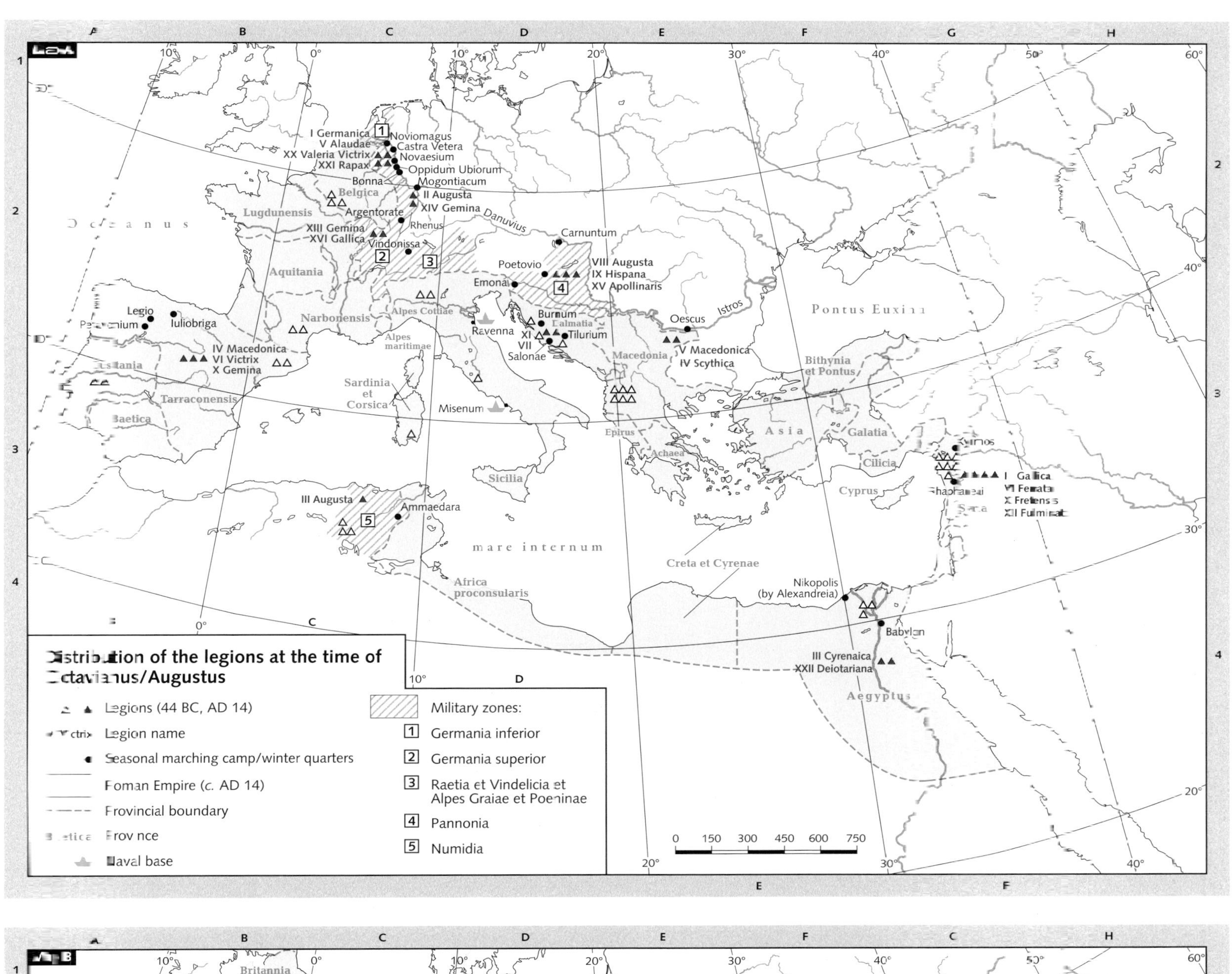

Distribution of the legions at the time of Octavianus/Augustus

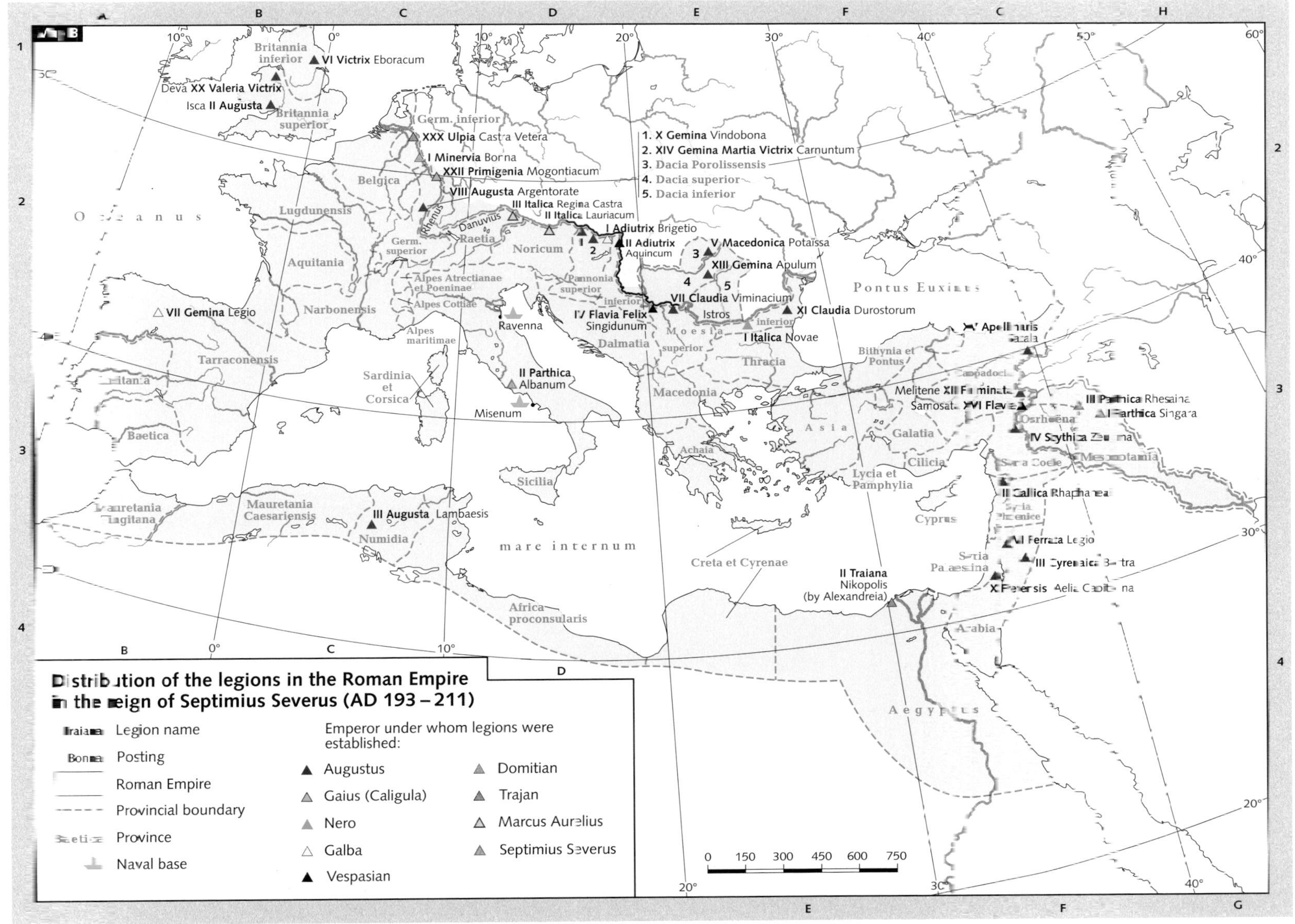

Distribution of the legions in the Roman Empire in the reign of Septimius Severus (AD 193–211)

The *limites* in Britannia, Germania and Raetia

In principle, the frontiers of Roman influence were as flexible as any particular neighbour permitted. Hence, the Roman Republic never had any external borders fixed by military installations. During the Principate, as the standing professional army developed, so also did an Empire-wide frontier control system. But this system, too, varied according to geographical, ethnic and political conditions. Some examples illustrate its diversity:

- the *Vallum Hadriani* (Hadrian's Wall) in Britannia, a defensive rampart and wall with military road (Stanegate) running alongside, 118 km long and protected by 14 forts
- the *Vallum Antonini* (Antonine Wall) in Britannia, a defensive rampart with military road running alongside, 60 km long
- the Upper Germanic-Rhaetian *limes*, a system of embankments, ditches and walls 548 km long, secured by approx. 900 turrets and 120 forts
- the Danube (river frontier), secured by forts
- the *limes Arabicus*, a frontier line controlled from forts, watch-towers and fortified towns by patrols along roads
- the Mauretanian border region in the northern Sahara, secured by an eclectic system of ditches and embankments.

It appears that the securing of frontiers was not planned centrally from Rome, but was arranged by the particular provincial governors. At times, the Romans would respond to threats on particular sections of frontier by setting up unusual concentrations of garrison bases; these would be abandoned again once the crisis had died down.

The sources

The literary sources fail us almost entirely on the subject of Roman frontier security. However, this informational void is filled by numerous inscriptions – soldiers' gravestones, military diplomas and milestones. The real basis for our knowledge of Roman frontier security is provincial archaeology. It has unearthed the ground plans of entire camps, revealing the construction of their buildings and fortifications and of their attendant camp villages. It is particularly revealing of the frontier walls, showing their chronology, structure and functions.

The map

The maps of the British and Germanic *limites* clearly demonstrate the ability of the Roman army command to respond flexibly to crisis situations as they arose. The construction of the *Vallum Hadriani* took into account the problem that adversaries might approach not only from the north, but also from the south, having first bypassed the wall's defences by sea. Particularly critical sections of the Rhine Delta and the region north of the lower Main had to be secured by unusually high concentrations of forts. Not the least of the reasons behind the several advances of the frontier in the Rhine-Danube region was the shortening of the militarily important link between the middle Rhine and the middle Danube. This took place in five phases:

I. 15 BC – AD 9 – Campaigns in the territory east of the Rhine and north of the Alps. The legionary camps of Castra Vetera (from 13 BC), Novaesium (from 16 BC), Mogontiacum (from 10 BC) and Augusta Vindelicum (*c.* 8 BC –AD 16) were established for these. The defeat in the Teutoburg Forest in AD 9 led to the abandonment of further plans, e.g. the conquest of Germany as far as the Elbe.

II. AD 9–50 – Withdrawal to the Rhine and Danube. Legionary camps were now set up at Colonia Agrippinensis (AD 10–43), Bonna (from AD 41–54), Argentorate (from AD 9) and Vindonissa (from AD 16/17), and these were linked together by roads and a chain of cohort forts. The link between the two river frontiers was shortened in the reigns of Tiberius and Claudius (*c.* AD 25–50) by the construction of the road via the forts at Riegel and Hüfingen.

III. AD 50–90 – This link was moved north in the 70s. It now went through the Kinzig Valley, and was secured by the construction of the Arae Flaviae military camp. At this time, the Romans also in places advanced east across the Upper Rhine, e.g. in the fertile Wetterau tract around Friedberg. The two military districts of Germania inferior and Germania superior were changed into provinces around AD 85.

IV. AD 90–120 – A new line of forts and a *limes* (in the sense of a frontier line secured by watch-towers and palisades) was built from the Wetterau along the Main and through the Odenwald, to protect a new link from Mogontiacum to Augusta Vindelicum via Lopodunum/Ladenburg and Stuttgart-Bad Cannstatt.

V. *c.* AD 160 – The *limes* was moved forward to the Miltenberg – Lorch line and expanded into a prestigious structure of frontier control.

→ Maps pp. 208 f., 212 f.

Sources

Britannia: SHA Hadr. 11,2; 12,6; SHA Pius 5,4; Flor. Epit. 2,30,26; A.K. Bowman, Life and Letters on the Roman Frontier. Vindolanda and its People, 1994; Id., J.D. Thomas, The Roman Writing Tablets from Vindolanda, 2 vols., 1983/1994.

Literature

D. Baatz, Der römische Limes, [3]1993; D.J. Breeze, The Northern Frontiers of Roman Britain, 1982; B. Campbell, The Roman Army 31 BC–AD 337. A Sourcebook, 1994; W.S. Hanson, G.S. Maxwell, Rome' s North West Frontier. The Antonine Wall, 1983; J. Heiligmann, Der 'Alb-Limes' , 1990; B. Jones, D. Mattingly, An Atlas of Roman Britain, 1993; L. Keppie, The Making of the Roman Army from Republic to Empire, 1984; D. Kortüm, Zur Datierung der römischen Militäranlagen im obergermanisch-rätischen Grenzgebiet, in: Saalburg-Jb. 49, 1998, 5–65; G. MacDonald, The Roman Wall in Scotland, [2]1934; A.S. Robertson, The Antonine Wall, [2]1973; M. Schmidt, Der römische Limes im Hohen Odenwald, am Mittleren Neckar und in der Schwäbischen Alb, 2005; C.E. Stevens, The Building of Hadrian' s Wall, 1966; C.R. Whittaker, Frontiers of the Roman Empire. A Social and Economic Study, 1994; Id., Rome and Its Frontiers. The Dynamics of Empire, 2004.

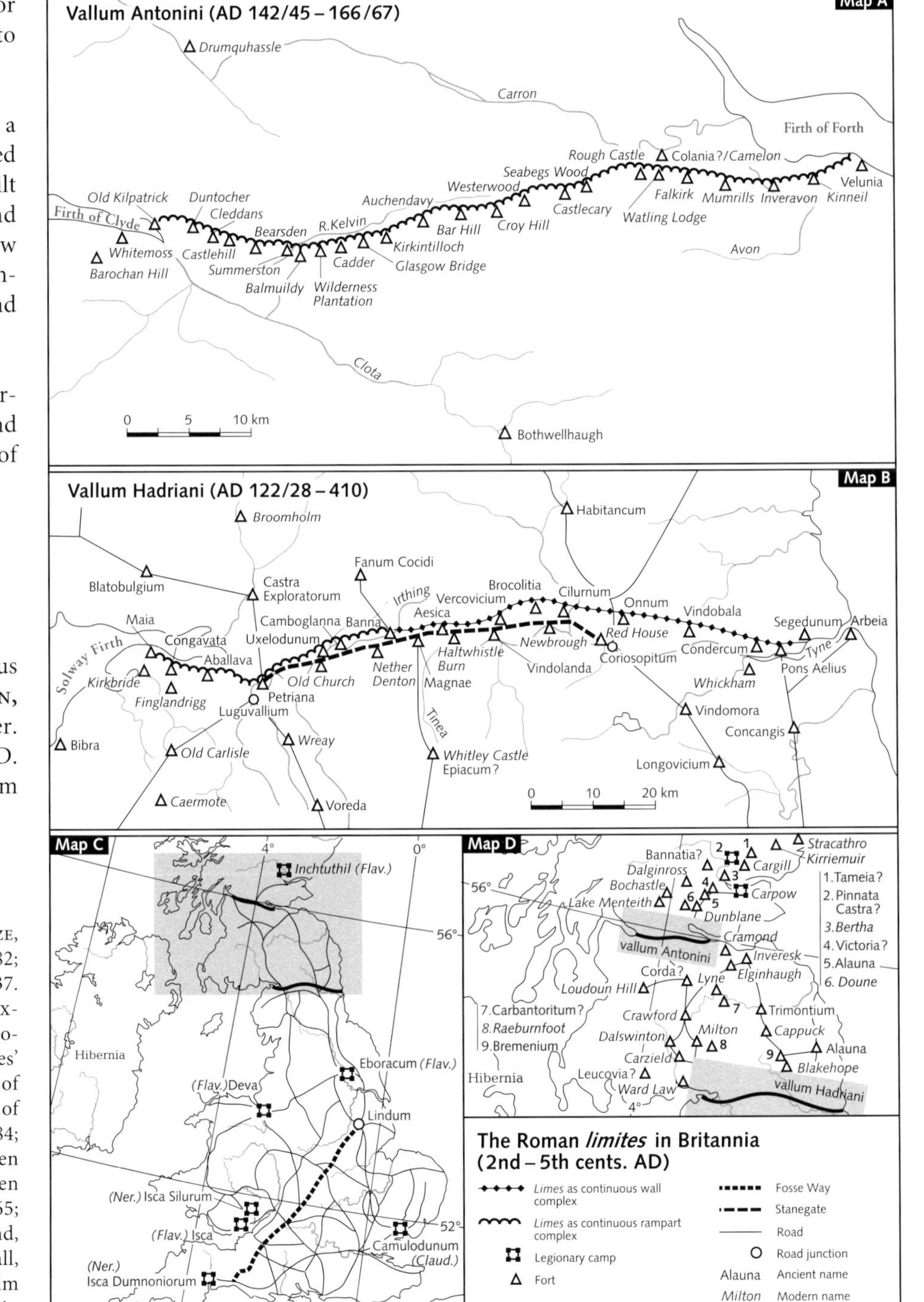

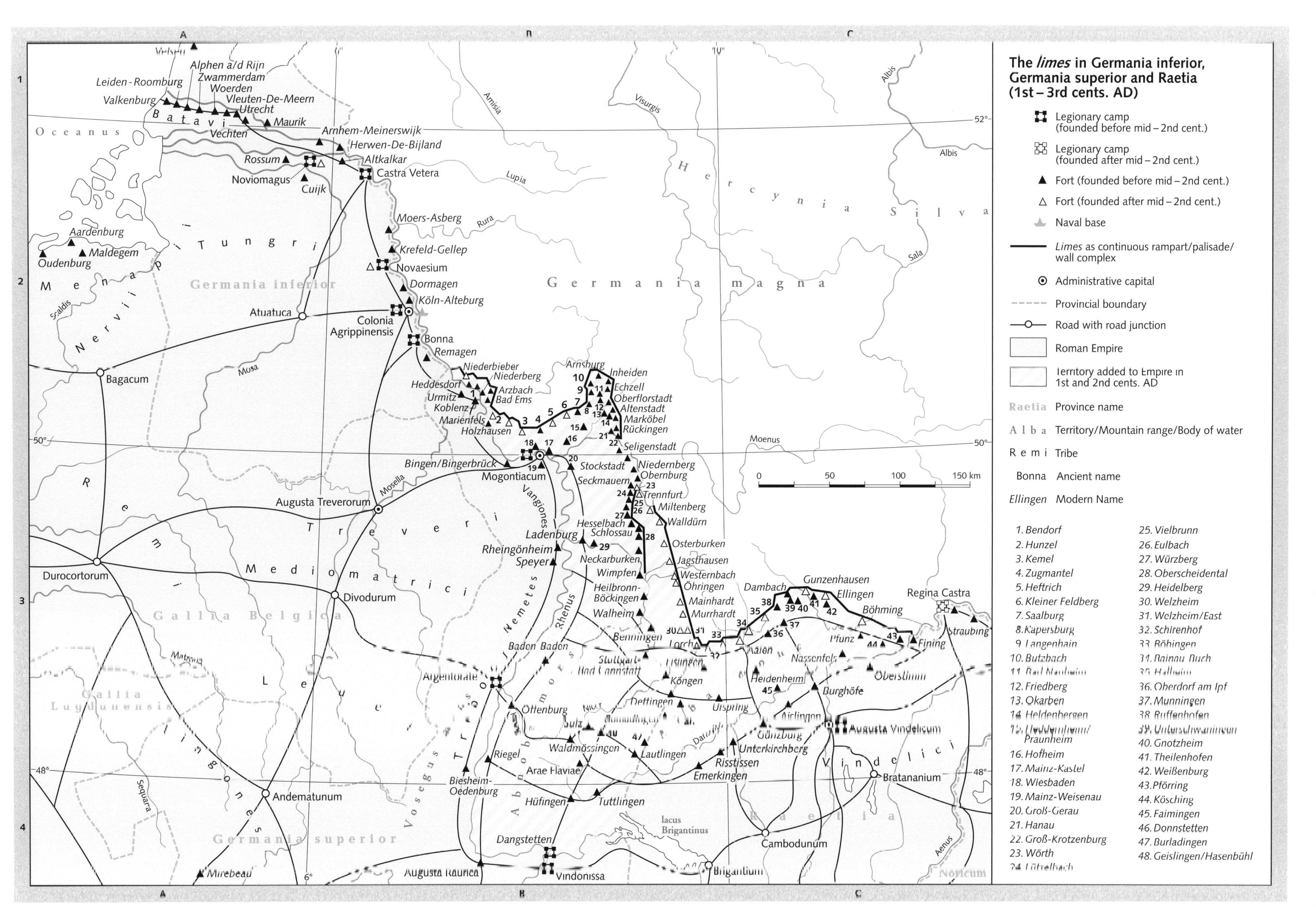

The limes in Germania inferior, Germania superior and Raetia (1st – 3rd cents. AD)
Legionary camp (founded before mid – 2nd cent.)
Legionary camp (founded after mid – 2nd cent.)
Fort (founded before mid – 2nd cent.)
Fort (founded after mid – 2nd cent.)
Naval base
Limes as continuous rampart/palisade/wall complex
Administrative capital
Provincial boundary
Road with road junction
Roman Empire
Territory added to Empire in 1st and 2nd cents. AD
Raetia Province name
A l b a Territory/Mountain range/Body of water
R e m i Tribe
Bonna Ancient name
Ellingen Modern Name
1. Bendorf
2. Hunzel
3. Kemel
4. Zugmantel
5. Heftrich
6. Kleiner Feldberg
7. Saalburg
8. Kapersburg
9. Langenhain
10. Butzbach
12. Friedberg
13. Okarben
14. Heldenbergen
16. Hofheim
17. Mainz-Kastel
18. Wiesbaden
19. Mainz-Weisenau
20. Groß-Gerau
21. Hanau
22. Groß-Krotzenburg
23. Wörth
25. Vielbrunn
26. Eulbach
27. Würzberg
28. Oberscheidental
29. Heidelberg
30. Welzheim
31. Welzheim/East
32. Schirenhof
33. Böhingen
34. Rainau Buch
36. Oberdorf am Ipf
37. Munningen
38. Ruffenhofen
40. Gnotzheim
41. Theilenhofen
42. Weißenburg
43. Pförring
44. Kösching
45. Faimingen
46. Donnstetten
47. Burladingen
48. Geislingen/Hasenbühl
Germania magna
Germania inferior
Germania superior
Gallia Belgica
Raetia
Noricum
Hercynia Silva
Oceanus
Rhenus
Moenus
Albis
Visurgis
Amisia
Lupia
Mosella
Mosa
lacus Brigantinus
Castra Vetera
Colonia Agrippinensis
Bonna
Mogontiacum
Augusta Treverorum
Augusta Vindelicum
Regina Castra
Argentorate
Vindonissa
Cambodunum
Brigantium
Durocortorum
Divodurum
Noviomagus
Atuatuca
Bagacum
Andematunum
150 km

The eastern and southern frontiers of the Roman Empire, 1st–3rd cents. AD

I. The eastern frontier of the Roman Empire, 2nd and 3rd cents. AD

Whereas on other imperial frontiers Rome had to defend itself against constantly shifting opponents, the foe in the region covered by this map was clear. Until AD 224 it was the Parthian Arsacids, and thereafter it was the Sassanids (→ map p. 217). This section of frontier was not protected with major fortification structures as in Britain and Germany, and only in places was it marked by river frontiers (the upper reaches of the Euphrates and Tigris and the Chaboras/Ḫabur) (*limes* of Asia Minor and Mesopotamia). North of the headwaters of the Euphrates, in the eastern Anatolian mountains (so-called Pontic *limes*), it was guarded only by a chain of forts and watch-towers.

In the north-east, the province of Cappadocia ran out into a narrow coastal strip along the western slopes of the Caucasus, which formed the northern projection of the Pontic *limes* from Pityus/Bicvinta and Sebastopolis/Suhumi via the as yet unlocalized city of Phasis on the river of the same name (nowadays called the Rioni), and Apsarus/Gonio as far as Trapezus/Trabzon – all towns secured with forts. The task of this cordon of forts was firstly to keep watch over the mountain tribes, and secondly to protect shipping in this part of the Black Sea. The Pontic fleet had been based at Trapezus since AD 64. South of Trapezus, the frontier crossed the Pontic Mountains at Mount Theches by the Zigana Pass (Zigana Geçidi) at an altitude of 2,025 m, before descending on to the high plain around Satala/Sadağ. This was the location of the most northerly legionary camp on the Pontic *limes*, and from here, a chain of auxiliary forts linked the legionary camps of Melitene/Eski Malatya, Samosata/Samsat (today submerged by the reservoir) and Zeugma/Belkis (also submerged) which lay along a military road, following the upper reaches of the Euphrates southwards. The military road followed the course of the river as far as Sura/al-Surīyya, where it turned south away from the Euphrates and joined the Syrian *limes* after Palmyra (→ map p. 181). Septimius Severus pushed the frontier eastwards in two campaigns, reaching a line that stretched from the Tigris at Balad to the Chaboras, and building legionary camps at Rhesaena (near Ras al-Ain) and Singara/Sinğār.

II. The *limes* in the south of the Roman Empire (1st–3rd cents. AD)

The mountain country of the Jebel Akhdar rises south of the narrow coastal strip around Cyrenae. Rainfall coming in off the sea on the northerly wind meant that the coastal region was always well-watered. To the west, this strip adjoins the Tripolitana, a littoral some 800 km long between Leptis Magna and Carthage, enclosed to the south like a theatre by the semi-circle of the Jebel Nafusa, which reaches 981 m in altitude. Here, too, the wet northerly winds dropped much rain, and the littoral strip was well-watered as a result. Farther still to the west is the chain of the Atlas Mountains, ascending westwards from Carthage to 4,165 m (Toubkal in Morocco). These mountain regions skirting the North African coast form a threshold between the moist coastal climate to the north and the extremely dry inland climate of the Anydros/Sahara to the south.

The southern frontier control systems in the African provinces were all linked by a network of roads connecting individual military installations. In places, there were systems of ramparts and ditches, sometimes also armed with palisades, so-called *fossae* or *fossata*. In sections, walls or at least walled barriers, so-called *clausurae*, prevented arbitrary northbound incursions. Lastly, there were also installations such as watch-towers (*speculae*), forts (*centenaria* and *burgi*) and strongly fortified camps such as Lambaesis/Tazoult-Lambèse. The purpose of these installations was not only to defend against a hostile force coming out of the desert, but also to control and direct transhumance migrations and to enable the regular collection of duties.

In the province of Cyrenae, whose southern frontiers have only been very inadequately researched, military coverage may have been organized by the cities of the Pentapolis around Cyrenae. Individual coastal forts such as Antipyrgus, Tauchira and Boreum controlled coastal shipping.

The coastal region of the Tripolitana, most of which is today only barely fertile, was profitably exploited for agriculture in Antiquity. The frontier region was defended with roads, ditch systems and, in places, walls.

The sole African legion, the *legio III Augusta*, had its permanent camp at Lambaesis from AD 129, in the hinterland immediately to the west of Carthage. This legion had previously been stationed at Ammaedara, then in Theveste. The region around the *mons Aurasius*/Massif de Aurès to the south of Lambaesis was systematically protected with roads and forts. Toll stations were set up at Lambaesis and several other towns along caravan routes to Carthage.

In Mauretania Caesariensis, the frontier control line through Auzia was moved south during the 2nd cent. AD as various forts were built. Castellum Dimmidi/Messad was now the southernmost point of this section of *limes*.

Difficulties began to arise immediately upon the foundation of the Mauretanian provinces in AD 40 requiring a reinforcement of the Roman military presence. Most troops were again withdrawn after the defeat of various revolts, and only a few remained in the country. We learn of these from a great number of military diplomas.

→ Map p. 209

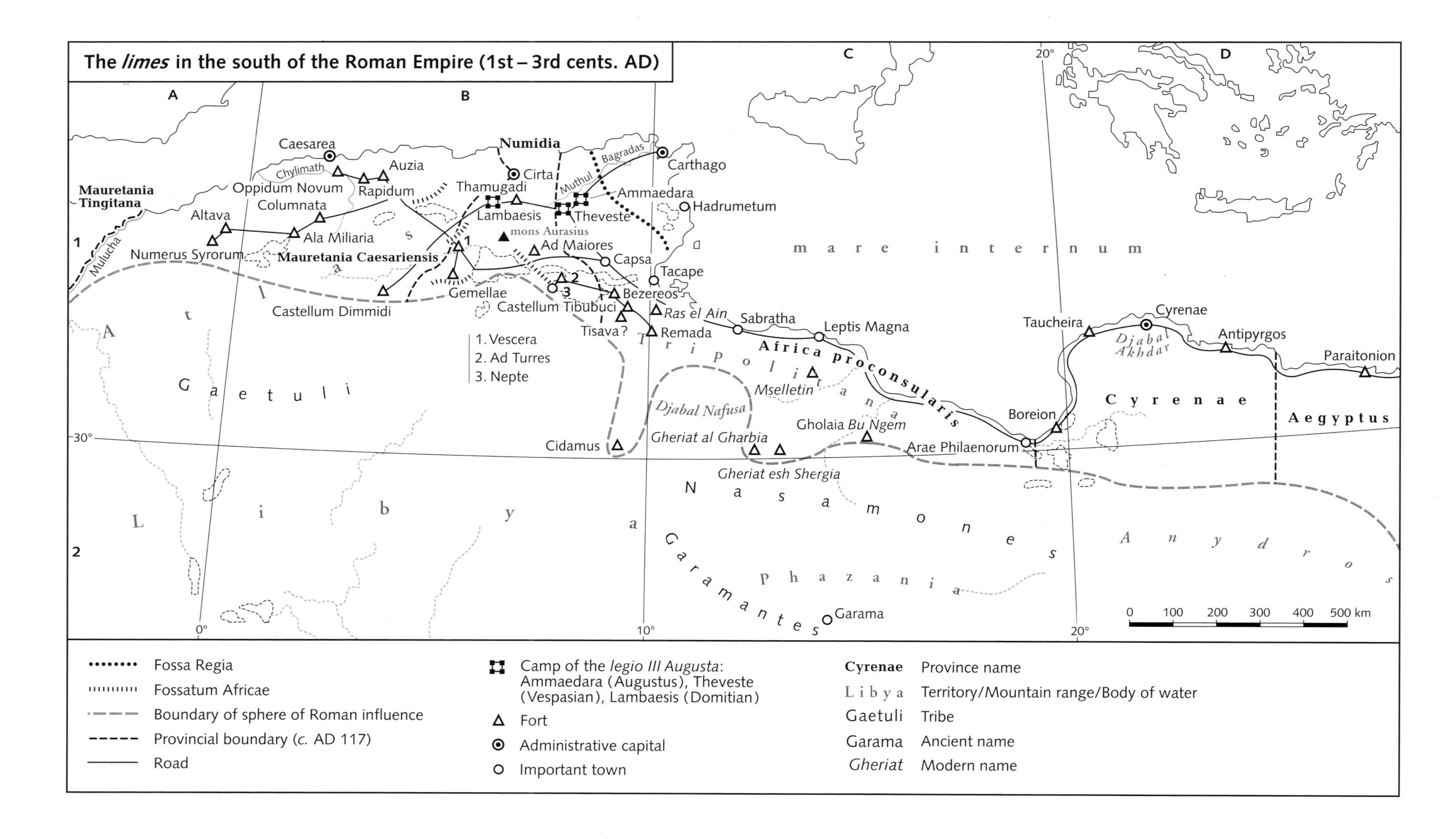

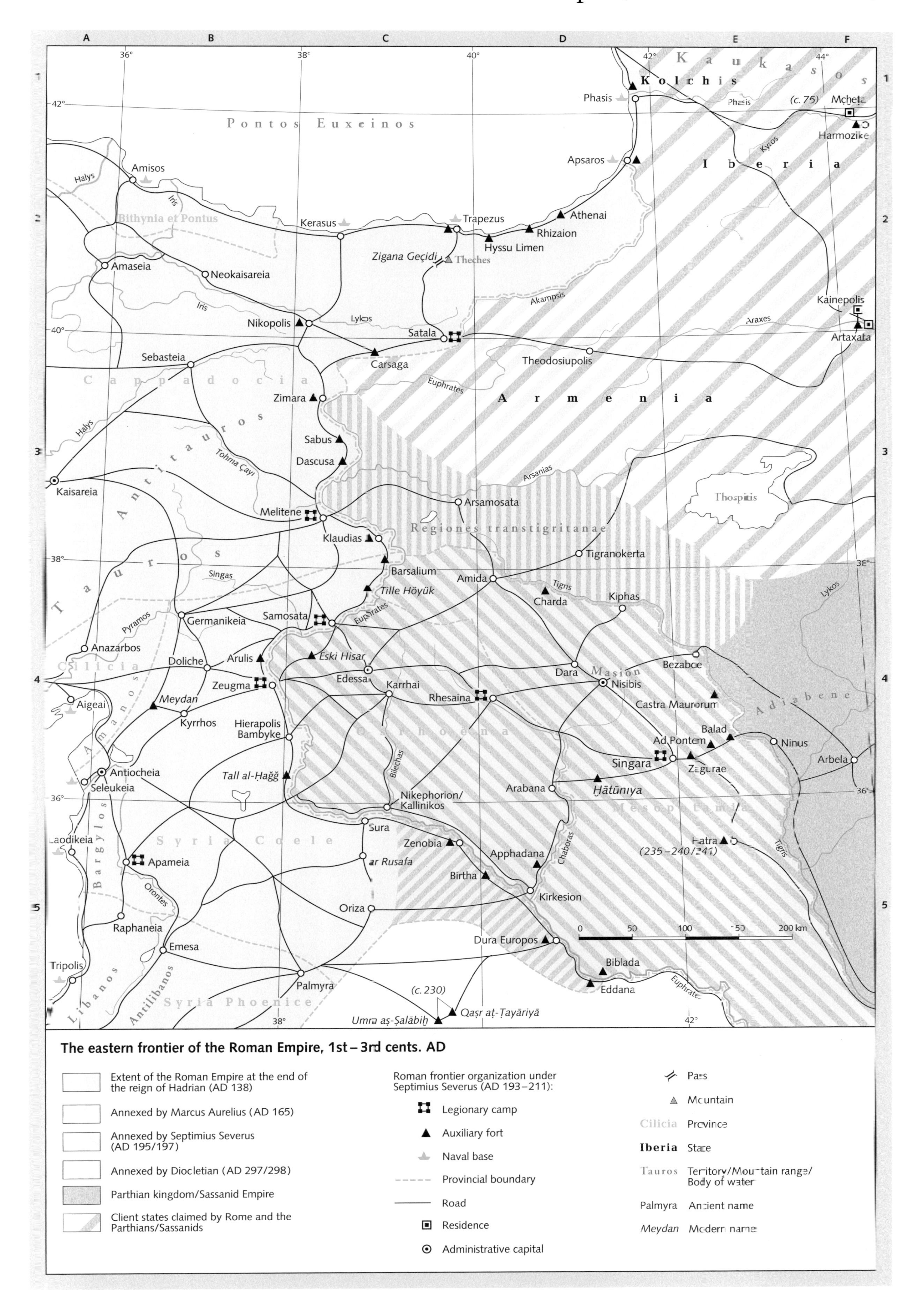

Kaukasos
Kolchis
Phasis
(c. 75)
Mcheta
Harmozike
Pontos Euxeinos
Apsaros
Iberia
Kyros
Amisos
Halys
Iris
Bithynia et Pontus
Kerasus
Trapezus
Athenai
Rhizaion
Hyssu Limen
Zigana Geçidi
Theches
Amaseia
Neokaisareia
Akampsis
Kainepolis
Araxes
Artaxata
Nikopolis
Lykos
Satala
Sebasteia
Carsaga
Theodosiupolis
Cappadocia
Euphrates
Armenia
Zimara
Antitauros
Halys
Sabus
Tohma Çayı
Dascusa
Arsanias
Kaisareia
Thospitis
Arsamosata
Melitene
Regiones transtigritanae
Klaudias
Tigranokerta
Tauros
Singas
Barsalium
Amida
Tigris
Tille Höyük
Charda
Kiphas
Lykos
Samosata
Germanikeia
Pyramos
Anazarbos
Arulis
Eski Hisar
Bezabde
Cilicia
Doliche
Edessa
Dara
Masion
Nisibis
Zeugma
Karrhai
Adiabene
Aigeai
Meydan
Rhesaina
Castra Maurorum
Kyrrhos
Hierapolis Bambyke
Osrhoene
Balad
Ad Pontem
Ninus
Amanos
Singara
Arbela
Antiocheia
Tall al-Ḥaǧǧ
Bilechas
Zagurae
Seleukeia
Arabana
Ḥātūnıya
Nikephorion/ Kallinikos
Mesopotamia
Laodikeia
Syria Coele
Sura
Hatra
(235–240/241)
Zenobia
Chaboras
Apameia
Apphadana
ar Rusafa
Bargylos
Orontes
Birtha
Kirkesion
Oriza
Raphaneia
Dura Europos
Emesa
Biblada
Tripolis
Libanos
Antilibanos
Palmyra
Eddana
(c. 230)
Syria Phoenice
Qaṣr aṭ-Ṭayāriyā
Umm aṣ-Ṣalābiḫ
0 50 100 150 200 km
36° 38° 40° 42° 44°
A B C D E F
1 2 3 4 5
The eastern frontier of the Roman Empire, 1st – 3rd cents. AD
Extent of the Roman Empire at the end of the reign of Hadrian (AD 138)
Annexed by Marcus Aurelius (AD 165)
Annexed by Septimius Severus (AD 195/197)
Annexed by Diocletian (AD 297/298)
Parthian kingdom/Sassanid Empire
Client states claimed by Rome and the Parthians/Sassanids
Roman frontier organization under Septimius Severus (AD 193–211):
Legionary camp
Auxiliary fort
Naval base
Provincial boundary
Road
Residence
Administrative capital
Pass
Mountain
Cilicia Province
Iberia State
Tauros Territory/Mountain range/ Body of water
Palmyra Ancient name
Meydan Modern name

The Arsacid kingdom in the 1st and 2nd cents. AD (to AD 224)

The Parthians, or to use the name of the ruling dynasty, the Arsacids, built a kingdom from the 3rd cent. BC in what is now Iran, taking the place of the Seleucids; it encompassed large parts of Mesopotamia, the south-eastern portion of Central Asia and some adjoining peripheral zones, and it formed the link between the Graeco-Roman world on the one hand and Central Asia (and China) on the other. Many details of its history remain obscure because of the poor state of the sources. The Parthians attracted the attention of Rome in the course of their involvement with the strategically important eastern Armenia/Armenia maior (where there were also copper and gold reserves) and their intermittent involvement with Syria and Asia Minor. Treaties of 69 and 66 BC set the Euphrates as the shared frontier between the Parthians and Rome. When the rump of the Seleucid kingdom was transformed into the Roman province of Syria in 64/3 BC, the neighbouring Parthian kingdom became Rome's rival for primacy in the East, a competition based not least on trading interests.

I. Historical development

The Parthians (Old Persian Parθava, Greek Parthoi, Latin Parthi) were an Iranian people. Originally they were the Parni, a Scythian-Massagetan branch of the Daai/Dahae tribe settled south-east of the Caspian Sea. According to linguistic findings, they spoke an eastern Middle Iranian dialect. Around 250 BC (Parthian era from 247 BC), they invaded the satrapy of Parthia under the command of Arsaces I. Here, they adopted the name 'Parthian' and, with time, the Northwest-Iranian language. They subsequently conquered part of the Iranian possessions of the Seleucid kingdom, making indirect contact with the Achaemenid empire in the process. They rapidly expanded their rule in all directions, taking western Iran and Mesopotamia and parts of Graeco-Bactria in the 2nd cent. (115 BC, references to the so-called Silk Road). Besides a few regions where the Greek influence was strong, especially in some cities (→ map p. 87), Hellenization of the conquered territories was generally superficial (e.g. Greek coinage, Greek as one of the administrative languages).

A number of 'civil wars' shook the Parthian kingdom in the 1st and 2nd cents. AD. There were setbacks in the conflict with the Roman Empire as well, but in the main, the Parthian kingdom was able to hold its own quite successfully against its western neighbour (AD 54–66: Nero's Armenian war; AD 114–117: Trajan's Parthian campaign; AD 162–166: Marcus Aurelius' Parthian War; AD 195, 197–199: Septimius Severus' Parthian War). In view of the numerous wars, the Euphrates frontier proved extraordinarily durable, surviving essentially unchanged until the fall of the Parthian kingdom – although at the latest Septimius Severus did conquer parts of northern Mesopotamia for Rome.

Despite the compromise on Armenia that had been found under Nero in 63, this territory remained contested for centuries. Nor would the situation change under the Sassanids, the region being of great strategic importance.

On the kingdom's north-eastern and eastern frontiers, the effort to hold off the tribes of the steppes (Sacae, Kuṣāna) was a constant drain on its powers, but here, too, the Parthians held their own until the surprising destruction of the Parthian kingdom in the early 3rd cent. AD at the hands of a vassal king from Persis, the Sassanid Ardashir (→ map p. 217).

From 140 at the latest, the Parthian kingdom was a polyethnic, multilingual empire (languages including Middle Persian (in Persis, coins), Parthian (language of court and administration), Sogdian (coins), Chorasmian (short inscriptions), Bactrian (inscriptions, coins), Armenian (toponyms), Babylonian (cf. Iamblichus; cuneiform until *c.* AD 75), Aramaic (inscriptions), Greek (inscriptions, coins); several scripts were also in use). It hosted a plethora of different cultural and religious traditions (the Parthians themselves were Zoroastrians), which for political (and personal) reasons, the Parthians not only tolerated but often even promoted. Conflicts with non-Iranian subject peoples had their roots in particular circumstances of internal and foreign policy, not in any principled xenophobia, although an intensifying recourse to Iranian cultural roots is perceptible from the 1st cent. AD, perhaps in a conscious effort to differentiate the Parthian kingdom from the Romans.

In view of the durability of Parthian rule, the obscurities of the political relations between the central Parthian authority and the regional centres of power must not be mistaken for signs of a 'weak kingdom'. These relations concern in particular the dependent vassal kingdoms (such as Adiabene, Media Atropatene, Characene, Elymais, Persis), and to a lesser extent the regions under direct Parthian rule administered by satraps or *strategoi* (e.g. Armenia) and frontier regions under 'margraves', which were governed centrally in terms of fiscal, political and military authority, but also had a sometimes extraordinary degree of autonomy (coinage prerogatives, other privileges). Rather, the slight Parthian presence in the material and written finds from the provinces should be seen as proof of the strength of the structure of the realm and the success of royal policies. For the most part, those cities that were of particular importance to the social and economic development of the kingdom, e.g. the royal residence of Ctesiphon on the Tigris and the 'mega-city' of Seleucia opposite, as well as Nisā, Ecbatana, Susa/Seleucia, Rhaga(e) and Hecatompylus (mints alongside the peripatetic court mint), enjoyed an economic and cultural florescence under the Parthians. As traders and intermediary traders, Parthians and Parthian subjects (Characenes) together with others (e.g. Palmyrenes) traded goods between China, India and Syria, e.g. silk, steel and spices westwards, fruit and horses eastwards.

II. The kingdom of the Kuṣāna

From the 1st–3rd cents. AD (until the 5th cent. locally in the Cophen valley), the Eastern Iranian dynasty of the Kuṣāna (Kushan), which had emerged from the nomadic tribes of the Yuezhi or Tochari, ruled a kingdom which at times reached from the Indus and Ganges to the Aral Sea (greatest extent between *c.* AD 100 and 250). Its heartland was in the former Bactria and Gandaritis (around Cophen). Territory was gradually won from the Greeks, Sacae, Parthians and others, and the kingdom expanded. The capitals of the 1st and 2nd cents. were Puruṣapura and Mathura (in northern India, west of the Ganges), while Capisa, the old capital of the Indo-Greek kings, served as a summer residence.

Art and religion, but also writing were strongly influenced by the Graeco-Bactrian heritage (Gandhāra art, Graeco-Buddhism, government inscriptions in an expanded Greek alphabet; coin minting). Contacts with the Roman Empire, the Parthians (later Sassanids), India and China gave the Kuṣāna a key position. Control of the 'Silk Road' and the routes to and from India (connections to maritime trading routes) ensured substantial earnings from trade. Roman coins and imports from the Mediterranean have been found in settlements of the Kuṣāna period (Capisa; also there, finds of Chinese lacquerwork, Indian ivories, imitations of Hellenistic art works). Yet Greek and Latin sources yield almost nothing about the Kuṣāna (Apollodorus of Artemita on the Yuezhi).

Sources

Although there is indigenous evidence in the form of inscriptions and coin finds (over 2,000 *ostraka* from Nisā and Šahr-i Qūmis: economic texts, naming of officials and titles; parchment finds from Avrōmān/western Persia and Dura Europus, papyrus finds from Dura Europus), and some cuneiform transmission (cf. Raḫimesu Archive at Babylon) and Armenian and Chinese texts (historiographies: Sima Qian, *c.* 98 BC; Ban Gu and Ban Zhao, Han Period; Fan Ye, 5th cent. AD), the state of the sources is not particularly revealing as far as the Parthians are concerned. Authors writing in Greek and Latin, such as Pompeius Trogus (whose work survives only in excerpts in Justin), Strabo, Tacitus, Flavius Josephus and Cassius Dio, are often biased. Other sources, such as the *History of the Parthians* by Apollodorus of Artemita, are entirely lost apart from a few quotations in other authors. The state of the sources is relatively fully described in the works indicated (e.g. Wiesehöfer 1998). There were pictorial representations of Parthians in the West, e.g. at Rome.

Matters of scholarly interest, aside from issues of the structure of the Parthian 'state' and the relations between the Great King and the nobility, include the origins of the kingdom and the relationship between the Parthian kings and individual parts of the kingdom. Much has yet to be clarified. As yet, there is insufficient evidence from literary, epigraphic, archaeological and numismatic sources to reconstruct satisfactorily Parthian history, the history of the so-called 'Indo-Parthian kingdom' or the kingdom of the Persidae.

→ Maps pp. 117, 129, 133, 161, 181, 183, 207, 213; India, trade with, BNP 6, 2005 (with map); Kushan, Kushanians, BNP 7, 2005; Yuezhi, BNP 15, 2009

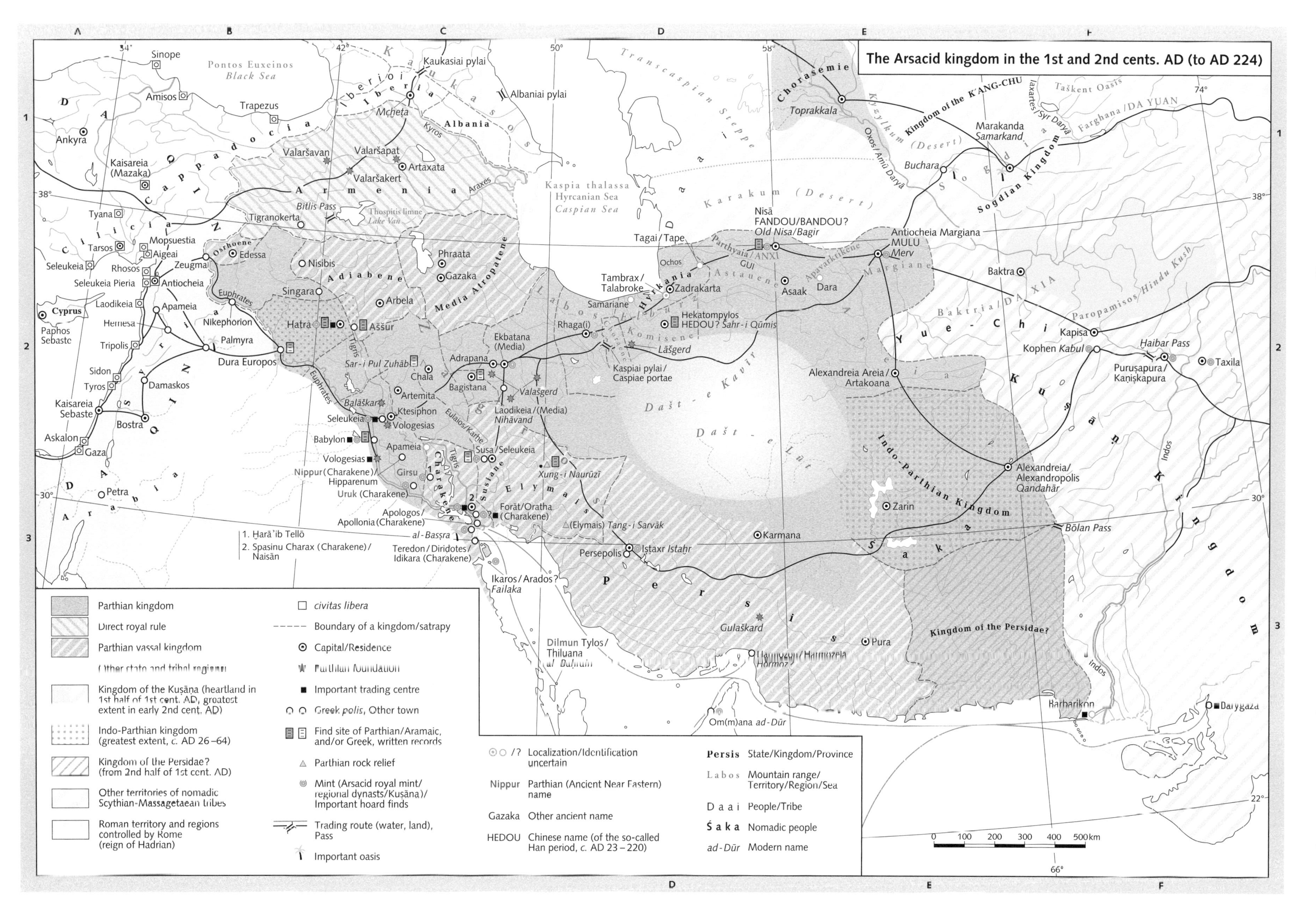
The Arsacid kingdom in the 1st and 2nd cents. AD (to AD 224)
Parthian kingdom
Direct royal rule
Parthian vassal kingdom
Kingdom of the Kušāṇa (heartland in 1st half of 1st cent. AD, greatest extent in early 2nd cent. AD)
Indo-Parthian kingdom (greatest extent, c. AD 26–64)
Kingdom of the Persidae? (from 2nd half of 1st cent. AD)
Other territories of nomadic Scythian-Massagetaean tribes
Roman territory and regions controlled by Rome (reign of Hadrian)
civitas libera
Boundary of a kingdom/satrapy
Capital/Residence
Parthian foundation
Important trading centre
Greek polis, Other town
Find site of Parthian/Aramaic, and/or Greek, written records
Parthian rock relief
Mint (Arsacid royal mint/regional dynasts/Kušāṇa)/Important hoard finds
Trading route (water, land), Pass
Important oasis
Localization/Identification uncertain
Nippur Parthian (Ancient Near Eastern) name
Gazaka Other ancient name
HEDOU Chinese name (of the so-called Han period, c. AD 23–220)
Persis State/Kingdom/Province
Labos Mountain range/Territory/Region/Sea
Daai People/Tribe
Śaka Nomadic people
ad-Dūr Modern name
1. Ḥarā'ib Tellō
2. Spasinu Charax (Charakene)/Naisān
Pontos Euxeinos
Black Sea
Kaspia thalassa
Hyrcanian Sea
Caspian Sea
Transcaspian Steppe
Karakum (Desert)
Kyzylkum (Desert)
Dašt-e Kavīr
Dašt-e Lūt
Chorasemie
Kingdom of the K'ANG-CHU
Sogdian Kingdom
Yue-Chi
Kušāṇa Kingdom
Indo-Parthian Kingdom
Kingdom of the Persidae?
Armenia
Albania
Iberia
Cappadocia
Cilicia
Osrhoene
Adiabene
Media Atropatene
Hyrkania
Komisene
Parthyaia
Astauene
Margiane
Baktria/DA XIA
Paropamisos/Hindu Kush
Elymais
Susiane
Charakene
Persis
Śaka
Sinope
Amisos
Trapezus
Ankyra
Kaisareia (Mazaka)
Tyana
Tarsos
Mopsuestia
Aigeai
Seleukeia
Rhosos
Zeugma
Seleukeia Pieria
Antiocheia
Laodikeia
Apameia
Cyprus
Paphos Sebaste
Hemesa
Tripolis
Sidon
Tyros
Damaskos
Kaisareia Sebaste
Bostra
Askalon
Gaza
Petra
Palmyra
Dura Europos
Nikephorion
Edessa
Nisibis
Singara
Hatra
Aššur
Arbela
Tigranokerta
Bitlis Pass
Thospitis limne
Lake Van
Valaršavan
Valaršapat
Valaršakert
Artaxata
Mcheta
Kaukasiai pylai
Albaniai pylai
Kyros
Araxes
Phraata
Gazaka
Ekbatana (Media)
Adrapana
Chala
Bagistana
Sar-i Pul Zuhāb
Artemita
Ktesiphon
Seleukeia
Vologesias
Balāškar
Babylon
Apameia
Nippur (Charakene)/Hipparenum
Uruk (Charakene)
Girsu
Apologos/Apollonia (Charakene)
al-Basra
Teredon/Diridotes/Idikara (Charakene)
Ikaros/Arados?
Failaka
Forāt/Oratha (Charakene)
Susa/Seleukeia
Xung-i Naurūzī
(Elymais) Tang-i Sarvāk
Laodikeia/(Media)
Nihāvand
Valašgerd
Eulaios/Karhe
Tigris
Euphrates
Rhaga(i)
Kaspiai pylai/Caspiae portae
Lāšgerd
Samariane
Tambrax/Talabroke
Zadrakarta
Hekatompylos
HEDOU? Šahr-i Qūmis
Ochos
Tagai/Tape
Nisā
FANDOU/BANDOU?
Old Nisa/Bagir
Parthyaia/ANXI
GUI
Asaak
Dara
Antiocheia Margiana
MULU
Merv
Oxos/Amū Daryā
Toprakkala
Buchara
Marakanda
Samarkand
Iaxartes/Syr Daryā
Taškent Oasis
Farghana/DA YUAN
Baktra
Kapisa
Kophen Kabul
Haibar Pass
Purusapura/Kaniṣkapura
Taxila
Alexandreia Areia/Artakoana
Alexandreia/Alexandropolis
Qandahār
Zarin
Bōlan Pass
Karmana
Persepolis
Istaxr Istahr
Gulaškard
Pura
Harmozeia/Harmozeia
Hormoz
Dilmun Tylos/Thiluana
al-Bahrain
Om(m)ana ad-Dūr
Barbarikon
Barygaza
Indos
0 100 200 300 400 500 km

The Sassanid Empire (AD 224–651)

I. Political developments

1. The relationship with Rome

The homeland of the Sassanids, who were named after their dynastic founder Sāsān (Sassan), was the territory of Pārsa (Greek Persis) north of the Persian Gulf, more precisely, the city of Staḫr. The connection between Sāsān and Papak, the father of the kingdom's founder, Ardashir I, is unclear. The royal history of the Sassanids, the dynasty as well as the Iranians and Persians living in its empire, really begins with Ardashir's elevation to Prince of Staḫr early in the 3rd cent. AD.

Ardashir asserted himself in a revolt against the Parthian King Artabanus IV (213–224). He inflicted a devastating defeat on Artabanus on 28 April 224, in a battle at Hormizdâgan (location unknown) in which Artabanus fell. In just a few years of fighting members of the Parthian royal house (e.g. Vologaeses VI) and its officials in Mesopotamia and Âdurbādagān (Azerbaijan), Ardashir succeeded in taking over the entire Parthian kingdom with the exception of Armenia. The conflicts with the Romans, which would continue until the end of Sassanid rule, are shown in the context of the maps on pp. 219 and 241. Three main phases of war can be defined in the course of Sassanid-Roman relations from AD 224 to 651:

- The first phase lasted from Ardashir's crossing of the Euphrates in AD 230 and the three great campaigns of Sapor I until the peace, sued for by the Sassanid king Narses after the crushing defeat at Satala and granted in negotiations with Diocletian and Galerius at Nisibis in 298.
- The second phase mainly encompasses the conflicts under Sapor II from 342 to 363, when the Roman Emperor Jovian was forced to solicit Sapor for peace.
- The third phase dates from 540, when Chosroes I broke the 'eternal peace' he himself had concluded with Justinian I in 532, and thereafter found himself permanently at war with the Romans. His grandson, Chosroes II, pursued this tradition with great success as far as the Nile and the gates of Constantinople, yet in the end he was unable to prevail against the Byzantine Emperor Heraclius. However, it was not the Byzantines who brought an end to the Sassanid Empire, but the Arabs, under their third caliph, 'Utmān ibn 'Affān, with the death of the last Sassanid ruler Isdigerdes (Yazdgird) III in AD 651 (→ maps pp. 237, 241 and 243).

2. The other frontiers

The Hephthalites of the north-eastern borderlands posed a great danger which also had a lasting influence on social developments within the Sassanid Empire. This Hunnish people (the 'White Huns' of the tradition) gathered a great number of nomadic tribes under their overlordship in a territory which, at the turn of the 5th/6th cents. AD, stretched from the Aral Sea to the Ganges. At a date probably after 469, they defeated the Sassanid king Perozes in battle and made the Sassanid Empire a tributary. They defeated Perozes a second time when he rebelled against this foreign rule in 484, and this time Perozes fell in battle. It was also the Hephthalites who put Cabades, Perozes' son, on the throne in the face of opposition from some Sassanid noble circles, Cabades having lived at the Hephthalite court as a hostage. Only in 560 did Chosroes I throw off the Hephthalite yoke, annihilating their kingdom. He received great help in this from the Western Turks, a nomadic tribal group who had migrated west from the Chinese interior and had come into conflict with the Hephthalites since the mid 6th cent. However, these Western Turks would themselves subsequently pose a constant threat to the Sassanid Empire in its eastern frontier zone in place of the Hephthalites.

II. Domestic political developments

A revolutionary religious popular movement led by Mazdak acquired extraordinary socio-political importance in the Sassanid Empire during the reign of Cabades. Espousing social egalitarianism, it aspired to establish a popular form of the hitherto absolutely elitist Zoroastrianism of the land-owning aristocracy. This movement was a threat to the nobility for as long as Cabades used it in his dispute with the followers of his brother, Zamasphes. When Cabades came to the throne for the second time, he gave up this alliance. His successor, Chosroes I, then persecuted the Mazdakites and put an end to the movement. Cabades and Chosroes used the weakness of the nobility which Mazdakism caused to bring about profound reforms in social, economic (taxation) and military (military zones) spheres.

III. Religious situation

The determinative religion of the Sassanid Empire was Zoroastrianism, which was particularly deeply rooted in aristocratic circles. Its founder, if he was a historical figure at all, is dated far back in Antiquity, between the 12th and 7th cents. BC. It is a matter of dispute, however, whether Zoroastrianism was the state religion of the Sassanid Empire. What is apparent is that religious dignitaries – in an orderly hierarchy led by the high priest, the so-called Kartēr – were consulted as experts not only on matters of religion but also on governmental issues of social order. Religious tensions in the Byzantine Empire drove some Christians (Monophysites, Nestorians) into exile in Persia, where they were welcomed. Jews, too, were able to live according to their beliefs, as initially were Manichaeans, although they migrated east after the death of Mani in AD 276. The preferred cult venue of Zoroastrianism was the fire sanctuary, which was realized in architectural terms as a fire temple or fire altar. Many such buildings can still be found in what was Sassanid territory – the map indicates their distribution. This was the focus of Zoroastrian ritual: fire in its various forms. It was regarded as a purifying force.

IV. The economy

The material basis of the Sassanid Empire was, as with all ancient states, agriculture, tax revenues from which were essential to the state. Various crafts played a part as well (textiles, glass, pottery), as did trading activity. The most important commodity was without doubt silk, and the intermediary trade between China and Constantinople brought substantial revenues. The Byzantines, who depended on their imports of raw silk, attempted to evade Sassanid control of the 'Silk Road' by using the (itself hardly straightforward) sea route through the Red Sea and across the Indian Ocean.

Sources

Of the primary sources on the history of the Sassanid Empire, the multilingual inscriptions discovered in Persis (Naqš-e Rustam, Naqš-e Rağab, Hāğğīābād, Bīšāhpuhr, Sar Mašhad, Tang-e Borāq) and Pāikūlī are of particular informative value. The trilingual (Parthian, Middle Persian, Greek) inscription, discovered in 1936 at Naqš-e Rustam on a tower-like Achaemenid building, the Ka'ba-i Zardušt, and containing a report of the deeds of Sapor I, is particularly impressive. Other primary sources for Sassanid culture are seal stones and clay *bullae*, Middle Persian papyri and *ostraka* and coins. Alongside these primary sources, we have the archaeological finds and findings, i.e. remains of houses, palace and city complexes, bridges, canals and dams, but also sculptures, reliefs and small finds of gold, silver and other metals, clay, glass and precious stones (gems and cameos). Little survives of Persian historical literature, but there is the universal history of the jurist aṭ-Ṭabarī (839–923), which is of particular value for its sometimes *verbatim* quotations of lost Iranian works including the Sassanid Book of Kings (*Ḫvadāy-nāmag*). Roman and Greek historians dealt with Sassanid history, generally from a hostile perspective. They include Cassius Dio of Nicaea in Bithynia (2nd/3rd cents.; on the foundation of the Sassanid Empire), the Greek (writing in Latin) Ammianus Marcellinus (4th cent.; on the conflicts of 353–371) and Procopius of Caesarea in Palestine (6th cent.; on the Roman-Sassanid wars in the reigns of the kings Cabades I and Chosroes I).

The map

The map particularly addresses conditions in the early Sassanid period of the 3rd cent. The central axes of infrastructure from the Central Asian highlands (the so-called Silk Road) and the Persian Gulf westwards make it clear why Antioch was repeatedly the target of the Sassanid kings' military expeditions – it was their window on the West. The map also shows why the sometimes successful Roman and Byzantine attempts to occupy the Sassanid residence of Tīsifōn (Ctesiphon) never led to the collapse of the kingdom: its real centre was in the hinterland.

The Sassanid Kings

Ardashir I 224–241/42
Sapor I 239/40–270/72
Hormisdas I 270/72–273
Vahram I 273–276
Vahram II 276–293
Vahram III 293
Narses 293–302
Hormisdas II 302–309
Sapor II 309–379
Ardashir II 379–383
Sapor III 383–388
Vahram IV 388–399
Yazdgird I 399–421
Vahram V 421–439
Yazdgird II 439–457
Hormisdas III 457–459
Perozes I 459–484
Vologaeses 484–488
Cabades I 488–496 and 499–531
Zamasphes 496–498
Chosroes I 531–579
Hormisdas IV 579–590
Chosroes II 590–628
Vahram VI 590–591
Cabades II 628
Ardashir III 628–630
Shahrbaraz 630
Chosroes III 630
Boran (female regent) 630–631
Azarmiducht (female regent) 631
Hormisdas V 631–632
Chosroes IV 631–633
Yazdgird III 633–651

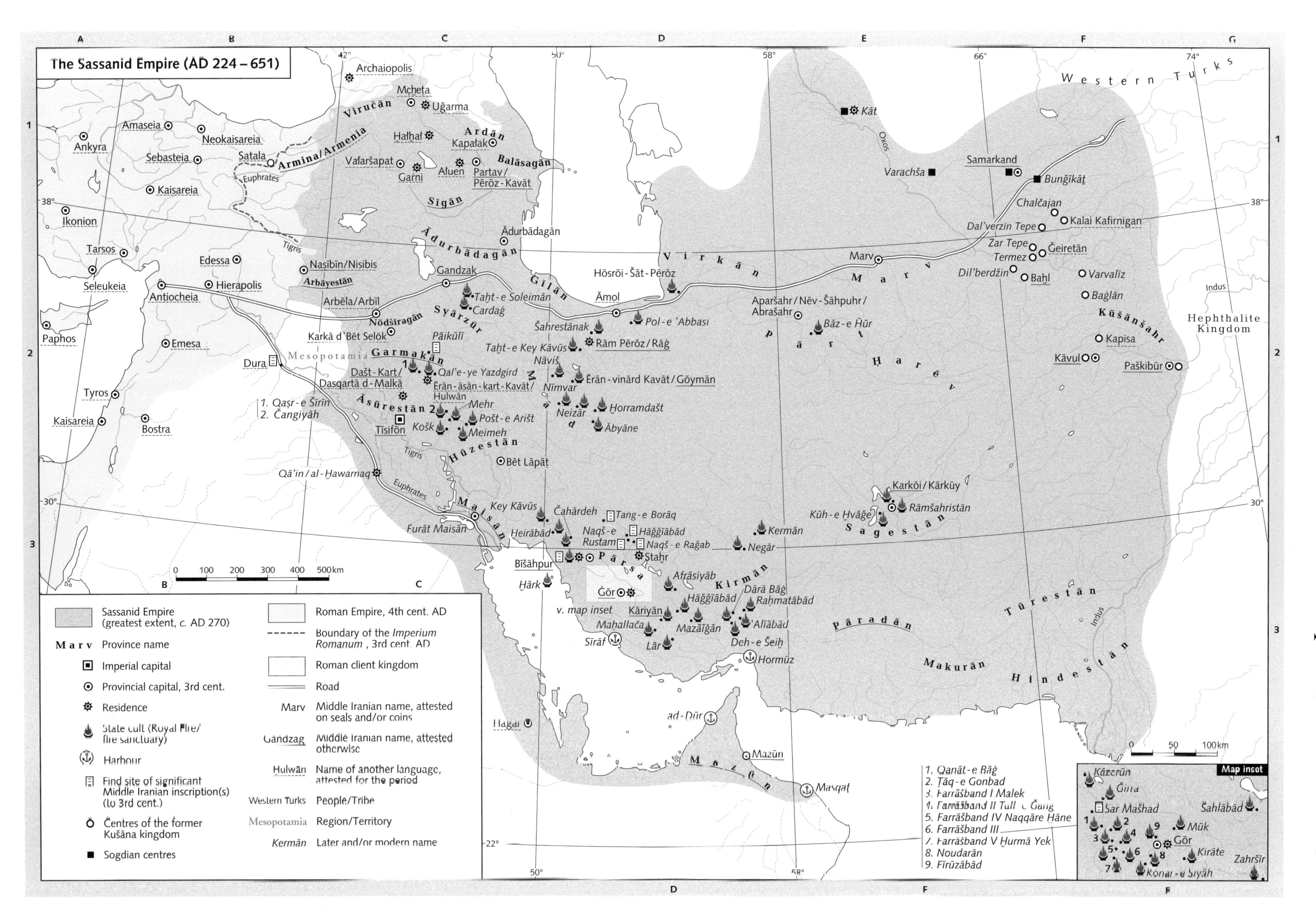
The Sassanid Empire (AD 224–651)
Sassanid Empire (greatest extent, c. AD 270)
Marv Province name
Imperial capital
Provincial capital, 3rd cent.
Residence
State cult (Royal Fire/ fire sanctuary)
Harbour
Find site of significant Middle Iranian inscription(s) (to 3rd cent.)
Centres of the former Kušāna kingdom
Sogdian centres
Roman Empire, 4th cent. AD
Boundary of the Imperium Romanum, 3rd cent. AD
Roman client kingdom
Road
Marv Middle Iranian name, attested on seals and/or coins
Gandzag Middle Iranian name, attested otherwise
Hulwān Name of another language, attested for the period
Western Turks People/Tribe
Mesopotamia Region/Territory
Kermān Later and/or modern name
Map inset
1. Qanāt-e Bāġ
2. Ṭāq-e Gonbad
3. Farrāšband I Malek
4. Farrāšband II Tull-e Ğunğ
5. Farrāšband IV Naqqāre Ḫāne
6. Farrāšband III
7. Farrāšband V Ḫurmā Yek
8. Noudarān
9. Fīrūzābād
1. Qaṣr-e Šīrīn
2. Čangiyāh
Western Turks
Hephthalite Kingdom
Kūšānšahr
Tūrestān
Hindestān
Makurān
Pāradān
Sagestān
Harēv
Marv
Pārs
Kirmān
Mazūn
Virkān
Gīlān
Ādurbādagān
Sīgān
Ardān
Balāsagān
Virūčān
Armina/Armenia
Syārzūr
Nōdšīragān
Arbāyestān
Garmakān
Āsūrestān
Hūzestān
Maišān
Mesopotamia
Archaiopolis
Mcheta
Uğarma
Halhal
Kapalak
Vałaršapat
Garni
Aluen
Partav/Pērōz-Kavāt
Amaseia
Neokaisareia
Ankyra
Sebasteia
Satala
Kaisareia
Ikonion
Tarsos
Seleukeia
Paphos
Edessa
Hierapolis
Antiocheia
Emesa
Tyros
Kaisareia
Bostra
Dura
Nasībīn/Nisibis
Arbēla/Arbīl
Karkā d'Bēt Selōk
Gandzak
Ādurbādagān
Taḫt-e Soleimān
Cardağ
Pāikūlī
Dašt-Kart/Dasqartā d-Malkā
Qal'e-ye Yazdgird
Ērān-āsān-kart-Kavāt/Hulwān
Tisifōn
Košk
Mehr
Pošt-e Arišt
Meimeh
Bēt Lāpāṭ
Qā'in/al-Ḥawarnaq
Furāt Maisān
Key Kāvūs
Heirābād
Čahārdeh
Tang-e Borāq
Naqš-e Rustam
Hāğğiābād
Naqš-e Rağab
Staḫr
Bīšāhpur
Hārk
Gōr
v. map inset
Kāriyān
Mahallača
Sīrāf
Lār
Afrāsiyāb
Hāğğiābād
Mazāiğān
Dārā Bāġ
Raḥmatābād
'Aliābād
Deh-e Šeiḫ
Hormūz
Negār
Kermān
Hāgar
ad-Dūr
Mazūn
Masqaṭ
Šahrestānak
Taḫt-e Key Kāvūs
Nāviš
Nīmvar
Neizār
Ērān-vinārd Kavāt/Gōymān
Ḫorramdašt
Ābyāne
Rām Pērōz/Rāğ
Pol-e 'Abbasī
Hōsrōi-Šāt-Pērōz
Āmol
Aparšahr/Nēv-Šāhpuhr/Abrašahr
Bāz-e Hūr
Marv
Karkōi/Kārkūy
Kūh-e Ḫvāğe
Rāmšahristān
Kāt
Oxos
Varachša
Samarkand
Bunğīkaṯ
Chalčajan
Kalai Kafirnigan
Dal'verzin Tepe
Zar Tepe
Termez
Čeiretān
Dil'berdžin
Baḫl
Varvalīz
Baġlān
Kapisa
Kāvul
Paškibūr
Indus
Tigris
Euphrates
Kāzerūn
Ğirra
Sar Mašhad
Šahlābād
Mūk
Gōr
Kirāte
Zahršīr
Konār-e Siyāh

Romans and Sassanids in the Soldier-Emperors period (3rd cent. AD)

In the strictest sense, the term 'Sassanids' denotes the Iranian dynasty descended from Sassan, and in a wider sense the inhabitants of the territory ruled by the Sassanid dynasty (cf. list of kings p. 216). The Sassanid Ardashir I rose up aginst Parthian rule and annihilated the forces of the Arsacid King Artabanus IV (213–224) on 28 April AD 224 at Hormizdâgan (not localized) in Media; Artabanus IV fell in the battle. The Sassanids went on to take over the entire Parthian kingdom, with the exception of Armenia, in a very short time.

In doing so, the Sassanids inherited the problem of relati ons with the Roman Empire, a problem whose history dated back to Sulla's Cilician proconsulship of 96 BC. Tensions had culminated in several annihilations of Roman armies (53 BC under Crassus, AD 161 under Sedatius Severianus), but also in several occupations of Ctesiphon (AD 165 under L. Verus, AD 198 under Septimius Severus). True, there had been periods of understanding regarding the status of the Armenian throne (62 BC) or the Euphrates frontier (under Hadrian). But the emperor Macrinus found himself with no choice other than to continue the war against the Parthians initiated in AD 216 by his predecessor Caracalla. He was defeated at Nisibis in the spring of 217 by Artabanus IV, and forced to accept an unfavourable peace with swingeing reparation payments.

Once the Sassanid Ardashir I had eliminated the last Parthian resistance in Iran, he crossed the Euphrates in AD 230. In the spring of 232, the Roman emperor Severus Alexander confronted him with his forces. The ensuing battle on the banks of the Euphrates, while not decisive, was enough to prevent Ardashir for the moment from advancing farther into Roman territory. This he only did in 236/38, taking Hatra, Nisibis and Carrhae.

Sapor I, in his account of his own exploits (preserved in inscription), describes the war of 242-244, in which he reacted to the huge expedition of the emperor Gordian III and his *praefectus praetorio* C. Furius Aquila Timesitheus, as his first campaign. The Romans had initial successes, advancing through the whole of Mesopotamia, winning a battle at Rhesaena and pushing forward into the vicinity of Ctesiphon. There, however, in the spring of 244, came the decisive battle, in which the Roman army was defeated at Misiche on the Euphrates. Gordian himself was killed. Philippus Arabs now had himself proclaimed emperor, having succeeded Timesitheus (who had died on the march to Ctesiphon) as *praefectus praetorio*. He immediately concluded a peace with Sapor, by which the Romans, in exchange for paying a vast war reparation, were permitted at least to retain parts of Armenia and Mesopotamia of the territory they had recently won. The emperor, however, undertook never again to intervene in the Armenian succession, which the Sassanid court now controlled. The treaty did not prevent the Sassanid king from annexing territories in the southern Caucasus to his empire.

The second campaign (252–256) of which Sapor tells in his report, was, as he says, caused by the Romans' breach of the treaty. Sapor began by invading Armenia, and his troops then moved through Mesopotamia and Syria. The Sassanids were defeated at Emesa, but they then annihilated a Roman army at Barbalissus and briefly even occupied Antioch. Sassanid forces also entered Cappadocia. Dura was conquered in 256.

The third campaign Sapor recounts culminated in the early summer of 260 in the battle of Edessa, in which the Romans were defeated. The emperor Valerian and his entire military staff were taken prisoner. The Sassanids occupied Antioch again, but also marched west through Cilicia and Cappadocia. Septimius Odaenathus, exarch of the oasis state of Palmyra and a *vir consularis* since 258, had already proved himself in battle against Sapor on the Euphrates in 260, and now sought an alliance with Sapor, but he was rebuffed. He therefore offered his services to Valerian's son and successor, Gallienus. Bearing the title *dux Romanorum* and hence supreme commander of all Roman forces in the East, Odaenathus conquered Carrhae and Nisibis in 262, and advanced to the vicinity of Ctesiphon. He repeated this campaign in 266/7, and this time actually took Ctesiphon. Evidently, it seemed with this success that the Sassanid front was secured. The Goths were posing more urgent problems in Asia Minor, and Odaenathus moved against them to Heraclea Pontica in 267. There, however, he was murdered. The safeguarding of the eastern front was now taken over, with distinctly self-serving motives, by his widow, Zenobia.

Zenobia acted as regent for her minor son Vaballathus, who inherited his father's title of *rex* in Palmyra and usurped those of *corrector* and *dux Romanorum*. The foundation of the fortress of Zenobia/Ḥalabīya on the right bank of the Euphrates upstream of Dura was probably intended to secure the Palmyrene territory against the Sassanids. Palmyrene troops moved through Bostra to Egypt in 270, but failed to occupy the Anatolian peninsula in 271, because Aurelian was already crossing the Bosporus and Asia Minor towards Antioch, where he annihilated the Palmyrene forces in 272 and put an end to Zenobia's self-proclaimed 'Palmyrene Empire'.

Our sources report no significant incidents in the Roman-Sassanid border country until AD 283. The absence of ventures in the east on the part of the Roman emperors from Claudius Gothicus to Probus is explained by the problems constantly developing in the Empire itself and on its northern frontiers. Scholars generally attribute Sapor's failure to exploit the lengthy crisis in the Roman Empire for further westward expansion to the problems presumably posed to the Sassanid Empire in the east by the Kushan rulers, whose realm extended from the Ganges to the Aral Sea until the 4th cent. At all events, they are mentioned in Sapor's inscription.

Not until the Roman emperor Carus was there a new campaign against the Sassanid Empire, led by the emperor and his son, Numerianus. The emperor brought his forces as far as Ctesiphon, won a victory and even occupied the city in 283, but died in the late summer of the same year, so that his victory came to nothing. Diocletian turned his attention east once the situation on the northern frontier had stabilized to a reasonable degree, and in 290 he restored the exiled Armenian king Tiridates III in place of the Sassanid governor. This was ample provocation for the Sassanid king Narses to invade Armenia in 296. Diocletian then ordered his *Caesar*, Galerius, to move against Narses. Galerius however suffered a defeat south of Carrhae, probably in the autumn of 296. Still, he was able in turn to defeat Narses at Satala in Armenia, and to take possession of his camp and harem. A peace was then signed at Nisibis, restoring the rule of Rome's clients, the Arsacids, in Armenia and promising Rome five Armenian provinces and a protectorate over Virčan (Iberia). The Arch of Galerius at Thessalonica bears a reference to this peace, which would remain in force for the next forty years.

Sources

Ancient literature is particularly poor as a source for the period covered by the map. We have the emperors' biographies from Maximinus Thrax to Carinus in the *Historia Augusta*, which is largely unreliable and difficult to date (4th–6th cents.), and there are also breviaries (Aurelius Victor, 4th cent.; Eutropius, *c.* 370) and very brief prose histories (Zosimus, early 6th cent.; Zonaras, 11th cent.). Coins give some important information, and valuable information can also be gleaned from inscriptions. For instance, north of Persepolis, rock tombs, rock reliefs and inscriptions have been found dating from various periods of Persian history from Darius I (522-486 BC) to the Sassanids; these include the inscription from Naqš-e Rustam on the so-called Kaʿba-i Zardušt, which glorifies Sapor's victories over the Roman Emperors Gordian, Philippus Arabs and Valerian. This was also the location of Sapor's autobiographical inscription, the Middle Persian-Parthian-Greek trilingual inscription of the so-called *Res gestae divi Saporis*.

The map

The map shows the expeditionary movements which can be localized on the basis of the testimony of Sapor's autobiographical inscription. They show that these conflicts mostly took place in the Roman-Sassanid frontier zone up to Ctesiphon, and that these decades of grappling ultimately did not bring results of any great consequence to either side.

→ Maps pp. 213, 217, 221

The Roman Emperors

Septimius Severus 193–211
Caracalla 211–217
Macrinus 217–218
Elagabalus 218–222
Severus Alexander 222–235
Maximinus Thrax 235–238
Gordian I 238
Gordian II 238
Pupienus 238
Balbinus 238
Gordian III 238–244
Philippus Arabs 244–249
Decius 249–251
Trebonianus Gallus 251–253
Valerian 253–260
Gallienus 253–268
Claudius II Gothicus 268–270
Quintillus 270
Aurelian 270–275
Tacitus 275–276
Florian 276
Probus 276–282
Carus 282–283
Carinus 283–285
Diocletian 284–305
Galerius 305–311

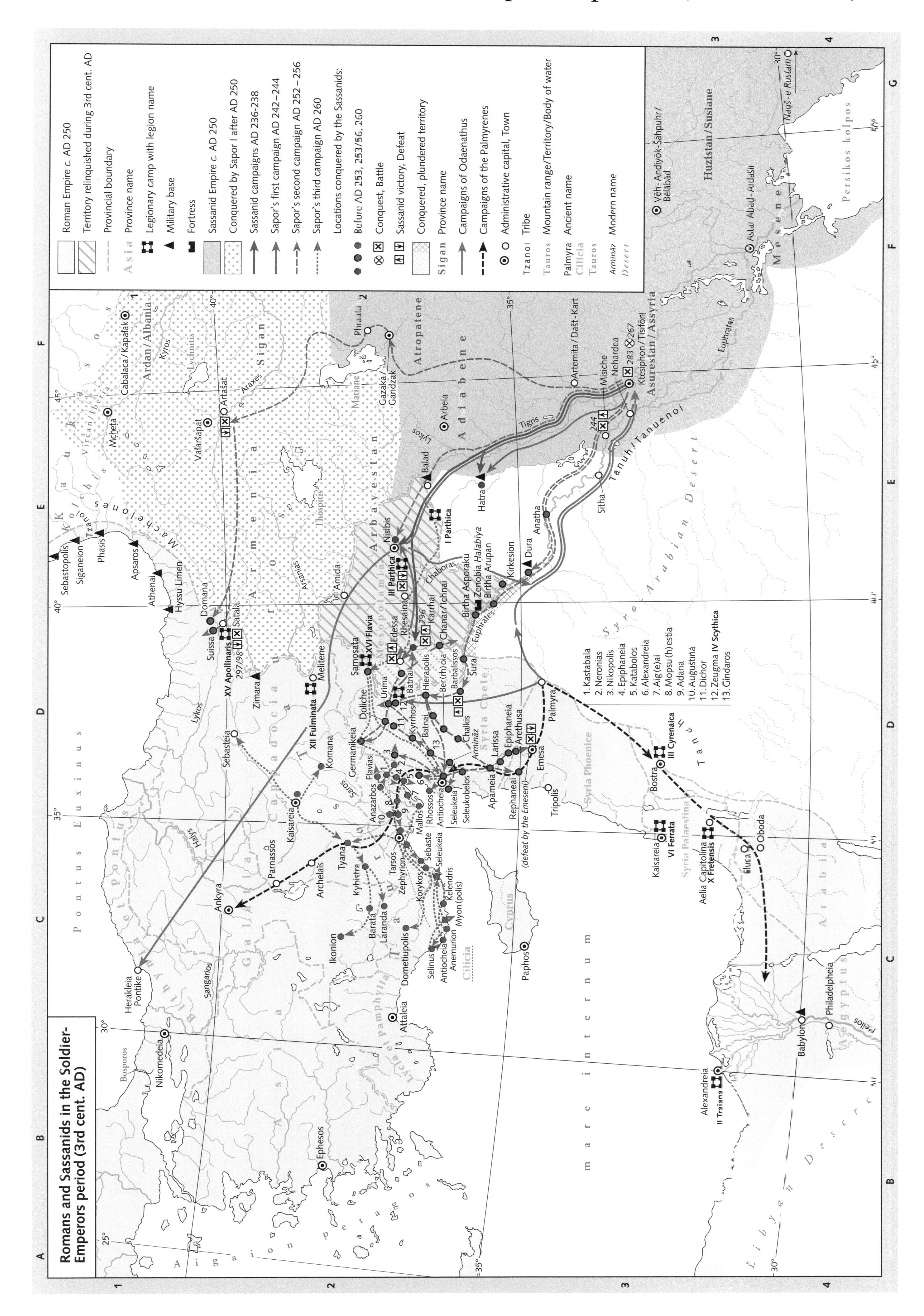
Romans and Sassanids in the Soldier-Emperors period (3rd cent. AD)
Roman Empire c. AD 250
Territory relinquished during 3rd cent. AD
Provincial boundary
Asia Province name
Legionary camp with legion name
Military base
Fortress
Sassanid Empire c. AD 250
Conquered by Sapor I after AD 250
Sassanid campaigns AD 236-238
Sapor's first campaign AD 242-244
Sapor's second campaign AD 252-256
Sapor's third campaign AD 260
Locations conquered by the Sassanids:
Before AD 253, 253/56, 260
Conquest, Battle
Sassanid victory, Defeat
Conquered, plundered territory
Sigan Province name
Campaigns of Odaenathus
Campaigns of the Palmyrenes
Administrative capital, Town
Tzanoi Tribe
Tauros Mountain range/Territory/Body of water
Palmyra Cilicia Tauros Ancient name
Arminaz Desert Modern name
1. Kastabala
2. Neronias
3. Nikopolis
4. Epiphaneia
5. Katabolos
6. Alexandreia
7. Aig(e)ai
8. Mopsu(h)estia
9. Adana
10. Augustina
11. Dichor
12. Zeugma IV Scythica
13. Gindaros

The 'Palmyrene Empire' (AD 250–272)

The Greeks adopted an Egyptian term denoting a depression with fertile soil in the desert, calling it an *oasis*. Palmyra was such an Oasis. It lies in the heart of the so-called Fertile Crescent (a term coined by the Orientalist James Henry Breasted, 1865–1935), the zone of steppe in what is now Jordan, Israel, the Lebanon, Syria and Iraq, surrounded by highlands (Libanus, Antilibanus, Antitaurus, Zagrus) and suitable for rain-fed agriculture (i.e. without artificial irrigation). This zone forms a semicircular fringe around the northern edge of the Syro-Arabian Desert, from Jordan to the mountain foothills of northern Iraq, and links the irrigation-dependent regions of Egypt in the west with those of Lower Mesopotamia in the east. Palmyra is at the eastern foot of the Palmyrene Mountains (Ǧabal as-Silsila at-Tadmurīya), where they meet the western edge of the Ǧabal al-Abyat highlands. It thus lies in a fracture zone in which a number of springs rise. This ensured the water supply to the oasis. Palmyra enjoyed excellent routes to the Euphrates and (via the river and the Euphrates Valley road) to Ctesiphon and the Persian Gulf, via Emesa to Antioch and Seleucia and from there to the Mediterranean, and via Damascus south to Egypt and the Red Sea. Caravan routes connected the city, which lay on the edge of the desert, with the Arabian hinterland to the south-east. Beyond regional limits, this advantageous location linked Palmyra into trade with India, Arabia and the Mediterranean as a trans-shipment centre. Through the first three centuries of the Roman Imperial period, it therefore grew into one of the richest and most important cities of Syria.

Palmyra had probably been part of the province of Syria since the reign of Tiberius (AD 14–37). It may have been a free city (*civitas libera*) since the time of Hadrian (117–138), and under Caracalla (211–217) it first became a Latin colony, then in 212 a Roman colony. The Palmyrene area, i.e. the *territorium* of the city of Palmyra, stretched north-eastwards as far as the Euphrates. A Roman garrison was stationed at Palmyra from the 160s, providing frontier protection, alongside the Palmyrene militia (recruited from the 1st cent. AD).

Palmyra's trading power also explains the wealth exhibited by the family of the exarch of Palmyra, Septimius Odaenathus. This family, which is recorded in inscriptions from AD 251, was one of only two senatorial families in the city. The family had possessed Roman citizenship since the reign of Alexander Severus at the latest (222-235), and Odaenathus became a member of the Palmyrene *curia*. The Palmyrenes subsequently named him tribal chieftain (*rš'*) and exarch, both titles of which Odaenathus is the first attested holder at Palmyra and which clearly indicate a position of social and military competence. His career was particularly determined by the precarious situation in which Palmyra found itself in the course of the anti-Roman campaigns of the Sassanid Sapor I. This situation demanded a senior official with executive powers, and the incumbent needed to be highly competent in military matters. Rome accepted this local empowerment when Odaenathus was adlected into the Roman Senate around AD 250 and named a *vir consularis* in 258. The not improbable inference that Odaenathus was handed the administration of the whole province of Syria Phoenice at this time is debated.

In this capacity, Odaenathus became involved in the events of the campaign, a campaign which would culminate in Sapor's capture of Valerian in 260. Odaenathus now made contact with the Sassanid king. It is conceivable that he was trying to prevent Sapor from marching through the province. The negotiations clearly failed: Sapor now took his army into the province. When he turned his troops, laden with spoils, for home and sought to cross the Euphrates (at Zeugma or Samosata), he was attacked by the forces of Odaenathus – regular Roman soldiers and Palmyrene militia – and heavily defeated.

After Valerian's catastrophe, events took the turn constantly feared during this period. Two brothers, Macrianus and Quietus, sons of the physically handicapped (and hence perceived to be unsuited to the dignity of *Augustus*) officer M. Fulvius Macrianus, usurped the imperial throne. The elder fell alongside his father in Illyricum in late 261, fighting Aureolus, while the younger was defeated by Odaenathus at Emesa and blockaded in the city, where he died. Gallienus rewarded Odaenathus for this achievement by naming him *dux Romanorum* and *corrector totius Orientis* in 261/2.

Odaenathus undertook a first campaign against the Sassanids on Gallienus' behalf in the autumn of 261. He began with a campaign in northern Mesopotamia, crossing the Euphrates at Zeugma and conquering Carrhae and Nisibis. From there, he moved down the Chaboras/Ḫābūr to its confluence with the Euphrates, then proceeded down the left bank of the latter to Nehardea. After taking Nehardea, he marched to Ctesiphon and tried to take the city. Finding no success and encountering ever greater difficulties of supply in enemy territory, he broke off the siege. After the end of this enterprise, in 263, Odaenathus and his son Herodianus took the Sassanid title of 'King of Kings' (*rex regum*). He named his wife, Zenobia, *regina* (Queen).

Odaenathus set off on a second campaign against the Sassanids in 267. This campaign, which is particularly poorly documented, led to the occupation of Ctesiphon. From there, however, Odaenathus was recalled to fight the Goths, who had invaded Bithynia by sea and were laying waste to the coastal region. He reached Heraclea Pontica still in 268, but the Goths had already put to sea and made for home on the northern shore of the Black Sea. While at Heraclea, Odaenathus and his son Herodianus were murdered by a relative, perhaps on the initiative of Gallienus (September 267/August 268).

After Odaenathus' death, his son Vaballathus took his father's inheritance: the titles and status of a *consularis*, of *corrector totius Orientis, dux Romanorum* and *rex regum*. Zenobia, his mother, acted on behalf of the boy, who was only just ten years of age. Until 270, nothing happened within her sphere of action that would indicate a usurpation. The titles now held by Vaballathus (*vir clarissimus, rex, consul, imperator, dux Romanorum*) still proclaimed him the emperor's representative in the East, and nothing more. Joint datings according to the reigns of the emperor Aurelian (270–275) and Vaballathus, always in this order – from December 270 to April 272 – make this particularly clear, as do the coins minted in Vaballathus' name at Alexandria. Zenobia, too, bore only the epithet *pia*, reserved for the wives of emperors, alongside her title of *regina*. She did not bear the title of *Augusta*. But in spite of her ostensible subordination to the *Augustus*, it seems that neither Gallienus, Claudius II nor Aurelian approved of the grandiose self-presentation of the Palmyrene rulers. Nor, however, did they act against it.

All this changed when Zenobia took possession of the provinces of Arabia and Aegyptus (with Alexandria), as if she were the successor to Odaenathus. In Arabia, the Roman garrison (*legio III Cyrenaica*) did not long resist the Palmyrene occupation. Late in the summer of 270, Alexandria, too, was occupied by Zenobia's forces after brief hostilities, in the absence of the *praefectus Aegypti* Tenagino Probus, who had been ordered away to fight Gothic pirates in the eastern Mediterranean. But when Probus found out about this, he hastened back and drove the Palmyrenes out of the country. The Palmyrenes marched back into the Nile Delta in the autumn of the same year, but they were again repulsed. This time, Probus tried to block the Romans' escape route. A battle ensued on the right bank of the Nile at Babylon. Probus' forces were annihilated; Probus himself was captured and chose to take his own life. After this success, Zenobia planned for 271 to take the lands formerly under her husband's administration in western Asia Minor as well. But since the cities in this territory had come out against Zenobia in the light of Aurelian's assumption of power, she abandoned this plan after a few feeble forays. Her troops may have reached Ancyra.

Aurelian was quite determined to reclaim the little empire that had developed around Palmyra for the Roman Empire. He therefore crossed the Bosporus into Asia Minor with a powerful army in the spring of 272 to proceed against the Palmyrenes. In all likelihood, it seems that this was the situation in which Zenobia and Vaballathus openly moved to claim power in the Roman Empire, by assuming the titles of Augusta and Augustus. Marching through Asia Minor, Aurelian was only challenged at Tyana; after a brief siege, the city was taken. He continued via Tarsus, taking the coast road towards Antioch. In what seems to have been an evasive manoeuvre, he reached the Orontes some 20 km east of Antioch: battle was joined here in May 272, and the Palmyrenes were utterly defeated. The emperor entered Antioch, encountering no resistance. The Palmyrenes had given up the city and fled to Emesa. It was here, in the high summer of 272, that the decisive battle of the war took place. In spite of their superiority in cavalry, the Palmyrenes were the losers. Zenobia fled to Palmyra, but because the city was not walled, it could not be held against Aurelian's army. Zenobia probably fled again, towards Dura, where she was captured with Vaballathus as she tried to cross the river.

Sources

The structural remains of Palmyra are informative (religion and economy). A number of inscriptions (Greek and Palmyrene) illustrate the history of the city, as do coins and, for Egypt, papyri. There is little that is reliable in the literary sources. There are the Latin breviaries (Aurelius Victor, Festus, Eutropius, *Historia Augusta*, *Epitome de Caesaribus*) with their spare and often confused information, the basis of which the historian Alexander Enmann (1884) deduced to be an imperial history (the so-called '*Enmannsche Kaisergeschichte*') up to the reign of Constantine I (composed between 337 and 361). The *New History* of the Greek scholar Zosimus is of a higher order, and it provides some information on Palmyrene history, esp. the occupation of Egypt. Also worthy of attention is the *Thirteenth Sibylline Oracle*, which offers revealing details of historical interpretation in the form of a *vaticinium ex eventu*.

→ Map p. 219; map s.v. Palmyra, BNP 10, 2007

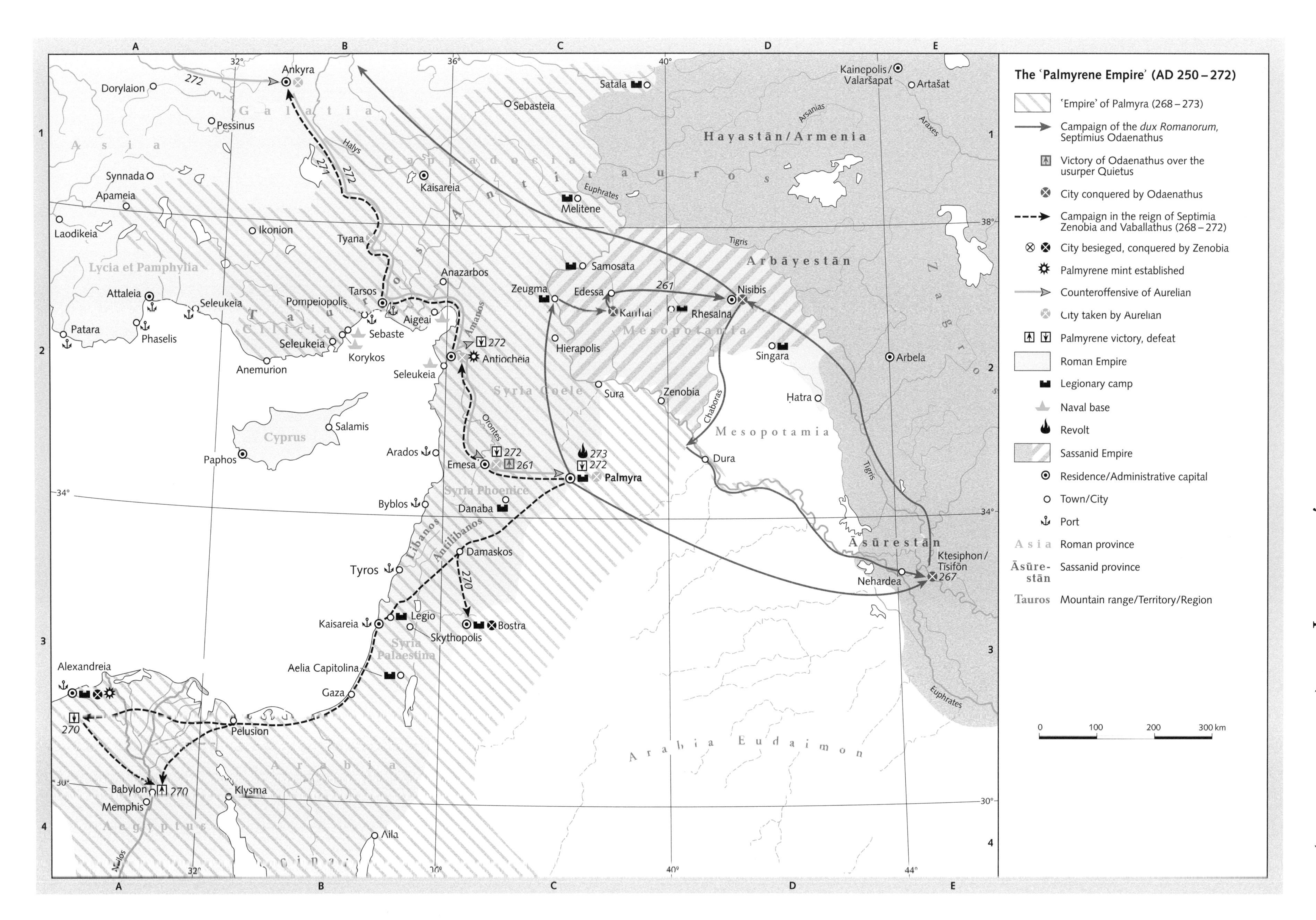
The ‘Palmyrene Empire’ (AD 250–272)
‘Empire’ of Palmyra (268–273)
Campaign of the *dux Romanorum*, Septimius Odaenathus
Victory of Odaenathus over the usurper Quietus
City conquered by Odaenathus
Campaign in the reign of Septimia Zenobia and Vaballathus (268–272)
City besieged, conquered by Zenobia
Palmyrene mint established
Counteroffensive of Aurelian
City taken by Aurelian
Palmyrene victory, defeat
Roman Empire
Legionary camp
Naval base
Revolt
Sassanid Empire
Residence/Administrative capital
Town/City
Port
Asia Roman province
Āsūre-stān Sassanid province
Tauros Mountain range/Territory/Region
0 100 200 300 km
Ankyra
Dorylaion
Pessinus
Synnada
Apameia
Laodikeia
Ikonion
Tyana
Tarsos
Pompeiopolis
Seleukeia
Attaleia
Patara
Phaselis
Sebaste
Korykos
Anemurion
Aigeai
Anazarbos
Kaisareia
Sebasteia
Satala
Melitene
Samosata
Zeugma
Edessa
Hierapolis
Rhesaina
Nisibis
Singara
Arbela
Hatra
Sura
Zenobia
Dura
Antiocheia
Seleukeia
Salamis
Paphos
Cyprus
Arados
Emesa
Palmyra
Byblos
Danaba
Damaskos
Tyros
Kaisareia
Legio
Skythopolis
Bostra
Aelia Capitolina
Gaza
Alexandreia
Pelusion
Babylon
Memphis
Klysma
Aila
Kainepolis/Valaršapat
Artašat
Ktesiphon/Tisifōn
Nehardea
Hayastān/Armenia
Arbāyestān
Āsūrestān
Mesopotamia
Arabia Eudaimon
Lycia et Pamphylia
Galatia
Asia
Cappadocia
Cilicia
Syria Coele
Syria Phoenice
Syria Palaestina
Arabia
Aegyptus
Tauros
Zagros
Amanos
Libanos
Antilibanos
Halys
Euphrates
Tigris
Arsanias
Araxes
Chaboras
Orontes
Nilos
261
267
270
271
272
273

Imperium Galliarum – the 'Gallic Empire' (AD 260–274)

Contemporaries had no conception of what we know as the 'Gallic Empire' or 'Imperium Galliarum' as a separate political entity. They saw the 'emperors' at Colonia Agrippinensis/Cologne and Augusta Treverorum/Trier as men who claimed dominion (with whatever, if any, justification) over the entire Roman Empire, but who were at present only able to implement this dominion in the western part of the Empire. The coins of Postumus, the founder of the 'Gallic Empire', make this claim with the correct imperial title, but he appears never to have attempted to enforce his claim against the imperial court at Rome. Gallienus (the legitimacy of whose rule historical tradition has never contested to this day), conversely tried on several occasions to eliminate Postumus as a usurper.

It is generally recognized in surveys of this field that the Gallic 'emperors' very effectively and successfully accomplished tasks that Rome either did not or could not accomplish, in particular the securing of the Rhine frontier, which was under threat from Germanic tribes to the east. It may have been at this time that the Franks, among others, pushed ahead to the Rhine. They are first mentioned in our sources in AD 287. Inscriptions reveal Postumus' rule to have been initially recognized from Spain to Britain, via Gaul to the east. In addition, an inscription recently discovered at Augusta Vindelicum/Augsburg attests to his acknowledgment as emperor in the province of Raetia in AD 260. It is a matter of dispute to what extent he was able to hold Germania superior west of the Rhine – unambiguous evidence for this is lacking. It is striking that the Spanish, Gaulish and British provinces had already once before found themselves in the same constellation with Raetia: at the beginning of the Year of the Four Emperors, AD 69, after Vitellius' victory over Otho.

Marius, Postumus' successor, lost Spain to Claudius II Gothicus, the emperor at Rome. Raetia had already been lost to Gallienus in 265.

The residence under Postumus was at Colonia Agrippinensis, and Victorinus moved it to Augusta Treverorum. Within their dominions, Postumus and his successors brought peace and order, creating the infrastructure for a prosperous economy and some degree of legal security. The reincorporation of the territory into the Imperium Romanum took place in association with a battle, which Tetricus I from the start gave up for lost, on the Catalaunian Plains in the spring of AD 274.

The 'Emperors' of the Imperium Galliarum
Postumus 260–269
Laelianus June 269 (usurper against Postumus)
Marius July/August 269
Victorinus October/November 269–271
Tetricus I 271–274
Tetricus II 273–274 (son and Caesar of Tetricus I)
Faustinus 274 (usurper against Tetricus I)

The sources

The state of the sources for the history of the Gallic Empire of 260-274 is distinctly poor. The literary sources give little information. There are the imaginative emperor biographies of the *Historia Augusta*, the historical sketches of such as Aurelius Victor, Eutropius and Zosimus and the chronicles of Zonaras and Polemius Silvius. No more informative, but more reliable are the inscriptions made under the Gallic emperors. The same is true of the coins of the Imperium Galliarum, which along with other, exceedingly sparse information at least make clear that the Gallic emperors maintained intensive contacts with Germania magna. Understandably, the Egyptian papyri give no insight at all into conditions in Gaul, although they do provide some indication of developments of relevance throughout the Empire, concerning the period before and after the Gallic emperors' seizure of power. Finally, there is the archaeological evidence. However, this most productively provides historians with details when amplified by literary references, and this is where the Gallic Empire is so ill-served.

The map

The map clearly shows the importance of inscriptions and coins to the history of the 'Gallic Empire' in view of the threadbare nature of the literary sources. The inscriptions set up under Postumus show the 'Empire' at its full extent, from the Spanish provinces to the Gaulish and British, and eastwards as far as Raetia. Raetia and the Spanish provinces were lost either while Postumus still reigned or under Marius or Victorinus, but Tetricus still ruled as far as the frontier of Hadrian's Wall in Britain. The evidence of the imperial gold coins found in Germania magna as far east as the Bay of Gdańsk shows that trading links from Gaul reached far to the east. The coinage attesting to Postumus at Mediolanum, however, is the product of a brief episode when Aureolus, a cavalry officer of Gallienus, defected from the latter and declared for Postumus.

Sources

Literary sources: Aur. Vict. Caes. 33–35; (Ps.-)Aur. Vict. Epit. Caes. 32–35; Eutr. 9,9–13; SHA Tyr. Trig.; SHA Aurelian.; SHA Clod.; Zon. 12,23–27; Zos. 1,29–62.
Epigraphic sources: CIL II 4919; 4943; 5736; VII 287; 802; 820; 822 f.; 1150 f.; 1160–1162; XIII 633; 3035; 3163; 3679; 5868; 6779; 8879; 8882 f.; 8925; 8927; 8955–8964; 8970; 8972; 8975; 8977; 8999 f.; 9006; 9012; 9023; 9040 f.; 9092; 11311; 11976; 12090; 12241; AE 1890, 154; 1930, 35; 1960, 175; 1969–1970, 415; 1971, 23; 279; RIB 2232; 2238; 2241; 2255; 2287; 2296; P. Wuilleumier, Inscriptions Latines des Trois Gaules, 1963, 465; JRS 50, 1960, 238 no. 13; 52, 1962, 195 nos. 23 f.; Germania 21, 1937, 29; Annales de Bretagne 85, 1978, 349–360; E. Espérandieu, Inscriptions Latines de la Gaule Narbonnaise, 1929, 656; *on the inscription from Augsburg*: L. Bakker, Rätien unter Postumus – das Siegesdenkmal einer Juthungenschlacht im Jahr 260 n. Chr., in: Germania 71, 1993, 369–386.

Numismatic sources: J. De Witte, Recherches sur les empereurs, qui ont régné dans les Gaules au IIIe siècle de l' ère chrétienne, 1868; G. Elmer, Die Münzprägung der gallischen Kaiser in Köln, Trier und Mailand, in: BJ 146, 1941, 1–106; P.H. Webb, The Roman Imperial Coinage 5,2, 1933.

Literature

J.F. Drinkwater, The Gallic Empire. Separatism and Continuity in the North-Western Provinces of the Roman Empire A.D. 260–274, 1987; I. König, Die gallischen Usurpatoren von Postumus bis Tetricus, 1981; J. Lafaurie, L' empire Gaulois. Apport de la numismatique, in: ANRW II.2, 1975, 853–1012.

Eboracum
Britannia inferior
Camulodunum
Britannia superior
Londinium
Germania magna
Colonia Agrippinensis
Franci
Germania inferior
Chatti
Bucinobantes
Augusta Treverorum
Mogontiacum
Vandali
Iuthungi
Raetovarii
Danuvius
Belgica
274
Campi Catalauni
Augusta Vindelicum
Lentienses
Brisgavi
Rhenus
Lugdunensis
Germania superior
Raetia
Augustodunum
Lemonum
Forum Claudii Vallensium
Mediolanum Santonum
Lugdunum
Alpes Graiae et Poeninae
Mediolanum
Segusio
Aquitania
Burdigala
Alpes Cottiae
Gallia transalpina
Alpes maritimae
Cemenelum
Narbo
Roma
Hispania citerior
Tarraco
Hispania ulterior
Augusta Emerita
Palma
Corduba
Hispania Baetica
Vistula
Imperium Galliarum – the 'Gallic Empire' (AD 260 – 274)
Domains of Postumus and his successors
Domains of the Emperor at Rome
Provincial boundary
Raetia Province name
Residence
Administrative capital
Mints of the 'Gallic Emperors'
Inscriptions of the 'Gallic Emperors':
Inscriptions of Postumus
Inscriptions of Victorinus
Inscriptions of Tetricus
Gold coins of the 'Gallic Emperors' outside Gaul:
Postumus
Victorinus
Tetricus
Battle
Major pass
Chatti Tribe/People
Campi Catalauni Region/Territory
0 100 200 300 400 500 km

The provincial administration of the Roman Empire in the 4th cent. AD

The provincial administration of the Roman Empire in the 4th cent. AD

The entire Roman Empire was administratively reshaped under the rule of the emperors Diocletian (284–305) and Constantine I (306–337) and their successors. One of the reforms implemented by these emperors concerned the subdivision of the civil administration of the Empire into *praefecturae* under *praefecti praetorio*, *dioeceses* under *vicarii* and *provinciae* under *praesides*. The division of the Empire into *dioeceses* began under Diocletian; the *praefecturae* were a creation of Constantine's. He removed the military competences of the *praefecti praetorio* – these passed to the office of the *magister militum*, also newly created – and assembled the most important civil competences, such as jurisprudence, taxation and the superintendence of public order, in their hands. The Empire was divided into four *praefecturae (Galliae, Italia, Illyricum, oriens)*, each under a praetorian prefect. The borders of these prefectures ran more or less parallel to the meridians and across the Mediterranean Sea. The particular importance of the praetorian prefectures was as a link between territorial administration and the imperial court. Rome itself had a special status in this system (excepted from the administration of the rest of Italy) under the *praefectus urbi*, as, from AD 359, did Constantinople. The intricately subdivided and strictly organized bureaucracy, which gradually came to span the entire Empire, and by means of which the emperors pursued the goal of better control over the huge Empire with its plethora of social, legal, economic and political problems, did not meet with universal approval. The levelling of Italy's status to that of the rest of the Empire (the process is known as 'provincialization') seems particularly to have evoked opposition. Italy was divided into two *dioeceses*, Italia annonaria and Italia suburbicaria, with six and seven provinces respectively. The Christian writer Lactantius (3rd/4th cents.), described such reforms under Diocletian in these terms: 'In order to fill everyone with fear, the provinces, too, were divided into pieces. Many governors and numerous subordinate officials laid their heavy oppression upon every region, almost indeed on every town' (Lact. De mort. pers. 7,4).

These profound changes to the Roman administrative and provincial system were not made according to any kind of integral concept, much less at a single stroke, but gradually, as and when a favourable opportunity for such restructuring presented itself.

List of *praefecturae*, *dioeceses* and *provinciae* (with administrative capitals of *praefecturae* and *dioeceses*)

I Praefectura oriens (Constantinople)
I 1 Dioecesis oriens (Antioch)
1 Libya superior
2 Libya inferior
3 Thebaïs
4 Aegyptus Iovia
5 Aegyptus Herculia
6 Arabia
7 Arabia nova
8 Augusta Libanensis
9 Palaestina
10 Phoenice
11 Syria Coele
12 Augusta Euphratensis
13 Cilicia
14 Isauria
15 Cyprus
16 Mesopotamia
17 Osrhoëna

I 2 Dioecesis Pontica (Nicomedia)
1 Bithynia
2 Cappadocia
3 Galatia
4 Paphlagonia
5 Diospontus
6 Pontus Polemoniacus
7 Armenia minor

I 3 Dioecesis Asiana (Ephesus)
1 Lycia et Pamphylia
2 Phrygia prima
3 Phrygia secunda
4 Asia
5 Lydia
6 Caria
7 Pisidia
8 Hellespontus
9 Insulae

I 4 Dioecesis Thraciae (Philippopolis)
1 Europa
2 Rhodope
3 Thracia
4 Haemimontus
5 Scythia
6 Moesia inferior

II Praefectura Illyricum (Thessalonica)
II 1 Dioecesis Moesiae (Thessalonica)
1 Dacia mediterranea
2 Dacia ripensis
3 Moesia superior/Margensis
4 Dardania
5 Macedonia
6 Thessalia
7 Achaea
8 Praevalitana
9 Epirus nova
10 Epirus vetus
11 Creta

III Praefectura Italia (Mediolanum)
III 1 Dioecesis Pannoniae (Aquincum)
1 Pannonia inferior
2 Savensis/Savia
3 Dalmatia
4 Valeria
5 Pannonia superior
6 Noricum ripense
7 Noricum mediterraneum

III 2.1 Dioecesis Italia annonaria (Mediolanum)
1 Venetia et Histria
2 Aemilia et Liguria
3 Flaminia et Picenum
4 Tuscia et Umbria
5 Alpes Cottiae
6 Raetia

III 2.2 Dioecesis Italia suburbicaria (Roma)
1 Campania
2 Apulia et Calabria
3 Lucania et Brutii
4 Samnium
5 Sicilia
6 Sardinia
7 Corsica

III 3 Dioecesis Africa (Carthago)
1 Africa proconsularis
2 Byzacena
3 Numidia Cirtensis
4 Numidia militiana
5 Mauretania Caesariensis
6 Mauretania Sitifensis
7 Tripolitana

IV Praefectura Galliae (Augusta Treverorum)
IV 1 Dioecesis Britanniae (Londinium)
1 Britannia prima
2 Britannia secunda
3 Maxima Caesariensis
4 Flavia Caesariensis

IV 2 Dioecesis Galliae (Augusta Treverorum)
1 Belgica prima
2 Belgica secunda
3 Germania prima
4 Germania secunda
5 Sequania
6 Lugdunensis prima
7 Lugdunensis secunda
8 Alpes Graiae et Poeninae

IV 3 Dioecesis Viennensis (Vienna/Constantina)
1 Viennensis
2 Narbonensis prima
3 Narbonensis secunda
4 Novempopulana
5 Aquitania prima
6 Aquitania secunda
7 Alpes maritimae

IV 4 Dioecesis Hispaniae (Augusta Emerita)
1 Baetica
2 Lusitania
3 Carthaginiensis
4 Gallaecia
5 Tarraconensis
6 Mauretania Tingitana
7 Baleares

Sources

The state of the sources is exceedingly poor. The earliest evidence for the nascent imperial reforms, and the most important for the adjacent map, is the *Laterculus Veronensis*, a list of twelve *dioeceses* each with a different number of constituent provinces (totalling 95). This text is named after the single extant MS from Verona (7th cent.), and it shows the state of provincial administration in the western and eastern halves of the Empire at different times (AD 303/14 and 314/24 respectively). – Rufius Festus was a high official of the Roman Empire in the reign of the emperor Valens (364–378). His summary of Roman history shows its author's particular interest in administrative history, esp. where it describes the creation of the provinces in the east of the Empire. – The *Laterculus* written by Polemius Silvius in 448/9 is a chronological register which also contains a list of 113 provinces of the Roman Empire. The work reveals itself to be a compilation from a variety of sources, none of which has survived. – The *Notitia Dignitatum* is a vademecum for the official use of Roman functionaries. It was written around AD 430, but it does not consistently reproduce the *status quo* of that date – earlier versions from the second half of the 4th cent. can be detected in it. It lists the offices of civil and military administration; the imperial administrative structures can be deduced from it. – In the 6th cent., Hierocles composed a *Synecdemus* (a 'travel guide' and vademecum in one) for private use. In it, he lists the 64 provinces of the Eastern Roman Empire with their 923 cities, probably basing his work on a handbook from the reign of Theodosius II (408–450). This register has likewise not been updated to represent the *status quo* consistently. – Joannes Laurentius Lydus, who held high civil and military office for many years under Justinian (527–565), wrote a thorough introduction to the bureaucracy of the state of Late Antiquity and its traditions in his three-volume treatise *De Magistratibus reipublicae Romanae* ('On the Magisterial Offices of the Roman Commonwealth'). Here, too, we learn details of the provincial administration of the 4th cent. in particular. – Also worthy of attention in this regard are the lists of participants appended to the various protocols of the Councils held at Nicaea, Serdica and Constantinople, in 325, 343/4 and 381 respectively.

Literature

T.D. Barnes, The New Empire of Diocletian and Constantine, 1982; Id., Emperors, Panegyrics, Prefects, Provinces and Palaces (284–317), in: Journal of Roman Archaeology, 9, 1996, 532–552; E.S. Dula bahn, Studies on the Laterculus of Polemius Silvius, 1987; J.W. Fadie, The Breviarium of Festus, 1967; K.L. Noethlichs, Zur Entstehung der Diözesen als Mittelinstanz des spätrömischen Verwaltungssystems, in: Historia 31, 1982, 70–81.

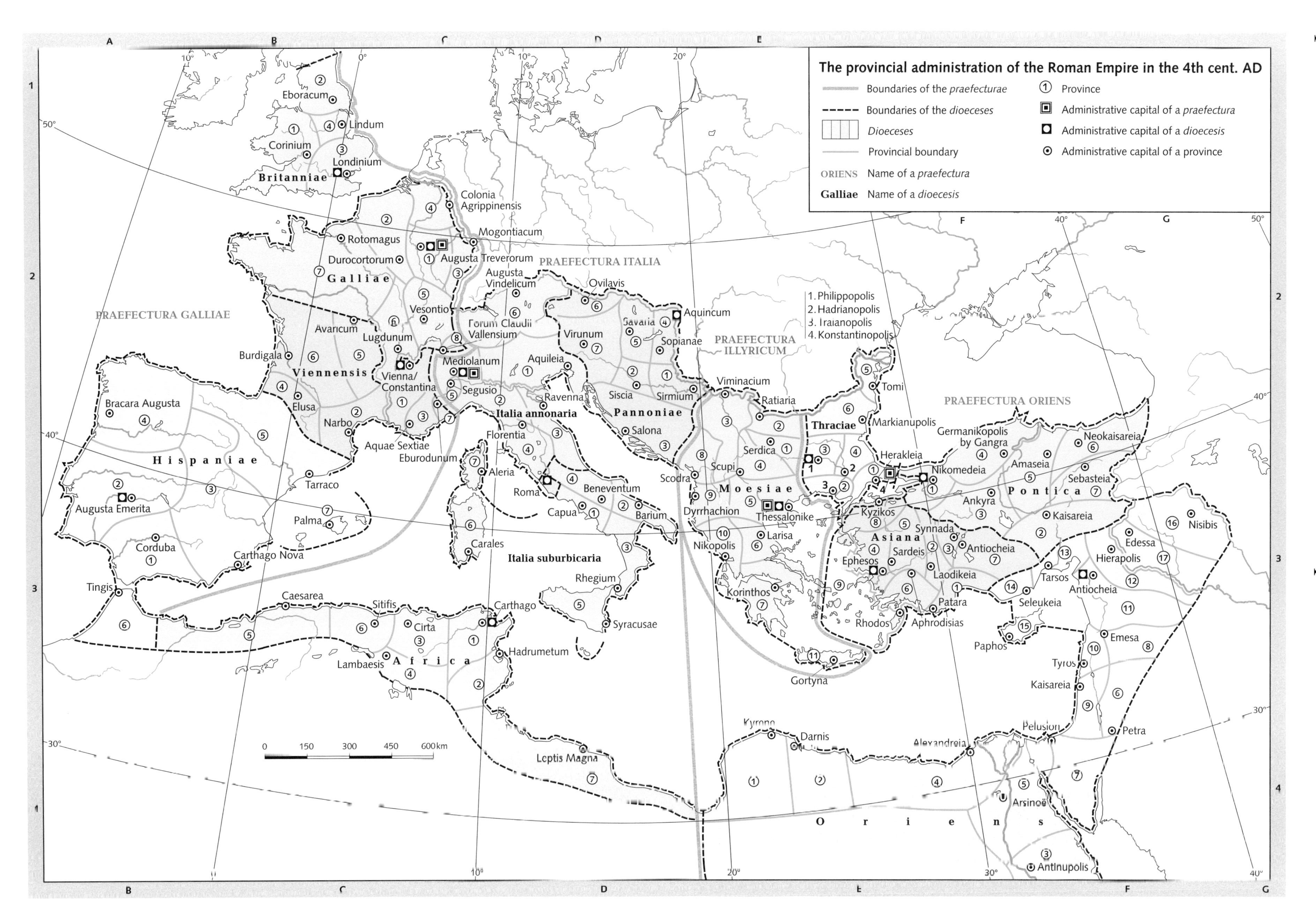
The provincial administration of the Roman Empire in the 4th cent. AD
Boundaries of the praefecturae
Boundaries of the dioeceses
Dioeceses
Provincial boundary
ORIENS Name of a praefectura
Galliae Name of a dioecesis
Province
Administrative capital of a praefectura
Administrative capital of a dioecesis
Administrative capital of a province
PRAEFECTURA GALLIAE
PRAEFECTURA ITALIA
PRAEFECTURA ILLYRICUM
PRAEFECTURA ORIENS
Britanniae
Galliae
Viennensis
Hispaniae
Italia annonaria
Italia suburbicaria
Pannoniae
Moesiae
Thraciae
Asiana
Pontica
Africa
Oriens
1. Philippopolis
2. Hadrianopolis
3. Traianopolis
4. Konstantinopolis
Eboracum
Lindum
Corinium
Londinium
Rotomagus
Durocortorum
Augusta Treverorum
Colonia Agrippinensis
Mogontiacum
Augusta Vindelicum
Avaricum
Lugdunum
Vesontio
Burdigala
Elusa
Narbo
Vienna/Constantina
Aquae Sextiae
Eburodunum
Segusio
Mediolanum
Aquileia
Forum Claudii Vallensium
Ovilavis
Virunum
Savaria
Sopianae
Aquincum
Siscia
Sirmium
Salona
Ravenna
Florentia
Roma
Aleria
Carales
Capua
Beneventum
Barium
Rhegium
Syracusae
Viminacium
Ratiaria
Serdica
Scupi
Scodra
Dyrrhachion
Thessalonike
Larisa
Nikopolis
Korinthos
Gortyna
Tomi
Markianupolis
Herakleia
Nikomedeia
Kyzikos
Ephesos
Sardeis
Synnada
Laodikeia
Patara
Aphrodisias
Rhodos
Germanikopolis by Gangra
Ankyra
Amaseia
Neokaisareia
Sebasteia
Kaisareia
Antiocheia
Tarsos
Seleukeia
Paphos
Edessa
Hierapolis
Nisibis
Emesa
Tyros
Kaisareia
Petra
Pelusion
Arsinoë
Antinupolis
Alexandreia
Darnis
Kyrene
Leptis Magna
Hadrumetum
Carthago
Cirta
Sitifis
Lambaesis
Caesarea
Tingis
Bracara Augusta
Augusta Emerita
Corduba
Carthago Nova
Tarraco
Palma
0 150 300 450 600 km

The growth of Christian congregations, 1st-4th cents. AD

I. Mission

From its very earliest beginnings, Christianity was a missionary religion, in accordance with the 'Great Commission' of the New Testament (Mt 28: 18–20; Mk 16: 15f.; Jo 20: 21). Paul already planned to expand his missionary activities to Spain (Rom 11,25f.). In the mid 40s, the so-called Convention of the Apostles took place at Jerusalem, and from it came important stimuli for the sending out of missionaries, including Paul (Acts 15; Gal 2: 1–10). However, this does not mean that the worldwide mission was organized according to a plan. Synagogues were often the centres of such missionary work. But it also often emerged from the campaigning power of individual Christians and itinerant preachers. It was an 'incidental' missionary phenomenon. Planned missionary activity can only be demonstrated from the 3rd cent., and that mainly in those regions where Christianity had already gained a substantial foothold, e.g. North Africa, Egypt and Asia Minor. Here, an ecclesiastical organizational structure was developing as new episcopal sees were founded; this made planned missionary involvement in the *territoria* of the cities and across the countryside possible.

Thus developed the first Christian congregations in the Greek cities of the western coast of Asia Minor and Greece itself (Ephesus, Thessalonica, Corinth). At first, it appears that it was the lower urban classes who were won over to Christianity – the poor, slaves, freedmen and foreigners, artisans and farmers in the urban surroundings. The new creed began to spread into higher strata of society in the late 2nd cent., appearing e.g. among the members of the Senate and the equestrian order, war veterans and the staff (slaves and freedmen) of the imperial court at Rome. By the mid 3rd cent., it is estimated that there were some sixty bishops in Italy.

Increasing numbers began to throng to the Christian communities with the emperor Galerius' (AD 305-311) Edict of Toleration of April 311 (Lact. De mort. pers. 34; Euseb. Hist. eccl. 8,17,1–11) and its adoption by the emperors Constantine (306–337) and Licinius (308–324) in the Edict of Milan of 313. By his law of 28 February 380 (Cod. Theod. 16,1,2), the emperor Theodosius I (379–395) issued a number of provisions (together with Gratian in the west of the Empire) by which Christianity, from a position of merely being tolerated, became the state religion, all other cultic activities being prohibited.

Although the attendance registers of the Synod of Nicaea (325) show that there was a Christian congregation in almost every city of the Roman Empire (well-attested for Alexandria, Antioch, Edessa in the Osrhoëne, Eumeneia and Orcistus in Phrygia, Maiuma, Bostra), Christianity had certainly not penetrated all rural areas. In some cities and regions of the Empire, the Christian mission was met with reserve or even hostility (Carrhae, Heliopolis and Emesa, Philae in Egypt). Northern Italy, Gaul and Britain (Druidic tradition) were only Christianized at a very late stage, i.e. in the 4th and 5th cents.

Christianity spread beyond the borders of the Roman Empire in the East in particular, where it probably became the state religion of Armenia under the Arsacid Tiridates III as early as AD 301. It also received state recognition in Iberia (337 or 356). There was successful missionary work in the Germanic north as well.

II. The reactions of the Roman authority to the spread of Christianity

Paul, one of the earliest Christian missionaries (→ map p. 228), already met resistance of various kinds (e.g. religious or economic in motivation) in various cities (Iconium, Ephesus, Philippi, Thessalonica). The renowned tolerance of the Roman authorities towards the countless cults of the Empire found its limit where public peace and order were disturbed by the followers of particular cults. To this extent, the resistance which Christians encountered could have been politically motivated on the part of the authorities.

Under Claudius, in AD 50, the Jews, with their leader Chrestus, were expelled from Rome as troublemakers (Suet. Claudius 25,4). There have been attempts to link this 'Chrestus' with 'Christus', but they are undermined by chronological and geographical objections.

The earliest evidence of both the existence and official proceedings against Christians (*Chrestiani*) dates from the reign of Nero. To divert the suspicion that he might himself have ordered the fire of Rome in AD 64, the emperor acted against the Christians, who were generally disliked 'for their abominations' (Tac. Ann. 15,44,2). They were accused, convicted and executed. The crime of which they were accused was probably arson. But Tacitus speculates that their fate was sealed by their fundamental misanthropy.

Little credence can be attached to a remark in Hegesippus, a church historian active in the late 2nd cent., according to which Domitian (AD 81–96) ordered a persecution of the Christians, but then retracted his order on becoming convinced of the harmlessness of the Christian faith. There are more trustworthy testimonies, however, for the reign of the emperor Trajan (98–117): a letter to the emperor from the younger Pliny, the governor of Bithynia et Pontus, and the emperor's reply (Plin. Epist. 10,96f.). However, these do not in fact concern the persecution of Christians, but the means of dealing with them according to the fundamental principles of the rule of law. The emperor instructed his governor here not to hunt out Christians, and not to act on any anonymous denunciations. But if they were properly accused and convicted, they were to be punished. These instructions leave much unresolved – as the emperor himself was the first to admit – but there is no question of this being a planned persecution of Christians.

There were repeated pogroms against Christians, e.g. in 165/68 in Smyrna, where Bishop Polycarpius died and was proclaimed a martyr. In 177, Christians were killed at Lugdunum. There were occasional persecutions, perhaps as early as the reign of Commodus (180-192) but certainly under Septimius Severus (193-211), at Rome, Carthage, Alexandria and various places in Cappadocia. But these, again, were not organized actions.

Even the so-called 'anti-Christian persecution' under Decius (249–251) was not solely directed at Christians or any other group, but was a systematically organized action across the entire Empire requiring all citizens on pain of death to make a declaration of loyalty to the Empire: it was a general command to make sacrifices (*supplicatio*). From the emperor's perspective, it concerned the unity of the Empire and its populace. The sacrifice was to the state gods, and had to be certified by the local authorities. Some fifty such certificates (*libelli*) have been preserved on papyrus. As many Christians refused to make this sacrifice, the action led to a persecution of Christians across the Empire, but the death of the emperor in June 251 brought this to an end. Two laws issued by Valerian in 257 and 258 were particularly directed against Christians in the Empire. His son Gallienus repealed these laws, but the sources do not reveal the reasons for this.

On 23 February 303, the church at Nicomedia was destroyed by imperial order. This was the beginning of the only true organized persecution of Christians across the whole Empire on the basis of an edict. The emperor Diocletian issued the edict the next day. The purpose of the decree seems to have been to eliminate the Christian ecclesiastical organization in the context of a renewal of society and the entire state. The edict ordered the incarceration of presbyters and bishops, forbade Christian services and ordered the destruction of churches and the burning of Christian writings. It ordered the imprisonment of Christian public officials, and Christians were henceforth banned from taking public office. Refusal to sacrifice to the emperor was punishable by death. This edict remained in force until AD 311, and was enforced with varying degrees of strictness in the various regions of the Empire – leniently by Constantius Chlorus and Constantine at Trier, very much more restrictively by Galerius in Nicomedia. Yet it was Galerius of all people who finally retracted Diocletian's edict in 311, stopping the persecutions and making Christians discharge their duties for the state.

The sources and the map

Reliable evidence of the demographic spread of Christianity in the Roman Empire is difficult to obtain. One revealing work for the social history of the early Christians calls itself *The Shepherd of Hirmas* and is a Christian prophetic vision, written at Rome in the face of the threat of persecution, i.e. probably in the early 4th cent. The lists of signatories at the synods give important information, as do the so-called 'Acts of the Martyrs' and other reports of martyrdom. There are also inscriptions, and the writings of Christian authors such as Tertullian, Minucius Felix and Eusebius. The letters of the emperor Julian ('the Apostate') (361–363) contain indications of the organization of Christian congregations. – The map is restricted to general information on the spread of Christianity, with a selection of attested congregations, organizational structures (patriarchates, archiepiscopal and episcopal sees) and instances of communication (synods). Centres of persecution are likewise shown only selectively.

Sources

Christian funerary inscriptions: ILCV 56; 276; 277; 427; 349; 705A; 3032; 3872; 3315; 3332; 3915; CIL II 2, 5458; VI 31749a; *Claudius:* Suet. Claudius 25,4; Acts 18,1; Tac. Ann. 15,44,3; *Nero:* Tac. Ann. 15,44,2; Suet. Nero 16; *Domitian:* Euseb. Hist. eccl. 3,20,1–7; *Trajan:* Plin. Epist. 10,96f.; *Septimius Severus:* SHA Septimius Severus 17,1; *Commodus/Septimius Severus:* Euseb. Hist. eccl. 5,28,8–12; 6,1; 6,2,2; 6,2,12; 6,6; 6,3,1; 6,3–5; 6,11,5; Tertullian, Epist. ad Scapulam 3,1; 3,5; Tert. Apol. 35; cf. 50,13; Passio SS. Perpetuae et Felicitatis; but cf. SHA Septimius Severus 17,1; *Smyrna:* Martyrium Polycarpi; *Lugdunum:* Euseb. Hist. eccl. 5,1; *Decius:* Euseb. Hist. eccl. 6,41,9f.; Lact. De mort. pers. 4,2; *Valerian:* Euseb. Hist. eccl. 7,10f.; Cypr. Epist. 80,1f.; *Gallienus:* Euseb. Hist. eccl. 7,23; *Diocletian:* Lact. De mort. pers. 12,2; *Galerius:* Lact. De mort. pers. 34; Euseb. Hist. eccl. 8,17.

Literature

A. Bludau, Die ägyptischen Libelli und die Christenverfolgung des Kaisers Decius, 1931; E. Ferguson, Backgrounds of Early Christianity, 2003; W.H.C. Frend, Martyrdom and Persecution in the Early Church, 1965; A. v. Harnack, Die Mission und Ausbreitung des Christentums in den ersten drei Jahrhunderten, 2 vols., 1924; E.J. Hunt, Christianity in the Second Century, 2003; H. Jedin, K.S. Latourette, J. Martin (eds.), Atlas zur Kirchengeschichte, 1987 (maps 2, 4f. and 13); P. Keresztes, The Decian *libelli* and Contemporary Literature, in: Latomus 34, 1975, 761–781; P.M. Meyer, Die Libelli aus der decianischen Christenverfolgung, 1910; J. Molthagen, Der römische Staat und die Christen im zweiten und dritten Jahrhundert, ²1975; J. Moreau, La persécution du Christianisme dans l'Empire Romain, 1956; H.A. Pohlsander, The Religious Policy of Decius, in: ANRW II,16.3 (1986), 1826–1842; J.B. Rives, The Decree of Decius and the Religion of Empire, in: JRS 89, 1999, 135–154; G. Schönaich, Die Libelli und ihre Bedeutung für die Christenverfolgung des Kaisers Decius, 1910; K.H. Schwarte, Das angebliche Christengesetz des Septimius Severus, in: Historia 12, 1963, 185–208.

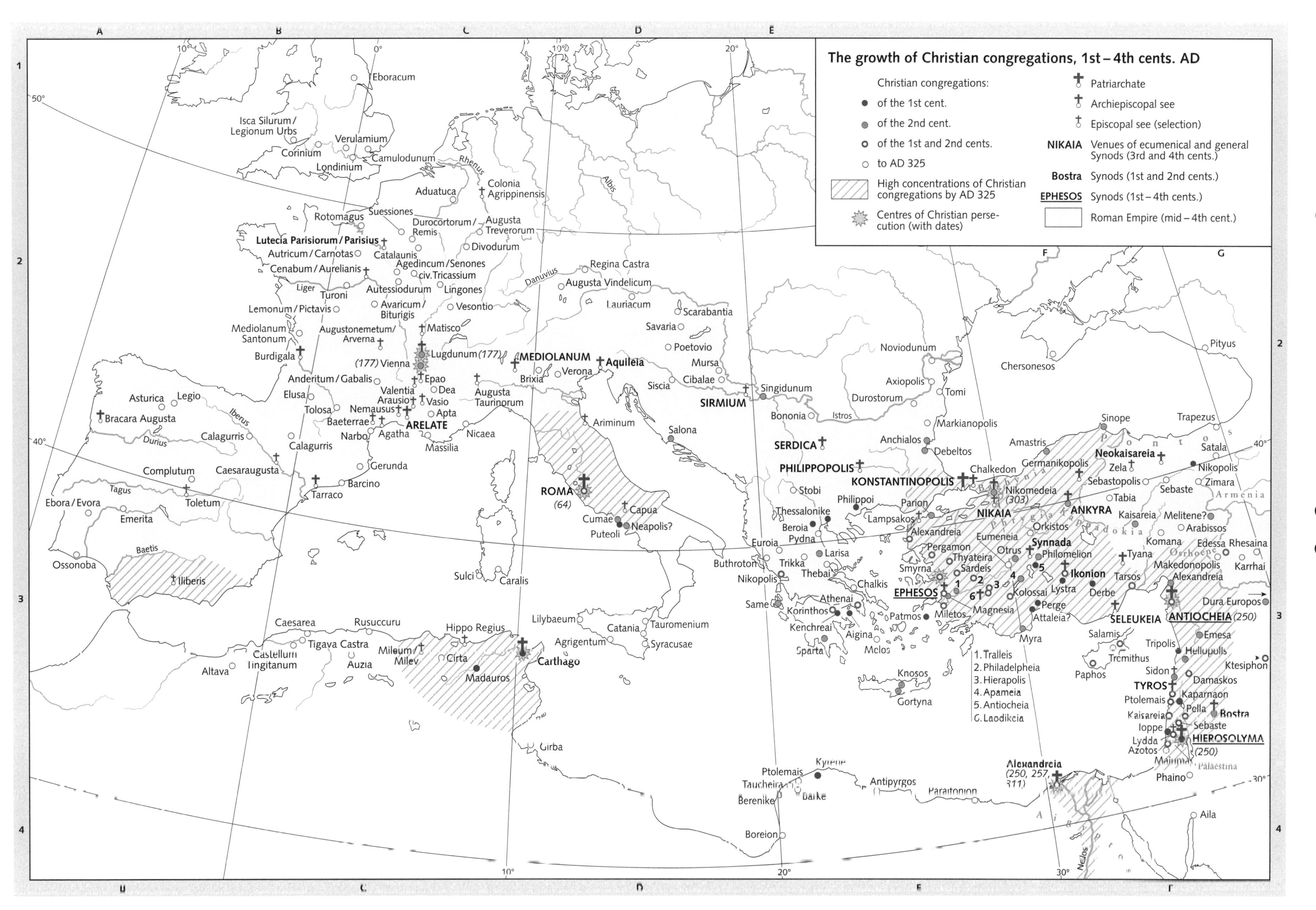
The growth of Christian congregations, 1st – 4th cents. AD
Christian congregations:
of the 1st cent.
of the 2nd cent.
of the 1st and 2nd cents.
to AD 325
High concentrations of Christian congregations by AD 325
Centres of Christian persecution (with dates)
Patriarchate
Archiepiscopal see
Episcopal see (selection)
NIKAIA Venues of ecumenical and general Synods (3rd and 4th cents.)
Bostra Synods (1st and 2nd cents.)
EPHESOS Synods (1st – 4th cents.)
Roman Empire (mid – 4th cent.)
1. Tralleis
2. Philadelpheia
3. Hierapolis
4. Apameia
5. Antiocheia
6. Laodikeia
ROMA (64)
MEDIOLANUM
Aquileia
ARELATE
Lugdunum (177)
(177) Vienna
Lutecia Parisiorum / Parisius
Carthago
SIRMIUM
SERDICA
PHILIPPOPOLIS
KONSTANTINOPOLIS
NIKAIA
Nikomedeia (303)
ANKYRA
Neokaisareia
Synnada
Ikonion
EPHESOS
ANTIOCHEIA (250)
SELEUKEIA
TYROS
HIEROSOLYMA (250)
Bostra
Alexandreia (250, 257, 311)
Dura Europos

The organization of the Christian church after the Synod of Chalcedon (AD 451)

In the 2nd cent., the Christian congregations had moved from collegial leadership by the presbyters – modelled on the structure of authority in the Jewish diaspora communities – to the sole authority of one representative of the congregation, a bishop, who held office by virtue of his *auctoritas* and *dignitas* in succession to the Apostles. His sphere of authority usually included a city and the villages and individual farmsteads in its *territorium*. The first ecumenical synod, at Nicaea (19 June – *c.* 25 August 325) already acknowledged what had developed since the end of the Diocletianic persecutions: the mirroring of the political administrative structure in the ecclesiastical. The *provinciae/eparchiae* and church provinces were equivalent, as were the *caput provinciae/metropolis* and the ecclesiastical administrative centre, the *consularis/praeses* and the so-called metropolitan, who was entrusted with the spiritual and administrative leadership of the bishops in his ecclesiastical province.

Depending on where his province was, the metropolitan himself was subordinate to the bishop

- of Rome (with the *dioeceses* Galliae, Viennensis, Hispaniae, Italia I and II, Africa, Pannoniae and Moesiae),
- of Alexandria (part of the *dioecesis Oriens*: Libya I and II, Thebaïs, Aegyptus I and II) or
- of Antioch (*dioecesis Oriens* except the provinces dependent on Alexandria).

The second ecumenical council at Constantinople (381) also raised

- the imperial residence, Constantinople, to the seat of a bishop superordinate to the metropolitans of the three *dioeceses* Thracia, Asiana and Pontica.

At the fourth ecumenical council at Chalcedon (8 October–1 November 451),

- Jerusalem was also made the seat of a bishop, superordinate to the metropolitans of the provinces of Arabia nova and Palaestina.

The hierarchy of sees was also laid down: Rome (founded by the Apostle Peter), Constantinople (founded by the Apostle Andrew), Alexandria (founded by the Evangelist Mark), Antioch (founded by the Apostles Peter and Paul) and Jerusalem (founded by all the Apostles). Although the Church historian Socrates (*Church History* 5,8; *c.* 380–after 439) refers to these bishops as 'patriarchs' in the context of the 325 Synod of Nicaea, the term only took root under Justinian I (527–565). It was later also granted to metropolitans on an honorific basis.

Sources

The most important sources for the ecclesiastical organization sketched in this map are the attendance and signatory lists of the synods, esp. the Synod of Chalcedon in 451, and the *Notitiae episcopatuum*, i.e. lists of metropolitans and autocephalous and suffragan sees compiled for purposes of protocol.

Routes of Christian pilgrims (4th–6th cents.)

Christian pilgrimages to the holy places became a general custom with the toleration and promotion of Christianity by Constantine I, who architecturally developed or built anew the favourite pilgrimage sites in Palestine (Church of the Holy Sepulchre, memorial church on the Mount of Olives, Basilica of the Nativity at Bethlehem), Rome (basilica over the tomb of Peter) and Constantinople (Church of the Holy Apostles). Destinations were martyrs' tombs and relics of all kinds, but also formerly pagan cult sites now transformed into Christian pilgrimage sites, e.g. the Asclepius sanctuaries at Athens and Epidaurus. Some descriptions of pilgrimages from the 4th-6th cents. have survived: the map records four.

1. The *Itinerarium Hierosolymitanum sive Burdigalense* (AD 333) is the oldest. This notes the stations (towns/*civitates*, accommodation/*mansiones*, horse/rest stations/*mutationes*) on a female pilgrim's journey from Burdigala to Jerusalem/Hierosolyma (595) via Mediolanum/Milan (557), Constantinople (570), Ancyra (575), Antioch (581) and Caesarea (585). This is followed by a tour of the Holy Land (596–599). The homeward journey uses the same route as the outward as far as Ancyra, but it then continues from Ancyra to Heraclea, and from there presumably by ship through the Bosporus to Heraeum, where the land journey continues via Thessalonica (605) on the *via Egnatia* to Aulon (608). The journey then takes the Adriatic passage from Hydruntum (609) via Rome to Mediolanum (617), where it ends.

2. The *Peregrinatio ad loca sancta* is an epistolary report of the Christian pilgrim Egeria – other traditions call her Aetheria, Etheria or Eucheria – of her journey (probably accomplished in 381–384) to the Holy Land, where she travelled around on the basis of the Bible and the advice of local holy men. She describes her journey through the Sinai and Egypt, into the land east of the Jordan and southern Syria, into upper Mesopotamia and through Asia Minor to Constantinople, from where she sent her report home (to Spain or Gaul).

3. Paula (347–404), a woman of the social elite of Rome, widowed since 379, decided after meeting Jerome in 382 to leave Rome and live a life dedicated solely to piety in Palestine. In 385, with her daughter Eustochium, she boarded a ship which took her through the Straits of Messina, via Methone, Rhodes and Salamis on Cyprus to Seleucia. She then continued her journey along the coast to Jerusalem, and even to Egypt and the monasteries of Nitria. Her account of her journey is found in a letter of Jerome, written to Eustochium after Paula's death (Jer. Epist. 108).

4. The *Itinerarium Antonini Placentini*, the travel accont of Antoninus of Piacenza, written around 570, begins in Constantinople and goes by way of Salamis on Cyprus and Antaradus to Jerusalem, continuing from there to make a trip into Lower Egypt as far as Alexandria. The text does not explain how the pilgrim returned from Alexandria to Jerusalem. From there, a new journey begins, this time north via Damascus to Antioch, continuing into the Syrian Desert via Carrhae to Sergiupolis. Here, the account breaks off abruptly.

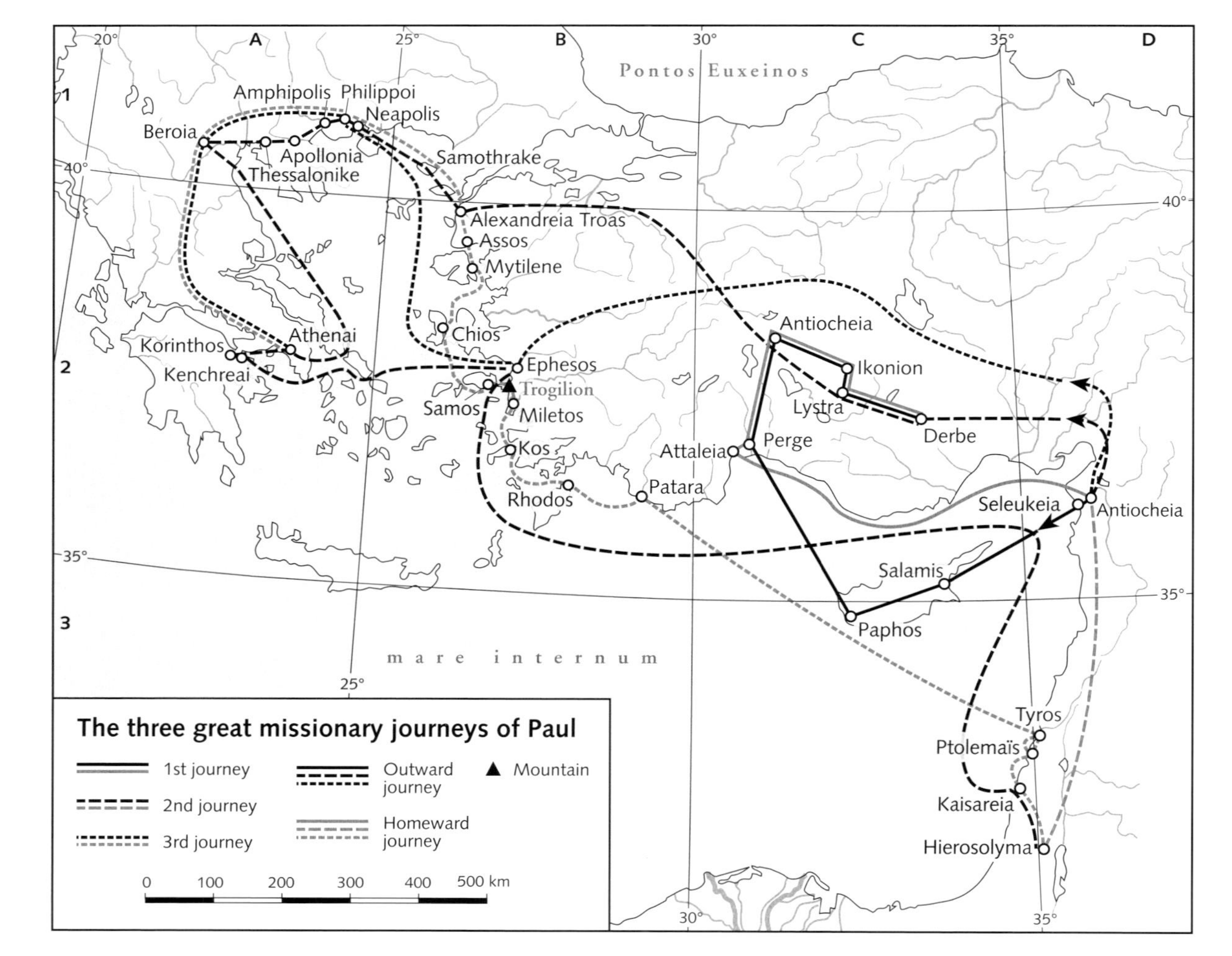

The three great missionary journeys of Paul

The Acts of the Apostles, the only near-contemporary (*c.* AD 90) biography of Paul, tells only allusively of Paul's missionary activities among the Nabataeans and in Syria and Cilicia, but more comprehensively of three missions Paul undertook to Asia Minor and from there to Macedonia and Greece.

The first missionary journey went via Cyprus, Pamphylia, Pisidia, Lycaonia and Cilicia back to Antioch.

The second missionary journey took him via Syria and Cilicia through Asia Minor to Alexandria Troas, from where he crossed to Macedonia. He continued to Athens and Corinth. A long sea voyage ensued, by way of Ephesus, past the Trogilium promontory to Caesarea, from where he returned to Antioch by land after three years.

The third missionary journey took him through Asia Minor to Ephesus, across to Macedonia and Achaia, then back to Alexandria, to sea via Tyre and on to Jerusalem, where Paul was arrested and deported to Rome. This journey, too, was said to have lasted three years, as Paul spent two whole years at Ephesus.

The Acts of the Apostles allow a relative chronology of these journeys to be established, while a decree of the emperor Claudius for Delphi, which mentions the proconsul L. Iunius Annaeus Gallio, a brother of the philosopher Seneca, permits the absolute dating of the second missionary journey and Paul's eighteen-month stay at Corinth to the years AD 51–53. Paul may therefore have undertaken these journeys between AD 45 and 60.

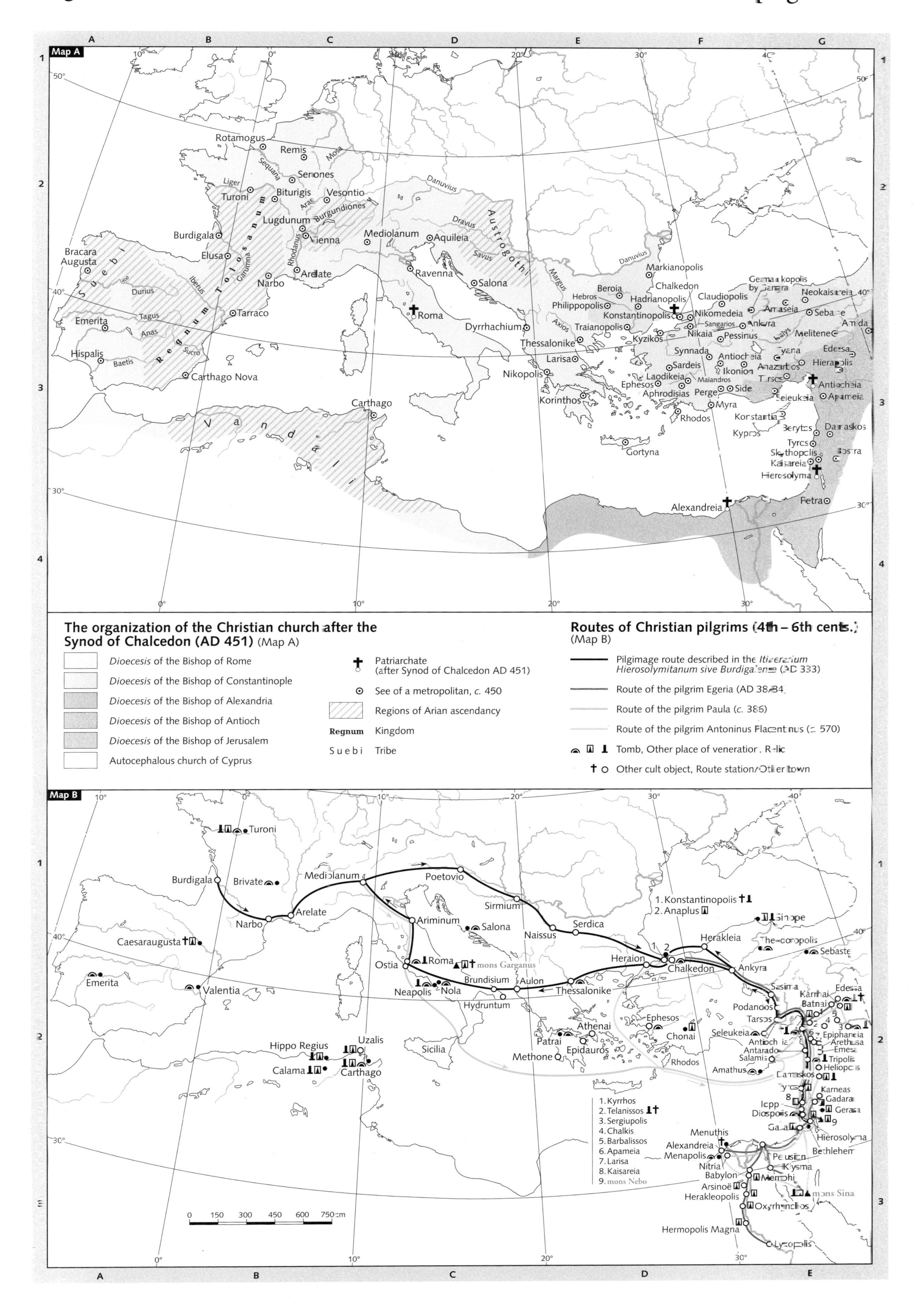
Map A
Rotamogus
Remis
Senones
Turoni
Biturigis
Vesontio
Lugdunum
Vienna
Burdigala
Elusa
Narbo
Arelate
Bracara Augusta
Emerita
Hispalis
Tarraco
Carthago Nova
Mediolanum
Aquileia
Ravenna
Salona
Roma
Dyrrhachium
Carthago
Thessalonike
Larisa
Nikopolis
Korinthos
Markianopolis
Chalkedon
Hadrianopolis
Philippopolis
Konstantinopolis
Nikomedeia
Traianopolis
Kyzikos
Nikaia
Pessinus
Synnada
Sardeis
Ephesos
Laodikeia
Aphrodisias
Perge
Side
Myra
Rhodos
Gortyna
Alexandreia
Petra
Suebi
Regnum Tolosanum
Austrogothi
Burgundiones
The organization of the Christian church after the Synod of Chalcedon (AD 451) (Map A)
Dioecesis of the Bishop of Rome
Dioecesis of the Bishop of Constantinople
Dioecesis of the Bishop of Alexandria
Dioecesis of the Bishop of Antioch
Dioecesis of the Bishop of Jerusalem
Autocephalous church of Cyprus
Patriarchate (after Synod of Chalcedon AD 451)
See of a metropolitan, c. 450
Regions of Arian ascendancy
Regnum Kingdom
Suebi Tribe
Routes of Christian pilgrims
(Map B)
Route of the pilgrim Paula (c. 386)
Map B
Turoni
Burdigala
Brivate
Mediolanum
Poetovio
Sirmium
Narbo
Arelate
Ariminum
Salona
Naissus
Serdica
1. Konstantinopolis
2. Anaplus
Herakleia
Caesaraugusta
Ostia
Roma
mons Garganus
Heraion
Chalkedon
Ankyra
Emerita
Valentia
Neapolis
Nola
Brundisium
Aulon
Thessalonike
Hydruntum
Ephesos
Athenai
Patrai
Epidauros
Methone
Chonai
Rhodos
Tarsos
Seleukeia
Sebaste
Hippo Regius
Uzalis
Calama
Carthago
Sicilia
Amathus
1. Kyrrhos
2. Telanissos
3. Sergiupolis
4. Chalkis
5. Barbalissos
6. Apameia
7. Larisa
8. Kaisareia
9. mons Nebo
Menuthis
Alexandreia
Menapolis
Nitria
Babylon
Arsinoë
Herakleopolis
Hermopolis Magna
Karneas
Gadara
Hierosolyma
Bethlehem
mons Sina
0 150 300 450 600

Cultural developments in the regions of Germanic settlement

I. Language

The Germanic languages are defined as a group of Indo-European languages, distinct from other Indo-European languages by reason of peculiarities developed through the course of time. Such peculiarities include

- phenomena arising from the First Germanic Sound Shift (Grimm's Law; this sound shift has recently been dated not to 500 BC, but only to the 1st cent. BC)
- the shift of accent to the first syllable
- the reduction of end syllables
- the expansion of the inherited vowel gradation system.

Surviving linguistic evidence allows the inference of a lost Proto-Germanic language – the formerly usual distinction of languages supposedly including Germanic according to the Centum-Satem isogloss, on the other hand, is probably no longer tenable. The transition from the Indo-European protolanguage to this Proto-Germanic, as a step towards the development of the individual Germanic languages, was probably complete for the most part after 500 BC. The individual Germanic languages then developed before AD 500. The origins of Germanic are sought in southern Scandinavia and the region of the lower Elbe. There are important references to Germanic in the Latin sources: there is name material in the writings of ancient authors such as Caesar (*De bello Gallico*) and Tacitus (*Germania*). Loan-words in Finnish and the Slavic languages into which Germanic merged, provide further help.

The Germanic languages are generally divided into:

1. Eastern Germanic – the languages of eastern Germanic peoples such as the Goths, Burgundians and Vandals. The most important source for Gothic is the earliest recorded text, the Gothic Bible translation of Ulfilas (d. 382/3). The eastern Germanic languages had already died out by the early Middle Ages, except for so-called Crimean Gothic, which probably survived until the 18th cent. A pupil of Erasmus of Rotterdam, the Flemish scholar Ogier Ghiselin de Busbecq (1522–1592), whom the Emperor Ferdinand I sent to Turkey in 1554 as ambassador to Sultan Süleyman I, met two people at Constantinople who still knew this language. The notes he made of his conversations and his collection of vocabulary are our only source for Crimean Gothic.

2. Northern Germanic – the language which, in the final phase of development, divided into the main individual languages of Swedish, Danish, Norwegian, Faroese and Icelandic. The closest to a Northern Germanic protolanguage was Old Norse (12th–14th cents.), and the linguistic material recorded in runic script by the Vikings (8th-11th cents.) and others. The earliest runic inscriptions date from around AD 200.

3. Western Germanic – the language which, in the final phase of development, divided into the main individual languages of English, High and Low German, Frisian and Dutch. Hypothetical protolanguages are divided into North Sea, Rhine-Weser and Elbe Germanic, and these probably did not separate from a single protolanguage as dialects, but amalgamated out of various dialect groups.

II. Other cultural characteristics

The material culture of the Germanic peoples is primarily described with the help of archaeological sources and methods. Evidence of linguistic history and written sources also contributes. The focus of research is on the ethnogenesis of the Germanic peoples in the pre-literate period, and the nature of their settlements, in terms of everyday culture, economy, cult of the dead and other cultural characteristics. The problem with which archaeologists are constantly faced is here, as elsewhere, to what extent the commonalities of archaeological finds can be equated to ethnic groupings. For the period in which the illiterate Germanic and literate Latin cultures developed in parallel, there are chronological points of reference.

Pre-Roman Iron Age (5th-1st cents. BC)

The Jastorf Culture, named after the urn grave field at Jastorf (Lower Saxony), is regarded as typical of the early Germanic culture, profoundly agricultural in nature and the first to use iron. Evidence of this culture (burials, jewellery, e.g. pins, *fibulae* and neck rings, clay vessels, weapons) is spread across an area from west of the Albis/Elbe to the Vidua/Oder and from the edge of the Mid-German Highlands to Jutland, and regionally differentiated groups can be distinguished within this area. Continuity of grave layout is variously attested into historically documented times. The people of this culture had close contacts with the Celts in southern Central Europe. The migrations of the late 2nd and 1st cents. BC are discernible more easily and in greater diversity in archaeology than in the earliest literary references to the Germanic peoples.

Cultural groups of the early and middle Roman Imperial Period (1st/2nd cents. AD)

These migrations are already clearly evident in the early 1st cent. BC in the growing distribution of the archaeological evidence of the Jastorf Culture (grave finds). Our knowledge of these finds is ever more subtle, and enables the various groups to be ever better defined. North Sea and Rhine-Weser Germans are clearly recognizable in the west of the particularly strongly characterized Elbe Germanic culture. The proximity of the Roman Empire influenced the Germanic region in very different ways: activities in the Roman frontier zone such as the organizational reform of the Germanic military districts into provinces around AD 85 and the continual construction of the *limes* up to the end of the 2nd cent. are barely perceptible here. Contrary to usual experience, according to which frontier zones remain economically underdeveloped, a distinctly prosperous picture emerges to both sides of the *limes*, characterized by intensive trade in goods of all kinds, including luxury goods, and the exchange of all manner of expertise. So-called 'princely graves' and hoard finds like the Hildesheim silver hoard attest to this. The beginning of the great migrations from the 4th cent. cannot be read with certainty from the archaeological finds. However, the legacy which tribes such as the Franks, Goths and Saxons left as they passed through, is clear enough.

According to the report of Tacitus (Tac. Germ. 40), there was a cult association of seven Germanic tribes dedicated to the goddess Nerthus and generally known as the 'Nerthus League'. The individual tribes, however, are difficult to identify. Their localization to western Mecklenburg is purely hypothetical, and has been included in the map solely because of the ongoing discussion.

→Maps pp. 233, 235

Literature

Language: H. Beck (ed.), Germanische Rest- und Trümmersprachen, 1989; H. Birkhan, Germanen und Kelten bis zum Ausgang der Römerzeit, 1970; E. Haugen, Die skandinavischen Sprachen. Eine Einführung in ihre Geschichte, 1984; C.J. Hutterer, Die germanischen Sprachen, 1975; H. Krahe, W. Meid, Germanische Sprachwissenschaft, 3 vols., 1965–1969; W. Krause, Runen, 1970; S. Krogh, Die Stellung des Altsächsischen im Rahmen der germanischen Sprachen, 1996; P. Ramat, Einführung in das Germanische, 1981; O.W. Robinson, Old English and its Closest Relatives. A Survey of the Earliest Germanic Languages, 1992; M. Stearns, Crimean Gothic. Analysis and Etymology of the Corpus, 1978; J.B. Voyles, Early Germanic Grammar: Pre-, Proto-, and Post-Germanic Languages, 1992; *culture in general*: H.J. Eggers, Zur absoluten Chronologie der römischen Kaiserzeit im freien Germanien, in: ANRW II 5, 1976, 3–64; H. Jankuhn, Siedlung, Wirtschaft und Gesellschaftsordnung der germanischen Stämme in der Zeit der römischen Angriffskriege, in: ANRW II 5, 1976, 65–126; W. Künnemann, Jastorf. Geschichte und Inhalt eines archäologischen Kulturbegriffs, in: Die Kunde 46, 1995, 61–122; A.A. Lund, Die ersten Germanen. Ethnizität und Ethnogenese, 1998; K. Motykova, Die ältere römische Kaiserzeit in Böhmen, in: ANRW II 5, 1976, 143–199; V. Pingel, Germanische Archäologie, in: DNP 4, 1998, 967–971; S. Rieckhoff, Süddeutschland im Spannungsfeld von Kelten, Germanen und Römern, 1995; D. Timpe, Germanen, Germania, Germanische Altertumskunde, in: RGA 11, 1998, 181–245; *Nerthus:* L. Motz, The Goddess Nerthus. A New Approach, in: Amsterdamer Beiträge zur älteren Germanistik 36, 1992, 1–19; R. Simek, Religion und Mythologie der Germanen, 2003; D. Timpe, Tacitus' Germania als religionsgeschichtliche Quelle, in: Romano-Germanica. Gesammelte Studien zur Germania des Tacitus, 1995, 93–143 (Original 1992); Id., Die Söhne des Mannus, in: Chiron 21, 1991, 69–125.

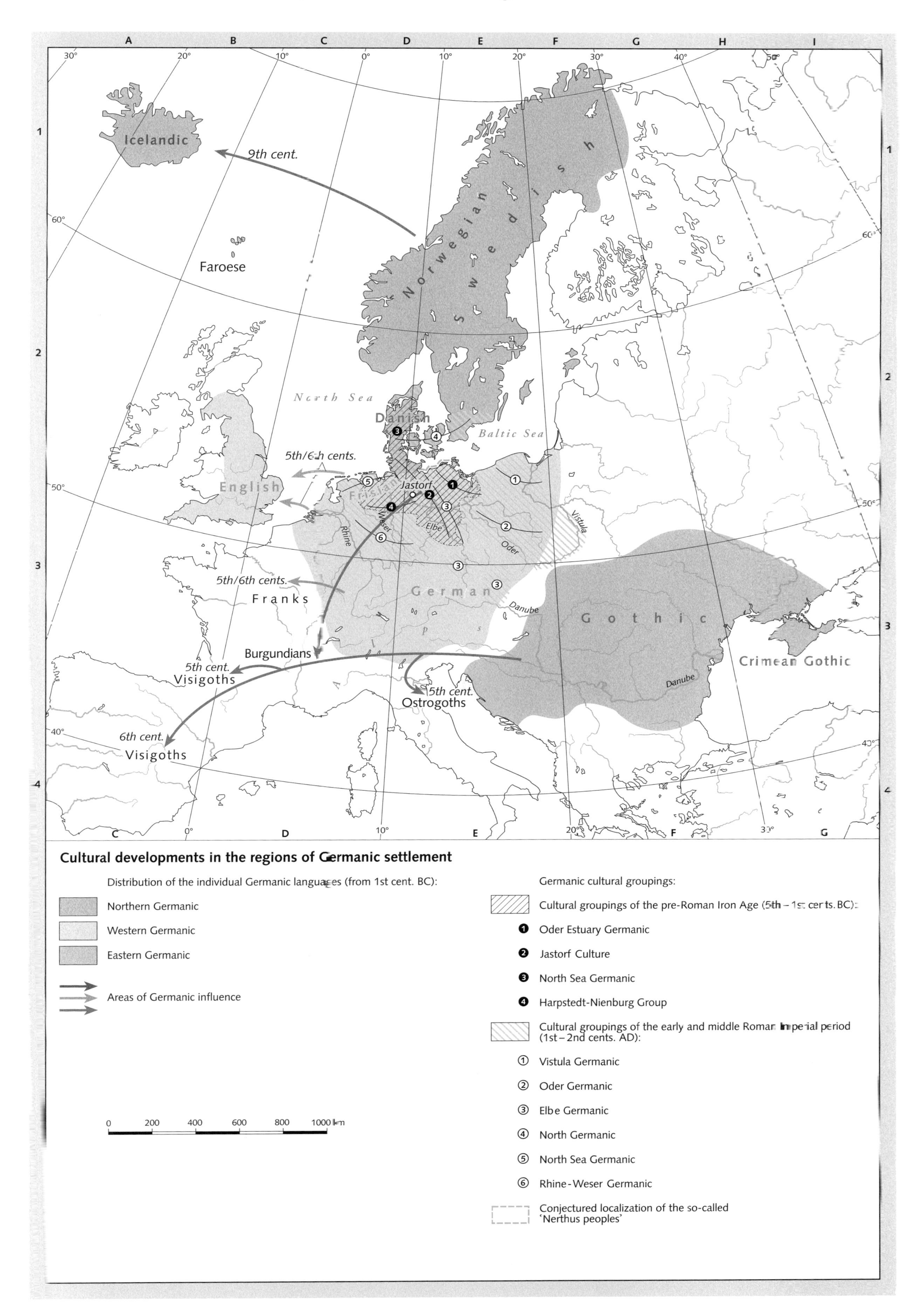
Icelandic
9th cent.
Faroese
Norwegian
Swedish
North Sea
Danish
Baltic Sea
5th/6th cents.
English
Frisian
Jastorf
Rhine
Weser
Elbe
Oder
Vistula
5th/6th cents.
Franks
German
Danube
Gothic
Crimean Gothic
Burgundians
5th cent.
Visigoths
5th cent.
Ostrogoths
6th cent.
Visigoths
Cultural developments in the regions of Germanic settlement
Distribution of the individual Germanic languages (from 1st cent. BC):
Northern Germanic
Western Germanic
Eastern Germanic
Areas of Germanic influence
Germanic cultural groupings:
Cultural groupings of the pre-Roman Iron Age (5th – 1st cents. BC):
Oder Estuary Germanic
Jastorf Culture
North Sea Germanic
Harpstedt-Nienburg Group
Cultural groupings of the early and middle Roman Imperial period (1st – 2nd cents. AD):
Vistula Germanic
Oder Germanic
Elbe Germanic
North Germanic
North Sea Germanic
Rhine-Weser Germanic
Conjectured localization of the so-called 'Nerthus peoples'
0 200 400 600 800 1000 km

Germanic migrations and invasions of the Roman Empire

Peoples have migrated for all kinds of reasons and at all periods of history. Problems occur when they encounter each other during their migrations, and especially when migrating people come into contact with sedentary populations. At first, the various assaults on the Roman frontier lands were exclusively campaigns of pillage. Later, however, tribes close to the frontier increasingly began to covet settlement land on Roman soil, hoping to integrate into the *imperium*. Roman history through Late Antiquity provides numerous examples. Not least in reaction to the great Hunnic incursion from the Caspian Sea to the Catalaunian Plains, returning eastwards via Aquileia (376–453), permanent states were then established in the territory of the Western Roman Empire: by the Burgundians (413), Swabians (411), Visigoths (418), Vandals (439), Franks (482) and Ostrogoths (493). Finally there emerged the kingdom of the Langobards (568) (→ maps p. 234f.).

I. Germanic migrations and invasions of the Roman Empire (3rd cent. AD)

Lower Danube, Greece and Asia Minor

The Heruli are a typical example of the Germanic peoples who broke into the Roman Empire ahead of the Age of Migrations. They are first recorded living along the Maeotis. In AD 267, they voyaged out of the Maeotis across the Black Sea, attacking Byzantium and plundering Athens, Corinth and Argos before carrying out raids along the south coast of the Peloponnese. Roman naval units forced them to cross Greece by land, by way of Sparta. Gallienus (253-268) decisively defeated them at Nestus in Thrace in 268 – of the survivors, some made their way back to their homeland and others joined the Roman army. The following year, the Heruli joined forces with other Germanic tribes, including Goths, for a land campaign through the upper Balkans, where they were annihilated at Naissus by Claudius II (268–270). Other Heruli put to sea from Tyras, attacking Thessalonica and continuing to Crete, Rhodes, Cyprus and Pamphylia. After battles with Roman naval units under Probus (276–282), their forces dissolved.

Goths migrated from Pomerania into the lands east of the middle reaches of the Vistula between AD 150 and 230, finally settling in the Ukraine around the mid 3rd cent. From there, beginning in 238, they terrorized the Balkan peninsula for some forty years. They were defeated by Decius (249–251) at Nicopolis in 250, but subsequently inflicted heavy defeats on him at Beroea and especially at Abrittus in 251. In 254, they even attacked Thessalonica. Setting out from the northern shore of the Black Sea (Tyras was in their hands from 268), they made naval forays to the shores of Asia Minor in 257 and 276/7. They repeatedly revisited these raids, in spite of the heavy losses inflicted on them by the cities they attacked and by Roman naval units.

The Marcomanni and Quadi repeatedly came into conflict with the Romans in Raetia, Noricum and Pannonia from the first years of the Principate. This development culminated in the three wars fought against them by Marcus Aurelius in the period AD 167-182. The domestic political problems which preoccupied the Roman Empire during the period of the soldier emperors emboldened many tribes of Germania magna to attack imperial territory – these once more included the Marcomanni, who crossed the Danube in 253 and pushed through the north of Greece as far as Thessalonica. As a result, Gallienus felt compelled to allow these Marcomanni to settle in Pannonia superior.

On the Rhine and upper Danube

Around 260, Germanic tribes in this area began to develop a dynamic that would extend far into the Roman territories. In the 3rd cent., the Saxons, a seafaring people, repeatedly laid waste to parts of the lower Rhine and, accompanied by Franks, the southern coast of Britannia. The Franks crossed the Rhine, taking Augusta Treverorum/Trier in 275/6, and their raids took them as far as the Iberian Peninsula. Various tribes, mainly Germanic, gathered in the section between the Main and the Raetian *limes*, including the Semnones, Iuthungi and Suebi, who made repeated incursions into the Empire between 258 and 275, even entering northern Italy. Others moved through Gaulish territory as far as Arelate/Arles. Much remains obscure about these events because of the lack of sources, but there are occasional glimpses, such as that afforded by the inscription telling us that 'Semnones or Iuthungi' returning from Italy were defeated by the Romans at Augusta Vindelicum/Augsburg on 24/25 April, probably in 260 (AE 1993, 1231). The Iuthungi were back in Italy in 271, where they were defeated by Aurelian (270–275) at the Ticinus.

II. Germanic migrations (4th-6th cents. AD)

The Age of Migrations is dated from the crossing of the Rha/Volga by the Huns in AD 376. The discussion of map A shows the precarious nature of this dating, but the end date is less problematic: it is AD 568, when the Langobards settled in Italy, the last of the tribes to become permanently sedentary.

Eastern Germanic tribes were moving south from Scandinavia and the southern shore of the Baltic long before 376, migrating in several waves and along different routes, with several interludes of inactivity. They reached as far south as the Black Sea. Here, the Goths also appeared, a polyethnic group of tribes whose division into Tervingi and Greuthungi (Visigoths and Ostrogoths, or West and East Goths) is attested from AD 291. While the Ostrogoths under Ermanaric ruled a large expanse north of the Black Sea in the 4th cent., the Visigoths moved west, deeper into Roman frontier regions.

Visigoths

Under their leader Alaric, elected in 391, the Visigoths moved out of Moesia and Thrace, where they had been stationed as *foederati* since 382. By order of the emperor at Constantinople, they advanced south to the Balkan peninsula, then to Italy, where they moved against the government at Ravenna, finally conquering Rome in 410. They then moved north into Gaul under Athaulf. Here they settled, on the basis of a treaty agreed by Vallia with the *patricius* Constantius (later Constantius III), founding a kingdom centring on the residence of Tolosa/Toulouse. Frankish pressure forced them to relinquish it and move to the Iberian Peninsula in 507.

Ostrogoths

After the death of Attila in 453, the Ostrogoths were still living near the Maeotis. Soon afterwards, they are found as *foederati* of the emperor Marcian (450–457) in Pannonia, then in Macedonia from 474 as *foederati* of the emperor Zeno (474–491). By order of the latter emperor, they moved on Italy in the summer of 488, under the Amal Theoderic, to oppose the *rex Italiae* Odoacer. After Odoacer's removal in 493, Theoderic founded an Ostrogothic state in Italy, formally as a vassal of the Byzantine emperor, but *de facto* independent.

Burgundians

The Eastern Germanic tribe of the Burgundians migrated out of Further Pomerania into the region between the Vistula/Weichsel and Vidua/Oder late in the 2nd cent. In the 3rd cent., allied with the Vandals, they moved into Roman territory, where they were defeated in 280 by Probus (276–282) and withdrew into the region between the Neckar and the Taunus. Some of the tribe crossed the Rhine at Mogontiacum/Mainz in 406/7, initially settling in the region around Borbetomagus/Worms under the emperor Honorius (395-423).

Franks

The Franks, a group of small tribes from the right bank of the Rhine, began repeated incursions from *c.* AD 260, penetrating as far as the Iberian Peninsula. However, they also integrated into the society, army and government of the Romans. For instance, they also took part in Carausius' British Usurpation. In 360, Julian 'the Apostate' (361–363) settled part of the tribe, the Salian Franks, in Toxandria as *foederati*, and Aetius settled others around Turnacum/Tournai in 445.

Vandals

Settled between the Vidua and Vistula in the 1st cent. BC, the Vandals were subsequently recognizable as two groups differing in culture and organization: the Hasdingi (on the borders of Dacia) and the Silingi (in Silesia). The Silingi stayed where they were, while the Hasdingi joined the Marcomanni to confront the Romans during the reign of Marcus Aurelius (161–180), later invading Pannonia in 270 and beaten back by the emperor Aurelian (270–275). The two Vandal groups later reunited, and together they migrated from Silesia with other Germanic groups (e.g. the Alani) towards the end of the 4th cent., following the westward movement of the Huns. Like the Burgundians, they crossed the Rhine in 406/7 between Mogontiacum and Borbetomagus, entering Gaul, but they continued through 409, crossing the Pyrenees on to the Iberian Peninsula. There, the Hasdingi settled in the north-west with the Suebi, while the Silingi settled in the south-west. In 417, on the orders of Honorius (395–423), the Silingi were almost completely annihilated by the Visigoths. The Hasdingi detached themselves from the Suebi, and crossed to Africa in 429 under Geiseric.

Suebi

The earliest archaeological evidence dates from the 1st cent. AD in the region between the Rhine, Danube and Elbe, but the Suebi are attested in literature from the 1st cent. BC, thanks to their encounter with Caesar. Their eventful tribal history is characterized by the repeated outward migrations of parts of the tribe, and by their absorption of other Germanic tribal groups. In the winter of 406/7, they crossed the Rhine with the Alani and Vandals, moved through Gaul, crossed the Pyrenees and spent four years moving through the Iberian Peninsula before they settled in 411, at first in the south-west, then in the north-west.

Langobards

The Germanic Langobards may have originated in southern Sweden, from where they migrated in the 1st cent. BC into the region west of the Gutalus/Neman, absorbing other Germanic groups. However, archaeological attestation only begins rather later, by the lower reaches of the Albis/Elbe. After crossing the river in AD 5, they were forced back to the right bank by Tiberius. They now broke away south-eastwards, entering Pannonia on a raid in *c.* AD 166. Around 488, they settled in northern Noricum, where they encountered the Heruli and defeated them in 508. In 526, they expanded their settlement area to Pannonia. Justinian (527–565) allocated settlement land to them in Pannonia and Noricum in 546. In 568, they entered Italy.

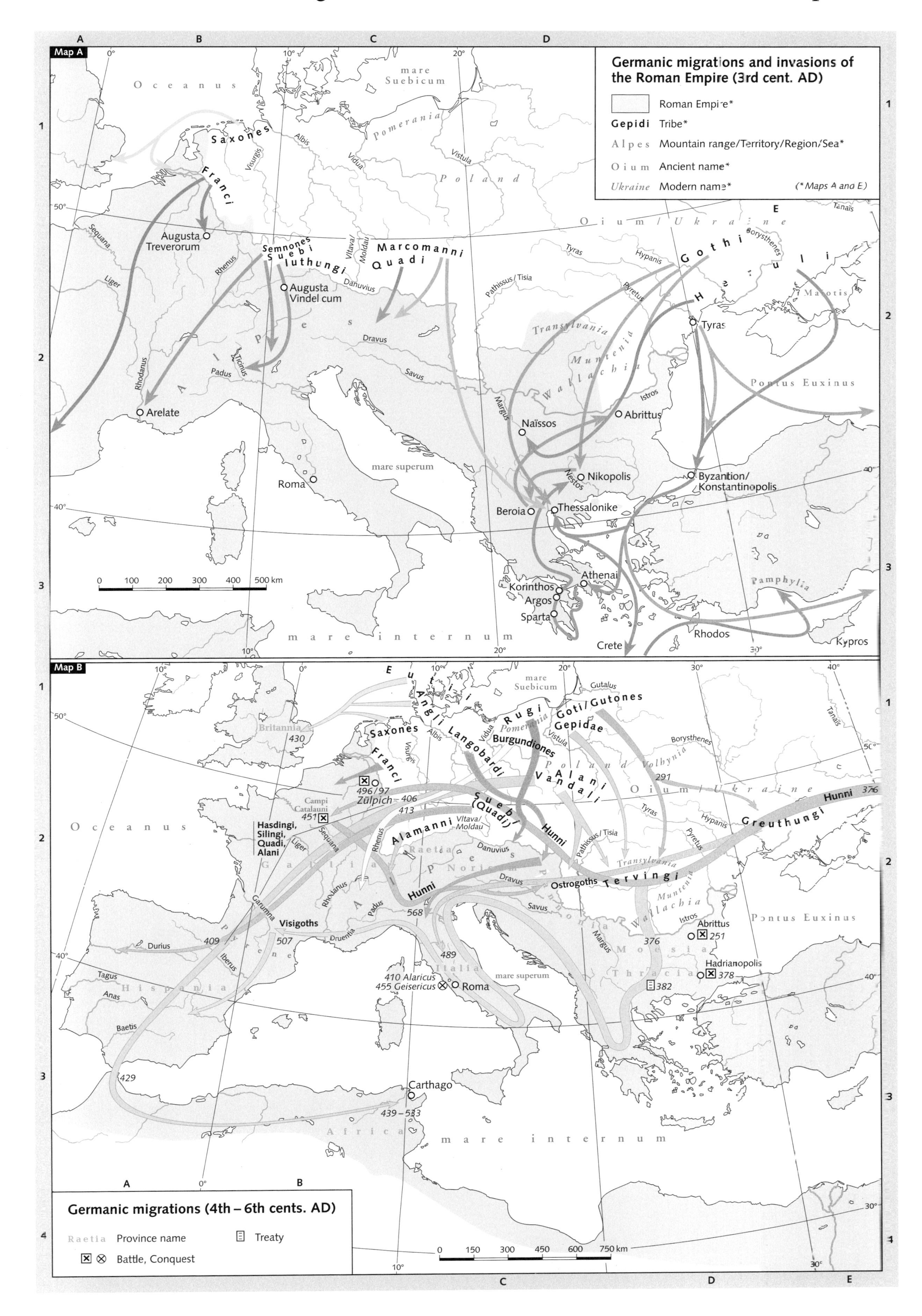
Map A
Germanic migrations and invasions of the Roman Empire (3rd cent. AD)
Roman Empire*
Gepidi Tribe*
Alpes Mountain range/Territory/Region/Sea*
Oium Ancient name*
Ukraine Modern name*
(*Maps A and E)
Oceanus
mare Suebicum
Saxones
Franci
Pomerania
Poland
Augusta Treverorum
Semnones
Suebi
Iuthungi
Marcomanni
Quadi
Augusta Vindelicum
Gothi
Heruli
Tyras
Transylvania
Muntenia
Wallachia
Arelate
Naïssos
Abrittus
Nikopolis
Byzantion/ Konstantinopolis
Roma
mare superum
Beroia
Thessalonike
Athenai
Korinthos
Argos
Sparta
Crete
Rhodos
Kypros
Pamphylia
Pontus Euxinus
mare internum
Map B
Germanic migrations (4th – 6th cents. AD)
Raetia Province name
Treaty
Battle, Conquest
Britannia
430
Saxones
Angli
Langobardi
Rugi
Goti/Gutones
Gepidae
Burgundiones
Alani
Vandali
Suebi (Quadi)
Alamanni
Hunni
Greuthungi
Tervingi
Ostrogoths
496/97
Zülpich
406
413
Campi Catalauni 451
Hasdingi, Silingi, Quadi, Alani
Visigoths
409
507
568
489
410 Alaricus
455 Geisericus
Roma
Hispania
429
Carthago
439 – 533
Africa
Abrittus 251
Hadrianopolis 378
382
376
291
Hunni 376
Pontus Euxinus
mare internum

The Germanic successor-states to the Western Roman Empire

First and Second Burgundian kingdoms
Around 413, the Burgundians settled around Borbetomagus/Worms. They retreated before the Huns into Belgica, where they were defeated in 436 by the Roman general Aetius with the help of Hunnic *foederati*. Aetius settled the survivors as *foederati* in Sapaudia (modern Savoy) between the Rhône and Lake Geneva in 438. They extended their kingdom into modern Burgundy, but were not in the end able to withstand the Franks. Their last king, Godomar, was defeated in 532 at Divio/Dijon and his realm incorporated into the Frankish kingdom.

The Visigothic kingdoms
The Visigoths were an effective weapon for the emperor at Constantinople in his dispute with the court at Ravenna. Led by their king, Alaric, they invaded Italy as far as the south, conquering Rome on 24 August AD 410. After Alaric's death, Athaulf led the Goths into Gaul in 412. In 418, his successor Vallia, who had fought the Vandals, Alans and Suebi on the Iberian Peninsula for the emperor Honorius, was rewarded with a treaty by which the Goths were granted settlement land in Aquitania. Their greatest king, Theoderic I (418–451), ruled from the Pyrenees to the Loire, and from the Atlantic to the Rhône. In the end, however, the Visigoths were no match for the Franks and Burgundians, and they were pushed back on to the Iberian Peninsula. Succession disputes initially weakened the Iberian Visigothic kingdom, giving Justinian the opportunity to gain a foothold in the south-east of the peninsula in 552. The kingdom stabilized when the Suebi were subjugated in 585, but it was no match for the Muslim troops under Tariq Abu Zara who crossed to Spain from Africa. The defeat at the Guadalete on 23 June 711 sealed the fate of the Visigothic kingdom.

The Ostrogothic kingdom
Theoderic rose to power over the Goths settled as *foederati* in Pannonia by 475 at the latest. Relations between the emperor Zeno and the Goths were tense. Zeno therefore took the opportunity to send them to Italy against Odoacer in 488. Amidst the tense interplay among the Vandals, Visigoths, Burgundians, Franks and the emperor, Theoderic succeeded for a time in bringing a vast complex of territories under his control. For instance, in 511, he was also elected king of the Visigoths, as a kind of proxy for his grandson Amalaric (son of the Visigothic King Alaric II), who was not yet of age. However, the *regnum Hesperiae* did not outlive Theoderic, who died in 526. Both parts of the kingdom went their own way once more – Amalaric now being elected king of the Visigoths, while the king's daughter Amalasuntha assumed the regency for her underage son Athalaric at Ravenna. Her murder in 535 then provided a pretext for the emperor Justinian to reconquer Italy.

The Frankish kingdom under the Merovingians
In 360, the emperor Julian settled Salian Franks as *foederati* in Toxandria. Salian Frankish groups then settled around Turnacum/Tournai in the mid 5th cent. Clovis I annexed the last rump of the Western Roman Empire, Syagrius' kingdom, in 486/7. In doing so, he came into conflict with the Visigoths, whom he defeated at Vouillé in 507 in an alliance with the Burgundians. He then expelled them from Gaul, with the exception of a small coastal strip along the Mediterranean.

The Vandal kingdom
Some 80,000 Vandals crossed from Spain to Africa in 429, led by their king, Geiseric. Both Ravenna and Constantinople sought to stop the Vandal advance. However, following the conquest of Carthage in 439, a peace treaty was signed in 442, by which the Vandals were permitted possession of the North African littoral from Africa proconsularis to Mauretania Tingitana. With campaigns of plunder on the coasts of Italy (conquest of Rome in 455) and the islands, Geiseric won possession of Sicily, Sardinia, Corsica and the Balearics, which was confirmed by treaty in 474. The Vandals were only able to secure their kingdom with individual bases – in Africa, the frontiers with the Moorish tribes were barely defensible, Sicily was ceded to Odoacer in exchange for tribute in 476, and Theoderic took almost the whole island in 491. The remaining territories were finally won back for the emperor at Constantinople by Belisarius.

The kingdom of the Suebi
The Suebi spent four years plundering the Iberian Peninsula with the Vandals and Alani before settling in 411, initially in the south-west and finally in the north-west, around Braga. They were unable to survive in the face of the numerically far superior Visigoths. Braga was taken and its kingdom occupied. The Suebi set up a second kingdom at Braga, but it was defeated in battle against the Visigothic King Leowigild in 585 and was relinquished.

The Roman kingdom of Syagrius
In 457, the emperor Majorian (457-461) appointed the Gaulish aristocrat Aegidius his *magister militum utriusque militiae per Gallias*. After the same emperor's execution by order of the *magister militum* Ricimer, Aegidius withheld recognition from his successor Libius Severus (461–465), and gradually distanced his domain around Augusta Suessionum/Soissons from the central Western government at Ravenna. After his death, his successor, the *comes* Paulus, devoted himself to assuring the military security of this Roman enclave. Following Paulus' death, Aegidius' son Syagrius sought to have his power base, as created by his father, recognized by the emperor Zeno at Constantinople, but his efforts were in vain. In 486/7, he was defeated by the Frankish king Clovis I, and his territories were incorporated into the Frankish realm.

The Langobardic kingdom
The Langobardi under their king Alboin (561-572) left their settlement lands in Pannonia and Noricum for Italy after 568, taking substantial parts of the peninsula from the Byzantine exarchate at Ravenna. The two Langobardic duchies of Beneventum and Spoletium, which had detached themselves from the central authority of the kingdom, were reincorporated by Luitprand (712–744). Under his successors, the kingdom fragmented in conflicts with the Franks, Constantinople and the Papacy. Apart from the Duchy of Benevento, it vanished in 774, when Charlemagne conquered the residence of Pavia and had himself proclaimed King of the Langobardi.

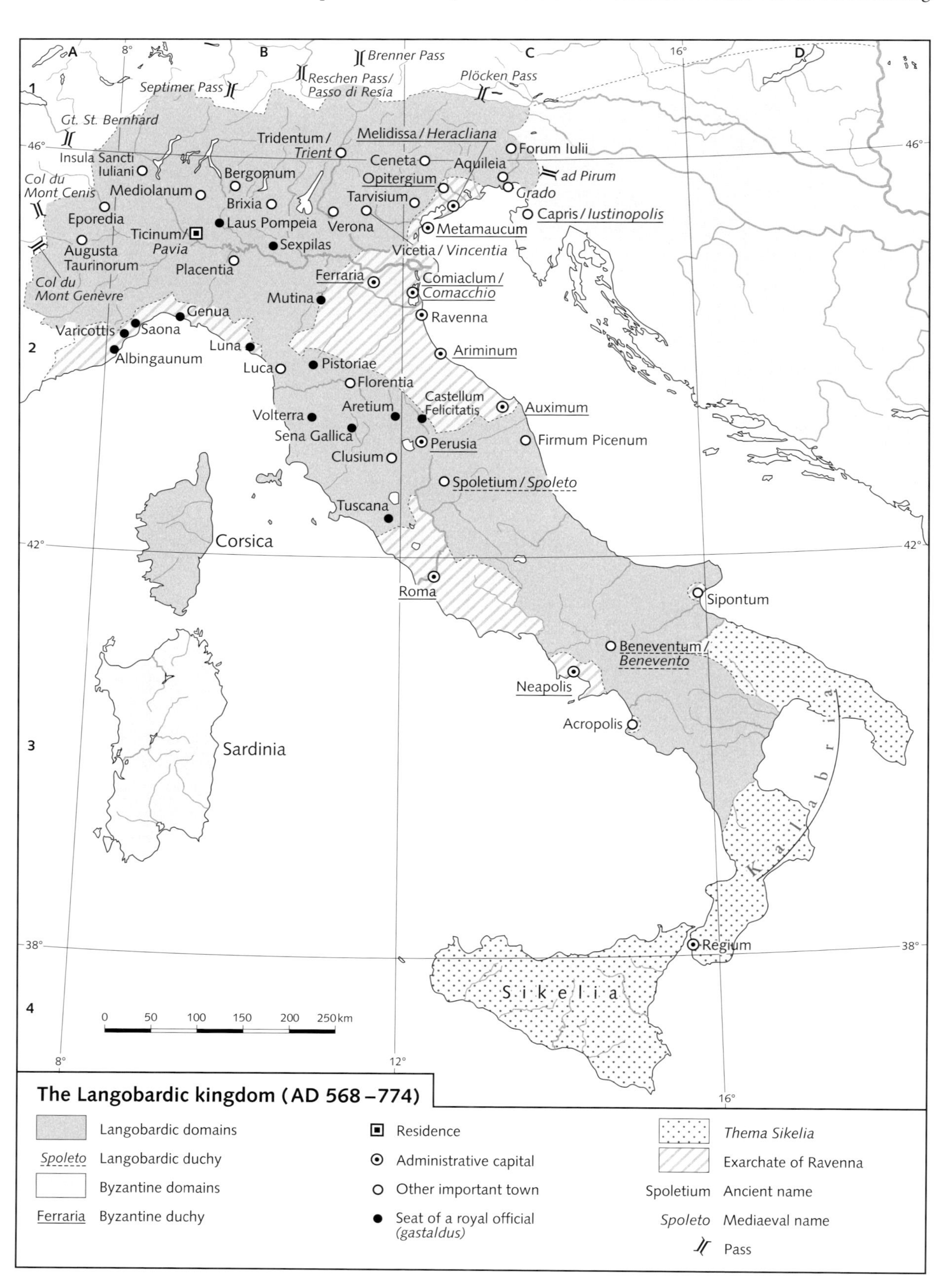

The Langobardic kingdom (AD 568–774)

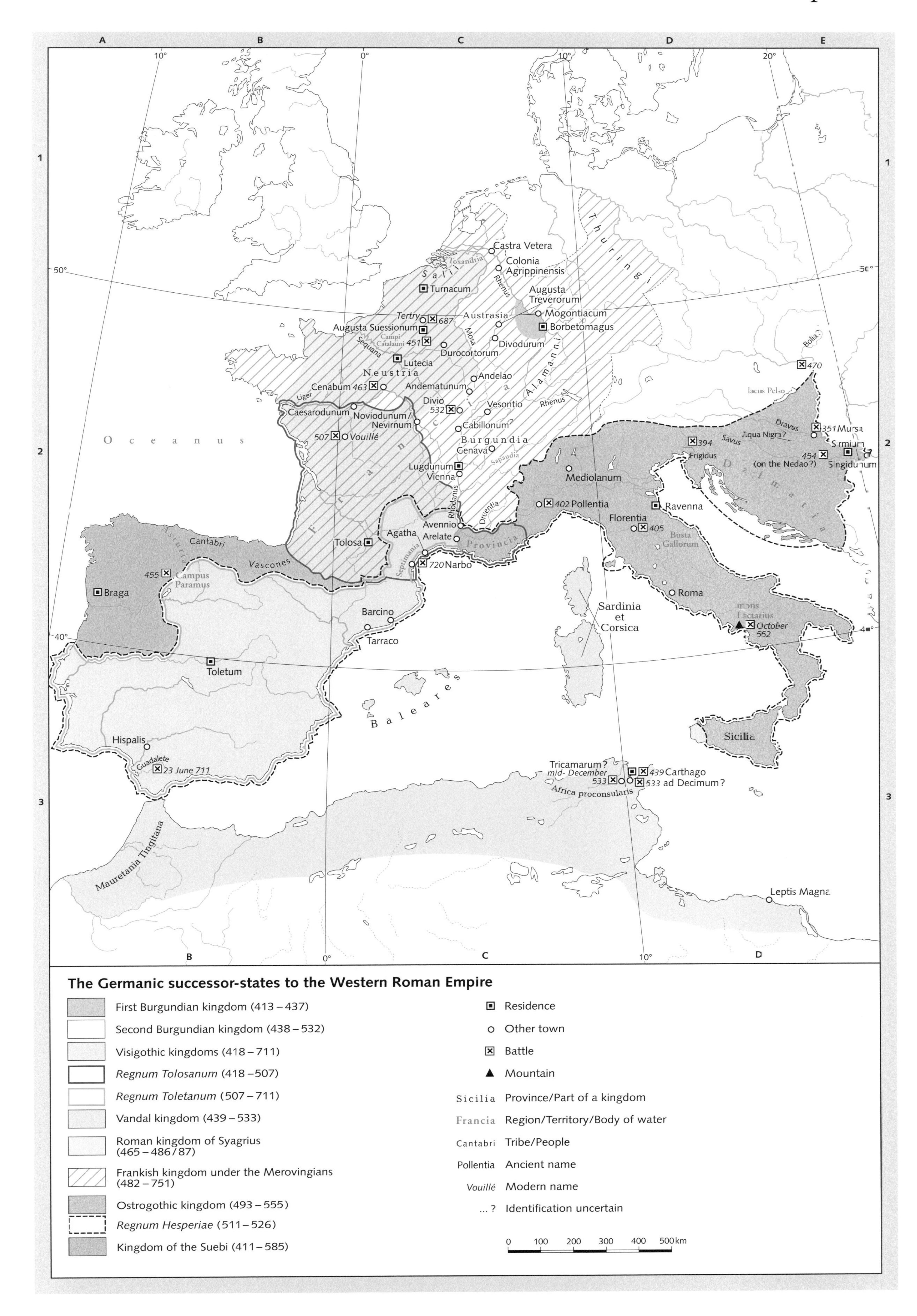
Castra Vetera
Colonia Agrippinensis
Augusta Treverorum
Mogontiacum
Borbetomagus
Turnacum
Austrasia
Augusta Suessionum
Durocortorum
Divodurum
Lutecia
Neustria
Cenabum 463
Andelao
Andematunum
Divio 532
Vesontio
Caesarodunum
Noviodunum/Nevirnum
507 Vouillé
Cabillonum
Burgundia
Genava
Lugdunum
Vienna
Oceanus
Mediolanum
402 Pollentia
Ravenna
Florentia
405
Avennio
Arelate
Agatha
Provincia
Tolosa
720 Narbo
Cantabri
Vascones
455
Campus Paramus
Braga
Barcino
Tarraco
Toletum
Sardinia et Corsica
Roma
October 552
Baleares
Sicilia
Hispalis
23 June 711
Tricamarum?
mid- December 533
439 Carthago
533 ad Decimum?
Africa proconsularis
Mauretania Tingitana
Leptis Magna
470
351 Mursa
394
454
Sirmium
Singidunum
(on the Nedao?)
Frigidus
Savus
Dravus
Tertry 687
451
Rhenus
Mosa
Sequana
Liger
Rhodanus
Druentia
The Germanic successor-states to the Western Roman Empire
First Burgundian kingdom (413 – 437)
Second Burgundian kingdom (438 – 532)
Visigothic kingdoms (418 – 711)
Regnum Tolosanum (418 – 507)
Regnum Toletanum (507 – 711)
Vandal kingdom (439 – 533)
Roman kingdom of Syagrius (465 – 486/87)
Frankish kingdom under the Merovingians (482 – 751)
Ostrogothic kingdom (493 – 555)
Regnum Hesperiae (511 – 526)
Kingdom of the Suebi (411 – 585)
Residence
Other town
Battle
Mountain
Sicilia Province/Part of a kingdom
Francia Region/Territory/Body of water
Cantabri Tribe/People
Pollentia Ancient name
Vouillé Modern name
... ? Identification uncertain
0 100 200 300 400 500 km

The Roman Empire under Justinian (527–565)

Justin I (518-527) prepared his nephew Justinian well for government, and he duly acceded on his uncle's death. Justinian (527–565) made the most of his long reign with energy and stamina.

Justinian is one of many to suffer from the tendency of the Thucydidean historiographical tradition to focus on foreign policy and the military aspects of government, while doing insufficient justice to achievements in internal government. His efforts in pursuit of an orderly fiscal policy, administrative and legal reforms (*Corpus iuris civilis*) and the unification of the Orthodox Church (Synod of Constantinople 553) all bore fruit. One particular feature of the emperor's reign is his building activities, of which we already know much through archaeology, but which Procopius also reports in his treatise *De aedificiis* ('On the Buildings [sc. of Justinian]').

In foreign policy, Justinian set himself an ambitious goal: the restoration of the *imperium Romanum*. To him, this was a divine mission (Procop. Aed. 2,6,6). The task required particular determination in the West. Here, Germanic successor states had arisen covering entire regions of the Empire (Franks, Visigoths, Ostrogoths, Vandals; → map p. 235). Justinian's approach to his aim was seldom pursued by diplomatic means – he tried to solve problems militarily, with the support of talented officers (Belisarius, Mundus, Narses). The Balkans, too, were threatened, by Huns, Slavs and Avars. Finally, the emperor had to confront the Sassanid king Chosroes I (531-579) in the East (→ map p. 241).

I. The African campaign of 533

The Vandals had ruled Carthage since 439, initially under Geiseric (d. 477), shutting off Rome from its traditional 'bread-basket'. Justinian sent Belisarius with naval and army forces against the Vandal king Gelimer (530–534), a nephew of Geiseric, to recover this territory for the Empire. The maritime campaign can be traced from Constantinople through the Aegean, round the southern tip of the Peloponnese and across to Sicily. Belisarius landed at Caput Vada on 31 August 533. After victories at Ad Decimum and Tricamarum, Belisarius captured the king, winning North Africa from Tripolis to Ad Septem, Sardinia, Corsica and the Balearics for the emperor.

II. Conflicts with the Ostrogoths in Italy, 535–555

A pretext welcomed by both sides facilitated the confrontation between Constantinople and the Ostrogoths in Italy. Amalasuntha, who had acted as regent for her son Athalaric since 526, and who after Athalaric's death (534) accepted her cousin Theodahat as co-regent, was murdered at the latter's instigation in 535. Whereas the queen was said to have occasionally considered the possibility of restoring her realm to the emperor, Theodahat openly opposed him. War was the result in 535, the emperor's declared aim being revenge for the death of the rightful queen.

In 535, Belisarius succeeded in winning back Sicily. In 536, he crossed to Regium and moved on Naples, where the strong Gothic garrison was forced to capitulate after several weeks of siege. From there, he moved on to Rome, entering the city unopposed on the night of the 9th-10th December. Witigis (536–540), Theodahat's successor, had meanwhile withdrawn to Ravenna to assemble more troops. His attempt to regain Rome failed after a siege lasting more than a year (March 538). Byzantine forces, augmented by troops now landing in Italy under Narses' command to support Belisarius, fought with varying success in the environs of Auximum, Faesulae, Ariminum and Mediolanum, accompanied by negotiations often shrouded in obscurity. Some kind of treaty was agreed by which Belisarius entered Ravenna without a fight in March 540 and took Witigis prisoner. For the time being, Italy was now restored to the emperor, except for a few Gothic bases north of the Padus.

Witigis' successor Hildebad was followed to the throne in 541 by his nephew Totila (541–552), who resumed the war against the emperor in an attempt to regain Italy for the Goths. During the years that followed, Totila and his small army (which grew day by day) pushed Byzantine rule far back through Italy. In 543, he conquered Naples, and took Rome late in 546 after a year of siege – but lost the city to Belisarius again the following year. Totila took Rome for a second time in 550, and thereafter shifted the theatre of war to southern Italy, crossing from there to Sicily. He enjoyed repeated successes everywhere, but was unable to win territory permanently. Narses crossed from Salona to Italy in 552 with an army and came up against Totila at Tadinae. Battle was joined at Busta Gallorum near Tadinae in June 552: Totila was decisively defeated, and died as he fled. Theia, who succeeded him, tried in October 552 to relieve the town of Cumae, which Narses had under siege. He was brought to battle at the *mons Lactarius* , and his army was heavily defeated. Theia himself fell in the battle. The Goths elected no successor to the unoccupied throne, and the fate of the Ostrogothic kingdom in Italy was thereby sealed, although resistance to Byzantine rule continued to flare up repeatedly, in northern Italy in particular, until 555.

III. The Spanish expedition of 552–555

Justinian planned to reconquer the western outpost of the old Roman Empire with a landing on the south-eastern coast of the Iberian Peninsula. Succession disputes surrounding the Visigothic kingdom provided him with the pretext for this enterprise. Athanagild (551/555–567), a prominent nobleman at the court of the Visigothic King Agila, rebelled against the luckless king in 551, and sought imperial support. In the spring of 552, Petrus Marcellinus Felix Liberius, the *praefectus Augustalis* at the court in Constantinople, was given command over this operation. He was a man with an extraordinarily eventful past (under Odoacer and Theoderic), and had by now reached the advanced age of 87 – Procopius calls him *eschatogeron*. The Byzantine forces did help Athanagild to the throne in 555, whereupon they were asked to return home, but stayed put. They held the littoral around Malaca and Cartagena, inland as far as Cordoba, for the Empire. The Byzantines minted imperial coins in this region until the reign of Heraclius (610–641).

IV. The three Persian wars

The sacrifices demanded of the Empire's populace become clear when it is recalled that not only was the emperor conducting wars in Africa, Italy and Spain, but also a concurrent long and attritional war with the Sassanids in the East, and that a plague epidemic struck the entire Empire in 542. There was already a long tradition of tension between Rome and the Persians. The catastrophe of Carrhae, the Roman defeat and the death of Crassus in 53 BC, was not even the beginning. At that time, Rome's conflict of interests was with the Parthian Arsacid Dynasty, and it can no longer be established which interests were at stake to whom exactly. It is certain that the intermediary trade between the Far East and the Mediterranean world along the so-called Silk Road through Antioch was one element capable of causing conflict. The cheaper but riskier sea voyage had lost much of its terror to Mediterranean sailors since the discovery of the monsoon cycle in the late 2nd cent. BC, but the Arsacids and their successors from AD 224, the Sassanids, could still play politics by blocking overland trade just as the Romans could by closing Antioch or its harbour at Seleucia. The Persians conquered and relinquished Antioch countless times – a process that still continued in Justinian's time. Conflicts with the Sassanid king Chosroes I (531-579) took place in three phases.

530–532: This conflict broke out over the kingdom of Lazica, whose king had become a vassal of Constantinople following his baptism in 522. The blocking of what had hitherto been free access to the Black Sea was the trigger of a war which broke out while Cabades I (488–531) was still king, and which his son Chosroes I brought to an end in 532 with a peace treaty.

540–545: Pretext arose not only from Chosroes' breach of the 532 treaty, but also from a dispute which arose in 540 between two Arab tribes, the emperor's Ghassanid allies and the Lachimids. A climax of this war was the brief occupation of Antioch by Chosroes in 540; Sassanid troops occupied Lazica. Mutual exhaustion brought a five-year truce from 545.

549–557: This war broke out again over Lazica, when the emperor acceded to the Laz request to be made a protectorate. The conflict that now ensued lasted until 557. A truce was then agreed, and a peace treaty was signed in 561, by which Lazica was given to the emperor.

Sources

There is good documentation of Justinian and the history of his times. Most sources are contemporary, the most prominent being the *Corpus Iuris Civilis*, a four-volume legal corpus. Commissioned in 528 from a committee of jurists, this collection of Latin imperial law from Hadrian to Justinian (*Codex Iustinianus*) was published in 529, with a second, expanded edition in 534. The revision of the codex was made necessary by the publication in 533 of the *Institutiones*, a juristic manual with force of law, and the *Digestae*, a collection of excerpts from important juristic treatises. The section of the *Iuris Civilis* termed the *Novellae* was planned as a collection of the imperial laws enacted since 534, but was not realized. Incomplete private collections of these laws have survived. – Another important source for Justinian's reign is the historical oeuvre of Procopius. Procopius was an eye-witness to the military history of his age, as a member of the staff of the imperial general Belisarius from the 530s, and accompanying him on campaigns against the Persians, Vandals and Goths; this lends authority to his work *De bellis* ('On the Wars [sc. of Justinian]'). His 'Unpublished' (*Anecdota*) treatise known as the 'Secret History' was published anonymously and probably not during Justinian's lifetime: in it, he excoriates the imperial couple. Conversely, his description of the buildings created under Justinian (*De aedificiis*) has a panegyric character. – Another contemporary of the emperor was Agathias, who wrote a history of the years 552-559, following on from Procopius' *De bellis*.

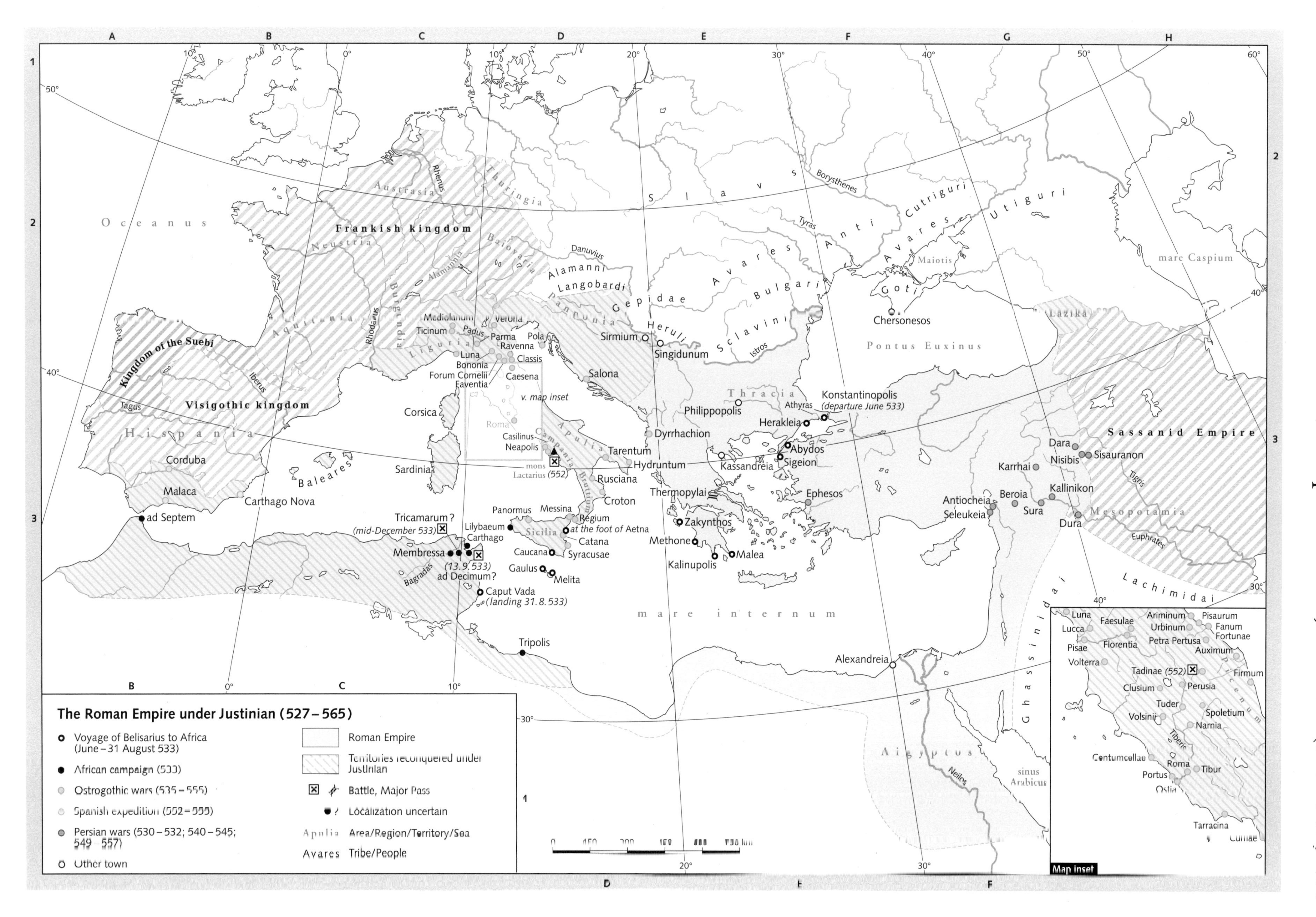
The Roman Empire under Justinian (527–565)
Voyage of Belisarius to Africa (June–31 August 533)
African campaign (533)
Ostrogothic wars (535–555)
Spanish expedition (552–555)
Persian wars (530–532; 540–545; 549–557)
Other town
Roman Empire
Territories reconquered under Justinian
Battle, Major Pass
Localization uncertain
Apulia Area/Region/Territory/Sea
Avares Tribe/People
Map inset
Oceanus
Kingdom of the Suebi
Visigothic kingdom
Hispania
Frankish kingdom
Neustria
Austrasia
Aquitania
Burgundia
Thuringia
Baiovaria
Alamanni
Langobardi
Gepidae
Heruli
Sclavini
Bulgari
Avares
Anti
Slavs
Cutriguri
Utiguri
Goti
Lazika
Sassanid Empire
Mesopotamia
Lachimidai
Ghassinidai
Aigyptus
Pontus Euxinus
mare Caspium
mare internum
Maiotis
sinus Arabicus
Thracia
Pannonia
Liguria
Apulia
Campania
Bruttium
Sicilia
Baleares
Rhenus
Rhodanus
Iberus
Tagus
Padus
Danuvius
Istros
Tyras
Borysthenes
Tigris
Euphrates
Neilos
ad Septem
Malaca
Corduba
Carthago Nova
Corsica
Sardinia
Tricamarum? (mid-December 533)
Membressa
ad Decimum? (13.9.533)
Carthago
Lilybaeum
Caput Vada (landing 31.8.533)
Bagradas
Panormus
Messina
Regium
at the foot of Aetna
Catana
Syracusae
Caucana
Gaulus
Melita
Tripolis
Mediolanum
Ticinum
Verona
Parma
Luna
Bononia
Forum Cornelii
Faventia
Ravenna
Classis
Caesena
Pola
v. map inset
Roma
Casilinus
Neapolis
mons Lactarius (552)
Tarentum
Hydruntum
Rusciana
Croton
Salona
Sirmium
Singidunum
Dyrrhachion
Philippopolis
Thermopylai
Zakynthos
Methone
Kalinupolis
Malea
Kassandreia
Herakleia
Athyras
Abydos
Sigeion
Konstantinopolis (departure June 533)
Ephesos
Chersonesos
Antiocheia
Seleukeia
Beroia
Sura
Kallinikon
Karrhai
Dara
Nisibis
Sisauranon
Dura
Alexandreia
Luna
Lucca
Pisae
Volterra
Faesulae
Florentia
Ariminum
Urbinum
Pisaurum
Fanum
Fortunae
Petra Pertusa
Auximum
Firmum
Picenum
Tadinae (552)
Perusia
Clusium
Tuder
Spoletium
Volsinii
Narnia
Tiberis
Centumcellae
Roma
Tibur
Portus
Ostia
Tarracina
Cumae

The Byzantine theme system (7th–9th cents. AD)

Heraclius (610-641), who acceded on the death of Phocas, would be another emperor who tightened the structures of the Empire and its society anew. And he it was (probably more so than his grandson Constans II (641–668)) who undertook a drastic administrative reorganization: the division of the Empire into military provinces, so-called *themata* or 'themes'.

The actual meaning of the word *thema* is disputed. It may have meant a 'sphere of operations' for particular army divisions that were transferred to Asia Minor in response to the loss of frontier territories. The themes were governed by *strategoi*, who assumed all civil powers in addition to their military responsibilities in the 8th cent. In the 7th cent., the Empire was divided into five large themes (the *themata* of *Armeniakon, Anatolikon, Opsikion, Kibyrrhaioton, Thrakesion*); as the emperors won back lost territories, more themes were added, especially in the West. For reasons of security, large *themata* were also divided, so that by the reign of Constantine VII (913-959), the Empire comprised 17 Asian themes and 12 European ones.

So-called *kleisourarchiai* were separated out of the *themata* in regions where defence was difficult. *Kleisoura*, derived from the Latin *clausura*, was originally a geographical expression denoting a mountain pass, later transferred to the defensive fortifications and garrison located at such passes. Under kleisourarchs, these *kleisourai* developed into administrative districts in frontier regions (Cappadocia, Charsianum, Seleucia). The volatility of the situation in the East is also shown by the existence of military bases that were set up under the command of a *dux*, the governor of a small region or city with military and civil powers, or of a *droungarios* ('drungary'), who led the military formations of a theme.

Constantinople

In the territory of Byzantium on the western shore of the Bosporus, Constantine I laid out his residence of Constantinople on a promontory between the Golden Horn (*Chryson Keras*) and the Propontis.

A wall had protected the landward side of Byzantium since its foundation in the 7th cent. BC. The location of this wall can only be guessed at (no. 1). The course of the land wall enclosing the enlarged city from the reign of Constantine can also only be approximately shown on the map (no. 2). However, the land wall built under Theodosius II from AD 404 to 447 (no. 4) survives. Inscriptions on this wall, its towers and gates document the history of these defensive fortifications until the 18th cent. The wall system is 5.7 km long, stretching from the Propontis to Blachernae, and is up to 55 m deep. Ahead of the main wall (4.8 m thick, 11 m high), which has 96 towers of an average height of 24 m, a second wall (4 m thick, 8 m high) stands at a distance of 14.5 m, with 90 towers placed in the gaps between the towers of the main wall. In front of this wall was a moat (with transverse walls; 18 m wide, 7 m deep). – A wall had also guarded the city to the seaward side since the reign of Constantine I (no. 3). Under Theodosius II this was extended at both ends to meet the land walls (on the Propontis by 8.5 km, on the Golden Horn by 5.4 km) and greatly strengthened.

The city, which had been rebuilt by Septimius Severus, was quadrupled in size by Constantine I, and Theodosius II again increased its size by more than a third. The expanded area inside the land walls fanned out from the acropolis along a main thoroughfare, the Mese, which forked to the north-west and south-west at the Philadelphion after about 3 km. Under Theodosius II, the urban territory was divided into 14 *regiones*. Four squares marked the main stations of the imperial triumphal procession, which entered the city in the south-west on the *via Egnatia* via the Porta Aurea and led eastwards to the Hagia Sophia (no. 5a). These were:

1. the Forum Arcadii, a square surrounded by *stoai*, built around an honorary column dedicated to Arcadius (395–408);
2. the Forum Tauri or Forum of Theodosius (no. 9), a square completed under Theodosius I (379–395) and also surrounded by *stoai*, at its centre an honorary column dedicated to Theodosius. The Philadelphion was a short way away to the west;
3. the Forum of Constantine built under Constantine I (no. 8), the market place and centre of urban life, a square lined with arcades, with two gates, the senate building in the north-west and the Column of Constantine in the middle (no. 8a);
4. the Tetrastoon with Augusteum, imperial palace (no. 5b), another senate building (no. 5c) and basilica (no. 5d), as well as the Hagia Sophia (no. 5a) and the Patriarch's residence.

The Hippodrome (no. 6) was an important link between emperor and people. This was a venue for festivals of all kinds, but also for political demonstrations. Three monuments from the *spina* survive:

1. in the south, a painted obelisk (32 m high; no. 6c); its age is not precisely known (before 390). According to an inscription, it was renovated under Constantine VI (913–959) and covered with gilt bronze plates (the dowel holes are still visible), which remained until the systematic pillage of the city by the Crusaders in 1204;
2. farther to the north, the Serpent Column (no. 6b), originally a votive gift to Delphi from the Greeks on the occasion of their victory over the Persians at Plataeae in 479 BC. The column found its way to Constantinople as Constantine I adorned his capital with art works from all over the known world;
3. in the north, the Egyptian obelisk (no. 6a), created by order of Thutmosis III (1490–1436 BC) at Karnak, earmarked for Constantinople by Constantine I and subsequently Julian 'Apostata', but only transported there under Theodosius I and put up in the Hippodrome in 390.

As the city had no springs, conduits were needed to bring water from the mountains to the north-west. The aqueduct built in the reign of Valens formed part of the conduit system whose destination was the Nymphaeum in the Forum of Theodosius. Many cisterns also provided a water supply; some covered, some open, some private, some public. The map shows the largest open cisterns. –Many of the countless churches of the city have survived (even with their internal architecture) thanks to their conversion into mosques. Some are included on the map by way of example.

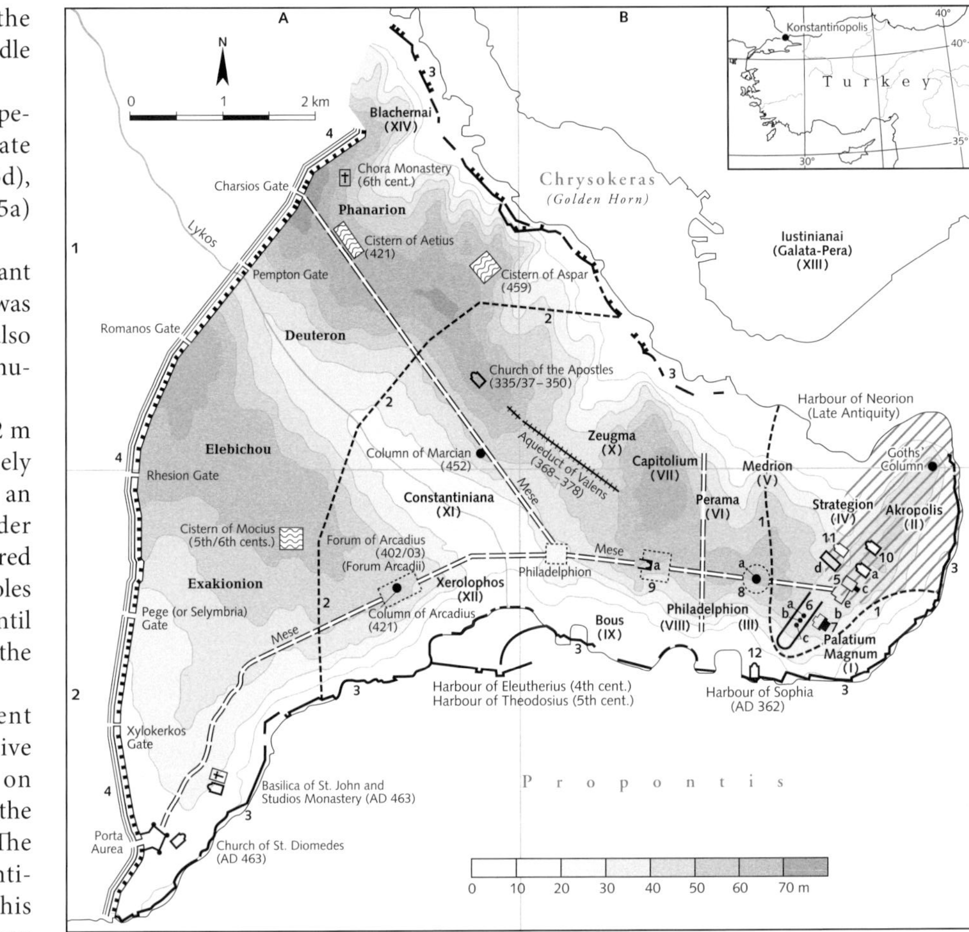

Constantinopolis: archaeological site-map

1. Land wall at the time of Emperor Septimius Severus (AD 193–211); conjectured course
2. Land wall at the time of Emperor Constantine I (AD 306–337); conjectured course
3. Sea wall at the time of Emperor Theodosius II (AD 408–450)
4. Land wall at the time of Emperor Theodosius II (AD 408–450)
5. Augusteion, a square surrounded by four stoai
   a: Hagia Sophia
   b: Imperial palaces
   c: Senate
   d: Administrative basilica
   e: Baths of Zeuxippus
6. Hippodrome
   a: Obelisk from Karnak in Egypt (15th cent. BC)
   b: Serpent Column from Delphi (479 BC)
   c: Obelisk in stonework (before AD 390)
7. Palace of Lausus (5th/6th cents. AD)
8. Forum Constantini (early 4th cent. AD)
   a: Column of Constantine (AD 328)
9. Forum Tauri (Forum of Theodosius; AD 372–393)
   a: Remains of the honorary arch
10. Hagia Eirene (4th cent. AD)
11. Church of the Theotokos Chalkoprateia (5th cent. AD)
12. Church of Sts Sergius and Bacchus (AD 527–536)

Zeugma (X) Theodosian regio
Elebichou District
Main streets
Acropolis of Byzantium

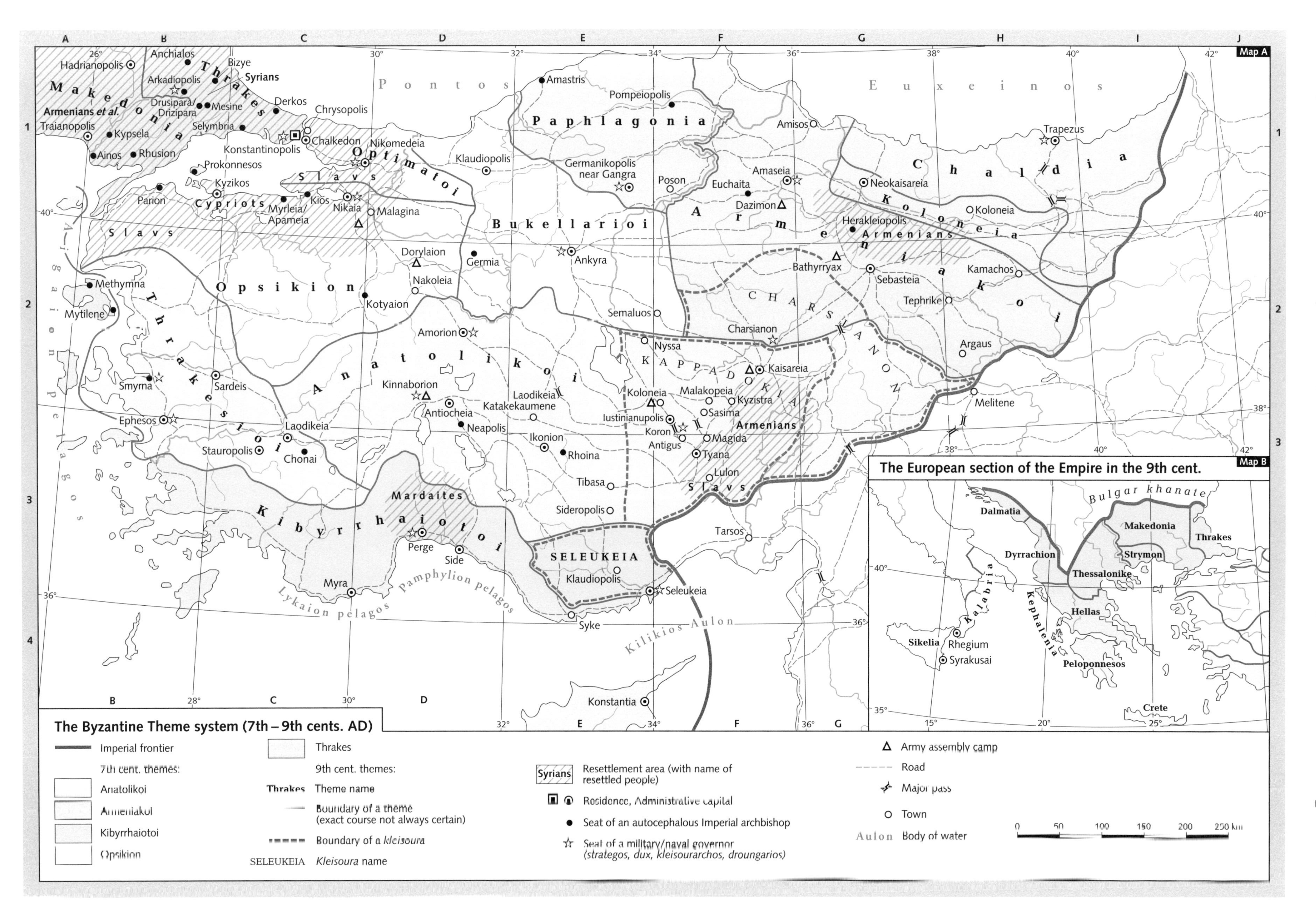
Map A
Pontos
Euxeinos
Makedonia
Thrakes
Armenians et al.
Syrians
Hadrianopolis
Anchialos
Bizye
Arkadiopolis
Drusipara/Drizipara
Mesine
Derkos
Chrysopolis
Traianopolis
Kypsela
Selymbria
Chalkedon
Ainos
Rhusion
Konstantinopolis
Nikomedeia
Optimatoi
Prokonnesos
Kyzikos
Slavs
Parion
Cypriots
Kios
Nikaia
Myrleia/Apameia
Malagina
Klaudiopolis
Amastris
Pompeiopolis
Paphlagonia
Germanikopolis near Gangra
Poson
Amisos
Euchaita
Amaseia
Dazimon
Armeniakoi
Bukellarioi
Ankyra
Dorylaion
Germia
Nakoleia
Opsikion
Kotyaion
Methymna
Mytilene
Aigaion pelagos
Thrakesioi
Amorion
Semaluos
Smyrna
Sardeis
Anatolikoi
Kinnaborion
Antiocheia
Laodikeia Katakekaumene
Ephesos
Laodikeia
Neapolis
Stauropolis
Chonai
Ikonion
Rhoina
Tibasa
Sideropolis
Mardaites
Kibyrrhaiotoi
Perge
Side
Myra
Lykaion pelagos
Pamphylion pelagos
SELEUKEIA
Klaudiopolis
Seleukeia
Syke
Kilikios Aulon
Konstantia
Tarsos
Nyssa
Kappadokia
Kaisareia
Koloneia
Malakopeia
Kyzistra
Iustinianupolis
Sasima
Koron
Antigus
Magida
Tyana
Armenians
Lulon
Charsianon
Bathyrryax
Sebasteia
Herakleiopolis
Neokaisareia
Chaldia
Trapezus
Koloneia
Kamachos
Tephrike
Argaus
Melitene
Map B
The European section of the Empire in the 9th cent.
Bulgar khanate
Dalmatia
Makedonia
Thrakes
Dyrrachion
Strymon
Thessalonike
Kalabria
Kephalenia
Hellas
Sikelia
Rhegium
Syrakusai
Peloponnesos
Crete
The Byzantine Theme system (7th – 9th cents. AD)
Imperial frontier
7th cent. themes:
Anatolikoi
Armeniakoi
Kibyrrhaiotoi
Opsikion
Thrakes
9th cent. themes:
Thrakes Theme name
Boundary of a theme (exact course not always certain)
Boundary of a kleisoura
SELEUKEIA Kleisoura name
Syrians Resettlement area (with name of resettled people)
Residence, Administrative capital
Seat of an autocephalous Imperial archbishop
Seat of a military/naval governor (strategos, dux, kleisourarchos, droungarios)
Army assembly camp
Road
Major pass
Town
Aulon Body of water
0 50 100 150 200 250 km

Byzantine-Sassanid conflicts in the 6th and 7th cents. AD

On the Sassanid kings, cf. the list on p. 216; on the Byzantine emperors, p. 264.

The reign of Sapor II, a great-grandson of Sapor I (240–272), was chiefly characterized by two themes: the war with the Romans and the persecution of Christians, who were feared in the Sassanid Empire for their contacts with Christians in the Roman Empire. In AD 344, the Persians suffered a heavy defeat at Singara at the hands of the emperor Constantius II; they also failed to take Nisibis from the Romans, besieging it in vain in 338 and 346. They did, however, succeed in conquering Amida in 359. The emperor Julian 'the Apostate' led a victorious campaign against the Persians to the gates of Ctesiphon, but he failed to take the Sassanid capital. Retreating up the Tigris, Julian was wounded in battle and died. His successor, Jovian, agreed a peace greatly to Rome's disadvantage, allowing Sapor huge territorial gains. Nisibis now also fell to the Sassanids once more.

Although the tensions between the Sassanids and Constantinople flared up in military conflict on a regular basis, the central problem in the years that followed for each of the Sassanid kings from Vahram V (421–438) to Chosroes I (531–579) were the Hunnic Hephthalites, who were pressing forward against the north-east of the Sassanid Empire. Vahram V and Yazdgird II had several successes against them, but in 484 they defeated Perozes, who fell in battle. With that, the Sassanids became tributaries to the Hephthalites. The Hephthalites also played an important part in the domestic politics of the Sassanid court. For instance, it was with their help that Cabades, the son of Perozes, returned to power in AD 499, against the court nobility under Zamasphes.

Only under Chosroes I did the Sassanid Empire succeed in extricating itself from the embarrassment in which it found itself with the Hephthalites, a situation which badly affected state and society for a century. Chosroes I reformed the Persian state and its society. A reorganization of the army also enabled him to pursue the war with Byzantium with success. Shortly after acceding to the throne, in 532, he concluded an 'eternal peace' with Justinian (Procop. Pers. 1,22,3), which he then breached in 540, even succeeding in occupying Antioch. At this point, Byzantine negotiators bought a truce which was again fragile, and which the conflict over Lazica caused to be interrupted and resumed several times. In 561, a fifty-year peace was agreed. By this, Chosroes relinquished his claim to Lazica.

Byzantium broke this peace agreement in 572, in the reign of Justin II, invading Mesopotamia. The Romans did not succeed in conquering Nisibis, and following an abortive skirmish in 573 were forced to retreat to the city of Dara and finally to capitulate. The Byzantine-Sassanid war nonetheless continued in northern Mesopotamia. For instance, the Sassanid army under Chosroes' personal command was defeated at Melitene in 575 – yet this battle still brought no outcome to the war. When Chosroes died in 579, he bequeathed the legacy of this war to his son Hormisdas IV.

Meanwhile, however, Chosroes I enjoyed success on other fronts. He was victorious in the south of the Arabian Peninsula, winning Mazūn on the south-east coast and al-Yamān on the south-west, both important trading locations. Around 560, he succeeded in destroying the Hephthalite kingdom with the help of the Göktürks ('Celestial Turks'). It would become clear, however, that in winning these 'allies', he had visited upon his empire a dangerous future opponent.

Indeed, Chosroes' son Hormisdas IV was only able to pursue the conflict with Byzantium half-heartedly, because the Göktürks in the north-east were threatening to invade his empire. In 589, his general Vahram Chobin defeated them in the vicinity of Balḫ, keeping them at bay for now.

The son of Hormisdas IV, Chosroes II, acceded to the throne in a palace revolution in 590, but was deposed by supporters of Vahram Chobin and fled to Constantinople, where the emperor Mauricius granted him asylum – only to take him back to Ctesiphon in the spring of 591 and restore him to the throne with military assistance. In return for this service, Chosroes gave Mauricius the cities of Dara and Martyroupolis, as well as part of Armenia as far as Lake Thospitis.

When Mauricius was murdered in 602 on the instigation of one of his officers, Phocas, Chosroes appointed himself his avenger. Sassanid armies moved westwards into Mesopotamia along the Euphrates as far as Circesium, and into Asia Minor through Armenia, as far as Amasea and Caesarea. Some cities, such as Edessa, offered resistance, but in the field no Byzantine army successfully confronted the Sassanid king, who undertook several campaigns in person in 606-610. Nor did this change when Phocas was deposed by Heraclius – even this capable emperor was unable to assemble an effective force in the field in such short order. Chosroes and his general Shahrbaraz advanced through the whole of Syria, and the Sassanids took the cities of Antioch (611), Damascus (613) and Jerusalem (614), where Chosroes destroyed the Constantinian Church of the Holy Sepulchre and had the 'True Cross' removed to Ctesiphon. The Sassanids invaded Egypt in 619, and moved up the Nile into the Sudan. They also pushed farther west, perhaps as far as Tripolis. Meanwhile, in Asia Minor in 615, even Chalcedon on the Bosporus, within sight of the Byzantine capital, was conquered by the Sassanids under Chosroes' general Shahin.

After completing his wide-ranging reforms of society and army, Heraclius from AD 622 was once more in a position to respond to the Sassanid challenge. A first campaign, which he conducted against a Sassanid army under Shahrbaraz, and which appears to have led both opposing armies hither and thither across Asia Minor, ended in the spring of 623 with a decisive Byzantine victory, probably at Caesarea in Cappadocia. At a stroke, the emperor had won back the Anatolian Peninsula. Heraclius entered northern Mesopotamia in 623–625, but won no decisive victories. In the summer of 626, in confederation with the Turkic Avars who had settled in the Carpathian Basin since 567 and conquered Sirmium in 582, Chosroes tried to force a victory at Constantinople with an army commanded by Shahrbaraz, while tying Heraclius and his army down in Asia Minor using lesser forces. This was the first of many sieges of the Byzantine capital. Its defence was led by the *magister militum praesentalis* Bonus and Patriarch Sergius. After a first defeat in the Golden Horn, the Avars retreated. The Sassanids, who were unable to cross to the European shore because of the Byzantine maritime superiority, remained at Chalcedon into 627 before finally withdrawing eastwards. Meanwhile, Heraclius was able to pursue his campaign plans from Armenia, in the course of which he decisively defeated a Sassanid army under Razates at Mosil near Niniveh in December 627. The Byzantine army moved on towards the Sassanid capital of Ctesiphon, but Heraclius forebore to assault it, in view of the anticipated fierce Persian resistance.

The Sassanid defeat at Niniveh and the loss of Chosroes' residence at Dastagird near Ctesiphon in 628 led to unrest around the throne, during which the king was deposed and murdered by his son Siroe. The peace treaty which Siroe (who took the regnal name Cabades II) immediately negotiated with Heraclius and which Boran, a sister of Cabades', ratified as regent in 630, required the return of all formerly Byzantine territories, from Egypt to the Black Sea, excepting only eastern Armenia. The 'True Cross' was also to be returned; the emperor raised it back into position at Jerusalem on 21 March 630.

There were several more successions to the throne before Yazdgird III, a grandson of Chosroes, acceded. He no longer had the Byzantine Empire to contend with: his opponents were the Arabs under the first three caliphs. After defeats in 636 at al-Qādisīya on the Euphrates and in 642 at Nihāwand in Media (→ map p. 243), the life of the last Sassanid ruler became one of ceaseless flight. Yazdgird was murdered at Marv in 651, and with his death, the Sassanid Empire fell in its entirety to the Arabs.

Sources

Procopius' *De bellis* is an important historical work. Its books 1 and 2 cover the wars of the Byzantine Romans against the Sassanids, providing information in particular about relations between Justinian and Chosroes I. Georgios Pisides (*c.* 600 – *c.* 630) was a cleric at Hagia Sophia under Patriarch Sergius. A contemporary of Heraclius', he extolled the military exploits of that emperor, e.g. in *De expeditione Heraclii imperatoris contra Persas* (622/3) and *Bellum Avaricum*. He also wrote the *Heraclias*, an epic hymn on Heraclius' victory over the Sassanids, as well as a song of praise to Bonus, who had led the defence of Constantinople alongside Patriarch Sergius in 626. Another contemporary of Heraclius was Theodore Syncellus, one of whose sermons, from the first anniversary of the city's liberation on 7 August 627, is preserved. The anonymous *Easter Chronicle* (from Adam to AD 627) is a contemporary document that has survived. Also of incalculable value for the period covered by this map are the histories of the monk Theophanes (*c.* 760–817/8) and the patriarch Nicephorus (758–828), respectively a world chronicle following on from Georgius Syncellus and a *Historia syntomos* of the years 602–769. For an Arab perspective on the Byzantine-Sassanid conflicts, we have the *Annals* (*Tarikh al-Tabari*) of the Persian Muhammad ibn Jarir al-Ṭabarī (839–923), which extend from the beginning of the world to the year 915.

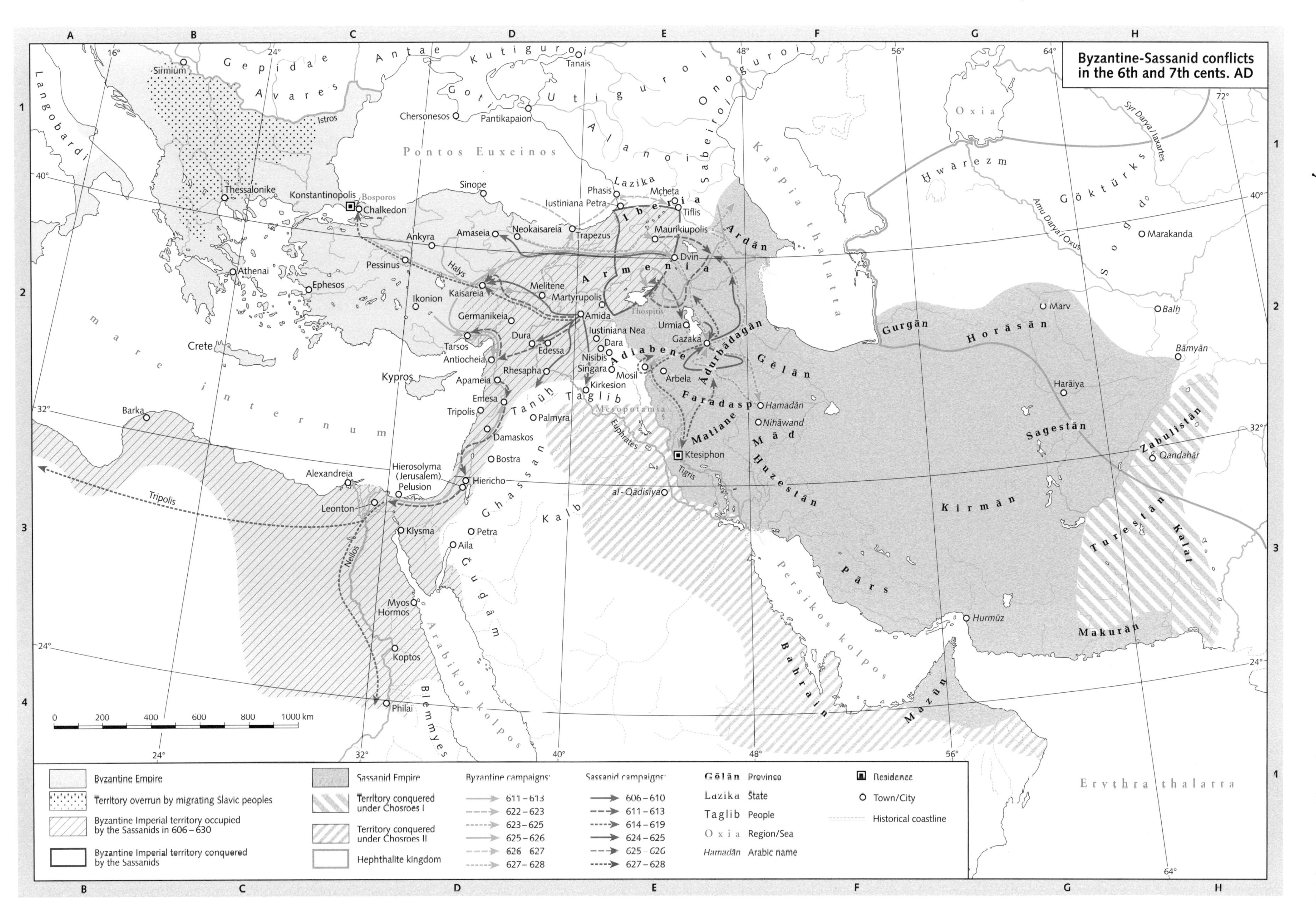

Byzantine-Sassanid conflicts in the 6th and 7th cents. AD
Pontos Euxeinos
Kaspia thalatta
mare internum
Persikos kolpos
Arabikos kolpos
Erythra thalatta
Oxia
Konstantinopolis
Chalkedon
Thessalonike
Athenai
Ephesos
Ankyra
Sinope
Trapezus
Kaisareia
Melitene
Amida
Dara
Nisibis
Edessa
Antiocheia
Damaskos
Hierosolyma (Jerusalem)
Alexandreia
Ktesiphon
Dvin
Tiflis
Gazaka
Hamadān
Marv
Balḫ
Hurmūz
Byzantine Empire
Territory overrun by migrating Slavic peoples
Byzantine Imperial territory occupied by the Sassanids in 606–630
Byzantine Imperial territory conquered by the Sassanids
Sassanid Empire
Territory conquered under Chosroes I
Territory conquered under Chosroes II
Hephthalite kingdom
Byzantine campaigns
611–613
622–623
623–625
625–626
626 627
627–628
Sassanid campaigns
606–610
611–613
614–619
624–625
625 626
627–628
Gēlān Province
Lazika State
Taglib People
Oxia Region/Sea
Hamadān Arabic name
Residence
Town/City
Historical coastline
0 200 400 600 800 1000 km

The unification and expansion of the Arabs under the first four caliphs (632–661)

The first four caliphs, successors of Muḥammad (d. 632):
'Abdallāh Abū Bakr (632–634)
'Umar ibn al-Ḫaṭṭāb (634–644)
'Utmān ibn 'Affān (644–656)
'Alī ibn Abī Ṭālib (656–661)

Muḥammad met such resistance to his doctrine in Makka/Mecca, the town of his birth, that he emigrated with his followers in 622, after long preparations, to the oasis city of Yaṯrīb, subsequently al-Madīna ('City [of the Prophet]'), where he had engendered great hopes as a mediator in the conflicts between various Arab tribes of the region.

To bring some degree of economic independence, he carried out a number of raids (*ghazawat*, 'razzias') in 623 and 624 against caravans from Mecca, and finally a *ghazi* against a large caravan on its way home to Mecca from Syria. This battle, which took place at Badr in March 624, was of a different order from the earlier skirmishes. It was now clear that the conflicts between Muḥammad and Mecca had grown into a full-scale war. A year or so later, an army left Mecca bent on revenge for the attack at Badr. Battle was joined at Mount Uḥud, and the followers of Muḥammad were defeated. However, the defeat was not decisive. In 627, an army from Mecca besieged Muḥammad at Yaṯrīb. Muḥammad withstood the siege. Over subsequent years, he succeeded in winning over more and more followers, not only in Yaṯrīb but also in Mecca, so that in 630 he finally entered Mecca and took the city almost without bloodshed. In the same year, in reaction to the fall of Mecca, two Arab tribes, including women, children and herds of cattle, went to Mecca to make use of the trading routes at aṭ-Ṭā'if, as heirs to the trading town. Muḥammad thwarted this intent, confronting them at Ḥunain with an army and forcing them to flee. – It was inevitable that, sooner or later, Muḥammad and his religious-political movement (which would soon unite the entire Arabian Peninsula) would come into conflict with the neighbouring states of Byzantium and the Persians. Muḥammad conducted a first campaign as early as 629, to Palestine, but was defeated by an imperial army at Mu'ṭa.

Following the death of Muḥammad on 8 June 632, a priority of his successor, 'Abdallāh Abū Bakr, was to keep together the Arab states which had been tied to Muḥammad by separate treaties and might no longer feel bound by them following the prophet's death. Ḫālid ibn al-Walīd for the most part led the so-called *ridda* wars (632–633) against the 'apostate' Bedouin tribes, esp. in the south of the Arabian Peninsula, by order of the caliph. Various Arab tribes here were refusing to pay the alms tax (*zakāt*). The most influential leader of the *ridda* was Ibn Ḥabib al-Ḥanefi, whom Ḫālid ibn al-Walīd decisively defeated in 633 at al-'Aqrabā'.

In spite of these conflicts within the Muslim world, 'Abdallāh Abū Bakr and his capable general Ḫālid ibn al-Walīd found the time in 634 both to take al-Ḥīra on the Euphrates and to defeat the army of Heraclius I at Ağnadain, a victory which won them south-eastern Palestine.

Under the caliph 'Umar ibn al-Ḫaṭṭāb, the Arabs won a victory over the imperial forces at the Yarmouk in 636. Jerusalem was taken two years later, under Ḫālid ibn T-ābit al-Fahmī. The Arabs then also advanced through Egypt, where they defeated an imperial army at Babylon in July 640. Alexandria surrendered without a fight in 642. In 643, the Arabs also occupied Barce in the Cyrenaica. There were also further battles with the Persian Sassanid Empire. For instance, Muslim troops were defeated in 634 at Kūfa by the Euphrates. In 636, a Sassanid contingent was defeated at al-Qādisīya, as was a larger army at Galūlā'. In 639, al-Ahwāz came into Arab hands. They defeated the Persian army under the last Sassanid king Yazdgird III at Nihāwand in 642, and shortly afterwards, the Sassanid capital of Ctesiphon was overrun.

Under the Caliph 'Utmān ibn 'Affān, the Arabs pushed into North Africa as far as Tripoli in 647. In 649, they occupied Cyprus, and in 655 briefly Rhodes. Arabic armies reached Dwīn, Tiflis and Marv in 645-653.

Social tensions which developed under 'Utmān, the third caliph, led to revolts in 655 and 656 in Kūfa and in Egypt respectively. In their wake, the caliph was murdered on 17 June 656. Members of the oldest following of Muḥammad at al-Madīna, who saw themselves disadvantaged in relation to the Meccan mercantile community in the light of the Umayyad clan's influence which 'Utmān had systematically promoted, chose the prophet's cousin and son-in-law, 'Alī ibn Abī Ṭālib as 'Utmān's successor. His opponents contested the legality of his succession, leading to armed conflict – 'Alī was victorious at al-Bṣra on 9 December 656. However, the governor of Syria, the Umayyad Mu'āwiya ibn Abī Sufyan, still withheld his recognition, and the result was an inconclusive battle in the summer of 657, on the right bank of the Euphrates at Ṣiffīn. A court of arbitration which was supposed to rule of the legality of 'Alī's caliphate, was unable to reach consensus. Consequently, Mu'āwiya had himself proclaimed caliph in 660 at Jerusalem by his army. 'Alī was murdered on 24 January 661, but his followers (the so-called Shi'ites) did not disband. Mu'āwiya was now generally recognized as caliph, and the caliphate remained the province of the Umayyads for the next hundred years.

Sources

The monk Theophanes Homologetes (*c.* 760–817/8) produced a strictly annalistic continuation of the Chronicle of George Syncellus (covering the years AD 284-813) in a simple style often close to the spoken language. For the years after 602, his chronicle is the best source for historical events, and often the only one. Another chronicle (*Historia syntomos*), which survives only in a fragmentary state and which is available for the period covered by this map, was written by the patriarch Nicephorus (*c.* 758–828). The period 591-661 is covered by the Armenian bishop Sebuos (7th cent.) in his *History of Heraclius*. Finally, the comprehensive *World Chronicle* of Michael the Syrian, patriarch of Antioch (1126–1199), is particularly important. Among the Arabic sources, the history by the scholar aṭ-Ṭabarī (839–923) is outstanding in terms of quality of information.

The map

The map uses colours to distinguish battles between the Arabs and the Sassanids and Byzantine Romans on the one hand and battles among the Arabs on the other, the latter generally regarded as civil wars. The campaigns of the Arab Abū Bakr in the so-called *ridda* wars are specially highlighted, because they clarify the particular achievement of Muḥammad: the political and religious unification of all the Arabs.

→Map p. 241

Literature

M.A. Cook, Muhammad, 1983; F.M. Donner, Narratives of Islamic Origins. The Beginnings of Islamic Historical Writing, 1998; Id., The Early Islamic Conquests, 1981; U. Haarmann, Geschichte der Arabischen Welt, [4]2001; W.E. Kaegi, Heraclius. Emperor of Byzantium, 2003; T. Nagel, Staat und Glaubensgemeinschaft im Islam, 2 vols., 1981; Id., The Frontier: Barrier or Bridge?, in: Congrès International des Études Byzantines 17, 1986, 288–293; R. Paret, Mohammed und der Koran, [9]2005; G.J. Reinink, B.H. Stolte (eds.), The Reign of Heraclius (610–641). Crisis and Confrontation, 2002; G. Rotter, Die Umayyaden und der zweite Bürgerkrieg, 1982.

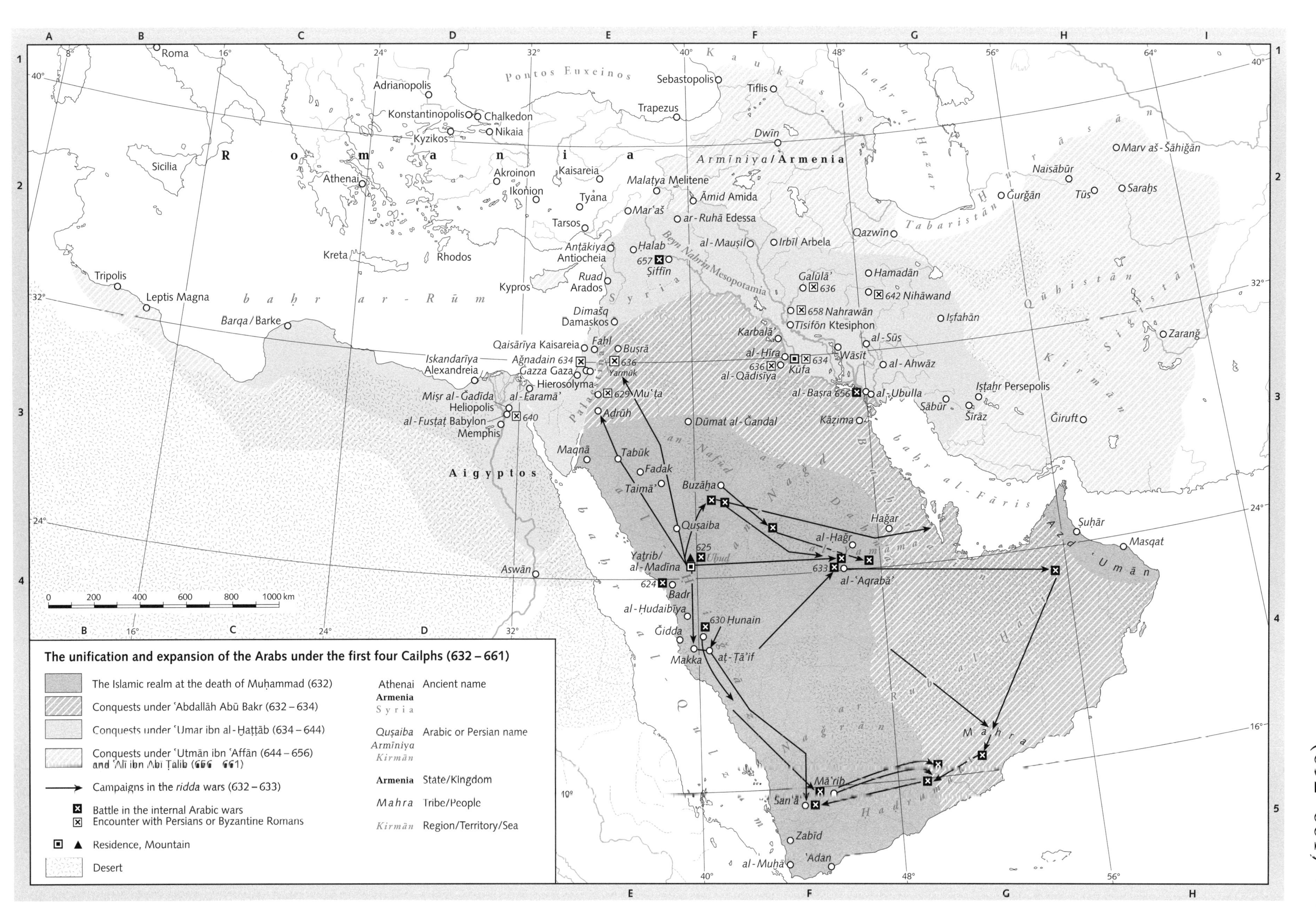
The unification and expansion of the Arabs under the first four Cailphs (632 – 661)
The Islamic realm at the death of Muḥammad (632)
Conquests under ʿAbdallāh Abū Bakr (632 – 634)
Conquests under ʿUmar ibn al-Ḫaṭṭāb (634 – 644)
Conquests under ʿUṯmān ibn ʿAffān (644 – 656) and ʿAlī ibn Abī Ṭālib (656 – 661)
Campaigns in the ridda wars (632 – 633)
Battle in the internal Arabic wars
Encounter with Persians or Byzantine Romans
Residence, Mountain
Desert
Athenai Ancient name
Armenia
Syria
Quṣaiba Arabic or Persian name
Armīniya
Kirmān
Armenia State/Kingdom
Mahra Tribe/People
Kirmān Region/Territory/Sea
0 200 400 600 800 1000 km
Roma
Sicilia
Tripolis
Leptis Magna
Barqa / Barke
Athenai
Kreta
Rhodos
Adrianopolis
Konstantinopolis
Chalkedon
Nikaia
Kyzikos
Akroinon
Ikonion
Kaisareia
Tyana
Tarsos
Kypros
Pontos Euxeinos
Sebastopolis
Trapezus
Tiflis
Dwīn
Armīniya / Armenia
Malaṭya Melitene
Āmid Amida
Marʿaš
ar-Ruhā Edessa
Anṭākiya Antiocheia
Ḥalab
657 Ṣiffīn
Ruad Arados
Dimašq Damaskos
Iskandarīya Alexandreia
Qaisārīya Kaisareia
Faḥl
Buṣrā
Aǧnadain 634
636 Yarmūk
Ġazza Gaza
Hierosolyma
629 Muʾta
al-Faramāʾ
Miṣr al-Ǧadīda
Heliopolis
640
al-Fusṭāṭ Babylon
Memphis
Adrūḥ
Aigyptos
Aswān
Maqnā
Tabūk
Fadak
Taimāʾ
Buzāḫa
Quṣaiba
Yaṯrib / al-Madīna
625 Uḥud
624 Badr
al-Ḥudaibīya
Ǧidda
630 Ḥunain
Makka
aṭ-Ṭāʾif
al-Mauṣil
Irbīl Arbela
Beyn Nahrīn Mesopotamia
Galūlāʾ 636
658 Nahrawān
Ṭīsifūn Ktesiphon
Karbalāʾ
al-Ḥīra
636 al-Qādisīya
634 Kūfa
Wāsiṭ
al-Baṣra 656
al-Ubulla
Kāẓima
Dūmat al-Ǧandal
al-Ahwāz
al-Sūs
Hamadān
642 Nihāwand
Iṣfahān
Qazwīn
Iṣṭaḫr Persepolis
Sābūr
Šīrāz
Ǧīruft
Ṭabaristān
Ǧurǧān
Naisābūr
Ṭūs
Saraḫs
Marv aš-Šāhiǧān
Zaranǧ
Qūhistān
Kirmān
baḥr ar-Rūm
baḥr al-Ḫazar
baḥr al-Fāris
baḥr al-Qulzum
Hağar
al-Ḥaǧr
633
al-ʿAqrabāʾ
Ṣuḥār
Masqat
ʿUmān
Mahra
Māʾrib
Ṣanʿāʾ
Zabīd
ʿAdan
al-Muḫāʾ
Ḥaḍramūt

The Byzantine Empire under Basil II (976–1025)

Under Basil II, a son of Romanus II and an emperor of the Macedonian Dynasty which had ruled since 867, the Byzantine Empire attained a degree of domestic political consolidation that enabled an exercise of power abroad the like of which had not been seen since the reign of Justinian I. Like his brother Constantine VIII, Basil II lived in the shadow of extra-dynastic emperors, Nicephorus II Phocas (963–969) and John I Tzimisces (969–976), following the early death of his father Romanus II in 963. After John Tzimiskes' death, Basil's eunuch great-uncle, the *parakoimomenos* (head of the imperial bodyguard), took control of government – the chief task being to ward off Bardas Skleros, a pretender to the throne from the rural aristocracy of Asia Minor. In the ensuing civil war (976–979), Skleros was defeated on 24 May 979 at Amorium, withdrawing into exile at Cairo with the Fatimid caliph al-Aziz (975–995). Yet even now, the great-uncle did not grant his great-nephew the helm of state: the latter therefore had him eliminated without further ado in 985. Bardas Skleros now returned from exile and renewed his claim to the throne – but another member of the Bardas family, Bardas Phokas, was by now also vying for power. In the civil war that duly erupted, Basil was finally able to prevail, but only with the help of mercenaries sent to him by prince Vladimir of the Kievan Rus'. The decisive battles were at Chrysopolis in 988 and at Abydus on 13 April 989.

When the civil war and the conflicts on the Syrian and Balkan fronts with the Fatimids and Bulgars respectively permitted him time, the emperor devoted himself to the social and economic conditions within the empire. His administrative and legal reforms worked to reinforce central power against the feudalizing tendencies of the major Anatolian landowners, whose ranks had spawned his opponents in the civil wars. He made use of agrarian reforms, which mostly benefited the estates of small farmers and soldiers. The consistent implementation of the theme system founded under Heraclius (610–641), a strongly hierarchical defence system encompassing the whole empire, also served this purpose. The empire's defensive powers were also enhanced by an army reform which included the creation of a heavily-armed cavalry.

Since the victory of iconodulism in 843, the emperor had increased his influence with the Church, and Basil took great pains to secure and expand upon the ground thereby won. He achieved particular success in his reform of ecclesiastical organization in the Balkans after his victory over the Bulgar khan Samuel. But the hand of the emperor was also at work in the missionary activity that began to spread on a large scale outside the empire, in particular among the Slavic peoples. The Byzantine Slavic mission had grown out of the desperation of civil war. In return for the military support provided by prince Vladimir, the emperor gave him the hand of his sister Anna. At the same time, however, he made Vladimir promise that he would have himself and his people baptized.

The emperor was personally averse to cultural matters. Nonetheless, his reign saw a cultural heyday which is generally termed the 'Macedonian Renaissance' – the reception of Antiquity was notably resumed, while creative innovation was clearly overshadowed.

Even when Basil had further advanced the empire's frontiers to the east in the Caucasus, to the detriment of the Armenian Bagratids (*thema* of *Basprakania*) and the princely house of Ani, the two major frontier regions to which the emperor had to devote most attention remained Syria, against the Fatimids of Cairo, and the west, against the Bulgars under Samuel.

In Syria, the Arabs attacked the Byzantine *dux* Michael Burtzes at the Orontes near Antioch on 15 September 994, under Fatimid command. They inflicted a harsh defeat. Basil II responded with a campaign that ended in 995 with a victory at Beroea (Aleppo) and the occupation of Rhaphanea and Emesa. In 998, the Arabs marched on Antioch again, annihilating the forces of Damianos Dalassenos, the *dux* of Antioch, at Apamea in July of that year. Basil found himself compelled to resume his Syrian campaign, and conquered Emesa, Byblus and Berytus. The result of this campaign was an offer of peace from the caliph al-Hakim (995–1021).

Meanwhile, the Bulgars in the Balkans were presenting more serious problems. After the death of emperor John I in 976, they had risen up in Macedonia, led by the sons of an imperial *comes*, Nikola: the ensuing revolt, ultimately led by the youngest of these sons, Samuel, would engulf the entire Balkan Peninsula. Samuel resided at first in Prespa, later in Ochrida. The centre of his realm between the Istrus/Danube and the Haemus/Balkans was Macedonia. In addition, he gained parts of Thessaly (Larisa 986). In 986, therefore, after his great-uncle's death, Basil undertook a Balkan campaign, but it came to abject failure at the pass of Succorum Angustiae on 16/17 August of the same year. It was this reverse that emboldened Bardas Skleros to return to the empire and claim the throne once more.

Protected by the fog of civil war which kept the Byzantine emperor in Asia Minor, Samuel determinedly pursued his plans to form a greater Bulgarian kingdom on the model of the Bulgar tsars Symeon (893–927) and Peter (927–969). He succeeded in advancing by way of Epirus as far as the Gulf of Corinth and on to the Peloponnese. From 991, the emperor, assisted by his *duces*, fought to wear Samuel down, and indeed the *domesticus* Nikephoros Uranos succeeded in defeating the Bulgars under Samuel's command in the Spercheius Valley. Samuel, though wounded, escaped. Undaunted by this defeat, he conquered Dyrrhachium and won Rascia and Diocleia in 998. From 1001, however, the emperor's victories became more frequent. He conquered Pliska, Megale Presthlaba (Preslav) and Mikra Presthlaba/Pereyaslavets (Little Preslav) and occupied Beroea and Servia, then took Voden. After an eight-month siege, Bidine (Vidin) on the Danube fell. Samuel, though, was not idle, conquering Adrianople. After an imperial victory on the Axius (Vardar) in 1004, Scopia (Skopje) capitulated, and the following year Dyrrhachium, too, returned to the Byzantine fold. The sources give no detailed information about subsequent years, but there were battles in each of them. Finally, the decisive imperial victory came, on 29 July 1014 at Clidium at a pass in the Belasitsa Mountains, on the upper reaches of the Strymon (Struma). Here too, Samuel escaped. No account of this war fails to describe the revenge Basil visited upon the Bulgars: 14,000 Bulgars are said to have been taken captive. The emperor had every one of them blinded. Every hundredth captive was left with the sight of one eye, and led the group. Thus the prisoners were released to return to Ochrida. Upon seeing them, Samuel is said to have collapsed, dying a few days later (6 October 1014). The emperor thus acquired the epithet Bulgaroktonos, the Bulgaricide. After the death of the Bulgar khan, his son Gabriel Radomir (1014/15) continued the fight against the emperor, as after his death in turn did Ivan Vladislav, a putative nephew of Samuel's (1015–1018). By 1019, however, the entire Balkan Peninsula was once more undisputedly part of the Byzantine Empire, from the Danube to the Peloponnese.

Sources

The Byzantine writer John Skylitzes (11th/12th cents.), author of a chronicle of the years 811-1057, consciously distanced his work from traditional chronicles by presenting an account in an artistic style, overrunning the strict year-by-year structure. Georgios Kedrenos Later copied this chronicle almost word for word. In John Zonaras, by contrast, Scylitzes and his work found a worthy successor, continuing the world chronicle for the years up to 1118.

The map

The map shows the great extent acquired by the Byzantine Empire under Basil II. Under the Comnenid emperors (1057–1185), as is shown here by a line boundary, large parts of Asia Minor in the east were lost to the Seljuks of Rum with their residence at Iconium, while new territory was won for a time in the west, in Sicily.

→Map p. 239

Literature

P. Carelos, Bemerkungen zur Herrschaft Basileios' II. Bulgaroktonos, in: Byzantino-Slavica 53, 1992, 1–16; W. Felix, Byzanz und die islamische Welt im frühen 11. Jahrhundert, 1981; C. Holmes, 'How the East was won' in the Reign of Basil II, in: A. Eastmond (ed.), Eastern Approaches to Byzantium, 2001, 41–56; P.M. Strässle, Krieg und Kriegführung in Byzanz. Die Kriege Kaiser Basileios' II. gegen die Bulgaren (976–1019), 2006; Id., Raum und Kriegführung in Byzanz – eine militärgeographische Beurteilung der Taktik und Strategie der Operation am Fluß Spercheios (996), in: Byzantinische Forschungen 26, 2000, 231–254; P. Stephenson, Byzantium' s Balkan Frontier 900–1204. A Political Overview, in: Acta Byzantina Fennica 10, 1999/2000, 153–167; C. Toumanoff, Caucasia and Byzantium, in: Traditio 27, 1971, 111–158.

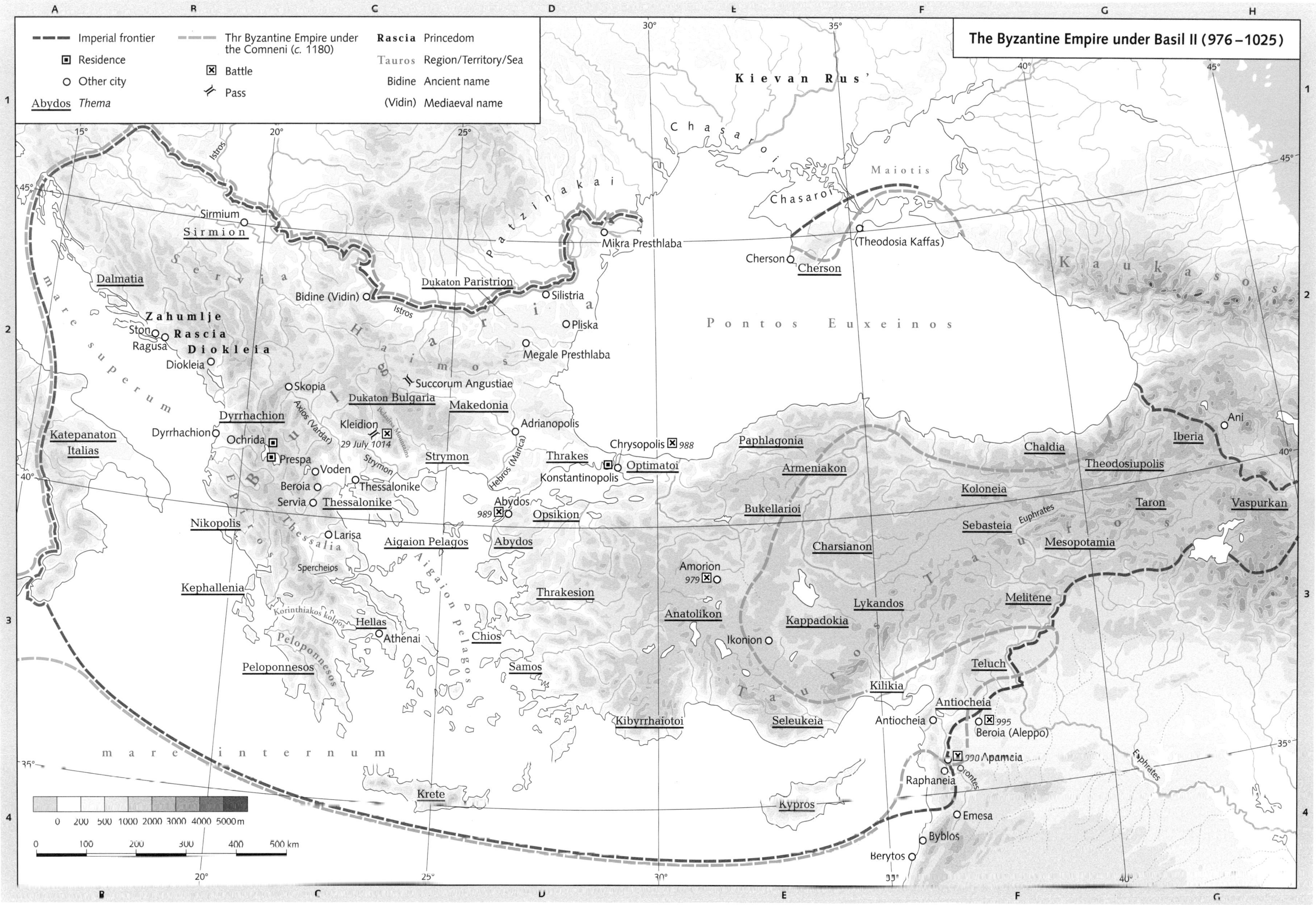
The Byzantine Empire under Basil II (976–1025)
Imperial frontier
Residence
Other city
Abydos Thema
Thr Byzantine Empire under the Comneni (c. 1180)
Battle
Pass
Rascia Princedom
Tauros Region/Territory/Sea
Bidine Ancient name
(Vidin) Mediaeval name
0 200 500 1000 2000 3000 4000 5000 m
0 100 200 300 400 500 km
Kievan Rus'
Chasaroi
Patzinakai
Maiotis
Chasaroi
Chersonȯ
Cherson
(Theodosia Kaffas)
Pontos Euxeinos
Kaukasos
Mikra Presthlaba
Silistria
Pliska
Megale Presthlaba
Dukaton Paristrion
Istros
Bidine (Vidin)
Sirmium
Sirmion
Servia
Dalmatia
Zahumlje
Ston
Ragusa
Rascia
Diokleia
Diokleia
mare superum
Haimos
Bulgaria
Skopia
Succorum Angustiae
Dukaton Bulgaria
Makedonia
Dyrrhachion
Dyrrhachion
Ochrida
Kleidion
29 July 1014
Belasitsa Mountains
Axios (Vardar)
Prespa
Voden
Strymon
Strymon
Beroia
Thessalonike
Servia
Thessalonike
Katepanaton Italias
Adrianopolis
Hebros (Marica)
Thrakes
Konstantinopolis
Chrysopolis 988
Optimatoi
Abydos
989
Opsikion
Abydos
Epiros
Nikopolis
Thessalia
Larisa
Aigaion Pelagos
Aigaion pelagos
Spercheios
Kephallenia
Korinthiakos kolpos
Hellas
Athenai
Peloponnesos
Peloponnesos
Chios
Samos
Thrakesion
mare internum
Krete
Paphlagonia
Armeniakon
Bukellarioi
Charsianon
Amorion
979
Anatolikon
Ikonion
Kappadokia
Lykandos
Tauros
Kibyrrhaiotoi
Seleukeia
Kilikia
Kypros
Chaldia
Theodosiupolis
Iberia
Ani
Koloneia
Taron
Vaspurkan
Sebasteia
Euphrates
Mesopotamia
Melitene
Teluch
Antiocheia
Antiocheia
995
Beroia (Aleppo)
990 Apameia
Orontes
Raphaneia
Emesa
Byblos
Berytos
Euphrates

The economy of the Middle Byzantine Empire

This map presents political and economic conditions at the beginning of the 11th cent., at the zenith of the so-called Macedonian Renaissance. It depicts neither the crises of the 7th-9th cents nor the slow disintegration of the Empire from the mid 11th cent.

From the 7th to the 9th cents., the Byzantine Empire found itself in a profound political, military, demographic and economic crisis. Wide expanses of territory from India to Spain had passed into the hands of the Arabs, who had advanced even as far as Sicily and southern Italy. The Bulgarian khanate and the Kievan Rus' also posed constant threats for long periods. The Empire at times lost control over the Balkans and southern Russia. Dire epidemics of plague caused a dramatic loss of population: cities shrank and commercial and agricultural production endured deep slumps.

However, the tide began to turn from the mid-9th cent. With the Arabs, Bulgars and Kievan Rus' hampering trade with the Orient, the Balkans and across the Black Sea respectively, the Byzantine Empire was thrown back on its own resources, and compelled to try to regain influence in the Balkans and southern Russia. In the end, it would largely succeed in the latter enterprise in the late 9th and 10th cents. Several indicators show the Byzantine economy taking wing at the end of the 9th cent. Firstly, the population clearly increased. It is also evident that the cities began to grow markedly once more in this phase. The consequent intensification of demand breathed life into crafts, trade and agriculture. The latter also profited from the end of labour shortage.

A free peasantry had emerged as the dominant rural group during the crisis period. These peasant village communities formed the backbone of the Byzantine defence and fiscal systems, which the Macedonian emperors initially also recognized, protecting the peasantry from the dangers posed by the growth of large-scale land ownership. When this imperial protection of the peasantry was abandoned in the mid 10th cent., large-scale land ownership began to spread again. The emerging large estates, as well as the large, long-established imperial crown estates, were now managed in an increasingly rational and profit-minded way. Overall, the expanse of land in agricultural use increased. New crops were introduced, e.g. oats and rye. This, together with the arrival of new technologies, such as the water mill, led to significant increases in agricultural production.

Production of other goods was mostly concentrated in the cities. It, too, enjoyed a clear recovery from the late 9th cent., in spite of state regulation. Manufacturing production increased dramatically, in particular in the cities of western Asia Minor and Greece, in Thessalonica and especially at Constantinople, the capital. Admittedly there were important areas of production that were subject to state monopoly, or at least to very strict state regulation – examples include the extraction and refinement of precious metals and the manufacture and processing of silk. The circulation of money began to flow once more during the 9th cent. Small coins were once more in circulation in the Byzantine provinces, something not seen in the preceding centuries. Even through the crisis period, the Byzantine gold *solidus* had preserved its value, and it became one of the most important currencies in the Mediterranean trade.

Finally, trade, too, enjoyed considerable growth. Interregional and international trade grew stronger after the crisis period. It received new impetus, not least through the opening up of trade into the western Mediterranean facilitated by relations with the trading republics of Venice, Genoa and Pisa (*c.* 1000: state protection for Italian traders; 1081 trading privileges motivate the monopoly accorded to Venice in trade between Byzantium and the western Mediterranean). Admittedly, the state restricted trade as well. For instance, the export of grain, iron, weapons, wine, olive oil, salt and gold was either forbidden or subject to state privilege. The silk trade was almost entirely in state hands, as was the greater part of the grain trade. Long-distance trade went mostly by sea. Road transport (the most important road connections are shown schematically on the map) tended to serve the regional exchange of goods.

Hence, the spatial economic order of the Middle Byzantine Empire had the following structure: in the countryside, the peasant communities produced small surpluses and the large estates bigger surpluses, to support the cities. The countryside was characterized by crafts, primarily for self-supply, and local or, at best, regional trade. Non-agricultural production and trade were concentrated in the cities. The most important economic centre was the capital. The heart of non-agricultural production was there, and it was the destination of the flows of surplus from across the empire. Constantinople was also the venue for most long-distance trade. By comparison, the other cities of the empire had a relatively auxiliary function. They served regional exchange and collected surpluses for onward transfer to Constantinople. For example, this is clearly evident in the case of Thessalonica, the second city of the empire after Constantinople. The resources of the Balkan region and Bulgaria were collected there, to be passed on to Constantinople by road or ship. By contrast, Thessalonica's links to Mediterranean trade were marginal.

Sources

Some descriptive sources supplement the knowledge of the economy of the Byzantine Empire obtained from the results of archaeological research. These narrative, hagiographic and juristic texts are listed fully in the bibliography.

Literature

A. Guillou, Byzantinisches Reich C. Sozial- und Wirtschaftsgeschichte, in: LMA 2, 1983, 1268–1275; A. Kazhdan, Economy, in: ODB 1, 1991, 674f.; A.E. Laiou (ed.), The Economic History of Byzantium from the Seventh through the Fifteenth Century, 2002 (in which esp. A. Avramea, Land and Sea Communications, Fourth-Fifteenth Centuries, 57–90; J. Lefort, The Rural Economy, Seventh-Twelfth Centuries, 231–310; G. Dragon, The Urban Economy, Seventh-Twelfth Centuries, 393–461; A.E. Laiou, Exchange and Trade, Seventh-Twelfth Centuries, 697–770).

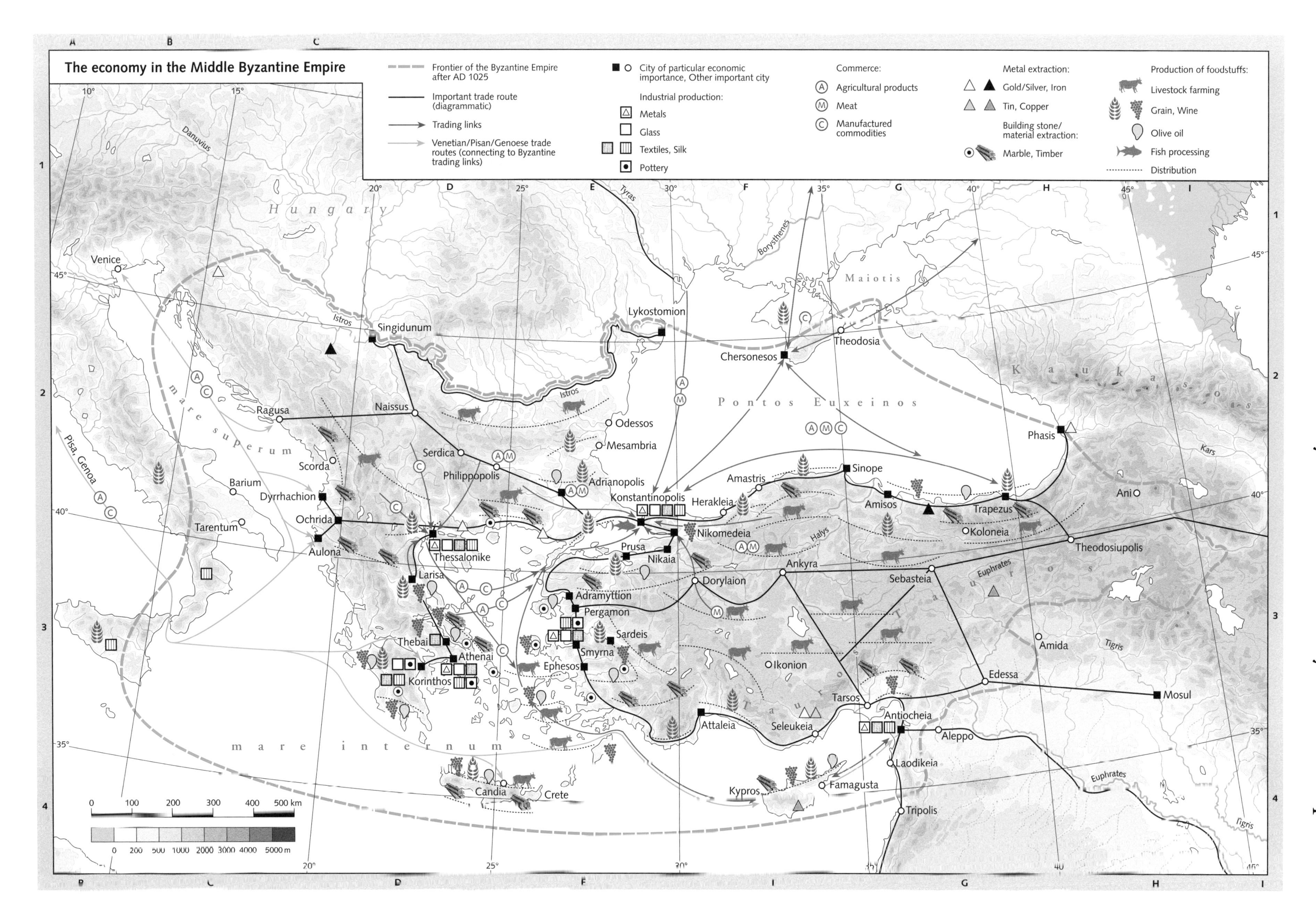
The economy in the Middle Byzantine Empire
Frontier of the Byzantine Empire after AD 1025
Important trade route (diagrammatic)
Trading links
Venetian/Pisan/Genoese trade routes (connecting to Byzantine trading links)
City of particular economic importance, Other important city
Industrial production:
Metals
Glass
Textiles, Silk
Pottery
Commerce:
Agricultural products
Meat
Manufactured commodities
Metal extraction:
Gold/Silver, Iron
Tin, Copper
Building stone/material extraction:
Marble, Timber
Production of foodstuffs:
Livestock farming
Grain, Wine
Olive oil
Fish processing
Distribution
Hungary
Danuvius
Istros
Tyras
Borysthenes
Maiotis
Pontos Euxeinos
Kaukasos
Kars
Halys
Euphrates
Tauros
Tigris
mare superum
mare internum
Venice
Pisa, Genoa
Singidunum
Lykostomion
Theodosia
Chersonesos
Ragusa
Naissus
Odessos
Mesambria
Phasis
Serdica
Philippopolis
Scorda
Barium
Dyrrhachion
Adrianopolis
Konstantinopolis
Amastris
Sinope
Herakleia
Amisos
Trapezus
Ani
Koloneia
Nikomedeia
Tarentum
Ochrida
Aulona
Thessalonike
Prusa
Nikaia
Theodosiupolis
Larisa
Ankyra
Dorylaion
Sebasteia
Adramyttion
Pergamon
Sardeis
Thebai
Athenai
Smyrna
Amida
Korinthos
Ephesos
Ikonion
Edessa
Tarsos
Mosul
Attaleia
Seleukeia
Antiocheia
Aleppo
Laodikeia
Candia
Crete
Kypros
Famagusta
Tripolis
0 100 200 300 400 500 km
0 200 500 1000 2000 3000 4000 5000 m

The first three Crusades (1096–1192)

A 'Crusade' is defined in the narrower sense as a military campaign initiated by the medieval Church against 'infidels', i.e. 'heathens' and 'heretics', but historically it refers to the campaigns of the western Christians aimed at regaining the 'Holy Land' between the 11th and 15th cents. Of the numerous campaigns in this latter sense, the map shows those counted by canonical tradition as the first three. However, this numbering should not obscure the fact that there were various other armed pilgrimages (some of which did not even reach their geographical destination) before, between and after these three and the other four 'canonical' Crusades, until the last Crusade, which ended on 10 November 1444 with an appalling disaster for the Crusader army in the battle against the Turks under sultan Murad II at Varna.

The Byzantine emperor Alexios I Komnenos (1081–1118) faced wars on several fronts: against the Normans under Robert Guiscard in the west (conquest of Dyrrhachium in October 1081), the Patzinaks/Bisseni in the north (siege of Constantinople 1090/91) and the Turkic Seljuks in the east (Byzantine defeat at Mantzikert, 19 August 1071). In view of such crises, the Byzantine emperor had already turned to Western powers (including the Pope in Rome) with appeals for military support on several occasions. Possible negotiating leverage was to be found in the unity discussions to abolish the schism between the Eastern and Western Churches that had endured since 1054. However, it would become clear that, for all the Pope's willingness to help, he was not the correct addressee of the emperor's appeals for assistance. What the emperor needed was soldiers, but what he got was armed pilgrims, who would in the end undertake the protection of Christians in the East on their own account.

I. The First Crusade

Pope Urban II (1088-1099) called up the French knights for the First Great Crusade (1096–1099) on 27 November 1095, at the Synod of Clermont, then again at Tours, Nîmes and elsewhere. Apart from the call for help from the Byzantine embassy at the Synod of Piacenza in March 1095, no direct cause for this call is known. Jerusalem had, after all, not been part of the Byzantine Empire since 636; thus the fact that the entire plan of the Crusade focused on the 'liberation of Jerusalem' seems a little strange.

This Crusade was led by Godfrey of Bouillon (*c.* 1060–1100), Duke of Lower Lorraine from 1087–1096, by Baldwin of Boulogne/Bouillon (1058–1118), Godfrey's brother, by their cousin Baldwin of Bourcq (d. 1131), by Robert II (*c.* 1054–1134), Duke of Normandy, by Raymond of Toulouse and Saint-Gilles (1041/42–1105), Margrave of Provence, and by Bohemond I of Taranto (1050/58–1111), Duke of Apulia, a son of Robert Guiscard. With no joint planning, they took their various forces (mostly French, Flemish and Norman knights from southern Italy and France) along various routes to Constantinople, arriving between late 1096 and the spring of 1097. More or less voluntarily, they all took oaths of fealty, giving the assurance that all lands they conquered which had formerly been Byzantine would be restored to the emperor. It took approximately eight weeks to take Nicaea (9 June 1097), which had since 1077 been the capital of the Seljuk sultanate of Rūm. Continuing their advance across Asia Minor, they encountered the Seljuks under the command of sultan Kilij Arslan on 1 July 1097 at Dorylaeum, finally forcing them to flee after fierce fighting.

The Crusaders continued without significant interruption to Heraclea, where Baldwin separated from the main Crusader force, moved through the Porta de Ferre by way of Tarsus and eastwards to Edessa, where in the spring of 1098 – against his oath of fealty – he founded the first Crusader state: the County of Edessa.

Meanwhile, the main army under Godfrey had bypassed the Porta de Ferre, going southwards via Caesarea to Antioch, where Bohemond, who had accompanied Baldwin for some distance, rejoined them. A seven-month siege ensued, and the arrival of a Seljuk relief force made for a dramatic situation, but the Crusaders won a complete victory on 28 June 1098. Bohemond, the first to enter the city, proclaimed himself regent of the Principality of Antioch, the second Crusader state. Protests from the emperor, whose embassy met the main army at Archis as it marched on Jerusalem, fell on deaf ears.

The Crusaders reached Jerusalem on 7 June 1099, and they conquered the city on 15 July. An army of the Fatimid sultan al-Musta'lī attempted to wrest back control of the city, but was defeated on 12 August by the Crusaders under Godfrey's command at Ascalon.

II. The Second Crusade

Pope Eugene III (1145-1153) called the Second Great Crusade (1147/48) on 24 December 1144, in response to the conquest of Edessa by Nūraddīn Zangī, the atabeg of Mosul and Aleppo. This Crusade was led by the French king Louis VII (1137–1180), the German king Conrad III (1138–1151) and bishop Otto of Freising (1138–1158), a half-brother of Conrad. The emperor at Constantinople, in support of whom the Crusaders nominally took the field, was at this time Manuel I Comnenus/Komnenos (1143–1180).

Conrad started out from Regensburg in May 1147, and Louis left from Metz a few weeks later. Their real destination was Edessa. Conrad, whose relations with the emperor could not have been more strained, did not remain long at Constantinople in September, but took his force to Asia Minor, crossing the *Brachium sancti Georgii* (Bosporus). At Nicaea, he broke away from Otto of Freising, who took with him most of the train via Lopadium and Sardis to Laodicea, where he was heavily defeated by Seljuk forces. The remains of his army were entirely wiped out at Istānūs – only the bishop managed to escape by sea to Jerusalem. The main force under Conrad did not fare conspicuously better – it was heavily defeated in October by a Seljuk army at Cedrea, south of Dorylaeum, and Conrad narrowly managed to struggle to safety behind the walls of Nicaea.

Louis had arrived at Constantinople on 4 October. He remained in the city for a time, then, under pressure from the emperor, who feared the French soldiers in his capital, crossed to Asia Minor. The French joined up with the remains of Conrad's German army at Nicaea, and took the coastal route by way of Smyrna and Ephesus. However, early in 1148, they were overwhelmed by the Seljuks at Laodicea. Anyone escaping the battle and reaching the coast at Attaleia was killed in battle with the Seljuks there. Louis nevertheless succeeded in escaping by sea to the port of St. Simeon near Antioch, while Conrad, who had fallen ill on the way at Ephesus, had for some time been enjoying the emperor's hospitality at Constantinople.

III. The Third Crusade

The Third Great Crusade (1187–1192) was called by Pope Gregory VIII in 1187. His *casus belli* was the defeat suffered by the army of the Kingdom of Jerusalem on 4 July 1187 at the hands of the Ayyubid sultan Saladin (1171/74–1193) at Ḥaṭṭīn, following which Saladin had taken Jerusalem on 2 October of the same year. This Crusade was commanded by the German emperor Frederick I, 'Barbarossa' (1155–1190), and his third son, Frederick, Duke of Swabia (1167–1191). They left Regensburg on 11 May 1189 with a large army. When they were denied access to the route via Constantinople, relations between Barbarossa and the Byzantine emperor Isaac II Angelus soured dramatically. After taking Philippopolis and Adrianople, Barbarossa briefly contemplated storming Constantinople, but was in the end able to negotiate the formalities of the territorial crossing. The Crusader army crossed to Asia Minor at Callipolis on 25 April 1190. As far as Laodicea, Barbarossa followed in the footsteps of Otto of Freising's 1147 campaign, but at Laodicea he turned east, and met the Seljuk army at Iconium on 18 May, annihilating it. Unchallenged by further enemy activity, Barbarossa reached the sea at Seleucia, only to drown there while swimming the Calycadnus (Saleph) on 10 June.

The great army now dispersed, wherever ships could be found at the harbours on the coast. Barbarossa's son Frederick took the remains of the force through Tarsus and Antioch bound for Acre, where they arrived on 7 October 1190. His few remaining troops joined the cordon besieging the city, which Sultan Saladin's forces had held against Guido of Lusignan (1186–1192), the king of Jerusalem, since August 1189. Meanwhile, the French king Philippe II (1180–1223) and the English king Richard I Lionheart (1189–1199) arrived at Tyre in June 1191, reaching Acre a few days later. The city fell on 12 July 1191.

Sources

First Crusade: Anna Comnena (1083–*c.* 1153) describes the reign of her father, Alexius I Comnenus, from 1069 in her *Alexiad*, combining elements of narrative history, biography and encomium.

Second Crusade: John Kinnamos (after 1143–after 1185) describes the years 1118-1176, following on from Anna Comnena's *Alexiad*. Niketas Choniates (*c.* 1150–1215), in his *Chronike Di(h)egesis,* deals with Byzantine history in a self-consciously objective style, covering the period 1118-1206. William of Tyre (*c.* 1130–1186) wrote a Latin history of the kings of Jerusalem, an excellent source of information for the first two Crusades. Odo of Deuil (*c.* 1110–1162) took part in the Second Crusade in the retinue of the French King Louis VII, and wrote a chronicle of it, also in Latin and entitled *De profectione Ludovici in orientem.*

Third Crusade: Ansbert (12th cent.) took part in the Third Crusade as a follower of Barbarossa; he compiled his own notes together with the diary records of another Crusader, the cathedral dean Tageno of Passau, in the *Historia de expeditione Friderici*, a vivid eye-witness account. Roger of Hoveden (*c.* 1150–1201) took part in the Third Crusade from 1190 to 1192 with Richard Lionheart: his English history (*Chronica*) contains valuable eye-witness information for these years.

→ Map p. 251

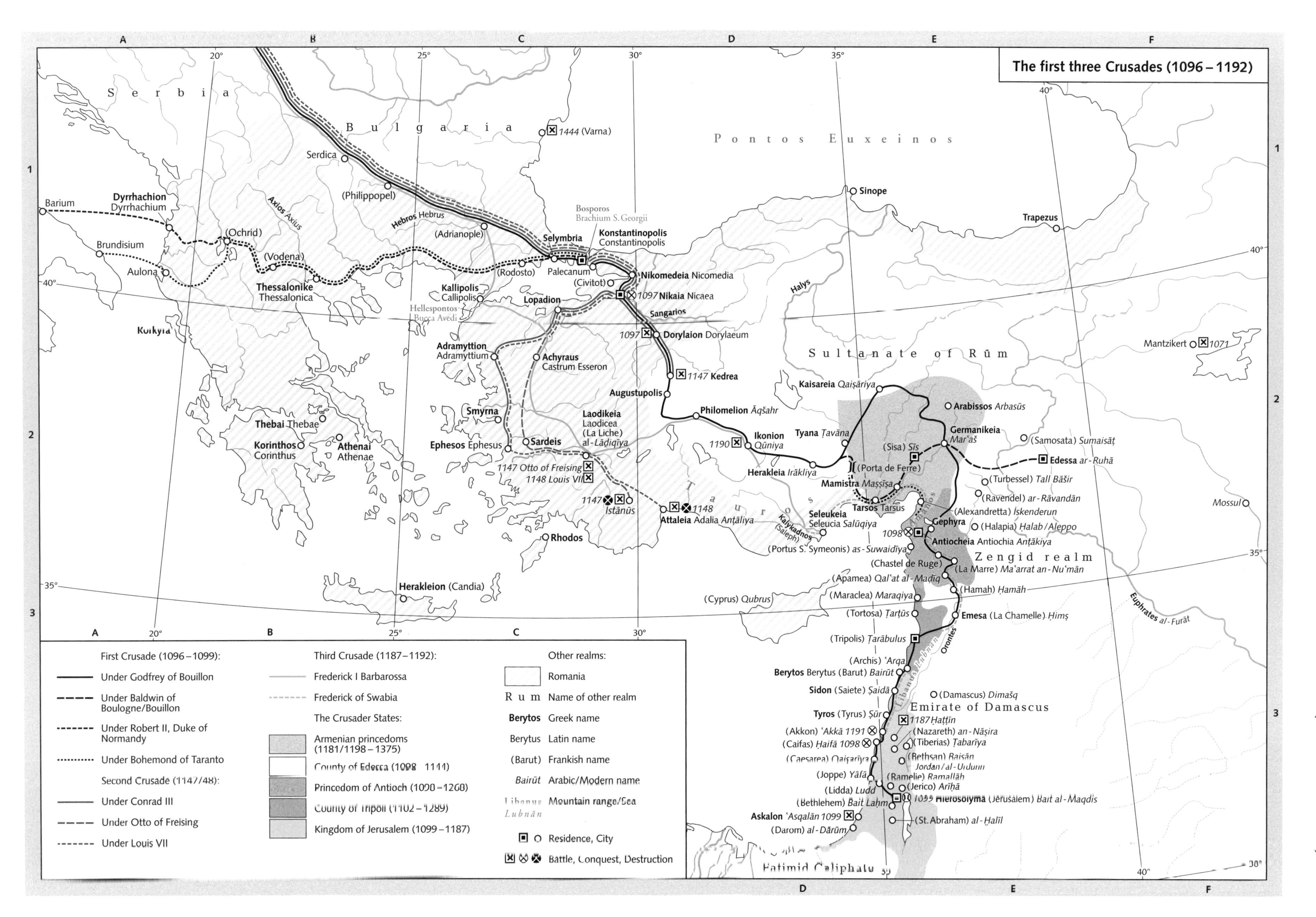
The first three Crusades (1096 – 1192)
First Crusade (1096 – 1099):
Under Godfrey of Bouillon
Under Baldwin of Boulogne/Bouillon
Under Robert II, Duke of Normandy
Under Bohemond of Taranto
Second Crusade (1147/48):
Under Conrad III
Under Otto of Freising
Under Louis VII
Third Crusade (1187 – 1192):
Frederick I Barbarossa
Frederick of Swabia
The Crusader States:
Armenian princedoms (1181/1198 – 1375)
Princedom of Antioch (1098 – 1268)
Kingdom of Jerusalem (1099 – 1187)
Other realms:
Romania
R u m Name of other realm
Berytos Greek name
Berytus Latin name
(Barut) Frankish name
Bairūt Arabic/Modern name
Mountain range/Sea
Residence, City
Battle, Conquest, Destruction
Pontos Euxeinos
Bulgaria
Serbia
Sultanate of Rūm
Zengid realm
Emirate of Damascus
Fatimid Caliphate
Barium
Dyrrhachion Dyrrhachium
Brundisium
Aulona
(Ochrid)
(Vodena)
Thessalonike Thessalonica
Serdica
(Philippopel)
Hebros Hebrus
Axios Axius
(Adrianople)
1444 (Varna)
Selymbria
Konstantinopolis Constantinopolis
Bosporos Brachium S. Georgii
Palecanum
(Rodosto)
(Civitot)
Nikomedeia Nicomedia
1097 Nikaia Nicaea
Sangarios
Kallipolis Callipolis
Lopadion
1097 Dorylaion Dorylaeum
Adramyttion Adramyttium
Achyraus Castrum Esseron
1147 Kedrea
Augustupolis
Philomelion Āqšahr
Smyrna
Ephesos Ephesus
Sardeis
Laodikeia Laodicea (La Liche) al-Lādiqīya
1147 Otto of Freising
1148 Louis VII
Istānūs
1148
Attaleia Adalia Anṭāliya
Rhodos
Thebai Thebae
Korinthos Corinthus
Athenai Athenae
Herakleion (Candia)
Korkyra
1190
Ikonion Qūniya
Tyana Ṭavāna
Herakleia Irākliya
Kaisareia Qaiṣāriya
Halys
Sinope
Trapezus
Mantzikert 1071
Arabissos Arbasūs
Germanikeia Mar'aš
(Samosata) Sumaisāṭ
Edessa ar-Ruhā
(Sisa) Sīs
(Porta de Ferre)
Mamistra Maṣṣīṣa
Tarsos Tarsus
Seleukeia Seleucia Salūqiya
(Turbessel) Tall Bāšir
(Ravendel) ar-Rāvandān
(Alexandretta) Iskenderun
Gephyra
(Halapia) Ḥalab/Aleppo
1098
Antiocheia Antiochia Anṭākiya
(Portus S. Symeonis) as-Suwaidīya
(Chastel de Ruge)
(La Marre) Ma'arrat an-Nu'mān
(Apamea) Qal'at al-Maḍīq
(Hamah) Ḥamāh
(Maraclea) Maraqiya
(Tortosa) Ṭarṭūs
Emesa (La Chamelle) Ḥimṣ
(Cyprus) Qubrus
Mossul
Euphrates al-Furāt
Orontes
(Tripolis) Ṭarābulus
(Archis) 'Arqa
Berytos Berytus (Barut) Bairūt
Sidon (Saiete) Ṣaidā
(Damascus) Dimašq
Tyros (Tyrus) Ṣūr
1187 Ḥaṭṭīn
(Akkon) 'Akkā 1191
(Nazareth) an-Nāṣira
(Caifas) Ḥaifā 1098
(Tiberias) Ṭabarīya
(Joppe) Yāfā
(Ramelie) Ramallāh
(Lidda) Ludd
(Jerico) Arīḥā
(Bethlehem) Bait Laḥm
Hierosolyma (Jerusalem) Bait al-Maqdis
Askalon 'Asqalān 1099
(St. Abraham) al-Ḫalīl
(Darom) al-Dārūm

The eastern Mediterranean at the time of the *Imperium Romaniae* (1204–1261)

Reigns of the Latin emperors, the Grand Comneni of Trebizond, the despots of Epirus and the emperors of Nicaea → p. 264

I. The Fourth Crusade and the Latin Empire

In August 1198, Pope Innocent III (1198–1216) issued the first call for a new Crusade – according to the official reckoning, it would be the Fourth. An army of French and Lombard knights eventually assembled at Venice, led by Count Theobald of Champagne, Louis of Blois, Baldwin of Flanders and Hainault, and Hugo IV of Saint-Pol. However, their passage to Egypt indebted the knights so deeply to the Venetians that the real leader of the Crusade became the doge, Enrico Dandolo. In November 1202, he ordered the taking of the port city of Iader (Italian: Zara), which had been lost to Venetian rule since 1181 and was a protectorate of the Hungarian king Béla. Venetian interests, linked to the widespread western antipathy to Byzantium, determined the further course of the Crusade, which was now diverted to Constantinople itself. The city was conquered and sacked on 13 April 1204. According to a plan already agreed between the knights and Enrico Dandolo on 15 March, the territory of the Byzantine state was essentially organized as a community of fiefdoms and divided up as follows: two eighths of the imperial territory went to the emperor of the new state, three eighths to various knights as imperial fiefdoms, and three eighths to Venice. Venice reserved for itself the Hagia Sophia, and hence the seat of the new Latin patriarch, the Venetian Tomaso Morosini.

1. Romania – the Latin Empire
The emperor, Baldwin I of Flanders, chosen by the Crusaders and the Venetians, was given Thrace, north-western Asia Minor and the islands of Lesbos, Chios and Samos. But taking the capital by no means won the Crusaders the entire Byzantine Empire. Another difficulty was that in falling heir to Constantinople, the Latins had also inherited the enemies of the Byzantine emperor, specifically the Bulgars and the Rum Seljuks. There were also the three Byzantine successor states at Arta (Principality of Epirus), Nicaea and Trebizond, which in view of the evident weakness of Latin rule found the regaining of Constantinople a distinctly feasible proposition.

2. Imperial fiefdoms
The kingdom of Thessalonica was short-lived. After the death of its ruler, Boniface of Montferrat, in 1207, it was reclaimed by its liege, the emperor Henry. – The Barony of Athens (attested from 1280 as a duchy) was initially in the hands of the Burgundian de la Roche family (1204–1311), before being taken over by mercenaries of the Catalan Company fighting on their own account (1311–1386). The Florentine Acciaiuoli family finally lost it to the Ottomans in 1456. – The Peloponnese had initially passed to the Venetians as an imperial fiefdom. They, however, kept only Methone, Corone and Nauplia, and the islands were occupied by Frankish knights. The Principality of Achaea was divided into twelve baronies. The Franks and Greeks co-existed better here than in other parts of the country.

3. Venice
The trading republic did not occupy all the territories nominally allotted to it as imperial fiefdoms (only Crete was Venetian state property) – it had no particular interest in doing so, and to do so would moreover have overstretched its administrative capacities. In many cases, therefore, the doge passed territories (Lemnos, Skopelos, Skyros, Skiathos) to Venetian nobles, who exercised power in their own right over various islands (Duchy of Naxos with twelve islands in total). The Venetians claimed and developed very specific locations of importance to their trading policies: for instance, they secured control of the sea passage from the Aegean into the Black Sea – the Hellespont with Lampsacus, Callipolis and Abydus, the Propontis with Rhaedestus and Heraclea, the Bosporus with their now further-expanded quarter on the Golden Horn. They secured their influence in the Black Sea in a similar way, with the foundation of the colony of Sugdia on the Crimea. – Also striking is the chain of Venetian bases on the route from Venice into the eastern Mediterranean and Aegean: Iader (Italian: Zara), Dyrrhachium (Italian: Durazzo), Corcyra (Italian: Corfù), Cephallenia (Italian: Cefalonia), Methone (Italian: Modon), Corone (Italian: Coron). – However, actual possession of the islands of Euboea (Italian: Negroponte; capital Chalcis, Italian: Negropolis) and Crete (capital Heracleum, Italian: Candia) was particularly important to the doge. Here, the Venetians set up full-scale military colonies in support of their projects of war.

II. The Byzantine exile empires

Meanwhile, the Fourth Crusade also enabled or brought about the emergence of states which nurtured and preserved the Byzantine heritage in opposition to Latin rule.

1. The Empire of Trebizond
The Empire of Trebizond was founded by Alexius and David Comnenus, grandsons of the Constantinopolitan emperor Andronicus I Comnenus (1183–1185). They had fled Constantinople at the time of its first conquest by the Crusaders on 17 July 1203, taking refuge at the court of the Georgian queen Tamar (d. 1213). With her military support, they had taken Trebizond/Trapezus without bloodshed. Theodoros Gabras, the Byzantine *dux* of the theme of Chaldia, had ruled here since 1075, almost independent of Constantinople, founding a dynasty which the Comneni extinguished in April 1204. David advanced westwards from 1205, crossing the Sangarius, but could not hold these territorial gains against Theodoros I Laskaris, the emperor at Nicaea (1205–1221). By 1208, Sinope was the westernmost outpost of the Empire of Trebizond. In 1214, this important port fell to the Seljuks of Rum, who, after taking Amisus, confined the Grand Comneni to the littoral strip from the mouth of the Iris to the Acampsis. The Comneni nonetheless succeeded in gaining another foothold on the southern shore of the Crimea, around the commercially important Chersonesus. Through the 13th cent., Trebizond even developed to become a commercial interface between Persia, Central Asia, China and the west, a process in which the Genoese (from 1280) and Venetian (from 1319) trading posts at Trebizond played leading roles. The Empire survived the Mongol invasions without harm, thanks to skilful diplomacy and matrimonial policies. However, it could not withstand the siege of the Osmanlı Turks under Mehmet II in 1461.

2. The Principality of Epirus
Michael Komnenos Dukas was a cousin of the Byzantine emperor Isaak II Angelos (1185–1195; 1203/04); after the catastrophe at Constantinople, he set up his own realm in the *thema* of Nicopolis, based at Arta, ancient Ambracia. Over the following ten years, he was able to expand his rule as far as Naupactus in the south and Dyrrhachium in the north. As he had with Dyrrhachium, he succeeded in winning Corcyra from the Venetians. His successor (from 1215) and brother Theodoros expanded the Epirote Empire against the Latins and Bulgars, taking the Latin emperor Peter of Courtenay prisoner in 1217 and entering Thessalonica in 1224, where he had himself crowned emperor by the archbishop of Ohrida. In 1230, though, he was heavily defeated by the Bulgars at Klokotnica: he was taken prisoner and blinded. His successor, Michael II, was defeated by the emperor of Nicaea, Michael VIII, at Pelagonia in 1259, and was forced to retreat to Epirus.

3. The Empire of Nicaea
Theodoros Laskaris, who proclaimed himself emperor at Nicaea, was a son-in-law of the Byzantine emperor Alexios III Angelos (1195–1203). His successor, John III, expanded his realm against the Seljuks in the east and the Latins to the north. After Theodoros I of Epirus, his one serious competitor in the effort to win back Constantinople, had failed against the Bulgars at Klokotnica in 1230, Theodoros Laskaris crossed to Europe and decisively defeated the former's successor, Michael II, at Pelagonia. In July 1261, an army division of the Nicene emperor Michael VIII succeeded in forcing its way into Constantinople and restoring the Byzantine Empire.

Sources

There are a number of eye-witness accounts of the conquest of Constantinople, the subsequent division of the Byzantine Empire and the history of the Latin Empire. A particularly important one is the chronicle of Niketas Choniates (*c.* 1155–1215/16). The history of the Empire of Nicaea is the main theme of the chronicle (*Annales*) of Georgios Akropolites (1217–1282), and Nikephoros Gregoras (1295–1359) reported on the same theme rather later in his *Historia Rhōmaikē*. The *Rhōmaikē historia* of George Pachymeres (1242–1310) is devoted to the first of the Palaeologi emperors. The eye-witness account of the Frenchman Geoffroy de Villehardouin (1160–1213), *La conquête de Constantinople*, is also highly informative. Important for events on the Peloponnese is the anonymous *Chronicle of Morea* (early 14th cent.). The document *Partitio Romaniae* provides a detailed account of the treaty of partition negotiated between the Crusader knights and Venice and published prior to the conquest of Constantinople in April/May or after it in September/October 1204.

The map

The map ought really to depict the world of the Crusaders from their point of departure in the Franco-Lombardic West by way of Venice, and Iader (Italian: Zara) to Egypt, the intended destination of the Fourth Crusade, and from there as far north as the Crimea, where the Venetians set up a trading-post at Sugdia (modern Sudak). However, for the sake of clarity, it restricts itself to the main crucible of events around Greece and the west of Asia Minor.

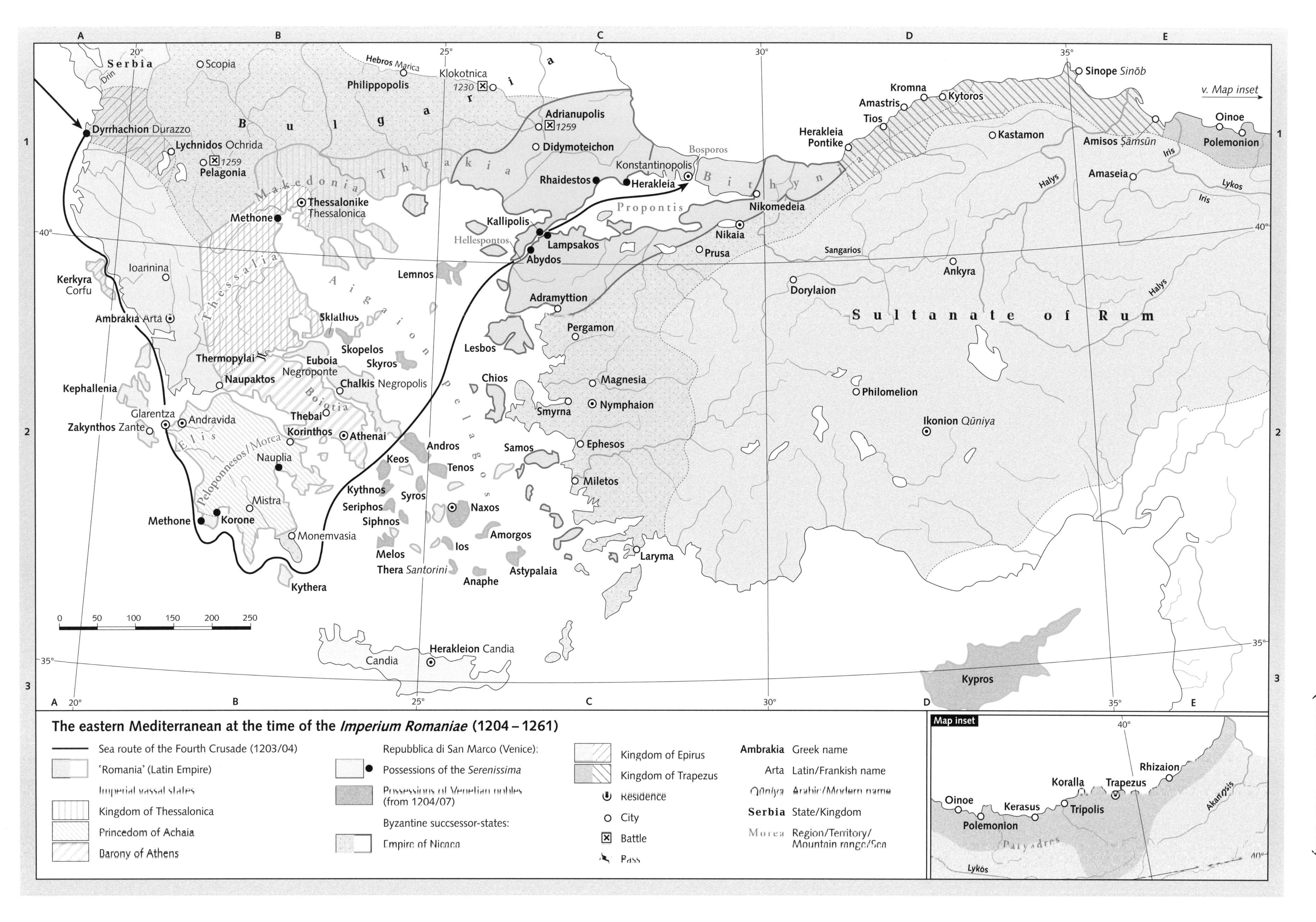
The eastern Mediterranean at the time of the *Imperium Romaniae* (1204 – 1261)
Sea route of the Fourth Crusade (1203/04)
'Romania' (Latin Empire)
Imperial vassal states
Kingdom of Thessalonica
Princedom of Achaia
Barony of Athens
Repubblica di San Marco (Venice):
Possessions of the *Serenissima*
Possessions of Venetian nobles (from 1204/07)
Byzantine successor-states:
Empire of Nicaea
Kingdom of Epirus
Kingdom of Trapezus
Residence
City
Battle
Pass
Ambrakia Greek name
Arta Latin/Frankish name
Qūniya Arabic/Modern name
Serbia State/Kingdom
Morea Region/Territory/Mountain range/Sea
Map inset
v. Map inset
Sultanate of Rum
Bulgaria
Serbia
Makedonia
Thessalia
Thrakia
Bithynia
Propontis
Hellespontos
Bosporos
Aigaion Pelagos
Peloponnesos/Morea
Elis
Boiotia
Kypros
Konstantinopolis
Herakleia
Rhaidestos
Adrianupolis
Didymoteichon
Philippopolis
Klokotnica 1230
Kallipolis
Lampsakos
Abydos
Adramyttion
Pergamon
Magnesia
Nymphaion
Smyrna
Ephesos
Miletos
Laryma
Nikomedeia
Nikaia
Prusa
Dorylaion
Philomelion
Ikonion *Qūniya*
Ankyra
Kastamon
Kytoros
Kromna
Amastris
Tios
Herakleia Pontike
Sinope *Sinōb*
Amisos *Sāmsūn*
Amaseia
Oinoe
Polemonion
Kerasus
Koralla
Tripolis
Trapezus
Rhizaion
Thessalonike Thessalonica
Methone
Scopia
Lychnidos Ochrida
Pelagonia 1259
Dyrrhachion Durazzo
Ioannina
Ambrakia Arta
Kerkyra Corfu
Kephallenia
Zakynthos Zante
Glarentza
Andravida
Naupaktos
Thermopylai
Thebai
Chalkis Negropolis
Euboia Negroponte
Athenai
Korinthos
Nauplia
Mistra
Monemvasia
Korone
Kythera
Skiathos
Skopelos
Skyros
Lemnos
Lesbos
Chios
Samos
Andros
Tenos
Keos
Syros
Kythnos
Seriphos
Siphnos
Melos
Naxos
Ios
Amorgos
Astypalaia
Anaphe
Thera *Santorini*
Herakleion Candia
Candia
Halys
Iris
Lykos
Sangarios
Hebros *Marica*
Drin
Paryadres
Akampsis

The Byzantine Empire under the Palaeologi (1261–1453)

Reigns of the Palaeologi emperors and the rulers of the House of Osman → p. 264

I. Constantinople under Michael VIII and Andronikos II

By winning back Constantinople on 26 July 1261, the Byzantine state of Nicaea (→ map p. 251) gained the coveted jewel for its crown. The expulsion of the Latin emperor Baldwin II, however, did not eliminate the danger from the west. Charles of Anjou, king of Naples and Sicily, made an alliance with the exiled Baldwin (27 May 1267), to restore him to Constantinople and divide the empire. However, the 'Sicilian Vespers' rebellion in Sicily (31 March 1282) robbed Charles of the throne he had only recently won, and forced him to abandon his plans. They were taken up by Charles of Valois, a brother of Philip the Fair, who in 1301 married the titular Latin empress Catherine of Courtenay, a granddaughter of Baldwin II, and took the Catalan Company (see below) into his service. But when Catherine died in the same year and his legal claim to the Latin throne lapsed, the Catalans abandoned him without ceremony and hired themselves out to Walter of Brienne, Duke of Athens. Thus the danger which had first threatened with the Crusade under Frederick I Barbarossa and which had materialized with the Fourth Crusade, had now entirely cleared away.

II. The Catalan Company (1303–1311)

The Catalan Company was a troop of mercenaries of Frederick III who had been without employment since the 1302 Peace of Caltabellotta between the Kingdom of Naples and Sicily. Under the command of Roger de Flor, it went into battle against the Turks in 1303, in the service of Andronikos II Palaiologos. In 1304, Roger de Flor crossed to Asia Minor with over 6,000 men. The company forces liberated Philadelphia from its Turkish besiegers, but troubled less and less to distinguish between Turkish and Byzantine booty, for which reason Roger was ordered back to the court. The mercenary army entered winter quarters on the Callipolis Peninsula (modern Gallipoli). When Roger demanded and received wide-ranging concessions from the emperor in return for his services, he became so dangerous a competitor in the eyes of the emperor's son and co-regent Michael that the latter had Roger killed (30 April 1305). There was now a dispute between the mercenaries and the court at Constantinople, leading to open warfare: battle was joined at Aprus (10 July 1305) and the Catalans were victorious. While the plains were defenceless and exposed to the mercenaries' pillage, the cities, such as Adrianople and Bizye, defended themselves successfully. Under Bernat de Rocafort, the mercenaries entered the service of the titular Latin emperor of Constantinople, Charles of Valois: they overwintered at Cassandria in 1307/08 and at the foot of Olympus in 1308/09. They then entered the service of Walter of Brienne, Duke of Athens, taking Demetrias, Zetunium and Halmyrus on his behalf. A dispute arose over outstanding pay, and there was a battle by the Cephissus in Boeotia in which the knights were annihilated (15 March 1311); Walter fell in the battle. The Catalans duly occupied the Duchy of Athens, and they and their descendants remained there until 1387/88.

III. Constantinople in civil war (1321–1357)

Civil wars shook the Byzantine state in the years 1321–1328 (Andronikos II against his grandson Andronikos III), 1341–1350 (John Cantacuzene against John V Palaeologus, 'Hesychasm Dispute', the zealots at Thessalonica) and 1352–1357 (between John V Palaeologus and Matthew Cantacuzene).

Asia Minor was already almost completely in the hands of Turkish emirates by 1300. Only a few cities were still Byzantine, including Heraclia on the Black Sea, Nicaea, Nicomedia, Prusa, Philadelphia, Magnesia on the Sipylus, Sardis, Phocaea and Smyrna. Participants in the civil wars repeatedly sought Turkish military support, from the Seljuk emirates and the Osmanlı (House of Osman, 'Ottomans') in Bithynia. But those very Osmanlı, under Orhan, caught the slipstream of these civil wars to claim base after base in western Asia Minor from the Seljuks and Byzantines alike. Prusa fell to them in 1326 and was made the Osmanlı residence. Nicaea followed in 1331, and the Osmanlı defeated a Byzantine army at Philocrene on 2 March 1331, taking the fortress. Nicomedia fell in 1337. They occupied the Bithynian Black Sea coast, and from it they terrorized the seas with pirate raids. Finally, they crossed the Hellespont to Europe, conquering Tzympe/Cinbi on the Callipolis in 1352, and the city of Callipolis itself in 1354.

In the Aegean, after Andronikos II had disbanded the navy as too costly (*c.* 1290), the Byzantines were entirely dependent on the aid of the Genoese, who pursued their own interests (occupation of Chios in 1346). Andronikos III therefore laid the keels of a new fleet in 1328, but it was destroyed off Constantinople in the spring of 1349 – by the Genoese.

After the Bulgars (who were allied with the Byzantines) suffered a devastating defeat in their conflict with the Serbs at Pautalia (Belebustion/Velbužd, modern Kyustendil) on 28 July 1330, and following the resolution of internal succession disputes joined forces with them under a peace treaty, the Serbs pushed south into Byzantine imperial territory, as far as Celetrum and Ochrida (1334). The occasional successes which brought the Byzantines to Thessaly (1333) and to Epirus on the Adriatic (1337) were short-lived. By 1348, the Serbs were already pushing back past Epirus to the Gulf of Corinth, led by Stefan Dušan.

IV. Constantinople and the Osmanlı (1357–1453)

By his occupation of the peninsula of Callipolis, Orhan had made it clear that he intended to establish a permanent kingdom not only in Asia Minor but also in Europe. John V therefore sought the support of the Pope and western courts, and promised to convert the court and populace of Constantinople to the Catholic faith. Attempts to overcome the 1054 schism between the two Churches had been made on several occasions – the Fourth Crusade, the Latin Empire and the Council of Lyons (1274) had all taken place under the banner of the 'union of the Churches'. The Byzantine emperors had repeatedly undertaken negotiations with the Pope, proffering promises of union (honestly meant or feigned). John V thus travelled to the Hungarian court in 1365 and to Rome in 1369/71, but in spite of many concessions his efforts were in vain.

Meanwhile, the Turks were consolidating their positions in Europe. Didymoteichon fell in 1361. Philippopolis followed in 1363, and so, in 1369, did Adrianople: Murad moved his residence there. The Byzantines were not alone in their efforts to arrest the Turkish invasion. The Serbs opposed them, too, but were utterly defeated in 1371 at Zeirinia (Tzernomianon/Černomen) on the Hebrus/Marica. In 1383, the Turks took Serrhae from the Serbs and in 1387 Thessalonica. The Serbian prince Lazar massed one final array of forces, which was destroyed in a bloody battle at the 'Kosovo Polje' ('Field of the Blackbirds') on 15 June 1389.

John V had already (in 1373) submitted to the obligation to provide Murad with all forms of military service. As a Turkish vassal, therefore, his son Manuel was compelled to support the Turks in 1390 in their conquest of Philadelphia, the last Byzantine city on the soil of Asia Minor.

Another Crusader army now assembled, led by the Hungarian king Sigismund, to fight the Turks. The Osmanlı annihilated it at Nicopolis on the lower Danube on 25 September 1396. For the moment, however, the Mongol invasion led by Timur Lenk ('Tamberlane') saved the capital.

Timur, the ruler of a Turko-Mongol empire in Central and Western Asia (1370–1405), undertook a westward campaign in 1400-1403 against the Mamluks in Egypt and their ally, the Osmanlı sultan Beyazid. After destroying the residence of the Mongol Jalayirids at Baghdad (July 1401), Timur entered Asia Minor in 1402, defeating Beyazid at Ancyra (28 July 1402). Beyazid himself was taken prisoner. Timur then advanced to the Aegean, but soon pulled back eastwards through Lesser Armenia. Brief though this Mongol invasion was, it nonetheless presented the Byzantine Empire with a quite unexpected opportunity to free itself from the threat to its existence posed by the Turks. A war among the sons of Beyazid, who had died in Mongol captivity near Ancyra, kindled hopes of a possibile peaceful co-existence between the Turks and Byzantines, especially under the rule of the last of Beyazid's sons, Mehmet I (1413–1421).

Regardless, Manuel II ordered the construction of a wall over the Isthmus of Corinth (the 'Hexamilion Wall'), to secure the Palaeologi's last Greek province on the Peloponnese, the Despotate of Morea. In 1430, Thessalonica, no longer in Byzantine hands after its sale to the Venetians, fell to the Turks.

In 1440, Pope Eugene called a Crusade, led by Władysław III of Poland and Hungary. This Crusader army reached the lower Danube before being annihilated by Murad at Varna (10 November 1444).

When Murad's son, Mehmet II, came to the throne with the avowed intent of conquering the Byzantine capital, all hope for Constantinople was lost. The siege began on 6 April, and ended on 29 May 1453 with the fall of the city.

Sources

Byzantine historians give accounts, some as eye-witnesses, of the history of the Palaeologi: George Pachymeres (1242–*c.* 1310), Nikephoros Gregoras (1290/94–1358/61) and emperor John VI Cantacuzene (*c.* 1295–1354) in his autobiographical history, also Laonikos Chalkokondyles (1423–1490), who made a thorough examination of Ottoman history, and a Greek in Genoese service, Dukas, who portrays the events of his time very clearly and reliably, in a language coloured by demotic idiom. George Phrantzes (1401–1478) gives insights into the official positions of the last three Palaeologi in particular, by virtue of the high office he held at the Constantinopolitan court and his friendship with the imperial house. Also worthy of mention as a historical source for the period of the Palaeologi is the substantial opus (letters, rhetorical writings) of the philosopher Demetrios Kydones (*c.* 1323–1398).

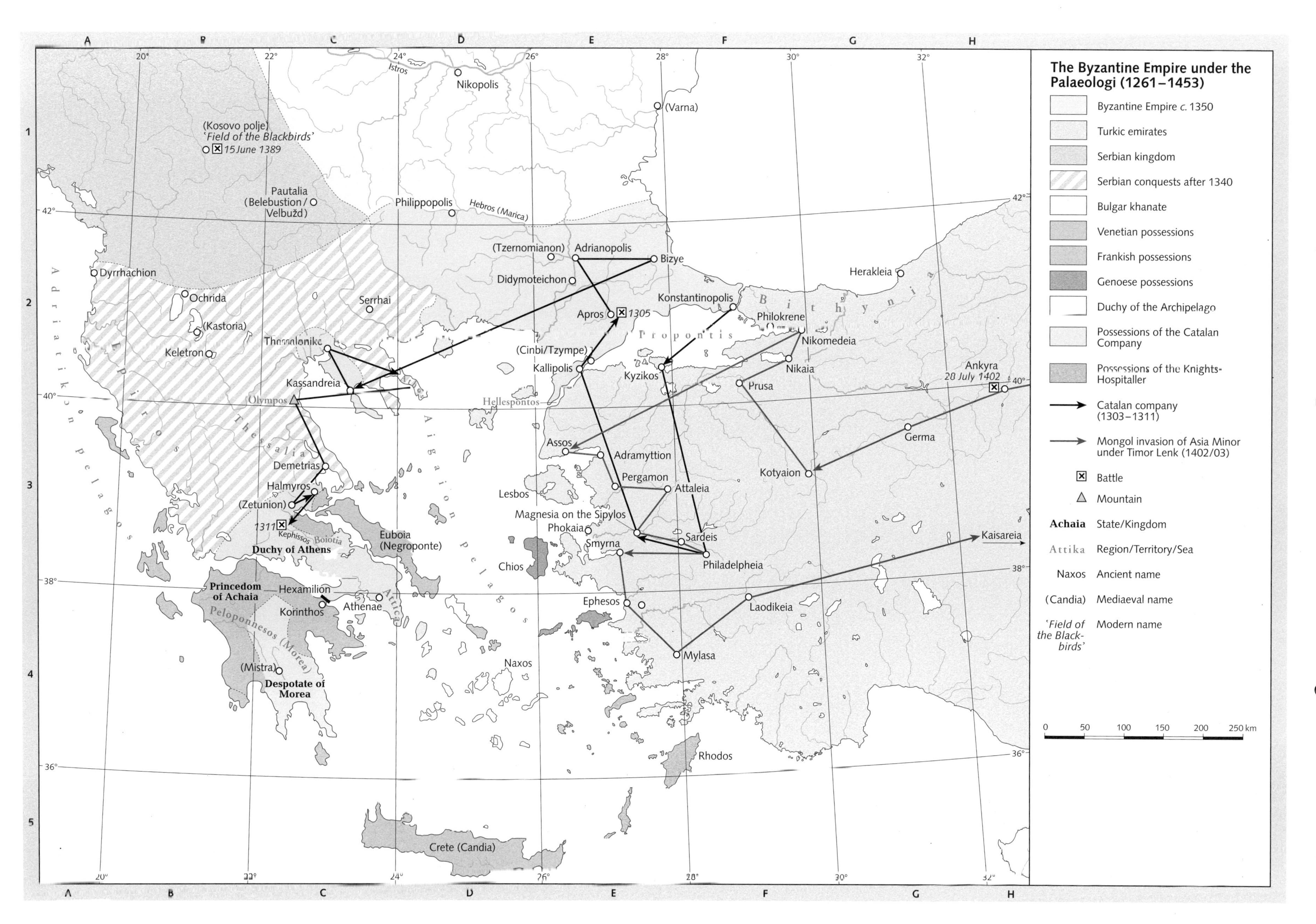
The Byzantine Empire under the Palaeologi (1261–1453)
Byzantine Empire c. 1350
Turkic emirates
Serbian kingdom
Serbian conquests after 1340
Bulgar khanate
Venetian possessions
Frankish possessions
Genoese possessions
Duchy of the Archipelago
Possessions of the Catalan Company
Possessions of the Knights-Hospitaller
Catalan company (1303–1311)
Mongol invasion of Asia Minor under Timor Lenk (1402/03)
Battle
Mountain
Achaia State/Kingdom
Attika Region/Territory/Sea
Naxos Ancient name
(Candia) Mediaeval name
'Field of the Black-birds' Modern name
0 50 100 150 200 250 km
Nikopolis
Istros
(Varna)
(Kosovo polje)
'Field of the Blackbirds'
15 June 1389
Pautalia (Belebustion / Velbužd)
Philippopolis
Hebros (Marica)
(Tzernomianon)
Adrianopolis
Bizye
Didymoteichon
Dyrrhachion
Ochrida
(Kastoria)
Keletron
Serrhai
Thessalonike
Kassandreia
Olympos
Apros
1305
Konstantinopolis
Philokrene
Bithynia
Herakleia
Propontis
Nikomedeia
Nikaia
Prusa
(Cinbi/Tzympe)
Kallipolis
Kyzikos
Hellespontos
Ankyra
20 July 1402
Germa
Kotyaion
Assos
Adramyttion
Pergamon
Attaleia
Lesbos
Magnesia on the Sipylos
Phokaia
Sardeis
Smyrna
Philadelpheia
Chios
Kaisareia
Ephesos
Laodikeia
Mylasa
Naxos
Rhodos
Crete (Candia)
Demetrias
Halmyros
(Zetunion)
1311
Kephissos
Boiotia
Duchy of Athens
Euboia (Negroponte)
Princedom of Achaia
Hexamilion
Korinthos
Athenae
Attica
Peloponnesos (Morea)
(Mistra)
Despotate of Morea
Thessalia
Epiros
Adriatikon pelagos
Aigaion pelagos

Addenda to commentaries

(sources, literature, tables)

Page 4

Sources

Hecataeus: Agathemerus, GGM 2, 1882, 471 § 1; Diog. Laert. 2,2; Eratosth. fr. 1B 5 Berger; Favorinus, FHG 581; Hecat. FGrH 1 F 18a; Ptol. 8,1,2 f.
Herodotus: Aristot. Mete. 2,2,354a; Hdt. 1,202; 2,31–34; 4,36; 4,42; 4,45; 4,48–50
Eratosthenes: Agathemerus, GGM 2, 1882, 472 § 5; Eratosth. fr. 3B 6 p. 224 ff.; 3B 11
Claudius Ptolemaeus: Ptol. Geographike Hyphegesis 1,6–14 *et passim*

Literature

Hecataeus: K. Brodersen, Terra Cognita (Spudasmata 59), 1995, 78 f.; K. von Fritz, Die griechische Geschichtsschreibung 1, 1967, 48–76; 2, 1967, 32–53; F. Jacoby, s.v. Hekataios [3], RE 7, 1912, 2667–2750, esp. 2700–2734
Herodotus: J. Lacarrière, Hérodote et la découverte de la terre, 1968; C. van Paassen, The Classical Tradition of Geography, 1957, 71–211
Eratosthenes: A. Thalamas, La géographie d'Eratosthène, 1921; W. Thonke, Die Karte des Eratosthenes und die Züge Alexanders, thesis, Strasbourg 1914
Claudius Ptolemaeus: Klaudios Ptolemaios. Handbuch der Geographie, 2 vols., ed. A. Stückelberger, G. Grasshoff, F. Mittenhuber, 2006; J.L. Berggren, A. Jones, Ptolemy's Geography. An annotated translation of the theoretical chapters, 2000; O.A. Dilke, The culmina- tion of Greek cartography in Ptolemy, in: J.B. Harley, D. Woodward (ed.), Cartography in Prehistoric, Ancient, and Medieval Europe and the Mediterranean (The History of Cartography 1), 1987, 177–200; E. Honigmann, s.v. Marinos [2], RE 14,2, 1930, 1767–1796, here 1771–1789; U. Lindgren, Die Geographie des Claudius Ptolemaeus, in: Archives internationales d'histoire des sciences 35, 1985, 148–239; E. Polaschek, s.v. Ptolemaios [66], RE S 10, 1965, 680–833; R. Wieber, Marinos von Tyros in der arabischen Überlieferung, in: M. Weinmann-Walser (ed.), Historische Interpretationen. FS G. Walser (Historia Einzelschriften 100), 1995, 161–190; F. Mittenhuber, Text- und Kartentradition in der Geographie des Klaudios Ptolemaios, 2009.

Page 10

Literature

W. Eder, J. Renger (ed.), Herrscherchronologien der antiken Welt, 2004; H.J. Nissen, Geschichte Alt-Vorder- asiens, 1998; W. Orthmann (ed.), Der Alte Orient (PropKG, vol. 14), 1975; RLA, 1928 ff.

Page 12

Literature

W. Eder, J. Renger (ed.), Herrscherchronologien der antiken Welt, 2004; H.J. Nissen, Geschichte Alt-Vorderasiens, 1998; W. Orthmann (ed.), Der Alte Orient, PropKg vol. 14, 1975; RLA, 1928ff.; G. Wilhelm, Grundzüge der Geschichte und Kultur der Hurriter, 1982.

Page 22

Literature

E. Akurgal, Die Kunst der Hethiter, 1976; T. Bryce, The Kingdom of the Hittites, 1998; Id., Life and Society in the Hittite World, 2002; V. Haas, Geschichte der hethitischen Religion, 1994; Id., Die hethitische Literatur, 2006; J.D. Hawkins, Corpus of Hieroglyphic Luwian Inscriptions, 2000; M. Hutter, S. Hutter-Braunsar (ed.), Offizielle Religion, lokale Kulte und individuelle Religiosität, 2004; Kunst- und Ausstellungshalle der BRD GmbH (ed.), Die Hethiter und ihr Reich, exhibition catalogue, Bonn, 2002; H.C. Melchers (ed.), The Luwians, 2003; M. Sommer, Der Untergang des hethitischen Reiches: Anatolien und der östliche Mittelmeerraum um 1200 v. Chr., in: Saeculum 52, 2001, 157–176; F. Starke, Zur "Regierung" des hethitischen Staates, in: Zeitschrift für Altorientalische und Biblische Rechtsgeschichte 2, 1996, 140–182.

Page 24

Literature

O. Brinna, König Minos und sein Volk, 1997; A. Chaniotis, Das antike Kreta, 2004; E.H. Cline, Sailing the Wine-Dark Sea. International Trade in the Late Bronze Age, 1994; J.L. Fitton, Die Minoer, 2004; S.W. Manning, The Absolute Chronology of Aegean Early Bronze Age, 1995; Id. et al., Chronology for the Aegean Late Bronze Age 1700–1400 B.C., Science 312, 2006, 565–569; F. Matz et al. (ed.), Corpus der minoischen und mykenischen Siegel, 1964 ff.; H. Siebenmorgen (ed.), Im Labyrinth des Minos, exhibition catalogue, 2000.

Page 26

Literature

K. Branigan (ed.), Cemetery and Society in the Aegean Bronze Age, 1998; H.-G. Buchholz, Ägäische Bronzezeit, 1987; S. Deger-Jalkotzky, O. Panagl (ed.), Die neuen Linear-B-Texte aus Theben. Ihr Aufschlußwert für die Mykenische Sprache und Kultur. Papers of the international research congress at the Austrian Academy of Sciences, 5 and 6 December 2002, 2006; S. Gitin, A. Mazar, E. Stern, Mediterranean Peoples in Transition, 1998; R. Laffineur, W.- D. Niemeier (ed.), Politeia: Society and State in the Aegean Bronze Age, 1995; F. Matz et al. (ed.), Corpus der minoischen und mykenischen Siegel, 1964 ff.; P.A. Mountjoy, Regional Mycenaean Decorated Pottery, 1999.

Page 28

Literature

E.H. Cline, Sailing the Wine-Dark Sea. International Trade in the Late Bronze Age, 1994; Id., D. Harris-Cline, The Aegean and the Orient in the Second Millennium, 1998; E. Edel (†), M. Görg, Die Ortsnamenlisten im nördlichen Säulenhof des Totempels Amenophis' III., 2005; H. Genz, Eine mykenische Scherbe aus Boğazköy, in: Archäologischer Anzeiger 2004, 77–84; S. Hiller, O. Panagl, Die frühgriechischen Texte aus mykenischer Zeit, ²1986; F. Matz et al. (ed.), Corpus der minoischen und mykenischen Siegel, 1964 ff.; W.H. van Soldt, The Topography of the City-State of Ugarit, 2005.

Page 30

Literature

M. Bettelli, Italia meridionale e mondo Miceneo, 2002; H.-G. Buchholz, Ägäische Bronzezeit, 1987; D. Cocchi Genick (ed.), Aspetti culturali della media età del bronzo nell' Italia centro-meridionale, 1995; O. Dickinson, The Aegean Bronze Age, 1996; B. Hänsel, B. Teržan, K. Mihovilić, Herrschaftseliten und Baumeister, in: Antike Welt 4, 2006, 55–60; F.-W. v. Hase, Ägäische, griechische und vorderorientalische Einflüsse auf das tyrrhenische Mittelitalien, in: Beiträge zur Urnenfelderzeit nördlich und südlich der Alpen, 1995, 239–286; H. Hencken, Tarquinia, Villanovians and Early Etruscans, 1968; M. Pallottino, Genti e culture dell' Italia preromana, Rome 1981; Id., Italien vor der Römerzeit, 1987; Die Picener. Ein Volk Europas, exhibition catalogue 1999, 36–54; R. Ross Holloway, Italy and the Aegean 3000–700 B.C., 1981; Id., The Archaeology of Ancient Sicily, 1991; F. Lo Schiavo, E. Macnamara, L. Vagnetti, Late Cypriot Imports to Italy and their Influence on Local Bronze Work, in: PBSR 53, 1985, 1–71; S. Tusa, La società siciliana e il "contatto" con il Mediterraneo centro-orientale dal II millenio a.C. agli inizi del primo millenio a.C., in: Sicilia Archeologica 98, 2000, 9–39; L. Vagnetti (ed.), Magna Grecia e mondo miceneo. Colloquium, Taranto 1982, Atti Taranto 22, 1983; K. v. Welck, R. Stupperich (ed.), Italien vor den Römern, 1996.

Page 32

Literature

G. Bunnens (ed.), Essays on Syria in the Iron Age 2000; H.-J. Gehrke, H. Schneider (ed.), Geschichte der Antike, ²2006; S. Gitin, A. Mazar, E. Stern, Mediterranean Peoples in Transition, 1998; J.D. Hawkins, Corpus of Hieroglyphic Luwian Inscriptions, 2000; K. Jansen-Winkeln, Der Beginn der libyschen Herrschaft in Ägypten, in: Biblische Notizen 71, 1994, 78–97; Id., Ägyptische Geschichte im Zeitalter der Wanderungen von Seevölkern und Libyern, in: E.A. Braun-Holzinger, H. Matthäus (ed.), Die nahöstlichen Kulturen und Griechenland an der Wende vom 2. zum 1. Jahrtausend v. Chr., 2002, 123–142; A. Leahy (ed.), Libya and Egypt c1300–750 BC, 1990; E. Lipinski, The Aramaeans, 2000; H.J. Nissen, Geschichte Alt-Vorderasiens, 1998; E.D. Oren (ed.), The Sea Peoples and Their World, 2000; C.G. Thomas, C. Conant, Citadel to City-State, 1999.

Page 36

Literature

J. Boardman, Kolonien und Handel der Griechen, 1981; W. Eder, K.-J. Hölkeskamp (ed.), Volk und Verfassung im vorhellenistischen Griechenland, 1997; H.-J. Gehrke, H. Schneider (ed.), Geschichte der Antike, ²2006; M.H. Hansen, T. Heine Nielsen (ed.), An Inventory of Archaic and Classical Poleis, 2004; F. Kolb, Die Stadt im Altertum, 1984; F. Krinzinger (ed.), Die Ägäis und das westliche Mittelmeer. Beziehungen und Wechselwirkungen 8. bis 5. Jahrhundert v. Chr., 2000; F. Lang, Archaische Siedlungen in Griechenland, 1996; L.G. Mitchell, P.J. Rhodes (ed.), The Development of the Polis in Archaic Greece, 1997; C. Özgünel, Karia geometrik seramiği, 2006; R. Rollinger, C. Ulf (ed.), Griechische Archaik: Interne Entwicklungen – Externe Impulse, 2004; W. Schuller, Griechische Geschichte, ⁴1995; P. Siewert, L. Aigner-Foresti, Föderalismus in der griechischen und römischen Antike, 2005; C.G. Thomas, C. Conant, Citadel to City-State. The Transformation of Greece, 1200–700 B.C.E., 1999; K.-W. Welwei, Die griechische Polis, ²1998.

Page 38

Literature

H. Blum et al. (ed.), Brückenland Anatolien?, 2002; C. Brixhe, M. Lejeune, Corpus des inscriptions paléophrygiennes, 1984; N. Ehrhardt, Die Ionier und ihr Verhältnis zu den Phrygern und Lydern, in: Neue Forschungen zu Ionien, 2005, 93–111; A. Fuchs, Die Inschriften Sargons II. aus Khorsabad, 1994; R. Gusmani, Lydisches Wörterbuch, 1964, with additional vols., 1980, 1982; Id. et al. (ed.), Frigi e Frigio, 1997; J.D. Hawkins, Corpus of Hieroglyphic Luwian Inscriptions, 2000; A. Heubeck, Schrift, Arch-Hom III, ch. X, 1979; H. Klinkott (ed.), Anatolien im Lichte kultureller Wechselwirkungen, 2001; G.B. Lanfranchi, Dinastie e tradizione regie d'Anatolia: Frigia, Cimmeri e Lidia nelle fonti, in: Universita di Trento (ed.), Dall' Indo a Thule 1996, 89–111; Id., The Ideological and Political Impact of the Neo-Assyrian Imperial Expansion on the Greek World in the 8th and 7th Centuries B.C., in: S. Aro, R. Whiting (ed.), The Heirs of Assyria, 2000, 7–34; H.C. Melchers (ed.), The Luwians, 2003; M. Novák et al. (ed.), Die Außenwirkung des späthethitischen Kulturraumes, 2004; F. Prayon, A.-M. Wittke, Kleinasien vom 12. bis 6. Jahrhundert v. Chr., TAVO Beih. B 82, 1994; M.M. Voigt, R.C. Henrickson, Formation of the Phrygian State: The Early Iron Age at Gordion, in: Anatolian Studies 50, 2000, 37–54; M.M. Voigt, T.C. Young, Jr., From Phrygian Capital to Achaemenid Entrepot: Middle and Late Phrygian Gordion, in: Iranica Antiqua 34, 1999, 191–241; A.-M. Wittke, Mušker und Phryger, (thesis, Stuttgart) 2004; L. Zgusta, Kleinasiatische Ortsnamen, 1984.

Page 44

Literature

A. Berlejung, Geschichte und Religionsgeschichte des antiken Israel, in: J.C. Gertz (ed.), Grundinformationen Altes Testament, [2]2007, 55–185; K. Bieberstein, H. Bloedhorn, Jerusalem I–III, 1994; H. Donner, Geschichte des Volkes Israel und seiner Nachbarn in Grundzügen, vol. 1, [4]2007; vol. 2, [3]2001; E.A. Knauf, Israel II., Geschichte 1. Allgemein und biblisch, RGG 4, [4]2001, 284–293; M. Liverani, Israel's History and the History of Israel, 2005.

Page 52

Literature

R. Borger, Die Inschriften Asarhaddons, König von Assyrien, repr. 1967; Id., Beiträge zum Inschriftenwerk Assurbanipals, 1996: E. Cancik-Kirschbaum, Die Assyrer, 2003; E. Ebeling et al. (ed.), RLA, 1928 ff.; W. Eder, J. Renger (ed.), Herrscherchronologien der antiken Welt, 2004; D.O. Edzard, Geschichte Mesopotamiens, 2004; J.D. Hawkins, Corpus of Hieroglyphic Luwian Inscriptions, 2000; M. Jursa, Die Babylonier, 2004; F.W. König, Handbuch der chaldischen Inschriften, 1955–1957; E. Lipinski, The Aramaeans, 2000; H.C. Melchers (ed.), The Luwians, 2003; H.-J. Nissen, Geschichte Altvorderasiens, 1998; M. Salvini, Geschichte und Kultur der Urartäer, 1995; H. Sauter, Studien zum Kimmerierproblem, 2000; M.W. Stolper, Elam, 1984; M. Streck, Assurbanipal und die letzten assyrischen Könige bis zum Untergang Ninivehs, repr. 1975; M.W. Waters, A Survey of Neo-Elamite History, 2002.

Page 56

Literature

W. Allinger-Csollich, Die Größe und Lage der Heiligtümer im Esagil von Babylon, in: R. Rollinger (ed.), Von Sumer bis Homer, FS M. Schretter, 2004, 7–19; R. Koldewey, Das wieder erstehende Babylon, 1990; S. Maul, Die altorientalische Hauptstadt: Nabel und Abbild der Welt, in: G. Wilhelm (ed.), Die orientalische Stadt: Kontinuität, Wandel, Bruch, 1997, 109–124; M. Novák, Die orientalische Residenzstadt: Funktion, Entwicklung und Form, in: Id., 169–199; Id., Herrschafts- form und Stadtbaukunst. Programmatik im mesopotamischen Residenzstadtbau von Agade bis Surra man ra'a, 1999; J. Renger (ed.), Babylon: Focus mesopotamischer Geschichte, Wiege früher Gelehrsamkeit, Mythos der Moderne, 1999; M. van de Mieroop, Reading Babylon, in: AJA 107, 2003, 257–276.

Page 58

Literature

Map A: F. Gomaà, Die libyschen Fürstentümer des Deltas vom Tod Osorkons II. bis zur Wiedervereinigung Ägyptens durch Psametik I., 1974; *Bubastis*: C. Tietze, M. Abd El Maksoud, Tell Basta. Ein Führer über das Grabungsgelände, 2004; *Buto*: U. Hartung, P. French, J. Bourriau, Spätdynastische Zeit, in: U. Hartung et al., Tell el-Fara'in – Buto. 8. Vorbericht, in: MDAI(K) 59, 2003, 209–233; *al-Gīza*: C.M. Zivie-Coche, Giza au premier millénaire autour du temple d' Isis dame des pyramides, 1991; *Heracleopolis*: M. del Carmen Perez-Die, P. Vernus, Excavaciones en Ehnasya el Medina (Heracleópolis Magna), 1992; *Hermupolis*: A.J. Spencer, Excavations at el-Ashmunein. III. The Town, 1993; *al-Kurru*: D. Dunham, El Kurru, 1950; *Mendes*: D.B. Redford, Excavations at Mendes. I: The Royal Necropolis, 2004; *Nūrī*: D. Dunham, Nuri, 1955; *Qaṣr Ibrīm*: M. Horton, Africa in . Egypt: New Evidence from Qasr Ibrim, in: W.V. Davies (ed.), Egypt and Africa. Nubia from Prehistory to Islam, 1991, 264–277; *Map sketches*: N.-C. Grimal, La stèle triomphale de Pi(cankh)y au Musée du Caire JE 48862 et 47086–47089, 1981, 221f. 231. 245; *Maps*: J. Baines, J. Málek, Weltatlas der alten Kulturen. Ägypten, 1980, 43. 47; I. Gamer-Wallert, Ägypten in der Spätzeit (21. bis sog. 31. Dynastie), TAVO B IV 1, 1993; F. Gomaà, Nildelta (Ägypten) Libysche Fürstentümer, TAVO B IV 2, 1977; K. Zibelius, Nubien und Sudan. Von der 25. Dynastie bis in die Ptolemäisch-Römische Zeit, TAVO B IV 3, 1981. *Maps B, C*: F. Grieshaber, Lexikographie einer Landschaft. Beiträge zur historischen Topographie Oberägyptens zwischen Theben und Gabal as-Silsila anhand demotischer und griechischer Quellen, 2004; *Aḫmīm*: K.P. Kuhlmann, Materialien zur Archäologie und Geschichte des Raumes von Achmim, 1983; *Athribis*: P. Vernus, Athribis. Textes et documents relatifs à la geographie, aux cultes, et à l' histoire d' une ville du Delta égyptien à l' époque pharaonique, 1978; *Bubastis*: C. Tietze, M. Abd El Maksoud, Tell Basta. Ein Führer über das Grabungsgelände, 2004; *Buto*: U. Hartung, P. French, J. Bourriau, Spätdynastische Zeit, in: U. Hartung et al., Tell el-Fara'in – Buto. 8. Vorbericht, in: MDAI(K) 59, 2003, 209–233; *Diospolis Kato*: A.J. Spencer, Excavations at Tell el-Balamun, 3 vols., 1996–2003; *Elephantine*: C. von Pilgrim, Textzeugnis und archäologischer Befund: Zur Topographie Elephantines in der 27. Dynastie, in: H. Guksch, D. Polz (ed.), Stationen. Beiträge zur Kulturgeschichte Ägyptens, 1998, 485–497; *al-Gīza*: C.M. Zivie-Coche, Giza au premier millénaire autour du temple d' Isis dame des pyramides, 1991; *Hermupolis*: A.J. Spencer, Excavations at el-Ashmunein. II. The Temple Area, 1989; *Koptos*: C. Traunecker, Coptus. Hommes et dieux sur le parvis de Geb, 1992; *Memphis*: D.G. Jeffreys, The Survey of Memphis. Part One: The Archaeological Report, 1985; *Mendes*: D.B. Redford, Excavations at Mendes. I: The Royal Necropolis, 2004; *Moiris*: O. Kimball Armayor, Herodotus' Autopsy of the Fayoum: Lake Moeris and the Labyrinth of Egypt, 1985; *Naucratis*: W.D.E. Coulson, Ancient Naukratis II,I: The Survey at Naukratis, 1996; *Pelusion*: H. Jaritz et al., Pelusium. Prospection archéologique et topographique de la région de Tell el-Kana'is 1993 et 1994, 1996; *Phakussa*: P. Davoli, Saft el-Henna. Archeologia e storia di una città del Delta orientale, 2001; *Saqqāra*: H.S. Smith, D.G. Jeffreys, The North Saqqâra Temple-Town Survey: Preliminary Report for 1976/77, in: JEA 64, 1978, 10–21; *Tall al-Masḫūṭa*: J.S. Holladay Jr., Tell el-Maskhuṭa. Preliminary Report on the Wadi Tumilat Project 1978–1979, 1982; *Wādī aṭ-Ṭumilāt*: C.A. Redmount, The Wadi Tumilat and the 'Canal of the Pharaohs', in: JNES 54, 1995, 127–135; *Maps*: J. Baines, J. Málek, Weltatlas der alten Kulturen. Ägypten, 1980, 49; I. Gamer-Wallert, Ägypten in der Spätzeit (21. bis sogenannte 31. Dynastie), TAVO B IV 1, 1993; K. Zibelius, Nubien. Neues Reich und Spätzeit vor Napata, TAVO B III 2, 1981; Id., Nubien und Sudan. Von der 25. Dynastie bis in die Ptolemäisch-Römische Zeit, TAVO B IV 3, 1981. **Map inset**: redrawn after: D.J. Thompson, Memphis under the Ptolemies, 1988, fig. 3 (with additions and revisions by the author).

Page 76

Literature

L. Aigner-Foresti, Die Etrusker und das frühe Rom, 2003; Id. (ed.), Etrusker nördlich von Etrurien. Etruskische Präsenz in Norditalien sowie ihre Einflüsse auf die einheimischen Kulturen, 1992; Id., P. Siewert (ed.), Entstehung von Staat und Stadt bei den Etruskern, 2006; H. Berve, Die Tyrannis bei den Griechen, 2 vols., 1967; M. Cristofani (ed.), La grande Roma dei Tarquini, exhibition catalogue, 1990; S. Haynes, Kulturgeschichte der Etrusker, 2005; E. Manni, Sicilia e Magna Grecia nel V secolo, in: Kokalos 14/15, 1968/69, 95–111; D. Mertens, Städte und Bauten der Westgriechen, 2006; S. Moscati, Die Karthager, 1996; M. Pallottino, Italien vor der Römerzeit, 1987; L. Pearson, The Greek Historians of the West. Timaeus and His Predecessors, 1987; F. Prayon, Die Etrusker, [3]2003; G. Pugliese Carratelli (ed.), Italia omnium terrarum alumna. La civiltà dei Veneti, Reti, Liguri, Celti, Piceni, Umbri, Latini, Campani e Iapigi, 1988; Id., Italia omnium terrarum parens. La civiltà degli Enotri, Choni, Ausoni, Sanniti, Lucani, Brettii, Sicani, Siculi, Elimi, 1989; P. Siewert, L. Aigner-Foresti, Föderalismus in der griechischen und römischen Antike, 2005; M. Sommer, Phönizier, 2005.

Page 78

Literature

L. Aigner-Foresti (ed.), Etrusker nördlich von Etrurien. Papers of the symposium of Vienna – Schloß Neuwaldegg 2–5 October 1989, 1992; J. Biel, D. Krausse (ed.),Frühkeltische Fürstensitze. Älteste Städte und Herrschaftszentren nördlich der Alpen?, 2005; K. Bittel et al. (ed.), Die Kelten in Baden-Württemberg, 1981; J. Bofinger et al., Die Keltenfürsten. Glanz und Gloria, 2006; R. Cordie-Hackenberg, Hunsrück-Eifel-Kultur, RGA 15, 2000, 266–271; H. Dannheimer, R. Gebhard (ed.), Das keltische Jahrtausend, exhibition catalogue Prähistorische Staatssammlung Munich, 1993; M. Egg, D. Kramer, Krieger, Feste, Totenopfer. Der letzte Hallstattfürst von Kleinklein in der Steiermark, 2005; F. Fischer, KEIMELIA. Bemerkungen zur kulturgeschichtlichen Interpretation des sogenannten Südimportes in der späten Hallstatt- und frühen Latène-Kultur des westlichen Mitteleuropas, in: Germania 51, 1973, 436–459; Id., Frühkeltische Fürstengräber in Mitteleuropa. Antike Welt special issue, 1982; Die Hallstattkultur. Frühform europäischer Einheit, exhibition catalogue Steyr 1980; Die Hallstattkultur. Report on Steyr Symposium 1980; K. Kromer, Das Gräberfeld von Hallstatt, 1959; M. Kuckenburg, Die Kelten in Mitteleuropa, 2004; I.R. Metzger, P. Gleirscher (ed.), Die Räter,1992; S. Moscati et al. (ed.), The Celts, exhibition catalogue Venice, 1993 (Italian 1991); L. Pauli, Studien zur Golasecca-Kultur. MDAI(R) Erg.-H. 19, 1971; S. Rieckhoff, J. Biel, Die Kelten in Deutschland, 2001; K. Spindler, Die frühen Kelten, [3]1991; T. Stöllner, Die Hallstattzeit und der Beginn der Latènezeit im Inn-Salzach-Gebiet. Archäologie in Salzburg vol. 3/I-II, 1996, 2002; Id., Hallstattkultur und Hallstattzeit, RGA 13, 1999, 446–453; Trésors des princes celtes, exhibition catalogue Paris 1987; C. Wilms, Der Keltenfürst von Frankfurt. Macht und Totenkult um 700 v. Chr., 2002.

Page 80

Literature

H. Birkhan, Die Kelten. Versuch einer Gesamtdarstellung ihrer Kultur, 1997; Id., Die Kelten. Bilder ihrer Kultur, 1999; K. Bittel et al. (ed.), Die Kelten in Baden-Württemberg, 1981; J. Collis, Oppida. Earliest Towns North of the Alps, 1984; H. Dannheimer, R. Gebhard (ed.), Das keltische Jahrtausend, exhibition catalogue Prähistorische Staatssammlung Munich, [2]1993; P.-M. Duval, V. Kruta (ed.), Les mouvements celtiques du Ve au Ier siècle avant notre ère. Actes du XXVIIIe colloque organisé à l' occasion du XIe Congrès International des Sciences Préhistoriques et Protohistoriques, Nice, 1976, 1979; J. Fries-Knoblach, Die Kelten. 3000 Jahre europäischer Kultur und Geschichte, 2002; V. Kruta, Die Kelten. Aufstieg und Niedergang einer Kultur, 2000; M. Kuckenburg, Die Kelten in Mitteleuropa, 2004; I.R. Metzger, P. Gleirscher (ed.), Die Räter, 1992; S. Moscati et al. (ed.), The Celts. Exhibition catalogue Venice, 1993 (Italian 1993); R. Müller, Latènekultur und Latènezeit, RGA 18, 2001, 118–124; L. Pauli, Die Alpen in Frühzeit und Mittelalter. Die archäologische Entdeckung einer Kulturlandschaft, 1980; Id. (ed.), Die Kelten in Mitteleuropa. Kunst – Kultur – Wirtschaft, exhibition catalogue Keltenmuseum Hallein, 1980; S. Rieckhoff, J. Biel, Die Kelten in Deutschland, 2001; Keltenmuseum Hochdorf/Enz (ed.), Schätze aus der Keltenzeit in Ungarn, exhibition catalogue Keltenmuseum Hochdorf/Enz, 1998; S. Sievers, Manching – Die Keltenstadt, 2003; T. Stöllner, Die Hallstattzeit und der Beginn der Latènezeit im Inn-Salzach-Gebiet. Archäologie in Salzburg vol. 3/I–II, 1996, 2002; Keltenmuseum Hochdorf/Enz (ed.), Thraker und Kelten beiderseits der Karpaten, exhibition catalogue Keltenmuseum Hochdorf/Enz, 2000; P. Vouga, La Tène. Monographie de la station, 1923.

Page 86

Literature

T. Bakir-Akbaşoğlu (ed.), Anatolia in the Achaimenid Period, 1999; P. Briant, Darius. Les Perses et l' Empire, 1992; Id., From Cyrus to Alexander. A History of the Persian Empire, 2002; Id., A. Kuhrt, M.C. Root, H. Sancisi-Weerdenburg, J. Wiesehöfer (ed.), Achaemenid History. vols. 1 ff., 1987 ff.; M. Brosius, The Persians, 2006; W. Hinz, Darius und die Perser. Eine Kulturgeschichte der Achämeniden, 2 vols., 1976 (evaluation of the Elamite clay tablets); H. Klinkott, Der Satrap, 2005; M.J. Olbrycht, Die Kultur der Steppengebiete, in: J. Wiesehöfer (ed.), Das Partherreich und seine Zeugnisse, 11–43; R. Schmitt, Persien, Perser. A. Sprache, RLA 10, 2004, 412–416; M. Stausberg, Persien, Perser. C. Religion, RLA 10, 2004 422–424; G. Summers, F. Summers, The Kerkenes Project, Anatolian Archaeology 11, 2005, 34–36; L. Trümpelmann, Zwischen Persepolis und Firuzabad, 1992; W.J. Vogelsang, Rise and Organisation of the Achaemenid Empire, 1992; J. Wiesehöfer, Das frühe Persien, Munich 1999, 11–73; Id., Das antike Persien. Von 550 v. Chr. bis 650 n. Chr., 2005.

Page 88

Literature

A.R. Burn, Persia and the Greeks: the Defence of the West, c. 546–478 B.C., [2]1984; G. Cawkwell, The Greek Wars. The Failure of Persia, 2005; M. Dreher, Athen und Sparta, 2001; M.H. Hansen, T. Heine Nielsen (ed.), An Inventory of Archaic and Classical Poleis, 2004; E. Hutz- feldt, Das Bild der Perser in der griechischen Dichtung des 5. vorchristlichen Jahrhunderts, 1999; M. Jung, Marathon und Plataiai, 2006; J.F. Lazenby, The Defence of Greece 490–479 B.C., 1993; M.C. Miller, Athens and Persia in the Fifth Century B.C., 1997; C. Schubert, Athen und Sparta in klassischer Zeit, 2003; M. Stahl, Gesellschaft und Staat bei den Griechen, vol. 2, 2003; L. Thommen, Sparta. Verfassungs- und Sozialgeschichte einer griechischen Polis, 2003; K.-W. Welwei, Das klassische Athen, 1999; Id., Sparta, 2004; J. Wiesehöfer, Das antike Persien, [4]2005.

Page 92

Literature

P. Barceló, Basileia, Monarchia, Tyrannis. Untersuchungen zu Entwicklung und Beurteilung von Alleinherrschaft im vorhellenistischen Griechenland, 1993; S. Berger, Revolution and Society in Greek Sicily and Southern Italy, 1992; H. Berve, Die Tyrannis bei den Griechen, 2 vols., 1967; B. Caven, Dionysios I., 1990; A. Demandt, Antike Staatsformen, 1995, 163–189; M. Hofer, Tyrannen Aristokraten, Demokraten. Untersuchungen zu Staat und Herrschaft des griechischen Sizilien von Phalaris bis zum Aufstieg des Dionysios I., 2000; E. Kinzl, Betrachtungen zur älteren Tyrannis, in: Id. (ed.), Die ältere Tyrannis bis zu den Perserkriegen, 1979, 298–325; J. Labarbe, L' apparition de la notion de tyrannie dans la Grèce archaique, in: L' Antiquité Classique 40, 1971, 479–504; L. de Libero, Archaische Tyrannis, 1996; N. Luraghi, Tirannidi arcaiche in Sicilia e Magna Grecia, 1994; D. Mertens, Griechen und Punier. Selinunt nach 409 v. Chr., in: RhM 104, 1997, 301–320; V. Parker, Tyrannos. The Semantics of a Political Concept from Archilochos to Aristotle, in: Hermes 126, 1998, 145–172; H.G. Plass, Die Tyrannis in ihren beiden Perioden bei den Griechen, 1852, 21859; H.W. Pleket, The Archaic Tyranny, in: Talanta 1, 1969, 19–61; H. Sancisi-Weerdenburg (ed.), Peisistratos and the Tyranny. A Reappraisal of the Evidence, 2000; M. Stahl, Aristokraten und Tyrannen im archaischen Athen, 1987; K.F. Stroheker, Dionysios I., 1958.

Page 98

Literature

H. Beck, Polis und Koinon. Untersuchungen zu Geschichte und Struktur der griechischen Bundesstaaten im 4. Jahrhundert v. Chr., 1997; E.N. Borza, In the Shadow of Olympus. The Emergence of Macedon, 1990; J. Buckler, The Theban Hegemony, 371–362 B.C., 1980 Id., Philip II. and the Sacred War, 1989; J. Cargill, The Second Athenian League. Empire or Free Alliance?, 1981; M. Dreher, Hegemon und Symmachoi. Untersuchungen zum Zweiten Attischen Seebund, 1995; P. Funke, Homonoia und Arche. Athen und die griechische Staatenwelt vom Ende des Peloponnesischen Krieges bis zum Königsfrieden (404/03–387/86 v. Chr.), 1980; C. Habicht, Athen. Die Geschichte der Stadt in hellenistischer Zeit, 1995; F. Jacoby, FGrH, 15 vols., 1923–1958; M. Jehne, Koine Eirene. Untersuchungen zu den Befriedungs- und Stabilisierungsbemühungen in der griechischen Poliswelt des 4. Jahrhunderts v. Chr., 1994; J.A.O. Larsen, Greek Federal States. Their Institutions and History, 1968; W.G. Runciman, Doomed to Extinction: The Polis as an Evolutionary Dead-End, in: O. Murray, S. Price (ed.), The Greek City. From Homer to Alexander, 1990, 347–367; T.T.B. Ryder, Koine Eirene. General Peace and Local Independence in Ancient Greece, 1965; W. Schmitz, Wirtschaftliche Prosperität, soziale Integration und die Seebundpolitik Athens, 1988; P. Siewert, L. Aigner-Foresti (ed.), Föderalismus in der griechischen und römischen Antike, 2005; M. Sordi, La lega tessala fino al Alessandro Magno, 1958; L. Thommen, Sparta. Verfassungs- und Sozialgeschichte einer griechischen Polis, 2003; L.A. Tritle (ed.), The Greek World in the Fourth Century. From the Fall of the Athenian Empire to the Successors of Alexander, 1997; R. Urban, Der Königsfrieden von 387/86 v. Chr., 1991; M. Zahrnt, Olynth und die Chalkidier, 1971.

Page 108

Literature

T.J. Cornell, The Beginnings of Rome, 1995; H. Galsterer Herrschaft und Verwaltung im republikanischen Italien, 1976; R.R. Holloway, The Archaeology of Early Rome and Latium, 1994; G. Pugliese Carratelli (ed.), Italia omnium terrarum alumna. La civiltà dei Veneti, Reti, Liguri, Celti, Piceni, Umbri, Latini, Campani e Iapigi, 1988; P. Siewert, L. Aigner-Foresti (ed.), Föderalismus in der griechischen und römischen Antike, 2005.

Page 110

Literature

G. Bradley, Ancient Umbria, State, Culture, and Identity in Central Italy from the Iron Age to the Augustan Era, 2000; F.E. Brown, Cosa, 1980; G. Camporeale, Die Etrusker, 2003 (Italian 2004); A. Carandini (ed.), La romanizzazione dell' Etruria: il territorio di Vulci, 1985; T.J. Cornell, The Beginnings of Rome, 1995; S.L. Dyson, Settlement Patterns in the Ager Cosanus: The Wesleyan University Survey, 1974–1976, in: Journal of Field Archaeology 5, 1978, 251–268; W. Eder, Staat und Staatlichkeit in der frühen römischen Republik, 1990; F. Enei, Progetto Ager Caeretanus: Il litorale di Alsium, 2001; P.W.M. Freeman, 'Romanisation' and Roman Material Culture, in: Journal of Roman Archaeology 6, 1993, 438–445; H. Galsterer, Herrschaft und Verwaltung im republikanischen Italien, 1976; T. Hantos, Das römische Bundesgenossensystem in Italien, 1983; W.V. Harris, Rome in Etruria and Umbria, 1971; S. Haynes, Kulturgeschichte der Etrusker, 2005; P. Hemphill, The Civitella Cesi Survey, Archaeological Investigations in Southern Etruria I, 2000; K.-J. Hölkeskamp, Die Entstehung der Nobilität, 1987; F. Perkins, Etruscan Settlement, Society and Material Culture in Central Coastal Etruria, 1999; E.T. Salmon, Roman Colonization under the Republic, 1969; Id., The Making of Roman Italy, 1982; P. Siewert, L. Aigner-Foresti (ed.), Föderalismus in der griechischen und römischen Antike, 2005; M. Sordi, I rapporti romano-ceriti e l' origine della civitas sine suffragio, 1960; N. Terrenato, Tam firmum municipium. The Romanization of Volaterrae and its Cultural Implications, in: JRS 88, 1998, 94–114.

Page 120

Literature

P. Bilde et al. (ed.), Ethnicity in Hellenistic Egypt, 1992; D.N. Edwards, The Archaeology of the Meroitic State, 1996; H.-J. Gehrke, Geschichte des Hellenismus, [3]2003; F. Hoffmann, Ägypten Kultur und Lebenswelt in griechisch-römischer Zeit. Eine Darstellung nach demotischen Quellen, 2000; W. Huss, Ägypten in hellenistischer Zeit: 332–30 v. Chr., 2001; C.A. La'da, Foreign Ethnics in Hellenistic Egypt, 2002; A. Lampela, Rome and the Ptolemies of Egypt The Development of Their Political Relations 273–80 B.C., 1998; C. Préaux, L' économie royale des Lagides, 1939; C. Schäfer, Kleopatra, 2006; T. Schrapel, Das Reich der Kleopatra. Quellenkritische Untersuchungen zu den "Landschenkungen" Mark Antons 1996; F. Uebel, Die Kleruchen Ägyptens unter den ersten sechs Ptolemäern, 1968.

Page 124

Literature

R.E. Allen, The Attalid Kingdom, 1983; K. Bringmann, H. v. Steuben (ed.), Schenkungen hellenistischer Herrscher an griechische Städte und Heiligtümer, I, 1995, II, 2000; H.-J. Gehrke, Geschichte des Hellenismus, [3]2003; J. Hopp, Untersuchungen zur Geschichte der letzten Attaliden, 1977; J. Kobes, "Kleine Könige". Untersuchungen zu den Lokaldynasten im hellenistischen Kleinasien (323–188 v. Chr.), 1996; I. Nielsen, Hellenistic Palaces, 1995; F. Queyrel, Les portraits des Attalides, 2003; W. Radt, Pergamon. Geschichte und Bauten einer antiken Metropole, 2005; H.-J. Schalles, Untersuchungen zur Kulturpolitik der pergamenischen Herrscher im 3. Jahrhundert v. Chr., 1985; K. Strobel, Die Galater. vol. 1, 1996.

Page 126

Literature

P.G. Bilde, V.F. Stolba (ed.), Surveying the Greek Chora: The Black Sea Region in a Comparative Perspective, 2006; J. Fornasier, B. Böttger (ed.), Das Bosporanische Reich, 2002; N.A. Frolova, The Coinage of the Kingdom of Bosporus: A.D. 69–238, 1979; Id., The Coinage of the Kingdom of Bosporus A.D. 242–341/42, 1983; V.F. Gajdukevič, Das Bosporanische Reich, [2]1971; M. Oppermann, Die Westpontischen Poleis und ihr indigenes Umfeld in vorrömischer Zeit.

2004; V.F. Stolba, L. Hannestad (ed.), Chronologies of the Black Sea Area in the Period, c. 400–100 B.C., 2005; G.R. Tsetskhladze, The Greek Colonisation of the Black Sea Area. Historical Interpretation of Archaeology, 1998; Id., North Pontic Archaeology: Recent Discoveries and Studies, 2001; Id. (ed.), Greek Settlements in the Eastern Mediterranean and the Black Sea, 2002; Id., The Black Sea Region in the Greek, Roman, and Byzantine Periods, 2002.

Page 128

Literature

A.E. Astin (ed.), Rome and the Mediterranean to 133 B.C., 1989; R.S. Bagnall, P. Derow (ed.), The Hellenistic Period. Historical Sources in Translation. New Edition, 2004; P. Bilde et al. (ed.), Centre and Periphery in the Hellenistic World, 1993; G.R. Bugh, The Cambridge Companion to the Hellenistic World, 2006; P. Cartledge et al. (ed.), Hellenistic Constructs. Essays in Culture, History, and Historiography, 1997; H.-J. Gehrke, Geschichte des Hellenismus, [3]2003; E.S. Gruen, The Hellenistic World and the Coming of Rome, 2 vols., 1984; J. Kobes, "Kleine Könige". Untersuchungen zu den Lokaldynasten im hellenistischen Kleinasien (323–188 v. Chr.), 1996; A. Lampela, Rome and the Ptolemies of Egypt. The Development of Their Political Relations 273–80 B.C., 1998; C. Vial, Délos indépendante (314–167 avant J.-C.), 1984; F.W. Walbank (ed.), The Hellenistic World, 1984; H.-U. Wiemer, Krieg, Handel und Piraterie. Untersuchungen zur Geschichte des hellenistischen Rhodos, 2002; J. Wiesehöfer (ed.), Das Partherreich und seine Zeugnisse, 1998.

Page 134

Literature

G.G. Aperghis, The Seleukid Royal Economy, 2004; J. Bleicken, Geschichte der römischen Republik, [6]2004; M.I. Finley, Die antike Wirtschaft, 1977; H.-J. Gehrke, Geschichte des Hellenismus, [3]2003; H. Kloft, Die Wirtschaft der griechisch-römischen Welt, 1992; H. Kreissig, Wirtschaft und Gesellschaft im Seleukidenreich, 1978; F. de Martino, Wirtschaftsgeschichte des alten Rom, [2]1991; F. Meijer, O. van Nijf, Trade, Transport, and Society in the Ancient World, 1992; J. Oelsner, Gesellschaft und Wirtschaft des seleukidischen Babylonien, in: Klio 63, 1981, 39ff.; T. Pekáry, Die Wirtschaft der griechisch-römischen Antike, 1979; C. Préaux, L' économie royale des Lagides, 1939; Id., L' économie Lagide: 1933–1958, in: Proceedings of the IX. International Congress of Papyrology, Oslo, 19th-22nd August 1958, 1961, 200ff.; M. Rostovtzeff, Gesellschafts- und Wirtschaftsgeschichte der hellenistischen Welt, vols. 1–3, 1998 (repr. of 1955 edition); J.-F. Salles, The Arab-Persian Gulf under the Seleucids, in: A. Kuhrt, S. Sherwin-White (ed.), Hellenism in the East, 1987, 75–109; W. Scheidel, S. v. Reden, The Ancient Economy, 2002; H. Schneider (ed.), Zur Sozial- und Wirtschaftsgeschichte der späten römischen Republik, 1976; M. Schuol, Die Charakene, 2000; C. Vial, Délos indépendante (314–167 avant J.-C.), 1984; H.-U. Wiemer, Krieg, Handel und Piraterie. Untersuchungen zur Geschichte des hellenistischen Rhodos, 2002.

Page 136

Literature

J. Beloch, Der italische Bund unter Roms Hegemonie, 1880; K. Bringmann, Geschichte der Römischen Republik, 2002; P.A. Brunt, Italian Manpower 225 B.C.–A.D. 14, 1971, repr. with additions 1987; G. Forsythe, A Critical History of Early Rome. From Prehistory to the First Punic War, 2005; H. Galsterer, Herrschaft und Verwaltung im republikanischen Italien, 1976; T. Hantos, Das römische Bundesgenossensystem in Italien, 1983; K.-H. Schwarte, Zum Ausbruch des zweiten Samnitenkrieges (326–304 v. Chr.), in: Historia 20, 1971, 368–376.

Page 138

Literature

N. Bagnall, Rom und Karthago, 1995; P. Barceló, Hannibal, [2]2003; H. Bellen, J. Bleicken, Geschichte der Römischen Republik, [6]2004; K. Christ, Hannibal, 2003; M. Gerhold, Rom und Karthago zwischen Krieg und Frieden: Rechtshistorische Untersuchung zu den römisch-karthagischen Beziehungen zwischen 241 v. Chr. und 149 v. Chr. (thesis, Vienna), 2002; U. Händel-Siegawe, Der Beginn des Zweiten Punischen Krieges, 1995; H. Heftner, Der Aufstieg Roms, 1997; B.D. Hoyos, Unplanned Wars.The Origins of the First and Second Punic War, 1998; W. Huss, Karthago, 1995; S. Lancel, Carthage, 1995; K.-H. Schwarte, Der Ausbruch des Zweiten Punischen Krieges,1983; J. Seibert, Hannibal (Antike Welt, special edition), 1997; K. Zimmermann, Rom und Karthago, 2005.

Page 140

Cf. table below, p. 265

Sources and literature

Where they are not listed here, the sources for the respective periods are found in the commentary to another map, to which the text makes reference.

219: Sources: First Illyrian War: Pol. 2,4,7–6,7; 2,8–9; 2,12,3; Flor. Epit. 1,21; App. Ill. 7,18; Cass. Dio fr. 49,1–7; 53; *Second Illyrian War*: Pol. 3,18,2 ff.; *Literature:* N.G.L. Hammond, The Illyrian Atintani, the Epirotic Atintanes and the Roman Protectorate, in: JRS 79, 1989, 11–25; D. Vollmer, Symploke. Das Übergreifen der römischen Ex- pansion auf den griechischen Osten. Untersuchungen zur römischen Außenpolitik am Ende des 3. Jahrhunderts, 1990.
146 (1): Sources: Pol. 38,19–22; Diod. Sic. 32,23 f.; Liv. Per. 51; Flor. Epit. 1,31,13–18; App. Lib. 127–135; *Literature:* N. Ferchiou, Nouvelles données sur un fossé inconnu en Afrique proconsulaire et sur la Fossa Regia, in: Histoire et archéologie de l'Afrique du Nord, 1986, 351–365; F. Gschnitzer, Die Stellung Karthagos nach dem Frieden von 201, in: WS 79, 1966, 276–289; W. Hoffmann, Die römische Politik des 2. Jahrhunderts und das Ende Karthagos, in: Historia 9, 1960, 309–344.
123 (1): Sources: Liv. Per. 60 f.; App. Civ. 1,34; Diod. Sic. 34,23; Vell. Pat. 2,6,4; 2,10,2 f.; 2,39,1; Str. 4,1,5–11; Cic. Font. 36; *Literature:* F. Benoit, Entremont, capitale celto-ligure de la Provence, in: Rhodania 20/22, 1938/1946, 58–60; D. Roman, M. Fulvius Flaccus et la frontière de la Gaule Transalpine, in: Y. Roman (ed.), La frontière. Séminaire de recherche, 1993, 57–66.
123 (2): Sources: Liv. Per. 60; Str. 3,5,1; Flor. Epit. 1,43; Oros. 5,13,1; *Literature:* A. Schulten, Iberische Landeskunde. Geographie des antiken Spanien 2, 1957, 251–256.
75: Sources: Cic. De or. 1,82; 2,2; Leg. Man. 33; Liv. Per. 68; Pomp. Trog. prol. 39; Plut. Pomp. 24,6; Sall. Hist. 2,87; Str. 14,3,3–14,5,7; *Literature:* A. Avidov, Were the Cilicians a Nation of Pirates?, in: Mediterranean Historical Review 10, 1997, 5–55; M. Benabou, Rome et la police des mers au 1er siècle avant J. C.: la répression de la piraterie Cilicienne, in: M. Galley, L. Sébai (ed.), L' homme méditerranéen et la mer, 1985, 60–69; P. de Souza, Piracy in the Graeco-Roman World, 1999; H. A. Ormerod, Piracy in the Ancient World, 21978.
74: Sources: Liv. Per. 100; Plut. Pompeius 29; agri Apionis: Cic. Leg. agr. 2,5,1; Sall. Hist. 2,42; App. Mithr. 600; *Literature:* A. Laronde, La Cyrénaïque romaine, des origines à la fin des Sévères, in: ANRW II 10.1, 1006–1064; G. Perl, Die römischen Provinzbeamten in Cyrenae und Creta zur Zeit der Republik, in: Klio 52, 1970, 319–354; P. Romanelli, La Cirenaica romana, 1943; I.F. Sanders, Roman Crete, 1982, 3–15.
58: Sources: Vell. Pat. 2,45,4 f.; Liv. Per. 104; Flor. Epit. 1,44; Plut. Cato minor 34–39; Str. 14,6,6; App. Civ. 2,85; Cass.Dio 38,30,5; *Literature:* E. Badian, M. Porcius Cato and the Annexation and Early Administration of Cyprus, in: JRS 55, 1965, 110–121.
146: Sources: App. Civ. 4,233; Plin. HN 5,22; Cass. Dio 43,3,1–4; Bell. Afr. 25,2 f.; *Literature:* J. Kolendo, C. Sallustius Crispus, premier gouverneur de l'Africa Nova et la dispersion géographique du gentilice Sallustius en Afrique, in: Arheoloski vestnik 28, 1977, 255–277.
33: Sources: App. Ill. 12–28; Liv. Per. 132 f.; Cass. Dio 49,34–43; 51,25,2; ILS 77; *Literature:* P.-S.G. Freber, Der hellenistische Osten und das Illyricum unter Caesar, 1993.
30: Sources: Vell. Pat. 2,87,1; Tac. Ann. 2,59; Tac. Hist. 1,11; Cass. Dio 51,17,1–3; *Literature:* G. Geraci, Genesi della provincia romana d' Egitto, 1983; Id., La formazione della provincia romana d' Egitto, in: Egitto e società antica, 1985, 163–180; E.G. Huzar, Augustus, Heir of the Ptolemies, in: ANRW II 10,1 (1988), 343–382.

Page 142

Cf. tables below, pp. 266 f.

Page 146

Sources

Development of provinces – Fossa Regia: Plin. HN 5,25; *146–46 BC*: Cic. Agr. 2,41; Sall. Iug. 5; 11; 13 ff.; 77 f.; Diod. Sic. 32; App. Pun. 106; 135; App. Libyca *passim*; *46–40/39 BC*: Plin. HN 5,27; 5,38; Ptol. 4,3,12 f.; ILS 5955, 1–3; *40/39 BC – AD 284*: Tac. Ann. 2,52; Tac. Hist. 1,11; 2,97; 4,48 f.; Cass. Dio 53,14; 55,23; 59,20; Ptol. 4,3; Amm. Marc. 28,6,7; Laterculus Veronensis 12; Not Dign. Occ. *passim*.
War against Jugurtha – Massinissa in Carthage: Liv. 34,49; App. Lib. 37; *Jugurtha's remark* (*urbem venalem et mature perituram, si emptorem invenerit*): Sall. Iug. 35,10; Liv. Per. 64; App. Num. 1; *Jugurthan War*: Sall. Iug. *passim*; Cic. Off. 3,20; Diod. Sic. 34,31; 34,35; 34,39; Liv. Per. 66; Plut. Marius 10–12; App. Num. 4 f.

Literature

Development of provinces: A. Beschaouch, Le territoire de Sicca Veneria (El-Kef), nouvelle Cirta, en Numidie Proconsulaire (Tunisie), in: CRAI 1981, 105–122; A. Chausa, Modelos de reservas de indígenas en el África romana, in: Gerión 12, 1994, 95–101; F. Decret, M. Fantar, L' Afrique du Nord dans l' Antiquité, 1981; J. Desanges, Mauretania Ulterior Tingitana, in: Bulletin d' Archéologie Marocaine 4, 1960, 437–441; D. Fishwick, B.D. Shaw, The Formation of Africa proconsularis, in: Hermes 105, 1977, 369–380; Id., On the Origins of Africa Proconsularis, in: AntAfr 29, 1993, 53–62; 30, 1994, 57–80; 32, 1996, 13–36; D. Fushöller, Tunesien und Ostalgerien in der Römerzeit, 1979; H. Halm, Eine Inschrift des Magister militum Solomon in arabischer Überlieferung. Zur Restitution der Mauretania Caesariensis unter Justinian, in: Historia 36, 250–256; F. Kolb, Der Aufstand der Provinz Africa Proconsularis im Jahre 238 n. Chr. Die wirtschaftlichen und sozialen Hintergründe, in: Historia 26, 1977, 440–478; Y. Le Bohec, Notes prosopographiques sur la Legio III Augusta (2–5), in: ZPE 36, 1979, 82 f.; 150; 206 f.; 226 f.; R.A. Markus, Additional Note on Justin II' s Privileges for the Ecclesiastical Province of Byzacena, in: Byzantion 49, 1979, 303–306; P. Maymó y Capdevila, Maximiano en campaña: matizaciones cronológicas a las expediciones hispanas y africanas del augusto hercúleo, in: Polis 12, 2000, 229–257; A. Paki, Quelques remarques sur l' inscription CIL VIII 18085, in: F. Koenig, S. Rebetez (ed.), Arculiana. FS H. Bögli, 1995, 493–498; P. Romanelli, Storia delle province romane dell' Africa, 1959; B.E. Thomasson, Zur Verwaltungsgeschichte der römischen Provinzen Nordafrikas (Proconsularis, Numidia, Mauretaniae), in: ANRW II 10.2, 3–61; TIR H/I 33, 1954.

War against Jugurtha: J.-M. Claassen, Sallust's Jugurtha – Rebel or Freedom Fighter?, in: CW 86, 1992/93, 273–297; J. Hellegouarc' h, Urbem uenalem... (Sall. Iug. 35,10), in: Bulletin de l' Association G. Budé 1990, 163–174; C.S. Mackay, Sulla and the Monuments. Studies in his Public Persona, in: Historia 49, 2000, 161–210; R. Morstein-Marx, The Alleged 'Massacre' at Cirta and its Consequences (Sallust Bellum Iugurthinum 26–27), in: CPh 95, 2000, 468–476; V. Parker, Romae omnia venalia esse. Sallust' s Development of a Thesis and Prehistory of the Jugurthine War, in: Historia 53, 2004, 408–423; H.W. Ritter, Rom und Numidien, 1987; C. Saumagne, La Numidie et Rome, 1966.

Page 148

Cf. table below, p. 267

Sources

Rome's wars in the west: Cato: Liv. 33,43; 34,8–21; Val. Max. 4,3,11; Plut. Cato Maior 10 f.; *Viriatus*: Diod. Sic. 33,1; 33,7; 33,19; 33,21a; Liv. Per. 54; Flor. Epit. 1,33,15; Cass. Dio fr. 73; 75; 77 f.; Oros. 5,4,1–14; *Numantia*: App. Ib. 323–424; Liv. Per. 53–69.
Repulsing of the Cimbri and Teutoni: Rhine crossing: Vell. Pat. 2,8,3; App. Celt. 1,11; *defeat of Silanus*: Liv. Per. 65; Vell. Pat. 2,12; Ascon. 68,15; 78,15; 80,19–25; Flor. Epit. 3,38; *defeat of Cassius Longinus*: Caes. B Gall. 1,7; 12; 30; Liv. Per. 65; App. Celt. 1,3; Oros. 5,15,23 ff.; *defeat at Arausio*: Liv. Per. 67; Vell. Pat. 2,12; Ascon. 89; Tac. Germ. 37; Plut. Marius 19; Plut. Lucullus 27; Granius Licinianus 16; 20; Eutr. 5,1; Oros. 5,16; Sall. Iug. 114; Cic. Balb. 28; Val. Max. 4,7,3; Gell. NA 3,9,7; *Cimbri in Hispania*: Liv. Per. 67; Iul. Obsequens 43; Plut. Marius 14; Seneca ad Helviam 7,2; Caes. B Gall. 1,33; 7,77; *over the Alps to Italy*: Plut. Marius 15; 19; Str. 7,2,2; *Tridentina iuga*: Flor. Epit. 1,38; *Aquae Sextiae*: Liv. Per. 68; Vell. Pat. 2,12; Plut. Marius 18 ff.; Flor. Epit. 3,3; Eutr. 5,1; Oros. 5,16; *at Vercellae in the Raudian Fields*: Liv. Per. 68; Vell. Pat. 2,12; Plut. Marius 24 ff.; Flor. Epit. 3,3; Eutr. 5,2; Oros. 5,16.

Literature

Rome's wars in the west: R.W. Bane, The Development of Roman Imperial Attitudes and the Iberian Wars, in: Emerita 44, 1976, 409–420; H. Gundel, Viriatus, RE 24A, 1961, 203–230; H.J. Hildebrandt, Die Römerlager von Numantia, in: Madrider Mitteilungen 20, 1979, 238–271; R. López Melero, Viriatus Hispaniae Romulus. FS E. Ripoll Perello 2, 1988, 247–261; A. Schulten, Iberische Landeskunde I, ²1974; H. Simon, Roms Kriege in Spanien 154–133 v. Chr. (Frankfurter Wissenschaftliche Beiträge 11), 1962; G.V. Sumner, Proconsuls and Provinciae in Spain, 218/17–196/95, in: Arethusa 3/4, 1970/71, 85–112; Id., Notes on Provinciae in Spain (197–133 B.C.), in: CPh 72, 1977, 126–130; A. Tovar, Iberische Landeskunde II.,1–3, 1974/76/89.
Repulsion of the Cimbri and Teutoni: A. Degrassi, Per quale via i Cimbri calarono nella Val Padana? in: Id., Scritti vari di antichità, 1962, 991–992; K.-P. Johne, Die Römer an der Elbe, 2006; E. Koestermann, Der Zug der Kimbern, in: Gymnasium 76, 1969, 310–329; R.G. Lewis, Catulus and the Cimbri 102 B.C., in: Hermes 102, 1974, 90–109; R. Loose, Kimbern am Brenner? Ein Beitrag zur Diskussion des Alpenübergangs der Kimbern 102–101 v. Chr., in: Chiron 2, 1972, 231–252; B. Luiselli, Quam de Roma imaginem gentes ad septentrionem spectantes sibi finxerint, in: Romanobarbarica 15, 1998, 75–95; T. Trazaska-Richter, Furor Teutonicus, 1991.

Page 150

Sources

Rome's rise to 'world' power: Pol. 1,1,5; 1,2,7; 3,1,10; *First Illyrian War*: Pol. 2,4,7–6,7; 2,8–9; App. Ill. 7,18; *Second Illyrian War*: Pol. 3,16–19; 7,9,14; *First Macedonian War*: alliance with Hannibal: StV III 528; alliance with the Aetolians: 536; Treaty of Phoenice: StV III 543; *Second Macedonian War*: Liv. 31,6,3; 31,8,1; Liv. 32,3,2–7; StV III 547; '*three fetters of Greece*': Pol. 18,11,4–5; Liv. 32,37,3–4; Str. 9,4,15; *Third Macedonian War*: Delphic Amphictyony: Syll.³ 636; Achaean, Aetolian, Boeotian Leagues: Liv. 42,12,1–2; 5–7; App. Mac. 11,2; *truce*: Liv. 42,38,8–43,3; *Macedonian Revolt under Philip VI*: Liv. Per. 49–50; Diod. Sic. 31,40a; 32,15; Zon. 9,28; *Pseudophilippus*: Pol. 36,9,1 f.; 36,17,13; *Achaean War*: Liv. Per. 51–52; Diod. Sic. 32,26,3–5; Vell. Pat. 1,13,1; Paus. 7,16; Iust. 34,2,1–6; Zon. 9,28; 9,31; *Achaea added to province of Macedonia*: Syll.³ 683; SEG 3,378; IGR 1,118.

Literature

First and Second Illyrian Wars: M.A. Levi, Le cause della guerra romana contro gli Illiri, in: PdP 28, 1973, 317–325; W. Pajakowski, Niedergang des Staates der Päonen. Umstände, Chronologie und Folgen, in: Eos 72, 1984, 269–283; *First Macedonian War*: J. Bibauw, La paix de Phoinikè, dernière KOINH EIRHNH de l' histoire grecque?, in: J. Bibauw (ed.), FS M. Renard 2 (Collection Latomus 102), 1969, 83–90; R.J. Sklenár, Sources and Individuality in Two Passages of Livy, in: Historia 53, 2004, 302–310; *Second Macedonian War*: R.M. Errington, The Alleged Syro-Macedonian Pact and the Origins of the Second Macedonian War, in: Athenaeum 49, 1971, 336–354; N.G.L. Hammond, The Opening Campaigns and the Battle of the Aoi Stena in the Second Macedonian War, in: JRS 56, 1966, 39–54; L. Raditsa, Bella Macedonica, in: ANRW I 1, 1972, 564–589, here 564–576; *Third Macedonian War*: A.M. Eckstein, Rome, the War with Perseus, and Third Party Mediation, in: Historia 37, 1988, 414–444; L. Raditsa, Bella Macedonica, in: ANRW I 1, 1972, 564–589, here 576–589; F.W. Walbank, The Causes of the Third Macedonian War: Recent Views, in: Archaia Makedonia 2, 1977, 81–94; H.-U. Wiemer, Der Beginn des 3. Makedonischen Krieges, in: Historia 53, 2004, 22–37; A. Ziolkowski, The Plundering of Epirus in 167 B.C. Economic Considerations, in: PBSR 54, 1986, 69–80; *Macedonian Revolt under Philip VI*: E.S. Gruen, The Hellenistic World and the Coming of Rome, 1984, 431–433; J.M. Helliesen, Andriscus and the Revolt of the Macedonians 149–148 B.C., dissertation, Wisconsin 1968; P.A. McKay, Studies in the History of the Republican Macedonia, dissertation, Berkeley 1964; M.G. Morgan, Metellus Macedonicus and the Province Macedonia, in: Historia 18, 1969, 422–446; K. Rosen, Andriskos. Milesische Geschichten und makedonische Geschichte, in: W. Will (ed.), Alexander der Grosse, 1998, 117–130; *Achaean War*: E. Badian Cicero and the Commission of 146 B.C., in: J. Bibauw (ed.), FS M. Renard 1 (Collection Latomus 101), 1969, 54–65; A. Fuks, The bellum Achaicum and its Social Aspects, in: JRS 90, 1970, 78–89; D. Knoepfler, L. Mummius Achaicus et les cités du golfe euboïque: à propos d' une nouvelle inscription d' Érétrie, in: Museum Helveticum 48, 1991, 252–280; Y.Z. Tzifopoulos, Mummius' Dedications at Olympia and Pausanias' Attitude to the Romans, in: Greek Roman and Byzantine Studies 34, 1993, 93–100; J. Wiseman, Corinth and Rome I. 228 B.C. – A.D. 267, in: ANRW II 7,1, 1979, 438–548, here 491–496.

Page 152

Cf. table below, p. 268

Sources

Syrian War: Liv. 37,1–7; 37,33–44; Pol. 21,4f.; 21,11–15; Diod. Sic. 25,5–10; Val. Max. 5,5,1; Frontin. Str. 4,7,30; Flor. Epit. 1,24,14–18; App. Syr. 23; 28f.; Iust. 31,7f.; Eutr. 4,4; Zon. 9,20; Oros. 4,20,22; *Raid of Cn. Manlius Vulso*: Pol. 21,33–39; Liv. 38,12–27; Diod. Sic. 29,12f.; *Rome's war with Eumenes III*: Liv. Per. 58f.; Str. 14,1,38; Vell. Pat. 2,4,1; Iust. 36,4–37,1; Flor. Epit. 1,35; *resolution of the people of the city of Pergamum (133 BC)*: OGIS 338; *honorific inscription for Menodorus*: M. Wörrle, Pergamon um 133 v. Chr., in: Chiron 30, 2000, 543–576; *honorific inscription for Apollonius*: P. Briant, P. Brun, E. Varinlioglu, Une inscription inédite de Carie et la guerre d' Aristonicos, in: A. Bresson, R. Descat (ed.), Les cités d' Asie Mineure occidentale au IIᵉ siècle a.C., 2001, 241–259; *honorific inscription for Andronicus*: F. Canali De Rossi, Attalo III e la fine della dinastia pergamena. Due note epigrafiche, in: Les Études Anciennes 31, 1999, 83–93; *Senatus Consultum Popillianum (132 BC)*: F. Daubner, Bellum Asiaticum, 2003, 275f.; Senatus Consultum de agro Pergameno (129 BC): R.K. Sherk, Documents from the Greek East. Senatus Consulta and Epistulae to the Age of Augustus, 1969, 12; *L. Tremelius Scrofa*: Liv. Per. 53; Eutr. 4,15.

Literature

Syrian War: J.P.V.D. Balsdon, L. Cornelius Scipio: A Salvage Operation, in: Historia 21, 1972, 224–234; J.D. Grainger, Antiochos III in Thrace, in: Historia 45, 1996, 329–343; É. Will, Histoire politique du monde hellénistique 2, ²1982, 178–240; *Campaign of the Roman consul Cn. Manlius Vulso against the Galatians*: J.G. Grainger, The Campaign of Cn. Manlius Vulso in Asia Minor, in: AS 45, 1995, 23–42; H.K. Kandlbinder, Die historische Bedeutung von Gnaeus Manlius Vulso, dem römischen Consul des Jahres 189 vor Chr., 1956; B. Pagnon, Le récit de l' expédition de Cn. Manlius Vulso contre les Gallo-Grecs et de ses prolongements dans le livre 38 de Tite-Live, in: Les Études Classiques 50, 1982, 115–128; G. Zecchini, Cn. Manlio Vulsone e l' inizio della corruzione a Roma, in: Contributi dell' Istituto di storia Antica dell' Università di Sacro Cuore 8, 1982, 159–178; *Rome's war with Eumenes III*: J.P. Adams, Aristonikos and the Cistophoroi, in: Historia 29, 1980, 302–314; M. Basile, Le città greche ed Aristonico, in: Seia 2, 1988, 104–116; T.R.S. Broughton, Stratonicea and Aristonicus, in: CPh 29, 1934, 252–254; K. Buraselis, Colophon and the War of Aristonicus, in: I. Velissaropoulou-Karakosta et al. (ed.), FS I. Trantafyllopoulou, 2000, 181–207; F. Carrata Thomes, La rivolta di Aristonico e le origini della provincia Romana d' Asia, 1968; F. Collins III, The Revolt of Aristonicus, dissertation, Univ. of Virginia 1978; F. Daubner, Bellum Asiaticum. Der Krieg der Römer gegen Aristonikos von Pergamon und die Einrichtung der Provinz Asia (Quellen und Forschungen zur antiken Welt 41), ²2006; J. Hopp, Untersuchungen zur Geschichte des letzten Attaliden (Vestigia 25), 1977; M. Kampmann, Aristonicos à Thyatire, in: RN 20, 1978, 38–42; D. Magie, Roman Rule in Asia Minor, 1950; E.S.G. Robinson, Cistophori in the Name of King Eumenes, in: NC 14, 1954, 1–8; M.L. Sanchez Leon, Aristónico: Basileus Eumenes III, in: Hispania Antiqua 13, 1986–1989, 135–157; F. Scardigli, Scharlatane, Hochstapler und Prätendenten in der römischen Republik, in: Antiquité Vivante 55, 2005, 149–171; É. Will, Histoire politique du monde hellénistique II, ²1982.

Page 154

Sources

Definition of the colonia: Serv. Aen. 1,12; Cic. Leg agr. passim; Gell. NA 16,13; F. Blume, K. Lachmann, T. Mommsen, A. Rudorff (ed.), Die Schriften der römischen Feldmesser, 2 vols., 1848/52; B. Campbell (ed.), The Writings of the Roman Land Surveyors. Introduction, Text, Translation and Commentary, 2000, 164–203 (*liber coloniarum*); M.H. Crawford (ed.), Roman Statutes 1, 1996, no. 25 (*lex Coloniae Genetivae Iuliae*).

Literature

P.A. Brunt, Italian Manpower 225 B.C.–A.D. 14, 1971; H. Galsterer, Herrschaft und Verwaltung im republikanischen Italien, 1976; Id., Die Kolonisation der hohen Republik und die römische Feldmesskunst, in: O. Behrends, L. Capogrossi Colognesi (eds.), Die römische Feldmesskunst, 1992, 412–443; A. Gonzales, J.-Y. Guillaumin (ed.), Autour des "Libri coloniarum". Colonisation et colonies dans le monde romain. Actes du Colloque International (Besançon, 16–18 octobre 2003), 2006; L. Keppie, Colonisation and Veteran Settlement in Italy 47–14 B.C., 1983; E. Kornemann, Coloniae, RE 4 1900, 510–588.

U. Laffi, La colonizzazione romana nell' età della Repubblica, in: Il fenomeno coloniale dall' Antichità ad oggi (Atti dei Convegni Lincei 189), 2003, 37–52; Id., La colonizzazione romana tra la guerra latina e l' età dei Gracchi: aspetti istituzionali, in: Id., Studi di storia romana e di diritto (Storia e Letteratura, Raccolta di Studi e Testi 206), 2001, 85–111; B. Levick, Roman Colonies in Southern Asia Minor, 1967; H. Papageorgiadou-Bani, The Numismatic Iconography of the Roman Colonies in Greece, 2004; E.T. Salmon, Roman Colonization under the Republic, 1969; Id., The Making of Roman Italy, 1982; A.N. Sherwin-White, The Roman Citizenship, 1973; F. Vittinghoff, Römische Kolonisation und Bürgerrechtspolitik unter Caesar und Augustus, 1950.

Page 158

Sources

First Mithridatic War: Laodicea on the Lycus: Str. 12,8,16; App. Mithr. 78; *Magnesia on the Sipylus*: App. Mithr. 82; *Rhodes*: Memnon FGrH 434 F 1,22,8; App. Mithr. 102 ff.; *'Asiatic Vespers'*: Memnon FGrH 434 F 1,22,9; App. Mithr. 85 f.; Plut. Sulla 24; *Sparta and Thespiae*: Memnon FGrH 434 F 1,22,10; App. Mithr. 112; Athens: App. Mithr. 116–155; Plut. Sulla 12–14; Str. 9,1,20; Paus. 1,20,6 f.; 9,7,4 f.; Liv. Per. 81; Vell. Pat. 2,23,4–6; Eutr. 5,6,2; Oros. 6,2,5; *Chaeronea*: App. Mithr. 161–176; Plut. Sulla 16–19; 23; *Orchomenus*: App. Mithr. 194–202; Plut. Sulla 21; Polyaenus, Strat. 8,9,2; Liv. Per. 82; *Treaty of Dardanus*: App. Mithr. 228–240; Str. 13,1,27; Plut. Sulla 22; 24; Sertorius 23; Memnon FGrH 434 F 25,2; Liv. Per. 83.
Second Mithridatic War: Murena: Cic. Manil. 9; App. Mithr. 264–267; *cessation*: App. Mithr. 280; cf. Cic. Fam. 15,4,6.
Third Mithridatic War – first phase: App. Mithr. 305–325; Plut. Lucullus 9–11; Diod. Sic. 37,22b; Str. 12,8,11; Sall. Hist. 3,26–37; Frontin. Strat. 3,13,6; Oros. 6,2,14–24; *Second phase*: App. Mithr. 360–368; Plut. Lucullus 15–17; Memnon FGrH 434 F 1,30,1; Cic. Manil. 22; Sall. Hist. 4,12; Liv. Per. 97; *confinement*: Memnon FGrH 434 F 1,38,1; *third phase – Tigranocerta*: App. Mithr. 382–388; Plut. Lucullus 27 f.; Memnon FGrH 434 F 1,38,4 f.; Frontin. Strat. 2,1,14; 2,4; *Artaxata*: Plut. Lucullus 31; App. Mithr. 397; *fourth phase*: App. Mithr. 414; Plut. Lucullus 35; Cass. Dio 36,9,1; Cic. Manil. 5; *fifth phase – Nicopolis*: App. Mithr. 458–462; Plut. Pompeius 32; Cass. Dio 36,48 ff.
Inscriptions and coins: J.G.C. Anderson, F. Cumont, H. Grégoire, Recueil des inscriptions grecques et latines du Pont et de l' Arménie 3,1, 1910; W.H. Waddington, E. Babelon, T. Reinach, Recueil générale des monnaies grecques d' Asie Mineure 1,1, ^{2}1925.

Literature

L. Ballesteros Pastor, Mitrídates Eupátor, 1996; Id., L' an 88 av. J.-C. Présages apocalyptiques et propagande idéologique, in: Dialogues d' Histoire Ancienne 25, 1999, 83–90; D. Magie, Roman Rule in Asia Minor, 1951; A. Mastrocinque, Studi sulle guerre Mitridatiche (Historia monograph 124), 1999; B.C. McGing, The Foreign Policy of Mithridates VI Eupator, King of Pontus, 1986; E. Olshausen, Pontos (2), RE Suppl. 15, 396–442; Id., Mithradates VI. und Rom, in: ANRW I 1, 1972, 806–815; Id., J. Wagner, TAVO B V 6; T. Reinach, Mithradates Eupator, 1895 (ND 1973).

Page 160

Sources

Bithynia et Pontus: Amm. Marc. 14,8,10; 14,8,12; App. Mithr. 470; 490; 494–501; 555–561; 565; 576; App. Syr. 250–254; 307; Caes. B Civ. 3,4,5; Caes. Bell. Alex. 65,4; Cass. Dio 36,50; 37,7; 37,14–16; Cic. Ad Q. fr. 1,1,3 f.; 2,11,2; Cic. Att. 2,16,2; Cic. Balb. 13; Cic. Fam. 15,1,2; Cic. Flacc. 67; 69; Cic. Leg. agr. 2,40; Cic. Prov. 10; Cic. Sest. 58; Diod. Sic. 37,7a; 37,14; 37,20; 37,40,2–4; Eutr. 6,14,1 f.; Jos. Ant. Iud. 13,395; 14,35–59; 14,72–88; 14,156; Jos. Bl. 1,132–137; 1,143; 1,153–160; 1,166; 1,201; 5,396; Iust. 40,2,3 f.; Memnon 18,10; 29,4; Oros. 6,4,7; 6,6,2; Plin. HN 37,16; Plin. Epist. 10,79 f.; 10,112,1; 10,114 f.; Plut. Pompeius 33 f.; 36–42; 45; 83; Plut. Crassus 21; Porph. FGrH 260 F 32,27; Ps.-Sall. in Tull. 2,2; 8,19; Str. 11,2,10–19; 11,8,4–6; 12,1,2–12,5,4; 14,5,17–21; 14,5,9–12; 16,1,28–16,2,42; Tac. Hist. 5,5,9; Vell. Pat. 2,40,1.
Syria: Amm. Marc. 14,8,12; App. Mithr. 553–560; App. Syr. 252–255; Caes. Bell. Alex. 65,4; Cass. Dio 37,14–16; Cic. Ad Q. fr. 2,11,2; Cic. Balb. 13; Cic. Flacc. 67; 69; Cic. Mur. 34; Cic. Prov. 10; Diod. Sic. 34,1; 40,2; Eutr. 6,14,2; Jos. Ant. Iud. 14,35–88; Jos. Bl. 1,132–166; 5,106; 5,392–396; Jos. Ap. 2,80; 2,91; Plin. HN 7,97; 37,14; Plut. Pompeius 39–42; Porph. FGrH 260 F 32,27; Str. 16,2,8–16; 16,40 f.; Tac. Hist. 5,5,5; 5,9,1.
Cilicia: App. Mithr. 96; 115; 117; Cass. Dio 36,37,6; Cic. Att. 5,15 ff.; Flor. Epit. 1,41,12 f.; Plut. Pompeius 25; 28; 45; Str. 14,3,3; 14,5,2; Vell. Pat. 2,31,2; 2,32,4.
Pirates' War 67 BC: App. Civ. 5,336; 5,400; App. Mithr. 434–445; Cass. Dio 36,16; 36,24–37; 36,45–52; Cic. Flacc. 29; 31; Cic. Manil. 35; 44; 52; Lucan. 1,346; 2,578 f.; 2,635 f.; 3,228; 8,26; Plin. HN 7,97; Plut. Pompeius 25–29; Str. 8,7,7; 14,3,3; 14,5,2; 14,5,8; Vell. Pat. 2,32,4 f.

Literature

A. Avidov, Were the Cilicians a nation of pirates?, in: Mediterranean Historical Review 10, 1997, 5–55; M. Benabou, Rome et la police de mers au 1er siècle avant J.C. La répression de la piraterie Cilicienne, in: M. Galley, L. Sébai (ed.), L' homme méditerranéen et la mer. Actes du troisième Congrès international d' études des cultures de la Méditerranée occidentale (Jerba, avril 1981), 1985, 60–69; A. Dreizehnter, Pompeius als Städtegründer, in: Chiron 5, 1975, 213–245; P. Freeman, The province of Cilicia and its origins, in: Id., D. Kennedy, The Defence of the Roman and Byzantine East, Proceedings of a Colloquium held at the University of Sheffield in April 1986, 1986, 253–275; K.M. Girardet, Imperium und provinciae des Pompeius seit 67 v. Chr., in: Cahiers du Centre G. Glotz 3, 1992, 177–188; M. Heil, Einige Bemerkungen zum Zollgesetz aus Ephesos, in: EA 17, 1991, 9–18; W. Leschhorn, Antike Ären, 1993, 78–200; F. Lewis, A History of Bithynia under Roman Rule 74 B.C. – 14 A.D., 1973; C. Marek, Stadt, Ära und Territorium in Pontus-Bithynia und Nord-Galatia (Istanbuler Forschungen 39), 1993, 26–46, 27 Anm. 219 (Literatur); E. Maróti, Die Piraterie vor dem römischen Bürgerkrieg, 1972 (Hungarian); T.B. Mitford, Roman Rough Cilicia, in: ANRW II 7.2, 1980, 1230–1261; J.-P. Rey-Coquais, Syrie Romaine de Pompée Dioclétien, in: JRS 68, 1978, 44–73; K. Strobel, Mithradates VI., in: Ktema 21. Hommages ed. Frézouls 2, 1996 (1998), 55–94; J. Wagner, W. Stahl, TAVO B V 7 (1983); G. Wirth, Pompeius – Armenien – Parther. Mutmaßungen zur Bewältigung einer Krisensituation, in: BJ 183, 1983, 1–60; Id., Pompeius im Osten, in: Klio 66, 1984, 574–580.

Page 162

Sources

Pentapolitana regio: Plin. HN 5,31; imperium *of Pompey*: Vell. Pat. 2,31,2–4; *dispute between Metellus and Pompey*: Liv. Per. 99.

Literature

F. Chamoux, La Cyrénaïque, des origines à 321 a.C., d' après les fouilles et les travaux récents, in: Society for Libyan Studies. Annual Reports 20, 1989, 63–70; R.G. Goodchild, Tabula Imperii Romani, Blatt H.I. 34, Cyrene, 1954; R. Haensch, Capita provinciarum. Statthaltersitze und Provinzialverwaltung in der römischen Kaiserzeit, 1997; A. Laronde, La Cyrénaïque romaine, des origines à la fin des Sévères (96 av. J.-C. – 235 ap. J.C.), in: ANRW II 10,1, 1988, 1006–1064; A. Lintott, Imperium Romanum, 1993; E. Malamut, Les îles de l' Empire Byzantin: VIIIe-XIIe siècles, 2 vols., 1988; P. Romanelli, La Cirenaica Romana (96 a.C. – 642 d.C.), 1943; I.F. Sanders, Roman Crete, 1982.

Page 164

Cf. table below, p. 268

Page 166

Cf. table below, p. 265

Sources

Via Domitia: CIL 17,2, 294 & pp. 75–106; Cic. Font. 18; *Narbo Martius*: Cic. Font. 13; Vell. Pat. 1,15,5; 2,7,8; Eutr. 4,23; *Act of State of 13 Jan. 27 BC*: Cass. Dio 53,12,5–7; 13,1; Str. 17,3,25; Suet. Aug. 28; 47.

Literature

T.D. Barnes, The New Empire of Diocletian and Constantine, 1982; R. Chevallier, G. Narbonensis, in: ANRW II 3, 1975, 686–828; Id., Gallia Lugdunensis, in: ANRW II 3, 1975, 860–1060; C. Delaplace, J. France, Histoire des Gaules, 1995; P.M. Duval, Gallien, 1979; R. Haensch, Capita provinciarum, 1997; J.J. Hatt, Histoire de la Gaule romaine, 1970; M. Pichon, Le transport par voie navigable – l' exemple du site de Tendu (Indre), in: Gallia 59, 2002, 83–88; J. Prieur, L' histoire des régions alpestres (Alpes Maritimes, Cottiennes, Graies et Pennines) sous le haut-empire romain (Ier–IIIe siècle après J.C.), in: ANRW II 5, 1975, 630–656; M.-T. Raepsaet-Charlier, G. Raepsaet, Gallia Belgica et Germania Inferior, in: ANRW II 4, 1975, 3–299; D. Roman, Y. Roman, Histoire de la Gaule, 1997; C.-M. Ternes, Die Provincia Germania Superior im Bilde der jüngeren Forschung, in: ANRW II 5, 1975, 721–1260; F. Zinn, Untersuchungen zu Wagenfahrtdarstellungen auf provinzialrömischen Grabdenkmälern, in: Kölner Jahrbuch für Vor- und Frühgeschichte 34, 2001, 141–266; TIR M 31, 1975.

Page 168

Cf. table below, p. 269

Page 170

Literature

A. Alföldi, J.B. Giard, Guerre civile et propagande politique. L' émission d' Octave au nom du Divus Julius (41–40 av. J.C.), in: Numismatica e Antichità classiche 13, 1984, 147–161; E. Badian, M. Lepidus and the Second triumvirate, in: Arctos 25, 1991, 5–16; J. Bleicken, Zwischen Republik und Prinzipat, 1990, 27–85; A.B. Bosworth, Asinius Pollio and Augustus, in: Historia 21, 1972, 441–473; H. Buchheim, Die Orientpolitik des Triumvirn M. Antonius, 1960; J. Ermatinger, ILS 77 and 78. The End of the Second Triumvirate, in: Historia 42, 1993, 109–110; D. Fishwick, On the Origins of Africa Proconsularis, in: Antiquités Africaines 30, 1994, 57–80; B. Flory Marleen, A Note on Octavian' s felicitas, in: RhM 135, 1992, 283–289; A.M. Gowing, Lepidus, the Proscriptions and the Laudatio Turiae, in: Historia 41, 1992, 283–296; P. Jal, La guerre civile à Rome. Étude littéraire et morale (Publ. Fac. des Lettres de Paris Sér. Recherches; VI), 1963; L. Keppie, Colonisation and Veteran Settlement in Italy 47–14 B.C., 1983; L. Labruna, Civitas misera, 1996; E. Maróti, Die Rolle der Seeräuber unter den Anhängern des Sextus Pompeius, in: H. Diesner et al. (ed.), Sozialökonomische Verhältnisse im Alten Orient und klassischen Altertum, 1961, 208–216; F. Millar, Triumvirate and Principate, in: JRS 63, 1973, 50–67; M. Pani, Lotta politica repubblicana e principato. Schemi di analisi, in: Quaderni di Storia 33, 1991, 177–185; C.B.R. Pelling, Plutarch, Life of Antony, 1988; M. Sordi, La guerra di Perugia e la fonte del libro V dei Bella civilia di Appiano, in: Latomus 44, 1985, 301–316; P. Wallmann, Münzpropaganda in den Anfängen des Zweiten Triumvirats (43/42 v. Chr.), 1977; Id., Triumviri Rei Publicae Constituendae. Untersuchungen zur politischen Propaganda im zweiten Triumvirat (43–30 v. Chr.), 1989; R.D. Weigel, Lepidus, The Tarnished Triumvir, 1992; Id., The Coins Issued by Lepidus as triumvir monetalis, in: Journal of the Society of Ancient Numismatics 5, 1973/74, 51–52.

Page 172

Literature

G.S. ALDRETE, Floods of the Tiber in ancient Rome, 2007; J.C. ANDERSON Jr., The Historical Topography of the Imperial Fora, 1984; B. BRIZZI (ed.), Mura e porte di Roma antica, 1995 ; A. CARANDINI, La nascita di Roma, 1997; F. CASTAGNOLI, Topografia e urbanistica di Roma antica, 1969; F. COARELLI, Il Foro Romano, 2 vols., [2]1986–1992; Id., Campo Marzio, 1997; D. FAVRO, The Urban Image of Augustan Rome, 1996; A.P. FRUTAZ, Le piante di Roma, 1962; E. GJERSTAD, Early Rome, 6 vols., 1953–1973; L. HASELBERGER, J. HUMPHREY (ed.), Imaging Ancient Rome. Documentation, visualization, imagination, 2006; M. HEINZELMANN, Rom. Topographie und Archäologie der Stadt, in: DNP 10, 2001, 1083–1106; H. JORDAN, T. ASHBY, A Topographical Dictionary of Ancient Rome, 1929; M.-J. KARDOS, Topographie de Rome: les sources littéraires latines, 2000; F. KOLB, Rom: die Geschichte der Stadt in der Antike, [2]2002; E. LA ROCCA, I Fori Imperiali, 1995; R. ROSS HOLLOWAY, The Archaeology of Early Rome and Latium, 1996; D. PALOMBI, Tra Palatino ed Esquilino: Velia, Carinae, Fagutal, 1997; J.R. PATTERSON, The City of Rome: From Republic to Empire, in: JRS 82, 1992, 186–215; L. RICHARDSON (jr.), A New Topographical Dictionary of Ancient Rome, 1992; E.M. STEINBY (ed.), Lexicon topographicum Urbis Romae, vols. 1–6, 1993–2000; U. VENTRIGLIA, La geologia della città di Roma, 1971.

Pages 176, 180, 182, 184

Cf. tables below, pp. 270 ff.

Page 188

Cf. tables below, p. 273

Sources

Raetia: L. BAKKER, Der Siegesaltar aus Augusta Vindelicum/Augsburg von 260 n. Chr., in: E. SCHALLMAYER (ed.), Niederbieber, Postumus und der Limesfall, 1996, 7–13; S. SCHUMACHER, Die rätischen Inschriften, 1992; Vell. Pat. 2,39,104; *Amber Road*: Plin. HN 37,45; *Noricum – regnum Noricum*: Vell. Pat. 2,109,5; Suet. Tib. 16,2; *ferrum Noricum*: Ov. Met. 14,712.

Literature

Raetia: K. DIETZ, Okkupation und Frühzeit, in: W. CZYSZ (ed.), Die Römer in Bayern, 1995, 18–99; P. GLEIR- SCHER, Die Räter, 1991; P. KOS, Sub principe Gallieno ... amissa Raetia? Numismatische Quellen zum Datum 259/260 n. Chr. in Raetien, in: Germania 73, 1995, 131–144; R. HAENSCH, Capita provinciarum, 1997; R. ROLLINGER, Raetiam autem et Vindelicos ... subiunxit provinciae. Oder: Wann wurde Raetien ... als römische Provinz eingerichtet?, in: Id., P. W. HAIDER (ed.), Althistorische Studien im Spannungsfeld zwischen Universalgeschichte und Wissenschaftsgeschichte, 2001, 267–315; F. SCHÖN, Der Beginn der römischen Herrschaft in Rätien, 1986; G. WALSER, Die römischen Straßen und Meilensteine in Raetien, 1983; J. WILKES, Les provinces danubiennes, in: Rome et l' intégration de l' Empire, 44 av. J.-C.–260 ap. J.-C. 2, in: Nouvelle Clio, 1998, 53–57, 231–297.
Noricum: G. ALFÖLDY, Noricum, 1974; A. BETZ, E. WEBER, Aus Österreichs römischer Vergangenheit, [2]1990; R. BRATOŽ, Severinus von Noricum und seine Zeit, 1983; R. GÖBL, Die Münzprägung der norischen Fürsten, in: J. GRAENMAYER (ed.), Die Kultur der Kelten, 1989, 54–66; H. GRASSL, Norisches Eisen aus dem Burgenland?, in Römisches Österreich 15/6, 1987/8, 83–88; Id., Zur Problematik des Ferrum Noricum, in: Berichte des 17. Öster- reichischen Historikertags 1989, 54–57; Id., Noricum im Bürgerkrieg des Jahres 196–197 n. Chr., in Römisches Österreich 2, 1974, 7–10; R. HAENSCH, Capita provinciarum, 1997; M. LOVENJAK, Die neuen Meilensteine von Celje (Celeia, Noricum) aus den Jahren 161, 214 und 218 n. Chr., in: ZPE 146, 2004, 205–210; W. LUGS, "Ripa" – der römische Grenzschutz an der Donau in Noricum von Augustus bis zu den Markomannenkriegen, 2002; J. OTT, Die norischen Militärdiplome und ein neuer Statthalter der Provinz, in: Rivista di Storia Antica 25, 1995, 91–110; M. ŠAŠEL KOS, From the Tauriscan Gold Mine to the Goldenhorn, in: Studia mythologica Slavica 1, 1998, 169–182; G. SPERL, Zum Stand der Erforschung des Ferrum Noricum, in: H. HEFTNER, K. TOMASCHITZ (ed.), Ad Fontes! FS G. Dobesch, 2004, 961–976; H. STRAUBE, Ferrum Noricum und die Stadt auf dem Magdalensberg, 1996; G. WINKLER, Die römischen Straßen und Meilensteine in Noricum, 1985; Id., Noricum und Rom, in: ANRW II 6, 1977, 183–262.
Pannonia: G. ALFÖLDY, B. LÖRINCZ, Ein neues Militärdiplomfragment und ein neuer Statthalter der Provinz Pannonia Superior, in: ZPE 139, 2002, 211–218; L. BARKÓCZI (ed.), Die römischen Inschriften Ungarns 1–4, 1972–1991; S. DUSANIC, Fragment of a Severan Auxiliary Diploma. Notes on a Variety of the 'Two-Province' Diplomata, in: ZPE 122, 1998, 219–228; J. FITZ, Der römische Limes in Ungarn, 1976; Id., Die Verwaltung Pannoniens in der Römerzeit 1–4, 1993–1995; R. HAENSCH, Capita provinciarum, 1997; B. LÖRINCZ, Ein neues Militärdiplom aus Pannonia Inferior, in: Tyche 14, 1999, 173–175; M. MIRKOVIC, Euphrata et Romano consulibus auf einem neuen Militärdiplom von der unteren Sava, in: ZPE 133, 2000, 286–290; A. MÓCSY, Pannonia, RE Suppl. 9, 516–776; Id., Pannonia and Upper Moesia, 1974; P. OLIVA, Pannonia and the Onset of Crisis in the Roman Empire, 1962; B. PFERDEHIRT, Ein neues Militärdiplom für Pannonia inferior vom 11.8.193 n. Chr., in: Archäologisches Korrespondenzblatt 32, 2002, 247–260; Id., Vier neue Militärdiplome im Besitz des Römisch-Germanischen Zentralmuseums, in: Archäologisches Korrespondenzblatt 31, 2001, 261–279; M.M. ROXAN, Two Complete Diplomas of Pannonia Inferior – 19 May 135 and 7 Aug. 143, in: ZPE 127, 1999, 249–273; M. ŠAŠEL KOS, The Defensive Policy of Valentinian I in Pannonia. A Reminiscence of Marcus Aurelius?, in: R. BRATOŽ (ed.), Westillyricum und Nordostitalien in der spätrömischen Zeit, 1996, 145–175; S. VISY, Der pannonische Limes in Ungarn, 1988; P. WEISS, Zwei vollständige Konstitutionen für die Truppen in Noricum (8. Sept. 79) und Pannonia inferior (27. Sept. 154), in: ZPE 146, 2004, 239–254. Also TIR L 34 Budapest, 1968; TIR L 33 Tergeste, 1961; TIR M 33 Praha, 1986.

Page 190

Sources

Tac. Hist. 2,69; 2,79; 4,12–37; 4,54–79; 5,14–26; Jos. Bl. 1,5; 7,75–88; Cass. Dio 66,3; cf. Pliny HRR 2,109–112; Plut. Galba; Plut. Otho; of the numerous inscriptions relevant to the history of the Rhine legions, the following are listed by way of example: ILS 983 (tombstone of C. Dillius Vocula), cf. ILS 1006; C.B. RÜGER, Ein Siegesdenkmal der Legio VI Victrix, in: BJ 179, 1979, 187 ff.

Literature

L. BESSONE, La rivolta batavica e la crisi del 69 d.C., 1972; P.A. BRUNT, Tacitus on the Batavian revolt, in: Latomus 19, 1960, 494–517 (= Roman Imperial Themes, 1990); A. W. BYVANCK, Nederland in den Romeinschen tijd 1, 1943, 220–279; H.-W. GOETZ, K.-W. WELWEI (ed.), Altes Germanien 2 (Ausgewählte Quellen zur deutschen Geschichte des Mittelalters vol. 1a), 1995, 170–261; F. MÜNZER, Die Quelle des Tacitus für die Germanenkriege, in: BJ 4, 1899, 67–111; O. SCHMITT, Anmerkungen zum Bataveraufstand, in: BJ 193, 1993, 141–160; D. TIMPE, Tacitus und der Bataveraufstand, in: T. SCHMITT, W. SCHMITZ, A. WINTERLING (ed.), Gegenwärtige Antike – antike Gegenwarten. FS R. Rilinger, 2005, 151–187; R. URBAN, Der 'Bataveraufstand' und die Erhebung des Iulius Classicus, 1985; G. WALSER, Rom, das Reich und die fremden Völker in der Geschichtsschreibung der frühen Kaiserzeit. Studien zur Glaubwürdigkeit des Tacitus, 1951, 86–128.

Page 194

Literature

General: H.E. HERZIG, Probleme des römischen Straßenwesens, in: ANRW II 1, 1974, 593–648; E. OLSHAUSEN, H. SONNABEND (ed.), Zu Wasser und zu Land. Verkehrswege in der antiken Welt, 2002; M. RATHMANN, Viae Publicae, in: DNP 12.2, 2002, 164–171; Id., Untersuchungen zu den Reichsstraßen in den westlichen Provinzen des Imperium Romanum, 2003; H.C. SCHNEIDER, Altstraßenforschung, 1982; T.P. WISEMAN, Roman Republican Road-Building, in: PBSR 38, 1970, 122–135.
Italien: R. CHEVALLIER, Les voies romaines, 1997; R. LAURENCE, The Roads of Roman Italy, 1999; L. QUILICI, S. QUILICI GIGLI (ed.), Strade romane, ponti e viadotti, 1996.
Corsica, Sardinia and Sicilia: P. MELONI, I miliari Sardi e le strade romane in Sardegna, in: Epigraphica 15, 1953, 20–50; R. REBUFFAT, Les stations Corses de l' itinéraire Antonin, in: Annales de la Faculté des Lettres et Sciences Humaine d' Aix 43, 1967, 217–227; G.P. VERBRUGGHE, Sicilia, 1976.
Balkans: A. MÓCSY, Pannonia VI. Straßen und Verkehr, in: RE Suppl. 9, 1962, 653–667; G. STADTMÜLLER, Das römische Straßennetz der Provinzen Epirus Nova und Epirus Vetus, in: Historia 3, 1954, 236–251; E. WEBER, Die römischen Meilensteine aus dem österreichischen Pannonien, in: JÖAI 49, 1968–1971, 121–145.
Greece: J. CHRISTIEN, Les liaisons entre Sparte et son territoire malgré l' encadrement montagneux, in: J.-F. BERGIER (ed.), Montagnes, Fleuves Forêts dans l' Histoire, 1989, 18–44; P. FUNKE, Grenzfestungen und Verkehrsverbindungen in Nordost-Attika, in: P. FLENSTED-JENSEN et al. (ed.), Polis & Politics. FS M.H. Hansen, 2000, 121–131; F. GSCHNITZER, Straßen, Wege und Märsche in Xenophons Hellenika, in: E. OLSHAUSEN, H. SONNABEND (ed.), Zu Wasser und zu Land. Verkehrswege in der antiken Welt, 2002, 202–208; E.W. KÄSE, Mycenaean Roads in Phocis, in: AJA 77, 1973, 74–77; H. LOHMANN, Antike Straßen und Saumpfade in Attika und der Megaris, in: E. OLSHAUSEN, H. SONNABEND (ed.), Zu Wasser und zu Land. Verkehrswege in der antiken Welt, 2002, 109–147; A. MÜLLER, Megarika, in: BCH 108, 1984, 249–266; K. TAUSEND, Ein antiker Weg über den Chelmos? in: JÖAI 63, 1994, Suppl., 41–52; Id., Von Artemis zu Artemis? Der antike Weg von Lousoi nach Pheneos in: JÖAI 64, 1995, Suppl., 1–20; Id., Der antike Weg von Pheneos nach Orchomenos, in: JÖAI 67, 1998, 109–116; Id., Verkehrswege der Argolis, 2006; E. VANDERPOOL, Roads and Forts in Northwestern Attica, in: Classical Antiquity 11, 1978, 227–245.
Alps: H.U. INSTINSKY, Septimius Severus und der Ausbau des raetischen Straßennetzes, in: Klio 31, 1938, 33–50; G. WALSER, Summus Poeninus, 1984; Id. Die römischen Straßen und Meilensteine in Raetien, 1983; G. WINKLER, Die römischen Straßen und Meilensteine in Noricum: Österreich, 1985.
Asia Minor: T. BEKKER-NIELSEN, The Roads of Ancient Cyprus, 2004; D.H. FRENCH, The Roman Roads and Milestones of Asia Minor, 2 fasc., 1981/88; Id., The Roman Road System of Asia Minor, in: ANRW II 7,2, 1980, 698–729; I.W. MACPHERESON, Roman Roads and Milestones of Galatia, in: AS 4, 1954, 111–120; M.H. SAYAR, Straßenbau in Kilikien unter den Flaviern, in: EA 20, 1992, 57–62; Id., Antike Straßenverbindungen Kilikiens in der römischen Kaiserzeit, in: E. OLSHAUSEN, H. SONNABEND (ed.), Zu Wasser und zu Land. Verkehrswege in der antiken Welt, 2002, 452–473; E. OLSHAUSEN, Pontica 4. Das römische Straßennetz in Pontos, in: Orbis Terrarum 5, 1999, 93–113.
Iberian Peninsula: E. ALVAREZ, Vías romanas de Galicia, in: Zephyrus 11, 1960, 5–103; G. CASTELLVI et al. (ed.), Voies romaines du Rhône à l' Èbre, 1997; J. LOSTAL PROS, Los miliarios de la Provincia Tarraconense, 1992; R.C. KNAPP, La Via Heraclea en el occidente, in: Emerita 54, 1986, 103–122; P. SILLIERES, Les voies de communication de l' Hispanie méridionale, 1990.
North Africa: S. AURIGEMMA, Pietre Miliari Tripolitane, in: Rivista della Tripolitania 2, 1925, 3–21, 135–150; M. EUZENNAT, Les voies romaines du Maroc dans l' Itinéraire Antonin, in: M. RENARD (ed.), FS

A. Grenier 2, 1962, 595–610; R.G. Goodchild, The Roman Roads and Milestones in Tripolitania, 1948; Id., Roman Milestones in Cyrenaica, in: PBSR 18, 1950, 83–91; Id., Roman Roads of Libya and their Milestones, in: F.F. Gadallah (ed.), Libya in History, 1971, 155–171; J. Marcillet-Jaubert, Bornes milliaires de Numidie, in: Antiquités africaines 16, 1980, 161–184; P. Salama, Les voies de l' Afrique du Nord, 1951; Id., Bornes milliaires de l' Afrique proconsulaire, 1987.
Egypt and Syria: M. Avi-Yonah, The development of the Roman road system in Palaestine, in: IEJ 1, 1950/51, 54–60; T. Bauzou, Les routes romaines de Syrie, in: Id., W. Orthmann (ed.), Archéologie et histoire de la Syrie 2, 1989, 205–221; D.A. Dorsey, The Roads and Highways of Ancient Israel, 1991; B.H. Isaac, Milestones in Judaea, in: Palestine Exploration Quarterly 110, 1978, 47–60; Id., I. Roll, Roman roads in Judaea, 2 vols., 1982/96; D. Kennedy, Roman roads and routes in north-east Jordan, in: Levant 29, 1997, 71–93; P. Thomsen, Die römischen Meilensteine der Provinzen Syria, Arabia und Palaestina, in: Zeitschrift des Deutschen Palästina-Vereins 40, 1917, 1–103.
Britain: I.D. Margary, Roman roads in Britain, ²1967; A.L.F. Rivet, The British Section of the Antonine Itinerary, Britannia 1, 1970, 34–82; J. P. Sedgley, The Roman Milestones of Britain, 1975; M. Zahrnt, Die frühesten Meilensteine Britanniens und ihre Deutung, in: ZPE 73, 1988, 195–199.
Gaul: G. Castellvi et al. (ed.), Voies romaines du Rhône à l' Èbre, 1997; I. König, Die Meilensteine der Gallia Narbonensis, 1970; M. Rathmann, Untersuchungen zu den Reichsstraßen in den westlichen Provinzen des Imperium Romanum, 2003.

Page 202

Sources

Route information of the *itineraria* and the *Tabula Peutingeriana* is collated in: K. Miller, Itineraria Romana, 1916; *Diocletian's Price Edict*: edition and commentary: S. Lauffer, Diokletians Preisedikt, 1971.

Literature

D.J. Blackman, Ancient Harbours in the Mediterranean, in: International Journal of Nautical Archaeology 11, 1982, 79–104, 185–211; L. Casson, Ships and Seamanship in the Ancient World, 1972; U. Fellmeth, Eine wohlhabende Stadt sei nahe… Die Standortfaktoren in der römischen Agrarökonomie im Zusammenhang mit den Verkehrs-und Raumordnungsstrukturen im römischen Italien, 2002, 81–146; Id., Handelsgeographie, in: Historicum 2, 2004, 20–25; T. Frank, An Economic Survey of Ancient Rome 1–6, 1933 ff.; P. Garnsey, K. Hopkins, C.R. Whittacker (ed.), Trade in the Ancient Economy, 1983; F.M. Heichelheim, Wirtschaftsgeschichte des Altertums 2, 1938, 458–550, 690–722, 796–813; D. Höckmann, Antike Seefahrt, 1985; K. Lehmann-Hartleben, Die antiken Hafenanlagen des Mittelmeeres (Klio suppl. 14) 1923; F. De Martino, Wirtschaftsgeschichte des alten Rom, 1991; F. Meijer, O. von Nijf, Trade, Transport, and Society in the Ancient World. A Sourcebook, 1992; E. Olshausen, Einführung in die Historische Geographie der Alten Welt, 1991; M.I. Rostovtzeff, Gesellschaft und Wirtschaft im römischen Kaiserreich, 1931; J. Rougé, Recherches sur l' organisation du commerce maritime en Méditerranée sous l' empire romain, 1966; Id., La marine dans l' Antiquité, 1975; O. Schlippschuh, Die Händler im Römischen Kaiserreich in Gallien, Germanien und den Donauprovinzen Rätien, Noricum und Pannonien, 1974; F. Vittinghoff (ed.), Handbuch der Europäischen Wirtschafts- und Sozialgeschichte 1. Römische Kaiserzeit, 1990, esp. E. Frézouls, 459–471 (Gaul), J.M. Blázquez, 529–533 (Iberian Peninsula), J. Sasel, 560–562, 565–566 (Alpine regions), J. Sasel, 576 (Dalmatia), A. Mócsy, 591–594 (Dalmatia); H. Warnecke, Schifffahrt, Schifffahrtswege, Schiffbarkeit, in: H. Sonnabend (ed.), Mensch und Landschaft in der Antike, Lexikon zur Historischen Geographie der Alten Welt, 1999, 438–449; L. Wierschowski, Die regionale Mobilität in Gallien nach den Inschriften des 1. bis 3. Jahrhunderts n. Chr. Quantitative Studien zur Sozial- und Wirtschaftsgeschichte der westlichen Provinzen des Römischen Reiches, 1995.

Page 204

Sources

Hdt. 3,97–106; Str. 2,3,4; 2,5,12; 15,1,1–73; Peripl. m.r.; Ptol. 7,1–4; Plin. HN 6,56–106; Tab. Peut. 12,2–5; Geogr. Rav. 2,1–4; Papyrus Vindobonensis G 40822.

Literature

V. Begley, R.D. De Puma (ed.), Rome and India. The Ancient Sea Trade, 1991; V. Begley, Ceramic Evidence for Pre-Periplus Trade on the Indian Coasts, in: Id., R.D. De Puma (ed.), Rome and India. The Ancient Sea Trade, 1991, 157–196; S. Bianchetti, Die Seerouten nach Indien in hellenistischer und römischer Zeit, in: E. Olshausen, H. Sonnabend (ed.), Zu Wasser und zu Land. Verkehrswege in der antiken Welt, 2002, 280–292; J. Carswell, The Port of Mantai, Sri Lanka, in: Rome and India, 1991, 197–203; L. Casson, Ancient Naval Technology and the Route to India, in: Rome and India, 1991, 8–11; R.M. Cimino (ed.), Ancient Rome and India. Commercial and Cultural Contacts between the Roman World and India, 1994; F. De Romanis, Hypalos. Distanze e venti tra Arabia e India nella scienza ellenistica, in: Topoi 7, 1997, 671–692; S.B. Deo, Roman Trade, in: Rome and India, 1991, 39–45; A. Dihle, Die entdeckungsgeschichtlichen Voraussetzungen des Indienhandels der römischen Kaiserzeit, in: ANRW II 9,2, 1978, 546–580; R. Drexhage, Untersuchungen zum römischen Osthandel, 1988; P.H.L. Eggermont, Hippalus and the Discovery of the Monsoons, in: A. Théodorides et al. (ed.), Humour, travail et science en Orient, 1988, 343–364; S. Faller, Taprobane im Wandel der Zeit. Das Sri-Lanka-Bild in griechischen und lateinischen Quellen zwischen Alexanderzug und Spätantike, 2000; L. Guasti, Le rotte oceaniche per l' India in età imperiale. Tempi e percorsi, in: Klio 85, 2003, 370–383; R. Gurukkal, D. Whittaker, In Search of Muziris, in: JRA 14, 2001, 334–350; K. Karttunen, Early Roman Trade with South India, in: Arctos 29, 1995, 81–91; J.I. Miller, The Spice Trade of the Roman Empire, 1969; I. Pill-Rademacher et al., Vorderer Orient. Römer und Parther (14–138 n. Chr.), TAVO B V 8, 1988; K.V. Raman, Further Evidence of Roman Trade from Coastal Sites in Tamil Nadu, in: Rome and India, 125–133; K.P. Shajan, R. Tomber, V. Selvakumar, P.J. Cherian, Locating the Ancient Port of Muziris. Fresh Findings from Pattanam, in: JRA 17, 2004, 312–320; S.E. Sidebotham, Ports of the Red Sea and the Arabia-India Trade, in: Rome and India, 1991, 12–38; E.M. Stern, Early Roman Export Glass in India, in: Rome and India, 1991, 113–124; A. Tchernia, Winds and Coins: From the Supposed Discovery of the Monsoon to the Denarii of Tiberius, in: F. De Romanis, A. Tchernia (ed.), Crossings. Early Mediterranean Contacts with India, 1997, 250–284; R. Tomber, Indo-Roman Trade. The Ceramic Evidence from Egypt, in: Antiquity 74, 2000, 624–631; P.J. Turner, Roman Coins from India, 1989; E.L. Will, The Mediterranean Shipping Amphoras from Arikamedu, in: Rome and India, 1991, 151–156.

Page 208

Literature

Y. Le Bohec (ed.), Les légions de Rome sous le Haut-Empire. 2 vols., 2000; F.M.C. Bruun, Pericula Alexandrina. The Adventures of a Recently Discovered Centurion of the legio II Parthica, in: Arctos 29, 1995, 9–27; B. Campbell, The Roman Army, 31 BC–AD 337. A Sourcebook, 1994; J. C.N. Coulston, Military identity and personal self-identity in the Roman army, in: L. De Ligt, E.A. Hemelrijk, H.W. Singor (ed.), Roman rule and civic life: local and regional perspectives, 2004, 133–152; G. De la Bédoyère, Eagles over Britannia. The Roman Army in Britain, 2001; G. Forni, Il reclutamento delle legioni da Augusto a Diocleziano, 1953; A.K. Goldsworthy, The Roman Army at War 100 BC–AD 200, 1996; Id., The Complete Roman Army, 2003; B.H. Isaac, The Limits of Empire. The Roman Army in the East, 1992; L.J.F. Keppie, The Making of the Roman Army, ²1998; Id., The changing face of the Roman legions (49 BC–AD 69), in: PBSR 65, 1997, 89–102; Id., Legions and Veterans. Roman Army Papers 1971–2000, 2000; E. Kornemann, Die unsichtbaren Grenzen des Römischen Kaiserreiches, in: Gestalten und Reiche. Essays zur Alten Geschichte, 1943, 323–338 (first published as a monograph 1934); J.C. Mann, Legionary Recruitment and Veteran Settlement During the Principate, 1983; E. Ritterling, Legio Nr.1, in: RE 12,1 (1924), 1186–1328; P. Southern, K.R. Dixon, The Late Roman Army, 1996; O. Stoll, Römisches Heer und Gesellschaft. Gesammelte Beiträge 1991–1999, 2001; K. Strobel, Ein neues Zeugnis für die Truppengeschichte der Partherkriege Trajans, in: EA 12, 1988, 39–42; G. Webster, The Roman Imperial Army of the First and Second Centuries A.D., ³1985; R. Wiegels, "Legio": Emil Ritterling und sein Beitrag zur Truppengeschichte der römischen Kaiserzeit, in: Y. Le Bohec (ed.), Les légions de Rome sous le Haut-Empire, 2000, 9–20; J.J. Wilkes, Army and Society in Roman Dalmatia, in: G. Alföldy, B. Dobson, W. Eck (ed.), FS E. Birley, 2000, 327–341.

Page 212

Literature

Eastern frontier: E. Dabrowa (ed.), The Roman and Byzantine Army in the East, 1994; D.H. French, C.S. Lightfoot (ed.), The Eastern Frontier of the Roman Empire, 2 vols., 1989; H. Hellenkemper, Der Limes am nordsyrischen Euphrat, in: D. Haupt, H.G. Horn (ed.), Studien zu den Militärgrenzen Roms 2, 1977, 461–471; D.L. Kennedy (ed.), The Roman Army in the East, 1996; T.B. Mitford, Cappadocia and Armenia Minor, in: ANRW II 7,2, 1980, 1169–1228; D.B. Saddington, The Roman Naval Presence in the East, the Classis Syriaca and the Roman Approach to the Euphrates, in: AKB 31, 2001, 581–586; F. Stark, Rom am Euphrat. Geschichte einer Grenze, 1969; J. Wagner, Die Ostgrenze des Römischen Reiches (1.–5. Jahrhundert n. Chr.) (TAVO B V 13), 1992; Id., Östlicher Mittelmeerraum und Mesopotamien. Die Neuordnung des Orients von Pompeius bis Augustus (TAVO B V 7), 1983; *Cyrenae*: M. Le Glay, A propos d' une inscription de Cyrène, in: ZPE 59, 1985, 120–122; P.C. Jones, A Constitution of Hadrian Concerning Cyrene, in: Chiron 28, 1998, 255–266; A. Laronde, Septime Sévère et Cyrène, in: BSAF 1983 59–70; Id., De Cyrène à Timgad. P. Flauius Pudens Pomponianus et sa famille, in: AFLM 18, 1985, 47–69; Id., P. Flauius Pudens Pomponianus et Cyrène, in: BCTH(B) 18, 1982, 188; Id., Claude et l' extension de la cité romaine à Cyrène, in: Y. Burnand, Y. Le Bohec, J.-P. Martin (ed.), Claude de Lyon, empereur romain, 1998, 333–339; J.H. Oliver, Antoninus Pius to Ptolemais Barca about the Capitolia, in: GRBS 20, 1979, 157–159; J.M. Reynolds, Hadrian, Antoninus Pius and the Cyrenaican Cities, in: JRS 68, 1978, 111–121; Id., Senators Originating in the Provinces of Egypt and of Crete and Cyrene, in: Tituli 5, 1982, 671–683; Id., A Cyrenaican Milestone Re-Read, in: S. Follet (ed.), L' hellénisme d' époque romaine, 2004, 183–186; E.M. Ruprechtsberger, Die römische Limeszone in Tripolitanien und der Kyrenaika. Tunesien-Libyen, 1993; S.E.C. Walker, Hadrian and the Renewal of Cyrene, in: LibStud 33, 2002, 45–56; *Africa Proconsularis*: Y. le Bohec, La troisième légion Auguste, 1989; D. Fishwick, B.D. Shaw, The Formation of Africa Proconsularis, in: Hermes 1977, 369–380; D. Fishwick, On the Origins of Africa Proconsularis I. The Amalgamation of Africa Vetus and Africa Nova, in: AntAfr 29, 1993, 53–62; Id., On the Origins of Africa Proconsularis II. The Administration of Lepidus and the Commission of M. Caelius Phileros, in: AntAfr 30, 1994, 57–80; Id., Zur Verwaltungsgeschichte der römischen Provinzen Nordafrikas (Proconsularis, Numidia, Mauretaniae), in: ANRW II 10,2, 1982,

3–61; Id., On the Origins of Africa Proconsularis III. The Era of the Cereres Again, in: AntAfr 32, 1996, 32, 13–36; F. Kolb, Der Aufstand der Provinz Africa Proconsularis im Jahre 238 n. Chr., in: Historia 26, 1977, 440–478; D.J. Mattingly, Tripolitania, 1995; B. Thomas, Praesides provinciarum Africae proconsularis Numidiae Mauretaniarum qui fuerint ab Augusti aetate usque ad Diocletianum, in: ORom 7, 1969, 163–211; B.E. Thomasson, Africa Proconsularis, in: RE Suppl. 13, 1973, 1–11; Id., Zum Problem der Diözesen in Africa Proconsularis, in: Eranos 62, 1964, 176–178; *Numidia*: A.R. Birley, A Persecuting Praeses of Numidia under Valerian, in: JThS 92, 1991, 598–610; E. Birley, The Governors of Numidia, A.D. 193–268, in: JRS 1950, 60–68; D. Cherry, Soldiers' Marriages and Recruitment in Upper Germany and Numidia, in: AHB 3, 1989, 128–130; R.M. Cid López, El cu to al emperador en Numidia de Augusto a Diocleciano, 1986; G.W. Clarke, Barbarian Disturbances in North Africa in the Mid-Third Century, in: Antichthon 4, 1970, 78–85; E.B. Fentress, Numidia and the Roman Army, 1979; M.A. Levi, Le iscrizioni di Lambaesis e l' esercito di Adriano, in: RAL 5, 1994, 711–723; M.G. Manna, Le formazioni ausiliarie di guarnigione nella provincia di Numidia da Augusto a Gallieno, 1970; F.J. Navarro, P. Stertinius Quartus, Governatore di Numidia? Epigraphica 61, 1999, 67–79; B.D. Shaw, Soldiers and Society. The army in Numidia, in: Opus 2, 1983, 133–159; B.E. Thomasson, Zur Verwaltungsgeschichte der römischen Provinzen Nordafrikas (Proconsularis, Numidia, Mauretaniae), in: ANRW II 10,2, 1982, 3–61; Id., Numidia. Die Legionskommandanten und Statthalter der Principatsepoche, in: RE Suppl. 13 (1973), 315–322; P.I. Wilkins, Legates of Numidia as Municipal Patrons, in: Chiron 23, 1993, 189–206; *Mauretania Caesariensis*: J. Deininger, CIL VIII, 9040. Eine Inschrift der Provinz Mauretania Caesariensis für Aurelian?, in: Hermes 98, 1970, 121–124; J. Desanges, Mauretania ulterior Tingitana, in: BAM 4, 1960, 437–441; D. Fishwick, The Institution of the Provincial Cult in Roman Mauretania, in: Historia 21, 1972, 698–711; R. Lawless, Romanization and Berber Resistance in Mauretania Caesariensis (Western Algeria), in: M. Galley (ed.), Actes du 2. Congrès international d' étude des cultures de la Méditerranée Occidentale 2, 1978, 161–167; P. Weiss, Ausgewählte neue Militärdiplome. Seltene Provinzen (Africa, Mauretania Caesariensis), späte Urkunden für Prätorianer (Caracalla, Philippus), in: Chiron 32, 2002, 491–543; *Mauretania Tingitana*: J. Desanges, Mauretania Ulterior Tingitana, in: BAM 4, 1960, 437–441; W. Eck, A. Pangerl, Neue Militärdiplome für die Truppen der mauretanischen Provinzen, in: ZPE 153, 2005, 187–206; E. Gozalbes, La conquista romana de la Mauritania, in: StudMagr 20, 1988, 1–43; Id., Tumultos y resistencia indígena en Mauretania Tingitana (siglo II), in: Gerión 20, 2002, 451–485; C. Hamdoune, Géographie et administration de la Maurétanie Tingitane: ad fines imperii Romani, Mauretania Tingitana, in: IH 53, 1991, 127–133; P.A. Holder, A Diploma for Mauretania Tingitana of 22 December 144, in: ZPE 149, 2004, 275–281; Id., Diploma militare da "Thamusida" ("Mauretania Tingitana"): 103/104, in: ZPE 146, 2004, 255–258; H. Nesselhauf, Zur Militärgeschichte der Provinz Mauretania Tingitana, in: Epigraphica 12, 1950, 34–48; E. Papi, Diploma militare da Thamusida (Mauretania Tingitana): 31 dicembre 133/134, in: ZPE 142, 2003, 257–266; R. Rebuffat, L' implantation militaire romaine en Mauretanie Tingitane, in: L' Africa romana 4, 1986, 31–78; M. Roxan, The auxilia of Mauretania Tingitana, in: Latomus 32, 1973, 838–855; M.C. Sigman, The Role of Indigenous Tribes in the Roman Occupation of Mauritania Tingitana, 1976; J.E.H. Spaul, Governors of Tingitana, in: AntAfr 30, 1994, 235–260.

Page 214/215

Literature

M.R. Altaweel, S.R. Hauser, Trade Routes to Hatra According to Evidence from Ancient Sources and Modern Satellite Imagery, in: BaM 35, 2004, 59–86; C. Baumer, Die südliche Seidenstraße – Inseln im Sandmeer, 2002; M. Brosius, The Persians, 2006; V.S. Curtis, Parther. B. Parthische Kunst, RLA 10, 2004, 346–350; W. Eder, J. Renger (ed.), Herrscherchronologien der antiken Welt, 2004; J. Harmatta et al. (ed.), History of Civilizations of Central Asia. Vol. II: The Development of Sedentary and Nomadic Civilizations. 700 B.C. to A.D. 250, 1994; S. Hauser, Die ewigen Nomaden? Bemerkungen zu Herkunft, Militär, Staatsaufbau und nomadischen Traditionen der Arsakiden, in: B. Meissner et al. (ed.), Krieg, Gesellschaft, Institutionen, 2005, 163–205; M. Karras-Klapproth, Prosopographische Studien zur Geschichte des Partherreiches auf der Grundlage antiker literarischer Überlieferung, 1988; A. Landskron, Parther und Sasaniden. Das Bild der Orientalen in der römischen Kaiserzeit, 2005 (summary of written sources 177–183); M.J. Olbrycht, Parthia et ulteriores gentes, 1998; W. Posch, Baktrien zwischen Griechen und Kuschan, 1995; K. Schippmann, Grundzüge der parthischen Geschichte, 1980; Id., Arsacids II. The Arsacid Dynasty, EncIr 2, 1986, 525–536; R. Schmitt, Mitteliranische Sprachen im Überblick, in: Id. (ed.), Compendium Linguarum Iranicarum, 1989, 95–105; Id., Persien, Perser. A. Sprache, RLA 10, 2004, 412–416; R.M. Schneider, Die Faszination des Feindes. Bilder der Parther und des Orients in Rom, in: J. Wiesehöfer (ed.), Das Partherreich und seine Zeugnisse, 1998, 95–146; M. Schuol, Die Charakene, 2000; M. Sommer, Hatra, 2003; H. Sonnabend, Fremdenbild und Politik: Vorstellungen der Römer von Ägypten und dem Partherreich in der späten Republik und der frühen Kaiserzeit, 1986; M. Stausberg, Persien, Perser. C. Religion, RLA 10, 2004, 422–424; M.P. Streck, Parther. A. In der schriftlichen Überlieferung, RLA 10, 2003–2004, 343–346; F. Thierry, Yuezhi et Kouchans: pièges et dangers des sources chinoises, in: Afghanistan ancien carrefour entre l' est et l' ouest, Actes du colloque international, Musée Archéologique Henri-Prades-Lattes 2003, 2005; L. Van den Berghe, Reliefs rupestres de L' Iran ancien, exhibition cat. Brussels 1984, 41–54; A. Verstandig, Histoire de l'Empire parthe, 2001; J. Wiesehöfer, Persien, Perser. B. Geschichte, RLA 10, 2004, 416–422; Id., Das frühe Persien, [3]2006; Id., Das antike Persien, 1994; Id. (ed.), Das Partherreich und seine Zeugnisse, 1998; J. Wolski, L' empire des Arsacides, 1993; K.-H. Ziegler, Die Beziehungen zwischen Rom und dem Parther-Reich, [2]1964.

Page 216

Sources

M. Back, Die sāsānidischen Staatsinschriften, 1978; A. Christensen, L' Iran sous les Sassanides, [2]1944; P. Gignoux, Les quatre inscriptions du mage Kirdīr, 1991; H. Humbach, P.O. Skjærvø, The Sassanian Inscription of Paikuli, 3 vols., 1978–1983; P. Huyse, Die dreisprachige Inschrift Šābuhrs I. an der Kaba-i Zardušt (ŠKZ), 2 vols., 1999; E. Kettenhofen, Tirdād und die Inschrift von Paikuli, 1995; D. Weber, Ostraca, Papyri und Pergamente (Corpus inscriptionum Iranicarum 3,4/5), 1992.

Literature

F. Altheim, R. Stiehl, Ein asiatischer Staat, 1954; R.C. Blockley, East Roman Foreign Policy, 1992; H. Börm, Prokop und die Perser, 2007; A. Christensen, L' Iran sous les Sassanides, [2]1944; R.N. Frye, The History of Ancient Iran, 1984; R. Gyselen, La géographie administrative de l' empire Sassanide, 1989; E. Kettenhofen, Das Sāsānidenreich (TAVO B VI 3), 1993; K. Schippmann, Grundzüge der Geschichte des sasanidischen Reiches, 1990; J. Wagner, Die Ostgrenze des Römischen Reiches (TAVO B V 13), 1992; G. Widengren, Iran, der große Gegner Roms, in: ANRW II 9,1, 1979, 219–306; J. Wiesehöfer, Das antike Persien [2]2005; Id., P. Huyse (ed.), Eran und Aneran. Studien zu den Beziehungen zwischen dem Sasanidenreich und der Mittelmeerwelt, 2006.

Page 218

Sources

P. Huyse, Die dreisprachige Inschrift Šābuhrs I. an der Kaba-i Zardušt (ŠKZ), 2 vols., 1999; E. Kettenhofen, Die römisch-persischen Kriege des 3. Jahrhunderts n. Chr. Nach der Inschrift Šāhpuhrs I. an der Ka be-ye Zartošt, 1982; H.D. Malek, A Survey of Research on Sasanian Numismatics, in: NC 153, 1993, 227–269.

Literature

A. Christensen, L' Iran sous les Sassanides, [2]1944; E. Dabrowa (ed.), The Roman and Byzantine army in the East, 1994; M.H. Dodgeon, S.N.C. Lieu (ed.), The Roman Eastern Frontier and the Persian Wars (AD 226–363) 1, 1991; U. Hartmann, Das Palmyrenische Teilreich, 2001; D.L. Kennedy (ed.), The Roman Army in the East (Journal of Roman Archaeology Suppl. 18), 1996; E. Kettenhofen, The Persian Campaign of Gordian III and the Inscription of Sahpuhr at the Ka' be-ye Zartost, in: S. Mitchell (ed.), Armies and Frontiers in Roman and Byzantine Anatolia, 1983, 151–171; Id., Die Einforderung der achaimenidischen Territorien durch die Sāsāniden – eine Bilanz, in: S. Kurz (ed.), Yādnāme-ye Iradj Khalifeh-Soltani. FS Iradj Khalifeh-Soltani, 2002, 49–75; Id., Vorderer Orient. Römer und Sasaniden in der Zeit der Reichskrise (224–284 n. Chr.) (TAVO B V 11), 1982; Id., Östlicher Mittelmeerraum und Mesopotamien (235–284 n. Chr.) (TAVO B V 12), 1983; C. Körner, Philippus Arabs, 2002; D. MacDonald, The Death of Gordian III – another Tradition, in: Historia 30, 1981, 502–508; M.J. Olbrycht, Parthian Military Strategy at Wars against Rome, in: Military Archaeology, 1998, 138–141; E. Winter, Die sasanidisch-römischen Friedensverträge des 3. Jahrhunderts n. Chr., 1988; Id., B. Dignas, Rom und das Perserreich. Zwei Weltmächte zwischen Konfrontation und Koexistenz, 2001.

Page 220

Sources

J. Cantineau (ed.), Inventaire des inscriptions de Palmyre, vols. 1–12, 1930–70; M. Peachin, Roman Imperial Titulature and Chronology, A.D. 235–284, 1990; D.S. Potter, Prophecy and History in the Crisis of the Roman Empire. A Historical Commentary on the Thirteenth Sibylline Oracle, 1990; J. Vogt, Die alexandrinischen Münzen 1, 1924, 213–15; 2 1924, 160–61.

Literature

G.W. Bowersock, Roman Arabia, 1983; S. Dirven, Bel at Palmyra and Elsewhere in the Parthian Period, in: Aram 7, 1995, 137–151; W. Denk, TAVO-Map A I 5, 1991; L. Dirven, The Nature of the Trade Connection between Dura-Europos and Palmyra, in: Aram 8, 1996, 39–54; H. J.W. Drijvers, Hatra, Palmyra und Edessa, in: ANRW II 8, 1977, 799–906; A. Enmann, Eine verlorene Geschichte der römischen Kaiser, in: Philologus Suppl. 4, 1884, 337–501; M. Gawlikowski, Palmyra and its Caravan Trade, in: Annales Archéologiques Arabes Syriennes 42, 1996, 139–145; U. Hartmann, Das Palmyrenische Teilreich, 2001; J. Healey, Palmyra and the Arabian Gulf Trade, in: Aram 8, 1996, 33–37; E. Kettenhofen, Vorderer Orient. Römer und Sasaniden in der Zeit der Reichskrise (224–284), TAVO B V 11, 1982; S.F. Kowalski, P. Slawomir, The Camp of Legio I Illyricorum in Palmyra, in: P. Dyczek (ed.), Novae and the Romans on Rhine, Danube, Black Sea and beyond the Frontiers of the Empire, 1998, 189–209; W.H. Mare, Abila and Palmyra. Ancient Trade and Trade Routes from Southern Syria into Mesopotamia, in: Aram 7, 1995, 189–215; D. Piacentini, Palmyra' s Springs in the Epigraphic Sources, in: Aram 13/14, 2001/02, 525–534; R. Stoneman, Palmyra and its Empire, 1992

Page 228

Sources

Organization of the Christian Church after the Synod of Chalcedon (AD 451) – Council of Elders: Acts 11,30; 15,2; 15,4; 21,18; *presbyter and bishop synonymous*: Acts 20,17; 20,28; Tit 1,5,7; cf. Phil 1,1; 1 Tim 3,1; 3,8; *office of bishop*: Didache 15; patriarchate: Socr. HE 5,8; hierarchy of episcopal sees: Nov. 131,2.
The three great missionary journeys of Paul – Nabataeans: Gal. 1,17; 2 Cor 11,26; *Syria and Cilicia*: Gal 1,21; *commission of the apostle*: Gal 1,15; Rom 1,5; 1,13 f.; 15,16; 15,19–24; 15,28 f.; *Gallio*: Syll.[3] 801D; Acts 18,12; *First missionary journey*: Acts 13–14,2; *Second journey*: Acts 15,41–18,22; *Stay at Corinth*: Acts 18,11; *Third journey*: Acts 18,23–21,14; *Stay at Ephesus*: Acts 19,10; *Jerusalem*: Acts 21–26; *Rome*: Acts 27 f.; *Death of Paul*: 1 Clement 5,5.

Literature

Organization of the Christian Church after the Synod of Chalcedon (AD 451): H.-G. Beck, Kirche und theologische Literatur im byzantinischen Reich, 1959, 44–60, 148–156; J. Darrouzès, Notitiae episcopatuum Ecclesiae Constantinopolitanae, 1981; Ps.-Epiphanius (ed. H. Gelzer, unprinted and inadequately published texts of the Notitiae episcopatuum, in: ABAW, phil. hist. Kl. 21, 1901, 534–542); E. Honigmann, The Original Lists of the Members of the Council of Nicaea, the Robber-Synod and the Council of Chalcedon, in: Byzantion 16, 1944, 20–80; H. Jedin, K.S. Latourette, J. Martin (ed.), Atlas zur Kirchengeschichte, 1987 (8); G. Parthey, Hierocles Synecdemus, 1866; E. Schwartz, Über die Bischofslisten der Synoden von Chalkedon, Nicaea und Konstantinopel, 1937; F.M. Young, From Nicaea to Chalcedon, 1983.
Routes of Christian pilgrims: H. Donner (ed.), Pilgerfahrt ins Heilige Land. Die ältesten Berichte christlicher Palästinapilger (4.–7. Jahrhundert), [2]2002 in German translation, 35–67 (*Itinerarium Hierosolymitanum*), 68–133 (*Peregrinatio ad loca sancta*), 226–295 (*Itinerarium Antonini Placentini*), 134–163 (Jer. Ep. 108); M. Giebel, Friedensbrief und Pilgerflasche: frühchristliche Reisen ins Heilige Land, in: Anregung 46, 2000, 400–408; M. Mulzer, Mit der Bibel in der Hand? Egeria und ihr 'Codex', in: Zeitschrift des Deutschen Palästina-Vereins 112, 1996, 156–164.
The three great missionary journeys of Paul: Y. Aharoni, M. Avi-Yonah, Die Geschichte des Heiligen Landes 3000 Jahre v. Chr. bis 200 Jahre n. Chr., 1998 (maps 240–248); E.-M. Becker (ed.), Biographie und Persönlichkeit des Paulus, 2005; J. Becker, Paulus, der Apostel der Völker, [2]1992; M.D. Goulder, Paul and the Competing Mission in Corinth, 2001; A. v. Harnack, Die Mission und Ausbreitung des Christentums in den ersten drei Jahrhunderten, 1924; H. Jedin, K.S. Latourette, J. Martin (ed.), Atlas zur Kirchengeschichte, 1987 (2); W.H. Ollrog, Paulus und seine Mitarbeiter. Untersuchungen zu Theorie und Praxis der paulinischen Mission, 1979; E.J. Schnabel, Urchristliche Mission, 2002; A. Suhl, Der Beginn der selbständigen Mission des Paulus. Ein Beitrag zur Geschichte des Urchristentums, in: New Testament Studies 38, 1992, 430–447; H. Warnecke, T. Schirrmacher, War Paulus wirklich auf Malta?, 1992.

Page 232

Literature

E. Demougeot, La formation de l' Europe et les invasions barbares, 1979; A. Krause, Die Geschichte der Germanen, 2002; S. Krautschick, Hunnensturm und Germanenflut: 375 – der Beginn der Völkerwanderung? in: ByzZ 92, 1999, 10–67; M. Maczynska, Die Völkerwanderung. Geschichte einer ruhelosen Epoche im 4. und 5. Jahrhundert, 1993; J. Martin, Spätantike und Völkerwanderung, [3]1995; E. Olshausen, Die Anfänge der grossen Völkerwanderung im 3. Jahrhundert n. Chr. am Schwarzen Meer: zur Frage der inneren Struktur wandernder Stämme, in: R. Pillinger, A. Pülz, H. Vetters (ed.), Die Schwarzmeerküste in der Spätantike und im Frühen Mittelalter, 1992, 9–12; W. Pohl, Die Völkerwanderung. Eroberung und Integration, [2]2005; K. Rosen, Die Völkerwanderung, 2002; M.T. Schmid, Die römische Außenpolitik des 2. Jahrhunderts n. Chr., 1997; K. Tausend, Wanderungen vor der Wanderung. Migrationen und Ethnogenese im germanischen Raum, in: E. Olshausen, H. Sonnabend (ed.), "Troianer sind wir gewesen" – Migrationen in der antiken Welt, 2006, 393–401; M. Todd, Die Zeit der Völkerwanderung, 2002; J. Ulrich, Barbarische Gesellschaftsstruktur und römische Außenpolitik zu Beginn der Völkerwanderung: ein Versuch zu den Westgoten, 1995; H. Wolfram, Das Reich und die Germanen. Zwischen Antike und Mittelalter, 1990; Id., Die Schlacht von Adrianopel, in: Anzeiger der Österreichischen Akademie der Wissenschaften 114, 1977, 227–250.

Page 234

Literature

Burgundian kingdoms:
C. Crumley, W.H. Marquardt (ed.), Regional dynamics. Burgundian landscapes in historical perspectives, 1987; G. Domanski, Die Frage der sogenannten Burgundischen Kultur, in: EAZ 19, 1978, 413–444; R. Guichard, Essai sur l' histoire du peuple burgonde, 1965; R. Kaiser, Die Burgunder, 2004; V. Ondrouch, Auf den Spuren der geschichtlichen Anabasis der Anten, Burgunder und Langobarden, in: J. Češka, G. Hejzlar (ed.), Mnema V. Groh, 1964, 71–106; B. Saitta, I Burgundi (413–534), 1977.
Gothic kingdoms:
D. Claude, Geschichte der Westgoten, 1970; R. Collins, Visigothic Spain 409–711, 2006; Id., Early Medieval Spain, [2]1995; O. Devillers, Le conflit entre Romains et Wisigoths en 436–439 d' après les Getica de Jordanès – fortune et infortune de l' abréviateur, in: RPh 69, 1995, 111–126; R.M. Errington, Theodosius and the Goths, in: Chiron 26, 1996, 1–27; A. Ferreiro, The Visigoths in Gaul and Spain. A Bibliography, 1988; L. García Moreno, Historia de España Visigoda, 1989; W. Giese, Die Goten, 2004; P. Heather, The Goths, 1996; Id. (ed.), The Visigoths from the Migration Period to the Seventh Century, 1999; J. Moorhead, Cassiodorus on the Goths in Ostrogothic Italy, in: Romanobarbarica 16, 1999, 241–259; J. Orlandis, Historia de España 4, 1987; A. Schwarcz, Die Goten in Pannonien und auf dem Balkan, in: Mitteilungen des Instituts für Österreichische Geschichte 100, 1992, 50–83; A. Søby Christensen, Cassiodorus, Jordanes and the history of the Goths, 2002; E. A. Thompson, The Goths in Spain, 1969; H. Wolfram, Die Goten, [4]2001; Id., Gotische Studien. Volk und Herrschaft im frühen Mittelalter, 2005; I. Goti, exhibition catalogue, 2003.
Frankish kingdom:
W. Bleiber, Das Frankenreich der Merowinger, 1988; E. Demougeot, La formation de l' Europe et les invasions barbares, 2 vols., 1979; E. Ewig, Die Merowinger und das Frankenreich, [2]1993; P. Perin, L.-C. Feffer, Les Francs, 2 vols., 1987; D. Geuenich (ed.), Die Franken und die Alemannen bis zur 'Schlacht bei Zülpich' (496/97), 1998; R. Scharer, G. Scheibelreiter (ed.), Historiographie im frühen Mittelalter, 1994; F. Siegmund, Alemannen und Franken, 2000; Die Franken, exhibition catalogue, 1996.
Vandal kingdom:
J. Arce, Los vándalos en "Hispania" (409–429 A.D.), in: Antiquité Tardive 10, 2002, 75–85; F.M. Ausbüttel, Die Verträge zwischen den Vandalen und Römern, in: Romanobarbarica 11, 1991, 1–20; H. Castritius, Die Vandalen, 2007; F.M. Clover, The Late Roman West and the Vandals, 1993; C. Courtois, Les Vandales et l' Afrique, 1955; P. Delogu (ed.), Le invasioni barbariche nel meridione dell' impero: Visigoti, Vandali, Ostrogoti, 2001; L. D. Dossey, The Last Days of Vandal Africa, in: Journal of Theological Studies 54, 2003, 60–138; A. Hettinger, Migration und Integration: zu den Beziehungen von Vandalen und Romanen im Norden Afrikas, in: Frühmittelalterliche Studien 35, 2001, 121–143; T. Hodkin, Huns, Vandals and the Fall of the Roman Empire, 1996; P. Hulten, The True Story of the Vandals, 2001; J. Leclant (ed.), Regards sur la Méditerranée, 1997; A.H. Merrills (ed.), Vandals, Romans and Berbers. New Perspectives on Late Antique North Africa, 2004; Y. Modéran, L' établissement territorial des Vandales en Afrique, in: Antiquité Tardive 10, 2002, 87–122; M. Todd, Die Germanen, 2000, 172–176; Die Vandalen, exhibition catalogue, 2003.
Kingdom of the Suebi:
S. Hamann, Vorgeschichte und Geschichte der Sueben in Spanien, 1971; J.L. Quiroga, M.R. Lovelle, De los Vándalos a los Suevos en Galicia: una visión crítica sobre su instalación y organización territorial en el noroeste de la Península Ibérica en el siglo V, in: Studie Historica. Historia Antigua 13–14, 1995–1996, 421–436.
The Roman 'kingdom' of Syagrius:
D. Claude, Geschichte der Westgoten, 1970; A. Demandt, magister militum, RE Suppl. 12, 1970, 687–691; D. Henning, Periclitans res publica, 1999, 300–305; R. Schulz, Die Entwicklung des römischen Völkerrechtes im 4. und 5. Jahrhundert n. Chr., 1993; D. Zolotenki, Galia polnocna za rzadów wojskowych Komesa Pawla (464–469/470) (La Gaule septentrionale sous le gouvernement militaire du comte Paul), in: Eos 84, 1996, 143–153 (with French résumé).
Langobardic kingdom:
E. Bernareggi, Moneta Langobardorum, 1983; R. Busch (ed.), Die Langobarden, 1988; D.J. Diesner, Byzanz, Rom und die Langobarden, in: Jahrbücher der Österreichischen Byzantinistik 25, 1976, 31–45; Id., Zur langobardischen Sozialstruktur. Gasindii und Verwandtes, in: Klio 58, 1976, 141–186; H. Ebling, J. Jarnut, G. Kampers, Nomen et gens. Untersuchungen zu den Führungsschichten des Franken-, Langobarden- und Westgotenreiches im 6. und 7. Jahrhundert, in: Francia 8, 1980, 687–745; J. Jarnut, Geschichte der Langobarden, 1982; H. Kalex, Die Anwendung des römischen Rechts im Langobardenreich und das Prinzip der Personalität des Rechts, in: Klio 61, 1979, 111–123; W. Menghin, Die Langobarden, 1985; C. G. Menis (ed.), I Longobardi, exhibition catalogue, 1990; W. Pohl, P. Erhart (ed.), Die Langobarden, 2005; R. Schneider, Königswahl und Königserhebung im Frühmittelalter. Untersuchungen zur Herrschaftsnachfolge bei den Langobarden und Merowingern, 1972.

Page 236

Literature

P. Amory, People and Identity in Ostrogothic Italy, 1997; J.E. Atkinson, The plague of 542, in: Acta Classica 45, 2002, 1–18; G. Claude, Geschichte der Westgoten, 1970; L.M. Hartmann, Belisarios, in: RE 3,1, 1897, 209–240; M. Konrad, Römische Grenzpolitik und die Besiedlung in der Provinz Syria Euphratensis, in: B. Geyer (ed.), Conquête de la steppe et appropriation des terres sur les marges arides du Croissant fertile, 2001, 145–158; M. Meier, Justinian, 2004; J.J. O'Donnell, Liberius the Patrician, in: Traditio 37, 1981, 31–72; J. Prostko-Prostynski, Zum Datum der Einrichtung der afrikanischen Prätorianerpräfektur durch Kaiser Iustinian I., in: ByzZ 91, 1998, 423–434; H.N. Roisl, Totila und die Schlacht bei den Busta Gallorum Ende Juni/Anfang Juli 552, in: Jahrbuch der Österreichischen Byzantinistik 30, 1981, 25–50; K.F. Stroheker, Das spanische Westgotenreich und Byzanz, in: BJ 163, 1963, 252–274; E.A. Thomson, The Goths in Spain, 1969; Id., Romans and Barbarians, 1982; Z.V. Udalcova, La campagne de Narsès et l' écrasement de Totila, in: Corso di cultura sull' arte Ravennate e Byzantina 18, 1971, 557–564; Id., L' Italie et Byzance au VI siècle, in: Id., 547–555; H. Wolfram, Die Goten, [4]2001.

Page 238

Sources

Themata: Konstantinos VII. Porphyrogennetos, De thematibus. *konstantinopolis*: Notitia urbis Constantinopolitanae, in: A. Riese, Geographi Latini Minores, 1878, 133–139; *Serpent Column (inscription)*: StV 2, no. 130.

Literature

Byzantine theme system: F. Dölger, Zur Ableitung des byzantinischen Verwaltungsterminus *thema*, in: Historia 4, 1955, 189–198; J. Ferluga, Le clisure bizantine in Asia Minore, in: ZRVI 16, 1975, 9–23; I.E. Karayannopoulos, Die Entstehung der byzantinischen Themenordnung, 1959; Id., Die byzantinische Themenordnung, in: FS E. Volterra, 1971, 521–526; R.J. Lilie, Die zweihundertjährige Reform, in: BS 45, 1984, 27–39, 190–201; A. Pertusi, La formation des thèmes byzantins, 1958; T. Riplinger, H. Brenner, TAVO B VI 5, 1985.
Constantinople: N. Asutay-Effenberger, Die Landmauer von Konstantinopel, 2007; J. Bardill, The Golden Gate in Constantinople, in: AJA 103, 1999, 671–696; S.E. Bassett, The urban image of Late Antique Constantinople, 2004; F. A. Bauer, Ausstattung des öffentlichen Raums in den spätantiken Städten Rom, Konstantinopel und Ephesos, 1996, 143–268; H.G. Beck (ed.), Studien zur Frühgeschichte Konstantinopels, 1973; A. Berger, Streets and public spaces in Constantinople, in: Dumbarton Oaks Papers 54, 2000, 161–172; Id., Regionen und Straßen im frühen Konstantinopel, in: MDAI(I) 47, 1997, 349–414; K.R. Dark, F. Özgümüs, New Evidence for the Byzantine Church of the Holy Apostles from Fatih Camii, Istanbul, in: The Oxford Journal of Archaeology 21, 2002, 393–413; D. Feissel, Le "Philadelphion" de Constantinople, in: CRAI 2003, 495–523; R. Janin, Constantinople Byzantine, [2]1964; H. Kalkan, S. Sahin, Ein neues Bauepigramm der theodosianischen Landmauer von Konstantinopolis aus dem Jahr 447, in: EA 23, 1994, 145–155; B. Kiilerich, The Obelisk Base in Constantinople. Court Art and Imperial Ideology, 1998; W.D. Lebek, Die Landmauer von Konstantinopel und ein neues Bauepigramm, in: EA 25, 1996, 107–153; C. Mango, G. Dagron (ed.), Constantinople and its Hinterland, 1995; M.M. Mango, The Commercial Map of Constantinople, in: Dumbarton Oaks Papers 54, 2000, 189–207; W. Müller-Wiener, Bildlexikon zur Topographie Istanbuls, 1977; Id., Die Häfen von Byzantion-Konstantinupolis-Istanbul, 1994.

Page 240

Byzantine Emperors
Constantine I 306–337
Constantine II 337–361
Julian 361–363
Jovian 363–364
Valens 364–378
Theodosius I 379–395
Arcadius 395–408
Theodosius II 408–450
Marcian 450–457
Leo I 457–474
Leo II 474
Zeno 474–475, 476–491
Basiliscus 475–476
Anastasius I 491–518
Justin I 518–527
Justinian I 527–565
Justin II 565–578
Tiberius II 578–582
Maurice 582–602
Phocas 602–610
Heraclius 610–641
Constans II 641–668

Sources

R.N. Frye, The History of Ancient Iran, 1984, 287–291; R. Göbl, Sasanidische Numismatik, 1968; Id., Der sasanidische Siegelkanon, 1973; Id., Die Tonbullen von Tacht-e Suleiman, 1976; T. Riplinger, TAVO B VI 5–7, 1988; K. Schippmann, Grundzüge der Geschichte des sasanidischen Reiches, 1990, 3–9; D. Sellwood, P. Whitting, R. Williams, An Introduction to Sasanian Coins, 1985; B. Spuler, Iran in früh-islamischer Zeit, 1952; G. Widengren, in: The Cambridge History of Iran 3,2, 1983, 1269ff.; E. Yar-Shater (ed.), The History of al-Tabari. An Annotated Translation, 38 vols., 1985/98.

Literature

H. Börm, Prokop und die Perser, 2007; A. Christensen, L' Iran sous les Sassanides, [2]1944; E. Dabrowa (ed.), The Roman and Byzantine Army in the East, 1994; M.H. Dodgeon, S.N.C. Lieu (ed.), The Roman Eastern Frontier and the Persian Wars (AD 226–363), 1991; G.B. Greatrex, Rome and Persia at War, 502–532, 1998; Id., The Two Fifth-century Wars between Rome and Persia, in: Florilegium 12, 1993, 1–14; R.J. Herbert, The Destruction of the Hephthalites, in: North American Journal of Numismatics 7, 1968, 187–197; J. Howard-Johnston, Heraclius' Persian Campaigns and the Revival of the East Roman Empire, 622–630, in: War in History 6, 1999, 1–44; D.L. Kennedy (ed.), The Roman Army in the East (Journal of Roman Archaeology, Suppl. 18), 1996; E. Kettenhofen, Die Einforderung der achaimenidischen Territorien durch die Sāsāniden – eine Bilanz, in: S. Kurz (ed.), Yādnāme-ye Iradj Khalifeh-Soltani. FS Iradj Khalifeh-Soltani, 2002, 49–75; M.J. Olbrycht, Parthian Military Strategy at Wars against Rome, in: Military Archaeology, 1998, 138–141; T. Riplinger, H. Brenner, Der östliche Mittelmeerraum. Das frühbyzantinische Reich (527–563 n. Chr.) (TAVO B VI 5), 1984; Id., Kaukasus und Mesopotamien. Byzantiner und Perser (581–628 n. Chr.) (TAVO B VI 6), 1984; Id., Der Vordere Orient zur Zeit des Byzantinisch-Persischen Konflikts (6.–7. Jahrhundert n. Chr.) (TAVO B VI 7), 1984; G. Widengren, Iran, der große Gegner Roms, in: ANRW II.9,1, 1976, 219–306; Id., in The Cambridge History of Iran 3,2, 1983, 1269ff.

Page 248

Literature

General: J. Engel (ed.), Großer Historischer Weltatlas II: Mittelalter, 1979, Nr. 85a; R.-J. Lilie, Byzanz und die Kreuzzüge, 2004; H.E. Mayer, Geschichte der Kreuzzüge, [10]2005; S. Runciman, Geschichte der Kreuzzüge, [3]2001; *First Crusade*: M. Balard (ed.), Autour de la Première Croisade, 1995; H.-G. Becker, Urban II., 2 vols., 1964/68; R. Beyer, H. Halm, C. Naumann, Kleinasien im 12. und 13. Jahrhundert, TAVO B VII 10, 1987; J.F. France, Victory in the East. A Military History of the First Crusade, 1994; G. Lobrichon, Die Eroberung Jerusalems im Jahre 1099, 1998; *Second Crusade*: G. Constable, The Second Crusade as Seen by Contemporaries, in: Traditio 9, 1953, 213–280; J. Phillips, M. Hoch (ed.), The Second Crusade. Scope and Consequences, 2003; *Third Crusade*: E. Eickhoff, Friedrich Barbarossa im Orient. Kreuzzug und Tod Friedrichs I., 1977; R. Hiestand, Precipua tocius christianismi columpna. Barbarossa und der Kreuzzug, in: A. Haverkamp, Friedrich Barbarossa. Handlungsspielräume und Wirkungsweisen, 1992, 51–108; H. Möhring, Saladin und der Dritte Kreuzzug, 1980.

Page 250

The Emperors of 'Romania' 1204–1261
Baldwin I of Flanders 1204–1205
Henry of Flanders 1206–1216
Peter of Courtenay 1217
Yolanda 1217–1219
Interregnum 1219–1221
Robert of Courtenay 1221–1228
Baldwin II 1228–1261, with
John of Brienne 1231–1237
The Great Comneni of Trebizond
David I 1204–1214
Alexius I 1204–1222
Andronicus I 1222–1235
John I 1235–1238
Manuel I 1238–1263
Andronicus II 1263–1266
George Comnenus 1266–1280
John II 1280–1284; 1285–1297
Theodora 1284/85
Alexius II 1297–1330
Andronicus III Comnenus 1330–1332
Manuel II 1332
Basilius 1332–1340
Irene Palaeologina 1340/41
Anna Comnena 1341/42
John III 1342–1344
Michael 1341; 1344–1349
Alexius III Comnenus 1349–1390
Manuel III Comnenus 1390–1417
Alexius IV Comnenus 1417–1429
John IV Comnenus 1429–1459
David II Comnenus 1459–1461
The Despots of Epirus
Michael I Angelus 1204–1215
Theodore I Angelus 1215–1230
Michael II Angelus 1230–1271
The Emperors of Nicaea
Theodore I Laskaris 1204–1222
John III Ducas Vatatzes 1222–1254
Theodore II Laskaris 1254–1258
John IV Laskaris 1258–1261 and Michael VIII Ducas Comnenus Palaeologus 1259–1282

Sources

A. Carile (ed.), Partitio Terrarum Imperii Romanie, in: Studi Veneziani 7, 1965, 125–305; C. Morris (ed.), Geoffroy de Villehardouin and the Conquest of Constantinople, in: History 53, 1968, 24–34; G.L.F. Tafel, G.M. Thomas, Urkunden zur älteren Handels- und Staatsgeschichte der Republik Venedig, 2 vols., 1856.

Literature

I. Booth, Theodore Laskaris and Paphlagonia, 1204–1214, in: Archeion Pontou 2003/4, 151–224; A. Carile, Per una storia dell' impero di Constantinopoli (1204–1261), [2]1978; B. Hendrickx, Les institutions de l' empire latin de Constantinople (1204–1261), in: Byzantina 6, 1974, 85–154; Id. Regestes des empereurs latins de Constantinople (1204–1261/1272), in: Byzantina 14, 1988, 7–221; J. Longnon, L' Empire Latin de Constantinople et la Principauté de Morée, 1949; R.L. Wolff, Studies in the Latin Empire of Constantinople, 1976.

Page 252

The Palaeologi Emperors
Michael VIII Palaeologus 1259–1282
Andronicus II Palaeologus 1282–1328
Andronicus III Palaeologus 1328–1341
John V Palaeologus 1341–1391
John VI Cantacuzenus 1341; 1347–1354
Michael IX Palaeologus 1294–1320
Andronicus IV Palaeologus 1376–1379
John VII Palaeologus 1390
Manuel II Palaeologus 1391–1425
John VIII Palaeologus 1425–1448
Constantine XI Palaeologus 1449–1453
Rulers of the House of Osman
Osman I 1288–1326
Orhan I 1326–1362
Murad I 1362–1389
Beyazid I 1389–1402
Mehmet I 1402–1421
Suleiman 1402–1410
Musa 1411–1413
Murad II 1421–1444; 1446–1451
Mehmet II Fatih 1444–1446; 1451–1481

Literature

M. Alexandrescu-Dersca, La campagne de Timur en Anatolie (1402), [2]1977; J.W. Barker, Manuel II Palaeologus (1391–1425). A Study in Late Byzantine Statesmanship, 1969; A. Bon, La Morée franque, 2 vols., 1969; A. Carile, La Partitio Terrarum Imperii Romanie del 1204 nella tradizione storica dei Veneziani, in: Rivista di Studi Bizantini e Neoellenici 2/3, 1965/66, 167–179; M. Carroll, A Contemporary Greek Source for the Siege of Constantinople 1453: The Sphrantzes Chronicle, 1985; J. Engel (ed.),

Großer Historischer Weltatlas II: Mittelalter, 1979 (90b); C.A. Gauci, M. Mallat, The Palaiologos Family. A Genealogical Review, 1985; A.E. Lalou, Constantinople and the Latins, 1972; B.F. Manz, The Rise and Rule of Timur, 1989; D. Morgan, The Mongols, 1986; T. Nagel, Timur, der Eroberer, 1993; D.M. Nicol, The Last Centuries of Byzantium 1261–1453, 1972; Id., The Despotate of Epiros. A Contribution to the History of Greece in the Middle Ages, 1984; Id., The Immortal Emperor. The Life and Legend of Constantine Palaiologos, Last Emperor of the Romans, 1992; D.E. Pitcher, An Historical Geography of the Ottoman Empire, 1972; K.M. Setton, Catalan Dominion of Athens 1311–1388, 1975; B. Spuler, Die Goldene Horde, ²1965; E. Trapp et al. (ed.), Prosopographisches Lexikon der Palaiologenzeit, 15 vols., 1976–1996.

Table to page 140

Date	Province	Sources
241 BC	Sicilia, western part	Pol. 3,27,2; App. Sic. 2,4; Oros. 4,11,2
237	Sardinia et Corsica	Pol. 1,88,8–12; Liv. 21,1,5; 21,40,5; 22,54,11; Fest. 430,14–20
229	Illyrian protectorate	App. Ill. 21 f.
210	Kingdom of Syracuse incorporated into the province of Sicilia	Liv. 26,40
206	Hispaniae, all of Sicilia	App. Ib. 152; Liv.25,31,5; 25,40,4; cf. 31,31,8
148	Macedonia	Flor. Epit.1,30; 1,32
146	Achaea incorporated into province of Macedonia	Pol. 39,3,3; 39,8,6; Cic. Verr. 2,1,55; Liv. Per 52; Str. 8,6,23
146	Africa	Str. 14,1,38; Vell. Pat. 2,38,2
129	Asia	Vell. Pat. 2,38,5
120	Gallia transalpina	Caes. B Gall. 1,45,2
123	Baleares incorporated into province of Hispania citerior	Liv. Per. 60; Flor. Epit.1,43; Oros. 5,13,1; Str. 3,5,1
102–75	Cilicia	Cic. Verr. 2,3,211; Sall. Hist. 2,87; Str. 14,3,3–14,5,7; CIL I2 2954
74	Cyrenaica	Sall. Hist. fr. incerta 2,2; App. Civ. 1,517
67	Creta et Cyrenae	Liv. Per. 100; Plut. Pompeius 29; App. Civ. 1,517
64	Syria	Plut. Pompeius 39,3
63	Bithynia et Pontus	Str. 12,3,1
58	Cyprus incorporated into province of Cilicia	Cic. Dom. 20 f.; 52 f.; 65; Cic. Sest. 56 f.
51	All of Gaul incorporated into province of Gallia transalpine	Caes. B Civ. 8; Liv. Per. 108; Suet. Iul. 25,1; Cass. Dio 40,42 f.
49/45	Caesar adds most of the *territorium* of Massilia to province of Gallia transalpina	Caes. B Civ. 1,34–36; 56–58; 2,1–16; 22; Liv. Per. 110; Lucan 3,298–762; Cass. Dio 41,19; 21; 25
46	Africa Nova and a military district around Cirta	App. Civ. 4,233; Plin. HN 5,22
30	Aegyptus	Tac. Ann. 2,59; Tac. Hist. 1,11; Cass. Dio 51,17,1–3

Table to page 166

Date	Province	Event	Sources
121 BC	Gallia transalpina	Province established	cf. Caes. B Gall. 1,45,2; Vell. Pat. 2,39; Capital Narbo Martius: Cic. Font. 13; Vell. Pat. 1,15,5
50 BC		Caesar adds Gallia comata to Gallia transalpina	Suet. Iul. 25,1; Cass. Dio 40,44,1; Eutr. 6,17,2; Festus, Breviarium 6; cf. D. Iunius Brutus as governor of all Gallia 47 and 46 in App. Civ. 2,48; 111; A. Hirtius 45 in Cic. Att. 10,37,4; 14,9,3
shortly before 15 Mar. 44 BC	Narbonensis, Transalpina	Division of the province into two separate provinces	M. Aemilius Lepidus proconsul of Narbonensis in Cass. Dio 43,51,8; L. Munatius Plancus proconsul of Transalpina in Cass. Dio 46,29,6
27 BC	Narbonensis, adm. capital Narbo; Aquitania, adm. capital Burdigala; Lugdunensis, adm. capital Lugdunum; Belgica, adm. capital Durocortorum	Establishment of four separate provinces on Gaulish soil under imperial administration	Cass. Dio 53,12,5; Suet. Aug. 28; 47; cf. Plin. HN 4,105; Ptol. 2,7,1; Narbo: Liv. Per. 134; Cass. Dio 43,22,5; Burdigala: ILS 1906; Lugdunum: ILS 1152; Durocortorum: Plin. HN 4,106; Tac. Hist. 4,68 f.
22 BC	Narbonensis	Placed under senatorial administration	Str. 17,3,25; Cass. Dio 53,12,7; 54,4,1
c. AD 85	Germania superior and Germania inferior	Military districts turned into provinces; adm. capitals Mogontiacum and Colonia Agrippinensis	CIL 11, 5744; cf. 3, 9960; Mogontiacum: AE 1964, 148 with Haensch, 150; Colonia Agrippinensis: evidence in Haensch, 65–73
AD 260–274		Gaulish provinces secede from the Roman Empire under Postumus	Aur. Vict. Caes. 33–35; (Ps.-)Aur. Vict. Epit. Caes. 32–35; Eutr. 9,9–13; → map p. 223
after AD 284	8 provinces in Dioecesis Galliarum, 7 provinces in Dioecesis Viennensis	Division of Gaul; Dioecesis Viennensis including Alpes maritimae; Alpes Cottiae now province of Italia annonaria	Laterculus Veronensis 8–10
reign of Constantine I (306–337)		4 *dioeceses* under praefectus praetorio Galliarum: Hispaniae, Galliae, Viennensis, Britanniae	Zos. 2,33,2

Tables to Page 142

Date	Event	Sources
508/07 BC	First treaty between Rome and Carthage: islands of Sardinia and Corsica assigned to the Carthaginian sphere of interest	Pol. 3,22,8 f. (StV II 121)
259	Consul L. Cornelius Scipio conquers Corsica in battle against Carthaginians, fighting on Sardinia	ILS 3,5; Zon. 8,11
258	Scipio's triumph over Carthaginians, Sardinians and Corsicans	Fast. triumphales
258	Triumph of consul C. Sulpicius Paterculus over Carthaginians and Sardinians	Fast. triumphales
238/37	Romans, under the consul Ti Sempronius Gracchus, conquer Sardinia; Carthaginians relinquish the island	Fest. 430; Pol. 1,88,8–12; Liv. 21,40,5; 22,54,11; cf. supplement (StV III 497) to the peace treaty (StV III 493)
237	Annexation of Corsica	Sinnius Capito in Fest. 430,14–20
236	Consul C. Licinius Varus fights rebels on Corsica	Liv. Per. 20; Zon. 8,18
234	Triumph of consul T. Manlius Torquatus over Sardinians	Fast. triumphales
233	Triumph of consul Sp. Carvilius Maximus over Sardinians	Fast. triumphales
232	Triumph of consul M'. Pomponius Matho over Sardinians	Fast. triumphales
231	Consul M. Pomponius Matho fighting rebels on Sardinia, his colleague C. Papirius Maso likewise on Corsica	Zon. 8,18; Fest. 131; Val. Max. 3,6,5
230	Ovatio of consul C. Papirius Maso over Corsicans	Fast. triumphales
227	Creation of province Sardinia et Corsica; adm. capital Carales; first praetor M. Valerius Laevinus	Solin. 5,1; Liv. 23,24,4; Liv. Per. 20; Zon. 8,19
225	Consul C. Atilius Regulus fighting rebels on Sardinia	Pol. 2,23,6; 2,27,1; Zon. 8,20
215	Proconsul T. Manlius Torquatus fighting rebel Sardinians (who are allied to Carthage) led by Hampsicora	Liv. 23,32,10; 23,32,40 f.; Zon. 9,4
177/76	Consul Ti. Sempronius Gracchus fighting Ilienses and Balari on Sardinia	Liv. 41,12,5 f.; cf. Pol. 25,4,1; Liv. 41,8 f.; 41,15,6
174	Triumph of the proconsul Ti. Sempronius Gracchus over Sardinians	Fast. triumphales
173	Praetor C. Cicereius defeats rebels on Sardinia	Liv. 42,1,3; 42,7,1 f.; Val. Max. 3,5,1; 4,5,3
172	Ovatio of the propraetor C. Cicereius over Corsians	Fast. triumphales; Liv. 42,21,6 f.
163	Praetor M'. Iuventius Thalna defeats rebels on Corsica	Val. Max. 9,12,3; Plin. HN 7,182
122	Triumph of proconsul L. Aurelius Orestes over Sardinians	Fast. triumphales
115–111	M. Caecilius Metellus, as consul and subsequently (until 111) proconsul, fighting Sardinians; triumph	Fast. triumphales; Eutr. 4,25,1; ILS 5947,7
104	Propraetor T. Albucius puts down rebels on Sardinia	Cic. Prov. 15; Cic. Div. Caec. 63
c. 100	Marius founds colony of Mariana on Corsica	Plin. HN 3,80; Mela 2,122; Sen. Dial. 7,9; Solin. 3,3
82/80	Sulla founds a colony at Aleria on Corsica	Plin. HN 3,80; Mela 2,122; Solin. 3,3
13 Jan. 27 BC	Sardinia et Corsica placed under senatorial administration	Str. 17,3,25; Cass. Dio 53,12,4
AD 6	Two separate provinces: Sardinia under imperial administration, adm. capital Carales; Corsica under senatorial administration, adm. capital Aleria	Cass. Dio 55,28,1; Str. 5,2,7; ILS 105; AE 1921, 86; ILS 2684; 5947,6; CIL XII 2455: praefectus Corsicae; Carales: CIL X 7583; 7859; ILS 1359; Aleria: CIL X 8036; 8038
probably 67	Both islands under senatorial administration	Paus. 7,17,3; ILS 5947,2; 4: proconsul
from 73	Under imperial administration	CIL X 8023 f.: procurator Augusti et praefectus
reign of Trajan (98–117)	Under senatorial administration	ILS 1038: proconsul
reign of Commodus (176–192) or Septimius Severus (193–211)	Under imperial administration	ILS 1358: procurator Augusti et praefectus
reign of Diocletian (284–305)	Still two separate provinces, each under a praeses	Not. Dign. Occ. 19,12 f.
c. 455	Both islands occupied by Vandals, province with the Baleares	Notitia provinciarum et civitatum Africae 484; Procop. Vand. 1,10,26
534	Sardinia et Corsica still one province under Byzantine rule	Procop. Vand. 2,5,2–4
241 BC	Province of Sicilia established in west of island, annual dispatch of one of the praetors, adm. capital Lilybaeum	Cic. Verr. 2,2,2; App. Sic. 2,6
227	Permanent praetorship, adm. capital Lilybaeum	Liv. Per. 20
210	Roman conquest of Syracusae, reorganization of the province (details unknown); one praetor with 2 quaestors, adm. capital Syracusae	Liv. 26,40; Syracusae: CIL X 7127
132	End of slave revolt (138–132), reorganization of province (details unknown)	*lex Rupilia*; Cic. Verr. 2,2,32–44; 50; 90; 125; Val. Max. 5,9,8; Flor. Epit. 1,22
99	Reorganization of province by proconsul M'. Aquillius after ending of slave revolt (104–101)	Cic. Verr. 2,5,5; Liv. Per. 69
70	Trial of Verres, Cicero's speeches against Verres	Cic. Verr. 1 and 2 *passim*

Date	Event	Sources
44	All freeborn Siculi granted Roman citizenship; law scrapped by *senatus consultum*	*lex Iulia de Siculis* (M. Antonius, *cos.* 44): Cic. Att. 14,12,1; Cic. Phil. 12,12; 13,5
43–36	Under rule of Sex. Pompeius	App. Civ. 5,122; Cass. Dio 48,17,5; 49,11,1; Flor. Epit. 2,18,9
43	Sex. Pompeius implements the *lex Iulia de Siculis*	Diod. Sic. 13,35,3
from 36	Reforms of Caesar the Younger/Augustus	Plin. HN 3,88–93
13 Jan. 27 BC	Under senatorial administration	Str. 17,3,25
AD 258/63	Colonial revolt	SHA Gall. 4,9
reign of Diocletian (284–305)	Imperial reform: province allocated to the Dioecesis Italia suburbicaria under a *consularis*, the *vicarius* in Rome	Not. Dign. Occ. 1,60
474	Sicilia part of the Vandal kingdom	Malchus fragment 3; Procop. Vand. 1,7; Victor Vitensis, Historia persecutionis Africanae provinciae 1,14,51

Table to page 148

Year	Roman magistrate(s)	Opponent	Sources
205 BC	L. Cornelius Lentulus, L. Manlius Acidinus	Ilergetes under Indibilis and Mandonius	Liv. 29,2 f.; 29,13,7; App. Ib. 156
200	C. Cornelius Cethegus	Edetani	Liv. 31,49,7
197	M. Helvius: Hisp. ult.; C. Sempronius Tuditanus: Hisp. cit.	Revolt of the cities of Bardo (or Baldo, location unknown) and Carmo, of the Malacini, Sexetani, all of Baeturia	Liv. 33,21,7 f.
196	C. Sempronius Tuditanus: Hisp. cit.	Roman defeat	Liv. 33,25,8 f.; Oros. 4,20,10
195	M. Porcius Cato	Ausetani, Bergistani, Celtiberi, Edetani, Lacetani, Suessetani, Turdetani	Liv. 34,8,4–21,8; App. Ib. 160–170; Zon. 9,17,5–7
195	P. Manlius, Ap. Claudius Nero: Hisp. ult.	Turdetani, Celtiberi	Liv. 34,10,1–5; 34,17–19; 40,16,7
195	Q. Minucius Thermus: Hisp. cit.	Turdetani; Roman victory at Turda (not localized)	Liv. 33,44,4 f.; 34,10,5–7
194	P. Cornelius Scipio: Hisp. ult.	Roman victory by northern reaches of the Iberus	Liv. 35,1,3 f.
194	Sex. Digitius: Hisp. cit.	Several Roman defeats	Liv. 35,1,1 f.; Oros. 4,20,16
193	M. Fulvius Nobilior: Hisp. ult.	Victories over Carpetani, Lusitani, Oretani, Vaccaei, Vettones	Liv. 35,7,8; Oros. 4,20,16; 4,20,19
193	P. Cornelius Scipio Nasica: Hisp. ult.	Victory over Lusitani	Liv. 35,1,3–12
193	Sex. Digitius: Hisp. cit.	Roman losses	Liv. 35,2,1–5; Oros. 4,20,16
192	C. Flaminius: Hisp. cit.	Roman conquest of Licabrum (not localized)	Liv. 35,20,11; 35,22,5
192	M. Fulvius Nobilior: Hisp. ult.	Victories over Oretani and Vettones	Liv. 35,22,6–8; Oros. 4,20,16; 4,20,19
190	L. Aemilius Paullus: Hisp. ult.	Losses in battle against Lusitani	Liv. 37,2,11; 37,46,7 f.
189	L. Aemilius Paullus: Hisp. ult.	Victory over Lusitani	Liv. 37,57,6–7; 37,58,5
186	L. Manlius Acidinus Fulvianus: Hisp. cit.	Victory over Lusitani at Calagurris	Liv. 39,21,6–10
186	C. Atinius: Hisp. ult.	Battles with Celtiberi, victory over Lusitani; conquest of Hasta	Liv. 39,21,4
185	C. Calpurnius Piso: Hisp. ult.	Battles by the Tagus against Carpetani	Liv. 39,30 f.
184	A. Terentius Varro: Hisp. cit.	Victory over Suessetani	Liv. 39,42,1
183	A. Terentius Varro: Hisp. cit.	Victory over Ausetani and Celtiberi	Liv. 39,45,4; 39,56,1
182	Q. Fulvius Flaccus: Hisp. cit.	Battles with Celtiberi, conquest of Urbicna (or Urbicuam; not localized)	Liv. 40,16,7–10
181	Q. Fulvius Flaccus: Hisp. cit.	Victory over Celtiberi (Lusones)	Liv. 40,30,1–40,33,9; App. Ib. 171
181	P. Manlius Vulso: Hisp. ult.	Battles with Lusitani	Liv. 40,34,1
180	Ti. Sempronius Gracchus: Hisp. cit.	Battles with Celtiberi at Caravis	Liv. 40,40,15
180	Q. Fulvius Flaccus: Hisp. cit.	Victory over Celtiberi (Arevaci)	Liv. 40,39,1–40,15; 40,43,4–7; App. Ib. 175
178	L. Postumius Albinus: Hisp. ult.	Battles with Vaccaei and Lusitani	Liv. 41,6,4; Liv. Per. 41
178	Tib. Sempronius Gracchus: Hisp. cit.	Victory over Celtiberi	Liv. 41,26,1; Liv. Per. 41
170	L. Canuleius Dives: both Hispaniae	Revolt of Celtiberi	Liv. 43,4,1–4; Liv. Per. 43; Flor. Epit. 1,33,14
168	M. Claudius Marcellus: both Hispaniae	Capture of Margolica (or Marcolica; not localized)	Liv. 45,4,1
154	L. Calpurnius Piso: Hisp. ult.	Defeat by Lusitani	App. Ib. 56; Liv. Per. 47

Table to page 152

Author	Dating	First Macedonian War	Second Macedonian War	Third Macedonian War	Syrian War	Manlius Vulso	Andriscus/Achaeans	Eumenes III
Appian	2nd cent. AD	Macedonica	Macedonica	Macedonica	Syriaca; Macedonica	Syriaca		
Cassiodorus	6th cent. AD						x/–	
Cassius Dio	2nd/3rd. cents. AD						–/x	
Cicero	1st cent. BC						–/x	
Diodorus Siculus	1st cent. BC		x	x	x	x	x/x	x
Florus	1st/2nd cents. AD							x
Justin	4th cent. AD	x	x	x	x		–/x	x
Livy	59 BC – AD 17	x	x	x	x	x	x/x	x
Memnon	Imperial period				x			
Pausanias	2nd cent. AD	x					–/x	
Plutarch	1st/2nd cents. AD	Aratus; Philopoemen	Flamininus; Philopoemen	Aemilius Paullus	Flamininus; Philopoemen; Cato Maior			Ti. Gracchus
Polybius	2nd cent. BC	x	x	x	x	x	x/x	
Strabo	1st cent. BC/1st cent. AD						x/x	x
Zonaras	11th/12th cents. AD	x	x	x	x	x	x/x	

Table to page 154

Date		Event	Source
58	March	Caesar's arrival at Genava	Plut. Caesar 17
	Summer	Victory over Helvetii at Bibracte	Caes. B Gall. 1,7–29; Liv. Per. 103; Plut. Caesar 17 f.; Cass. Dio 38,31–33
	September	Victory over Suebi under Ariovistus at Vesontio	Caes. B Gall. 1,30–54; Liv. Per. 104; Plut. Caesar 19; Cass. Dio 38,34–50
57	July	Subjection of Remi, Belgic Suessiones, Bellovaci and Ambiani; victory over Nervii by the Sabis; subjection of Aduatuci	Caes. B Gall. 2,1–35; Liv. Per. 104; Cass. Dio 39,1–5; Plut. Caesar 20 f.
	Autumn	Caesar in Illyricum	Caes. B Gall. 3,7,1
		Senate declares fifteen days of thanksgiving (*supplicatio*)	Cic. Prov. cons. 27; Cic. Balb. 61; Caes. B Gall. 2,35,4; Cass. Dio 39,5,1; Plut. Caesar 21,1 f.
56		Land and sea war against Veneti	Caes. B Gall. 3,7–16; Liv. Per. 104; Cass. Dio 39,40–43
		Campaign against Belgic Morini and Menapii	Caes. B Gall. 3,28 f.; Cass. Dio 39,44; Flor. Epit. 1,45,6
	April	Negotiations in Ravenna and Luca: renewal of 'Triumvirate'	Cic. Fam. 1,9,9; Suet. Iul. 24; Plut. Caesar 21; Plut. Cato Minor 41; Plut. Crassus 14
55	Summer	Campaign against Usipetes and Tencteri	Caes. B Gall. 4,1–16; 6,35,5; Plut. Caesar 22,1–5; Cass. Dio 39,47,1–53,1
		Reconnaissance of Britain	Caes. B Gall. 4,20–36; 4,38,5; Liv. Per. 105; Vell. Pat. 2,46,1; Cass. Dio 39,50–52
	Autumn	First Rhine crossing (on pile-and-beam bridge)	Caes. B Gall. 4,1–19; Liv. Per. 105; Cass. Dio 39,47 f.; Plut. Caesar 22 f.; Flor. Epit. 1,45,14 f.
		M. Porcius Cato calls on senate to hand Caesar over to Usipetes	Plut. Caesar 22,4; Plut. Cato Minor 51,1 f.; App. Celt. 18
		Senate declares twenty days of thanksgiving (*supplicatio*)	Caes. B Gall. 4,38,5; Cass. Dio 39,53,2
54		Invasion of Britain	Caes. B Gall. 5,1–23; Cic. ad Q. fr. 2,13,1 f.; 2,15,4; 3,1,10; Cic. Att. 4,15,10; 4,16,7; 4,18,5; Cic. Fam. 7,6,2; 7,7,1; 7,17,3; Liv. Per. 105; Cass. Dio 40,1–3
		Campaign against Eburones, Nervii and Treveri	Caes. B Gall. 5,24–52; Cass. Dio 40,4–11; Liv. Per. 106
	September	Death of Julia, daughter of Caesar and wife of Pompey	Cic. ad Q. fr. 3,1; 17,25; Plut. Pompeius 53,5; Cass. Dio 39,64,1
53		Second Rhine crossing (on pile-and-beam bridge)	Caes. B Gall. 6,9 f.; 6,29; Liv. Per. 107; Flor. Epit. 1,45,15; 2,30,22; Cass. Dio 40,32,1 f.
		Defeat of the legate L. Aurunculeius Cotta in battle against Eburones under Ambiorix	Caes. B Gall. 5,24–52; Liv. Per. 106 f.; Cass. Dio 40,5–11; Caes. B Gall. 6,5; 29–34
	15 June	M. Licinius Crassus dies following Battle of Carrhae	Cic. Brut. 282; Plut. Crassus 25; Plut. Cicero 36; Liv. Per. 106; Cass. Dio 40,21,3
52	18 January	P. Clodius Pulcher murdered	Cic. Mil. 27; 45; Ascon. 27
	25 February	Pompey *consul sine collega*	Ascon. 31; Plut. Pompeius 54,5
		War against the Arvernian Vercingetorix: conquest of Avaricum	Caes. B Gall. 7,13–28
		Fruitless siege of Gergovia	Caes. B Gall. 7,34; Liv. Per. 107
		Conquest of Alesia	Caes. B Gall. 7,68–90
		Senate declares twenty days of thanksgiving (*supplicatio*)	Caes. B Gall. 7,90,8; Suet. Iul. 24; Cass. Dio 40,50,4 (60 days)
51		Campaign against Bellovaci	Caes. B Gall. 8,6–22
50		Caesar sets two legions aside for Parthian War	Plut. Caesar 29; Plut. Pompeius 56; App. Civ. 2,115
49	10/11 January	Crossing of the Rubicon, outbreak of Civil War	Suet. Iul. 31,2; 81,2; Vell. Pat. 2,49,4; Plut. Pompeius 60,2

Table to page 168

Date	Protagonist	Action
		49 BC
10/11 Jan.	Caesar	crosses Rubicon
11 Jan.	Caesar	in Ariminum
13 Jan.	Caesar	in Fanum
14 Jan.	Caesar	in Ancona
17 Jan.	Pompey	leaves Rome
18 Jan.	Consuls	leave Rome
20 Jan.	C. Scribonius Curio	in Iguvium
3 Feb.	Caesar	in Auximum
5 Feb.	Caesar	in Firmum
14 Feb.	Caesar	outside Corfinium
21 Feb.	Caesar	takes Corfinium (*clementia Corfiniensis*)
25 Feb.	Pompey	in Brundisium
4 Mar.	Consuls	embark at Brundisium to meet with the senate at Thessalonica
9 Mar.	Caesar	outside Brundisium
17 Mar.	Pompey	embarks at Brundisium
31 Mar.–6 Apr.	Caesar	in Rome
1 Apr.	Caesar	chairs a sitting of the rump senate
19 Apr.–5 Jun.	Caesar	outside Massilia
23 Apr.	C. Scribonius Curio	wins Sicily for Caesar
7 Jun.	Cicero	takes the side of Pompey
23 Jun.	Caesar	outside Ilerda in Hispania citerior
2 Aug.	Battle of Ilerda, victory for Caesar, capitulation of forces under L. Afranius and M. Petreius	
11 Aug.	C. Scribonius Curio	lands in Africa
20 Aug.	C. Scribonius Curio	defeated by Juba by the Bagradas, falls
7 Sept.	Caesar	in Corduba
17 Sept.	Caesar	in Gades
25 Sept.–1 Oct.	Caesar	in Tarraco
October	Caesar	appointed dictator
late October	Caesar	outside Massilia; Massilia capitulates
2–12 Dec.	Caesar	back in Rome
22 Dec. 49–4 Jan. 48	Caesar	in Brundisium
		48 BC
January	Caesar	relinquishes dictatorship
5 Feb.	Caesar	lands at Palaeste south of Oricum
15 Mar.	M. Calpurnius Bibulus	dies
6 Feb.–12 Jul.	Positional warfare in Illyricum; fluctuating fortunes	
6 Jul. or 17 Jul.	Pompey	defeats Caesar at Dyrrhachium
26 Jul. or 31 Jul.	Caesar	takes Gomphus
29 Jul. or 3 Aug.	Caesar	encamps at Pharsalus
9 Aug.	Battle of Pharsalus, Caesar victorious	
10 Aug.	Caesar	in Larisa
12 Aug.	Pompey	in Amphipolis
16 Aug.	Pompey	in Mytilene
19 Aug.	Caesar	in Ephesus
23 Aug.	Pompey	on Cyprus
28 Sep.	Pompey	murdered at Pelusium
October	Caesar	named dictator for one year
2 Oct.	Caesar	lands at Alexandria
		47 BC
27 Mar.	Caesar	victorious over Ptolemaic forces on the Nile
May	Cn. Domitius Calvinus	defeated by Pharnaces II at Nicopolis/Pontus
27 Jun.	Caesar	in Antioch
2 Aug	Caesar victorious over Pharnaces II at Zela	
24 Sep.	Caesar	arrives at Tarentum
October to December	Caesar	in Rome
17 Dec.	Caesar	in Lilybaeum
28 Dec.	Caesar	lands at Hadrumetum in Africa
		46 BC
4 Jan.	Indecisive battle at Ruspina between Caesar and T. Labienus	
March	P. Sittius	conquers Numidian capital of Cirta for Caesar
6 Apr.	Caesar victorious over army of Q. Caecilius Metellus Pius Scipio at Thapsus	
before mid-April	M. Porcius Cato	takes his own life
mid-April	Caesar	named dictator for ten years
15 Jun.	Caesar	in Caralis, Sardinia
25 Jul.	Caesar	again in Rome
August	Caesar	celebrates four triumphs (ex Gallia, ex Aegypto, ex Ponto, ex Africa)
2nd *mensis intercalaris*	Caesar	departs for Spain, arrives 26 days later at Obulco
		45 BC
19 Feb.	Caesar	accepts capitulation of Ategua, Hispania ulterior
17 Mar.	Caesar	victorious over Cn. Pompeius at Munda; Cn. Pompeius and T. Labienus fall, Sex. Pompeius escapes

Table to page 176 (cont'd)

Date	Province	Event	Sources
AD 74	Lycia et Pamphylia	Province established; administrative capital Perge or Patara	Suet. Vesp. 8,4; Cass. Dio 60,17,3; discussion of capitals in HAENSCH, 290–293
c. AD 85	Germania superior and Germania inferior	Provinces established; administrative capitals Mogontiacum and Colonia Agrippinensis	CIL 11, 5744; cf. 3, 9960; Mogontiacum: AE 196[illegible], 148 with HAENSCH, 150; Colonia Agrippinensis: evidence in HAENSCH, 65–73
AD 86	Moesia superior and Moesia inferior	Province divided (cf. AD 45/46); administrative capitals Viminacium and Tomi	ILS 1005; SHA Hadrian 2,3; Viminacium: ILS 2[illegible]78; CIL 3, 1651; Tomi: ILS 1074; CIL 3, 7529
reign of Trajan (98–117) at the latest	Epirus	Province established; adm. capital Nicopolis	CIL 3, 12299; Ptol. 3,14,1; Nicopolis: ILS 8849
AD 103/[illegible]	Pannonia inferior and Pannonia superior	Province of Pannonia divided (cf. mid-1st cent.), adm. capitals Aquincum and Carnuntum	Vell. Pat. 2,110,2; SHA Hadrian 3,9; Aquincum: ILS 2375; Carnuntum: ILS 2382
AD 105	Arabia	Province established; adm. capital Petra or Bostra	Cass. Dio 68,14,5; Chron. pasch. 472; Amm. Marc. 14,8,13; discussion of capitals in HAENSCH, 238–24[illegible]
AD 106	Dacia	Province established; adm. capital Apulum or Sarmizegetusa	Cass. Dio 68,14; Aur. Vict. Caes. 13; discussion of capital in HAENSCH, 338–345
AD 114 – 117	Armenia	Province; adm. capital unknown (Artaxata?)	Cass. Dio 68,19 f.; Eutr. 8,3,2; 8,3,62; Festus 14; [illegible]; RIC 2, 289 no. 642
	Mesopotamia	Province; adm. capital unknown (Nisibis?)	Cass. Dio 68,21,1; 68,23,2; 68,28,2; Eutr. 8,3,2; 8,[illegible],2; Rufius Festus, Breviarium 14,3; 20,3
	Assyria	Province; adm. capital unknown (Ctesiphon?)	Eutr. 8,3,2; 8,6,2; Rufius Festus, Breviarium 14,3; 20,3
AD 118/[illegible]	Dacia superior/Apulensis, Dacia inferior, Dacia Porolissensis	Province of Dacia divided (cf. 106); adm. capitals respectively Apulum, unknown, Porolissum	CIL 16, 75; Apulum: ILS 1098; CIL 3, [illegible]177; Porolissum: AE 1973, 459; tres Daciae: Apulum: CIL 3, 836; [illegible]741
AD 194	Syria Phoenice and Syria Coele	Province of Syria divided into Syria Phoenice (adm. capital Tyre) and Syria Coele (adm. capital Laodicea)	SHA Hadr. 14,1; Dig. 50,15,1; Tyre: Dig *l.c.*; Laodicea: Malalas 12, 294 BONN
AD 195	Osrhoëne	Province established	ILS 1353
AD 197	Britannia superior, Britannia inferior	Province of Britannia divided: Britannia superior in south, Britannia inferior in north; adm. capitals Londinium (B. superior), Eboracum (B. inferior)	Herodian. 3,8,2; Londinium, Eburacum: ILS 88[illegible]
between AD 198 and 203	Numidia	Separated from Africa proconsularis as province in its own right; adm. capital Lambaesis	AE 1959, 181; Lambaesis: ILS 2376; CIL 8, 2739 [illegible]8083
AD 197 – [illegible]	Mesopotamia	Province re-established; adm. capital unknown (Nisibis?)	ILS 1331; 1388; 8847; 9148

Table to page 180

Date	Event	Sources
Winter 6[illegible]/6[illegible] BC	Province of Syria established; adm. capital Antioch	Plut. Pompeius 39,3; 70; App. Syr. 250; App. Mithr. 492; Oros. 6,4; Eu[illegible] 6 14; Antioch: Tac. Ann. 2,69,2; Jos. Ant. Iud. 16,270; 17,132; city lists: Plin. NH 5,74; 5,79; Ptol. 5,7,14–17
36–30 BC	Chalcis, Phoenician coastal cities, Coele Syria, Jericho: separated from province, added to Ptolemaic kingdom	Plut. Antonius 36,3; Jos. Ant. Iud. 15,92–97; Jos. Bl. 1,361 f.; 439; Cass. Dio 49,32
27 BC	Under imperial administration	Cass. Dio 53,12
AD 6	Judaea added to province of Syria	Jos. Bl. 2,117; 2,167; Jos. Ant. Iud. 17,355; cf. Dig. 50,15,1; Antioch: [illegible] 16,2,5; Cass. Dio 69,2,1
AD 72	Separation and establishment of province of Judaea (Syria Palaestina); adm. capital Caesarea	Jos. Bl. 2,111; 2,117; 2,167; Jos. Ant. Iud. 17,344; 17,355; Caesarea: Tac. Hist. 2,78,4
after AD [illegible]	Parts of the Syrian Dekapolis added to province of Arabia	cf. Amm. Marc. 14,8,13
AD 195	Division of province: Syria Phoenice, adm. capital Heliopolis; Syria Coele, adm. capital Antioch	Cass. Dio 53,12; 55,23; 79,7; Dig. 50,15,1; Heliopolis: CIL 3, 14387f; Antioch: R.o. FINK, Roman Military Records on Papyrus, 1971, 99,3
under Diocletian (284–305)	Division of Syrian region into six provinces: Syria I, Syria II, Syria Phoenice, Augusta Libanensis, Palaestina, Arabia I	Laterculus Veronensis 1

Table to page 182 (cont'd)

Date	Province	Event	Sources
from 74 BC	Cilicia	Constant Roman presence; incorporation of Pamphylia	Sall. Hist. 2,87; Str. 14,3,3–5,7; Cic. Att. 5,21,9
44 BC		Awarded to Tarcondimotus	Cass. Dio 41,63,1; 47,26,2; Plut. Antonius 61,2
20 BC		Cilicia Trachea awarded to Archelaus Sisines	Str. 12,1,4; 14,5,6
AD 72		Again a Roman province	Suet. Vesp. 8,4; ILS 8971; IGR 3,840; F. Imhoof-Blumer, Kleinasiatische Münzen, 1901/02, 445
25/24 BC	Galatia	Province established including Pamphylia	Cass. Dio 53,26,3; Str. 12,5 f.; 12,7,3; cf. AE 1969/70, 601
6/5 BC		Paphlagonia incorporated into province	Ptol. 5,4,5 f.
3/2 BC		Amasea and Caranitis incorporated in province as Pontus Galaticus	Str. 12,3,37–39; SNG Aulock 17
AD 34/35		Priestly state of Comana Pontica incorporated	Ptol. 5,6,9
AD 43/69		Pamphylia (with parts of Pisidia) added to/reincorporated in new province of Lycia	Cass. Dio 60,17,3; Suet. Claud. 25/Cass. Dio 49,32,3
AD 55–66		United with Cappadocia	Tac. Ann. 13,35,4; 15,6,5; Cass. Dio 62,20,4; 62,22,3 f.
AD 64/65		Pontus Polemoniacus/Polemonianus incorporated, including Sebastea	Tac. Hist. 3,47,1; Suet. Nero 18; CIL III 6818; Polemonianus: ILS 1017; Sebastea: W. Leschhorn, Antike Ären, 1993, 140–149, 469 f.
AD 69		Pamphylia reincorporated	Tac. Hist. 2,9
AD 71/72		Abolition of kingdom of Aristoboulus, incorporation into province of Galatia	Coins of Nicopolis, cf. W. Leschhorn, Antike Ären, 1993, 144–149
AD 76		United with Cappadocia	ILS 8904
AD 113		Cappadocia, Pontus Polemoniacus and Pontus Galaticus divided	ILS 1038 f.
c. AD 162		Abonutichus/Ionopolis, Sinope, Amisus incorporated	Ptol. 5,6,1–3
AD 230		Separation of Pontus as a province in its own right (from 4th cent. Diospontus, Helenopontus)	Milestones, cf. D. French, EA 8, 1986, 75–77, 80; Id., ZPE 43, 1981, 151–153 nos. 4 f.
reign of Diocletian (284–305)		Phrygia Paroreius separated	Laterculus Veronensis 3
after AD 381		Division into Galatia I (adm. capital Ancyra) and Galatia II/Salutaris (adm. capital Pessinus)	Not. Dign. or. 2,42 and 51; 25,4 and 7
Early 4th cent. AD	Isauria	Province established	ILS 1017; Not. Dign. or. 29
AD 371	Lycaonia	Province established	Polemius Silvius 7; Not. Dign. or. 1,97
AD 43	Lycia	Province established	Suet. Claud. 25,3
AD 74	Lycia et Pamphylia	Province established	Suet. Vesp. 8,4; Cass. Dio 60,17,3; Eus. Chron. 188
mid-2nd cent. AD		Southern Pisidia incorporated	Ptol. 5,5,4; 54,9; CIL 3,6885
3rd/4th cents. AD		Lycia separated from Pamphylia	Not. Dign. or. 2,31; 38
AD 116–117	Mesopotamia	Provincial status	Cass. Dio 68,22; 68,29; Eutr. 8,3,2; 8,6,2; Rufius Festus, Breviarium 14; 20; RIC 2,289 no. 642
AD 197–227		Again provincial status	Cass. Dio 75,1; 9; Eutr. 8,18; Rufius Festus, Breviarium 14
AD 195	Osrhoëne	Provincial status	ILS 1353
6/5 BC	Paphlagonia	Province established	Coins of Paphlagonian cities, cf. W. Leschhorn, Antike Ären, 1993, 170–177, 481–484
reign of Diocletian (284–305)	Pontus	Separated from Bithynia	CIL III 307; 13643
before AD 313		Amasia again incorporated	Laterculus Veronensis 9
c. AD 325		Divided into Pontus Galaticus and Diospontus (Helenopontus before 337)	CIL III 14184,17; 31; 37; Cod. Theod. 13,11,2; Polemius Silvius 9

Tables to page 184

Date	Province	Event	Sources
from 58 BC at the latest	Illyricum	C. Iulius Caesar first governor of province; 45–43 BC P. Vatinius	Caesar: Caes. B Gall. 2,35,2; 5,1,5; Suet. Iul. 22,1; Cass. Dio 38,8,5; Vatinius: Cic. Fam. 5,1–10; App. Ill. 38
13 Jan. 27 BC		Under senatorial administration	Cass. Dio 53,12,4 (tò Delmatikón)
AD 9	Illyricum superius (later Dalmatia) and military district of Illyricum inferius (later Pannonia; cf. map p. 189)	Province of Illyricum divided	ILS 938; Vell. Pat. 2,110,2; 2,114,4; 2,116,2 (already province of Dalmatia in this year?)
reign of Nero (54–68)	Dalmatia	Province established (former Illyricum superius); adm. capital Salona	ILS 219 (?)
reign of Diocletian (284–305)		Province of the Dioecesis Pannoniae	Laterculus Veronensis 6,3; Pol. Silv. 5,1
late 4th cent.	Dalmatia (north); Praevalitana (south)	Province divided into Dalmatia (Dioecesis Pannoniae) and Praevalitana (Dioecesis Dacia)	Not. Dign. occ. 2,31; Not. Dign. or. 3,19

Date	Province	Event	Sources
28 BC		Military district of Moesia incorporated into province of Macedonia	Cass. Dio 51,23–27; 53,7,1; Liv. Per. 134 f.; Flor. Epit. 2,26
c. AD 46	Moesia	Province established	ILS 969 with Tac. Hist. 2,23
AD 86	Moesia superior, Moesia inferior	Province divided; adm. capitals Viminacium, Tomi	cf. Ptol. 2,16,1; 3,9,1

Date	Province	Event	Sources
AD 46	Thracia	Province established; adm. capital Perinthus	Cassiod. chronica 659; Jer. Chron. 2064; Sync. p. 630 Bonn; Perinthus: ILS 1093; CIL 3, 731
imperial reform of Diocletian (284–305)		Dioecesis Thraciae with provinces Europa, Rhodope, Thracia and Haemimontus	Laterculus Veronensis 4; Amm. Marc. 27,4,12 f.

Date	Province	Event	Sources
AD 106	Dacia	Province established, adm. capital Apulum or Sarmizegetusa	Cass. Dio 68,14,3; Aur. Vict. Caes. 13; Eutr. 8,6,2
AD 118	Dacia superior, Dacia inferior	Province divided, adm. capitals Apulum and Napoca	SHA Hadr. 4,1–10; Apulum: cf. CIL 3, 7794b; Napoca: CIL 3, 853
AD 120	tres Daciae: Dacia superior/Apulensis, Dacia inferior/Malvensis, Dacia Porolissensis	Tripartition of province, main adm. capital Apulum/Sarmizegetusa, other adm. capitals Romula/Malva, Napoca	Apulum: cf. CIL 3, 7794b; AE 1964, 193; Romula/Malva: ILS 1403; Napoca: CIL 3, 853
AD 271		Provinces abandoned	Eutr. 9,15,1; Iord. De summa tempore Romanorum 217

Tables to page 188

Date	Province	Event	Sources
Augustan-Tiberian period (early 1st cent. AD)	Military district of Raetia et Vindelicia et Alpes Graiae et Poeninae	Occupation following Alpine campaigns of Drusus and Tiberius	Plin. HN 3,136 f.; cf. CIL 5, 7817; 9, 3044
reign of Claudius (41–54)	Raetia	Province established; adm. capital at first Cambodunum (?), then Augusta Vindelicum Separation of Alpes Graiae et Poeninae as a procuratorial province	Vell. Pat. 2,39,3; Suet. Aug. 21,1; Cambodunum: Tac. Germ. 41,1; Augusta Vindelicum: ILS 2386; 3203; other evidence in Haensch, 146–149; Alpes Graiae: ILS 1368
260–265		Part of the Imperium Galliarum under Postumus	cf. the inscription published by L. Bakker
mid-3rd cent.		Withdrawal to Danube frontier	cf. the inscription published by L. Bakker
reign of Diocletian (284–305)		in Dioecesis Italiciana	Laterculus Veronensis 10,9
reign of Constantine I (306–337)		Province divided: Raetia I, adm. capital Curia, Raetia II, adm. capital Augusta Vindelicum	Not. Dign. occ. 1,43 f.; 92 f.; 2,22 f.

Date	Province	Event	Sources
170 BC	regnum Noricum	Conclusion of a *hospitium publicum* with Rome	Liv. 43,5,7; 10
15 BC		Administration in Roman hands	Vell. Pat. 2,39,3; Tac. Ann. 2,63
reign of Claudius (AD 41–54)	Noricum	Province established	cf. J. Ott on the military diplomas
167–182		Marcomannic Wars; *legio II Italica* stationed, under a *legatus Augusti pro praetore*	ILS 2419a
reign of Diocletian (284–305)		Province divided: Noricum ripense and Noricum mediterraneum; in Dioecesis Pannoniae	Laterculus Veronensis 6,6 f.; Pol. Silv. 5,10 f.
from reign of Constantine I (306–337)		both Norican provinces in Dioecesis Illyrici	Not. Dign. occ. 2,33 f.; 88 f.

Date	Province	Event	Sources
AD 9	Illyricum	Pacification; province established	R. Gest. div. Aug. 30; Cass. Dio 55,2,4
	Illyricum superius (later Dalmatia) and military district of Illyricum inferius (later Pannonia)	Province of Illyricum divided	ILS 938; Vell. Pat. 2,110,2; 2,114,4
mid-1st cent. AD	Pannonia	Province established (formerly Illyricum inferius); adm. capital Carnuntum or Poetovio or Savaria	Vell. Pat. 2,116,2; ILS 1991; 2737; discussion of the capital in Haensch, 349 note 155 with 693 f.
85–103		Area around Sirmium transferred to Moesia superior; then part of Pannonia inferior	Military diplomas CIL 16, 31; 164; ILS 39; 9094
103/07	Pannonia inferior; Pannonia superior	Province of Pannonia divided, adm. capitals Aquincum and Carnuntum	Vell. Pat. 2,110,2; SHA Hadrian 3,9; Aquincum: ILS 2375; Carnuntum: ILS 238
200		Region around Emona transferred to province of Venetia et Histria	Herodian. 8,1
214		Transfer of region around Brigetio to Pannonia inferior	cf. CIL 3, 1670
reign of Diocletian (284–305)	Pannonia I/ripariensis, Savia, Pannonia II, Valeria	Pannonia superior divided into Pannonia I/ripariensis (Savaria) and Savia (Siscia), Pannonia inferior into Pannonia II (Sirmium) and Valeria (Sopianae)	Amm. Marc. 19,11,4; Aur. Vict. Caes. 40,10; Laterculus Veronensis 6,5

Addenda to commentaries

Authors and Rights

Unless otherwise indicated, authors of maps and commentaries are identical.

p. 2: A. Fuchs after: W. Horowitz, Mesopotamian Cosmic Geography, 1998
p. 3, Map A: R. Müller-Wollermann
p. 3, Map B: A. Fuchs
p. 4, Map A: E. Olshausen after J.O. Thomson, History of Ancient Geography, 1948, fig. 11
p. 5, Map B: E. Olshausen after E.H. Bunbury, A History of Ancient Geography 1, 1879, pl. III, pp. 172 f.
p. 5, Map C: E. Olshausen after K. Miller, Mappae Mundi. Die ältesten Weltkarten 6, 1898, pl. VIII 1
p. 5, Map D: E. Olshausen after H. Kiepert, Atlas Antiquus, [12]1903, pl. 1 inset
p. 8 f.: E. Olshausen
p. 10, Maps B and C: M. Novák with K. Volk, Tübingen
p. 11: M. Novák with K. Volk, Tübingen
p. 12: M. Novák
p. 13, Maps A and B: M. Novák
p. 14: M. Novák
p. 15, Maps A and B: M. Novák
p. 17, Maps A and B: R. Müller-Wollermann
p. 17, Map C: R. Müller-Wollermann after R. Stadelmann, Die ägyptischen Pyramiden. Vom Ziegelbau zum Weltwunder, 1985, 6 (with revisions by the map author)
p. 19, Maps A and B: R. Müller-Wollermann
p. 21, Maps A and B: R. Müller-Wollermann
p. 21, Map C: R. Müller-Wollermann after J. Dorner, Die Topographie von Piramesse – Vorbericht, in: Ägypten und Levante 9, 1999, map 1 (with revisions by the map author)
p. 22: A.-M. Wittke based on N. Oettinger, DNP 6, 557 f. (top map)
p. 23: A.-M. Wittke based on F. Starke, DNP 5, 195 f.
p. 24: A.-M. Wittke based on W.-D. Niemeier/A.-M. Wittke, DNP 1, 145 f.
p. 25, Map A: A.-M. Wittke based on W.-D. Niemeier/A.-M. Wittke, DNP 1, 145 f.
p. 25, Map B: A.-M. Wittke based on W.-D. Niemeier/A.-M. Wittke, DNP 1, 147 f.
p. 26: A.-M. Wittke based on R. Plath, DNP 7, 245 f.
p. 27: A.-M. Wittke based on G. Hiesel/A.-M. Wittke, DNP 8, 581 f.
p. 28: A.-M. Wittke after W.H. van Soldt, The Topography of the City-State of Ugarit, 2005
p. 29: A.-M. Wittke based on A.-M. Wittke, DNP 1, 151 f.
p. 31: A.-M. Wittke with L. Simons M.A., Tübingen
p. 33, Maps A and B: A. Fuchs, J. Kamlah, R. Müller-Wollermann, M. Novák, A.-M. Wittke
p. 34: A.-M. Wittke based on L. García-Ramón, DNP 4, 1233 f.
p. 35: A.-M. Wittke based on S. Deger-Jalkotzy/A.-M. Wittke, DNP 3, 839 f.
p. 36: A.-M. Wittke after F. Lang, Archaische Siedlungen in Griechenland, 1996
p. 37: A.-M. Wittke with C. Drosihn M.A. and D. Piras M.A., Tübingen
p. 39: A.-M. Wittke
p. 41: W. Röllig based on W. Röllig/A. Fuchs, DNP 12.1, 1027 f.
p. 43: M. Novák; © for the map: M. Novák
p. 44: Jens Kamlah based on K. Bieberstein, DNP 5, 903 f. (Map A)
p. 45, Maps A and B: J. Kamlah
p. 47, Maps A and B: A. Fuchs
p. 49: A. Fuchs
p. 51: A. Fuchs
p. 53: Maps A and B: A. Fuchs
p. 55: Maps A and B: A. Fuchs
p. 57, Map A: M. Novák based on S. Maul/A.-M. Wittke, DNP 2, 385 f.
p. 57, Map inset: © M. Novák
p. 58: R. Müller-Wollermann after D.J. Thompson, Memphis under the Ptolemies, 1988, fig. 3 (with additions and revisions by the map author)
p. 59, Maps A-C: R. Müller-Wollermann
p. 61: A.-M. Wittke based on W. Röllig/A.-M. Wittke, DNP 11, 235 f.
p. 62: A.-M. Wittke based on J. Untermann, DNP 5, 625 f.
p. 63: A.-M. Wittke
p. 65: A.-M. Wittke with M. Lesky, Bietigheim and A. Naso, Rome
p. 66: G. Meiser based on G. Meiser, DNP 5, 1173 f.
p. 67: G. Meiser based on G. Meiser, DNP 5, 1171 f.
p. 68: A.-M. Wittke based on A.-M. Wittke, DNP 6, 657 f.
p. 69: A.-M. Wittke based on A.-M. Wittke/W. Eder/H.G. Niemeyer (†)/F. Prayon, DNP 6, 655 f.
p. 71: A.-M. Wittke/H.G. Niemeyer (†) with assistance of M. Sommer, based on H.G. Niemeyer (†)/A.-M. Wittke, DNP 9, 923 f.
p. 73, Map A: A.-M. Wittke/H.G. Niemeyer (†) based on H.G. Niemeyer (†), DNP 6, 299 f.
p. 73, Map B: A.-M. Wittke based on A.-M. Wittke/E. Olshausen/H.G. Niemeyer (†)
p. 75: A.-M. Wittke
p. 77: A.-M. Wittke based on A.-M. Wittke, DNP 4, 173 f.
p. 79: T. Hoppe
p. 80: T. Hoppe based on V. Pingel (†), DNP 6, 389 f.
p. 81: T. Hoppe
p. 83: M. Terp
p. 84: A.-M. Wittke based on R. Schneider, DNP 7, 929 f.
p. 85: A.-M. Wittke
p. 87: A.-M. Wittke with J. Wiesehöfer, Kiel, and H. Klinkott, Tübingen
p. 88: W. Eder/A.-M. Wittke
p. 89: A.-M. Wittke based on W. Eder/A.-M. Wittke, DNP 9, 607 f.
p. 91, Map A: A.-M. Wittke based on R. Goette, DNP 2, 171 f.; 173 f. (lower map), commentary: W. Eder
p. 91, Map B: A.-M. Wittke based on F. Weber, DNP 8, 177 f., commentary: W. Eder
p. 93, Maps A and B: A.-M.Wittke, commentary: W. Eder
p. 95: A.-M. Wittke based on W. Eder/A.-M. Wittke, DNP 2, 253 f.
p. 97: A.-M. Wittke based on W. Eder/A.-M. Wittke, DNP 9, 503 f.
p. 99: A.-M. Wittke, commentary: W. Eder
p. 100, Map C: A.-M. Wittke based on E. Olshausen/D. Strauch, DNP, 1, 393 f., commentary: K. Freitag
p. 101, Map A: A.-M. Wittke based on E. Olshausen/A.-M. Wittke, DNP 1, 373 f., commentary: K. Freitag
p. 101, Map B: A.-M. Wittke based on A.-M. Wittke/E. Olshausen, DNP 2, 735 f., commentary: K. Freitag
p. 102, Map E: A.-M. Wittke based on E. Olshausen/A.-M. Wittke, DNP 2, 1 f., commentary: K. Freitag
p. 103, Map D: A.-M. Wittke based on E. Olshausen/A.-M. Wittke, DNP 1, 63 f.
Commentary: K. Freitag
p. 105: A.-M. Wittke based on A.-M. Wittke/E. Olshausen, DNP 7, 729 f.
p. 107: C. Winkle based on A.-M. Wittke, DNP 12.1, 803 f., commentary: E. Olshausen
p. 108: A.-M. Wittke based on A.-M. Wittke/W. Eder/H. Galsterer, DNP 6, 1167 f.
p. 109: E. Olshausen based on C. Winkle, based on H. Galsterer/A.-M. Wittke, DNP 3, 77 f.
p. 111: M. Köder based on A.-M. Wittke, DNP 12.1, 987 f., commentary: A.-M. Wittke/M. Köder
p. 112: A.-M. Wittke based on A.-M. Wittke, DNP 1, 463 f.
p. 113: A.-M. Wittke based on W. Eder/A.-M. Wittke, DNP 1, 469 f.
p. 115: A.-M. Wittke based on A.-M. Wittke, DNP 3, 501 f.
p. 116: A.-M. Wittke based on N. Oettinger, DNP 6, 557 f. (lower map)
p. 117: A.-M. Wittke based on W. Eder/A.-M. Wittke, DNP 5, 319 f.
p. 118: A.-M. Wittke
p. 119: A.-M. Wittke based on A.-M. Wittke, DNP 2, 669 f.
p. 120: A.-M. Wittke
p. 121, Map A: A.-M. Wittke
p. 121, Map B: A.-M. Wittke based on A.-M. Wittke, DNP 1, 161 f.
p. 123: A.-M. Wittke
p. 124: K. Strobel based on K. Strobel, DNP 6, 395 f., commentary: A.-M. Wittke
p. 125: A.-M. Wittke based on W. Eder/A.-M. Wittke, DNP 9, 555 f. Text to 124 f.: Wittke
p. 126, Map B: A.-M. Wittke based on I. von Bredow/W. Eder/A.-M. Wittke, DNP 10, 839 (map inset)
p. 126, Map C: A.-M. Wittke based on R. Rolle, DNP 11, 647 f.
p. 127, Map A: A.-M. Wittke with R. Posamentir, Istanbul, based on I. von Bredow/W. Eder/A.-M. Wittke, DNP 10, 841
p. 129: A.-M. Wittke based on W. Eder/A.-M. Wittke, DNP 5, 321 f.
p. 131: A.-M. Wittke
p. 133, Maps A and B: A.-M. Wittke with J. Wiesehöfer, Kiel, based on K. Kartunnen/W. Eder/A.-M. Wittke, DNP 4, 1191–1194
p. 135: A.-M. Wittke
p. 137: A.-M. Wittke based on H. Galsterer/W. Eder/A.-M. Wittke, DNP 2, 847 f.
p. 138: A.-M. Wittke based on W. Eder/A.-M. Wittke, DNP 10, 591 f.
p. 139: A.-M. Wittke based on W. Eder/A.-M. Wittke, DNP 10, 595 f.
p. 141: E. Olshausen
p. 143: E. Olshausen
p. 145: E. Olshausen based on P. Barceló/A.-M. Wittke, DNP 5, 629 f.
p. 146: E. Olshausen
p. 147: E. Olshausen based on A.-M. Wittke, DNP 1, 221 f.
p. 148: E. Olshausen
p. 149: E. Olshausen
p. 151: E. Olshausen
p. 152: E. Olshausen
p. 153: E. Olshausen
p. 155: E. Olshausen with C. Winkle, based on H. Galsterer/A.-M. Wittke, DNP 3, 81 f.
p. 157: E. Olshausen based on W. Eder/A.-M. Wittke, DNP 2, 843 f.
p. 159: E. Olshausen after E. Olshausen, J. Wagner, TAVO B V 6 (© Dr. Ludwig Reichert Verlag, Wiesbaden)
p. 161: E. Olshausen based on W. Eder/A.-M. Wittke DNP 10, 101 f. (after J. Wagner, TAVO B V 7; © Dr Ludwig Reichert Verlag, Wiesbaden)
p. 163: E. Olshausen
p. 165: E. Olshausen
p. 166: E. Olshausen
p. 167: E. Olshausen based on A.-M. Wittke/E. Olshausen, DNP 4, 765 f.
p. 169: E. Olshausen based on W. Will/W. Eder/A.-M. Wittke, DNP 2, 917 f.
p. 171: E. Olshausen
p. 172: C. Winkle based on M. Heinzelmann, DNP 10, 1093 f., commentary: E. Olshausen
p. 173: C. Winkle based on A.-M. Wittke, DNP 10, 1087–1090, commentary: E. Olshausen
p. 175: E. Olshausen based on A.-M. Wittke, DNP 10, 833 f. and M. Heinzelmann, DNP 10, 1097 f.
p. 177: E. Olshausen based on W. Eder/A.-M. Wittke,

DNP 10, 1063 f.
p. 179: E. Olshausen
p. 181: E. Olshausen based on A.-M. Wittke, DNP 11, 1175–1178
p. 183: E. Olshausen based on K. Strobel, DNP 6, 543–546
p. 184: E. Olshausen
p. 185: E. Olshausen based on F. Schön/A.-M. Wittke, DNP 8, 329 f.
p. 186: E. Olshausen
p. 187: E. Olshausen based on A.-M. Wittke, DNP 7, 733 f.
p. 189: E. Olshausen
p. 191: E. Olshausen based on E. Olshausen, DNP 2, 489 f.
p. 193: E. Olshausen
p. 195: E. Olshausen based on M. Rathmann, DNP 12.2, 1135 f.
p. 196, Map A: E. Olshausen based on M. Rathmann, DNP 12.2, 1155 f.
p. 196, Map B: E. Olshausen
p. 197, Map A: E. Olshausen
p. 197, Map B: E. Olshausen based on M. Rathmann, DNP 12.2, 1153 f.
p. 198: E. Olshausen based on M. Rathmann, DNP 12.2, 1143 f. and 1149 f.
p. 199: E. Olshausen based on M. Rathmann, DNP 12.2, 1139–1142
p. 201: U. Fellmeth
p. 203: U. Fellmeth based on A.-M. Wittke/H.-J. Drexhage/H. Schneider, DNP 5, 119 f.
p. 205: E. Olshausen based on A.-M. Wittke/H. Schneider, DNP 5, 973 f.
p. 207: E. Olshausen
p. 208: E. Olshausen
p. 209: E. Olshausen based on A.-M. Wittke, DNP 7, 11–14
p. 210: E. Olshausen based on A.-M. Wittke/E. Olshausen, DNP 7, 197 f.
p. 211: E. Olshausen based on A.-M. Wittke/E. Olshausen, DNP 7, 205 f.
p. 212: E. Olshausen based on A.-M. Wittke/E. Olshausen, DNP 7, 225 f.
p. 213: E. Olshausen based on J. Wagner, DNP 7, 217 f.
p. 215: A.-M. Wittke
p. 217: E. Olshausen based on J. Wiesehöfer, DNP 11, 91 f. (after E. Kettenhofen, TAVO B VI 3; © Dr. Ludwig Reichert Verlag, Wiesbaden)
p. 219: E. Olshausen after E. Kettenhofen, TAVO B V 11 (© Dr. Ludwig Reichert Verlag, Wiesbaden)
p. 221: E. Olshausen based on A.-M. Wittke, DNP 12.2, 731 f.
p. 223: E. Olshausen
p. 225: E. Olshausen based on B. Bleckmann, DNP 3, 579–582
p. 227: E. Olshausen based on A.-M. Wittke, DNP 2, 1159 f.
p. 228: E. Olshausen
p. 229, Map A: E. Olshausen after H. Jedin, K.S. Latourette, J. Martin (ed.), Atlas zur Kirchengeschichte, 1987, map 8
p. 229, Map B: E. Olshausen based on A. Merkt/A.-M. Wittke, DNP 9, 1015 f.
p. 231: E. Olshausen based on V. Pingel (†), DNP 4, 969 f. and S. Ziegler, DNP 4, 973 f.
p. 233: E. Olshausen based on K. Tausend, DNP 12.2, 283–286
p. 234: E. Olshausen after Großer Historischer Weltatlas, Bayerischer Schulbuch-Verlag II, 1970, p. 62, map b
p. 235: E. Olshausen
p. 237: E. Olshausen
p. 238: C. Winkle based on A.-M. Wittke/E. Olshausen, DNP 6, 713 f., commentary: E. Olshausen
p. 239: E. Olshausen
p. 241: E. Olshausen after T. Riplinger, H. Brenner, TAVO B VI 6 (© Dr. Ludwig Reichert Verlag, Wiesbaden)
p. 243: E. Olshausen after U. Rebstock, TAVO B VII 2 (© Dr. Ludwig Reichert Verlag, Wiesbaden)
p. 245: E. Olshausen
p. 247: U. Fellmeth
p. 249: E. Olshausen after R. Beyer, H. Halm, C. Naumann, TAVO B VII 10 (© Dr. Ludwig Reichert Verlag, Wiesbaden)
p. 251: E. Olshausen after H. Halm, T. Riplinger, P. Thorau, TAVO B VIII 6 (© Dr. Ludwig Reichert Verlag, Wiesbaden)
p. 253: E. Olshausen

Index

Explanations regarding the index are found in the notes on use. Abbreviations used in the index are as follows:

dcs. *Dioecesis*
des. Desert
isl. Island(s) (no extra indication for: *insula/insulae* or *nesos/nesoi*)
kgd. 'Kingdom', 'Empire', etc.: political units of Graeco-Roman antiquity
MI Map inset
mtn. Mountain/range of mountains (no extra indication for: *mons/montes*)
oas. Oasis
p-d. Present-day
ppl. People, tribe, tribal group
prf. *Praefectura*
prv. Province (Roman unless otherwise stated)
reg. *Regio*
ter. Territory, region, area; esp. also political and administrative names of the Ancient Near East
thm. Theme (Byzantine)
trb. *Tribus*
wtr. Body of water (river, sea, lake; no extra indication for: *lacus, mare*)

A

E

H

M

Q

R

W

X

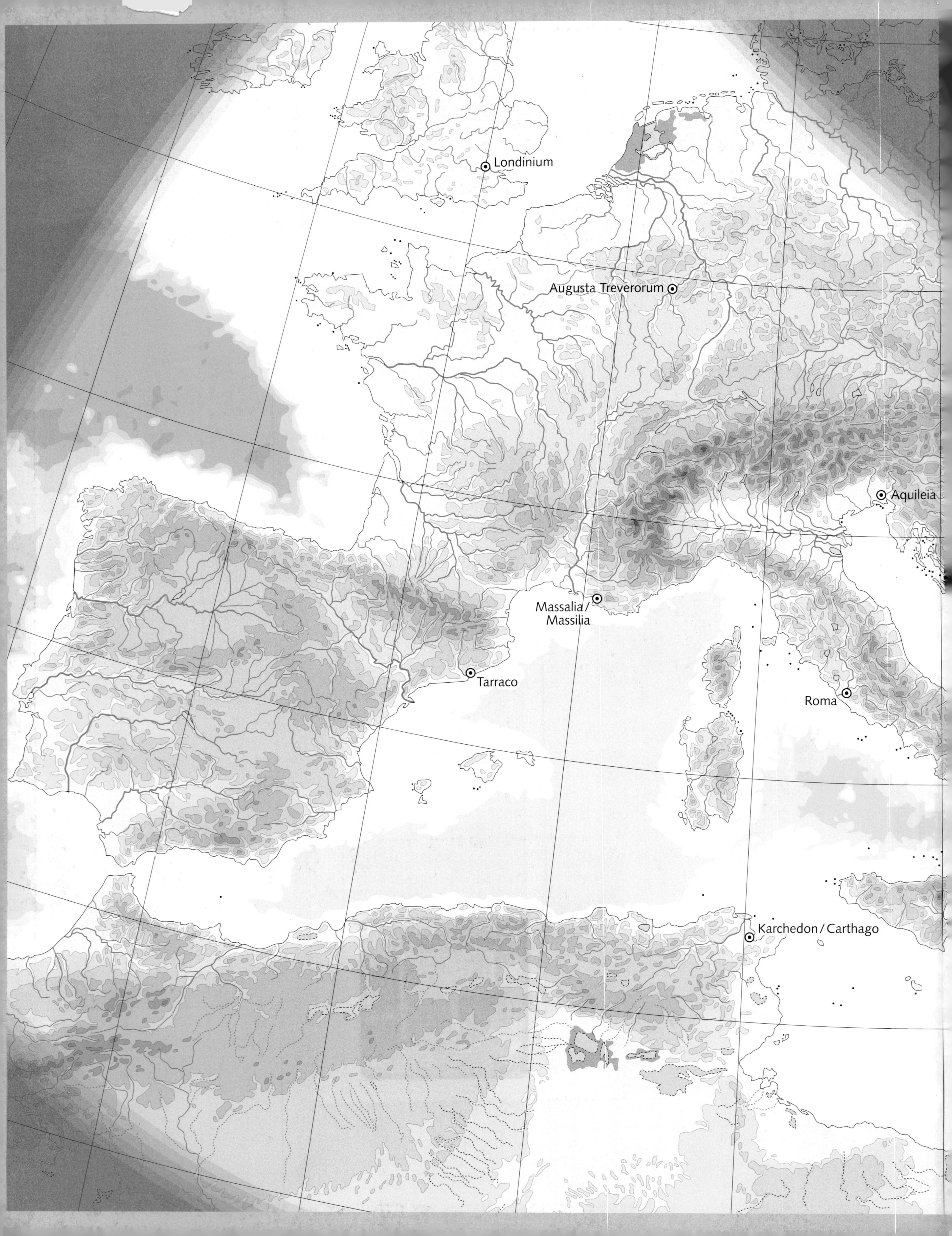

Londinium
Augusta Treverorum
Aquileia
Massalia / Massilia
Tarraco
Roma
Karchedon / Carthago